# Regression and Other Stories

Many textbooks on regression focus on theory and the simplest of examples. Real statistical problems, however, are complex and subtle. This is not a book about the theory of regression. It is a book about how to use regression to solve real problems of comparison, estimation, prediction, and causal inference. It focuses on practical issues such as sample size and missing data and a wide range of goals and techniques. It jumps right in to methods and computer code you can use fresh out of the box.

Key features:

- Real examples, real stories from the authors' real-world experience demonstrate what can be achieved by regression and what the limitations are
- Uses computation with the popular open-source programs R and Stan instead of deriving formulas, with all code available online
- Emphasis on using graphics and presentation to understand and check models that have been fit to data
- Practical advice for understanding assumptions and implementing methods for experiments and observational studies
- Smooth transition to logistic regression and generalized linear models
- Clear presentation of key ideas in data collection, sampling, generalization, and causal inference

The authors are experienced researchers who have published articles in hundreds of different scientific journals in fields including statistics, computer science, policy, public health, political science, economics, sociology, and engineering. They have also published articles in the *Washington Post, New York Times, Slate*, and other public venues. Their previous books include *Bayesian Data Analysis, Teaching Statistics: A Bag of Tricks*, and *Data Analysis and Regression Using Multilevel/Hierarchical Models*.

ANDREW GELMAN is Higgins Professor of Statistics and Professor of Political Science at Columbia University.

JENNIFER HILL is Professor of Applied Statistics at New York University.

AKI VEHTARI is Associate Professor in Computational Probabilistic Modeling at Aalto University.

## Analytical Methods for Social Research

Analytical Methods for Social Research presents texts on empirical and formal methods for the social sciences. Volumes in the series address both the theoretical underpinnings of analytical techniques as well as their application in social research. Some series volumes are broad in scope, cutting across a number of disciplines. Others focus mainly on methodological applications within specific fields such as political science, sociology, demography, and public health. The series serves a mix of students and researchers in the social sciences and statistics.



### Other Titles in the Series

*Maximum Likelihood for Social Science*, by Michael D. Ward and John S. Ahlquist
*Computational Social Science*, by R. Michael Alvarez
*Spatial Analysis for the Social Sciences*, by David Darmofal
*Counterfactuals and Causal Inference*, Second Edition, by Stephen L. Morgan and Christopher Winship
*Time Series Analysis for the Social Sciences*, by Janet M. Box-Steffensmeier, John R. Freeman, Matthew P. Hitt, and Jon C. W. Pevehouse
*Statistical Modeling and Inference for Social Science*, by Sean Gailmard
*Formal Models of Domestic Politics*, by Scott Gehlbach
*Data Analysis Using Regression and Multilevel/Hierarchical Models*, by Andrew Gelman and Jennifer Hill
*Political Game Theory: An Introduction*, by Nolan McCarty and Adam Meirowitz
*Essential Mathematics for Political and Social Research*, by Jeff Gill
*Spatial Models of Parliamentary Voting*, by Keith T. Poole
*Event History Modeling: A Guide for Social Scientists*, by Janet M. Box-Steffensmeier and Bradford S. Jones
*Ecological Inference: New Methodological Strategies*, edited by Gary King, Ori Rosen, and Martin A. Tanner

# Regression and Other Stories

ANDREW GELMAN
*Columbia University, New York*

JENNIFER HILL
*New York University*

AKI VEHTARI
*Aalto University, Finland*

**CAMBRIDGE**
**UNIVERSITY PRESS**

University Printing House, Cambridge CB2 8BS, United Kingdom

One Liberty Plaza, 20th Floor, New York, NY 10006, USA

477 Williamstown Road, Port Melbourne, VIC 3207, Australia

314–321, 3rd Floor, Plot 3, Splendor Forum, Jasola District Centre, New Delhi – 110025, India

79 Anson Road, #06–04/06, Singapore 079906

Cambridge University Press is part of the University of Cambridge.

It furthers the University's mission by disseminating knowledge in the pursuit of education, learning, and research at the highest international levels of excellence.

www.cambridge.org
Information on this title: www.cambridge.org/9781107023987
DOI:10.1017/9781139161879



First published 2021

Printed in the United Kingdom by TJ International Ltd, Padstow Cornwall

*A catalogue record for this publication is available from the British Library.*

ISBN 978-1-107-02398-7 Hardback
ISBN 978-1-107-67651-0 Paperback

# Contents

# Preface

Existing textbooks on regression typically have some mix of cookbook instruction and mathematical derivation. We wrote this book because we saw a new way forward, focusing on understanding regression models, applying them to real problems, and using simulations with fake data to understand how the models are fit. After reading this book and working through the exercises, you should be able to simulate regression models on the computer and build, critically evaluate, and use them for applied problems.

The other special feature of our book, in addition to its wide range of examples and its focus on computer simulation, is its broad coverage, including the basics of statistics and measurement, linear regression, multiple regression, Bayesian inference, logistic regression and generalized linear models, extrapolation from sample to population, and causal inference. Linear regression is the starting point, but it does not make sense to stop there: once you have the basic idea of statistical prediction, it can be best understood by applying it in many different ways and in many different contexts.

After completing Part 1 of this book, you should have access to the tools of mathematics, statistics, and computing that will allow you to work with regression models. These early chapters should serve as a bridge from the methods and ideas you may have learned in an introductory statistics course. Goals for Part 1 include displaying and exploring data, computing and graphing linear relations, understanding basic probability distributions and statistical inferences, and simulation of random processes to represent inferential and forecast uncertainty.

After completing Part 2, you should be able to build, fit, understand, use, and assess the fit of linear regression models. The chapters in this part of the book develop relevant statistical and computational tools in the context of several applied and simulated-data examples. After completing Part 3, you should be able to similarly work with logistic regression and other generalized linear models. Part 4 covers data collection and extrapolation from sample to population, and in Part 5 we cover causal inference, starting with basic methods using regression for controlled experiments and then considering more complicated approaches adjusting for imbalances in observational data or capitalizing on natural experiments. Part 6 introduces more advanced regression models, and the appendixes include some quick tips and an overview on software for model fitting.

## What you should be able to do after reading and working through this book

This text is structured through models and examples, with the intention that after each chapter you should have certain skills in fitting, understanding, and displaying models:

- *Part 1:* Review key tools and concepts in mathematics, statistics, and computing.
  - *Chapter 1:* Have a sense of the goals and challenges of regression.
  - *Chapter 2:* Explore data and be aware of issues of measurement and adjustment.
  - *Chapter 3:* Graph a straight line and know some basic mathematical tools and probability distributions.
  - *Chapter 4:* Understand statistical estimation and uncertainty assessment, along with the problems of hypothesis testing in applied statistics.
  - *Chapter 5:* Simulate probability models and uncertainty about inferences and predictions.

- *Part 2:* Build linear regression models, use them in real problems, and evaluate their assumptions and fit to data.
  - *Chapter 6:* Distinguish between descriptive and causal interpretations of regression, understanding these in historical context.
  - *Chapter 7:* Understand and work with simple linear regression with one predictor.
  - *Chapter 8:* Gain a conceptual understanding of least squares fitting and be able to perform these fits on the computer.
  - *Chapter 9:* Perform and understand probabilistic prediction and simple Bayesian information aggregation, and be introduced to prior distributions and Bayesian inference.
  - *Chapter 10:* Build, fit, and understand linear models with multiple predictors.
  - *Chapter 11:* Understand the relative importance of different assumptions of regression models and be able to check models and evaluate their fit to data.
  - *Chapter 12:* Apply linear regression more effectively by transforming and combining predictors.
- *Part 3:* Build and work with logistic regression and generalized linear models.
  - *Chapter 13:* Fit, understand, and display logistic regression models for binary data.
  - *Chapter 14:* Build, understand, and evaluate logistic regressions with interactions and other complexities.
  - *Chapter 15:* Fit, understand, and display generalized linear models, including the Poisson and negative binomial regression, ordered logistic regression, and other models.
- *Part 4:* Design studies and use data more effectively in applied settings.
  - *Chapter 16:* Use probability theory and simulation to guide data-collection decisions, without falling into the trap of demanding unrealistic levels of certainty.
  - *Chapter 17:* Use poststratification to generalize from sample to population, and use regression models to impute missing data.
- *Part 5:* Implement and understand basic statistical designs and analyses for causal inference.
  - *Chapter 18:* Understand assumptions underlying causal inference with a focus on randomized experiments.
  - *Chapter 19:* Perform causal inference in simple settings using regressions to estimate treatment effects and interactions.
  - *Chapter 20:* Understand the challenges of causal inference from observational data and statistical tools for adjusting for differences between treatment and control groups.
  - *Chapter 21:* Understand the assumptions underlying more advanced methods that use auxiliary variables or particular data structures to identify causal effects, and be able to fit these models to data.
- *Part 6:* Become aware of more advanced regression models.
  - *Chapter 22:* Get a sense of the directions in which linear and generalized linear models can be extended to attack various classes of applied problems.
- *Appendixes:*
  - *Appendix A:* Get started in the statistical software R, with a focus on data manipulation, statistical graphics, and fitting and using regressions.
  - *Appendix B:* Become aware of some important ideas in regression workflow.

After working through the book, you should be able to fit, graph, understand, and evaluate linear and generalized linear models and use these model fits to make predictions and inferences about quantities of interest, including causal effects of treatments and exposures.

## Fun chapter titles

The chapter titles in the book are descriptive. Here are more dramatic titles intended to evoke some of the surprise you should feel when working through this material:

- *Part 1:*
  - *Chapter 1:* Prediction as a unifying theme in statistics and causal inference.
  - *Chapter 2:* Data collection and visualization are important.
  - *Chapter 3:* Here's the math you actually need to know.
  - *Chapter 4:* Time to unlearn what you thought you knew about statistics.
  - *Chapter 5:* You don't understand your model until you can simulate from it.
- *Part 2:*
  - *Chapter 6:* Let's think deeply about regression.
  - *Chapter 7:* You can't just *do* regression, you have to *understand* regression.
  - *Chapter 8:* Least squares and all that.
  - *Chapter 9:* Let's be clear about our uncertainty and about our prior knowledge.
  - *Chapter 10:* You don't just *fit* models, you *build* models.
  - *Chapter 11:* Can you convince *me* to trust *your* model?
  - *Chapter 12:* Only fools work on the raw scale.
- *Part 3:*
  - *Chapter 13:* Modeling probabilities.
  - *Chapter 14:* Logistic regression pro tips.
  - *Chapter 15:* Building models from the inside out.
- *Part 4:*
  - *Chapter 16:* To understand the past, you must first know the future.
  - *Chapter 17:* Enough about your data. Tell me about the population.
- *Part 5:*
  - *Chapter 18:* How can flipping a coin help you estimate causal effects?
  - *Chapter 19:* Using correlation and assumptions to infer causation.
  - *Chapter 20:* Causal inference is just a kind of prediction.
  - *Chapter 21:* More assumptions, more problems.
- *Part 6:*
  - *Chapter 22:* Who's got next?
- *Appendixes:*
  - *Appendix A:* R quick start.
  - *Appendix B:* These are our favorite workflow tips; what are yours?

In this book we present many methods and illustrate their use in many applications; we also try to give a sense of where these methods can fail, and we try to convey the excitement the first time that we learned about these ideas and applied them to our own problems.

## Additional material for teaching and learning

### Data for the examples and homework assignments; other teaching resources

The website `www.stat.columbia.edu/~gelman/regression` contains pointers to data and code for the examples and homework problems in the book, along with some teaching materials.

## Prerequisites

This book does not require advanced mathematics. To understand the linear model in regression, you will need the algebra of the intercept and slope of a straight line, but it will not be necessary to follow the matrix algebra in the derivation of least squares computations. You will use exponents and logarithms at different points, especially in Chapters 12–15 in the context of nonlinear transformations and generalized linear models.

## Software

Previous knowledge of programming is not required. You will do a bit of programming in the general-purpose statistical environment R when fitting and using the models in this book, and some of these fits will be performed using the Bayesian inference program Stan, which, like R, is free and open source. Readers new to R or to programming should first work their way through Appendix A.

We fit regressions using the `stan_glm` function in the `rstanarm` package in R, performing Bayesian inference using simulation. This is a slight departure from usual treatments of regression (including our earlier book), which use least squares and maximum likelihood, for example using the `lm` and `glm` functions in R. We discuss differences between these different software options, and between these different modes of inference, in Sections 1.6, 8.4, and 9.5. From the user's perspective, switching to `stan_glm` doesn't matter much except in making it easier to obtain probabilistic predictions and to propagate inferential uncertainty, and in certain problems with collinearity or sparse data (in which case the Bayesian approach in `stan_glm` gives more stable estimates), and when we wish to include prior information in the analysis. For most of the computations done in this book, similar results could be obtained using classical regression software if so desired.

## Suggested courses

The material in this book can be broken up in several ways for one-semester courses. Here are some examples:

- *Basic linear regression*: Chapters 1–5 for review, then Chapters 6–9 (linear regression with one predictor) and Chapters 10–12 (multiple regression, diagnostics, and model building).
- *Applied linear regression*: Chapters 1–5 for review, then Chapters 6–12 (linear regression), Chapters 16–17 (design and poststratification), and selected material from Chapters 18–21 (causal inference) and Chapter 22 (advanced regression).
- *Applied regression and causal inference*: Quick review of Chapters 1–5, then Chapters 6–12 (linear regression), Chapter 13 (logistic regression), Chapters 16–17 (design and poststratification), and selected material from Chapters 18–21 (causal inference).
- *Causal inference*: Chapters 1, 7, 10, 11, and 13 for review of linear and logistic regression, then Chapters 18–21 in detail.
- *Generalized linear models*: Some review of Chapters 1–12, then Chapters 13–15 (logistic regression and generalized linear models), followed by selected material from Chapters 16–21 (design, poststratification, and causal inference) and Chapter 22 (advanced regression).

## Acknowledgments

We thank the many students and colleagues who have helped us understand and implement these ideas, including everyone thanked on pages xxi–xxii of our earlier book, *Data Analysis Using Regression and Multilevel/Hierarchical Models*. In addition, we thank Pablo Argote, Bill Behrman, Danilo Bzdok, Andres Castro, Devin Caughey, Zad Chow, Dick De Veaux, Vince Dorie, Sander Greenland, Daphna Harel, Merlin Heidemanns, Christian Hennig, David Kane, Katharine Khanna, Lydia Krasilnikova, Stefano Longo, Jenny Pham, Eric Potash, Phil Price, Malgorzata Roos, Michael

Sobel, Melinda Song, Scott Spencer, Mireia Triguero, Jasu Vehtari, Zane Wolf, Lizzie Wolkovich, Adam Zelizer, Shuli Zhang, and students and teaching assistants from several years of our classes for helpful comments and suggestions, Alan Chen for help with Chapter 20, Andrea Cornejo, Zarni Htet, and Rui Lu for helping to develop the simulation-based exercises for the causal chapters, Ben Silver for help with indexing, Beth Morel and Clare Dennison for copy editing, Luke Keele for the example in Section 21.3, Kaiser Fung for the example in Section 21.5, Mark Broadie for the golf data in Exercise 22.3, Michael Betancourt for the gravity-measuring demonstration in Exercise 22.4, Jerry Reiter for sharing ideas on teaching and presentation of the concepts of regression, Lauren Cowles for many helpful suggestions on the structure of this book, and especially Ben Goodrich and Jonah Gabry for developing the `rstanarm` package which allows regression models to be fit in Stan using familiar R notation.

We also thank the developers of R and Stan, and the U.S. National Science Foundation, Institute for Education Sciences, Office of Naval Research, Defense Advanced Research Projects Agency, Google, Facebook, YouGov, and the Sloan Foundation for financial support.

Above all, we thank our families for their love and support during the writing of this book.

# Part 1: Fundamentals

# Chapter 1

# Overview

This book explores the challenges of building, understanding, and using predictive models. It turns out there are many subtleties involved even with simple linear regression—straight-line fitting. After a review of fundamental ideas of data, measurement, and statistics in the first five chapters of the book, we cover linear regression with one predictor and multiple predictors, and then logistic regression and other generalized linear models. We next consider various applications of regression involving generalization from the data at hand to larger questions involving sampling and causal inference. The book concludes with a taste of more advanced modeling ideas and appendixes on quick tips and getting started with computing.

This introductory chapter lays out the key challenges of statistical inference in general and regression modeling in particular. We present a series of applied examples to show how complex and subtle regression can be, and why a book-length treatment is needed, not just on the mathematics of regression modeling but on how to apply and understand these methods.

## 1.1 The three challenges of statistics

The three challenges of statistical inference are:

1. *Generalizing from sample to population*, a problem that is associated with survey sampling but actually arises in nearly every application of statistical inference;
2. *Generalizing from treatment to control group*, a problem that is associated with causal inference, which is implicitly or explicitly part of the interpretation of most regressions we have seen; and
3. *Generalizing from observed measurements to the underlying constructs of interest*, as most of the time our data do not record exactly what we would ideally like to study.

All three of these challenges can be framed as problems of prediction (for new people or new items that are not in the sample, future outcomes under different potentially assigned treatments, and underlying constructs of interest, if they could be measured exactly).

The key skills you should learn from this book are:

- *Understanding regression models*. These are mathematical models for predicting an outcome variable from a set of predictors, starting with straight-line fits and moving to various nonlinear generalizations.
- *Constructing regression models*. The regression framework is open-ended, with many options involving the choice of what variables to include and how to transform and constrain them.
- *Fitting regression models to data*, which we do using the open-source software R and Stan.
- *Displaying and interpreting the results*, which requires additional programming skills and mathematical understanding.

A central subject of this book, as with most statistics books, is *inference*: using mathematical models to make general claims from particular data.

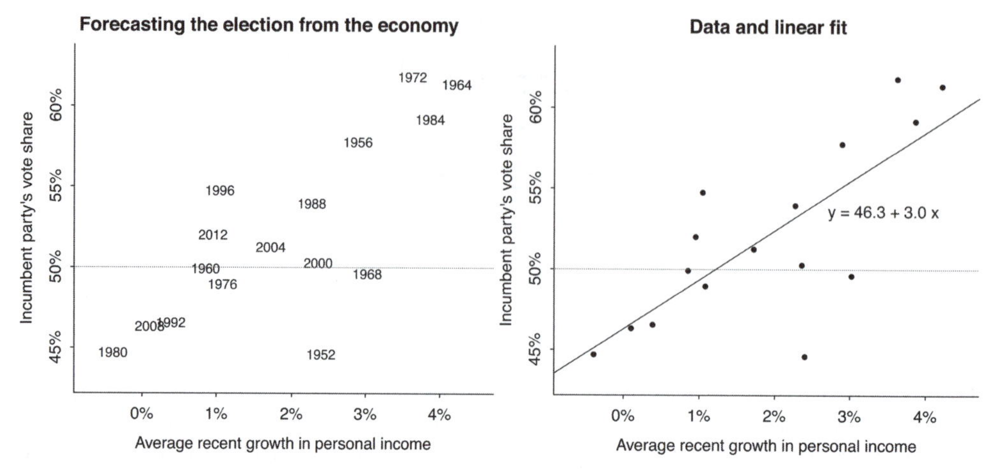


Figure 1.1: *Predicting elections from the economy: (a) the data, (b) the linear fit,* $y = 46.3 + 3.0x$.

## 1.2 Why learn regression?

Example: Elections and the economy

Regression is a method that allows researchers to summarize how predictions or average values of an *outcome* vary across individuals defined by a set of *predictors*. For example, Figure 1.1a shows the incumbent party's vote share in a series of U.S. presidential elections, plotted vs. a measure of economic growth in the period leading up to each election year. Figure 1.1b shows a linear regression fit to these data. The model allows us to predict the vote—with some uncertainty—given the economy and under the assumption that future elections are in some way like the past.

We do our computing in the open-source software R; see Appendix A for how to set up and use R on your computer. For this example, we first load in the data:[1]

```
hibbs <- read.table("hibbs.dat", header=TRUE)
```

Then we make a scatterplot:

```
plot(hibbs$growth, hibbs$vote, xlab="Average recent growth in personal income",
  ylab="Incumbent party's vote share")
```

Then we estimate the regression, $y = a + bx + \text{error}$:[2]

```
M1 <- stan_glm(vote ~ growth, data=hibbs)
```

And then we add the fitted line to our graph:

```
abline(coef(M1), col="gray")
```

This produces something similar to Figure 1.1b.

To display the fitted model, we type `print(M1)`, which gives the following output:

```
            Median MAD_SD
(Intercept) 46.3    1.7
growth       3.0    0.7

Auxiliary parameter(s):
      Median MAD_SD
sigma 3.9    0.7
```

[1]Data and code for all examples in this book are at `www.stat.columbia.edu/~gelman/regression/`. Information for this particular example is in the folder `ElectionsEconomy` at this website.

[2]Section 1.6 introduces R code for least squares and Bayesian regression.

The first column shows estimates: 46.3 and 3.0 are the coefficients in the fitted line, $y = 46.3 + 3.0x$ (see Figure 1.1b). The second column displays uncertainties in the estimates using median absolute deviations (see Section 5.3). The last line of output shows the estimate and uncertainty of $\sigma$, the scale of the variation in the data unexplained by the regression model (the scatter of the points above and below from the regression line). In Figure 1.1b, the linear model predicts vote share to roughly an accuracy of 3.9 percentage points. We explain all the above code and output starting in Chapter 6.

If desired we can also summarize the fit in different ways, such as plotting residuals (differences between data and fitted model) and computing $R^2$, the proportion of variance explained by the model, as discussed in Chapter 11.

Some of the most important uses of regression are:

- *Prediction*: Modeling existing observations or forecasting new data. Examples with continuous or approximately continuous outcomes include vote shares in an upcoming election, future sales of a product, and health status in a medical study. Examples with discrete or categorical outcomes (sometimes referred to as classification) include disease diagnosis, victory or defeat in a sporting event, and individual voting decisions.
- *Exploring associations*: Summarizing how well one variable, or set of variables, predicts the outcome. Examples include identifying risk factors for a disease, attitudes that predict voting, and characteristics that make someone more likely to be successful in a job. More generally, one can use a model to explore associations, stratifications, or structural relationships between variables. Examples include associations between pollution levels and disease incidence, differential police stop rates of suspects by ethnicity, and growth rates of different parts of the body.
- *Extrapolation*: Adjusting for known differences between the *sample* (that is, observed data) and a population of interest. A familiar example is polling: real-world samples are not completely representative and so it is necessary to perform some adjustment to extrapolate to the general population. Another example is the use of data from a self-selected sample of schools to make conclusions about all the schools in a state. Another example would be using experimental data from a drug trial, along with background characteristics from the full population, to estimate the average effect of the drug in the population.
- *Causal inference*: Perhaps the most important use of regression is for estimating *treatment effects*. We define causal inference more carefully in Part 5 of this book; for now we'll just talk about comparing outcomes under treatment or control, or under different levels of a treatment. For example, in an education study, the outcome could be scores on a standardized test, the control could be an existing method of teaching, and the treatment could be some new innovation. Or in public health, the outcome could be incidence of asthma and the continuous treatment could be exposure to some pollutant. A key challenge of causal inference is ensuring that treatment and control groups are similar, on average, before exposure to the treatment, or else adjusting for differences between these groups.

In all these settings, it is crucial that the regression model have enough complexity to carry the required information. For example, if most of the participants in a drug trial are healthy and under the age of 70, but there is interest in estimating an average effect among the general elderly population, then it is important to include age and prior health condition as predictors in the regression model. If these predictors are not included, the model will simply not have enough information to allow the adjustment that we want to do.

## 1.3 Some examples of regression

To give a sense of the difficulties involved in applied regression, we briefly discuss some examples involving sampling, prediction, and causal inference.

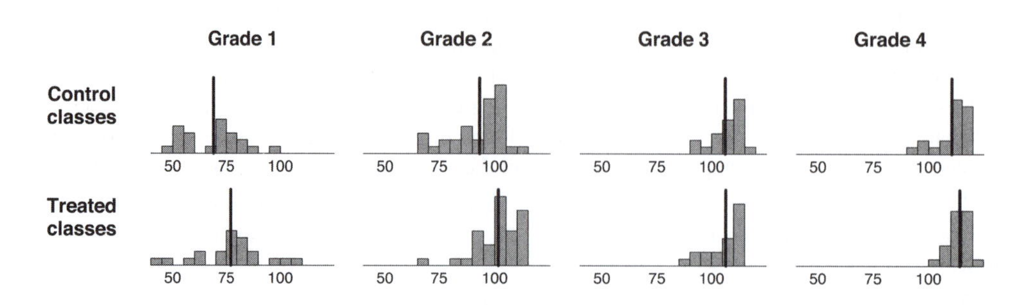


Figure 1.2 *Post-treatment classroom-average test scores from an experiment measuring the effect of an educational television program, The Electric Company, on children's reading abilities. The dark vertical line in each histogram shows the average for the corresponding group of classrooms.*

## Estimating public opinion from an opt-in internet survey

Example: Xbox survey

In a research project with colleagues at Microsoft Research, we used a regression model to adjust a convenience sample to obtain accurate opinion monitoring, at a sharper time scale and at less expense than traditional survey methods. The data were from a novel and highly non-representative survey dataset: a series of daily voter intention polls for the 2012 presidential election conducted on the Xbox gaming platform with a total sample size of 750 148 interviews from 345 858 unique respondents. This is a characteristic problem of big data: a very large sample, relatively inexpensive to collect, but not immediately representative of the larger population. After adjusting the Xbox responses via multilevel regression and poststratification (MRP), we obtained estimates in line with forecasts from leading poll analysts, which were based on aggregating hundreds of traditional polls conducted during the election cycle.

The purpose of the Xbox project was not to forecast individual survey responses, nor was it to identify important predictors or causal inference. Rather, the goal was to learn about nationwide trends in public opinion, and regression allowed us to adjust for differences between sample and population, as we describe in Section 17.1; this required *extrapolation*.

## A randomized experiment on the effect of an educational television program

Example: Electric Company experiment

A study was performed around 1970 to measure the effect of a new educational television program, The Electric Company, on children's reading abilities. An experiment was performed on children in grades 1–4 in two small cities in the United States. For each city and grade, the experimenters selected 10 to 20 schools, within each school selecting the two classes in the grade whose average reading test scores were lowest. For each pair, one of these classes was randomly assigned to continue with their regular reading course and the other was assigned to view the TV program. Each student was given a pre-test at the beginning of the school year and a post-test at the end.

Figure 1.2 shows post-test data for the control and treated classrooms in each grade.[3] Comparing the top and bottom row of graphs, we see what appears to be large beneficial effects in grades 1 and 2 with smaller effects for the higher grades, a plausible result given that most children in grades 3 and 4 already know how to read.

Further statistical analysis is required to adjust for differences in pre-treatment test scores between the two groups, and to assess uncertainty in the estimates. We return to this example in Section 19.2.

[3]Data and code for this example are in the folder `ElectricCompany`.

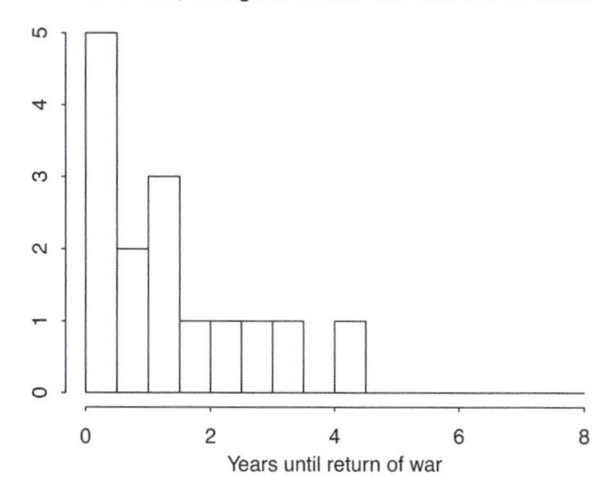


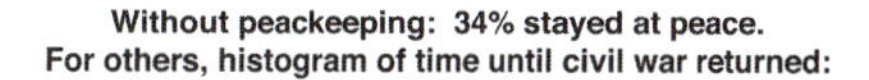


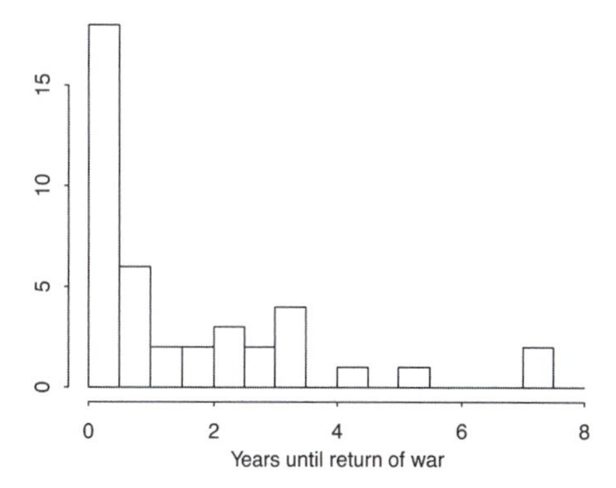


Figure 1.3 *Outcomes after civil war in countries with and without United Nations peacekeeping. The countries with peacekeeping were more likely to stay at peace and took on average about the same amount of time to return to war when that happened. However, there is a concern that countries with and without peacekeeping may differ in their pre-existing conditions; see Figure 1.4.*

## Estimating the effects of United Nations peacekeeping, using pre-treatment variables to adjust for differences between treatment and control groups

Example: United Nations peace-keeping

Several years ago, political scientist Page Fortna conducted a study on the effectiveness of international peacekeeping. She analyzed data from countries that had been involved in civil wars between 1989 and 1999, comparing countries with and without United Nations peacekeeping. The outcome measure was whether there was a return to civil war in the country and, if so, the length of time until that happened. Data collection ended in 2004, so any countries that had not returned to civil war by the end of that year were characterized as being still at peace. The subset of the data summarized here contains 96 ceasefires, corresponding to 64 different wars.[4]

A quick comparison found better outcomes after peacekeeping: 56% stayed at peace, compared to 34% of countries without peacekeeping. When civil war did return, it typically came soon: the average lag between ceasefire and revival of the fighting was 17 months in the presence of peacekeeping and 18 months without. Figure 1.3 shows the results.

There is, however, a concern about *selection bias*: perhaps peacekeepers chose the easy cases. Maybe the really bad civil wars were so dangerous that peacekeepers didn't go to those places, which would explain the difference in outcomes.

To put this in more general terms: in this study, the "treatment"—peacekeeping—was not randomly assigned. In statistics jargon, Fortna had an *observational study* rather than an *experiment*, and in an observational study we must do our best to adjust for pre-treatment differences between the treatment and control groups.

Fortna adjusted for how bad off the country was before the peacekeeping-or-no-peacekeeping decision was made, using some objective measures of conditions within the country. The analysis was further complicated because in some countries we know the time until return to civil war, whereas in other countries all we can say is that civil war had not yet returned during the period of data collection. In statistics, this sort of incomplete data process is called "censoring," which does not mean that someone has refused to provide the data but rather that, due to the process of data collection, certain ranges of data cannot be observed: in this case, the length of time until resumption of civil war is inherently unknowable for the countries that remained at peace through the date at which data collection had concluded. Fortna addressed this using a "survival model," a complexity that we will ignore here. For our purposes here we summarize the combination of pre-treatment predictors as a

[4]Data and code for this example are in the folder `Peacekeeping`.

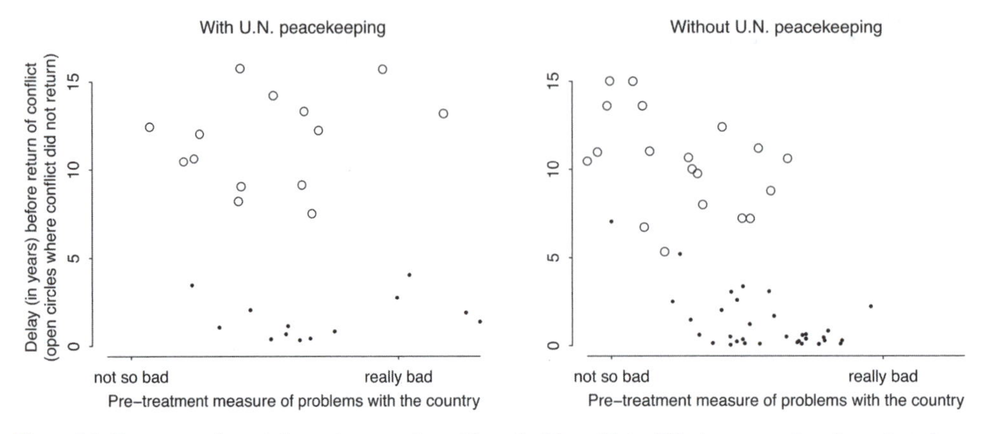


Figure 1.4 *Outcomes after civil war in countries with and without United Nations peacekeeping, plotted vs. a measure of how bad the situation was in the country. After adjusting for this pre-treatment variable, peacekeeping remains associated with longer periods without war.*

scalar "badness score," which ranges from 1.9 for the Yemen civil war in 1994 and 2.0 for India's Sikh rebellion in 1993, to the cases with the highest badness scores, 6.9 for Angola in 1991 and 6.5 for Liberia in 1990.

Figure 1.4 shows outcomes for treated and control countries as a function of badness score, with some missing cases where not all the variables were available to make that assessment. According to these data, peacekeeping was actually performed in tougher conditions, on average. As a result, adjusting for badness in the analysis (while recognizing that this adjustment is only as good as the data and model used to perform it) *increases* the estimated beneficial effects of peacekeeping, at least during the period of this study.

### Estimating the effect of gun laws, and the difficulty of inference using regression with a large number of predictors

Example: gun-control policies

A leading medical journal published an article purporting to estimate the effects of a set of gun-control policies:

> Of 25 firearm laws, nine were associated with reduced firearm mortality, nine were associated with increased firearm mortality, and seven had an inconclusive association.... Projected federal-level implementation of universal background checks for firearm purchase could reduce national firearm mortality from 10.35 to 4.46 deaths per 100 000 people, background checks for ammunition purchase could reduce it to 1.99 per 100 000, and firearm identification to 1.81 per 100 000.

This study attempted causal inference using regression on the treatment variables, adjusting for background variables to account for differences between treatment and control groups. The model was also used to make forecasts conditional on different values of the predictors corresponding to various hypothetical policy implementations.

But we believe these results are essentially useless, for two reasons: First, in this sort of regression with 50 data points and 30 predictors and no prior information to guide the inference, the coefficient estimates will be hopelessly noisy and compromised by dependence among the predictors. Second, the treatments were observational, not externally applied. To put it another way, there are systematic differences between states that have implemented different gun-control policies, differences which will not be captured in the model's other predictors (state-level covariates or background variables), and there is no reason to think that the big differences in gun-related deaths between states are mostly attributable to these particular policies.

### Comparing the peacekeeping and gun-control studies

Why do we feel satisfied with the conclusions drawn from the peacekeeping study but not with the gun-control study? In both cases, policy conclusions have been drawn from observational data, using regression modeling to adjust for differences between treatment and control groups. So what distinguishes these two projects?

One difference is that the peacekeeping study is focused, whereas the gun-control study is diffuse. It is more practical to perform adjustments when there is a single goal. In particular, in the peacekeeping study there was a particular concern that the United Nations might be more likely to step in when the situation on the ground was not so bad. The data analysis found the opposite, that peacekeeping appeared to be performed in slightly worse settings, on average. This conclusion is not airtight—in particular, the measure of badness is constructed based on particular measured variables and so it is possible that there are important unmeasured characteristics that would cause the adjustment to go the other way. Still, the pattern we see based on observed variables makes the larger story more convincing.

In contrast, it is hard to make much sense of the gun-control regression, for two reasons. First, the model adjusts for many potential causal variables at once: the effect of each law is estimated conditional on all the others being held constant, which is not realistic given that multiple laws can be changed at once, and there is no particular reason for their effects to add up in a simple manner. Second, the comparisons are between states, but states vary in many systematic ways, and it is not at all clear that a simple model can hope to adjust for the relevant differences. Yes, the comparisons in the peacekeeping project vary between countries, but the constructed badness measure seems more clearly relevant for the question being asked in that study.

We don't want to make too much of the differences between these studies, which ultimately are of degree and not of kind. Policies need to be evaluated in peacekeeping, gun control, and other areas, and it makes sense to use data and statistical analysis to aid in decision making. We see the peacekeeping study, for all its potential flaws, as a good example in that it starts with a direct comparison of data and then addresses a potential threat to validity in a focused way. In contrast, in the gun-control study the adjustments for pre-treatment variables seem less convincing, indeed fatally dependent on implausible model assumptions, which can happen when data are modeled in an unstructured way.

Indeed, statistical methods are part of the problem in that the gun-control claims would never have been publishable without the false sense of confidence supplied by regression analysis and statistical statements of uncertainty. Regression analysis was taken naively to be able to control for variation and give valid causal inference from observational data; and statistical significance and confidence intervals were taken naively to be able to screen out noise and deliver replicable statements about the world outside the data at hand. Put these together, and the result was that a respected medical journal was induced to publish strong and poorly supported conclusions taken from a messy set of aggregate trend data.

## 1.4 Challenges in building, understanding, and interpreting regressions

We can distinguish two different ways in which regression is used for causal inference: estimating a relationship and adjusting for background variables.

### Regression to estimate a relationship of interest

Start with the simplest scenario of comparability of treatment and control groups. This condition can be approximated by *randomization*, a design in which people—or, more generally, experimental units—are randomly assigned to treatment or control groups, or through some more complicated

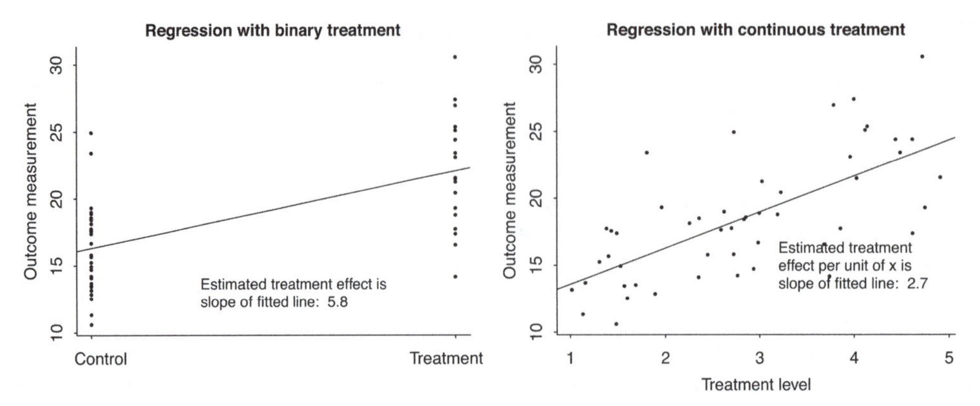


Figure 1.5 *Regression to estimate a causal effect with (a) simple comparison of treatment and control, or (b) a range of treatment levels.*

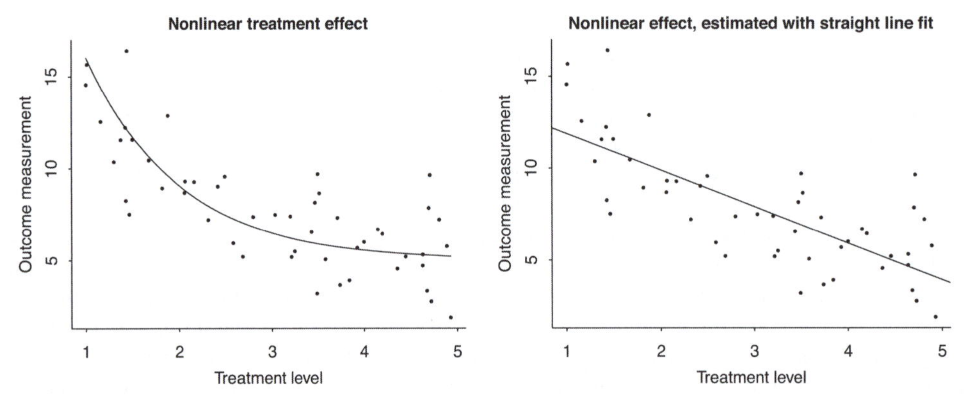


Figure 1.6 *(a) Hypothetical data in which the causal effect is a nonlinear function of the treatment level; (b) same data with a linear effect estimated. It is always possible to estimate a linear model, even if it does not fit the data.*

design that assures balance between the groups. In Part 5 of this book we discuss in detail the connections between treatment assignments, balance, and statistical analysis. For now, we just note that there are various ways to attain approximate comparability of treatment and control groups, and to adjust for known or modeled differences between the groups.

If we are interested in the effect of some treatment $x$ on an outcome $y$, and our data come from a randomized or otherwise balanced experiment, we can fit a regression—that is, a model that predicts $y$ from $x$, allowing for uncertainty.

Example: Hypothetical linear and nonlinear models

If $x$ is binary ($x = 0$ for control or $x = 1$ for treatment), then the regression is particularly simple; see Figure 1.5a. But the same idea holds for a continuous predictor, as shown in Figure 1.5b.[5]

In this setting, we are assuming comparability of the groups assigned to different treatments, so that a regression analysis predicting the outcome given the treatment gives us a direct estimate of the causal effect. Again, we defer to Part 5 a discussion of what assumptions, both mathematical and practical, are required for this simple model to make sense for causal inference.

But setting those qualms aside, we can continue by elaborating the model in various ways to better fit the data and make more accurate predictions. One direction is to consider nonlinear modeling of a continuous treatment effect. Figure 1.5b shows a linear estimate, Figure 1.6a shows an example of an underlying nonlinear effect, and Figure 1.6b shows what happens if this curve is fit by a straight line.

Another important direction is to model *interactions*—treatment effects that vary as a function of

[5]Code for this example is in the folder `SimpleCausal`.

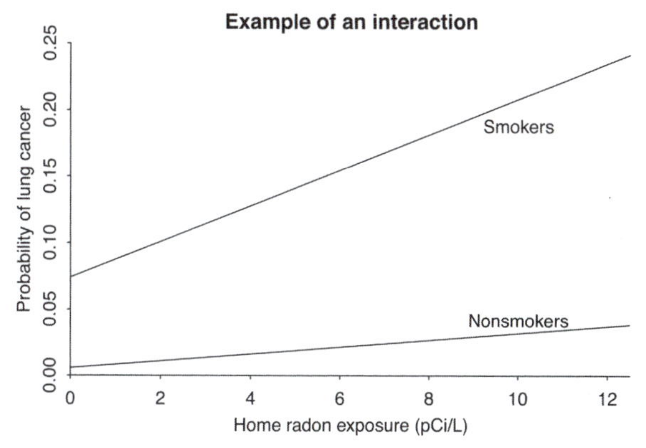


Figure 1.7 *Lifetime added risk of lung cancer for men, as a function of average radon exposure in picocuries per liter (pCi/L). The relation between cancer rate and radon is different for smokers and nonsmokers.*

xample: adon, noking, nd lung ancer

other predictors in the model. For example, Figure 1.7 shows the estimated effects of radon gas on lung cancer rates for men. Radon causes cancer (or, to be more precise, it increases the probability of cancer), with this effect being larger among smokers than nonsmokers. In this model (which is a summary of the literature and is not the result of fitting to any single dataset), the effect of radon is assumed to be linear but with an interaction with smoking.

Interactions can be important and we discuss them throughout the book. If we care about the effect of a treatment, then we also care about how this effect varies. Such variation can be important for practical reasons—for example, in deciding how to allocate some expensive medical procedure, or who is at most risk from some environmental hazard—or for the goal of scientific understanding.

## Regression to adjust for differences between treatment and control groups

In most real-world causal inference problems, there are systematic differences between experimental units that receive treatment and control. Perhaps the treated patients were sicker, on average, than those who received the control. Or, in an educational experiment, perhaps the classrooms that received the new teaching method had more highly motivated teachers than those that stuck with the old program. In such settings it is important to *adjust* for pre-treatment differences between the groups, and we can use regression to do this.

xample: ypothetical ausal djustment

Figure 1.8 shows some hypothetical data with a fitted linear regression.[6] A key difference compared to Figures 1.5 and 1.6 is that in this case the variable on the $x$-axis is a pre-treatment predictor, *not* the treatment level.

Adjusting for background variables is particularly important when there is *imbalance* so that the treated and control groups differ on key pre-treatment predictors. Such an adjustment will depend on some model—in the example of Figure 1.8, the key assumptions are linearity and additivity—and a good analysis will follow up with a clear explanation of the consequences of any adjustments.

For example, the hypothetical analysis of Figure 1.8 could be summarized as follows:

> On average, the treated units were 4.8 points higher than the controls, $\bar{y} = 31.7$ under the treatment and $\bar{y} = 25.5$ for the controls. But the two groups differed in their pre-treatment predictor: $\bar{x} = 0.4$ for the treated units and $\bar{x} = 1.2$ for the controls. After adjusting for this difference, we obtained an estimated treatment effect of 10.0.

This estimated effect is necessarily model based, but the point of this example is that when there is imbalance between treated and controls on a key predictor, some adjustment should be done.

[6]Code for this example is in the folder `SimpleCausal`.

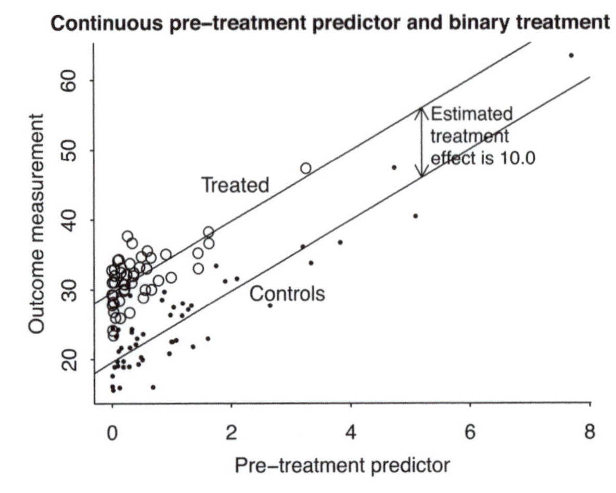


Figure 1.8 *Hypothetical data with a binary treatment and a continuous pre-treatment variable. Treated units are displayed with circles on the scatterplot, and controls are shown with dots. Overlaid is a fitted regression predicting the outcome given treatment and background variable, with the estimated treatment effect being the difference between the two lines.*

## Interpreting coefficients in a predictive model

There can be challenges in interpreting regression models, even in the simplest case of pure prediction.

Example: Earnings and height

Consider the following model fit to survey data: earnings = 11 000 + 1500 ∗ (height − 60) + error, where annual earnings are measured in dollars, height is measured in inches, and the errors are mostly in the range ±22 000 (in mathematical terms, the errors have mean 0 and standard deviation 22 000). This is a prediction model, but it is close to useless for *forecasting* because the errors from the model are so large: it is not particularly helpful to predict someone's earnings as 25 000 with uncertainty 22 000. The regression is, however, somewhat useful for *exploring an association* in that it shows that the estimated slope is positive (with an associated standard error conveying uncertainty in that slope). As *sampling inference*, the regression coefficients can be interpreted directly to the extent that the people in the survey are a representative sample of the population of interest (adult residents of the United States in 1990); otherwise, it is best to include additional predictors in the model to bridge the gap from sample to population. Interpreting the regression as a *causal inference*—each additional inch of height gives you another $1500 a year in earnings—may feel natural. But such an interpretation is questionable because tall people and short people may differ in many other ways: height is not a randomly assigned treatment. Moreover, height is a problematic variable to consider causally in other ways that will be discussed later in the book. Rather, the best fit for this example might be the *exploring associations* category. Observing a pattern in data might prompt a researcher to perform further research to study reasons that taller people earn more than shorter people.

## Building, interpreting, and checking regression models

Statistical analysis cycles through four steps:

- Model building, starting with simple linear models of the form, $y = a + bx$ + error and expanding through additional predictors, interactions, and transformations.
- Model fitting, which includes data manipulation, programming, and the use of algorithms to estimate regression coefficients and their uncertainties and to make probabilistic predictions.
- Understanding model fits, which involves graphics, more programming, and an active investigation of the (imperfect) connections between measurements, parameters, and the underlying objects of study.

- Criticism, which is not just about finding flaws and identifying questionable assumptions, but is also about considering directions for improvement of models. Or, if nothing else, limiting the claims that might be made by a naive reading of a fitted model.

The next step is return to the model-building step, possibly incorporating new data in this effort.

A challenge in serious applied work is how to be critical without being nihilistic, to accept that we can learn from statistical analysis—we can generalize from sample to population, from treatment to control, and from observed measurements to underlying constructs of interest—even while these inferences can be flawed.

A key step in criticizing research claims—and in understanding the limits to such criticisms—is to follow the steps that link the larger claims to the data and the statistical analysis. One weakness of the gun-control study discussed on page 8 is that conclusions were made regarding proposed changes in laws, but the comparisons were done across states, with no direct data on laws being implemented or removed. In contrast, the analysis of height and earnings was more clearly descriptive, not claiming or implying effects of policy changes. Another concern with the gun-control study was that the estimated effects were so large, up to fivefold reductions of the death rate. This is a sign of overinterpretation of noisy data, in this case taking existing variation among states and too eagerly attributing it to available factors. One might just as well try correlating firearm mortality with various laws on poultry processing and find similar correlations that could be given causal attributions. In contrast, the study of peacekeeping is more controlled—looking at one intervention rather than trying to consider 25 possibilities at once—and is more open about variation. The point of Figure 1.4 is not to claim that peacekeeping has some particular effect but rather to reveal that it was associated with a delay in return to civil war, in comparison to comparable situations in countries that did not have United Nations intervention.

No study is perfect. In the Xbox analysis, we used a non-representative sample to draw inference about the general population of voters. The Electric Company study was a controlled experiment, so that we have little worry about differences between treatment and control group, but one can be concerned about generalizing from an experimental setting to make claims about the effects of a national release of the television show. The common theme is that we should *recognize* challenges in extrapolation and then work to *adjust* for them. For the Xbox survey we used regression to model opinion as a function of demographic variables such as age, sex, and education where the sample differed from the population; the Electric Company data were analyzed separately for each grade, which gives some sense of variation in the treatment effect.

## 1.5 Classical and Bayesian inference

As statisticians, we spend much of our effort fitting models to data and using those models to make predictions. These steps can be performed under various methodological and philosophical frameworks. Common to all these approaches are three concerns: (1) what *information* is being used in the estimation process, (2) what *assumptions* are being made, and (3) how estimates and predictions are *interpreted*, in a classical or Bayesian framework. We investigate these in turn.

### Information

The starting point for any regression problem is data on an outcome variable $y$ and one or more predictors $x$. When data are continuous and there is a single predictor, the data can be displayed as a scatterplot, as in Figures 1.5 and 1.6. When there is one continuous predictor and one binary predictor, the data can be displayed as a scatterplot with two different symbols, as in Figure 1.8. More generally, it is not always possible to present all the data in a single display.

In addition to the data themselves, we typically know something about how they were collected. For example, in a survey, we can look at the survey questions, and we might know something about how they were asked and where and when the interviews took place. If data are laboratory assays, we might have knowledge of the biases and variation of the measurements, and so on.

Information should also be available on what data were observed at all. In a survey, respondents may be a random sample from a well-defined population (for example, sampled by extracting random names from a list) or they could be a convenience sample, in which case we should have some idea which sorts of people were more or less likely to be reached. In an experiment, treatments might be assigned at random or not, in which case we will typically have some information on how assignment was done. For example, if doctors are choosing which therapies to assign to individual patients, we might be able to find out which therapies were considered by each doctor, and which patient characteristics were relevant in the assignment decisions.

Finally, we typically have *prior knowledge* coming from sources other than the data at hand, based on experience with previous, similar studies. We have to be careful about how to include such information. For example, the published literature tends to overestimate effect sizes, as there is a selection by which researchers are under pressure to find large and "statistically significant" results; see Section 4.5. There are settings, however, where local data are weak and it would be foolish to draw conclusions without using prior knowledge. We give an example in Section 9.4 of the association between parental characteristics and the sexes of their children.

### Assumptions

There are three sorts of assumptions that are essential to any regression model of an outcome $y$ given predictors $x$. First is the functional form of the relation between $x$ and $y$: we typically assume linearity, but this is more flexible than it might seem, as we can perform transformations of predictors or outcomes, and we can also combine predictors in linear or nonlinear ways, as discussed in Chapter 12 and elsewhere in this book. Still, the choices of transformations, as well as the choice of which variables to include in the model in the first place, correspond to assumptions about the relations between the different variables being studied.

The second set of assumptions involves where the data came from: which potential observations are seen and which are not, who is surveyed and who does not respond, who gets which experimental treatment, and so on. These assumptions might be simple and strong—assuming random sampling or random treatment assignment—or weaker, for example allowing the probability of response in a survey to be different for men and women and to vary by ethnicity and education, or allowing the probability of assignment of a medical treatment to vary by age and previous health status. The strongest assumptions such as random assignment tend to be simple and easy to understand, whereas weaker assumptions, being more general, can also be more complicated.

The third set of assumptions required in any statistical model involves the real-world relevance of the measured data: are survey responses accurate, can behavior in a lab experiment be generalized to the outside world, are today's measurements predictive of what might happen tomorrow? These questions can be studied statistically by comparing the stability of observations conducted in different ways or at different times, but in the context of regression they are typically taken for granted. The interpretation of a regression of $y$ on $x$ depends also on the relation between the measured $x$ and the underlying predictors of interest, and on the relation between the measured $y$ and the underlying outcomes of interest.

### Classical inference

The traditional approach to statistical analysis is based on summarizing the information in the data, not using prior information, but getting estimates and predictions that have well-understood statistical properties, low bias and low variance. This attitude is sometimes called "frequentist," in that the classical statistician is interested in the long-run expectations of his or her methods—estimates should be correct on average (unbiasedness), confidence intervals should cover the true parameter value 95% of the time (coverage). An important principle of classical statistics is *conservatism*: sometimes data are weak and we can't make strong statements, but we'd like to be able to say, at least approximately, that our estimates are unbiased and our intervals have the advertised coverage. In classical statistics there should be a clear and unambiguous ("objective") path from data to inferences, which in turn should be checkable, at least in theory, based on their frequency properties.

Classical statistics has a lot to offer, and there's an appeal to summarizing the information from the data alone. The weaknesses of the classical approach arise when studies are small and data are indirect or highly variable. We illustrate with an example.

Example: Jamaica childhood intervention

In 2013, a study was released by a team of economists, reporting "large effects on the earnings of participants from a randomized intervention that gave psychosocial stimulation to stunted Jamaican toddlers living in poverty. The intervention consisted of one-hour weekly visits from community Jamaican health workers over a 2-year period . . . We re-interviewed the study participants 20 years after the intervention." The researchers estimated the intervention to have increased earnings by 42%, with a 95% confidence interval for the treatment effect which we reconstruct as [+2%, +98%]. That is, the estimate based on the data alone is that the treatment multiplies average earnings by a factor of 1.42, with a 95% interval of [1.02, 1.98] for this multiplicative factor; see Exercise 3.8.

The uncertainty here is wide, which is unavoidable given that the estimate is based on comparing earnings of only 127 children, who when they grow up have earnings that are highly variable. From the standpoint of classical inference, there's nothing wrong with that wide interval—if this same statistical procedure were applied over and over, to many different problems, the resulting 95% confidence intervals would contain the true parameter values 95% of the time (setting aside any imperfections in the data and experimental protocols). However, we know realistically that these intervals are more likely to be reported when they exclude zero, and therefore we would *not* expect them to have 95% coverage in the real world; see Exercises 5.8 and 5.9. And, perhaps more to the point, certain values in this interval are much more plausible than others: the treatment might well have an effect of 2% or even 0%, but it is highly unlikely for it to have an benefit of 98% and actually double people's earnings. We say this from prior knowledge, or general understanding. Indeed, we do not trust the estimate of 42%: if the study were to be replicated and we were offered a bet on whether the result would be greater or less than 42%, we would confidently bet on the "less than" side. This is not to say that the study is useless, just that not much can be learned about the effects of early childhood intervention from these data alone.

### Bayesian inference

Bayesian inference is an approach to statistics which incorporates prior information into inferences, going beyond the goal of merely summarizing existing data. In the early childhood intervention example, for instance, one might start with the assumption that the treatment could make a difference but that the average effect would most likely be less than 10% in a positive or negative direction. We can use this information as a *prior* distribution that the multiplicative treatment effect is likely to be in the range [0.9, 1.1]; combining this with the data and using the rules of Bayesian inference, we get a 95% posterior interval of [0.92, 1.28], which ranges from an 8% negative effect of the intervention to a possible 28% positive effect; see Exercise 9.6 for details. Based on this Bayesian analysis, our best guess of the observed difference in a future replication study is much lower than 42%.

This simple example illustrates both the strength and the weaknesses of Bayesian inference. On the plus side, the analysis gives more reasonable results and can be used to make direct predictions about future outcomes and about the results of future experiments. On the minus side, an additional piece of information is required—the "prior distribution," which in this case represents the perhaps contentious claim that the effect of the treatment on earnings is probably less than 10%. For better or worse, we can't have one without the other: in Bayesian inference, the prior distribution represents the arena over which any predictions will be evaluated. In a world in which the treatment could plausibly double average earnings, the raw estimate of 1.42 and interval of [1.02, 1.98] yield reasonable predictions. But in a world in which such huge effects are implausible, we must adjust our expectations and predictions accordingly.

So, in that sense, we have a choice: classical inference, leading to pure summaries of data which can have limited value as predictions; or Bayesian inference, which in theory can yield valid predictions even with weak data, but relies on additional assumptions. There is no universally correct answer here; we should just be aware of our options.

There is also a practical advantage of the Bayesian approach, which is that all its inferences are probabilistic and thus can be represented by random simulations. For this reason, whenever we want to summarize uncertainty in estimation beyond simple confidence intervals, and whenever we want to use regression models for predictions, we go Bayesian. As we discuss in Chapter 9, we can perform Bayesian inference using noninformative or weakly informative priors and obtain results similar to classical estimates, along with simulation draws that can be used to express predictive uncertainty, or we can use informative priors if so desired.

To the extent that we have relevant information that is *not* in our model (for example, awareness of bias, selection on unmeasured characteristics, prior information on effect sizes, etc), then we have a duty to account for this as well as we can when interpreting our data summaries.

## 1.6 Computing least squares and Bayesian regression

We write R code to make graphs and compute data summaries, fit statistical models, and simulate fake data from theoretical or fitted models. We introduce code in the book as needed, with background on R in Appendix A.

In general we recommend using Bayesian inference for regression: if prior information is available, you can use it, and, if not, Bayesian regression with weakly informative default priors still has the advantage of yielding stable estimates and producing simulations that enable you to express inferential and predictive uncertainty (that is, estimates with uncertainties and probabilistic predictions or forecasts). For example, in the election model presented in Section 1.2, to which we return in Chapter 7, simulations from the fitted Bayesian model capture uncertainty in the estimated regression coefficients and allow us to compute probabilistic predictions for future elections conditional on assumptions about the election-year economy.

You can fit Bayesian regression in R using commands of the form,

```
fit <- stan_glm(y ~ x, data=mydata)
```

But some users of statistics will be unfamiliar or uncomfortable with Bayesian inference. If you are one of these people, or if you need to communicate with people who are more comfortable with classical statistics, you can fit least squares regression:

```
fit <- lm(y ~ x, data=mydata)
```

Finally, another concern about `stan_glm` is that it can go slowly for large problems. We can make it faster by running it in optimizing mode:

```
fit <- stan_glm(y ~ x, data=mydata, algorithm="optimizing")
```

For the examples in this book, datasets are small and speed is not really a concern, but it is good to be aware of this option in larger applications. When run in optimizing mode, `stan_glm` performs an approximate fit, but it still produces simulations that can again be used to summarize inferential and predictive uncertainty.

In summary, if you would prefer to avoid Bayesian inference, you can replace most of the instances of `stan_glm` in this book with `lm` for linear regression or `glm` for logistic and generalized linear models and get nearly identical results. Differences show up in an example in Section 9.5 with a strong prior distribution, examples of logistic and ordered logistic regressions with complete separation in Sections 14.6 and 15.5, our implementation of cross validation in Section 11.8, and various examples throughout the book where we use simulations to express uncertainty in estimates or predictions. It is also possible to fit least squares regression and get Bayesian uncertainties by running `stan_glm` with flat prior distributions, as we discuss in Section 8.4.

Bayesian and simulation approaches become more important when fitting regularized regression and multilevel models. These topics are beyond the scope of the present book, but once you are comfortable using simulations to handle uncertainty, you will be well situated to learn and work with those more advanced models.

## 1.7 Bibliographic note

We return to the election forecasting example in Chapter 7. We discuss the height/earnings regression, the Xbox survey, and the Electric Company experiment in detail in Chapters 6, 17, and 19, respectively. Further references on these topics appear in the bibliographic notes to those chapters.

Fortna (2008) discusses the United Nations peacekeeping study and its implications. The gun-control study appears in Kalesan et al. (2016); see also Gelman (2016a, b); the last of these references links to a reply by the authors to criticisms of that controversial paper.

For more on radon and cancer risk, see Lin et al. (1999). The Jamaica childhood intervention experiment comes from Gertler et al. (2013) and is discussed further by Gelman (2013, 2018).

## 1.8 Exercises

Data for examples and assignments in this and other chapters are at `www.stat.columbia.edu/~gelman/regression/`. See Appendix A for an introduction to R, the software you will use for computing.

Example: Helicopter design

1.1 *From design to decision*: Figure 1.9 displays the prototype for a paper "helicopter." The goal of this assignment is to design a helicopter that takes as long as possible to reach the floor when dropped from a fixed height, for example 8 feet. The helicopters are restricted to have the general form shown in the sketch. No additional folds, creases, or perforations are allowed. The wing length and the wing width of the helicopter are the only two design parameters, that is, the only two aspects of the helicopter that can be changed. The body width and length must remain the same for all helicopters. A metal paper clip is attached to the bottom of the helicopter.
Here are some comments from previous students who were given this assignment:

> Rich creased the wings too much and the helicopters dropped like a rock, turned upside down, turned sideways, etc.
>
> Helis seem to react very positively to added length. Too much width seems to make the helis unstable. They flip-flop during flight.
>
> Andy proposes to use an index card to make a template for folding the base into thirds.
>
> After practicing, we decided to switch jobs. It worked better with Yee timing and John dropping. 3 – 2 – 1 – GO.

Your instructor will hand out 25 half-sheets of paper and 2 paper clips to each group of students. The body width will be one-third of the width of the sheets, so the wing width can be anywhere from $\frac{1}{6}$ to $\frac{1}{2}$ of the body width; see Figure 1.9a. The body length will be specified by the instructor. For example, if the sheets are U.S.-sized ($8.5 \times 5.5$ inches) and the body length is set to 3 inches, then the wing width could be anywhere from 0.91 to 2.75 inches and the wing length could be anywhere from 0 to 5.5 inches.
In this assignment you can experiment using your 25 half-sheets and 10 paper clips. You can make each half-sheet into only one helicopter. But you are allowed to design sequentially, setting the wing width and body length for each helicopter given the data you have already recorded. Take a few measurements using each helicopter, each time dropping it from the required height and timing how long it takes to land.

(a) Record the wing width and body length for each of your 25 helicopters along with your time measurements, all in a file in which each observation is in its own row, following the pattern of `helicopters.txt` in the folder `Helicopters`, also shown in Figure 1.9b.

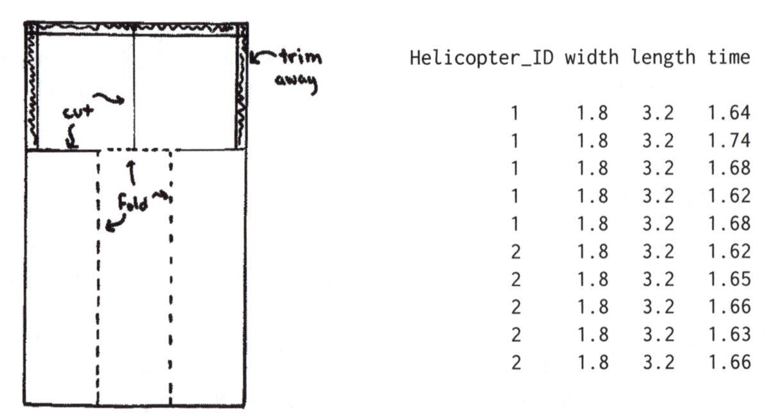


| Helicopter_ID | width | length | time |
|---|---|---|---|
| 1 | 1.8 | 3.2 | 1.64 |
| 1 | 1.8 | 3.2 | 1.74 |
| 1 | 1.8 | 3.2 | 1.68 |
| 1 | 1.8 | 3.2 | 1.62 |
| 1 | 1.8 | 3.2 | 1.68 |
| 2 | 1.8 | 3.2 | 1.62 |
| 2 | 1.8 | 3.2 | 1.65 |
| 2 | 1.8 | 3.2 | 1.66 |
| 2 | 1.8 | 3.2 | 1.63 |
| 2 | 1.8 | 3.2 | 1.66 |

Figure 1.9 *(a) Diagram for making a "helicopter" from half a sheet of paper and a paper clip. The long segments on the left and right are folded toward the middle, and the resulting long 3-ply strip is held together by a paper clip. One of the two segments at the top is folded forward and the other backward. The helicopter spins in the air when dropped. (b) Data file showing flight times, in seconds, for 5 flights each of two identical helicopters with wing width 1.8 inches and wing length 3.2 inches dropped from a height of approximately 8 feet. From Gelman and Nolan (2017).*

(b) Graph your data in a way that seems reasonable to you.

(c) Given your results, propose a design (wing width and length) that you think will maximize the helicopter's expected time aloft. It is not necessary for you to fit a formal regression model here, but you should think about the general concerns of regression.

The above description is adapted from Gelman and Nolan (2017, section 20.4). See Box, Hunter, and Hunter (2005) for a more advanced statistical treatment of this sort of problem.

1.2 *Sketching a regression model and data*: Figure 1.1b shows data corresponding to the fitted line $y = 46.3 + 3.0x$ with residual standard deviation 3.9, and values of $x$ ranging roughly from 0 to 4%.

(a) Sketch hypothetical data with the same range of $x$ but corresponding to the line $y = 30 + 10x$ with residual standard deviation 3.9.

(b) Sketch hypothetical data with the same range of $x$ but corresponding to the line $y = 30 + 10x$ with residual standard deviation 10.

1.3 *Goals of regression*: Download some data on a topic of interest to you. Without graphing the data or performing any statistical analysis, discuss how you might use these data to do the following things:

(a) Fit a regression to estimate a relationship of interest.

(b) Use regression to adjust for differences between treatment and control groups.

(c) Use a regression to make predictions.

1.4 *Problems of statistics*: Give examples of applied statistics problems of interest to you in which there are challenges in:

(a) Generalizing from sample to population.

(b) Generalizing from treatment to control group.

(c) Generalizing from observed measurements to the underlying constructs of interest.

Explain your answers.

1.5 *Goals of regression*: Give examples of applied statistics problems of interest to you in which the goals are:

(a) Forecasting/classification.

(b) Exploring associations.

(c) Extrapolation.

(d) Causal inference.

Explain your answers.

1.6 *Causal inference*: Find a real-world example of interest with a treatment group, control group, a pre-treatment predictor, and a post-treatment predictor. Make a graph like Figure 1.8 using the data from this example.

1.7 *Statistics as generalization*: Find a published paper on a topic of interest where you feel there has been insufficient attention to:

(a) Generalizing from sample to population.

(b) Generalizing from treatment to control group.

(c) Generalizing from observed measurements to the underlying constructs of interest.

Explain your answers.

1.8 *Statistics as generalization*: Find a published paper on a topic of interest where you feel the following issues *have* been addressed well:

(a) Generalizing from sample to population.

(b) Generalizing from treatment to control group.

(c) Generalizing from observed measurements to the underlying constructs of interest.

Explain your answers.

1.9 *A problem with linear models*: Consider the helicopter design experiment in Exercise 1.1. Suppose you were to construct 25 helicopters, measure their falling times, fit a linear model predicting that outcome given wing width and body length:

$$\text{time} = \beta_0 + \beta_1 * \text{width} + \beta_2 * \text{length} + \text{error},$$

and then use the fitted model time $= \beta_0 + \beta_1 * \text{width} + \beta_2 * \text{length}$ to estimate the values of wing width and body length that will maximize expected time aloft.

(a) Why will this approach fail?

(b) Suggest a better model to fit that would not have this problem.

1.10 *Working through your own example*: Download or collect some data on a topic of interest of to you. You can use this example to work though the concepts and methods covered in the book, so the example should be worth your time and should have some complexity. This assignment continues throughout the book as the final exercise of each chapter. For this first exercise, discuss your applied goals in studying this example and how the data can address these goals.

# Chapter 2

# Data and measurement

In this book, we'll be fitting lines (and some curves) to data, making comparisons and predictions and assessing our uncertainties in the resulting inferences. We'll discuss the assumptions underlying regression models, methods for checking these assumptions, and directions for improving fitted models. We'll discuss the challenges of extrapolating from available data to make causal inferences and predictions for new data, and we'll use computer simulations to summarize the uncertainties in our estimates and predictions.

Before fitting a model, though, it is a good idea to understand where your numbers are coming from. The present chapter demonstrates through examples how to use graphical tools to explore and understand data and measurements.

## 2.1 Examining where data come from

Example: Human Development Index

Figure 2.1 went viral on the web a few years ago. The map compares the 50 states and Washington, D.C., in something called the Human Development Index (HDI), which had previously been used to compare different countries in public health measures. The coding of the map is kind of goofy: the states with the three lowest values are Louisiana at 0.801, West Virginia at 0.800, and Mississippi at 0.799, but their shading scheme makes Mississippi stand out.

But we have bigger concerns than that. Is Alaska really so developed as all that? And what's up with Washington, D.C., which, according to the report, is ranked at #4, behind only Connecticut, Massachusetts, and New Jersey?

Time to look behind the numbers. From the published report, the HDI combines three basic dimensions:

- Life expectancy at birth, as an index of population health and longevity.
- Knowledge and education, as measured by the adult literacy rate (with two-thirds weighting) and the combined primary, secondary, and tertiary gross enrollment ratio (with one-third weighting).
- Standard of living, as measured by the natural logarithm of gross domestic product (GDP) per capita at purchasing power parity (PPP) in U.S. dollars.

Now we can see what's going on. There is not much variation by state in life expectancy, literacy, or school enrollment. Sure, Hawaiians live a few years longer than Mississippians, and there are some differences in who stays in school, but by far the biggest differences between states, from these measures, are in GDP. The average income in Connecticut is twice that of Mississippi. And Washington, D.C., ranks high because its residents have a high average income.

To check out the relation between HDI and income, we loaded in the tabulated HDI numbers and plotted them versus some historical data on average income by state.[1] Figure 2.2a shows the result. The pattern is strong but nonlinear. Figure 2.2b plots the ranks and reveals a clear pattern, with most of the states falling right on the 45-degree line and a high correlation between the two rankings. We were surprised the correlation isn't higher—and surprised the first scatterplot above is

[1] Data and code for this example are in the folder `HDI`.

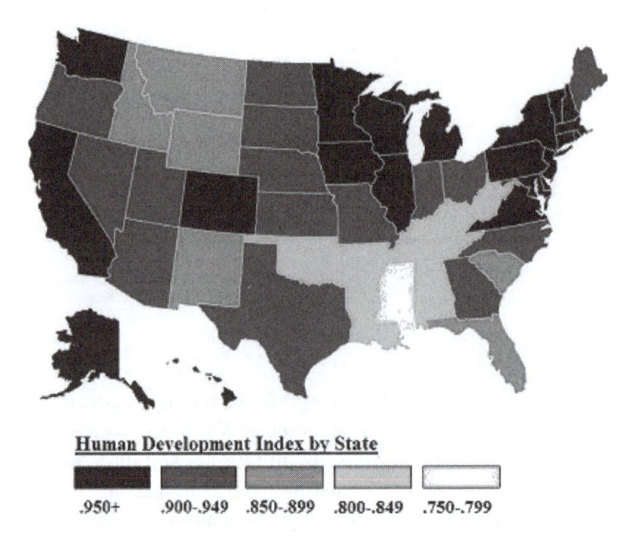


Figure 2.1 *Map that appeared on the internet of the so-called "Human Development Index," ranking the 50 states and Washington, D.C., from PlatypeanArchcow (2009).*

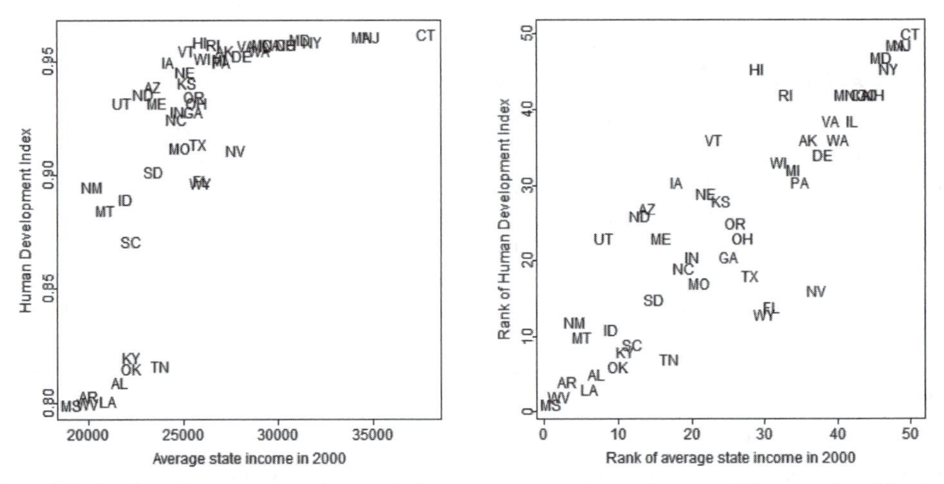


Figure 2.2 *Graphing the Human Development Index versus average income by state: (a) scatterplot of the data, (b) scatterplot of ranks.*

so nonlinear—but, then again, we're using state income rather than GDP, so maybe there's something going on with that. No, the logarithmic transformation is not what's doing this, at least not if you're logging income as is stated in the report. Logging stretches out the lower end of the scale a bit but does not change the overall pattern of the plot. The income values don't have enough dynamic range for the log transformation to have much effect.

Or maybe more is going on than we realize with those other components. If anyone is interested in following up on this, we suggest looking into South Carolina and Kentucky, which are so close in average income and so far apart on the HDI; see Figure 2.2a.

In any case, the map in Figure 2.1 is pretty much a map of state income with a mysterious transformation and a catchy name. The relevance of this example is that we were better able to understand the data by plotting them in different ways.

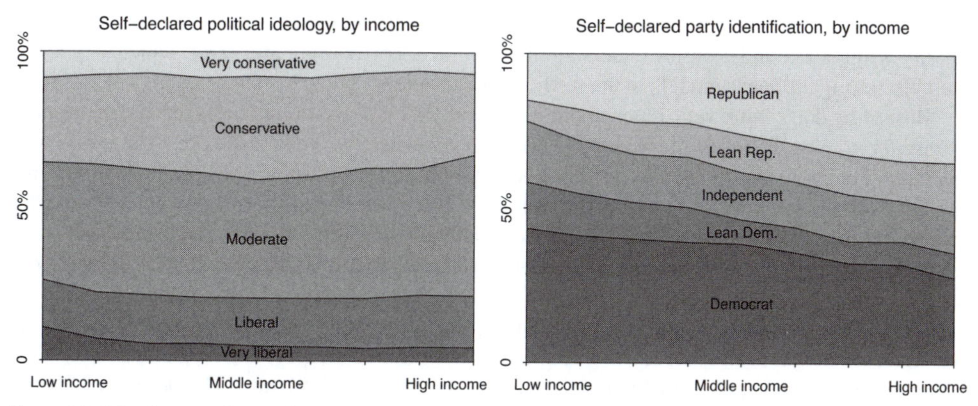


Figure 2.3 *Distribution of (a) political ideology and (b) party identification, by income, from a survey conducted during the 2008 U.S. election campaign.*

### Details of measurement can be important

Example: Political ideology and party identification

In American politics, there are two major parties, and most voters fall in an ideological spectrum ranging from left, or liberal, to right, or conservative. The split between Democrats and Republicans roughly aligns with the division between liberal and conservative.

But these two scales of partisanship and ideology are not identical.[2] Figure 2.3a shows that the proportion of political liberals, moderates, and conservatives is about the same for all income levels. In contrast, Figure 2.3b shows a strong relation between income and Republican partisanship, at least as of 2008, when these survey data were gathered. Party identification and political ideology were each measured on a five-point scale running from left to right, but, as the graphs show, there are clear differences between the two variables.

How does Figure 2.3 relate to the general themes of our book? Regression is a way to summarize and draw inferences from data. As such, conclusions from regressions will depend on the quality of the data being analyzed and the relevance of these data to questions of interest. The partisanship and ideology example is a reminder that even very similar measures can answer quite different questions.

Unfortunately, gaps between measurement and reality are a general problem in scientific research and communication. For example, Temple University's medical school issued a press release entitled "Extra-virgin olive oil preserves memory & protects brain against Alzheimer's"—but the actual study was performed on mice and had no direct connection with dementia or Alzheimer's disease. The claim thus lacks external validity (see Section 2.2). This sort of leap happens all the time. In some sense it is necessary—lab experimentation precedes clinical trials—but we should be open and aware of what we know.

## 2.2 Validity and reliability

We discuss the important issue of measurement for two reasons. First, we need to understand what our data actually mean. We have looked at ways to visualize data and extract information. But if we do not know what the data actually represent, then we cannot extract the right information.

Data analysis reaches a dead end if we have poor data. There are some measurement problems that no amount of fixing and adjusting can solve. In Section 1.3 we discussed how we made adjustments to the Xbox polling data to account for differences between sample and population. But if we had asked our respondents the wrong question, or had nor recorded key background variables that could be used for the adjustment, then there would have been no easy fix.

[2]Data and code for this example are in the folder Pew.

The second reason for discussing measurement is that learning about accuracy, reliability, and validity will set a foundation for understanding variance, correlation, and error, which will all be useful in setting up linear models in the forthcoming chapters.

Most of us don't think very much about measurement on a day-to-day basis, primarily because we take for granted the measures we work with, and even where we know there are some issues with precision, the precision we have is usually good enough for our purposes. So we have no trouble talking about the temperature outside, the weight of groceries, the speed of a car, etc. We take for granted the correspondence between the numbers and the "thing" that we are measuring. And we're usually not worried about the precision—we don't need temperature to the nearest half degree, or our car speed to six decimal places.

This is all dependent on what we are measuring and what our proposed inferences are. A scale that measures weight to an accuracy of 1 kilogram is fine for most purposes of weighing people, great for weighing elephants, and terrible for weighing medicine at a pharmacy. The property of being precise enough is a combination of the properties of the scale and what we are trying to use it for.

In social science, the way to measure what we are trying to measure is not as transparent as it is in everyday life. Sometimes this is because what we want to measure is "real" and well defined, but difficult to actually count. Examples include counting the number of immigrants, or measuring daily intake of food in uncontrolled conditions.

Other times, the thing we are trying to measure is pretty straightforward, but a little bit fuzzy, and the ways to tally it up aren't obvious, for example, counting the number of people in your neighborhood you know or trust, or counting the number of vocabulary words you know.

Sometimes we are trying to measure something that we all agree has meaning, but which is subjective for every person and does not correspond to a "thing" we can count or measure with a ruler. Examples include attitudes, beliefs, intentions to vote, and customer satisfaction. In all these cases, we share an understanding of what we are talking about; it is deeply embedded in our language and understanding that people have opinions about things and feelings. But attitudes are private; you can't just weigh them or measure their widths. And that also means that to probe them you have to invent some kind of measure such as, "Tell us on a scale of 0 to 100 how much you enjoyed the service you got today?" The relative answer matters, but we could have asked on a scale of 1 to 3, or for that matter 300 too 500. We just hope that people can be sincere when they answer and that they use the scale the same way. These concerns arise if you are designing your own study or when analyzing data collected by others.

It can be helpful to take multiple measurements on an underlying construct of interest. For example, in a class evaluation survey, students are typically asked several questions about the quality of an instructor and a course. And various health conditions are measured using standard batteries of questions. For example, the Beck Depression Inventory includes 21 items, each of which is given a score from 0 to 3, and then these are added to get a total from 0 to 63.

A measure can be useful for some purposes but not others. For example, in public health studies, a "never smoker" is typically defined as someone who has smoked fewer than 100 cigarettes in his or her lifetime, which generally seems like a reasonable definition when studying adult behavior and health. But in a study of adolescents, it would be mistaken to put a youth who has smoked 90 cigarettes in the same "never smoker" category as a youth who has smoked zero or one or two cigarettes.

### Validity

A measure is *valid* to the degree that it represents what you are trying to measure. It's easy to come up with negative examples. A written test is not a valid measure of musical ability. There is a vast gap between the evidence and what we want to make inferences about.

Similarly, asking people how satisfied they are with some government service might not be considered a valid measure of the effectiveness of that service. Valid measures are ones in which there is general agreement that the observations are closely related to the intended construct.

We can define the *validity* of a measuring process as the property of giving the right answer on average across a wide range of plausible scenarios. To study validity in an empirical way, ideally you want settings in which there is an observable true value and multiple measurements can be taken.

In social science, validity can be difficult to assess. When the truth is not available, measurements can be compared to expert opinion or another "gold standard" measurement. For instance, a set of survey questions designed to measure depression in a new population could be compared to the opinion of an experienced psychiatrist for a set of patients, and it can also be compared to a well-established depression inventory.

### Reliability

A *reliable* measure is one that is precise and stable. If we make a measurement, and then we have occasion to do it again, we would hope that the value would not move (much). Put another way, the variability in our sample is due to real differences among people or things, and not due to random error incurred during the measurement process.

For example, consider a test that is given twice to the same group of students. We could use the correlation between the scores across the two administrations of the test to help understand the extent to which the test *reliably* measures the given construct.

Another approach would be to have different raters administer the same measure in the same context. For instance, we could compare the responses on a measure of classroom quality across raters who observed the same classroom at the same time. Or we could compare judges' ratings of proficiency of gymnasts' performance of a given skill based on the same demonstration of that skill. This is referred to as inter-rater reliability.

### Sample selection

Yet another feature of data quality is *selection*, the idea that the data you see can be a nonrepresentative sample of a larger population that you will not see. For example, suppose you are interested in satisfaction with a city's public transit system, so you interview people who ride the buses and trains; maybe you even take some measurements such as travel times or percentage of time spent sitting or standing. But there is selection: you only include people who have chosen to ride the bus or train. Among those excluded are those who have chosen not to ride the bus or the train because they are unhappy with those services.

In addition to this sort of selection bias based on who is included in the dataset, there are also biases from nonresponse to particular survey items, partially observed measurements, and choices in coding and interpretation of data. We prefer to think about all these measurement issues, including validity, reliability, and selection, in the context of larger models connecting measurements to underlying relationships of interest.

## 2.3 All graphs are comparisons

As demonstrated throughout this book, we can learn a lot by looking at data with an open mind. We present three quick examples here. In the larger context of workflow, we go back and forth between data exploration, modeling, inference, and model building, and each step requires its own tools.

### Simple scatterplots

Example: Health spending and lifespan

Figure 2.4 shows some data on health spending and life expectancy, revealing that the United States spends much more per person than any other country without seeing any apparent benefit in lifespan.[3]

Here is R code to plot the data:

[3]Data and code for this example are in the folder `HealthExpenditure`.

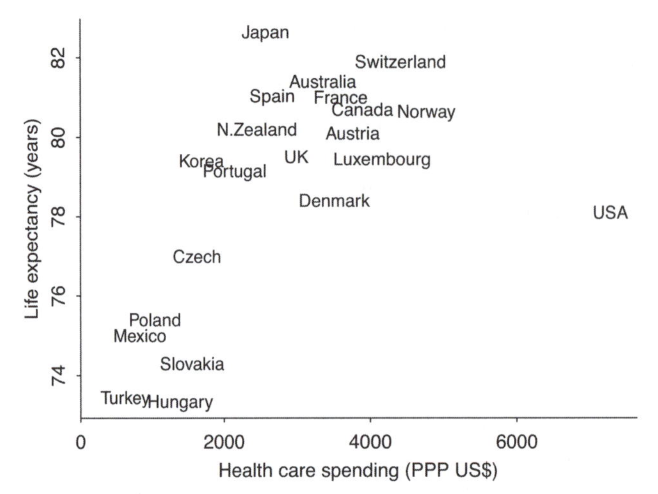


Figure 2.4 *Health care spending and life expectancy in several countries. This scatterplot shows two things: the generally positive correlation between spending and lifespan, and the extreme position of the United States.*

```
health <- read.table("healthdata.txt", header=TRUE)
country <- rownames(health)
plot(health$spending, health$lifespan, type="n")
text(health$spending, health$lifespan, country)
```

To make the graph just as displayed in Figure 2.4, further commands are required, and these are available on our website, but the code here gives the basic idea.

The graph shows the exceptional position of the United States and also shows the relation between spending and lifespan in the other countries.

## Displaying more information on a graph

You can make as many plots as you want (or as your patience allows), but it is useful to think a bit about each plot, just as it is useful to think a bit about each model you fit.

The points within a scatterplot correspond to the unit of analysis in your study. At least in theory, you can display five variables easily with a scatterplot: $x$ position, $y$ position, symbol, symbol size, and symbol color. A two-way grid of plots allows two more dimensions, bringing the total number of variables potentially displayed to seven.

Example: Redistricting and partisan bias

We demonstrate some of the virtues of a rich visual description of data and estimates with Figure 2.5, a graph from our applied research that was central to the discovery and presentation of our key finding. The scatterplot in question displays three variables, conveyed by $x$ position, $y$ position, and symbol, a comparison of treatments to control with a before and after measurement. In this case, the units are state legislative elections, and the plot displays estimated partisan bias (a measure of the extent to which the drawing of district boundaries favors one party or the other) in two successive election years. The "treatments" are different kinds of redistricting plans, and the "control" points (indicated by dots on the figure) represent pairs of elections with no intervening redistricting. We display all the data and also show the regression lines on the same scale. As a matter of fact, we did not at first think of fitting nonparallel regression lines; it was only after making the figure and displaying parallel lines that we realized that nonparallel lines (that is, an interaction between the treatment and the "before" measurement) are appropriate. The interaction is crucial to the interpretation of these data: (1) when there is no redistricting, partisan bias is not systematically changed; (2) the largest effect of any kind of redistricting is typically to bring partisan bias closer to zero. The lines and points together show this much more clearly than any numerical summary.

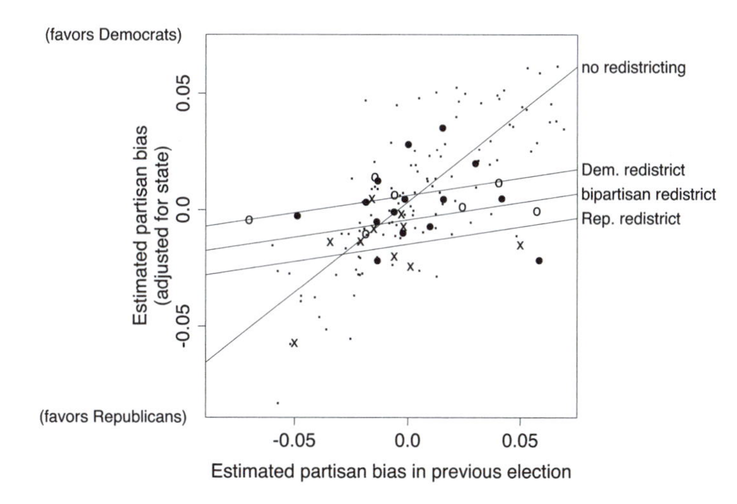


Figure 2.5 *Effect of redistricting on partisan bias in U.S. state legislative elections. Each symbol represents a state and election year, with solid circles, open circles, and crosses representing Democratic, bipartisan, and Republican redistricting, respectively. The small dots are the control cases—state-years that did not immediately follow a redistricting. Lines show fit from a regression model. Redistricting tends to make elections less biased, but small partisan biases remain based on the party controlling the redistricting.*

We sometimes have had success using descriptive symbol names such as two-letter state abbreviations. But if there are only two or three categories, we're happier with visually distinct symbols. For example, to distinguish men and women, we would not use M and W or M and F. In genealogical charts, men and women are often indicated by open squares and open circles, respectively, but even these symbols are hard to tell apart in a group. We prefer clearly distinguishable colors or symbols such as the open circles, solid circles, crosses, and dots in Figure 2.5. When a graph has multiple lines, label them directly, as in Figure 1.7.

These suggestions are based on our experience and attempts at logical reasoning; as far as we know, they have not been validated (or disproved) in any systematic study.

## Multiple plots

Example: Last letters of names

Looking at data in unexpected ways can lead to discovery. For example, Figure 2.6 displays the distribution of the last letters of boys' names in the United States in 1906. The most common names in that year included John, James, George, and Edward, for example.

We can learn by putting multiple related graphs in a single display. Figures 2.6 and 2.7 show the dramatic change in the distribution of last letters of boys' names during the twentieth century. In recent years, over a third of boys have been given names that end in "n," with the most common being Ethan, Jayden, Aiden, Mason, and Logan.

There is no single best way to display a dataset. For another view of the data just discussed, we created Figure 2.8, which shows time series of the percentage of boys' names recorded each year ending in each letter.[4] The graph has 26 lines, and we have labeled three of them. We played around with different representations but found the graphs hard to read when more than three lines were highlighted. There has been a steady increase in boys' names ending in "n" during the past 60 years.

Looking at names data another way, Figure 2.9 plots the proportion of boys' and girls' names each year that were in the top 10 names for each sex. Traditionally, boys' names were chosen from a narrower range than girls, with the top 10 names representing 30–40% of all boys, but in recent years,

[4]Data and code for this example are in the folder `Names`.

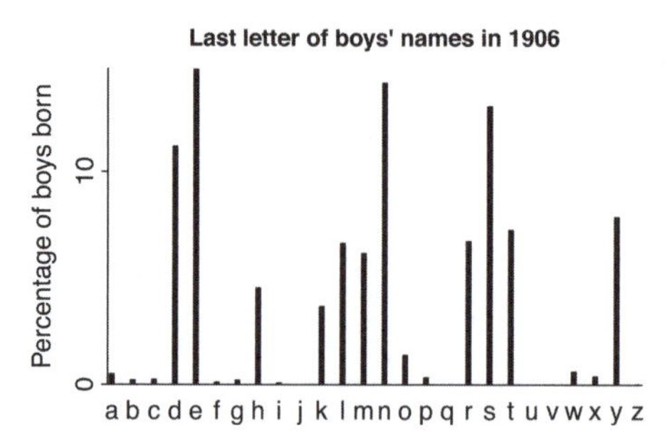


Figure 2.6 *Distribution of last letters of boys' names from a database of American babies born in 1906. Redrawn from a graph by Laura Wattenberg.*

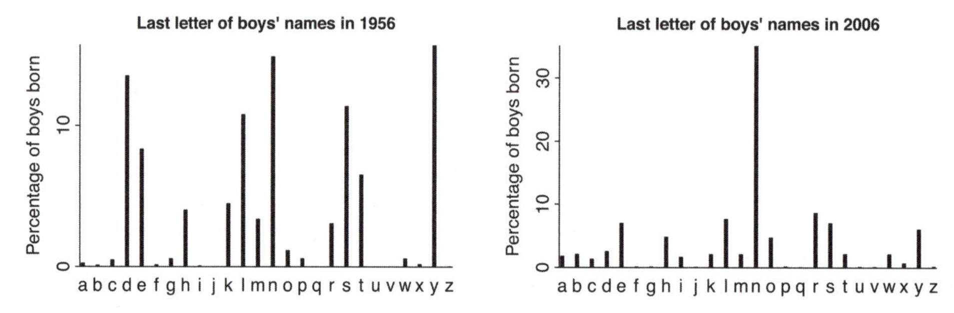


Figure 2.7 *Distribution of last letters of boys' names of American babies born in 1956 and 2006. Redrawn from graphs by Laura Wattenberg. Putting these plots together with the 1906 graph (Figure 2.6) shows a striking trend.*

name choice in the United States has become much more diverse. There are many more patterns to be found in this rich dataset.

## Grids of plots

A scatterplot displays two continuous variables, say $y$ vs. $x_1$. Coloring the dots enables us to plot a third variable, $x_2$, with some small number of discrete levels. Realistically it can be difficult to read a plot with more than two colors. We can then include two more discrete variables by constructing a two-way grid of plots representing discrete variables $x_3$ and $x_4$. This approach of *small multiples* can be more effective than trying to cram five variables onto a single plot.

Example: Swings in congressional elections

Figure 2.10 demonstrates with a grid relating to incumbency in U.S. congressional elections.[5] Each graph plots the swing toward the Democrats from one election to another, vs. the Democratic candidate's share of the vote in the first election, where each dot represents a different seat in the House of Representatives, colored gray for elections where incumbents are running for reelection, or black for open seats. Each row of the graph shows a different pair of national election years, and the four columns show data from different regions of the country.

Breaking up the data in this way allows us to see some patterns, such as increasing political polarization (going from the 1940s through the 1960s to the 1980s, we see a decreasing number of elections with vote shares near 50%), increasing volatility of elections (larger swings in the later periods than before), and a change in the South, which in the 1940s was overwhelmingly Democratic but by the 1980s had a more symmetric range of election results. It would be difficult to see all this in a single plot; in addition, the graph could be easily extended to additional rows (more years of data) or columns (smaller geographic subdivisions).

More generally, we can plot a continuous outcome $y$ vs. a continuous predictor $x_1$ and discrete

[5]Data and code for this example are in the folder `Congress`.

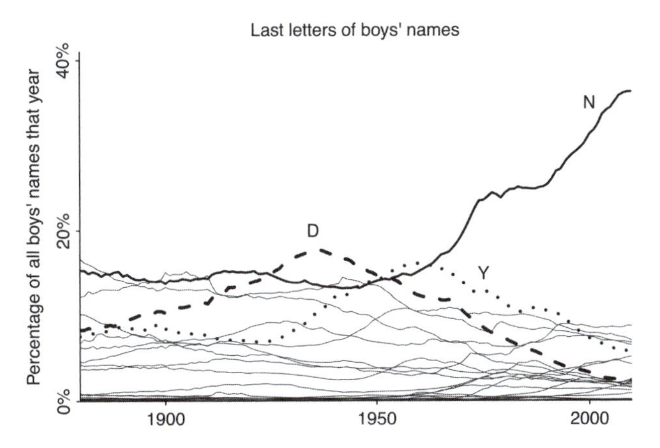


Figure 2.8 *Trends in percentage of boys' names ending in each letter. This graph has 26 lines, with the lines for N, D, and Y in bold to show the different trends in different-sounding names. Compare to Figures 2.6 and 2.7, which show snapshots of the last-letter distribution in 1906, 1956, and 2006.*

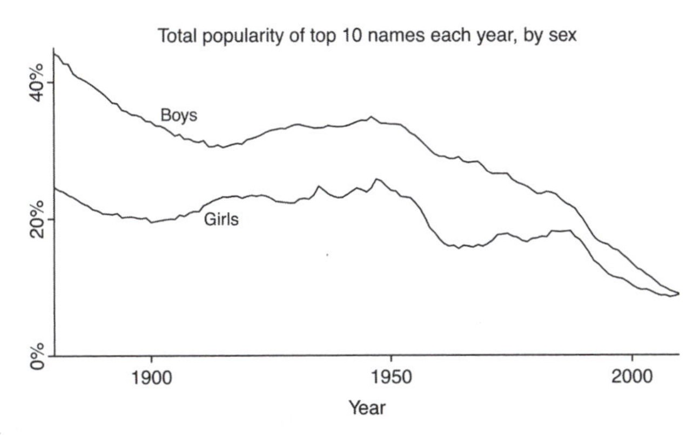


Figure 2.9 *Trends in concentration of boys' and girls' names. In the late 1800s, and then again at different periods since 1950, there have been steep declines in the percentage of babies given the most popular names, so that now the top 10 names of each sex represent only about 10% of baby names. Thus, even as the sounds of boys' names have become more uniform (as indicated by the pattern of last letters shown in Figure 2.6), the particular names chosen have become more varied.*

predictors $x_2$, $x_3$, and $x_4$. If there is interest, we can also plot fitted lines within each plot, showing the expected value of $y$ as a function of $x_1$ for different fixed values of the other three predictors.

The discrete variables can also represent continuous bins. For example, to display data from an experiment on blood-pressure medication, we could plot after vs. before measurements with different colors for treated and control students, with top and bottom rows of plots showing data from men and women, and rows corresponding to different age categories of patients. Age is a continuous variable, but it could be binned into categories for the graph.

## Applying graphical principles to numerical displays and communication more generally

When reporting data and analysis, you should always imagine yourself in the position of the reader of the report. Avoid overwhelming the reader with irrelevant material. For the simplest (but still important) example, consider the reporting of numerical results, either alone or in tables.

Do not report numbers to too many decimal places. There is no absolute standard for significant digits; rather, you should display precision in a way that respects the uncertainty and variability in the

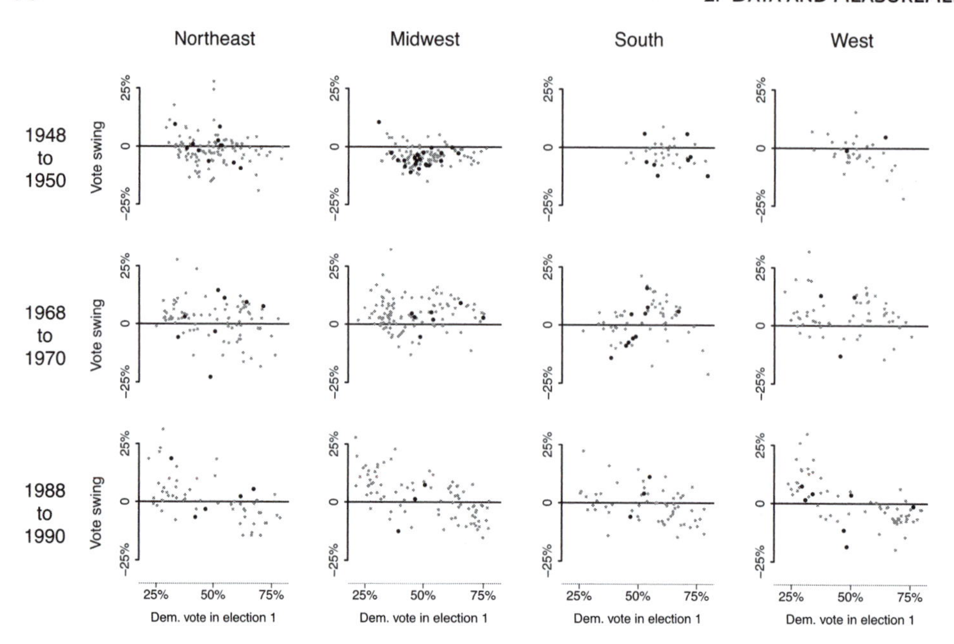


Figure 2.10 *Swings in U.S. congressional elections in three different periods. This grid of plots demonstrates how we can display an outcome (in this case, the swing toward the Democrats or Republicans between two elections in a congressional district) as a function of four predictors: previous Democratic vote share, incumbency status (gray for incumbents running for reelection, black for open seats), region of the country, and time period. Uncontested and landslide elections have been excluded.*

numbers being presented. For example, the uncertainty interval [3.276, 6.410] would be more clearly written as [3.3, 6.4]. (An exception is that it makes sense to save lots of extra digits for intermediate steps in computations, for example, 51.7643 – 51.7581.) A related issue is that you can often make a list or table of numbers more clear by first subtracting out the average (or for a table, row and column averages). The appropriate number of significant digits depends on the uncertainty. But in practice, three digits are usually enough because if more were necessary, we would subtract out the mean first.

The biggest source of too many significant digits may be default computer output. One solution is to set the rounding in the computer program (for example in R, `options(digits=2)`).

A graph can almost always be made smaller than you think and still be readable. This then leaves room for more plots on a grid, which then allows more patterns to be seen at once and compared.

Never display a graph you can't talk about. Give a full caption for every graph, as we try to do in this book. This explains, to yourself and others, what you are trying to show and what you have learned from each plot. Avoid displaying graphs that have been made simply because they are conventional.

## Graphics for understanding statistical models

We can consider three uses of graphics in statistical analysis:

1. Displays of raw data, often called "exploratory analysis." These don't have to look pretty; the goal is to see things you did not expect or even know to look for.
2. Graphs of fitted models and inferences, sometimes overlaying data plots in order to understand model fit, sometimes structuring or summarizing inference for many parameters to see a larger pattern. In addition, we can plot simulations of replicated data from fitted models and compare them to comparable plots of raw data.

3. Graphs presenting your final results—a communication tool. Often your most important audience here is yourself—in presenting all of your results clearly on the page, you'll suddenly understand the big picture.

The goal of any graph is communication to self or others. More immediately, graphs are comparisons: to zero, to other graphs, to horizontal lines, and so forth. We "read" a graph both by pulling out the expected (for example, the slope of a fitted regression line, the comparison of a series of uncertainty intervals to zero and each other) and the unexpected. In our experience, the unexpected is usually not an "outlier" or aberrant point but rather a systematic pattern in some part of the data.

Some of the most effective graphs simply show us what a fitted model is doing. See Figure 15.6 for an example.

### Graphs as comparisons

All graphical displays can be considered as comparisons. When making a graph, line things up so that the most important comparisons are clearest. Comparisons are clearest when scales are lined up. Creative thinking might be needed to display numerical data effectively, but your creativity can sometimes be enhanced by carefully considering your goals. Just as in writing, you sometimes have to rearrange your sentences to make yourself clear.

### Graphs of fitted models

It can be helpful to graph a fitted model and data on the same plot, as we do throughout the book. We also like to graph sets of estimated parameters; see, for example, in Figure 10.9. Graphs of parameter estimates can be thought of as proto-models in that the graph suggests a relation between the $y$-axis (the parameter estimates being displayed) and the $x$-axis (often time, or some other index of the different data subsets being fit by a model). These graphs contain an implicit model, or a comparison to an implicit model, the same way that any scatterplot contains the seed of a prediction model.

Another use of graphics with fitted models is to plot predicted datasets and compare them visually to actual data, as we discuss in Sections 11.4 and 11.5. For data structures more complicated than simple exchangeable batches or time series, plots can be tailored to specific aspects of the models being checked.

## 2.4 Data and adjustment: trends in mortality rates

Even when there are no questions of data quality or modeling, it can make sense to adjust measurements to answer real-world questions.

Example: Trends in mortality rates

In late 2015, economists Anne Case and Angus Deaton published a graph illustrating "a marked increase in the all-cause mortality of middle-aged white non-Hispanic men and women in the United States between 1999 and 2013." The authors stated that their numbers "are not age-adjusted within the 10-y 45–54 age group." They calculated the mortality rate each year by dividing the total number of deaths for the age group by the population as a whole, and they focused on this particular subgroup because it stood out with its increase: the death rates for other age and ethnic groups were declining during this period.

Suspecting an aggregation bias, we examined whether much of the increase in aggregate mortality rates for this age group could be due to the changing composition of the 45-to-54-year-old age group over the 1990 to 2013 time period. If this were the case, the change in the group mortality rate over time may not reflect a change in age-specific mortality rates. Adjusting for age confirmed this suspicion. Contrary to the original claim from the raw numbers, we find there is no longer a steady increase in mortality rates for this age group after adjusting for age composition. Instead, there is an increasing trend from 1999 to 2005 and a constant trend thereafter. Moreover, stratifying age-adjusted mortality rates by sex shows a marked increase only for women and not men.

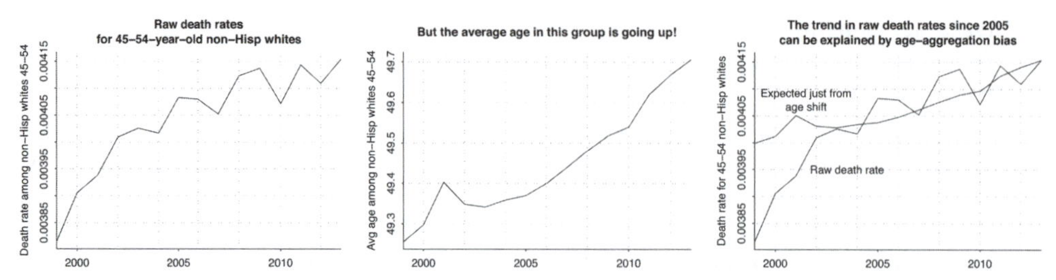


Figure 2.11 *(a) Observed increase in raw mortality rate among 45-to-54-year-old non-Hispanic whites, unadjusted for age; (b) increase in average age of this group as the baby boom generation moves through; (c) raw death rate, along with trend in death rate attributable by change in age distribution alone, had age-specific mortality rates been at the 2013 level throughout.*

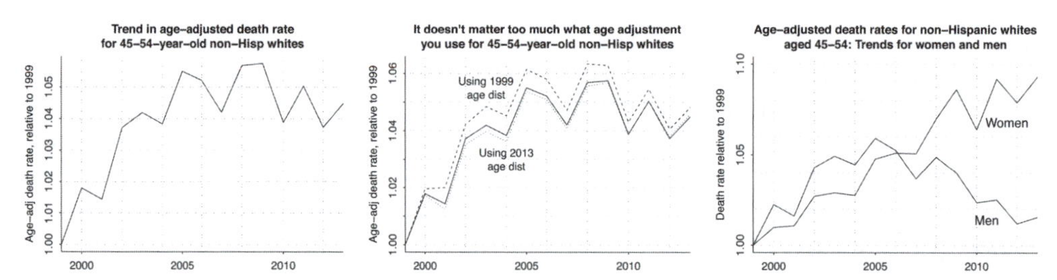


Figure 2.12 *(a) Age-adjusted death rates among 45-to-54-year-old non-Hispanic whites, showing an increase from 1999 to 2005 and a steady pattern since 2005; (b) comparison of two different age adjustments; (c) trends in age-adjusted death rates broken down by sex. The three graphs are on different scales.*

We demonstrate the necessity of the age adjustment in Figure 2.11.[6] The unadjusted numbers in Figure 2.11a show a steady increase in the mortality rate of 45-to-54-year-old non-Hispanic whites. During this period, however, the average age in this group increased as the baby boom generation passed through. Figure 2.11b shows this increase.

Suppose for the moment that mortality rates did not change for individuals in this age group from 1999 to 2013. In this case, we could calculate the change in the group mortality rate due solely to the change in the underlying age of the population. We do this by taking the 2013 mortality rates for each age and computing a weighted average rate each year using the number of individuals in each age group. Figure 2.11c shows the result. The changing composition in age explains about half the change in the mortality rate of this group since 1999 and all the change since 2005.

Having demonstrated the importance of age adjustment, we now perform an adjustment for the changing age composition. We ask what the data would look like if the age groups remained the same each year and only the individual mortality rates changed. Figure 2.12a shows the simplest such adjustment, normalizing each year to a hypothetical uniformly distributed population in which the number of people is equal at each age from 45 through 54. That is, we calculate the mortality rate each year by dividing the number of deaths for each age between 45 and 54 by the population of that age and then taking the average. This allows us to compare mortality rates across years. Consistent with Figure 2.11c, the resulting mortality rate increased from 1999 to 2005 and then stopped increasing.

We could just as easily use another age distribution to make valid comparisons across years. Checking, we find that age-adjusted trend is not sensitive to the age distribution used to normalize the mortality rates. Figure 2.12b shows the estimated changes in mortality rate under three options: first assuming a uniform distribution of ages 45–54; second using the distribution of ages that existed in 1999, which is skewed toward the younger end of the 45–54 group; and third using the 2013 age distribution, which is skewed older. The general pattern does not change.

[6]Data and code for this example are in the folder `AgePeriodCohort`.

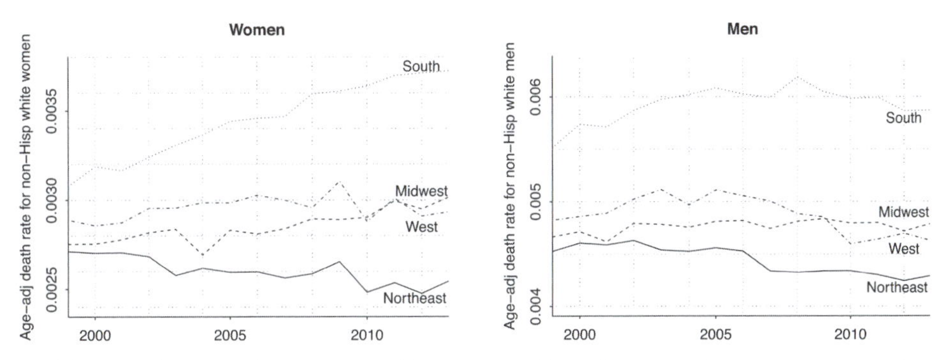


Figure 2.13 *Age-adjusted death rates among 45-to-54-year-old non-Hispanic white men and women, broken down by region of the country. The most notable pattern has been an increase in death rates among women in the South. In contrast, death rates for both sexes have been declining in the Northeast. The graphs are on different scales; as can be seen from the $y$-axes, death rates are lower for women than for men.*

Calculating the age-adjusted rates separately for each sex reveals a crucial result, which we display in Figure 2.12c. The mortality rate among white non-Hispanic American women increased from 1999 to 2013. Among the corresponding group of men, however, the mortality rate increase from 1999 to 2005 was nearly reversed from 2005 to 2013.

In summary, age adjustment is not merely an academic exercise. Due to the changing composition of the 45-to-54-year-old age group, adjusting for age changes the interpretation of the data in important ways. This does not change a key finding that had been seen in the unadjusted data: the comparison of non-Hispanic U.S. middle-aged whites to other countries and other ethnic groups. These comparisons hold up after our age adjustment. The aggregation bias in the published unadjusted numbers is on the order of 5% in the trend from 1999 to 2003, while mortality rates in other countries and other groups fell by around 20% during that period.

One can learn more by further decomposing these data. For example, Figure 2.13 breaks down the age-adjusted death rates in that group by U.S. region. The most notable pattern has been an increase in death rates among women in the South. In contrast, death rates for both sexes have been declining in the Northeast, the region where mortality rates were lowest to begin with. These graphs demonstrate the value of this sort of data exploration.

## 2.5 Bibliographic note

Some general references on data display and exploration include Cleveland (1985, 1993), Friendly and Kwan (2003), Chambers et al. (1983), Tukey (1977), Mosteller and Tukey (1977), Tufte (1983, 1990), Bertin (1967), and Wainer (1984, 1997). Gelman and Unwin (2013) discuss different goals of information visualization and statistical graphics.

For statistical graphics in R, the books by Healy (2018), Wickham (2016), and Murrell (2005) are good starting points. Fox (2002) is also helpful in that it focuses on regression models. An important topic not discussed in the present book is dynamic graphics; see Buja et al. (1988).

There are various systematic ways of studying statistical graphics. One useful approach is to interpret exploratory visualization as checks of explicit or implicit models. Another approach is to perform experiments to find out how well people can gather information from various graphical displays; see Hullman, Resnick, and Adar (2015) for an example of such research. More work is needed on both these approaches: relating to probability models is important for allowing us to understand graphs and devise graphs for new problems, and effective display is important for communicating to ourselves as well as others.

For some ideas on the connections between statistical theory, modeling, and graphics, see Buja et al. (2009), Wilkinson (2005), and, for our own perspective, Gelman (2004a). Unwin, Volinsky, and Winkler (2003), Urbanek (2004), and Wickham (2006) discuss exploratory model analysis, that is, visualization of different models fit to the same data.

For different perspectives on tabular displays, compare Tukey (1977); Ehrenberg (1978); Gelman, Pasarica, and Dodhia; and Wickham and Grolemund (2017, chapter 10).

For background on the Human Development Index, see Gelman (2009a). The graphs of political ideology, party identification, and income come from Gelman (2009b). The graph of health spending and life expectancy appears in Gelman (2009c). The graphs of baby names are adapted from Wattenberg (2007).

Validity and reliability are discussed in textbooks on psychometrics but have unfortunately been underemphasized in applied statistics; see Gelman (2015b). Rodu and Plurphanswat (2018) discuss a problem with the "never smoker" definition in a study of adolescent behavior. The middle-aged mortality rate example appears in Gelman (2015c) and Gelman and Auerbach (2016); see also Schmid (2016), Case and Deaton (2015, 2016), and Gelman (2017).

## 2.6 Exercises

2.1 *Composite measures*: Following the example of the Human Development Index in Section 2.1, find a composite measure on a topic of interest to you. Track down the individual components of the measure and use scatterplots to understand how the measure works, as was done for that example in the book.

2.2 *Significant digits*:

(a) Find a published article in a statistics or social science journal in which too many significant digits are used, that is, where numbers are presented or displayed to an inappropriate level of precision. Explain.

(b) Find an example of a published article in a statistics or social science journal in which there is *not* a problem with too many significant digits being used.

2.3 *Data processing*: Go to the folder `Names` and make a graph similar to Figure 2.8, but for girls.

2.4 *Data visualization*: Take any data analysis exercise from this book and present the *raw data* in several different ways. Discuss the advantages and disadvantages of each presentation.

2.5 *Visualization of fitted models*: Take any data analysis exercise from this book and present the *fitted model* in several different ways. Discuss the advantages and disadvantages of each presentation.

2.6 *Data visualization*: Take data from some problem of interest to you and make several plots to highlight different aspects of the data, as was done in Figures 2.6–2.8.

2.7 *Reliability and validity*:

(a) Give an example of a scenario of measurements that have *validity* but not *reliability*.

(b) Give an example of a scenario of measurements that have *reliability* but not *validity*.

2.8 *Reliability and validity*: Discuss validity, reliability, and selection in the context of measurements on a topic of interest to you. Be specific: make a pen-on-paper sketch of data from multiple measurements to demonstrate reliability, sketch true and measured values to demonstrate validity, and sketch observed and complete data to demonstrate selection.

2.9 *Graphing parallel time series*: The mortality data in Section 2.4 are accessible from this site at the U.S. Centers for Disease Control and Prevention: `wonder.cdc.gov`. Download mortality data from this source but choose just one particular cause of death, and then make graphs similar to those in Section 2.4, breaking down trends in death rate by age, sex, and region of the country.

2.10 *Working through your own example*: Continuing the example from Exercise 1.10, graph your data and discuss issues of validity and reliability. How could you gather additional data, at least in theory, to address these issues?

# Chapter 3

# Some basic methods in mathematics and probability

Simple methods from introductory mathematics and statistics have three important roles in regression modeling. First, linear algebra and simple probability distributions are the building blocks for elaborate models. Second, it is useful to understand the basic ideas of inference separately from the details of particular classes of model. Third, it is often useful in practice to construct quick estimates and comparisons for small parts of a problem—before fitting an elaborate model, or in understanding the output from such a model. This chapter provides a quick review of some of these basic ideas.

## 3.1 Weighted averages

In statistics it is common to reweight data or inferences so as to adapt to a target population.

Here is a simple example. In 2010 there were 456 million people living in North America: 310 million residents of the United States, 112 million Mexicans, and 34 million Canadians. The average age of people in each country in that year is displayed in Figure 3.1. The average age of all North Americans is a *weighted average*:

$$\begin{aligned}\text{average age} &= \frac{310\,000\,000 * 36.8 + 112\,000\,000 * 26.7 + 34\,000\,000 * 40.7}{310\,000\,000 + 112\,000\,000 + 34\,000\,000} \\ &= 34.6.\end{aligned}$$

This is a weighted average rather than a simple average because the numbers 36.8, 26.7, 40.7 are multiplied by "weights" proportional to the population of each country. The total population of North America was 310 + 112 + 34 = 456 million, and we can rewrite the above expression as

$$\begin{aligned}\text{average age} &= \frac{310\,000\,000}{456\,000\,000} * 36.8 + \frac{112\,000\,000}{456\,000\,000} * 26.7 + \frac{34\,000\,000}{456\,000\,000} * 40.7 \\ &= 0.6798 * 36.8 + 0.2456 * 26.7 + 0.0746 * 40.7 \\ &= 34.6.\end{aligned}$$

The above proportions 0.6798, 0.2456, and 0.0746 (which by necessity sum to 1) are the *weights* of the countries in this weighted average.

We can equivalently write a weighted average in summation notation:

$$\text{weighted average} = \frac{\sum_j N_j \bar{y}_j}{\sum_j N_j},$$

where $j$ indexes countries and the sum adds over all the *strata* (in this case, the three countries).

The choice of weights depends on context. For example, 51% of Americans are women and 49% are men. The average age of American women and men is 38.1 and 35.5, respectively. The average age of all Americans is thus 0.51 ∗ 38.1 + 0.49 ∗ 35.5 = 36.8 (which agrees with the U.S. average

| Stratum, $j$ | Label | Population, $N_j$ | Average age, $\bar{y}_j$ |
|---|---|---|---|
| 1 | United States | 310 million | 36.8 |
| 2 | Mexico | 112 million | 26.7 |
| 3 | Canada | 34 million | 40.7 |

Figure 3.1 *Populations and average ages of countries in North America. (Data from CIA World Factbook 2010.) The average age of all North Americans is a* weighted average *of the average ages within each country.*

in Figure 3.1). But now consider a slightly different problem: estimating the average salary of all teachers in the country. According to the Census in 2010, there were 5 700 000 female teachers and 1 500 000 male teachers (that is, the population of teachers was 79% female and 21% male) in the United States, with average incomes \$45 865 and \$49 207, respectively. The average income of all teachers was $0.79 * \$45\,865 + 0.21 * \$49\,207 = \$46\,567$, *not* $0.51 * \$45\,865 + 0.49 * \$49\,207 = \$47\,503$.

## 3.2 Vectors and matrices

A list of numbers is called a *vector*. A rectangular array of numbers is called a *matrix*. Vectors and matrices are useful in regression to represent predictions for many cases using a single model.

Example: Elections and the economy

In Section 1.2 we introduced a model for predicting the incumbent party's vote percentage in U.S. presidential elections from economic conditions in the years preceding the election:

$$\text{Predicted vote percentage} = 46.3 + 3.0 * (\text{growth rate of average personal income}),$$

which we shall write as

$$\hat{y} = 46.3 + 3.0x,$$

or, even more abstractly, as

$$\hat{y} = \hat{a} + \hat{b}x.$$

The expressions $\hat{a}$ and $\hat{b}$ denote estimates—the coefficients 46.3 and 3.0 were obtained by fitting a line to past data—and $\hat{y}$ denotes a predicted value. In this case, we would use $y$ to represent an actual election result, and $\hat{y}$ is the prediction from the model. Here we are focusing on the linear prediction, and so we work with $\hat{y}$.

Let's apply this model to a few special cases:

1. $x = -1$. A rate of growth of $-1\%$ (that is, a 1% decline in the economy) translates into an incumbent party vote share of $46.3 + 3.0 * (-1) = 43.3\%$.
2. $x = 0$. If there is zero economic growth in the year preceding the presidential election, the model predicts that the incumbent party's candidate will receive $46.3 + 3.0 * 0 = 46.3\%$ of the two-party vote; that is, he or she is predicted to lose the election.
3. $x = 3$. A 3% rate of economic growth translates to the incumbent party's candidate winning $46.3 + 3.0 * 3 = 55.3\%$ of the vote.

We can define $x$ as the vector that comprises these three cases, that is $x = (-1, 0, 3)$.

We can put these three predictions together:

$$\begin{aligned}\hat{y}_1 &= 43.3 = 46.3 + 3.0 * (-1),\\ \hat{y}_2 &= 46.3 = 46.3 + 3.0 * 0,\\ \hat{y}_3 &= 55.3 = 46.3 + 3.0 * 3,\end{aligned}$$

which can be written as vectors:

$$\hat{y} = \begin{pmatrix} 43.3 \\ 46.3 \\ 55.3 \end{pmatrix} = \begin{pmatrix} 46.3 + 3.0 * (-1) \\ 46.3 + 3.0 * 0 \\ 46.3 + 3.0 * 3 \end{pmatrix},$$

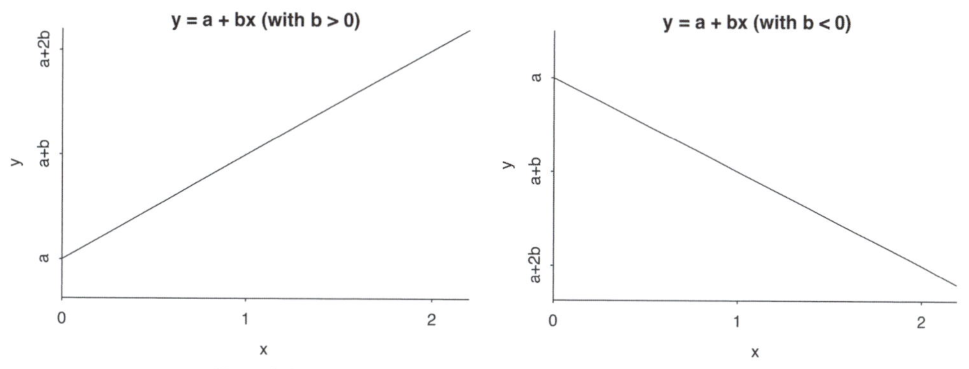


Figure 3.2: *Lines $y = a + bx$ with positive and negative slopes.*

or, in matrix form,

$$\hat{y} = \begin{pmatrix} 43.3 \\ 46.3 \\ 55.3 \end{pmatrix} = \begin{pmatrix} 1 & -1 \\ 1 & 0 \\ 1 & 3 \end{pmatrix} \begin{pmatrix} 46.3 \\ 3.0 \end{pmatrix},$$

or, more abstractly,

$$\hat{y} = X\hat{\beta}.$$

Here $y$ and $x$ are vectors of length 3, $X$ is a $3 \times 2$ matrix with a column of ones and a column equal to the vector $x$, and $\hat{\beta} = (46.3, 3.0)$ is the vector of estimated coefficients.

## 3.3 Graphing a line

To use linear regression effectively, you need to understand the algebra and geometry of straight lines, which we review briefly here.

Figure 3.2 shows the line $y = a + bx$. The *intercept*, $a$, is the value of $y$ when $x = 0$; the coefficient $b$ is the *slope* of the line. The line slopes upward if $b > 0$ (as in Figure 3.2a), slopes downward if $b < 0$ (as in Figure 3.2b), and is horizontal if $b = 0$. The larger $b$ is, in absolute value, the steeper the line will be.

Example: Mile run

Figure 3.3a shows a numerical example: $y = 1007 - 0.39x$. Thus, $y = 1007$ when $x = 0$, and $y$ decreases by 0.39 for each unit increase in $x$. This line approximates the trajectory of the world record time (in seconds) for the mile run from 1900 to 2000 (see Figure A.1). Figure 3.3b shows the graph of the line alone on this scale. In R it is easy to draw this line:[1]

```
curve(1007 - 0.393*x, from=1900, to=2000, xlab="Year", ylab="Time (seconds)",
  main="Approximate trend of world record times\nfor the mile run")
```

How would we draw it by hand? We cannot simply start with the intercept at $x = 0$ and go from there, as then the entire range of interest, from 1900 to 2000, would be crammed into a small corner of the plot. The value $x = 0$ is well outside the range of the data. Instead, we use the equation to calculate the value of $y$ at the two extremes of the plot:

$$\text{At } x = 1900,\ y = 1007 - 0.393 * 1900 = 260.$$
$$\text{At } x = 2000,\ y = 1007 - 0.393 * 2000 = 221.$$

These are the two endpoints in Figure 3.3, between which we can draw the straight line.

This example demonstrates the importance of location and scale. The lines in Figures 3.3a and

[1]Data and code for this example are in the folder `Mile`.

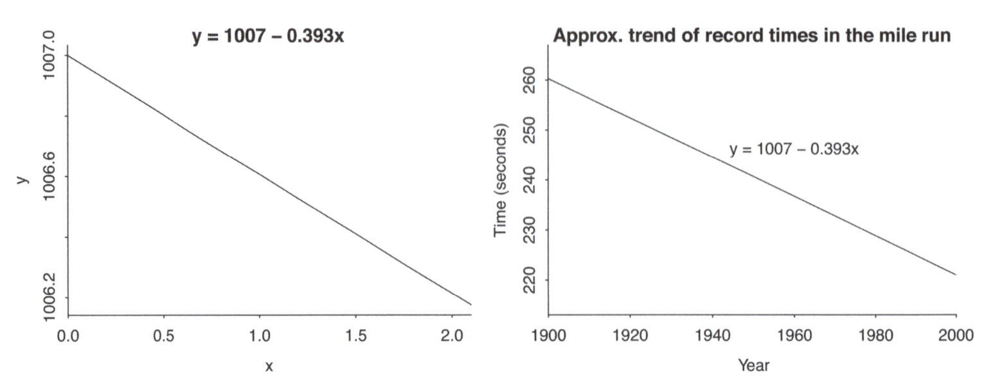


Figure 3.3 *(a) The line $y = 1007 - 0.393x$. (b) For x between 1900 and 2000, the line $y = 1007 - 0.393x$ approximates the trend of world record times in the mile run. Compare to Figure A.1.*

3.3b have the same algebraic equation but displayed at different ranges of $x$, and only the second graph serves any applied goal.

We also see the difficulty of interpreting the intercept (the $a$ in $y = a + bx$) in this example. In the equation $y = 1007 - 0.393x$, the intercept of 1007 seconds (equivalently, 16.8 minutes) represents the predicted world record time in the mile run in the year 0, an obviously inappropriate extrapolation. It could be better to express this model as $y = 260 - 0.393\,(x - 1900)$, for example, or perhaps $y = 241 - 0.393\,(x - 1950)$.

Finally we revisit the interpretation of the slope. The usual shorthand is that an increase of one year leads to a decrease in the world record times for the mile run by 0.393 seconds on average. However it is not so helpful to give this sort of implicit causal role to the passage of time. It's more precise to describe the result in a purely descriptive way, saying that when comparing any two years, we see a world record time that is, on average, 0.393 seconds per year less for the more recent year.

## 3.4 Exponential and power-law growth and decline; logarithmic and log-log relationships

The line $y = a + bx$ can be used to express a more general class of relationships by allowing logarithmic transformations.

The formula $\log y = a + bx$ represents exponential growth (if $b > 0$) or decline (if $b < 0$): $y = Ae^{bx}$, where $A = e^a$. The parameter $A$ is the value of $y$ when $x = 0$, and the parameter $b$ determines the rate of growth or decline. A one-unit difference in $x$ corresponds to an additive difference of $b$ in $\log y$ and thus a multiplicative factor of $e^b$ in $y$. Here are two examples:

- *Exponential growth.* Suppose that world population starts at 1.5 billion in the year 1900 and increases exponentially, doubling every 50 years (not an accurate description, just a crude approximation). We can write this as $y = A * 2^{(x-1900)/50}$, where $A = 1.5 * 10^9$. Equivalently, $y = A\,e^{(\log(2)/50))(x-1900)} = A\,e^{0.014(x-1900)}$. In statistics we use "log" to refer to the natural logarithm (log base $e$, not base 10) for reasons explained in Section 12.4.
  The model $y = A\,e^{0.014\,(x-1900)}$ is exponential growth with a rate of 0.014, which implies that $y$ increases by a factor of $e^{0.014} = 1.014$ per year, or $e^{0.14} = 1.15$ per ten years, or $e^{1.4} = 4.0$ per hundred years. We can take the log of both sides of the equation to get $\log y = 21.1 + 0.014(x-1900)$. Here, $\log A = \log(1.5 * 10^9) = 21.1$.
- *Exponential decline.* Consider an asset that is initially worth \$1000 and declines in value by 20% each year. Then its value at year $x$ can be written as $y = 1000 * 0.8^x$ or, equivalently, $y = 1000\,e^{\log(0.8)x} = 1000\,e^{-0.22x}$. Logging both sides yields $\log y = \log 1000 - 0.22x = 6.9 - 0.22x$.

The formula $\log y = a + b \log x$ represents power-law growth (if $b > 0$) or decline (if $b < 0$):

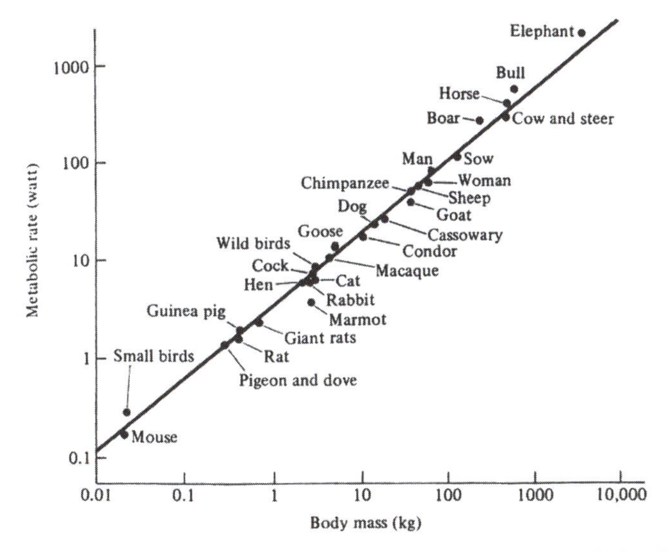


Figure 3.4 *Log metabolic rate vs. log body mass of animals, from Schmidt-Nielsen (1984). These data illustrate the log-log transformation. The fitted line has a slope of 0.74. See also Figure 3.5.*

$y = Ax^b$, where $A = e^a$. The parameter $A$ is the value of $y$ when $x = 1$, and the parameter $b$ determines the rate of growth or decline. A one-unit difference in $\log x$ corresponds to an additive difference of $b$ in $\log y$. Here are two examples:

- *Power law.* Let $y$ be the area of a square and $x$ be its perimeter. Then $y = (x/4)^2$, and we can take the log of both sides to get $\log y = 2(\log x - \log 4) = -2.8 + 2\log x$.
- *Non-integer power law.* Let $y$ be the surface area of a cube and $x$ be its volume. If $L$ is the length of a side of the cube, then $y = 6L^2$ and $x = L^3$, hence the relation between $x$ and $y$ is $y = 6x^{2/3}$; thus, $\log y = \log 6 + \frac{2}{3}\log x = 1.8 + \frac{2}{3}\log x$.

Example: Metabolic rates of animals

Here is an example of how to interpret a power law or log-log regression.[2] Figure 3.4 displays data on log metabolic rate vs. body mass indicating an approximate underlying linear relation. To give some context, the point labeled Man corresponds to a body mass of 70 kilogram man and a metabolism of 100 watts; thus, a classroom with 100 men is the equivalent of a 10 000-watt space heater. By comparison, you could compute the amount of heat given off by a single elephant (which weighs about 7000 kilograms according to the graph) or 10 000 rats (which together also weigh about 7000 kilograms). The answer is that the elephant gives off less heat than the equivalent weight of men, and the rats give off more. This corresponds to a slope of less than 1 on the log-log scale.

What is the equation of the line in Figure 3.4? The question is not quite as simple as it looks, since the graph is on the log-log scale, but the axes are labeled on the original scale. We start by relabeling the axes on the logarithmic (base $e$) scale, as shown in Figure 3.5a. We can then determine the equation of the line by identifying two points that it goes through: for example, when $\log x = -4$, $\log y = -1.6$, and when $\log x = 6$, $\log y = 5.8$. So, comparing two animals where $\log x$ differs by 10, the average difference in $\log y$ is $5.8 - (-1.6) = 7.4$. The slope of the line is then $7.4/10 = 7.4$. Since the line goes through the point (0, 1.4), its equation can be written as,

$$\log y = 1.4 + 0.74 \log x.$$

We can exponentiate both sides of (3.1) to see the relation between metabolic rate and body mass on

[2]Code for this example is in the folder `Metabolic`.

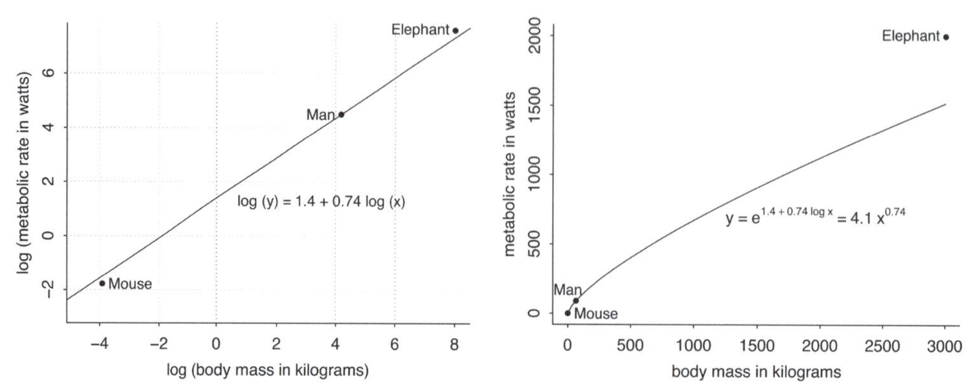


Figure 3.5 *Fitted curve (from data in Figure 3.4) of metabolic rate vs. body mass of animals, on the log-log and untransformed scales. The difference from the elephant's metabolic rate from its predictive value is relatively small on the logarithmic scale but large on the absolute scale.*

the untransformed scales:

$$\begin{aligned} e^{\log y} &= e^{1.4+0.74 \log x} \\ y &= 4.1\, x^{0.74}. \end{aligned} \quad (3.1)$$

This curve is plotted in Figure 3.5b. For example, when increasing body mass on this curve by a factor of 2, metabolic rate is multiplied by $2^{0.74} = 1.7$. Multiplying body mass by 10 corresponds to multiplying metabolic rate by $10^{0.74} = 5.5$, and so forth.

Now we return to the rats and the elephant. The relation between metabolic rate and body mass is less than linear (that is, the exponent 0.74 is less than 1.0, and the line in Fig. 3.5b is concave, not convex), which implies that the equivalent mass of rats gives off more heat, and the equivalent mass of elephant gives off less heat, than the men. This seems related to the general geometrical relation that surface area and volume are proportional to linear dimension to the second and third power, respectively, and thus surface area should be proportional to volume to the $\frac{2}{3}$ power. Heat produced by an animal is emitted from its surface, and it would thus be reasonable to suspect metabolic rate to be proportional to the $\frac{2}{3}$ power of body mass. Biologists have considered why the empirical slope is closer to $\frac{3}{4}$ than to $\frac{2}{3}$; our point here is not to discuss these issues but rather to show the way in which the coefficient in a log-log regression (equivalently, the exponent in power-law relation) can be interpreted.

See Section 12.4 for further discussion and examples of logarithmic transformations.

## 3.5 Probability distributions

In Section 3.3, we reviewed straight-line prediction, which is the deterministic part of linear regression and the key building block for regression modeling in general. Here we introduce probability distributions and random variables, which we need because our models do not fit our data exactly. Probability distributions represent the unmodeled aspects of reality—the *error term* $\epsilon$ in the expression $y = a + bx + \epsilon$—and it is randomness that greases the wheels of inference.

A probability distribution corresponds to an urn with a potentially infinite number of balls inside. When a ball is drawn at random, the "random variable" is what is written on this ball. Our treatment is not formal or axiomatic; rather, we mix conceptual definitions with mathematical formulas where we think these will be useful for using these distributions in practice.

Areas of application of probability distributions include:

- Distributions of data (for example, heights of men, incomes of women, political party preference), for which we use the notation $y_i,\ i = 1, \ldots, n$.

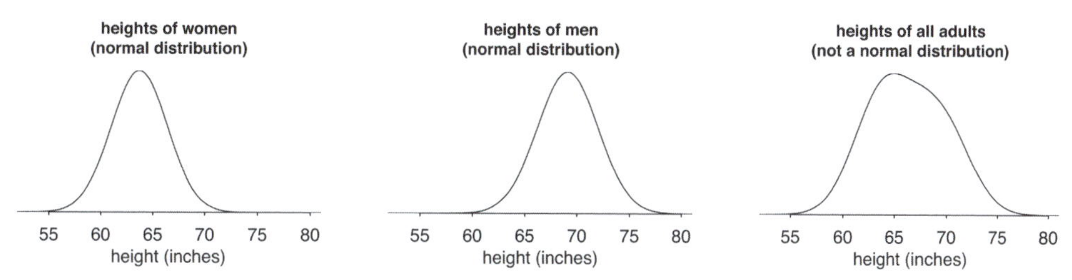


Figure 3.6 *(a) Heights of women, which approximately follow a normal distribution, as predicted from the Central Limit Theorem. The distribution has mean 63.7 and standard deviation 2.7, so about 68% of women have heights in the range* $63.7 \pm 2.7$. *(b) Heights of men, approximately following a normal distribution with mean 69.1 and standard deviation 2.9. (c) Heights of all adults in the United States, which have the form of a mixture of two normal distributions, one for each sex.*

- Distributions of error terms, which we write as $\epsilon_i,\ i = 1, \dots, n$.

A key component of regression modeling is to describe the typical range of values of the outcome variable, given the predictors. This is done in two steps that are conceptually separate but which in practice are performed at the same time. The first step is to predict the average value of the outcome given the predictors, and the second step is to summarize the variation in this prediction. Probabilistic distributions are used in regression modeling to help us characterize the variation that remains *after* predicting the average. These distributions allow us to get a handle on how uncertain our predictions are and, additionally, our uncertainty in the estimated parameters of the model.

## Mean and standard deviation of a probability distribution

A probability distribution of a random variable $z$ takes on some range of values (the numbers written on the balls in the urn). The *mean* of this distribution is the average of all these numbers or, equivalently, the value that would be obtained on average from a random sample from the distribution. The mean is also called the expectation or expected value and is written as $\mathrm{E}(z)$ or $\mu_z$. For example, Figure 3.6a shows the (approximate) distribution of heights of women in the United States. The mean of this distribution is 63.7 inches: this is the average height of all the women in the country and it is also the average value we would expect to see from sampling one woman at random.

The *variance* of the distribution of $z$ is $\mathrm{E}((z - \mu_z)^2)$, that is, the mean of the squared difference from the mean. To understand this expression, first consider the special case in which $z$ takes on only a single value. In that case, this single value is the mean, so $z - \mu_z = 0$ for all $z$ in the distribution, and the variance is 0. To the extent that the distribution has variation, so that sampled values of $z$ from the "urn" can be different, this will show up as values of $z$ that are higher and lower than $\mu_z$, and the variance of $z$ is nonzero.

The *standard deviation* is the square root of the variance. We typically work with the standard deviation rather than the variance because it is on the original scale of the distribution. Returning to Figure 3.6a, the standard deviation of women's heights is 2.7 inches: that is, if you randomly sample a woman from the population, observe her height $z$, and compute $(z - 63.7)^2$, then the average value you will get is 7.3; this is the variance, and the standard deviation is $\sqrt{7.3} = 2.7$ inches. The variance of 7.3 is on the uninterpretable scale of inches squared.

## Normal distribution; mean and standard deviation

The Central Limit Theorem of probability states that the sum of many small, independent random variables will be a random variable that approximates what is called a *normal distribution*. If we write this summation of independent components as $z = \sum_{i=1}^n z_i$, then the mean and variance of $z$ are the sums of the means and variances of the $z_i$'s: $\mu_z = \sum_{i=1}^n \mu_{z_i}$ and $\sigma_z = \sqrt{\sum_{i=1}^n \sigma_{z_i}^2}$. In statistical

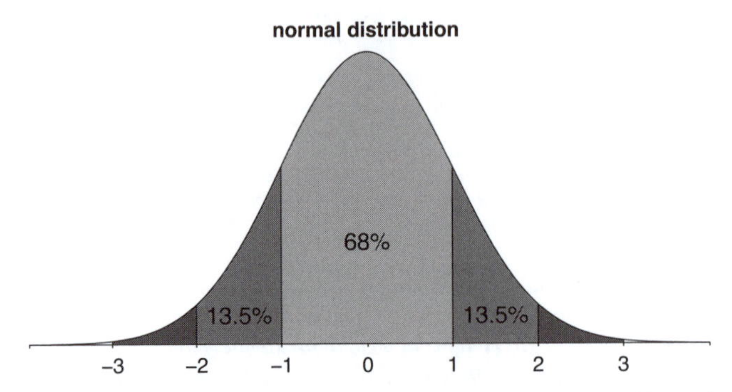


Figure 3.7 *Approximately 50% of the mass of the normal distribution falls within 0.67 standard deviations from the mean, 68% of the mass falls within 1 standard deviation from the mean, 95% within 2 standard deviations of the mean, and 99.7% within 3 standard deviations.*

notation, the normal distribution is written as $z \sim \mathrm{N}(\mu_z, \sigma_z^2)$. The Central Limit Theorem holds in practice—that is, $\sum_{i=1}^{n} z_i$ actually follows an approximate normal distribution—if the individual $\sigma_{z_i}$'s are small compared to the standard deviation $\sigma_z$ of the sum.

We write the normal distribution with mean $\mu$ and standard deviation $\sigma$ as $\mathrm{N}(\mu, \sigma^2)$. Approximately 50% of the mass of this distribution falls in the range $\mu \pm 0.67\sigma$, 68% in the range $\mu \pm \sigma$, 95% in the range $\mu \pm 2\sigma$, and 99.7% in the range $\mu \pm 3\sigma$. The 1 and 2 standard deviation ranges are shown in Figure 3.7. To put it another way, if you take a random draw from a normal distribution, there is a 50% chance it will fall within 0.67 standard deviations from the mean, a 68% chance it will fall within 1 standard deviation from the mean, and so forth.

Example: Heights of men and women

There's no reason to expect that a random variable representing direct measurements in the world will be normally distributed. Some examples exist, though. For example, the heights of women in the United States follow an approximate normal distribution. We can posit that the Central Limit Theorem applies here because women's height is affected by many small additive factors. In contrast, the distribution of heights of *all* adults in the United States is not so close to the normal curve. The Central Limit Theorem does not apply in this case because there is a single large factor—sex—that represents much of the total variation. See Figure 3.6c.[3]

The normal distribution is useful in part because summaries such as sums, differences, and estimated regression coefficients can be expressed mathematically as averages or weighted averages of data. There are many situations in which we can use the normal distribution to summarize uncertainty in estimated averages, differences, and regression coefficients, even when the underlying data do not follow a normal distribution.

### Linear transformations

Linearly transformed normal distributions are still normal. If $y$ is a variable representing men's heights in inches, with mean 69.1 and standard deviation 2.9, then 2.54 $y$ is height in centimeters, with mean $2.54 * 69 = 175$ and standard deviation $2.54 * 2.9 = 7.4$.

For a slightly more complicated example, suppose we take independent samples of 100 men and 100 women and compute the difference between the average heights of the men and the women. This difference will be approximately normally distributed with mean $69.1 - 63.7 = 5.4$ and standard deviation $\sqrt{2.9^2/100 + 2.7^2/100} = 0.4$; see Exercise 3.5.

[3]Code for this example is in the folder `CentralLimitTheorem`.

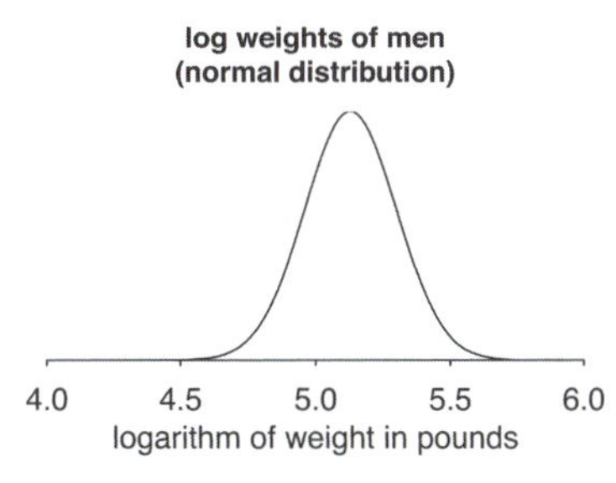


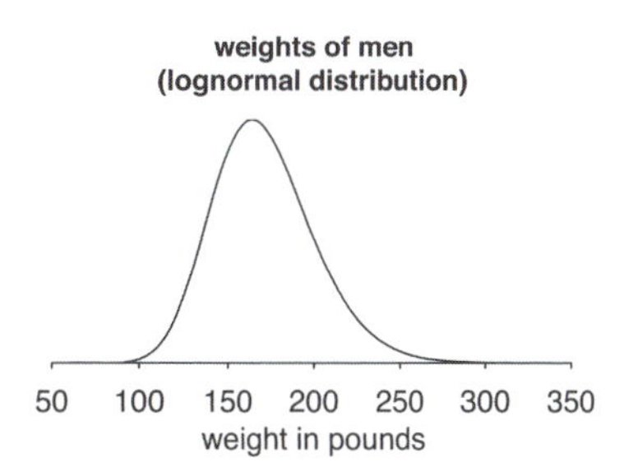


Figure 3.8 *Weights of men (which approximately follow a lognormal distribution, as predicted from the Central Limit Theorem from combining many small multiplicative factors), plotted on the logarithmic and original scales.*

### Mean and standard deviation of the sum of correlated random variables

If two random variables $u$ and $v$ have mean $\mu_u, \mu_v$ and standard deviations $\sigma_a, \sigma_b$, then their *correlation* is defined as $\rho_{uv} = \text{E}((u - \mu_u)(v - \mu_v))/(\sigma_a\sigma_b)$. It can be shown mathematically that the correlation must be in the range $[-1, 1]$, attaining the extremes only when $u$ and $v$ are linear functions of each other.

Knowing the correlation gives information about linear combinations of $u$ and $v$. Their sum $u + v$ has mean $\mu_u + \mu_v$ and standard deviation $\sqrt{\sigma_u^2 + \sigma_v^2 + 2\rho\sigma_u\sigma_v}$. More generally, the weighted sum $au + bv$ has mean $a\mu_u + b\mu_v$, and its standard deviation is $\sqrt{a^2\sigma_u^2 + b^2\sigma_v^2 + 2ab\rho\sigma_u\sigma_v}$. From this we can derive, for example, that $u-v$ has mean $\mu_u - \mu_v$ and standard deviation $\sqrt{\sigma_u^2 + \sigma_v^2 - 2\rho\sigma_u\sigma_v}$.

### Lognormal distribution

Example: Weights of men

It is often helpful to model all-positive random variables on the logarithmic scale because it does not allow for values that are 0 or negative. For example, the logarithms of men's weights (in pounds) have an approximate normal distribution with mean 5.13 and standard deviation 0.17. Figure 3.8 shows the distributions of log weights and weights among men in the United States. The exponential of the mean and standard deviations of log weights are called the *geometric mean* and *geometric standard deviation* of the weights; in this example, they are 169 pounds and 1.18, respectively.

The logarithmic transformation is nonlinear and, as illustrated in Figure 3.8, it pulls in the values at the high end, compressing the scale of the distribution. But in general, the reason we perform logarithmic transformations is *not* to get distributions closer to normal, but rather to transform multiplicative models into additive models, a point we discuss further in Section 12.4.

### Binomial distribution

If you take 20 shots in basketball, and each has 0.3 probability of succeeding, and if these shots are independent of each other (that is, success in one shot is not associated with an increase or decrease in the probability of success for any other shot), then the number of shots that succeed is said to have a *binomial distribution* with $n = 20$ and $p = 0.3$, for which we use the notation $y \sim \text{binomial}(n, p)$. Even in this simple example, the binomial model is typically only an approximation. In real data with multiple measurements (for example, repeated basketball shots), the probability $p$ of success can vary, and outcomes can be correlated. Nonetheless, the binomial model is a useful starting point for modeling such data. And in some settings—most notably, independent sampling with Yes/No responses—the binomial model generally is appropriate, or very close to appropriate. The binomial distribution with parameters $n$ and $p$ has mean $np$ and standard deviation $\sqrt{np(1-p)}$.

### Poisson distribution

The *Poisson distribution* is used for count data such as the number of cases of cancer in a county, or the number of hits to a website during a particular hour, or the number of people named Michael whom you know:

- If a county has a population of 100 000, and the average rate of a particular cancer is 45.2 per million people per year, then the number of cancers in this county could be modeled as Poisson with expectation 4.52.
- If hits are coming at random, with an average rate of 380 per hour, then the number of hits in any particular hour could be modeled as Poisson with expectation 380.
- If you know approximately 750 people, and 1% of all people in the population are named Michael, and you are as likely to know Michaels as anyone else, then the number of Michaels you know could be modeled as Poisson with expectation 7.5.

As with the binomial distribution, the Poisson model is almost always an idealization, with the first example ignoring systematic differences among counties, the second ignoring clustering or burstiness of the hits, and the third ignoring factors such as sex and age that distinguish Michaels, on average, from the general population.

Again, however, the Poisson distribution is a starting point—as long as its fit to data is checked. We generally recommend the Poisson model to be expanded to account for "overdispersion" in data, as we discuss in Section 15.2.

### Unclassified probability distributions

Real data will not in general correspond to any named probability distribution. For example, the distribution shown in Figure 3.6c of heights of all adults is not normal, or lognormal, or any other tabulated entry. Similarly, the distribution of incomes of all Americans has no standard form. For a discrete example, the distribution of the number of children of all U.S. adults does not follow the Poisson or binomial or any other named distribution. That is fine. Catalogs of named distributions are a starting point in understanding probability but they should not bound our thinking.

### Probability distributions for error

Above we have considered distributions for raw data. But in regression modeling we typically model as much of the data variation as possible with a *deterministic model*, with a probability distribution included to capture the *error*, or unexplained variation. A simple and often useful model for continuous data is $y$ = deterministic_part + error. A similar model for continuous positive data is $y$ = deterministic_part $*$ error.

For discrete data, the error cannot be so easily mathematically separated from the rest of the model. For example, with binary data, the deterministic model will give predicted probabilities, with the error model corresponding to the mapping of these probabilities to 0 and 1. Better models yield predicted probabilities closer to 0 or 1. The probability that a U.S. professional basketball team wins when playing at home is about 60%; thus we can model the game outcome ($y = 1$ if the home team wins, 0 otherwise) as a binomial distribution with $n = 1$ and $p = 0.6$. However, a better job of prediction can be done using available information on the matchups before each game. Given a reasonably good prediction model, the probability that the home team wins, as assessed for any particular game, might be some number between 0.3 and 0.8. In that case, each outcome $y_i$ is modeled as binomially distributed with $n = 1$ and a probability $p_i$ that is computed based on a fitted model.

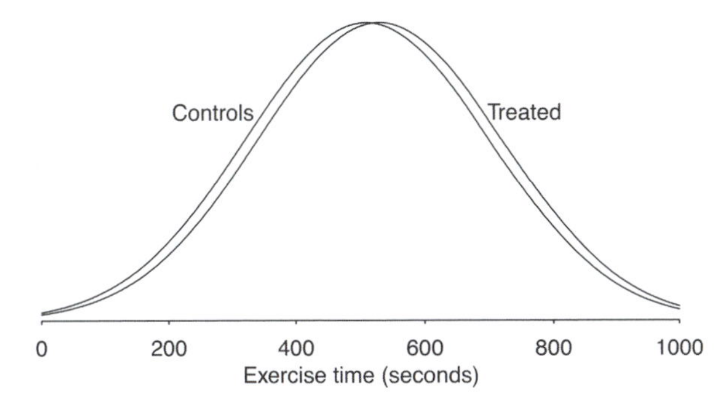


Figure 3.9 *Distributions of potential outcomes for patients given placebo or heart stents, using a normal approximation and assuming a treatment effect in which stents improve exercise time by 20 seconds, a shift which corresponds to taking a patient from the 50th to the 54th percentile of the distribution under the placebo.*

### Comparing distributions

We typically compare distributions using summaries such as the mean, but it can also make sense to look at shifts in quantiles.

Example: Stents

We demonstrate with data from a clinical trial of 200 heart patients in which half received percutaneous coronary intervention (stents) and half received placebo, and at follow-up various health measures were recorded. Here, we consider a particular outcome, the amount of time that a patient was able to exercise on a treadmill. They show a pre-randomization distribution (averaging the treatment and control groups) with a mean of 510 seconds and a standard deviation of 190 seconds.[4]

The treatment effect was estimated as 20 seconds. How does this map to the distribution? One way to make the connection is to suppose that the responses under the control group follow a normal distribution, and suppose that, under the treatment, the responses would increase by 20 seconds; see Figure 3.9. Take a person at the median of the distribution, with an exercise time of 510 seconds under the control and an expected 530 seconds under the treatment. This corresponds to a shift from the 50th to the 54th percentile of the distribution. We compute these probabilities in R like this: `pnorm(c(510, 530), 510, 190)`. Comparing to the quantiles gives some context to an estimated treatment effect of 20 seconds, a number that could otherwise be hard to interpret on its own.

## 3.6 Probability modeling

Example: Probability of a decisive vote

What is the probability that your vote is decisive in an election? No dice, coins, or other randomization devices appear here, so any answer to this question will require assumptions. We use this example to demonstrate the challenges of probability modeling, how it can work, and how it can go wrong.

Consider an election with two candidates and $n$ voters. An additional vote will be potentially decisive if all the others are divided equally (if $n$ is even) or if the preferred candidate is otherwise one vote short of a tie (if $n$ is odd).

We consider two ways of estimating this probability: an empirical forecasting approach that we recommend, and a binomial probability model that has serious problems.

### Using an empirical forecast

Suppose an election is forecast to be close. Let $y$ be the proportion of the vote that will be received by one of the candidates, and suppose we can summarize our uncertainty in $y$ by a normal distribution with mean 0.49 and standard deviation 0.04: that is, the candidate is predicted to lose, but there is enough uncertainty that the vote could go either way. The probability that the $n$ votes are exactly split

[4] Data and code for this example are in the folder `Stents`.

if $n$ is even, or that they are one vote less than being split if $n$ is odd, can then be approximated as $1/n$ times the forecast vote share density evaluated at 0.5. For example, in an election with 200 000 voters, we can compute this probability as `dnorm(0.5, 0.49, 0.04)/2e5`, which comes to `4.8e-5`, approximately 1 in 21 000.

This is a low probability for an individual voter; however it is not so low for a campaign. For example, if an additional 1000 voters could be persuaded to turn out for your candidate, the probability of *this* being decisive is the probability of the election being less than 1000 votes short of a tie, which in this case is approximately `1000*dnorm(0.5, 0.49, 0.04)/2e5`, or 1/21. A similar probability of decisiveness arises from convincing 10 000 people to vote in an election with 10 times as many other voters, and so on.

The key step in the above calculation is the probabilistic election forecast. The probability of a decisive vote can be substantial if the probability density is high at the 50% point, or very low if the forecast distribution is far from that midpoint. If no election-specific forecast is available, one can use a more generic probability model based on the distribution of vote shares in candidates in some large set of past elections. For example, the normal distribution with mean 0.5 and standard deviation 0.2 corresponds to a forecast that each candidate is likely to get somewhere between 30% and 70% of the vote. Under this very general model, we can calculate the probability of a tie from 200 000 voters as `dnorm(0.5, 0.5, 0.2)/2e5`, which comes to 1 in 100 000. This makes sense: if the election is not forecast to be particularly close, there is a smaller chance of it being exactly tied.

### Using an reasonable-seeming but inappropriate probability model

We next consider an alternative calculation that we have seen but which we do not think makes sense. Suppose there are $n$ voters, each with probability $p$ of voting for a particular candidate. Then the probability of an exact tie, or one vote short of a tie, can be computed using the binomial distribution. For example, with $n = 200\,000$ and $p = 0.5$, the probability of a tie election can be computed in R as `dbinom(1e5, 2e5, 0.5)`, which comes to 0.0018, or 1/560.

If we shift to $p = 0.49$, the probability of a tie—that is, exactly 10 000 votes for each candidate—becomes `dbinom(1e5, 2e5, 0.49)`, which is approximately $10^{-20}$, a number that is much too close to zero to make sense: unusual things happen in elections, and it is inappropriate to make a claim with such a high level of certainty about a future result

What went wrong? Most directly, the binomial model does not apply here. This distribution represents the number of successes in $n$ independent trials, each with probability $p$. But voters are not making their decisions independently—they are being affected by common factors such as advertising, candidates' statements, and other news. In addition, voters do not have a shared probability $p$. Some voters are almost certainly going to choose one candidate, others are almost certainly on the other side, and voters in the middle have varying leanings.

But the evident wrongness of the model is not enough, by itself, to dismiss its implications. After all, the normal distribution used earlier is only an approximation for actual forecast uncertainty. And other examples of the binomial distribution we have used, such as successes in basketball shots, can work well for many purposes even while being just approximate.

The key problem in the election example is that the binomial model does not do a good job at capturing uncertainty. Suppose we were to provisionally consider the $n$ voters as making independent decisions, and further assume (unrealistically) that they each have a common probability $p$ of voting for a certain candidate, or perhaps $p$ can be interpreted as an average probability across the electorate. Then where does $p$ come from? In any real election, we would not know this probability; we might have a sense that the election is close (so $p$ might be near 0.5), or that one side is dominant (so that $p$ will likely be far from 0.5), or maybe we have no election-specific information at all (so that $p$ could be in some wide range, corresponding to some prior distribution or reference set of historical elections). In any of these cases, $p$ is not known, and any reasonable probability model must average over a distribution for $p$, which returns us to the forecasting problem used in the earlier calculation.

To put it another way, if we were to apply the model that the number of votes received by one candidate in an election is binomial with $n = 200\,000$ and $p = 0.5$, this corresponds to *knowing* that the election is virtually tied, and this is not knowledge that would be available before any real election. Similarly, modeling the vote count as binomial with $n = 200\,000$ and $p = 0.49$ would imply an almost certain knowledge that the election outcome would be right around 0.49—not 0.485 and not 0.495—and that would not make sense for a real election. Those computed probabilities of 1/560 or $10^{-20}$ came from models that make very strong assumptions, as can be seen in this case by the extreme sensitivity of the output to the input.

### General lessons for probability modeling

In some settings, the binomial probability model can make sense. A casino can use this sort of model to estimate its distribution of winnings from a slot machine or roulette wheel, under the assumption that there is no cheating going on and the machinery has been built to tolerance and tested sufficiently that the probability assumptions are reasonable. Casino management can also compute alternatives by perturbing the probability model, for example, seeing what might happen if the probability of winning in some game is changed from 0.48 to 0.49.

Ultimately we should check our probability models by their empirical implications. When considering the probability of a tied election, we can compare to past data. Elections for the U.S. House of Representatives typically have about 200 000 votes and they are occasionally very close to tied. For example, during the twentieth century there were 6 elections decided by fewer than 10 votes and 49 elections decided by fewer than 100 votes, out of about 20 000 elections total, which suggests a probability on the order of $1/40\,000$ that a randomly chosen election will be tied.

When a probability model makes a nonsensical prediction (such as that $10^{-20}$ probability of a tied election), we can take that as an opportunity to interrogate the model and figure out which assumptions led us astray. For another example, suppose we were to mistakenly apply a normal distribution to the heights of all adults. A glance at Figure 3.6c reveals the inappropriateness of the model, which in turn leads us to realize that the normal distribution, which applies to the sum of many small factors, does not necessarily fit when there is one factor—in this case, sex—that is very predictive of the outcome. Probability modeling is a powerful tool in part because when it goes wrong we can use this failure to improve our understanding.

## 3.7 Bibliographic note

Ramsey and Schafer (2001) and Snedecor and Cochran (1989) are good sources for classical statistical methods. A quick summary of probability distributions appears in Appendix A of Gelman et al. (2013).

The data on teachers come from Tables 603 and 246 of the 2010 edition of the *Statistical Abstract of the United States*. The example of animals' body mass and metabolism comes from Gelman and Nolan (2017, section 3.8.2); for further background, see Schmidt-Nielsen (1984). The data on heights and weights of Americans come from Brainard and Burmaster (1992). See Swartz and Arce (2014) and J. F. (2015) for more on the home-court advantage in basketball. The stents example comes from Gelman, Carlin, and Nallamothu (2019). The probability of a tie vote in an election is discussed by Gelman, King, and Boscardin (1998), Mulligan and Hunter (2003), Gelman, Katz, and Bafumi (2004), and Gelman, Silver, and Edlin (2012).

## 3.8 Exercises

3.1 *Weighted averages*: A survey is conducted in a certain city regarding support for increased property taxes to fund schools. In this survey, higher taxes are supported by 50% of respondents aged 18–29, 60% of respondents aged 30–44, 40% of respondents aged 45–64, and 30% of respondents aged 65 and up. Assume there is no nonresponse.

Suppose the sample includes 200 respondents aged 18–29, 250 aged 30–44, 300 aged 45–64, and 250 aged 65+. Use the weighted average formula to compute the proportion of respondents in the *sample* who support higher taxes.

3.2 *Weighted averages*: Continuing the previous exercise, suppose you would like to estimate the proportion of all adults in the *population* who support higher taxes, so you take a weighted average as in Section 3.1. Give a set of weights for the four age categories so that the estimated proportion who support higher taxes for all adults in the city is 40%.

3.3 *Probability distributions*: Using R, graph probability densities for the normal distribution, plotting several different curves corresponding to different choices of mean and standard deviation parameters.

3.4 *Probability distributions*: Using a bar plot in R, graph the Poisson distribution with parameter 3.5.

3.5 *Probability distributions*: Using a bar plot in R, graph the binomial distribution with $n = 20$ and $p = 0.3$.

3.6 *Linear transformations*: A test is graded from 0 to 50, with an average score of 35 and a standard deviation of 10. For comparison to other tests, it would be convenient to rescale to a mean of 100 and standard deviation of 15.

(a) Labeling the original test scores as $x$ and the desired rescaled test score as $y$, come up with a linear transformation, that is, values of $a$ and $b$ so that the rescaled scores $y = a + bx$ have a mean of 100 and a standard deviation of 15.

(b) What is the range of possible values of this rescaled score $y$?

(c) Plot the line showing $y$ vs. $x$.

3.7 *Linear transformations*: Continuing the previous exercise, there is another linear transformation that also rescales the scores to have mean 100 and standard deviation 15. What is it, and why would you *not* want to use it for this purpose?

3.8 *Correlated random variables*: Suppose that the heights of husbands and wives have a correlation of 0.3, husbands' heights have a distribution with mean 69.1 and standard deviation 2.9 inches, and wives' heights have mean 63.7 and standard deviation 2.7 inches. Let $x$ and $y$ be the heights of a married couple chosen at random. What are the mean and standard deviation of the average height, $(x + y)/2$?

3.9 *Comparison of distributions*: Find an example in the scientific literature of the effect of treatment on some continuous outcome, and make a graph similar to Figure 3.9 showing the estimated population shift in the potential outcomes under a constant treatment effect.

3.10 *Working through your own example*: Continuing the example from Exercises 1.10 and 2.10, consider a deterministic model on the linear or logarithmic scale that would arise in this topic. Graph the model and discuss its relevance to your problem.

# Chapter 4

# Statistical inference

Statistical inference can be formulated as a set of operations on data that yield estimates and uncertainty statements about predictions and parameters of some underlying process or population. From a mathematical standpoint, these probabilistic uncertainty statements are derived based on some assumed probability model for observed data. In this chapter, we sketch the basics of probability modeling, estimation, bias and variance, and the interpretation of statistical inferences and statistical errors in applied work. We introduce the theme of uncertainty in statistical inference and discuss how it is a mistake to use hypothesis tests or statistical significance to attribute certainty from noisy data.

## 4.1 Sampling distributions and generative models

### Sampling, measurement error, and model error

Statistical inference is used to learn from incomplete or imperfect data. There are three standard paradigms for thinking about the role of inference:

- In the *sampling model*, we are interested in learning some characteristics of a population (for example, the mean and standard deviation of the heights of all women in the United States), which we must estimate from a sample, or subset, of that population.
- In the *measurement error model*, we are interested in learning aspects of some underlying pattern or law (for example, the coefficients $a$ and $b$ in the model $y_i = a + bx_i$), but the data are measured with error (most simply, $y_i = a + bx_i + \epsilon_i$, although one can also consider models with measurement error in $x$). Measurement error need not be additive: multiplicative models can make sense for positive data, and discrete distributions are needed for modeling discrete data.
- *Model error* refers to the inevitable imperfections of the models that we apply to real data.

These three paradigms are different: the sampling model makes no reference to measurements, the measurement model can apply even when complete data are observed, and model error can arise even with perfectly precise observations. In practice, we often consider all three issues when constructing and working with a statistical model.

For example, consider a regression model predicting students' grades from pre-test scores and other background variables. There is typically a sampling aspect to such a study, which is performed on some set of students with the goal of generalizing to a larger population. The model also includes measurement error, at least implicitly, because a student's test score is only an imperfect measure of his or her abilities, and also model error because any assumed functional form can only be approximate. In addition, any student's ability will vary by time and by circumstance; this variation can be thought of as measurement or model error.

This book follows the usual approach of setting up regression models in the measurement-error framework ($y_i = a + bx_i + \epsilon_i$), with the $\epsilon_i$'s also interpretable as model error, and with the sampling interpretation implicit in that the errors $\epsilon_1, \dots, \epsilon_n$ can be considered as a random sample from a distribution (for example, normal with mean 0 and standard deviation $\sigma$) that represents a hypothetical "superpopulation." We raise this issue only to clarify the connection between probability distributions (which are typically modeled as draws from an urn, or distribution) and the measurement or model errors used in regression.

### The sampling distribution

The *sampling distribution* is the set of possible datasets that could have been observed if the data collection process had been re-done, along with the probabilities of these possible values. The sampling distribution is determined by the data collection process, or the model being used to represent that process, which can include random sampling, treatment assignment, measurement error, model error, or some combination of all of these. The term "sampling distribution" is somewhat misleading, as this variation need not come from or be modeled by any sampling process—a more accurate term might be "probabilistic data model"—but for consistency with traditional terminology in statistics, we call it the sampling distribution even when no sampling is involved.

The simplest example of a sampling distribution is the pure random sampling model: if the data are a simple random sample of size $n$ from a population of size $N$, then the sampling distribution is the set of all samples of size $n$, all with equal probabilities. The next simplest example is pure measurement error: if observations $y_i$, $i = 1, \dots, n$, are generated from the model $y_i = a + bx_i + \epsilon_i$, with fixed coefficients $a$ and $b$, pre-specified values of the predictor $x_i$, and a specified distribution for the errors $\epsilon_i$ (for example, normal with mean 0 and standard deviation $\sigma$), then the sampling distribution is the set of possible datasets obtained from these values of $x_i$, drawing new errors $\epsilon_i$ from their assigned distribution.

As both these examples illustrate, the sampling distribution in general will not typically be known, as it depends on aspects of the population, not merely on the observed data. In the case of the simple random sample of size $n$, the sampling distribution depends on all $N$ datapoints. In practice, then, we will not *know* the sampling distribution; we can only *estimate* it. Similarly, for the measurement-error model, the sampling distribution depends on the parameters $a$, $b$, and $\sigma$, which in general will be estimated from the data, not known.

Even if data have not been collected by any random process, for statistical inference it is helpful to assume some probability model for the data. For example, in Section 7.1 we fit and interpret a regression predicting presidential election outcomes from the national economy. The 16 elections in our dataset are not a random sample from any larger set of elections, nor are elections the result of some random process. Nonetheless, we assume the model $y_i = a + bx_i + \epsilon_i$ and work out the sampling distribution implied from that.

The sampling distribution is said to be a *generative model* in that it represents a random process which, if known, could generate a new dataset. Next we discuss how to use the sampling distribution to define the statistical properties of estimates.

## 4.2 Estimates, standard errors, and confidence intervals

### Parameters, estimands, and estimates

In statistics jargon, *parameters* are the unknown numbers that determine a statistical model. For example, consider the model $y_i = a + bx_i + \epsilon_i$, in which the errors $\epsilon_i$ are normally distributed with mean 0 and standard deviation $\sigma$. The parameters in this model are $a$, $b$, and $\sigma$. The parameters $a$ and $b$ are called *coefficients*, and $\sigma$ is a called a *scale* or *variance parameter*. One way to think about estimated parameters is that they can be used to simulate new (hypothetical) data from the model.

An *estimand*, or *quantity of interest*, is some summary of parameters or data that somebody is interested in estimating. For example, in the regression model, $y = a + bx$ + error, the parameters $a$ and $b$ might be of interest—$a$ is the intercept of the model, the predicted value of $y$ when $x = 0$; and $b$ is the slope, the predicted difference in $y$, comparing two data points that differ by 1 in $x$. Other quantities of interest could be predicted outcomes for particular new data points, or combinations of predicted values such as sums, differences, averages, and ratios.

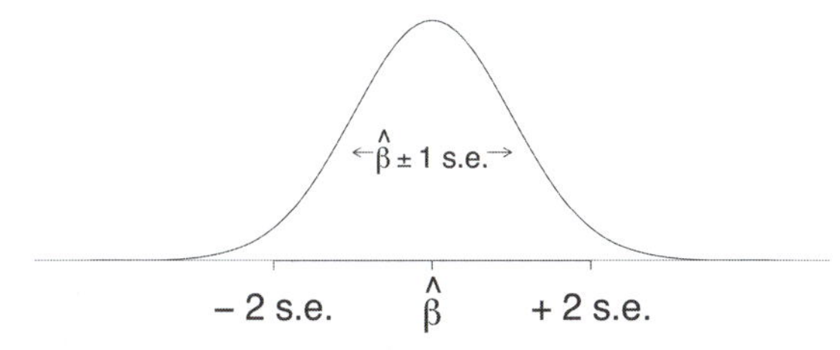


Figure 4.1 *Distribution representing uncertainty in the estimate of a quantity of interest $\beta$. The range of this distribution corresponds to the possible values of $\beta$ that are consistent with the data. In this case, we have assumed a normal distribution for this sampling distribution and therefore we have assigned an approximate 68% chance that $\beta$ will lie within 1 standard error (s.e.) of the point estimate, $\hat{\beta}$, and an approximate 95% chance that $\beta$ will lie within 2 standard errors. Assuming the model is correct, it should happen only about 5% of the time that the estimate, $\hat{\beta}$, falls more than 2 standard errors away from the true $\beta$.*

We use the data to construct *estimates* of parameters and other quantities of interest. The sampling distribution of an estimate is a byproduct of the sampling distribution of the data used to construct it. We evaluate the statistical properties of estimates analytically or by repeatedly simulating from the random sampling distribution on the computer, as discussed in Chapter 5.

## Standard errors, inferential uncertainty, and confidence intervals

The *standard error* is the estimated standard deviation of an estimate and can give us a sense of our uncertainty about the quantity of interest. Figure 4.1 illustrates in the context of a normal (bell-shaped) sampling distribution, which for mathematical reasons (the Central Limit Theorem; see page 41) can be expected to arise in many statistical contexts.

As discussed in Sections 5.3 and 9.1, in our current practice we usually summarize uncertainty using simulation, and we give the term "standard error" a looser meaning to cover any measure of uncertainty that is comparable to the posterior standard deviation.

However defined, the standard error is a measure of the variation in an estimate and gets smaller as sample size gets larger, converging on zero as the sample increases in size.

Example: Coverage of confidence intervals

The *confidence interval* represents a range of values of a parameter or quantity of interest that are roughly consistent with the data, given the assumed sampling distribution. If the model is correct, then in repeated applications the 50% and 95% confidence intervals will include the true value 50% and 95% of the time; see Figure 4.2 and Exercise 5.7.[1]

The usual 95% confidence interval for large samples, based on an assumption that the sampling distribution follows the normal distribution, is to take an estimate $\pm 2$ standard errors; see Figure 4.1. Also from the normal distribution, an estimate $\pm 1$ standard error is a 68% interval, and an estimate $\pm\frac{2}{3}$ of a standard error is a 50% interval. A 50% interval is particularly easy to interpret since the true value should be as likely to be inside as outside the interval. A 95% interval based on the normal distribution is about three times as wide as a 50% interval.

## Standard errors and confidence intervals for averages and proportions

When estimating the mean of an infinite population, given a simple random sample of size $n$, the standard error is $\sigma/\sqrt{n}$, where $\sigma$ is the standard deviation of the measurements in the population. This property holds regardless of any assumption about the shape of the sampling distribution, but the standard error might be less informative for sampling distributions that are far from normal.

A proportion is a special case of an average in which the data are 0's and 1's. Consider a survey of size $n$ with $y$ Yes responses and $n-y$ No responses. The estimated proportion of the population who would answer Yes to this survey is $\hat{p} = y/n$, and the standard error of this estimate is $\sqrt{\hat{p}(1-\hat{p})/n}$. If

[1]Code for this example is in folder `Coverage`.

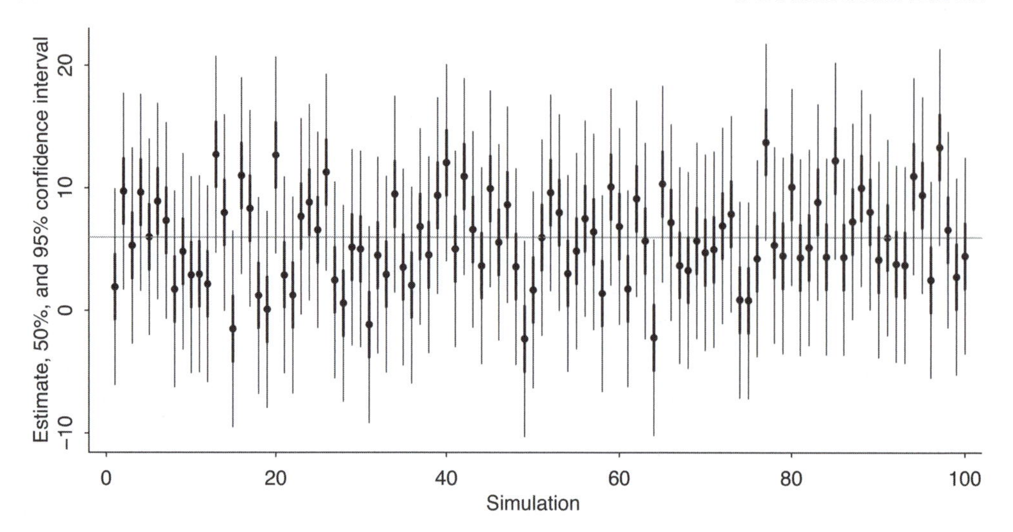


Figure 4.2 *Simulation of coverage of confidence intervals: the horizontal line shows the true parameter value, and dots and vertical lines show estimates and confidence intervals obtained from 100 random simulations from the sampling distribution. If the model is correct, 50% of the 50% intervals and 95% of the 95% intervals should contain the true parameter value, in the long run.*

$p$ is near 0.5, we can approximate this by $0.5/\sqrt{n}$. Consider that $\sqrt{0.5 * 0.5} = 0.5$, $\sqrt{0.4 * 0.6} = 0.49$, and $\sqrt{0.3 * 0.7} = 0.46$.)

Confidence intervals for proportions come directly from the standard-error formula. If 700 people in a random sample support the death penalty and 300 oppose it, then a 95% interval for the proportion of supporters in the population is simply $[0.7 \pm 2\sqrt{0.7 * 0.3/1000}] = [0.67, 0.73]$ or, in R,

```
estimate <- y/n
se <- sqrt(estimate*(1-estimate)/n)
int_95 <- estimate + qnorm(c(0.025, 0.975))*se
```

### Standard error and confidence interval for a proportion when $y = 0$ or $y = n$

The above estimate and standard error are usually reasonable unless the number of Yes or the number of No responses is close to zero. Conventionally, the approximation is considered acceptable if $y$ and $n - y$ are both at least 5. In the extreme case in which $y = 0$ or $n - y = 0$, there is an obvious problem in that the formula yields a standard error estimate of zero, and thus a zero-width confidence interval.

A standard and reasonable quick correction for constructing a 95% interval when $y$ or $n - y$ is near zero is to use the estimate $\hat{p} = \frac{y+2}{n+4}$ with standard error $\sqrt{\hat{p}(1-\hat{p})/(n+4)}$. For example, if $y = 0$ and $n = 75$, the 95% interval is $[\hat{p} \pm 2\,\text{s.e.}]$, where $\hat{p} = \frac{2}{79} = 0.025$ and s.e. $= \sqrt{(0.025)(1-0.025)/79} = 0.018$; this comes to $[-0.01, 0.06]$. It makes no sense for the interval for a proportion to contain negative values, so we truncate the interval to obtain $[0, 0.06]$. If $y = n$, we perform the same procedure but set the upper bound of the interval to 1.

### Standard error for a comparison

The standard error of the difference of two independent quantities is computed as,

$$\text{standard error of the difference} = \sqrt{\text{se}_1^2 + \text{se}_2^2}. \tag{4.1}$$

Consider a survey of 1000 people—400 men and 600 women—who are asked how they plan to vote in an upcoming election. Suppose that 57% of the men and 45% of the women plan to vote for the Republican candidate. The standard errors for these proportions are $\text{se}_{\text{men}} = \sqrt{0.57 * 0.43/400}$ and $\text{se}_{\text{women}} = \sqrt{0.45 * 0.55/600}$. The estimated gender gap in support for the Republican is $0.57 - 0.45 = 0.12$, with standard error $\sqrt{\text{se}^2_{\text{men}} + \text{se}^2_{\text{women}}} = 0.032$.

### Sampling distribution of the sample mean and standard deviation; normal and $\chi^2$ distributions

Suppose you draw $n$ data points $y_1, \dots, y_n$ from a normal distribution with mean $\mu$ and standard deviation $\sigma$, and then compute the sample mean, $\bar{y} = \frac{1}{n}\sum_{i=1}^n y_i$, and standard deviation, $s_y = \sqrt{\frac{1}{n-1}\sum_{i=1}^n (y_i - \bar{y})^2}$. These two statistics have a sampling distribution that can be derived mathematically from the properties of independent samples from the normal. The sample mean, $\bar{y}$, is normally distributed with mean $\mu$ and standard deviation $\sigma/\sqrt{n}$. The sample standard deviation has a distribution defined as follows: $s_y^2 * (n-1)/\sigma^2$ has a $\chi^2$ distribution with $n - 1$ degrees of freedom. We give neither the formulas nor the derivations of these distributions here, as they are not necessary for the applied methods in this book. But it is good to know the names of these distributions, as we can compute their quantiles and simulate from them in R to get a sense of what can be expected from data summaries. In addition, these distributions reappear in regression modeling when performing inference for coefficients and residual variation.

### Degrees of freedom

The concept of *degrees of freedom* arises with the $\chi^2$ distribution and several other places in probability and statistics. Without going into the technical details, we can briefly say that degrees of freedom relate to the need to correct for overfitting when estimating the error of future predictions from a fitted model. If we simply calculate predictive error on the same data that were used to fit the model, we will tend to be optimistic unless we adjust for the parameters estimated involved in the fitting. Roughly speaking, we can think of observed data as supplying $n$ "degrees of freedom" that can be used for parameter estimation, and a regression with $k$ coefficients is said to use up $k$ of these degrees of freedom. The fewer degrees of freedom that remain at the end, the larger the required adjustment for overfitting, as we discuss more carefully in Section 11.8. The degrees of freedom in the above-mentioned $\chi^2$ distribution corresponds to uncertainty in the estimation of the residual error in the model, which in turn can be shown to map to overfitting adjustments in prediction.

### Confidence intervals from the *t* distribution

The $t$ distribution is a family of symmetric distributions with heavier tails (that is, a greater frequency of extreme values) compared to the normal distribution. The $t$ is characterized by a center, a scale, and a degrees of freedom parameter that can range from 1 to $\infty$. Distributions in the $t$ family with low degrees of freedom have very heavy tails; in the other direction, in the limit as degrees of freedom approach infinity, the $t$ distribution approaches the normal.

When a standard error is estimated from $n$ data points, we can account for uncertainty using the $t$ distribution with $n - 1$ degrees of freedom, calcuated as $n$ data points minus 1 because of the mean is being estimated from the data. Suppose an object is weighed five times, with measurements $y = 35, 34, 38, 35, 37$, which have an average value of 35.8 and a standard deviation of 1.6. In R, we can create the 50% and 95% $t$ intervals based on 4 degrees of freedom as follows:

```
n <- length(y)
estimate <- mean(y)
se <- sd(y)/sqrt(n)
int_50 <- estimate + qt(c(0.25, 0.75), n-1)*se
int_95 <- estimate + qt(c(0.025, 0.975), n-1)*se
```

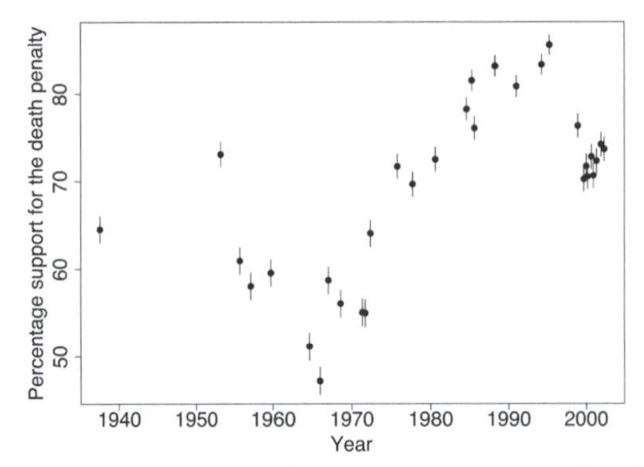


Figure 4.3 *Illustration of visual comparison of confidence or uncertainty intervals. Graph displays the proportion of respondents supporting the death penalty (estimates ±1 standard error—that is, 68% confidence intervals—under the simplifying assumption that each poll was a simple random sample of size 1000), from Gallup polls over time.*

The difference between the normal and $t$ intervals will only be apparent with low degrees of freedom.

### Inference for discrete data

For nonbinary discrete data, we can simply use the continuous formula for the standard error. Consider a hypothetical survey that asks 1000 randomly selected adults how many dogs they own, and suppose 600 have no dog, 300 have 1 dog, 50 have 2 dogs, 30 have 3 dogs, and 20 have 4 dogs. What is a 95% confidence interval for the average number of dogs in the population? If the data are not already specified in a file, we can quickly create the data as a vector of length 1000 in R:

```
y <- rep(c(0,1,2,3,4), c(600,300,50,30,20))
```

We can then continue by computing the mean, standard deviation, standard error, and confidence interval as shown with continuous data above.

### Linear transformations

To get confidence intervals for a linearly transformed parameter, simply transform the intervals. In the above example, the 95% interval for the number of dogs per person is [0.52, 0.62]. Suppose this (hypothetical) random sample were taken in a city of 1 million adults. The confidence interval for the total number of pet dogs in the city is then [520 000, 620 000].

### Comparisons, visual and numerical

Example: Death penalty opinions

Uncertainties can often be compared visually, as in Figure 4.2, which displays 68% confidence intervals for the proportion of American adults supporting the death penalty (among those with an opinion on the question), from a series of Gallup polls.[2] For an example of a formal comparison, consider a change in the estimated support for the death penalty from [80% ± 1.4%] to [74% ± 1.3%]. The estimated difference is 6%, with a standard error of $\sqrt{(1.4\%)^2 + (1.3\%)^2} = 1.9\%$.

### Weighted averages

Confidence intervals for other derived quantities can be determined by appropriately combining the separate means and variances. Suppose that separate surveys conducted in France, Germany, Italy, and

[2]Data and code for this example are in the folder `Death`.

other countries yield estimates of $0.55 \pm 0.02$, $0.61 \pm 0.03$, $0.38 \pm 0.03$, ..., for some opinion question. The estimated proportion for all adults in the European Union is $\frac{N_1}{N_{\rm tot}}0.55 + \frac{N_2}{N_{\rm tot}}0.61 + \frac{N_3}{N_{\rm tot}}0.38 + \cdots$, where $N_1, N_2, N_3, \ldots$ are the total number of adults in France, Germany, Italy, ..., and $N_{\rm tot}$ is the total number in the European Union. Put this all together, and the standard error of this weighted average becomes $\sqrt{(\frac{N_1}{N_{\rm tot}}0.02)^2 + (\frac{N_2}{N_{\rm tot}}0.03)^2 + (\frac{N_3}{N_{\rm tot}}0.03)^2 + \cdots}$.

Given N, p, se—the vectors of population sizes, estimated proportions of Yes responses, and standard errors—we can first compute the stratum weights and then compute the weighted average and its 95% confidence interval in R:

```
W <- N/sum(N)
weighted_avg <- sum(W*p)
se_weighted_avg <- sqrt(sum(W*se)^2)
interval_95 <- weighted_avg + c(-2,2)*se_weighted_avg
```

## 4.3 Bias and unmodeled uncertainty

The inferences discussed above are all consistent on the model being true, with unbiased measurements, random samples, and randomized experiments. But real data collection is imperfect, and where possible we should include the possibility of model error in our inferences and predictions.

### Bias in estimation

Roughly speaking, we say that an estimate is *unbiased* if it is correct on average. For a simple example, consider a survey, a simple random sample of adults in the United States, in which each respondent is asked the number of hours he or she spends watching television each day. Assuming responses are complete and accurate, the average response in the *sample* is an unbiased estimate of the average number of hours watched in the *population*. Now suppose that women are more likely than men to answer the survey, with nonresponse depending only on sex. In that case, the sample will, on average, overrepresent women, and women on average watch less television than men; hence, the average number of hours watched in the sample is now a *biased* estimate of the proportion in the population. It is possible to correct for this bias by reweighting the sample as in Section 3.1; recognizing the existence of the bias is the first step in fixing it.

In practice, it is typically impossible to construct estimates that are truly unbiased, because any bias correction will itself only be approximate. For example, we can correct the bias in a sample that overrepresents women and underrepresents men, but there will always be other biases: the sample might overrepresent white people, more educated people, older people, and so on.

To put it another way, bias depends on the sampling distribution of the data, which is almost never exactly known: random samples and randomized experiments are imperfect in reality, and any approximations become even more tenuous when applied to observational data. Nonetheless, a theoretically defined sampling distribution can still be a helpful reference point, and so we speak of the bias and variation of our estimates, recognizing that these are defined relative to some assumptions.

### Adjusting inferences to account for bias and unmodeled uncertainty

Consider the following example: a poll is conducted with 600 respondents to estimate the proportion of people who support some political candidate, and your estimate then has a standard error of approximately $0.5/\sqrt{600} = 0.02$, or 2 percentage points. Now suppose you could redo the survey with 60 000 respondents. With 100 times the sample size, the standard error will be divided by 10, thus the estimated proportion will have a standard error of 0.002, or 0.2 percentage points, much too low to use as a measure of uncertainty for the quantity of interest.

What is wrong with taking a survey with $n = 60\,000$ and saying that the support for the candidate is 52.5% ± 0.2%, or reporting a confidence interval of [52.1, 52.9]? The problem here is that 0.002 is the scale of the sampling error, but there are many other sources of uncertainty, including nonsampling error (coming from the sample not being representative, because different people can choose to answer the survey), systematic differences between survey respondents and voters, variation in opinion over time, and inaccurate survey responses. All these other sources of error represent unknown levels of bias and variance. From past election campaigns, we know that opinions can easily shift by a percentage point or two from day to day, so 0.2% would represent meaningless precision.

Survey respondents are not balls drawn from an urn, and the probabilities in the "urn" are changing over time. In other examples, there are problems with measurement or with the assumption that treatments are randomly assigned in an experiment or observational study. How can we account for sources of error that are not in our statistical model? In general, there are three ways to go: improve data collection, expand the model, and increase stated uncertainty.

Data collection can be improved using more careful measurement and sampling: in the polling example, this could be done by collecting data at different places and times: instead of a single massive survey of 60 000 people, perform a series of 600-person polls, which will allow you to estimate sources of variability other than sampling error.

Models can be expanded in many ways, often using regression modeling as described later in this book. For example, instead of assuming that survey respondents are a simple random sample, we can divide the population into demographic and geographic categories and assume a simple random sample within each category; see Section 17.1. This model is still far from perfect, but it allows us to reduce bias in estimation by adjusting for certain known differences between sample and population. Similar modeling can be performed to adjust for differences between treatment and control groups in a causal analysis, as discussed in Chapter 20.

Finally, when all else fails—which it will—you can increase your uncertainty to account for unmodeled sources of error. We typically assume errors are independent, and so we capture additional uncertainty by adding variances. The variance is the square of the standard deviation; see Section 3.5. For example, in polling for U.S. elections, nonsampling error has been estimated to have a standard error of approximately 2.5 percentage points. So if we want to account for total uncertainty in our survey of 600 people, we would use a standard error of $\sqrt{2^2 + 2.5^2} = 3.2$ percentage points, and for the survey of 60 000 people, the standard error would be $\sqrt{0.2^2 + 2.5^2} = 2.51$ percentage points. This formula shows that very little would be gained by increasing the sample size in this way.

More generally, we can think of total uncertainty as $\sqrt{S_1^2 + S_2^2}$, where $S_1$ is the standard error (that is, the standard deviation of an estimate) from the model, and $S_2$ represents the standard deviation of independent unmodeled uncertainty. The mathematics of this expression implies that it will typically be most effective to reduce the *larger* of these two quantities. For example, suppose that $S_1 = 2$ and $S_2 = 5$, so we start with a total uncertainty of $\sqrt{2^2 + 5^2} = 5.5$. If we reduce the first source of error from 2 to 1, that takes us down to $\sqrt{1^2 + 5^2} = 5.1$. But if we reduce the second source of error from 5 to 4, that reduces the total uncertainty to $\sqrt{1^2 + 4^2} = 4.1$. In this case, a 20% reduction in the larger error was more effective than a 50% reduction in the smaller error.

As noted above, unmodeled error can be decisive in many real problems. Why, then, in this book do we focus on quantification of error within models? The simple answer is that this is what we can do: modeled error is what statistical methods can handle most easily. In practice, we should be aware of sources of errors that are not in our models, we should design our data collection to minimize such errors, and we should set up suitably complex models to capture as much uncertainty and variation as we can. Indeed, this is a key role of regression: adding information to a model to improve prediction should also allow us to better capture uncertainty in generalization to new data. Beyond this, we recognize that our inferences depend on assumptions such as representativeness and balance (after adjusting for predictors) and accurate measurement, and it should always be possible to increase standard errors and widen interval estimates to account for additional sources of uncertainty.

## 4.4 Statistical significance, hypothesis testing, and statistical errors

One concern when performing data analysis is the possibility of mistakenly coming to strong conclusions that do not replicate or do not reflect real patterns in the underlying population. Statistical theories of hypothesis testing and error analysis have been developed to quantify these possibilities in the context of inference and decision making.

### Statistical significance

A commonly used decision rule that we do *not* recommend is to consider a result as stable or real if it is "statistically significant" and to take "non-significant" results to be noisy and to be treated with skepticism. For reasons discussed in this section and the next, we prefer not to focus on statistical significance, but the concept is important enough in applied statistics that we define it here.

Statistical significance is conventionally defined as a $p$-value less than 0.05, relative to some *null hypothesis* or prespecified value that would indicate no effect present, as discussed below in the context of hypothesis testing. For fitted regressions, this roughly corresponds to coefficient estimates being labeled as statistically significant if they are at least two standard errors from zero, or not statistically significant otherwise.

Speaking more generally, an estimate is said to be not statistically significant if the observed value could reasonably be explained by simple chance variation, much in the way that a sequence of 20 coin tosses might happen to come up 8 heads and 12 tails; we would say that this result is not statistically significantly different from chance. In that example, the observed proportion of heads is 0.40 but with a standard error of 0.11—thus, the data are less than two standard errors away from the null hypothesis of 50%.

### Hypothesis testing for simple comparisons

We shall review the key concepts of conventional hypothesis testing with a simple hypothetical example. A randomized experiment is performed to compare the effectiveness of two drugs for lowering cholesterol. The mean and standard deviation of the post-treatment cholesterol levels are $\bar{y}_T$ and $s_T$ for the $n_T$ people in the treatment group, and $\bar{y}_C$ and $s_C$ for the $n_C$ people in the control group.

**Estimate, standard error, and degrees of freedom.** The parameter of interest here is $\theta = \theta_T - \theta_C$, the expectation of the post-test difference in cholesterol between the two groups. Assuming the experiment has been done correctly, the estimate is $\hat{\theta} = \bar{y}_T - \bar{y}_C$ and the standard error is $\text{se}(\hat{\theta}) = \sqrt{s_C^2/n_C + s_T^2/n_T}$. The approximate 95% interval is then $[\hat{\theta} \pm t^{0.975}_{n_C+n_T-2} * \text{se}(\hat{\theta})]$, where $t^{0.975}_{df}$ is the 97.5$^{\text{th}}$ percentile of the unit $t$ distribution with $df$ degrees of freedom. In the limit as $df \to \infty$, this quantile approaches 1.96, corresponding to the normal-distribution 95% interval of $\pm 2$ standard errors.

**Null and alternative hypotheses.** To frame the above problem as a hypothesis test problem, one must define *null* and *alternative* hypotheses. The null hypothesis is $\theta = 0$, that is, $\theta_T = \theta_C$, and the alternative is $\theta \neq 0$, that is, $\theta_T \neq \theta_C$.

The hypothesis test is based on a *test statistic* that summarizes the deviation of the data from what would be expected under the null hypothesis. The conventional test statistic in this sort of problem is the absolute value of the $t$-score, $t = |\hat{\theta}|/\text{se}(\hat{\theta})$, with the absolute value representing a "two-sided test," so called because either positive or negative deviations from zero would be noteworthy.

**$p$-value.** In a hypothesis test, the deviation of the data from the null hypothesis is summarized by the *p-value*, the probability of observing something at least as extreme as the observed test statistic. For this problem, under the null hypothesis the test statistic has a unit $t$ distribution with $\nu$ degrees of freedom. In R, we can compute the $p$-value by doing this: `2*(1 - pt(abs(theta_hat)/se_theta, n_C+n_T, 2))`. If the standard deviation of $\theta$ is known, or if the sample size is large, we can use the normal distribution (also called the $z$-test) instead. The factor of 2 in the previous expression corresponds to a *two-sided test* in which the hypothesis is rejected if the observed difference is too much higher or too much lower than the comparison point of 0. In common practice, the null hypothesis is said to be "rejected" if the $p$-value is less than 0.05—that is, if the 95% confidence interval for the parameter excludes zero.

### Hypothesis testing: general formulation

In the simplest form of hypothesis testing, the null hypothesis $H_0$ represents a particular probability model, $p(y)$, with potential replication data $y^{\text{rep}}$. To perform a hypothesis test, we must define a test statistic $T$, which is a function of the data. For any given data $y$, the $p$-value is then $\Pr(T(y^{\text{rep}}) \geq T(y))$: the probability of observing, under the model, something as or more extreme than the data.

In regression modeling, testing is more complicated. The model to be fit can be written as $p(y|x, \theta)$, where $\theta$ represents a set of parameters including coefficients, residual standard deviation, and possibly other parameters, and the null hypothesis might be that some particular coefficient of interest equals zero. To fix ideas, consider the model $y_i = a + bx_i + \text{error}_i$, where the errors are normally distributed with mean 0 and standard deviation $\sigma$. Then $\theta$ is the vector $(a, b, \sigma)$. In such settings, one might test the hypothesis that $b = 0$; thus, the null hypothesis is *composite* and corresponds to the regression model with parameters $(a, 0, \sigma)$ for any values of $a$ and $\sigma$. The $p$-value, $\Pr(T(y^{\text{rep}}) \geq T(y))$, then depends on $a$ and $\sigma$, and what is typically done is to choose the maximum (that is, most conservative) $p$-value in this set. To put it another way, the hypothesis test is performed on the null distribution that is closest to the data.

Much more can be said about hypothesis testing. For our purposes here, all that is relevant is how to interpret and compute $p$-values for simple comparisons, and how to understand the general connections between statistical significance, $p$-values, and the null hypothesis.

### Comparisons of parameters to fixed values and each other: interpreting confidence intervals as hypothesis tests

The hypothesis that a parameter equals zero (or any other fixed value) can be directly tested by fitting the model that includes the parameter in question and examining the corresponding 95% interval. If the interval excludes zero (or the specified fixed value), then the hypothesis is said to be rejected at the 5% level.

Testing whether two parameters are equal is equivalent to testing whether their difference equals zero. We can do this by including both parameters in the model and then examining the 95% interval for their difference. As with inference for a single parameter, the confidence interval is commonly of more interest than the hypothesis test. For example, if support for the death penalty has decreased by $6 \pm 2$ percentage points, then the magnitude of this estimated difference is probably as important as that the confidence interval for the change excludes zero.

The hypothesis of whether a parameter is positive is directly assessed via its confidence interval. Testing whether one parameter is greater than the other is equivalent to examining the confidence interval for their difference and testing for whether it is entirely positive.

The possible outcomes of a hypothesis test are "reject" or "not reject." It is never possible to "accept" a statistical hypothesis, only to find that the data are not sufficient to reject it. This wording may feel cumbersome but we need to be careful, as it is a common mistake for researchers to act as if an effect is negligible or zero, just because this hypothesis cannot be rejected from data at hand.

### Type 1 and type 2 errors and why we don't like talking about them

Statistical tests are typically understood based on *type 1 error*—the probability of falsely rejecting a null hypothesis, if it is in fact true–and *type 2 error*—the probability of *not* rejecting a null hypothesis that is in fact false. But this paradigm does not match up well with much of social science, or science more generally.

A fundamental problem with type 1 and type 2 errors is that in many problems we do not think the null hypothesis can be true. For example, a change in law will produce *some* changes in behavior; the question is how these changes vary across people and situations. Similarly, a medical intervention will work differently for different people, and a political advertisement will change the opinions of some people but not others. In all these settings, one can imagine an average effect that is positive or negative, depending on whom is being averaged over, but there is no particular interest in a null hypothesis of no effect. The second concern is that, in practice, when a hypothesis test is rejected (that is, when a study is a success), researchers and practitioners report, and make decisions based on, the point estimate of the magnitude and sign of the underlying effect. So, in evaluating a statistical test, we should be interested in the properties of the associated effect-size estimate, conditional on it being statistically significantly different from zero.

The type 1 and 2 error framework is based on a deterministic approach to science that might be appropriate in the context of large effects (including, perhaps, some of the domains in which significance testing was developed in the early part of the last century), but it is much less relevant in modern social and biological sciences with highly variable effects.

### Type M (magnitude) and type S (sign) errors

With these concerns in mind, we prefer the concepts of type S ("sign") and type M ("magnitude") errors, both of which can occur when a researcher makes a *claim with confidence* (traditionally, a $p$-value of less than 0.05 or a confidence interval that excludes zero, but more generally any statement that is taken as strong evidence of a positive effect). A *type S error* occurs when the sign of the estimated effect is of the opposite direction as the true effect. A *type M error* occurs when the magnitude of the estimated effect is much different from the true effect. A statistical procedure can be characterized by its type S error rate—the probability of an estimate being of the opposite sign of the true effect, conditional on the estimate being statistically significant—and its expected exaggeration factor—the expected ratio of the magnitude of the estimated effect divided by the magnitude of the underlying effect.

When a statistical procedure is noisy, the type S error rate and the exaggeration factor can be large. In Section 16.1 we give an example where the type S error rate is 24% (so that a statistically significant estimate has a one-quarter chance of being in the wrong direction) and the expected exaggeration factor is 9.7.

In quantitative research we are particularly concerned with type M errors, or exaggeration factors, which can be understood in light of the "statistical significance filter." Consider any statistical estimate. For it to be *statistically significant*, it has to be at least two standard errors from zero: if an estimate has a standard error of $S$, any publishable estimate must be at least $2S$ in absolute value. Thus, the larger the standard error, the higher the estimate *must* be, if it is to be published and taken as serious evidence. No matter how large or small the underlying effect, the minimum statistically significant effect size *estimate* has this lower bound. This selection bias induces type M error.

### Hypothesis testing and statistical practice

We do not generally use null hypothesis significance testing in our own work. In the fields in which we work, we do not generally think null hypotheses can be true: in social science and public health, just about every treatment one might consider will have *some* effect, and no comparisons or regression coefficient of interest will be *exactly* zero. We do not find it particularly helpful to formulate and test null hypotheses that we know ahead of time cannot be true. Testing null hypotheses is just a matter of data collection: with sufficient sample size, any hypothesis can be rejected, and there is no real point to gathering a mountain of data just to reject a hypothesis that we did not believe in the first place.

That said, not all effects and comparisons are detectable from any given study. So, even though we do not ever have the research goal of rejecting a null hypothesis, we do see the value of checking the consistency of a particular dataset with a specified null model. The idea is that *non-rejection* tells us that there is not enough information in the data to move beyond the null hypothesis. We give an example in Section 4.6. Conversely, the point of *rejection* is not to disprove the null—in general, we disbelieve the null hypothesis even before the start of any study—but rather to indicate that there is information in the data to allow a more complex model to be fit.

A use of hypothesis testing that bothers us is when a researcher starts with hypothesis A (for example, that a certain treatment has a generally positive effect), then as a way of confirming hypothesis A, the researcher comes up with null hypothesis B (for example, that there is a zero correlation between treatment assignment and outcome). Data are found that reject B, and this is taken as evidence in support of A. The problem here is that a *statistical* hypothesis (for example, $\beta = 0$ or $\beta_1 = \beta_2$) is much more specific than a *scientific* hypothesis (for example, that a certain comparison averages to zero in the population, or that any net effects are too small to be detected). A rejection of the former does not necessarily tell you anything useful about the latter, because violations of technical assumptions of the statistical model can lead to high probability of rejection of the null hypothesis even in the absence of any real effect. What the rejection *can* do is motivate the next step of modeling the comparisons of interest.

## 4.5 Problems with the concept of statistical significance

A common statistical error is to summarize comparisons by statistical significance and to draw a sharp distinction between significant and nonsignificant results. The approach of summarizing by statistical significance has five pitfalls: two that are obvious and three that are less well understood.

### Statistical significance is not the same as practical importance

A result can be statistically significant—not easily explainable by chance alone—but without being large enough to be important in practice. For example, if a treatment is estimated to increase earnings by \$10 per year with a standard error of \$2, this would be statistically but not practically significant (in the U.S. context). Conversely, an estimate of \$10 000 with a standard error of \$10 000 would not be statistically significant, but it has the possibility of being important in practice (and is also consistent with zero or negative effects).

### Non-significance is not the same as zero

In Section 3.5 we discussed a study of the effectiveness of arterial stents for heart patients. For the primary outcome of interest, the treated group outperformed the control, but not statistically significantly so: the observed average difference in treadmill time was 16.6 seconds with a standard error of 9.8, corresponding to a 95% confidence interval that included zero and a $p$-value of 0.20. A fair summary here is that the results are uncertain: it is unclear whether the net treatment effect is positive or negative in the general population. It would be inappropriate to say that stents have no effect.

### The difference between "significant" and "not significant" is not itself statistically significant

Changes in statistical significance do not themselves necessarily achieve statistical significance. By this, we are not merely making the commonplace observation that any particular threshold is arbitrary—for example, only a small change is required to move an estimate from a 5.1% significance level to 4.9%, thus moving it into statistical significance. Rather, we are pointing out that even large changes in significance levels can correspond to small, nonsignificant changes in the underlying variables.

For example, consider two independent studies with effect estimates and standard errors of $25 \pm 10$ and $10 \pm 10$. The first study is statistically significant at the 1% level, and the second is not at all significant at 1 standard error away from zero. Thus it would be tempting to conclude that there is a large difference between the two studies. In fact, however, the difference is not even close to being statistically significant: the estimated difference is 15, with a standard error of $\sqrt{10^2 + 10^2} = 14$.

### Researcher degrees of freedom, $p$-hacking, and forking paths

Another problem with statistical significance is that it can be attained by multiple comparisons, or multiple potential comparisons. When there are many ways that data can be selected, excluded, and analyzed in a study, it is not difficult to attain a low $p$-value even in the absence of any true underlying pattern. The problem here is *not* just the "file-drawer effect" of leaving non-significant findings unpublished, but also that any given study can involve a large number of "degrees of freedom" available to the researcher when coding data, deciding which variables to include in the analysis, and deciding how to perform and summarize the statistical modeling. Even if a published article shows just a single regression table, there could well be thousands of possible alternative analyses of the same data that are equally consistent with the posited theory. Researchers can use this freedom to "$p$-hack" and achieve a low $p$-value (and thus statistical significance) from otherwise unpromising data. Indeed, with sufficient effort, statistically significant patterns can be found from just about any data at all, as researchers have demonstrated by finding patterns in pure noise, or in one memorable case by finding statistically significant results from a medical imaging study performed on a dead salmon.

Researcher degrees of freedom can lead to a multiple comparisons problem, even in settings where researchers perform only a single analysis on their data. The problem is there can be a large number of potential comparisons when the details of data analysis are highly contingent on data, without the researcher having to perform any conscious procedure of fishing or examining multiple $p$-values.

Consider the following testing procedures:

1. Simple classical test based on a unique test statistic, $T$, which when applied to the observed data yields $T(y)$.
2. Classical test pre-chosen from a set of possible tests: thus, $T(y; \phi)$, with preregistered $\phi$. Here, $\phi$ does *not* represent parameters in the model; rather, it represents choices in the analysis. For example, $\phi$ might correspond to choices of control variables in a regression, transformations, and data coding and excluding rules, as well as deciding on which main effect or interaction to focus.
3. Researcher degrees of freedom without fishing: computing a single test based on the data, but in an environment where a different test would have been performed given different data; thus $T(y; \phi(y))$, where the function $\phi(\cdot)$ is observed in the observed case.
4. "Fishing": computing $T(y; \phi(y))$, for $j = 1, \ldots, J$: that is, performing $J$ tests and then reporting the best result given the data, thus $T(y; \phi^{\text{best}}(y))$.

We believe that researchers are commonly doing #3, but the confusion is that, when this problem is pointed out to them, researchers think they are being accused of doing #4. To put it another way, researchers assert that they are not doing #4 and the implication is that they are doing #2. The problem with #3 is that, even without explicit fishing, a researcher can induce a huge number of researcher degrees of freedom and thus obtain statistical significance from noisy data, leading to apparently strong conclusions that do not truly represent the underlying population or target of study and that fail to reproduce in future controlled studies.

Our recommended solution to this problem of "forking paths" is not to compute adjusted $p$-values but rather to directly model the variation that is otherwise hidden in all these possible data coding and analysis choices, and to accept uncertainty and not demand statistical significance in our results.

### The statistical significance filter

A final concern is that statistically significant estimates tend to be overestimates. This is the type M, or magnitude, error problem discussed in Section 4.4. Any estimate with $p < 0.05$ is by necessity at least two standard errors from zero. If a study has a high noise level, standard errors will be high, and so statistically significant estimates will automatically be large, no matter how small the underlying effect. Thus, routine reliance on published, statistically significant results will lead to systematic overestimation of effect sizes and a distorted view of the world.

All the problems discussed above have led to what has been called a replication crisis, in which studies published in leading scientific journals and conducted by researchers at respected universities have failed to replicate. Many different problems in statistics and the culture of science have led to the replication crisis; for our purposes here, what is relevant is to understand how to avoid some statistical misconceptions associated with overcertainty.

### Example: A flawed study of ovulation and political attitudes

Example: Ovulation and voting

We demonstrate the last two problems mentioned above—multiple potential comparisons and the statistical significance filter—using the example of a research article published in a leading journal of psychology. The article begins:

> Each month many women experience an ovulatory cycle that regulates fertility. Whereas research finds that this cycle influences women's mating preferences, we propose that it might also change women's political and religious views. Building on theory suggesting that political and religious orientation are linked to reproductive goals, we tested how fertility influenced women's politics, religiosity, and voting in the 2012 U.S. presidential election. In two studies with large and diverse samples, ovulation had drastically different effects on single versus married women. Ovulation led single women to become more liberal, less religious, and more likely to vote for Barack Obama. In contrast, ovulation led married women to become more conservative, more religious, and more likely to vote for Mitt Romney. In addition, ovulatory-induced changes in political orientation mediated women's voting behavior. Overall, the ovulatory cycle not only influences women's politics, but appears to do so differently for single versus married women.

One problem here is that there are so many different things that could be compared, but all we see is some subset of the comparisons. Some of the choices available in this analysis include the days of the month characterized as peak fertility, the dividing line between single and married (in this particular study, unmarried but partnered women were counted as married), data exclusion rules based on reports of menstrual cycle length and timing, and the decision of which interactions to study. Given all these possibilities, it is no surprise at all that statistically significant comparisons turned up; this would be expected even were the data generated purely by noise.

In addition, relative to our understanding of the vast literature on voting behavior, the claimed effects seem implausibly large—a type M error. For example, the paper reports that, among women in relationships, 40% in the ovulation period supported Romney, compared to 23% in the non-fertile part of their cycle. Given that opinion polls find very few people switching their vote preferences during the campaign for any reason, these numbers seem unrealistic. The authors might respond that they don't care about the magnitude of the difference, just the sign, but (a) with a magnitude of this size, we are talking noise (not just sampling error but also errors in measurement), and (b) one could just as easily explain this as a differential nonresponse pattern: maybe liberal or conservative women in different parts of their cycle are more or less likely to participate in a survey. It would be easy enough to come up with a story about that.

As researchers and as evaluators of the research of others, we need to avoid the trap of considering

| | | | | | | |
|---|---|---|---|---|---|---|
| Clotelia Smith | 208 | 416 | 867 | 1259 | 1610 | 2020 |
| Earl Coppin | 55 | 106 | 215 | 313 | 401 | 505 |
| Clarissa Montes | 133 | 250 | 505 | 716 | 902 | 1129 |
| ... | ... | ... | ... | ... | ... | ... |

Figure 4.4 *Subset of results from the cooperative board election, with votes for each candidate (names altered for anonymity) tallied after 600, 1200, 2444, 3444, 4444, and 5553 votes. These data were viewed as suspicious because the proportion of votes for each candidate barely changed as the vote counting went on. (There were 27 candidates in total, and each voter was allowed to choose 6 candidates.)*

this sort of small study as providing definitive evidence—even if certain comparisons happen to be statistically significant.

## 4.6 Example of hypothesis testing: 55,000 residents need your help!

Example: Co-op election

We illustrate the application of statistical hypothesis testing with a story. One day several years ago, we received a fax, entitled HELP!, from a member of a residential organization:

> Last week we had an election for the Board of Directors. Many residents believe, as I do, that the election was rigged and what was supposed to be votes being cast by 5,553 of the 15,372 voting households is instead a fixed vote with fixed percentages being assigned to each and every candidate making it impossible to participate in an honest election.
>
> The unofficial election results I have faxed along with this letter represent the tallies. Tallies were given after 600 were counted. Then again at 1200, 2444, 3444, 4444, and final count at 5553.
>
> After close inspection we believe that there was nothing random about the count and tallies each time and that specific unnatural percentages or rigged percentages were being assigned to each and every candidate.
>
> Are we crazy? In a community this diverse and large, can candidates running on separate and opposite slates as well as independents receive similar vote percentage increases tally after tally, plus or minus three or four percent? Does this appear random to you? What do you think? HELP!

Figure 4.4 shows a subset of the data.[3] These vote tallies were deemed suspicious because the proportion of the votes received by each candidate barely changed throughout the tallying. For example, Clotelia Smith's vote share never went below 34.6% or above 36.6%. How can we HELP these people and test their hypothesis?

We start by plotting the data: for each candidate, the proportion of vote received after 600, 1200, ... votes; see Figure 4.5. These graphs are difficult to interpret, however, since the data points are not in any sense independent: the vote at any time point includes all the votes that came before. We handle this problem by subtraction to obtain the number of votes for each candidate in the intervals between the vote tallies: the first 600 votes, the next 600, the next 1244, then next 1000, then next 1000, and the final 1109, with the total representing all 5553 votes.

Figure 4.6 displays the results. Even after taking differences, these graphs are fairly stable—but how does this variation compare to what would be expected if votes were actually coming in at random? We formulate this as a hypothesis test and carry it out in five steps:

1. *The null hypothesis* is that the voters are coming to the polls at random. The fax writer believed the data contradicted the null hypothesis; this is what we want to check.
2. *The test statistic* is some summary of the data used to check the hypothesis. Because the concern was that the votes were unexpectedly stable as the count proceeded, we define a test statistic to summarize that variability. For each candidate $i$, we label $y_{i1}, \dots, y_{i6}$ to be the numbers of votes received by the candidates during each of the six recorded stages of the count. (For example, from

[3] Data and code for this example are in the folder Coop.

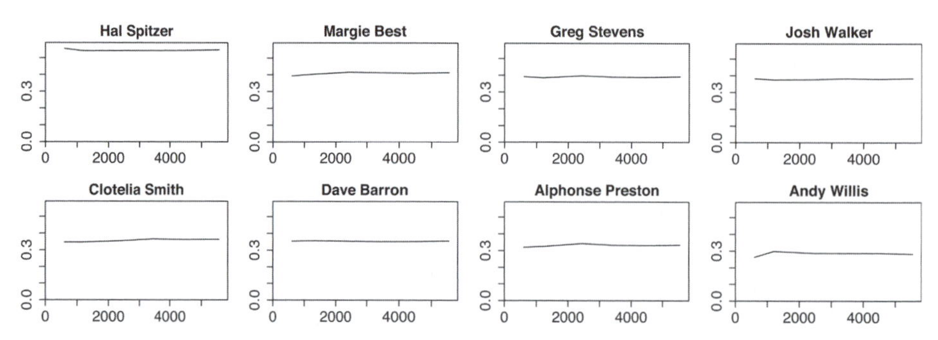


Figure 4.5 *Proportion of votes received by each candidate in the cooperative board election, after each stage of counting: 600, 1200, 2444, ..., 5553 votes. There were 27 candidates in total; for brevity we display just the leading 8 vote-getters here. The vote proportions appear to be extremely stable over time; this might be misleading, however, since the vote at any time point includes all the previous vote tallies. See Figure 4.6.*

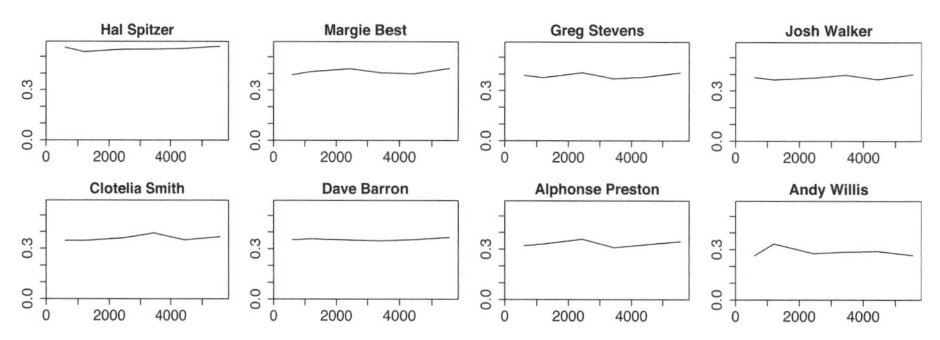


Figure 4.6 *Proportion of votes received by each of the 8 leading candidates in the cooperative board election, at each disjoint stage of voting: the first 600 votes, the next 600, the next 1244, then next 1000, then next 1000, and the final 1109, with the total representing all 5553 votes. The plots here and in Figure 4.5 have been put on a common scale which allows easy comparison of candidates, although at the cost of making it difficult to see details in the individual time series.*

Figure 4.4, the values of $y_{i1}, y_{i2}, \dots, y_{i6}$ for Earl Coppin are $55, 51, \dots, 104$.) We then compute $p_{it} = y_{it}/n_t$ for $t = 1, \dots, 6$, the proportion of the votes received by candidate $i$ during each stage. The test statistic for candidate $i$ is then the sample standard deviation of these six values $p_{i1}, \dots, p_{i6}$,

$$T_i = \text{sd}_{t=1}^{6} \, p_{it},$$

a measure of the variation in his or her support over time.

3. *The theoretical distribution of the data if the null hypothesis were true.* Under the null hypothesis, the six subsets of the election are simply six different random samples of the voters. If $\pi_i$ is the total proportion of voters who would vote for candidate $i$, then the proportion who vote for candidate $i$ during time period $t$, $p_{it}$, follows a distribution with mean $\pi_i$ and a variance of $\pi_i(1-\pi_i)/n_t$. Under the null hypothesis, the variance of the $p_{it}$'s across time should on average equal the average of the six corresponding theoretical variances. Therefore, the variance of the $p_{it}$'s—whose square root is our test statistic—should equal, on average, the theoretical value $\text{avg}_{t=1}^{6} \pi_i(1-\pi_i)/n_t$. The probabilities $\pi_i$ are not known, so we follow standard practice and insert the empirical probabilities, $p_i$, so that the expected value of the test statistic, for each candidate $i$, is

$$T_i^{\text{theory}} = \sqrt{p_i(1-p_i)\,\text{avg}_{t=1}^{6}(1/n_t)}.$$

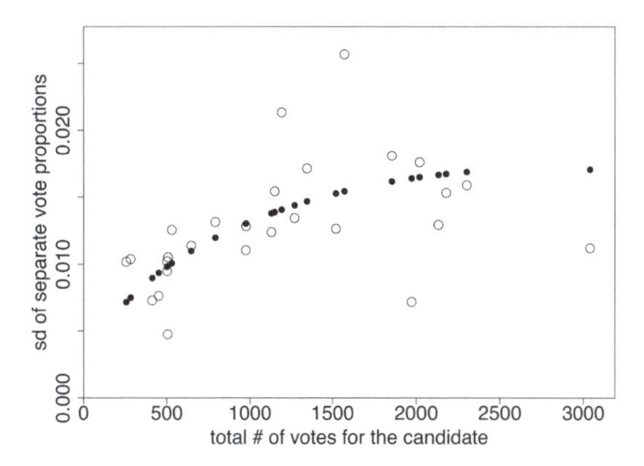


Figure 4.7 *The open circles show, for each of the 27 candidates in the cooperative board election, the standard deviation of the proportions of the vote received by the candidate across the six time points plotted versus the total number of votes received by the candidate. The solid dots show the expected standard deviation of the separate vote proportions for each candidate, based on the binomial model that would be appropriate if voters were coming to the polls at random. The actual standard deviations appear consistent with the theoretical model.*

4. *Comparing the test statistic to its theoretical distribution.* Figure 4.7 plots the observed and theoretical values of the test statistic for each of the 27 candidates, as a function of the total number of votes received by the candidate. The theoretical values follow a simple curve (which makes sense, since the total number of votes determines the empirical probabilities $p_i$, which determine $T_i^{\text{theory}}$), and the actual values appear to fit the theory fairly well, with some above and some below.
5. *Summary comparisons using $\chi^2$ tests.* We can also express the hypothesis tests numerically. Under the null hypothesis, the probability of a candidate receiving votes is independent of the time of each vote, and thus the $2 \times 6$ table of votes including or excluding each candidate would be consistent with the model of independence; see Figure 4.7 for an example. We can then compute for each candidate a summary, called a $\chi^2$ statistic, $\sum_{j=1}^{2} \sum_{t=1}^{6} (\text{observed}_{jt} - \text{expected}_{jt})^2/\text{expected}_{jt}$, and compare it to its theoretical distribution: under the null hypothesis, this statistic has what is called a $\chi^2$ distribution with $(6-1) * (2-1) = 5$ degrees of freedom.
   Unlike the usual applications of $\chi^2$ testing in statistics, in this case we are looking for unexpectedly *low* values of the $\chi^2$ statistic (and thus $p$-values close to 1), which would indicate vote proportions that have suspiciously little variation over time. In fact, however, the $\chi^2$ tests for the 27 candidates show no suspicious patterns: the $p$-values range from 0 to 1, with about 10% below 0.1, about 10% above 0.9, and no extreme $p$-values at either end.
   Another approach would be to perform a $\chi^2$ test on the entire $27 \times 6$ table of votes over time—that is, the table whose first row is the top row of the left table on Figure 4.4, then continues with the data from Earl Coppin, Clarissa Montes, and so forth. This test is somewhat suspect since it ignores that the votes come in batches (each voter can choose up to 6 candidates), but we can still perform the calculation. The value of the $\chi^2$ statistic is 115. Under the null hypothesis, this would be compared to a $\chi^2$ distribution with $(27 - 1) * (6 - 1) = 130$ degrees of freedom, which has a mean of 130 and standard deviation $\sqrt{2 * 130} = 16.1$. We would not be particularly surprised to see a $\chi^2$ statistic of 115 from this distribution.

We thus conclude that the intermediate vote tallies are consistent with random voting. As we explained to the writer of the fax, opinion polls of 1000 people are typically accurate to within 2%, and so, if voters really are arriving at random, it makes sense that batches of 1000 votes are highly stable. This does not rule out the possibility of fraud, but it shows that this aspect of the voting is consistent with the null hypothesis.

## 4.7 Moving beyond hypothesis testing

Null hypothesis significance testing has all sorts of problems, but it addresses a real concern in quantitative research: we want to be able to make conclusions without being misled by noisy data, and hypothesis testing provides a check on overinterpretation of noise. How can we get this benefit of statistical reasoning while avoiding the overconfidence and exaggerations that are associated with conventional reasoning based on statistical significance?

Here is some advice, presented in the context of the study of ovulation and political attitudes discussed on page 62 but applicable to any study requiring statistical analysis, following the principle noted above that the most important aspect of a statistical method is its ability to incorporate more information into the analysis:

- Analyze *all* your data. For most of their analyses, the authors threw out all the data from participants who were premenstrual or having their period. ("We also did not include women at the beginning of the ovulatory cycle (cycle days 1–6) or at the very end of the ovulatory cycle (cycle days 26–28) to avoid potential confounds due to premenstrual or menstrual symptoms.") That was a mistake. Instead of discarding one-third of their data, they should have included that other category in their analysis. This is true of any study in management or elsewhere: use all of your data to provide yourself and your readers with all the relevant information. Better to anticipate potential criticisms than to hide your data and fear for the eventual exposure.
- Present *all* your comparisons. The paper quoted on page 62 leads us through various comparisons and $p$-values that represent somewhat arbitrary decisions throughout of what to look for. It would be better to display and analyze more data, for example a comparison of respondents in different parts of their cycle on variables such as birth year, party identification, and marital status, along with seeing the distribution of reported days of the menstrual cycle. In this particular case we would not expect to find anything interesting, as any real underlying patterns will be much less than the variation, but speaking generally we recommend displaying more of your data rather than focusing on comparisons that happen to reach statistical significance. The point here is not to get an improved $p$-value via a multiple comparisons correction but rather to see the big picture of the data. We recognize that, compared to the usual deterministically framed summary, this might represent a larger burden of effort for the consumer of the research as well as the author of the paper.
- Make your data public (subject to any confidentiality restrictions). If the topic is worth studying, you should want others to be able to make rapid progress.

As we discuss starting in Chapter 10, regression models are helpful in allowing us to model varying treatment effects and situation-dependent phenomena. At the same time, good analysis is no substitute for good data collection. In small-sample studies of small effects, often all that a careful analysis will do is reveal the inability to learn much from the data at hand. In addition, we must move beyond the idea that effects are "there" or not, and the idea that the goal of a study is to reject a null hypothesis. As many observers have noted, these attitudes lead to trouble because they deny the variation inherent in the topics we study, and they deny the uncertainty inherent in statistical inference.

It is fine to design a narrow study to isolate some particular features of the world, but you should think about variation when generalizing your findings to other situations. Does $p < 0.05$ represent eternal truth or even a local truth? Quite possibly not, for two reasons. First, uncertainty: when studying small effects, it is very possible for a large proportion of statistically significant findings to be in the wrong direction as well as be gross overestimates of the magnitude of the underlying effect. Second, variation: even if a finding is "real" in the sense of having the same sign as the corresponding comparison in the population, things can easily be different in other populations and other scenarios. In short, an estimated large effect size is typically too good to be true, whereas a small effect could disappear in the noise.

Does this mean that quantitative research is hopeless? Not at all. We can study large differences, we can gather large samples, and we can design studies to isolate real and persistent effects. In such settings, regression modeling can help us estimate interactions and make predictions that more fully account for uncertainty. In settings with weaker data and smaller samples that may be required to study rare but important phenomena, Bayesian methods can reduce the now-common pattern of researchers getting jerked around by noise patterns that happen to exceed the statistical significance threshold. We can move forward by accepting uncertainty and embracing variation.

## 4.8 Bibliographic note

Agresti and Coull (1998) consider the effectiveness of various quick methods of inference for binomial proportions and propose the confidence interval based on the estimate $(y + 2)/(n + 4)$.

The death penalty example comes from Shirley and Gelman (2015). See Krantz-Kent (2018) for evidence that women watch less television than men. Shirani-Mehr et al. (2018) estimate the variation arising from nonsampling errors in political polls. The voting example in Section 4.6 comes from Gelman (2004b).

For further discussion and references on the problems with statistical significance discussed in Sections 4.4 and 4.5, see de Groot (1956), Meehl (1967, 1978, 1990), Browner and Newman (1987), Krantz (1999), Gelman and Stern (2006), McShane and Gal (2017), and McShane et al. (2019). Type M and type S errors were introduced by Gelman and Tuerlinckx (2000) and further discussed by Gelman and Carlin (2014). Gelman (2018) discusses failures of null hypothesis significance testing in many modern research settings. Simmons, Nelson, and Simonsohn (2011) introduced the terms "researcher degrees of freedom" and "$p$-hacking." Our discussion of forking paths is taken from Gelman and Loken (2014). Orben and Przybylski (2019) dissect a series of controversial research articles and find 604 million forking paths. The salmon scan example comes from Bennett et al. (2009). The discussion of the study of ovulation and political attitudes comes from Gelman (2015a).

Various discussions of the replication crisis in science include Vul et al. (2009), Nosek, Spies, and Motyl (2012), Button et al. (2013), Francis (2013), Open Science Collaboration (2015), and Gelman (2016c).

## 4.9 Exercises

4.1 *Comparison of proportions*: A randomized experiment is performed within a survey. 1000 people are contacted. Half the people contacted are promised a $5 incentive to participate, and half are not promised an incentive. The result is a 50% response rate among the treated group and 40% response rate among the control group. Give an estimate and standard error of the average treatment effect.

4.2 *Choosing sample size*: You are designing a survey to estimate the gender gap: the difference in support for a candidate among men and women. Assuming the respondents are a simple random sample of the voting population, how many people do you need to poll so that the standard error is less than 5 percentage points?

4.3 *Comparison of proportions*: You want to gather data to determine which of two students is a better basketball shooter. One of them shoots with 30% accuracy and the other is a 40% shooter. Each student takes 20 shots and you then compare their shooting percentages. What is the probability that the better shooter makes more shots in this small experiment?

4.4 *Designing an experiment*: You want to gather data to determine which of two students is a better basketball shooter. You plan to have each student take $N$ shots and then compare their shooting percentages. Roughly how large does $N$ have to be for you to have a good chance of distinguishing a 30% shooter from a 40% shooter?

4.5 *Sampling distribution*: Download a data file on a topic of interest to you. Read the file into R and order the data by one of the variables.

(a) Use the `sample` function in R to draw a simple random sample of size 20 from this population. What is the average value of the variable of interest in your sample?

(b) Repeat this exercise several times to get a sense of the sampling distribution of the sample mean for this example.

Example: Girl births

4.6 *Hypothesis testing*: The following are the proportions of girl births in Vienna for each month in 1908 and 1909 (out of an average of 3900 births per month):

.4777 .4875 .4859 .4754 .4874 .4864 .4813 .4787 .4895 .4797 .4876 .4859
.4857 .4907 .5010 .4903 .4860 .4911 .4871 .4725 .4822 .4870 .4823 .4973

The data are in the folder `Girls`. These proportions were used by von Mises (1957) to support a claim that that the sex ratios were less variable than would be expected under the binomial distribution. We think von Mises was mistaken in that he did not account for the possibility that this discrepancy could arise just by chance.

(a) Compute the standard deviation of these proportions and compare to the standard deviation that would be expected if the sexes of babies were independently decided with a constant probability over the 24-month period.

(b) The observed standard deviation of the 24 proportions will not be identical to its theoretical expectation. In this case, is this difference small enough to be explained by random variation? Under the randomness model, the actual variance should have a distribution with expected value equal to the theoretical variance, and proportional to a $\chi^2$ random variable with 23 degrees of freedom; see page 53.

4.7 *Inference from a proportion with $y = 0$*: Out of a random sample of 50 Americans, zero report having ever held political office. From this information, give a 95% confidence interval for the proportion of Americans who have ever held political office.

4.8 *Transformation of confidence or uncertainty intervals*: On page 15 there is a discussion of an experimental study of an education-related intervention in Jamaica. The point estimate of the multiplicative effect is 1.42 with a 95% confidence interval of [1.02, 1.98], on a scale for which 1.0 corresponds to a multiplying by 1, or no effect. Reconstruct the reasoning by which this is a symmetric interval on the log scale:

(a) What is the point estimate on the logarithmic scale? That is, what is the point estimate of the treatment effect on log earnings?

(b) What is the standard error on the logarithmic scale?

4.9 *Inference for a probability*: A multiple-choice test item has four options. Assume that a student taking this question either knows the answer or does a pure guess. A random sample of 100 students take the item, and 60% get it correct. Give an estimate and 95% confidence interval for the percentage in the population who know the answer.

4.10 *Survey weighting*: Compare two options for a national opinion survey: (a) a simple random sample of 1000 Americans, or (b) a survey that oversamples Latinos, with 300 randomly sampled Latinos and 700 others randomly sampled from the non-Latino population. One of these options will give more accurate comparisons between Latinos and others; the other will give more accurate estimates for the total population average.

(a) Which option gives more accurate comparisons and which option gives more accurate population estimates?

(b) Explain your answer above by computing standard errors for the Latino/other comparison and the national average under each design. Assume that the national population is 15% Latino, that the items of interest are yes/no questions with approximately equal proportions of each response, and (unrealistically) that the surveys have no problems with nonresponse.

4.11 *Working through your own example*: Continuing the example from the final exercises of the earlier chapters, perform some basic comparisons, confidence intervals, and hypothesis tests and discuss the relevance of these to your substantive questions of interest.

# Chapter 5

# Simulation

Simulation of random variables is important in applied statistics for several reasons. First, we use probability models to mimic variation in the world, and the tools of simulation can help us better understand how this variation plays out. Patterns of randomness are notoriously contrary to normal human thinking—our brains don't seem to be able to do a good job understanding that random swings will be present in the short term but average out in the long run—and in many cases simulation is a big help in training our intuitions about averages and variation. Second, we can use simulation to approximate the sampling distribution of data and propagate this to the sampling distribution of statistical estimates and procedures. Third, regression models are not deterministic; they produce probabilistic predictions. Simulation is the most convenient and general way to represent uncertainties in forecasts. Throughout this book and in our practice, we use simulation for all these reasons; in this chapter we introduce the basic ideas and the tools required to perform simulations in R.

## 5.1 Simulation of discrete probability models

### How many girls in 400 births?

Example: Simulation of births

The probability that a baby is a girl or boy is approximately 48.8% or 51.2%, respectively, and these do not vary much across the world. Suppose that 400 babies are born in a hospital in a given year. How many will be girls?

We can *simulate* the 400 births using the binomial distribution:[1]

```
n_girls <- rbinom(1, 400, 0.488)
print(n_girls)
```

which shows us what could happen in 400 births. To get a sense of the *distribution* of what could happen, we simulate the process 1000 times (after first creating the vector `n_girls` to store the simulations):

```
n_sims <- 1000
n_girls <- rep(NA, n_sims)
for (s in 1:n_sims){
  n_girls[s] <- rbinom(1, 400, 0.488)
}
hist(n_girls)
```

The 1000 simulations capture the uncertainty.

We performed these simulations in a loop. It would also be possible to simulate 1000 draws from the binomial distribution directly:

```
n_girls <- rbinom(n_sims, 400, 0.488)
```

In other settings one can write the simulation as a function and perform the looping implicitly using the `replicate` function in R, as we illustrate on page 72.

[1] Code for this example is in the folder `ProbabilitySimulation`.

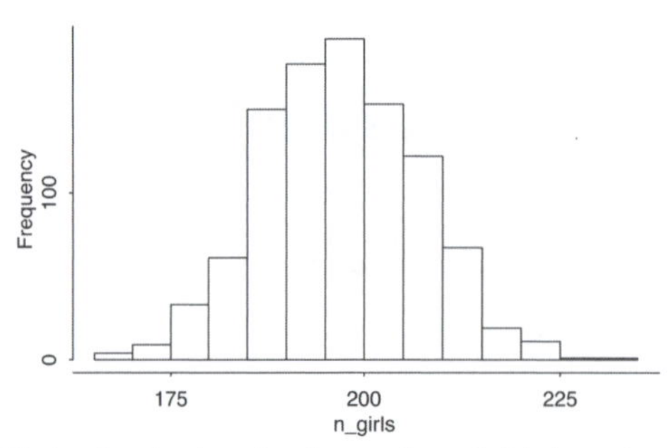


Figure 5.1 *Histogram of 1000 simulated values for the number of girls born in a hospital from 400 births, as simulated from the model that includes the possibility of twins.*

## Accounting for twins

We can complicate the model in various ways. For example, there is a 1/125 chance that a birth event results in fraternal twins, of which each has an approximate 49.5% chance of being a girl, and a 1/300 chance of identical twins, which have an approximate 49.5% chance of being a pair of girls. We can simulate 400 birth events as follows:

```
birth_type <- sample(c("fraternal twin","identical twin","single birth"),
  size=400, replace=TRUE, prob=c(1/125, 1/300, 1 - 1/125 - 1/300))
girls <- rep(NA, 400)
for (i in 1:400){
  if (birth_type[i]=="single birth") {
    girls[i] <- rbinom(1, 1, 0.488)
  } else if (birth_type[i]=="identical twin") {
    girls[i] <- 2*rbinom(1, 1, 0.495)
  } else if (birth_type[i]=="fraternal twin") {
    girls[i] <- rbinom(1, 2, 0.495)
  }
}
n_girls <- sum(girls)
```

Here, `girls` is a vector of length 400, of 0's, 1's, and 2's (mostly 0's and 1's) representing the number of girls in each birth event. Again, this calculation could also be performed without looping using vector operations in R:

```
girls <- ifelse(birth_type=="single birth", rbinom(400, 1, 0.488),
  ifelse(birth_type=="identical twin", 2*rbinom(400, 1, 0.495),
  rbinom(400, 2, 0.495)))
```

To approximate the *distribution* of the number of girls in 400 births, we put the simulation in a loop and repeat it 1000 times:

```
n_sims <- 1000
n_girls <- rep(NA, n_sims)
for (s in 1:n_sims){
  birth_type <- sample(c("fraternal twin","identical twin","single birth"),
    size=400, replace=TRUE, prob=c(1/125, 1/300, 1 - 1/125 - 1/300))
  girls <- rep(NA, 400)
  for (i in 1:400) {
    if (birth_type[i]=="single birth") {
      girls[i] <- rbinom(1, 1, 0.488)
    } else if (birth_type[i]=="identical twin") {
      girls[i] <- 2*rbinom(1, 1, 0.495)
    } else if (birth_type[i]=="fraternal twin") {
      girls[i] <- rbinom(1, 2, 0.495)
```

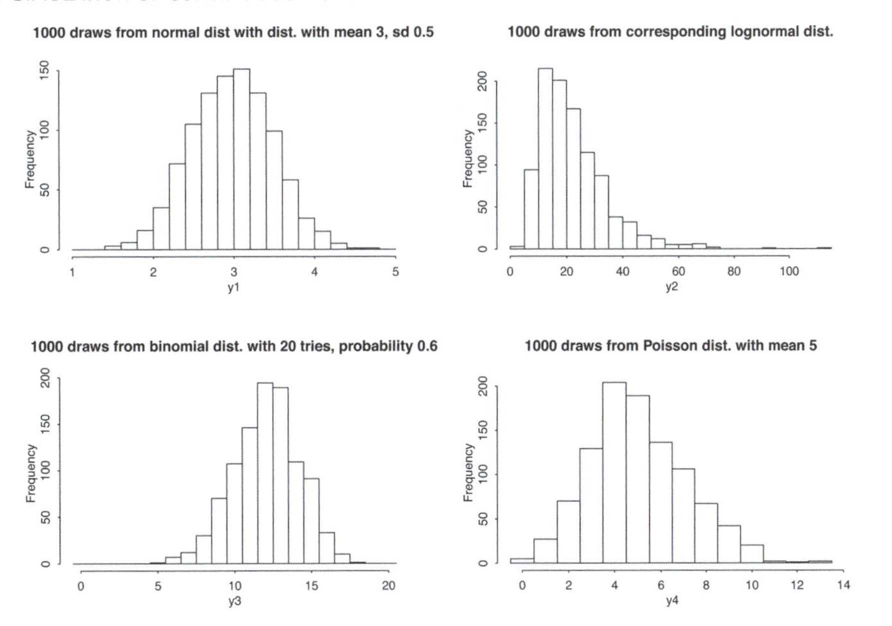


Figure 5.2 *Histograms of 1000 simulated values from four distributions, demonstrating the ability to draw from continuous and discrete random variables in R.*

```
      }
    }
    n_girls[s] <- sum(girls)
  }
```

This nested looping is characteristic of simulations of complex data structures and can also be implemented using custom R functions and the `replicate` function, as we discuss shortly.

Figure 5.1 shows the result, representing the probability distribution for the number of girl births under this model.

## 5.2 Simulation of continuous and mixed discrete/continuous models

Section 3.5 presented the normal, lognormal, binomial, and Poisson probability distributions. We can simulate all of these in R, as we demonstrate with the following code:

```
n_sims <- 1000
y1 <- rnorm(n_sims, 3, 0.5)
y2 <- exp(y1)
y3 <- rbinom(n_sims, 20, 0.6)
y4 <- rpois(n_sims, 5)
par(mfrow=c(2,2))
hist(y1, breaks=seq(floor(min(y1)), max(y1) + 0.2, 0.2),
  main="1000 draws from a normal dist. with mean 3, sd 0.5")
hist(y2, breaks=seq(0, max(y2) + 5, 5),
  main="1000 draws from the corresponding lognormal dist.")
hist(y3, breaks=seq(-0.5, 20.5, 1),
  main="1000 draws from the binomial dist. with 20 tries, probability 0.6")
hist(y4, breaks=seq(-0.5, max(y4) + 1, 1),
  main="1000 draws from the Poisson dist. with mean 5")
```

Figure 5.2 shows the results. As the code indicates, some care can be required to make such histograms clear, especially for discrete data.

Example: Heights of men and women

We can also incorporate continuous probability distributions with the sorts of simulations discussed in the previous section. Here is an example of a mixed discrete/continuous model: 52% of adults in the United States are women and 48% are men. The heights of the men are approximately normally distributed with mean 69.1 inches and standard deviation 2.9 inches; women with mean 63.7 and standard deviation 2.7. Here is code to generate the height of one randomly chosen adult:

```
male <- rbinom(1, 1, 0.48)
height <- ifelse(male==1, rnorm(1, 69.1, 2.9), rnorm(1, 63.7, 2.7))
```

Suppose we select 10 adults at random. What can we say about their average height?

```
N <- 10
male <- rbinom(N, 1, 0.48)
height <- ifelse(male==1, rnorm(N, 69.1, 2.9), rnorm(N, 63.7, 2.7))
avg_height <- mean(height)
print(avg_height)
```

To simulate the distribution of avg_height, we loop the simulation 1000 times:

```
n_sims <- 1000
avg_height <- rep(NA, n_sims)
for (s in 1:n_sims){
  N <- 10
  male <- rbinom(N, 1, 0.48)
  height <- ifelse(male==1, rnorm(N, 69.1, 2.9), rnorm(N, 63.7, 2.7))
  avg_height[s] <- mean(height)
}
hist(avg_height, main="Dist of avg height of 10 adults")
```

What about the maximum height of the 10 people? To determine this, just add the following line within the loop:

```
  max_height[s] <- max(height)
```

and before the loop, initialize max_height:

```
max_height <- rep(NA, n_sims)
```

Then, after the loop, you can make a histogram of max_height.

### Simulation in R using custom-made functions

The coding for simulations becomes cleaner if we express the steps for a single simulation as a function in R. We illustrate with the simulation of average heights. First, the function:

```
height_sim <- function(N){
  male <- rbinom(N, 1, 0.48)
  height <- ifelse(male==1, rnorm(N, 69.1, 2.9), rnorm(N, 63.7, 2.7))
  mean(height)
}
```

For simplicity we have "hard-coded" the proportion of women and the mean and standard deviation of men's and women's heights, but more generally these could be supplied as additional arguments to the function.

We then use the replicate function to call height_sim 1000 times:

```
avg_height <- replicate(1000, height_sim(N=10))
hist(avg_height)
```

## 5.3 Summarizing a set of simulations using median and median absolute deviation

There are many settings where it makes sense to use a set of simulation draws to summarize a distribution, which can represent a simulation from a probability model (as in the examples earlier in this chapter), a prediction for a future outcome from a fitted regression (as we shall illustrate in Figure 7.3 and many other examples in the book), or uncertainty about parameters in a fitted model (the probabilistic equivalent of estimates and standard errors).

When our distributions are constructed as simulations on the computer, it can be convenient to summarize them in some way. We typically summarize the location of a distribution by its mean or median, which we can directly compute in R using the functions `mean` and `median`.

The variation in the distribution is typically summarized by the standard deviation (in R, computed using the function `sd`), but we often prefer using the *median absolute deviation*. If the median of a set of simulations $z_1, \ldots, z_n$, is $M$, then the median absolute deviation is mad = $\text{median}_{i=1}^{n}|z_i - M|$. However, because we are so used to working with standard deviations, when we compute the median absolute deviation, we then rescale it by multiplying by 1.483, which reproduces the standard deviation in the special case of the normal distribution. We call this the "mad sd" and it is returned by the R function `mad` or, if you wanted to code it by hand in R, as `1.483*median(abs(y - median(z)))`.

We typically prefer median-based summaries because they are more computationally stable, and we rescale the median-based summary of variation as described above so as to be comparable to the standard deviation, which we already know how to interpret in usual statistical practice.

We demonstrate with 10 000 draws from the normal distribution with mean 5 and standard deviation 2:

```
z <- rnorm(1e4, 5, 2)
cat("mean =", mean(z), ", median =", median(z), ", sd =", sd(z), ", mad sd =", mad(z))
```

This returns:

```
mean = 4.97 , median = 4.98 , sd = 2.02 , mad sd = 2.02
```

These numbers would be exactly 5 and 2, except for sampling variation because of the finite number of simulations. We can get a sense of this sampling variation by looping the simulation many times, or by a simple analytic calculation. The standard deviation of the mean of $N$ draws from a distribution is simply the standard deviation of the distribution divided by $\sqrt{N}$; hence in the above example, the sampling standard deviation of `mean(z)` is $2/\sqrt{10\,000} = 0.02$, and so it is no surprise to see a sample mean of 4.97, which is approximately 0.02 from the true value of 5. A similar calculation for the variation in the sample standard deviation would use the $\chi^2_{N-1}$ distribution.

When a regression is fit using `stan_glm`, inference for each coefficient is summarized by its posterior median and mad sd, which roughly correspond to the classical estimate and standard error but are more stable summaries for low sample sizes or skewed distributions which can arise with logistic regression and generalized linear models. In general, we consider median and mad sd to be more reliable summaries, but in practice we treat them as point estimate and uncertainty, in the same way that one would typically interpret the classical estimate and standard error. We discuss this further in Section 9.1.

Finally, we can summarize any distribution by uncertainty intervals; for example, `quantile(z, 0.25, 0.75)` returns a central 50% interval and `quantile(z, 0.025, 0.975)` returns a central 95% interval. As discussed in later chapters, these intervals can be helpful in expressing inferential or predictive uncertainty.

## 5.4 Bootstrapping to simulate a sampling distribution

So far we have talked about simulation from prespecified probability models, and from Chapter 7 onward we shall sample parameters and predictions from fitted regressions. Here we briefly mention a different approach to assessing uncertainty, which is to resample the data, creating a set of simulated datasets that can be used to approximate some aspects of the sampling distribution, thus giving some sense of the variation that could be expected if the data collection process had been re-done.

Resampling data is called *bootstrapping*, based on the idea that the users of this procedure "pull themselves up by their own bootstraps" by obtaining an estimate of a sampling distribution without the need for any distributional assumptions or ever constructing a generative model of the form discussed in Section 4.1.

Bootstrap resampling is done *with replacement*; that is, the same data point can appear multiple times in a resampled dataset. This is necessary, as otherwise it would not be possible to get new datasets of the same size as the original.

We demonstrate with an example and then return to more general comments about bootstrapping.

In the example, the goal is to estimate a ratio: the median earnings of women in the United States divided by the median earnings of men. We use data from a national survey from 1990.[2]

We read in the data and compute the ratio of medians from the sample:

```
earnings <- read.csv("earnings.csv")
earn <- earnings$earn
male <- earnings$male
ratio <- median(earn[male==0]) / median(earn[male==1])
```

The value is 0.60: in this sample, the median earnings of women are 60% that of men. But what uncertainty can we attach to this? One way to approach the problem is to look at the variation in 100 random bootstrap samples from the data, which can be taken as an approximation to the sampling distribution of our survey, that is, possible data we could have seen, had the survey been conducted again in the same population.

To program the bootstrap, we first write the code to draw a single bootstrap sample and analyze it:

```
n <- nrow(earnings)
boot <- sample(n, replace=TRUE)
earn_boot <- earn[boot]
male_boot <- male[boot]
ratio_boot <- median(earn_boot[male_boot==0]) / median(earn_boot[male_boot==1])
```

That code constructs one resampled dataset. But to do the bootstrap we must repeat this many times to estimate a sampling distribution; thus, we wrap the code in a function so we can easily loop it:

```
boot_ratio <- function(data){
  n <- nrow(data)
  boot <- sample(n, replace=TRUE)
  earn_boot <- data$earn[boot]
  male_boot <- data$male[boot]
  median(earn_boot[male_boot==0]) / median(earn_boot[male_boot==1])
}
```

And then we write the code that gives the set of bootstrap simulations:

```
n_sims <- 10000
output <- replicate(n_sims, boot_ratio(data=earnings))
```

We conclude by summarizing the results graphically and numerically:

```
hist(output)
print(sd(output))
```

The resulting bootstrap distribution has standard deviation of 0.03; thus we can say that the estimated ratio of median earnings is 0.60 with a standard error of 0.03.

---

[2]Data and code for this example are in the folder Earnings.

### Choices in defining the bootstrap distribution

In the above example, the resampling distribution for the bootstrap was clear, redrawing from a set of independent observations. In a simple regression, one can simply resample data $(x, y)_i$. An alternative is to bootstrap the residuals: that is, to fit a model $y = X\beta + \text{error}$, compute the residuals $r_i = y_i - X_i\hat{\beta}$ from the fitted model, resample from the set of $n$ values of $r_i$, and then add them back in, to create bootstrapped data, $y_i^{\text{boot}} = X_i\hat{\beta} + r_i^{\text{boot}}$. Doing this once produces a bootstrapped dataset, and doing it 1000 times gives a simulated bootstrap sampling distribution.

With more structured models, one must think carefully to design the resampling procedure. Here are some examples:

**Time series.** Consider ordered observations $y_1, \ldots, y_n$, which can be written as $(t, y)_i, i = 1, \ldots, n$, with time measurements $t_i$. Simple resampling of $(t, y)$ results in datasets with multiple identical observations at some time points and no observations at others, which can be difficult to analyze. It can make sense to bootstrap the residuals, but this procedure has issues too. If, for example, you fit a model sequentially in time with bootstrapped residuals, each bootstrapped series will start out near the actual data but can look much different by the end. The point here is that the very concept of a "sampling distribution" is not so clearly defined once we move beyond explicit sampling models.

**Multilevel structure.** Suppose we have data from students within schools. Should we bootstrap by resampling students, resampling schools, or first resampling schools and then resampling students within schools? These three choices correspond to different sampling models and yield different bootstrap standard errors for estimates of interest. Similar questions arise when there are multiple observations on each person in a dataset: Should people be resampled, or observations, or both?

**Discrete data.** In logistic regression with binary data the simplest bootstrap is to resample the data $(x, y)_i$. But what about binomial logistic regression, as discussed in Section 15.3, where the data have the form $(x, n, y)_i$? One option is to bootstrap on the clusters, that is, to resample $(x, n, y)_i$. But another choice is to first expand each observation into $n$ separate data points: $(x_i, n_i, y_i)$ becomes $y_i$ observations of the form $(x_i, 1)$ and $n_i - y_i$ observations of the form $(x_i, 0)$, and then bundle these all into a logistic regression with $\sum_i n_i$ data points. The bootstrap can then be applied to this new dataset. As with multilevel structures more generally, neither of these two bootstrapping options is wrong here; they just correspond to two different sampling models.

### Limitations of bootstrapping

One of the appeals of the bootstrap is its generality. Any estimate can be bootstrapped; all that is needed are an estimate and a sampling distribution. The very generality of the bootstrap creates both opportunity and peril, allowing researchers to solve otherwise intractable problems but also sometimes leading to an answer with an inappropriately high level of certainty.

For example, in Chapters 13 and 14 we discuss logistic regression, a model for predicting binary outcomes. Section 14.6 has an example of a poll from the 1964 U.S. presidential election campaign, in which none of the black respondents in the sample supported the Republican candidate, Barry Goldwater. As a result, when presidential preference (the binary variable representing support for the Republican or Democratic candidate) was modeled using a logistic regression including several demographic variables, the usual default model follows the data and predicts that 0% of the African Americans in the population would support the Republican in that election. In the notation of logistic regression, the maximum likelihood for the coefficient of "black" is $-\infty$. This is a poor estimate, and the bootstrap does nothing to resolve it: there are no black Goldwater supporters in the data, so there are no black Goldwater supporters in any of the bootstrap samples. Thus the coefficient of "black" remains $-\infty$ in every bootstrap sample.

As this example illustrates, the effectiveness of the bootstrap does not depend only on the sampling protocol (in this case, resampling from the set of survey responses); it also depends both on the data and how they are analyzed. A main selling point of the bootstrap is that it can be performed algorithmically without requiring any formal probability model of the data: the implicit probability model is simply the discrete distribution corresponding to the observed data points. But, paradoxically, bootstrap estimates can be more effective when applied to estimates that have already been smoothed or regularized; in the above example, the problem arose with a simple unregularized logistic regression.

We will not consider the bootstrap further in this book; we discussed it here because it is often used in applications as an approximation to the sampling distribution. In Chapter 9 we consider the related idea of Bayesian simulation for representing uncertainty in estimation and prediction.

## 5.5 Fake-data simulation as a way of life

As discussed in Chapter 4, the sampling distribution is fundamental to the understanding and use of statistical methods. Now that you have the building blocks of simulation of random variables and random sampling, you can simulate generative models directly.

We demonstrate with a simple example in Section 7.2, simulating a regression that was used to predict election outcomes from the economy. The point of the fake-data simulation is not to provide insight into the data or the real-world problem being studied, but rather to evaluate the properties of the statistical methods being used, given an assumed generative model.

## 5.6 Bibliographic note

Random simulation for performing computations in probability and statistics was one of the first applications of computers, dating back to the 1940s. As computing power has become more dispersed since the 1970s, simulation has been used increasingly frequently for summarizing statistical inferences; Rubin (1981) is an early example.

The congressional election analysis in Section 10.6 uses a simplified version of the model in Gelman and King (1994).

The idea of resampling or bootstrap simulation comes from Efron (1979); see Efron and Tibshirani (1993) for a fuller discussion and many references. For some of our thoughts on the matter, see Gelman (2011a) and Gelman and Vehtari (2014).

Fake-data simulation is commonly used to validate statistical models and procedures. Some formal approaches to Bayesian simulation-based validation appear in Geweke (2004), Cook, Gelman, and Rubin (2006), and Talts et al. (2018). Predictive model checking is described in detail in Gelman et al. (2013, chapter 6) and Gelman, Meng, and Stern (1996), deriving from the ideas of Rubin (1984). Gelman (2004a) connects graphical model checks to exploratory data analysis (Tukey, 1977). Examples of simulation-based model checking appear throughout the statistical literature, especially for highly structured models; early examples include Bush and Mosteller (1955) and Ripley (1988).

## 5.7 Exercises

5.1 *Discrete probability simulation*: Suppose that a basketball player has a 60% chance of making a shot, and he keeps taking shots until he misses two in a row. Also assume his shots are independent (so that each shot has 60% probability of success, no matter what happened before).

  (a) Write an R function to simulate this process.

  (b) Put the R function in a loop to simulate the process 1000 times. Use the simulation to estimate the mean and standard deviation of the total number of shots that the player will take, and plot a histogram representing the distribution of this random variable.

  (c) Using your simulations, make a scatterplot of the number of shots the player will take and the proportion of shots that are successes.

5.2 *Continuous probability simulation*: The logarithms of weights (in pounds) of men in the United States are approximately normally distributed with mean 5.13 and standard deviation 0.17; women's log weights are approximately normally distributed with mean 4.96 and standard deviation 0.20. Suppose 10 adults selected at random step on an elevator with a capacity of 1750 pounds. What is the probability that their total weight exceeds this limit?

5.3 *Binomial distribution*: A player takes 10 basketball shots, with a 40% probability of making each shot. Assume the outcomes of the shots are independent.

(a) Write a line of R code to compute the probability that the player makes exactly 3 of the 10 shots.

(b) Write an R function to simulate the 10 shots. Loop this function 10 000 times and check that your simulated probability of making exactly 3 shots is close to the exact probability computed in (a).

5.4 *Demonstration of the Central Limit Theorem*: Let $x = x_1 + \cdots + x_{20}$, the sum of 20 independent uniform(0, 1) random variables. In R, create 1000 simulations of $x$ and plot their histogram. What is the normal approximation to this distribution provided by the Central Limit Theorem? Overlay a graph of the normal density on top of the histogram. Comment on any differences between the histogram and the curve.

5.5 *Distribution of averages and differences*: The heights of men in the United States are approximately normally distributed with mean 69.1 inches and standard deviation 2.9 inches. The heights of women are approximately normally distributed with mean 63.7 inches and standard deviation 2.7 inches. Let $x$ be the average height of 100 randomly sampled men, and $y$ be the average height of 100 randomly sampled women. In R, create 1000 simulations of $x - y$ and plot their histogram. Using the simulations, compute the mean and standard deviation of the distribution of $x - y$ and compare to their exact values.

5.6 *Propagation of uncertainty*: We use a highly idealized setting to illustrate the use of simulations in combining uncertainties. Suppose a company changes its technology for widget production, and a study estimates the cost savings at $5 per unit, but with a standard error of $4. Furthermore, a forecast estimates the size of the market (that is, the number of widgets that will be sold) at 40 000, with a standard error of 10 000. Assuming these two sources of uncertainty are independent, use simulation to estimate the total amount of money saved by the new product (that is, savings per unit, multiplied by size of the market).

5.7 *Coverage of confidence intervals*: Reconstruct the graph in Figure 4.2. The simulations are for an estimate whose sampling distribution has mean 6 and standard deviation 4.

5.8 *Coverage of confidence intervals*: On page 15 there is a discussion of an experimental study of an education-related intervention in Jamaica, in which the point estimate of the treatment effect, on the log scale, was 0.35 with a standard error of 0.17. Suppose the true effect is 0.10—this seems more realistic than the point estimate of 0.35—so that the treatment on average would increase earnings by 0.10 on the log scale. Use simulation to study the statistical properties of this experiment, assuming the standard error is 0.17.

(a) Simulate 1000 independent replications of the experiment assuming that the point estimate is normally distributed with mean 0.10 and standard deviation 0.17.

(b) For each replication, compute the 95% confidence interval. Check how many of these intervals include the true parameter value.

(c) Compute the average and standard deviation of the 1000 point estimates; these represent the mean and standard deviation of the sampling distribution of the estimated treatment effect.

5.9 *Coverage of confidence intervals after selection on statistical significance*: Take your 1000 simulations from Exercise 5.8, and select just the ones where the estimate is statistically significantly different from zero. Compute the average and standard deviation of the selected point estimates. Compare these to the result from Exercise 5.8.

5.10 *Inference for a ratio of parameters*: A (hypothetical) study compares the costs and effectiveness of two different medical treatments.

- In the first part of the study, the difference in costs between treatments A and B is estimated at $600 per patient, with a standard error of $400, based on a regression with 50 degrees of freedom.
- In the second part of the study, the difference in effectiveness is estimated at 3.0 (on some relevant measure), with a standard error of 1.0, based on a regression with 100 degrees of freedom.
- For simplicity, assume that the data from the two parts of the study were collected independently.

Inference is desired for the *incremental cost-effectiveness ratio*: the difference between the average costs of the two treatments, divided by the difference between their average effectiveness, a problem discussed further by Heitjan, Moskowitz, and Whang (1999).

(a) Create 1000 simulation draws of the cost difference and the effectiveness difference, and make a scatterplot of these draws.

(b) Use simulation to come up with an estimate, 50% interval, and 95% interval for the incremental cost-effectiveness ratio.

(c) Repeat, changing the standard error on the difference in effectiveness to 2.0.

5.11 *Predictive checks*: Using data of interest to you, fit a model of interest.

(a) Simulate replicated datasets and visually compare to the actual data.

(b) Summarize the data by a numerical test statistic, and compare to the values of the test statistic in the replicated datasets.

5.12 *Randomization*: Write a function in R to assign $n$ items to treatment and control conditions under the following assignment procedures:

- Independent random assignment. Each item is independently randomly assigned to treatment or control with probabilities $p$ and $1 - p$.
- Complete random assignment. The $n$ items are randomly partitioned into $np$ items that receive the treatment, and the other $n(1 - p)$ are assigned to control.
- Matched pairs. This is the simplest version of block random assignment. The $n$ items come sequentially in $n/2$ pairs. Within each pair, one item is randomly chosen for treatment and one for control. In other words, $p = 0.5$.

Write one R function, not three. Your function should have three inputs: $n$, $p$, and a code for which procedure to use. The output should be a vector of treatment assignments (1's and 0's). Hint: Your function likely won't work for all combinations of $n$ and $p$. If it doesn't, write it so that it throws an error and alerts the user.

5.13 *Working through your own example*: Continuing the example from the final exercises of the earlier chapters, construct a probability model that is relevant to your question at hand and use it to simulate some fake data. Graph your simulated data, compare to a graph of real data, and discuss the connections between your model and your larger substantive questions.

# Part 2: Linear regression

# Chapter 6

# Background on regression modeling

At a purely mathematical level, the methods described in this book have two purposes: prediction and comparison. We can use regression to predict an outcome variable, or more precisely the distribution of the outcome, given some set of inputs. And we can compare these predictions for different values of the inputs, to make simple comparisons between groups, or to estimate causal effects, a topic to which we shall return in Chapters 18–21. In this chapter we use our favored technique of fake-data simulation to understand a simple regression model, use a real-data example of height and earnings to warn against unwarranted causal interpretations, and discuss the historical origins of regression as it relates to comparisons and statistical adjustment.

## 6.1 Regression models

The simplest regression model is linear with a single predictor:

$$\text{Basic regression model: } y = a + bx + \text{error}.$$

The quantities $a$ and $b$ are called *coefficients* or, more generally, *parameters* of the model.

The simple linear model can be elaborated in various ways, including:

- Additional predictors: $y = \beta_0 + \beta_1 x_1 + \beta_2 x_2 + \cdots + \beta_k x_k + \text{error}$, written in vector-matrix notation as $y = X\beta + \text{error}$.
- Nonlinear models such as $\log y = a + b \log x + \text{error}$.
- Nonadditive models such as $y = \beta_0 + \beta_1 x_1 + \beta_2 x_2 + \beta_3 x_1 x_2 + \text{error}$, which contains an *interaction* between the input variables $x_1$ and $x_2$.
- *Generalized linear models*, which extend the linear regression model to work with discrete outcomes and other data that cannot be fit well with normally distributed additive errors, for example predicting support for the Republican or Democratic presidential candidate based on the age, sex, income, etc. of the survey respondent.
- *Nonparametric models*, which include large numbers of parameters to allow essentially arbitrary curves for the predicted value of $y$ given $x$.
- *Multilevel models*, in which coefficients in a regression can vary by group or by situation. For example, a model predicting college grades given admissions test scores can have coefficients that vary by college.
- *Measurement-error models*, in which predictors $x$ as well as outcomes $y$ are measured with error and there is a goal of estimating the relationship between the underlying quantities. An example is estimating the effect of a drug under partial compliance so that the dose taken by each individual patient is not exactly known.

In this book we shall focus on the first four of the generalizations above.

## 6.2 Fitting a simple regression to fake data

Example: Regression fit to simulated data

We demonstrate linear regression with a simple example in R.[1] First we load in the rstanarm package, which allows us to fit regression models using the statistical inference engine Stan:

```
library("rstanarm")
```

We then simulate 20 fake data points $y_i$ from the model, $y_i = a + bx_i + \epsilon_i$, where the predictor $x_i$ takes on the values from 1 to 20, the intercept is $a = 0.2$, the slope is $b = 0.3$, and the errors $\epsilon_i$ are normally distributed with mean 0 and standard deviation $\sigma = 0.5$, so that we expect roughly two-thirds of the points to fall within $\pm 1$ standard error of the line. Here is the code:[2]

```
x <- 1:20
n <- length(x)
a <- 0.2
b <- 0.3
sigma <- 0.5
y <- a + b*x + sigma*rnorm(n)
```

### Fitting a regression and displaying the results

To fit the regression we set up a *data frame* containing predictor and outcome. The data frame can have any name; here we call it fake to remind ourselves that this is a fake-data simulation:

```
fake <- data.frame(x, y)
```

And then we can fit the model using the stan_glm (using Stan to fit a generalized linear model) function in R;[3] we can save the fit using any name:

```
fit_1 <- stan_glm(y ~ x, data=fake)
```

And then we can display the result:

```
print(fit_1, digits=2)
```

which yields:

```
            Median MAD_SD
(Intercept) 0.40   0.23
x           0.28   0.02

Auxiliary parameter(s):
      Median MAD_SD
sigma 0.43   0.08
```

The first two rows of output tell us that the estimated intercept is 0.40 with uncertainty 0.23, and the estimated slope is 0.28 with uncertainty 0.02. The residual standard deviation $\sigma$ is estimated at 0.43 with an uncertainty of 0.08.

Under the hood, fitting the model in Stan produced a set of simulations summarizing our inferences about the parameters $a$, $b$, and $\sigma$, and the output on the screen shows the median and mad sd (see Section 5.3) to produce a point estimate and uncertainty for each parameter.

It can be helpful to plot the data and fitted regression line:

```
plot(fake$x, fake$y, main="Data and fitted regression line")
a_hat <- coef(fit_1)[1]
b_hat <- coef(fit_1)[2]
abline(a_hat, b_hat)
```

For convenience we can also put the formula on the graph:

[1]See Appendix A for instructions on how to set up and use R.
[2]Code for this example is in the folder Simplest.
[3]See Sections 1.6 and 8.4 for background on the stan_glm function.

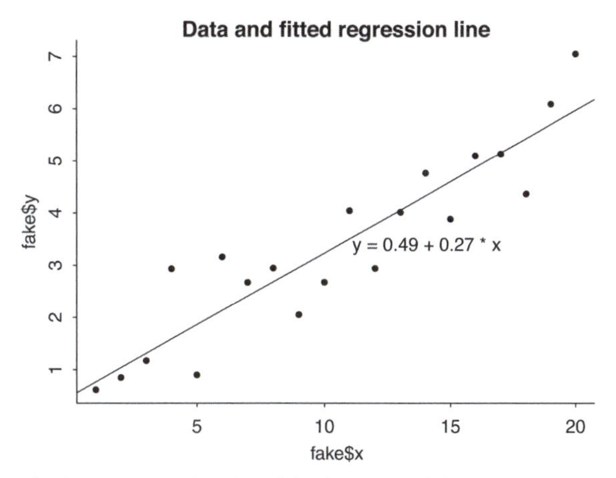


Figure 6.1 *Simple example of a regression line fit to fake data. The 20 data points were simulated from the model, $y = 0.2 + 0.3x$ + error, with errors that were independent and normally distributed with mean 0 and standard deviation 0.5.*

| Parameter | Assumed value | Estimate | Uncertainty |
|---|---|---|---|
| $a$ | 0.2 | 0.40 | 0.23 |
| $b$ | 0.3 | 0.28 | 0.02 |
| $\sigma$ | 0.5 | 0.43 | 0.08 |

Figure 6.2 *After simulating 20 fake data points from a simple linear regression, $y_i = a + bx_i + \epsilon_i$, with errors $\epsilon_i$ drawn from a normal distribution with mean 0 and standard deviation $\sigma$, we then fit a linear regression to these data and obtain estimates and uncertainties for the three parameters from the model. We can then see that the estimates are roughly consistent with the specified parameter values.*

```
x_bar <- mean(fake$x)
text(x_bar, a_hat + b_hat*x_bar,
  paste("y =", round(a_hat, 2), "+", round(b_hat, 2), "* x"), adj=0)
```

The result is shown in Figure 6.1.

## Comparing estimates to assumed parameter values

Having fit the model to fake data, we can now compare the parameter estimates to their assumed values. For simplicity we summarize the results in Figure 6.2, which simply repeats the results from the fitted regression model on page 82.

To read these results, start with the intercept $a$, which we set to 0.2 in the simulations. After fitting the model to fake data, the estimate is 0.40, which is much different from the assumed 0.2—but the uncertainty, or standard error, in the estimate is 0.23. Roughly speaking, we expect the difference between the estimate and the true value to be within 1 standard error 68% of the time, and within 2 standard errors 95% of the time; see Figure 4.1. So if the true value is 0.2, and the standard error is 0.23, it's no surprise for the estimate to happen to be 0.40. Similarly, the estimates for $b$ and $\sigma$ are approximately one standard error away from their true values.

As just illustrated, any given fake-data simulation with continuous data would not exactly reproduce the assumed parameter values. Under repeated simulations, though, we should see appropriate coverage, as illustrated in Figure 4.2. We demonstrate fake-data simulation for linear regression more fully in Section 7.2.

## 6.3 Interpret coefficients as comparisons, not effects

Example: Height and earnings

Regression coefficients are commonly called "effects," but this terminology can be misleading. We illustrate with an example of a regression model fit to survey data from 1816 respondents, predicting yearly earnings in thousands of dollars, given height in inches and sex, coded as male = 1 for men and 0 for women:[4]

```
earnings$earnk <- earnings$earn/1000
fit_2 <- stan_glm(earnk ~ height + male, data=earnings)
print(fit_2)
```

This yields,

```
            Median MAD_SD
(Intercept) -26.0   11.8
height        0.6    0.2
male         10.6    1.5

Auxiliary parameter(s):
      Median MAD_SD
sigma 21.4    0.3
```

The left column above shows the estimated parameters of the model, and the right column gives the uncertainties in these parameters. We focus here on the estimated model, putting off discussion of inferential uncertainty until the next chapter.

The table begins with the regression coefficients, which go into the fitted model:

$$\text{earnings} = -26.0 + 0.6 * \text{height} + 10.6 * \text{male} + \text{error}.$$

Then comes sigma, the residual standard deviation, estimated at 21.4, which indicates that earnings will be within $\pm$ 21 400 of the linear predictor for about 68% of the data points and will be within $\pm 2 *$ \$21 400 = \$42 800 of the linear predictor approximately 95% of the time. The 68% and 95% come from the properties of the normal distribution reviewed in Figure 3.7; even though the errors in this model are not even close to normally distributed, we can use these probabilities as a rough baseline when interpreting the residual standard deviation.

We can get a sense of this residual standard deviation by comparing it to the standard deviation of the data and then estimating the proportion of variance explained, which we compute as 1 minus the proportion of variance unexplained:

```
R2 <- 1 - sigma(fit_2)^2 / sd(earnings$earnk)^2
```

which returns the value $R^2 = 0.10$, meaning that the linear model accounts for only 9% of the variance in earnings in these data. This makes sense, given that people's earnings vary a lot, and most of this variation has nothing to do with height or sex. We discuss $R^2$ further in Section 11.6; for now, you can just think of it as a way of putting a scale on $\sigma$, the residual standard deviation.

We have to be careful not to overinterpret the fitted model. For example, it might seem natural to report that the estimated effect of height is \$600 and the estimated effect of sex is \$10 600.

Strictly speaking, though, it is inappropriate to label these as "effects"—at least, not without a lot of assumptions. We say this because we define an *effect* as the change associated with some *treatment*, or intervention. To say that "the effect of height on earnings" is \$600 is to suggest that, if we were to increase someone's height by one inch, his or her earnings would increase by an expected amount of \$600. But this is not really what's being estimated from the model. Rather, what is observed is an observational pattern, that taller people in the sample have higher earnings on average. These data allow between-person comparisons, but to speak of effect of height is to reference a hypothetical within-person comparison.

[4]The survey was conducted in 1990, and for the analyses in this book we exclude respondents with missing values of height or earnings. Data and code for this example are in the folder Earnings.

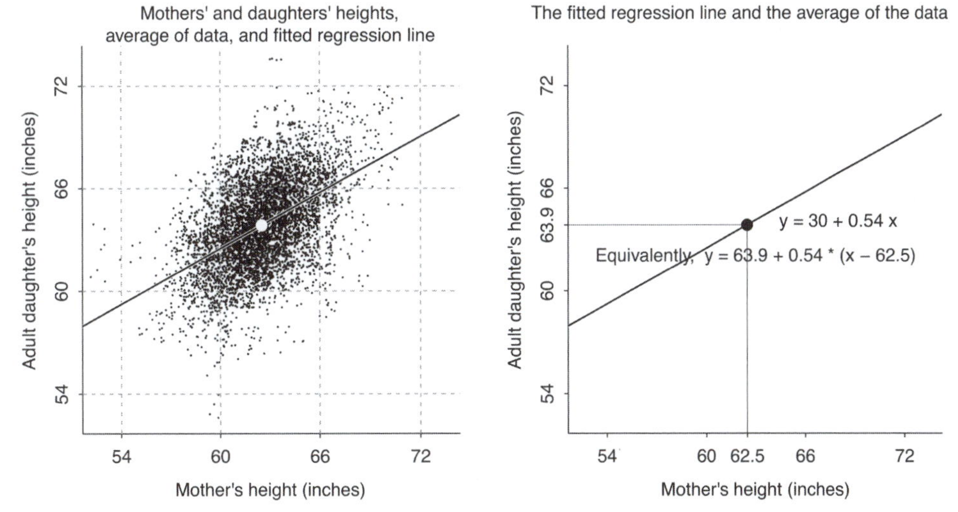


Figure 6.3 *(a) Scatterplot adapted from data from Pearson and Lee (1903) of the heights of mothers and their adult daughters, along with the* regression line *predicting daughters' from mothers' heights. (b) The regression line by itself, just to make the pattern easier to see. The line automatically goes through the mean of the data, and it has a slope of 0.54, implying that, on average, the difference of a daughter's height from the average (mean) of women's heights is only about half the difference of her mother's height from the average.*

How, then, can we think of the coefficient for height in the fitted model? We can say that, under the fitted model, the average difference in earnings, comparing two people of the same sex but one inch different in height, is \$600. *The safest interpretation of a regression is as a comparison.*

Similarly, it would be inappropriate to say that the estimated "effect of sex" is \$10 600. Better to say that, when comparing two people with the same height but different sex, the man's earnings will be, on average, \$10 600 more than the woman's in the fitted model.

Under some conditions, the between-person inferences from a regression analysis can be interpreted as causal effects—see Chapters 18–21—but as a starting point we recommend describing regression coefficients in predictive or descriptive, rather than causal, terms.

To summarize: regression is a mathematical tool for making predictions. Regression coefficients can sometimes be interpreted as effects, but they can always be interpreted as average comparisons.

## 6.4 Historical origins of regression

Example: Mothers' and daughters' heights

"Regression" is defined in the dictionary as "the process or an instance of regressing, as to a less perfect or less developed state." How did this term come to be used for statistical prediction? This connection comes from Francis Galton, one of the original quantitative social scientists, who fit linear models to understand the heredity of human height. Predicting children's heights from parent's heights, he noticed that children of tall parents tended to be taller than average but less tall than their parents. From the other direction, children of shorter parents tended to be shorter than average but less short than their parents. Thus, from one generation to the next, people's heights have "regressed" to the average or *mean*, in statistics jargon.

### Daughters' heights "regressing" to the mean

We illustrate with a classic study of the heredity of height, published in 1903 by Karl Pearson and Alice Lee.[5] Figure 6.3a shows the data of mothers' and daughters' heights along with the *regression*

[5]Data and code for this example are in the folder PearsonLee.

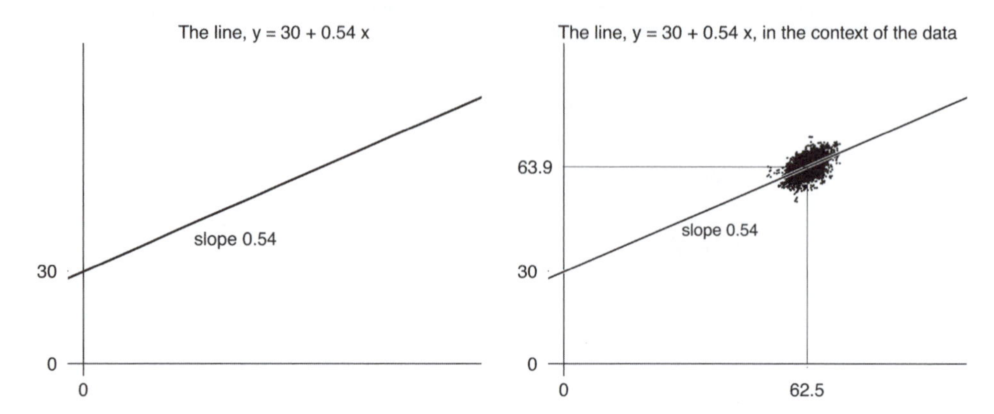


Figure 6.4 *(a) Fitted regression line, $y = 30 + 0.54\,x$, graphed using intercept and slope. (b) Difficulty of the intercept-slope formulation in the context of the data in the height example. The intercept of 30 inches corresponds to the predicted height of a daughter whose mother is a meaningless 0 inches tall.*

*line*—the best-fit line for predicting daughters' from mothers' heights. The line goes through the mean (average) of $x$ and $y$, shown with a large dot in the center of the graph.

Figure 6.3b shows the line by itself, the formula, $y = 30 + 0.54x$, which we can also write as,

$$y = 30 + 0.54x + \text{error}, \tag{6.1}$$

to emphasize that the model does not fit individual data points perfectly. We shall give R code for displaying the data and fitting the line, but first we briefly discuss the line itself.

The equation $y = 30 + 0.54x$ describes a line with intercept 30 and slope 0.54, as shown in Figure 6.4a. The intercept-slope formula is an easy way to visualize a line, but it can have problems in various real-world settings, as we demonstrate in Figure 6.4b. The line's slope of 0.54 is clearly interpretable in any case—adding one inch to mother's height corresponds to an increase of 0.54 inches in daughter's predicted height—but the intercept of 30 is hard to understand on its own: it corresponds to the predicted height of a daughter whose mother is a meaningless 0 inches tall.

Instead we can use a different expression of the regression line, centering it not at 0 but at the mean of the data. The equation $y = 30 + 0.54x$ can equivalently be written as,

$$y = 63.9 + 0.54(x - 62.5), \tag{6.2}$$

as shown in Figure 6.3b. This formula shows that when $x = 62.5$, $y$ is predicted to be 63.9.

To put this in the context of the example, if a mother has average height, her adult daughter is predicted to have average height. And then for each inch that a mother is taller (or shorter) than the average height, her daughter is expected to be about half an inch taller (or shorter) than the average for her generation.

## Fitting the model in R

The equation $y = 30 + 0.54x$ is the approximate best-fit line, where "best fit" is defined as minimizing the sum of squared errors; that is, an algorithm finds the values $a$ and $b$ that minimize $\sum_{i=1}^{n}(y_i - (a + bx_i))^2$. Section 8.1 discusses the formula by which this solution is calculated, but here we show how to get the answer in R.

We start by reading in the data and looking at the first five rows, just to check the numbers:

```
heights <- read.table("Heights.txt", header=TRUE)
print(heights[1:5,])
```

This is what appears:

```
  daughter_height mother_height
1            52.5          59.5
2            52.5          59.5
3            53.5          59.5
4            53.5          59.5
5            55.5          59.5
```

One can more easily look at the beginning of a matrix by typing head(data) but here we explicitly choose the first five rows to demonstrate R's indexing capabilities. We continue by fitting a regression to predict daughters' from mothers' heights:

```
fit_1 <- stan_glm(daughter_height ~ mother_height, data=heights)
print(fit_1)
```

And this is what we see:

```
              Median MAD_SD
(Intercept)   29.8    0.8
mother_height  0.5    0.0

Auxiliary parameter(s):
      Median MAD_SD
sigma 2.3    0.0
```

The data has heights from 5524 mother-daughter pairs, and the model has three parameters: an intercept, a coefficient for mothers' height, and a residual standard deviation.

Our next step is to graph the data. The numbers are reported discretely in one-inch bins (for example, "59.5" corresponds to a height between 59 and 60 inches); to be able to display them as a scatterplot, we impute random values within these ranges:

```
n <- nrow(heights)
mother_height_jitt <- heights$mother_height + runif(n, -0.5, 0.5)
daughter_height_jitt <- heights$daughter_height + runif(n, -0.5, 0.5)
```

We then make the scatterplot:

```
plot(mother_height_jitt, daughter_height_jitt, xlab="Mother's height (inches)",
  ylab="Adult daughter's height (inches)")
```

And we extract the coefficients from the regression and add the fitted line to the graph:

```
a_hat <- coef(fit_1)[1]
b_hat <- coef(fit_1)[2]
abline(a_hat, b_hat)
```

If you do these steps yourself, you'll get a graph that looks like Figure 6.3a but without some of the pretty features that require a few more lines of R to render.

## 6.5 The paradox of regression to the mean

Now that we have gone through the steps of fitting and graphing the line that predicts daughters' from mothers' heights, we can return to the question of heights "regressing to the mean."

When looked at a certain way, the regression slope of 0.54 in Figure 6.3—indeed, any slope other than 1—seems paradoxical. If tall mothers are likely to have daughters who are only tallish, and short mothers are likely to have shortish daughters, does this not imply that daughters will be more average than their mothers, and that if this continues, each generation will be more average than the last, until, after a few generations, everyone will be just about of average height? For example, a mother who is 8 inches taller than average is predicted to have a daughter 4 inches taller than average, whose

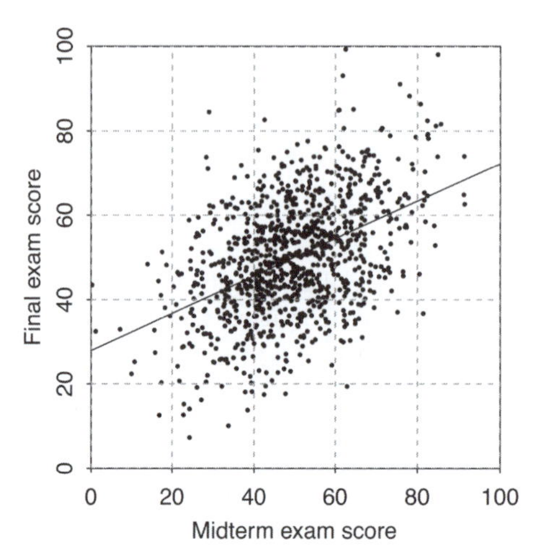


Figure 6.5 *Scatterplot of simulated midterm and final exam scores with fitted regression line, which has a slope of 0.45, implying that if a student performs well on the midterm, he or she is expected to do not so well on the final, and if a student performs poorly on the midterm, he or she is expected to improve on the final; thus, regression to the mean.*

daughter would be predicted to be only 2 inches taller than average, with *her* daughter predicted to be only an inch taller than average, and so forth.

But clearly this is not happening. We are already several generations after Pearson and Lee, and women's heights are as variable as ever.

The resolution of the apparent paradox is that yes, the *predicted* height of a woman is closer to the average, compared to her mother's height, but the actual height is not the same thing as the prediction, which has error; recall equation (6.1). The point predictions regress toward the mean—that's the coefficient less than 1—and this reduces variation. At the same time, though, the error in the model—the imperfection of the prediction—*adds* variation, just enough to keep the total variation in height roughly constant from one generation to the next.

Regression to the mean thus will always arise in some form whenever predictions are imperfect in a stable environment. The imperfection of the prediction induces variation, and regression in the point prediction is required in order to keep the total variation constant.

### How regression to the mean can confuse people about causal inference; demonstration using fake data

Example: Simulated midterm and final exams

Regression to the mean can be confusing and it has led people to mistakenly attribute causality. To see how this can happen, we move from heights of parents and children to the mathematically equivalent scenario of students who take two tests.

Figure 6.5 shows a hypothetical dataset of 1000 students' scores on a midterm and final exam. Rather than using real data, we have simulated exam scores using the following simple process representing signal and noise:[6]

1. Each student is assumed to have a true ability drawn from a distribution with mean 50 and standard deviation 10.
2. Each student's score on the midterm exam is the sum of two components: the student's true ability, and a random component with mean 0 and standard deviation 10, reflecting that performance on any given test will be unpredictable: a midterm exam is far from a perfect measuring instrument.

[6]Code for this example is in the folder `FakeMidtermFinal`.

3. Likewise, each student's score on the final exam is his or her true ability, plus another, independent, random component.

Here's our code for simulating the fake data:

```
n <- 1000
true_ability <- rnorm(n, 50, 10)
noise_1 <- rnorm(n, 0, 10)
noise_2 <- rnorm(n, 0, 10)
midterm <- true_ability + noise_1
final <- true_ability + noise_2
exams <- data.frame(midterm, final)
```

We then plot the data and the fitted regression line:

```
fit_1 <- stan_glm(final ~ midterm, data=exams)
plot(midterm, final, xlab="Midterm exam score", ylab="Final exam score")
abline(coef(fit_1))
```

And here's the regression output:

```
            Median MAD_SD
(Intercept) 24.8    1.4
midterm      0.5    0.0

Auxiliary parameter(s):
      Median MAD_SD
sigma 11.6   0.3
```

The estimated slope is 0.5 (see also Figure 6.5), which by being less than 1 is an example of regression to the mean: students who score high on the midterm tend to score only about half as high, compared to the average, on the final; students who score low on the midterm score low, but typically not as low, compared to the average, on the final. For example, on the far left of Figure 6.5 are two students who scored zero on the midterm and 33 and 42 on the final; on the far right of the graph are three students who scored 91 on the midterm and between 61 and 75 on the final.

It might seem natural to interpret this causally, to say that students who score well on the midterm have high ability but then they tend to get overconfident and goof off; hence, they typically don't do so well on the final. From the other direction, the appealing causal story is that poor-scoring students on the midterm are motivated to try extra hard, so they improve when the final exam comes along.

Actually, though, the data were simulated from a theoretical model that contained *no* motivational effects at all; both the midterm and the final were a function of true ability plus random noise. We know this because we created the simulation!

The pattern of regression to the mean—that is, the slope of the line in Figure 6.5 being less than 1—is a consequence of variation between the first and second observations: a student who scores very well on the midterm is likely to have been somewhat lucky and also to have a high level of skill, and so in the final exam it makes sense for the student to do better than the average but worse than on the midterm.

The point is that a naive interpretation of the data in Figure 6.5 could lead you to infer an effect (better-scoring students being lazy on the final; worse-scoring students studying harder) that is entirely spurious. This error is called the "regression fallacy."

Example: Flight school

A famous real-world example was reported by the psychologists Amos Tversky and Daniel Kahneman in 1973:

> The instructors in a flight school adopted a policy of consistent positive reinforcement recommended by psychologists. They verbally reinforced each successful execution of a flight maneuver. After some experience with this training approach, the instructors claimed that contrary to psychological doctrine, high praise for good execution of complex maneuvers typically results in a decrement of performance on the next try.

Actually, though, they explain:

> Regression is inevitable in flight maneuvers because performance is not perfectly reliable and progress between successive maneuvers is slow. Hence, pilots who did exceptionally well on one trial are likely to deteriorate on the next, regardless of the instructors' reaction to the initial success. The experienced flight instructors actually discovered the regression but attributed it to the detrimental effect of positive reinforcement. This true story illustrates a saddening aspect of the human condition. We normally reinforce others when their behavior is good and punish them when their behavior is bad. By regression alone, therefore, they are most likely to improve after being punished and most likely to deteriorate after being rewarded. Consequently, we are exposed to a lifetime schedule in which we are most often rewarded for punishing others, and punished for rewarding.

The point of this story is that a *quantitative* understanding of prediction clarifies a fundamental *qualitative* confusion about variation and causality. From purely mathematical considerations, it is expected that the best pilots will decline, relative to the others, while the worst will improve in their rankings, in the same way that we expect daughters of tall mothers to be, on average, tall but not quite as tall as their mothers, and so on.

### Relation of "regression to the mean" to the larger themes of the book

The regression fallacy described above is a particular example of a misinterpretation of a comparison. The key idea is that, for causal inference, you should compare like with like.

We can apply this idea to the examples of regression to the mean. In the test scores problem, the causal claim is that doing poorly on the midterm exam is a motivation for students to study hard for the final, while students who do well on the midterm are more likely to relax. In this comparison, the outcome $y$ is the final exam score, and the predictor $x$ is the midterm score. The striking result is that, comparing students who differ by 1 unit on $x$, their expected difference is only $\frac{1}{2}$ unit on $y$.

And why is this striking? Because it is being compared to the slope of 1. The observed pattern as shown in the regression table and in Figure 6.5 is being compared to an implicit default model in which midterm and final exam scores are the same. But the comparison between these two models is inappropriate because the default model is not correct—there is not, in fact, any reason to suspect that midterm and final exam scores would be identical in the absence of any motivational intervention.

Our point here is not that there is a simple analysis which would allow us to perform causal inference in this setting. Rather, we are demonstrating regression to the mean, along with a comparison to an implicit (but, upon reflection, inappropriate) model can lead to incorrect causal inferences.

Again, in the flight school example, a comparison is being made to an implicit model in which, absent any positive or negative reinforcement, individual performance would stay still. But such a model is inappropriate in the context of real variation from trial to trial.

## 6.6 Bibliographic note

For background on the height and earnings example, see Ross (1990) and the bibliographic note at the end of Chapter 12.

The data on mothers' and daughters' heights in Figure 6.3 come from Pearson and Lee (1903); see also Wachsmuth, Wilkinson, and Dallal (2003), and Pagano and Anoke (2013) for more on this example. The idea of regression coefficients as comparisons relates to the four basic statistical operations of Efron (1982).

The historical background of regression to the mean is covered by Stigler (1986), and some of its connections to other statistical ideas are discussed by Stigler (1983). Lord (1967, 1969) considers how regression to the mean can lead to confusion about causal inference. The story of the pilots' training comes from Kahneman and Tversky (1973).

## 6.7 Exercises

6.1 *Data and fitted regression line*: A teacher in a class of 50 students gives a midterm exam with possible scores ranging from 0 to 50 and a final exam with possible scores ranging from 0 to 100. A linear regression is fit, yielding the estimate $y = 30 + 1.2 * x$ with residual standard deviation 10. Sketch (by hand, not using the computer) the regression line, along with hypothetical data that could yield this fit.

6.2 *Programming fake-data simulation*: Write an R function to: (i) simulate $n$ data points from the model, $y = a + bx + \text{error}$, with data points $x$ uniformly sampled from the range $(0, 100)$ and with errors drawn independently from the normal distribution with mean 0 and standard deviation $\sigma$; (ii) fit a linear regression to the simulated data; and (iii) make a scatterplot of the data and fitted regression line. Your function should take as arguments, $a, b, n, \sigma$, and it should return the data, print out the fitted regression, and make the plot. Check your function by trying it out on some values of $a, b, n, \sigma$.

6.3 *Variation, uncertainty, and sample size*: Repeat the example in Section 6.2, varying the number of data points, $n$. What happens to the parameter estimates and uncertainties when you increase the number of observations?

6.4 *Simulation study*: Perform the previous exercise more systematically, trying out a sequence of values of $n$, for each simulating fake data and fitting the regression to obtain estimate and uncertainty (median and mad sd) for each parameter. Then plot each of these as a function of $n$ and report on what you find.

6.5 *Regression prediction and averages*: The heights and earnings data in Section 6.3 are in the folder `Earnings`. Download the data and compute the average height for men and women in the sample.

(a) Use these averages and fitted regression model displayed on page 84 to get a model-based estimate of the average earnings of men and of women in the population.

(b) Assuming 52% of adults are women, estimate the average earnings of adults in the population.

(c) Directly from the sample data compute the average earnings of men, women, and everyone. Compare these to the values calculated in parts (a) and (b).

6.6 *Selection on $x$ or $y$*:

(a) Repeat the analysis in Section 6.4 using the same data, but just analyzing the observations for *mothers'* heights less than the mean. Confirm that the estimated regression parameters are roughly the same as were obtained by fitting the model to all the data.

(b) Repeat the analysis in Section 6.4 using the same data, but just analyzing the observations for *daughters'* heights less than the mean. Compare the estimated regression parameters and discuss how they differ from what was obtained by fitting the model to all the data.

(c) Explain why selecting on daughters' heights had so much more of an effect on the fit than selecting on mothers' heights.

6.7 *Regression to the mean*: Gather before-after data with a structure similar to the mothers' and daughters' heights in Sections 6.4 and 6.5. These data could be performance of athletes or sports teams from one year to the next, or economic outcomes in states or countries in two successive years, or any other pair of measurements taken on a set of items. Standardize each of the two variables so it has a mean of 0 and standard deviation of 1.

(a) Following the steps of Section 6.4, read in the data, fit a linear regression, and plot the data and fitted regression line.

(b) Repeat the above steps with fake data that look similar to the data you have gathered.

6.8 *Regression to the mean with fake data*: Perform a fake-data simulation as in Section 6.5, but using the flight school example on page 89. Simulate data from 500 pilots, each of whom performs two maneuvers, with each maneuver scored continuously on a 0–10 scale, that each pilot has a true ability that is unchanged during the two tasks, and that the score for each test is equal to this true ability plus independent errors. Further suppose that when pilots score higher than 7 on the scale during the first maneuver, that they get praised, and that scores lower than 3 on the first maneuver result in negative reinforcement. Also suppose, though, that this feedback has no effect on performance on the second task.

(a) Make a scatterplot with one dot for each pilot, showing score on the second maneuver vs. score on the first maneuver. Color the dots blue for the pilots who got praised, red for those who got negative reinforcement, and black for the other cases.

(b) Compute the average change in scores for each group of pilots. If you did your simulation correctly, the pilots who were praised did worse, on average, and the pilots who got negative reinforcement improved, on average, for the second maneuver. Explain how this happened, given that your data were simulated under a model in which the positive and negative messages had no effects.

6.9 *Working through your own example*: Continuing the example from the final exercises of the earlier chapters, find two variables that represent before-and-after measurements of some sort. Make a scatterplot and discuss challenges of "regression to the mean" when interpeting before-after changes here.

# Chapter 7

# Linear regression with a single predictor

As discussed in Chapter 1, regression is fundamentally a technology for predicting an outcome $y$ from inputs $x_1, x_2, \ldots$. In this chapter we introduce regression in the simple (but not trivial) case of a linear model predicting a continuous $y$ from a single continuous $x$, thus fitting the model $y_i = a + bx_i + \text{error}$ to data $(x_i, y_i), i = 1, \ldots, n$. We demonstrate with an applied example that includes the steps of fitting the model, displaying the data and fitted line, and interpreting the fit. We then show how to check the fitting procedure using fake-data simulation, and the chapter concludes with an explanation of how linear regression includes simple comparison as a special case.

## 7.1 Example: predicting presidential vote share from the economy

xample: ections nd the conomy

Figure 7.1 tells a story of elections and the economy. It's based on the "bread and peace" model created by political scientist Douglas Hibbs to forecast elections based solely on economic growth (with corrections for wartime, notably Adlai Stevenson's exceptionally poor performance in 1952 and Hubert Humphrey's loss in 1968, years when Democratic presidents were presiding over unpopular wars). Better forecasts are possible using additional information (most notably, incumbency and opinion polls), but what is impressive here is that this simple model does pretty well.

### Fitting a linear model to data

Figure 7.2 displays the economy and elections data as a prediction problem, with a fitted line predicting $y$ (vote share) from $x$ (economic performance). The data are in the file `hibbs.dat`, with the incumbent party's vote percentage of the two-party vote coded as `vote` and average personal income growth in the previous years coded as `growth`. We first read in and display the data:[1]

```
hibbs <- read.table("hibbs.dat", header=TRUE)
plot(hibbs$growth, hibbs$vote, xlab="Economic growth",
  ylab="Incumbent party's vote share")
```

Now we fit and display the model:

```
M1 <- stan_glm(vote ~ growth, data=hibbs)
print(M1)
```

In Chapter 8 we discuss the fitting procedure. For now, we work with the fitted model assuming it summarizes the data we have.

### Understanding the fitted model

This is what we see in the R console after typing `print(M1)`:

[1]Data and code for this example are in the folder `ElectionsEconomy`.

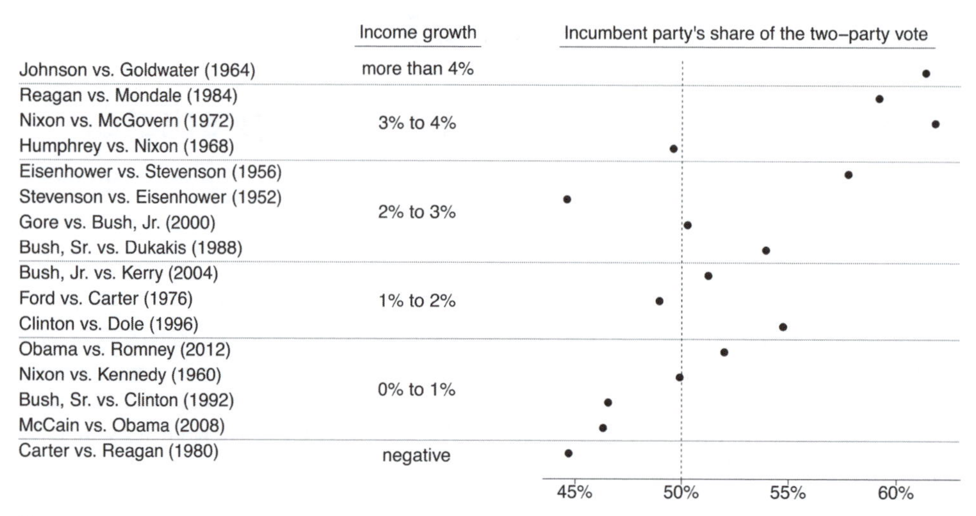


Figure 7.1 *Douglas Hibbs's "bread and peace" model of voting and the economy. Presidential elections since 1952 are listed in order of the economic performance at the end of the preceding administration (as measured by inflation-adjusted growth in average personal income). Matchups are listed as incumbent party's candidate versus other party's candidate. The better the economy was performing, the better the incumbent party's candidate did, with the biggest exceptions being 1952 (Korean War) and 1968 (Vietnam War).*

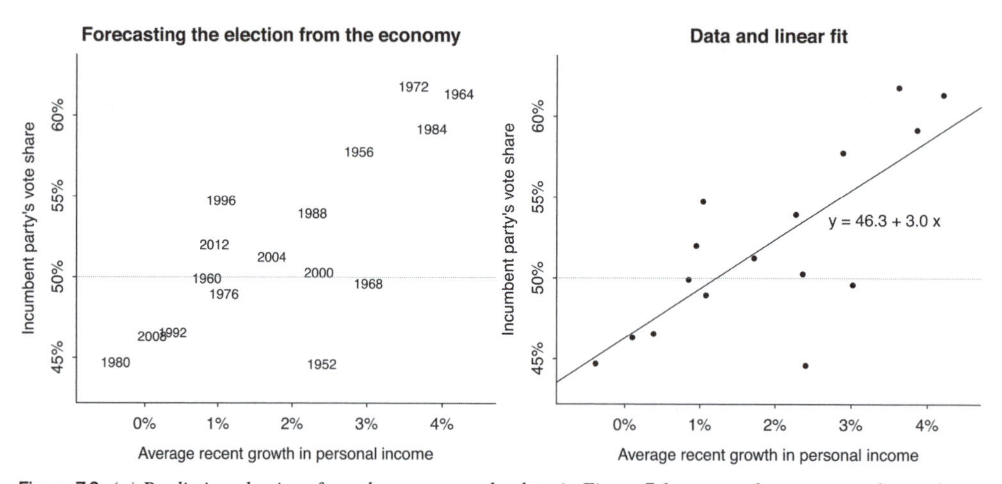


Figure 7.2 *(a) Predicting elections from the economy: the data in Figure 7.1 expressed as a scatterplot, with one data point for each year, (b) the data with the linear fit,* $y = 46.3 + 3.0x$. *Repeated from Figure 1.1.*

```
            Median MAD_SD
(Intercept) 46.3   1.7
growth       3.0   0.7

Auxiliary parameter(s):
      Median MAD_SD
sigma 3.9    0.7
```

In this example, vote is the outcome and growth is the predictor. By default the fit includes an intercept term: the line is $y = a + bx$, not simply $y = bx$.

To fit a line with zero intercept, $y = bx$, use the expression, `stan_glm(vote ~ -1 + growth)`; the `-1` tells R to exclude the intercept or constant term from the fit. In this example such a model would make no sense, but there are cases where it can be appropriate to fit a regression with no intercept, if you want the model to constrain the predicted value of $y$ to be zero when $x = 0$.

To continue with the model we have fit, next is a table showing the estimates (Median) and uncertainties (MAD_SD) for the two coefficients, Intercept and growth, and the residual standard deviation sigma.

From this information, the *fitted line* is $y = 46.3 + 3.0x$:

- At $x = 0$ (zero economic growth), the incumbent party's candidate is predicted to receive 46.3% of the vote (that is, to lose). This makes sense: in recent history, zero growth represents poor performance, and a party that presides over this poor economy will be punished.
- Each percentage point of economic growth that is larger than 0 corresponds to an expected vote share for the incumbent party that is 3.0 percentage points higher than 46.3%. This jibes with the range of data and the line on the graph.
- The standard errors for the coefficients are small. We don't usually care so much about the standard error on the intercept, but the standard error on the slope is a measure of the uncertainty in this estimate. In this case, the standard error is 0.7, so the data are roughly consistent with a slope in the range $3.0 \pm 0.7$, and the 95% interval is $[3.0 \pm 2 * 0.7] = [1.6, 4.4]$. The interval is well separated from zero, indicating that, if the data had been generated from a model whose true slope is 0, it would be very unlikely to get an estimated slope coefficient this large.
- The estimated *residual standard deviation*, $\sigma$, is 3.9. That is, the linear model predicts the election outcome to within about 3.9 percentage points. Roughly 68% of the outcomes will fall between $\pm 3.9$ of the fitted line. This makes sense. The predictions from the model are informative but not determinative. To put it another way: the model doesn't exactly predict each observation. Based on a linear model that only takes into account the economic growth, there is a limit to the accuracy with which you can predict an election.

One reason that Hibbs (and others) were confident in predicting a Democratic victory in 2008 could be seen from this linear model. The equation is $y = 46.3 + 3.0x$. In the years leading up to the 2008 election, the economic growth was approximately 0.1%, or $x = 0.1$. This linear model yields a prediction of $y = 46.6$, or an estimate of 46.6% of the popular vote going to the incumbent party, and thus a predicted 53.4% for Barack Obama, as the Republicans were the incumbent party in that year. The model fit when Obama was running for president was not identical to the model fit above, as it did not have the data points for 2008 and 2012. But it was similar, as can be seen by re-fitting the regression excluding those two elections.

## Graphing the fitted regression line

We complete Figure 7.2b by adding the fitted line to the scatterplot we have already made:

```
abline(coef(M1))
```

The coef function returns the estimated coefficients of the linear model. The first coefficient is the intercept and the second is the slope for growth, which can be seen by typing `coef(M1)`. By plugging in these values into the `abline()` function, we are able to add the fitted model to the graph. The vertical discrepancies between the points and the line indicate the imperfections of the fitted model.

## Using the model to predict

We fit the above model in advance of the 2016 presidential election of Democrat Hillary Clinton vs. Republican Donald Trump. At the time of fitting the model, the model's economic predictor—the

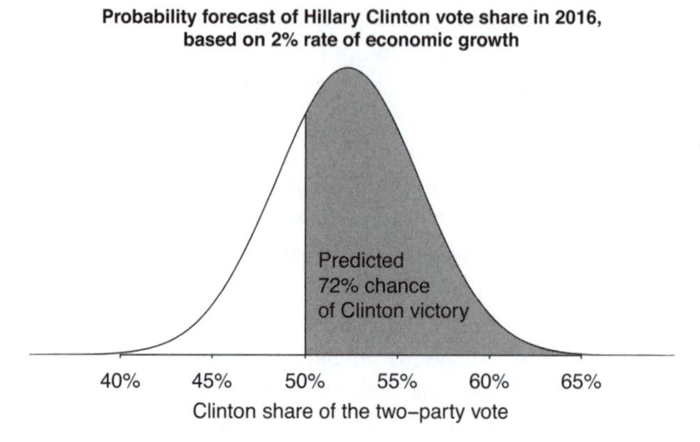


Figure 7.3 *Forecast distribution for Hillary Clinton's percentage of the two-party vote in 2016 based on an economic growth rate of 2%. The curve shows a normal distribution centered at the point forecast of* $46.3 + 3.0 * 2.0 = 52.3$ *and with standard deviation of 3.9, the estimated uncertainty based on the model fit. The shaded area indicates the probability (based on the model) that Clinton wins, which is* $1 - \text{pnorm}(50, 52.3, 3.9)$*, or 0.72.*

weighted average of economic growth of the previous four years—was approximately 2%. What, then, was Clinton's forecast vote percentage? The linear model predicted $46.3 + 3.0 * 2.0 = 52.3$. This forecast vote percentage does not alone tell us how likely it is that Clinton would win the popular vote in 2016. We also need some assessment of the uncertainty in that prediction.

Figure 7.3 shows the probability distribution of this forecast, which is centered on 52.3 with a standard deviation of 3.9. In constructing the regression model, we were assuming a normal approximation for the errors, which is conventional and also makes sense for the reasons discussed in Section 3.5 regarding the Central Limit Theorem, implicitly modeling errors as sums of many small independent components. We are very slightly understating uncertainty here by considering the fitted parameters of the regression model as known; in Chapter 9, we fix this by using simulation to combine the uncertainties in estimation and prediction.

To continue with the simple illustrative calculation, we can use the forecast of *vote share* to estimate the probability that Clinton would win the popular vote. She wins if her vote share exceeds 50%; this is the shaded area in the curve in Figure 7.3. This is a probability distribution, so the total area under the curve is 1. To find the shaded area in R, type:

```
1 - pnorm(50, 52.3, 3.9)
```

This returns 0.72. According to this model, there was a 72% chance that Clinton would win the popular vote, running in 2016 with economic growth of 2%.

If the goal is to predict the winner, why predict vote share at all? As shown in Figure 7.2, Hibbs's model predicts the election outcome as a continuous vote share, some number between 0 and 100% (in practice, between 40% and 60%). Why not instead use the economic data to directly predict who wins the election?

The reason for not simply predicting the winner can be seen by considering three sorts of elections:

- *Evenly divided elections* such as 1960 (Kennedy vs. Nixon) that were essentially tied. It would be unreasonable and meaningless to attempt to predict the winner of such an election. The best you can hope to do is to predict the vote percentages—that is, in these cases, to predict that the election is likely to be close.
- *Somewhat competitive elections* such as 2008 (Obama vs. McCain) in which one side was predicted to win but the other side was perceived as having some chance of an upset. Here you would like to predict the vote margin and the probability of either candidate winning, rather than trying to make a deterministic call.

- *Landslides* such as 1984 (Reagan vs. Mondale). Here there is nothing impressive about predicting the winner, but it is still a challenge to accurately predict the percentage of the vote received by the two candidates.

In any of these cases, a prediction of the vote share is more informative than simply predicting the winner.

In the event, Clinton received 51.1% of the popular vote, well within the forecast distribution, but she lost the electoral college, which is decided by which states are won by which candidates. To forecast the electoral-vote winner, one must predict the vote in each state, which can be done using a multilevel model, but this is beyond the scope of this book.

## 7.2 Checking the model-fitting procedure using fake-data simulation

The preceding example is simple enough that we can just make a graph and see that the line goes through it, but more generally it is good practice to check our fits by performing them under controlled conditions where we know the truth. We demonstrate here with the election model.[2]

### Step 1: Creating the pretend world

We start by assuming true values for all the parameters in the model. In this case, we have already fit a model to data so we shall proceed by pretending that these particular parameter values are the truth. That is, we assume the relation $y = 46.3 + 3.0x + \text{error}$ is true, with the errors drawn from a normal distribution with mean 0 and standard deviation 3.9. Then (using the predictors $x$ that we already have from our dataset) we examine whether these predictors create a distribution of $y$ that is consistent with our observed $y$.

```
a <- 46.3
b <- 3.0
sigma <- 3.9
x <- hibbs$growth
n <- length(x)
```

### Step 2: Simulating fake data

We then simulate a vector $y$ of fake data and put this all in a data frame:

```
y <- a + b*x + rnorm(n, 0, sigma)
fake <- data.frame(x, y)
```

### Step 3: Fitting the model and comparing fitted to assumed values

The next step is to fit a regression to these data. The fitting makes no use of the assumed true values of $\alpha$, $\beta$, and $\sigma$.

```
fit <- stan_glm(y ~ x, data=fake)
print(fit)
```

Here is the regression output:

```
            Median MAD_SD
(Intercept) 44.4    1.7
x            3.2    0.7
```

[2]Data and code for this example are in the folder `ElectionsEconomy`.

```
Auxiliary parameter(s):
      Median MAD_SD
sigma 4.0    0.8
```

Comparing the estimated coefficients to the assumed true values 46.3 and 3.0, the fit seems reasonable enough: the estimates are not exact but are within the margins of error.

We can perform this comparison more formally by extracting from the regression object the estimates and standard errors of the coefficients. Here for simplicity in programming we do this for just one of the coefficients, the slope $b$:

```
b_hat <- coef(fit)["x"]
b_se <- se(fit)["x"]
```

We then check whether the true value of $b$ falls within the estimated 68% and 95% confidence intervals obtained by taking the estimate $\pm 1$ or $\pm 2$ standard errors (recall Figure 4.1):

```
cover_68 <- abs(b - b_hat) < b_se
cover_95 <- abs(b - b_hat) < 2*b_se
cat(paste("68% coverage: ", cover_68, "\n"))
cat(paste("95% coverage: ", cover_95, "\n"))
```

### Step 4: Embedding the simulation in a loop

The confidence intervals worked once, but do they have the correct *coverage probabilities*—that is, do the intervals contain the true value the advertised percentage of the time, as in Figure 4.2? To check, we use the normal distribution; recall Figure 4.1. We can simulate the sampling distribution and thus compute the coverage of the confidence interval by embedding the data simulation, model fitting, and coverage checking in a loop and running 1000 times. This and other loops in this chapter could also be performed implicitly using the `replicate` function in R, as illustrated on pages 72 and 74. Here we code the loop directly:

```
n_fake <- 1000
cover_68 <- rep(NA, n_fake)
cover_95 <- rep(NA, n_fake)
for (s in 1:n_fake){
  y <- a + b*x + rnorm(n, 0, sigma)
  fake <- data.frame(x, y)
  fit <- stan_glm(y ~ x, data=fake, refresh=0)     # suppress output on console
  b_hat <- coef(fit)["x"]
  b_se <- se(fit)["x"]
  cover_68[s] <- abs(b - b_hat) < b_se
  cover_95[s] <- abs(b - b_hat) < 2*b_se
}
cat(paste("68% coverage: ", mean(cover_68), "\n"))
cat(paste("95% coverage: ", mean(cover_95), "\n"))
```

This takes a few minutes to run, and then the following appears on the console:

```
68% coverage:  0.628
95% coverage:  0.928
```

That is, `mean(cover_68)` = 63% and `mean(cover_95)` = 93%, not far from the nominal values of 0.68 and 0.95. The coverage is a bit low, in part because $\pm 1$ and $\pm 2$ are standard error bounds for the normal distribution, but with a sample size of only 16, our inferences should use the $t$ distribution with 14 degrees of freedom. This is not a big deal, but just to be careful we can redo by checking the coverage of the appropriate $t$ intervals:

```
n_fake <- 1000
cover_68 <- rep(NA, n_fake)
cover_95 <- rep(NA, n_fake)
t_68 <- qt(0.84, n - 2)
t_95 <- qt(0.975, n - 2)
for (s in 1:n_fake){
  y <- a + b*x + rnorm(n, 0, sigma)
  fake <- data.frame(x, y)
  fit <- stan_glm(y ~ x, data=fake, refresh=0)
  b_hat <- coef(fit)["x"]
  b_se <- se(fit)["x"]
  cover_68[s] <- abs(b - b_hat) < t_68 * b_se
  cover_95[s] <- abs(b - b_hat) < t_95 * b_se
}
cat(paste("68% coverage: ", mean(cover_68), "\n"))
cat(paste("95% coverage: ", mean(cover_95), "\n"))
```

This simulation gives the desired result that approximately 68% of the 68% intervals, and approximately 95% of the 95% intervals, contain the true parameter values.

## 7.3 Formulating comparisons as regression models

To connect to the basic statistical methods of Chapter 4, we show how simple averages and comparisons can be interpreted as special cases of linear regression. This more general formulation is helpful on a conceptual level to unify these seemingly different ideas, and it will help in practice as a key component in building larger models, as we show in the following chapters.

To express comparisons as regressions we need the concept of an *indicator variable*, which is a predictor that equals 1 or 0 to indicate whether a data point falls into a specified category. Indicator variables can be used for inputs with two categories (for example, an indicator for "male" that takes on the value 1 for men and 0 for women) or multiple categories (for example, indicators for "White," "Black," and "Hispanic," with each taking on the value of 1 for people in the specified category and zero otherwise).

For simplicity we here use fake data to demonstrate the connection between comparisons and regression on indicators.[3]

### Estimating the mean is the same as regressing on a constant term

Example: Comparison and simple regression

Let's simulate 20 observations from a population with mean 2.0 and standard deviation 5.0:

```
n_0 <- 20
y_0 <- rnorm(n_0, 2.0, 5.0)
fake_0 <- data.frame(y_0)
print(y_0)
```

These are the numbers that arose in one particular simulation:

```
 -0.3   4.1  -4.9   3.3   6.4   7.2  10.7  -4.6   4.7   6.0   1.1  -6.7  10.2
  9.7   5.6   1.7   1.3   6.2  -2.1   6.5
```

Considering these as a random sample, we can estimate the mean of the population as `mean(y_0)` with standard error `sd(y_0)/sqrt(n_0)`. The result: an estimate of 3.3 with standard error 1.1.

We get the identical result using least squares regression on a constant term:

```
fit_0 <- stan_glm(y_0 ~ 1, data=fake_0,
  prior_intercept=NULL, prior=NULL, prior_aux=NULL)
print(fit_0)
```

[3]Code for this example is in the folder `Simplest`.

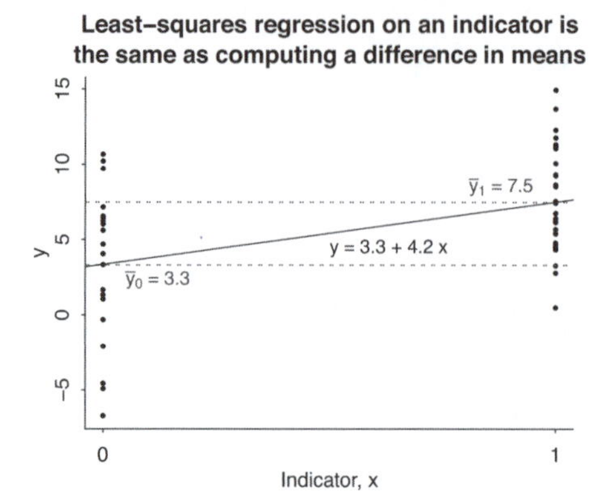


Figure 7.4 *Simulated-data example showing how regression on an indicator variable is the same as computing the difference in means between two groups.*

which yields:

```
            Median MAD_SD
(Intercept) 3.3    1.1

Auxiliary parameter(s):
      Median MAD_SD
sigma 5.3    0.9
```

As discussed in Chapters 8 and 9, the above settings assign a flat prior so that `stan_glm` will produce the classical least squares regression estimate. We discuss prior distributions more in Section 9.5; here we use the simple least-squares estimate to demonstrate the equivalence between the simple average and the intercept-only regression; in practice the default prior for `stan_glm` would give a very similar result.

From our simulations, we know that the true intercept for this regression is 2.0 and the residual standard deviation $\sigma$ is 5.0. But this small sample size yields noisy estimates. The estimates of 3.3 and 5.3 are consistent with the true parameter values, given the standard errors of the fit.

### Estimating a difference is the same as regressing on an indicator variable

Next add in a new group: 30 observations from a population with mean 8.0 and standard deviation 5.0:

```
n_1 <- 30
y_1 <- rnorm(n_1, 8.0, 5.0)
```

We can directly compare the averages in each group and compute the corresponding standard error:

```
diff <- mean(y_1) - mean(y_0)
se_0 <- sd(y_0)/sqrt(n_0)
se_1 <- sd(y_1)/sqrt(n_1)
se <- sqrt(se_0^2 + se_1^2)
```

In our particular simulation, these return the value 4.2 for the difference and 1.3 for its standard error. This result is consistent with the construction of our simulations in which the true population difference was 6.0.

Alternatively we can frame the problem as a regression by combining the data into a single vector, $y$, and then creating an *indicator variable* $x$, defined as,

$$x_i = \begin{cases} 0 & \text{if observation } i \text{ is in group 0} \\ 1 & \text{if observation } i \text{ is in group 1.} \end{cases}$$

In R:

```
n <- n_0 + n_1
y <- c(y_0, y_1)
x <- c(rep(0, n_0), rep(1, n_1))
fake <- data.frame(x, y)
fit <- stan_glm(y ~ x, data=fake, prior_intercept=NULL, prior=NULL, prior_aux=NULL)
print(fit)
```

The result:

```
            Median MAD_SD
(Intercept) 3.3    1.1
x           4.2    1.2

Auxiliary parameter(s):
      Median MAD_SD
sigma 4.7    0.5
```

The estimate of the slope, 4.2, is identical to the difference in means, $\bar{y}_1 - \bar{y}_0$, as it must be for this simple model. We discuss priors more fully in Chapter 9. The standard error is nearly identical but differs slightly because the regression model estimates a single residual standard deviation parameter, as compared to the difference calculation which uses separate values of $\text{se}_0$ and $\text{se}_1$.

Figure 7.4 visually displays the equivalence of the two estimates: the regression on an indicator variable (shown by the fitted line) and the comparison of the two means (shown by the difference between the two dotted lines). In this simple setting with no other predictors, the least squares line will go through the points $(0, \bar{y}_0)$ and $(1, \bar{y}_1)$, so the slope of the line is simply the difference, $\bar{y}_1 - \bar{y}_0$. In more complicated settings, we can use indicator variables to compare groups, adjusting for other predictors.

The point of doing all this using fake-data simulation is, first, to directly check that the direct comparison and the regression give the same answer and, second, to understand the properties of statistical fits using a general tool that will continue to be helpful in more complicated settings. In Section 10.4 we discuss the use of indicator variables in regression models with multiple predictors.

## 7.4 Bibliographic note

The material in this and the following chapters on linear regression is covered in many books which specialize in various aspects of the problem. Harrell (2001) covers applied and conceptual issues in regression. Fox (2002) is a good introductory text on fitting and displaying regressions in R.

For research on economic performance and incumbent party vote for president, see Hibbs, Rivers, and Vasilatos (1982) with updates by Hibbs (2000, 2012), and also Rosenstone (1983), Fair (1978), and Wlezien and Erikson (2004, 2005).

## 7.5 Exercises

7.1 *Regression predictors*: In the election forecasting example of Section 7.1, we used inflation-adjusted growth in average personal income as a predictor. From the standpoint of economics, it makes sense to adjust for inflation here. But suppose the model had used growth in average personal income, not adjusting for inflation. How would this have changed the resulting regression? How would this change have affected the fit and interpretation of the results?

7.2 *Fake-data simulation and regression*: Simulate 100 data points from the linear model, $y = a + bx + \text{error}$, with $a = 5$, $b = 7$, the values of $x$ being sampled at random from a uniform distribution on the range $[0, 50]$, and errors that are normally distributed with mean 0 and standard deviation 3.

(a) Fit a regression line to these data and display the output.

(b) Graph a scatterplot of the data and the regression line.

(c) Use the `text` function in R to add the formula of the fitted line to the graph.

7.3 *Fake-data simulation and fitting the wrong model*: Simulate 100 data points from the model, $y = a + bx + cx^2 + \text{error}$, with the values of $x$ being sampled at random from a uniform distribution on the range $[0, 50]$, errors that are normally distributed with mean 0 and standard deviation 3, and $a, b, c$ chosen so that a scatterplot of the data shows a clear nonlinear curve.

(a) Fit a regression line `stan_glm(y ~ x)` to these data and display the output.

(b) Graph a scatterplot of the data and the regression line. This is the best-fit linear regression. What does "best-fit" mean in this context?

7.4 *Prediction*: Following the template of Section 7.1, find data in which one variable can be used to predict the other, then fit a linear model and plot it along with the data, then display the fitted model and explain in words as on page 95. Use the model to obtain a probabilistic prediction for new data, and evaluate that prediction, as in the last part of Section 7.1.

7.5 *Convergence as sample size increases*: Set up a simulation study such as in Section 7.2, writing the entire simulation as a function, with one of the arguments being the number of data points, $n$. Compute the simulation for $n = 10, 30, 100, 300, 1000, 3000, 10\,000$, and $30\,000$, for each displaying the estimate and standard error. Graph these to show the increasing stability as $n$ increases.

7.6 *Formulating comparisons as regression models*: Take the election forecasting model and simplify it by creating a binary predictor defined as $x = 0$ if income growth is less than 2% and $x = 1$ if income growth is more than 2%.

(a) Compute the difference in incumbent party's vote share on average, comparing those two groups of elections, and determine the standard error for this difference.

(b) Regress incumbent party's vote share on the binary predictor of income growth and check that the resulting estimate and standard error are the same as above.

7.7 *Comparing simulated data to assumed parameter values*:

(a) Simulate 100 data points from the model, $y = 2 + 3x + \text{error}$, with predictors $x$ drawn from a uniform distribution from 0 to 20, and with independent errors drawn from the normal distribution with mean 0 and standard deviation 5. Save $x$ and $y$ into a data frame called `fake`. Fit the model, `stan_glm(y ~ x, data=fake)`. Plot the data and fitted regression line.

(b) Check that the estimated coefficients from the fitted model are reasonably close to the assumed true values. What does "reasonably close" mean in this context?

7.8 *Sampling distribution*: Repeat the steps of the previous exercise 1000 times (omitting the plotting). Check that the coefficient estimates are approximately unbiased, that their standard deviations in the sampling distribution are approximately equal to their standard errors, and that approximately 95% of the estimate $\pm$ 2 standard error intervals contain the true parameter values.

7.9 *Interpretation of regressions*: Redo the election forecasting example of Section 7.1, but switching $x$ and $y$, that is, predicting economic growth given the subsequent election outcome. Discuss the problems with giving a causal interpretation to the coefficients in this regression, and consider what this implies about any causal interpretations of the original regression fit in the chapter.

7.10 *Working through your own example*: Continuing the example from the final exercises of the earlier chapters, fit a linear regression with a single predictor, graph the data along with the fitted line, and interpret the estimated parameters and their uncertainties.

# Chapter 8

# Fitting regression models

Most of this book is devoted to examples and tools for the practical use and understanding of regression models, starting with linear regression with a single predictor and moving to multiple predictors, nonlinear models, and applications in prediction and causal inference. In this chapter we lay out some of the mathematical structure of inference for regression models and some algebra to help understand estimation for linear regression. We also explain the rationale for the use of the Bayesian fitting routine `stan_glm` and its connection to classical linear regression. This chapter thus provides background and motivation for the mathematical and computational tools used in the rest of the book.

## 8.1 Least squares, maximum likelihood, and Bayesian inference

We now step back and consider *inference*: the steps of estimating the regression model and assessing uncertainty in the fit. We start with *least squares*, which is the most direct approach to estimation, based on finding the values of the coefficients $a$ and $b$ that best fit the data. We then discuss *maximum likelihood*, a more general framework that includes least squares as a special case and to which we return in later chapters when we get to logistic regression and generalized linear models. Then we proceed to *Bayesian inference*, an even more general approach that allows the probabilistic expression of prior information and posterior uncertainty.

### Least squares

In the classical linear regression model, $y_i = a + bx_i + \epsilon_i$, the coefficients $a$ and $b$ are estimated so as to minimize the errors $\epsilon_i$. If the number of data points $n$ is greater than 2, it is not generally possible to find a line that gives a perfect fit (that would be $y_i = a + bx_i$, with no error, for all data points $i = 1, \dots, n$), and the usual estimation goal is to choose the estimate $(\hat{a}, \hat{b})$ that minimizes the sum of the squares of the residuals,

$$r_i = y_i - (\hat{a} + \hat{b}x_i).$$

We distinguish between the *residuals* $r_i = y_i - (\hat{a} + \hat{b}x_i)$ and the *errors* $\epsilon_i = y_i - (a + bx_i)$. The model is written in terms of the errors, but it is the residuals that we can work with: we cannot calculate the errors as to do so would require knowing $a$ and $b$.

The residual sum of squares is

$$\text{RSS} = \sum_{i=1}^{n} (y_i - (\hat{a} + \hat{b}x_i))^2. \tag{8.1}$$

The $(\hat{a}, \hat{b})$ that minimizes RSS is called the least squares or ordinary least squares or OLS estimate and can be written in matrix notation as,

$$\hat{\beta} = (X^t X)^{-1} X^t y, \tag{8.2}$$

where $\beta = (a, b)$ is the vector of coefficients and $X = (\mathbf{1}, x)$ is the matrix of predictors in the regression. In this notation, **1** represents a column of ones—the constant term in the regression—and must be included because we are fitting a model with an intercept as well as a slope. We show more general notation for linear regression with multiple predictors in Figure 10.8.

Expression (8.2) applies to least squares regression with any number of predictors. In the case of regression with just one predictor, we can write the solution as,

$$\hat{b} = \frac{\sum_{i=1}^{n}(x_i - \bar{x})y_i}{\sum_{i=1}^{n}(x_i - \bar{x})^2}, \tag{8.3}$$

$$\hat{a} = \bar{y} - \hat{b}\bar{x}. \tag{8.4}$$

We can then can write the least squares line as,

$$y_i = \bar{y} + \hat{b}\,(x_i - \bar{x}) + r_i;$$

thus, the line goes through the mean of the data, $(\bar{x}, \bar{y})$, as illustrated in Figure 6.3.

Formula (8.2) and the special case (8.3)–(8.4) can be directly derived using calculus as the solution to the problem of minimizing the residual sum of squares (8.1). In practice, these computations are done using efficient matrix solution algorithms in R or other software.

### Estimation of residual standard deviation $\sigma$

In the regression model, the errors $\epsilon_i$ come from a distribution with mean 0 and standard deviation $\sigma$: the mean is zero by definition (any nonzero mean is absorbed into the intercept, $a$), and the standard deviation of the errors can be estimated the from data. A natural way to estimate $\sigma$ would be to simply take the standard deviation of the residuals, $\sqrt{\frac{1}{n}\sum_{i=1}^{n} r_i^2} = \sqrt{\frac{1}{n}\sum_{i=1}^{n}(y_i - (\hat{a} + \hat{b}x_i))^2}$, but this would slightly underestimate $\sigma$ because of *overfitting*, as the coefficients $\hat{a}$ and $\hat{b}$ have been set based on the data to minimize the sum of squared residuals. The standard correction for this overfitting is to replace $n$ by $n - 2$ in the denominator (with the subtraction of 2 coming from the estimation of two coefficients in the model, the intercept and the slope); thus,

$$\hat{\sigma} = \sqrt{\frac{1}{n-2}\sum_{i=1}^{n}(y_i - (\hat{a} + \hat{b}x_i))^2}. \tag{8.5}$$

When $n = 1$ or 2 this expression is meaningless, which makes sense: with only two data points you can fit a line exactly and so there is no way of estimating error from the data alone.

More generally, in a regression with $k$ predictors (that is, $y = X\beta + \epsilon$, with an $n \times k$ predictor matrix $X$), expression (8.5) becomes $\hat{\sigma} = \sqrt{\frac{1}{n-k}\sum_{i=1}^{n}(y_i - (X_i\hat{\beta}))^2}$, with $n - k$ in the denominator rather than $n$, adjusting for the $k$ coefficients fit by least squares.

### Computing the sum of squares directly

The least squares estimates of the coefficients can be computed directly using the formula in (8.2). But to develop understanding it can be helpful to write an R function to compute the sum of squares and then play around with different values of $a$ and $b$.

First we write the function, which we call `rss` for "residual sum of squares":

```
rss <- function(x, y, a, b){     # x and y are vectors, a and b are scalars
  resid <- y - (a + b*x)
  return(sum(resid^2))
}
```

The above function is somewhat crude and is intended for home consumption and not production use, as we have not, for example, added lines to check that x and y are vectors of the same length, nor have we incorporated any methods for handling missing data (NA entries). Hence we are at the mercy of the R defaults if the function arguments are not specified exactly as desired. For exploration on our own, however, this simple function will do.

We can try it out: `rss(hibbs$growth, hibbs$vote, 46.3, 3.0)` evaluates the residual sum of squares at the least squares estimate, $(a, b) = (46.2, 3.1)$, and we can experiment with other values of $(a, b)$ to check that it's not possible to get any lower with these data; see Exercise 7.1.

### Maximum likelihood

If the errors from the linear model are independent and normally distributed, so that $y_i \sim \mathrm{N}(a+bx_i, \sigma^2)$ for each $i$, then the least squares estimate of $(a, b)$ is also the maximum likelihood estimate. The *likelihood function* in a regression model is defined as the probability density of the data given the parameters and predictors; thus, in this example,

$$p(y|a, b, \sigma, X) = \prod_{i=1}^{n} \mathrm{N}(y_i|a + bx_i, \sigma^2), \tag{8.6}$$

where $\mathrm{N}(\cdot|\cdot,\cdot)$ is the normal probability density function,

$$\mathrm{N}(y \mid m, \sigma^2) = \frac{1}{\sqrt{2\pi}\sigma} \exp\left(-\frac{1}{2}\left(\frac{y-m}{\sigma}\right)^2\right). \tag{8.7}$$

A careful study of (8.6) reveals that maximizing the likelihood requires minimizing the sum of squared residuals; hence the least squares estimate $\hat{\beta} = (\hat{a}, \hat{b})$ can be viewed as a maximum likelihood estimate under the normal model.

There is a small twist in fitting regression models, in that the maximum likelihood estimate of $\sigma$ is $\sqrt{\frac{1}{n}\sum_{i=1}^{n}(y_i - (\hat{a} + \hat{b}x_i))^2}$, without the $\frac{1}{n-2}$ adjustment given in (8.5).

In Bayesian inference, the uncertainty for each parameter in the model automatically accounts for the uncertainty in the other parameters. This property of Bayesian inference is particularly relevant for models with many predictors, and for advanced and hierarchical models.

### Where do the standard errors come from? Using the likelihood surface to assess uncertainty in the parameter estimates

In maximum likelihood estimation, the likelihood function can be viewed as a hill with its peak at the maximum likelihood estimate.

Figure 8.1a displays the likelihood for a simple example as a function of the coefficients $a$ and $b$. Strictly speaking, this model has three parameters—$a$, $b$, and $\sigma$—but for simplicity we display the likelihood of $a$ and $b$ conditional on the estimated $\hat{\sigma}$.

Figure 8.1b shows the maximum likelihood estimate $(\hat{a}, \hat{b}) = (46.2, 3.1)$. This is the value of the parameters where the likelihood function—the hill in Figure 8.1a—has its peak. Figure 8.1b also includes uncertainty bars showing $\pm 1$ standard error for each parameter. For example, the data are consistent with $a$ being roughly in the range $46.2 \pm 1.6$ and with $b$ being in the range $3.1 \pm 0.7$.

The likelihood function does not just have a maximum and a range; it also has a correlation. The area with highest likelihood surrounding the peak can be represented by an ellipse as is shown in Figure 8.1c. The shape of the uncertainty ellipse tells us something about the information in the data and model about the two parameters jointly. In this case the correlation is negative.

To understand this inferential correlation, see the scatterplot of the data from Figure 7.2 which we have reproduced in Figure 8.2a: the regression line goes through the cloud of points, most of which have positive values for $x$. Figure 8.2b shows a range of lines that are consistent with the data, with

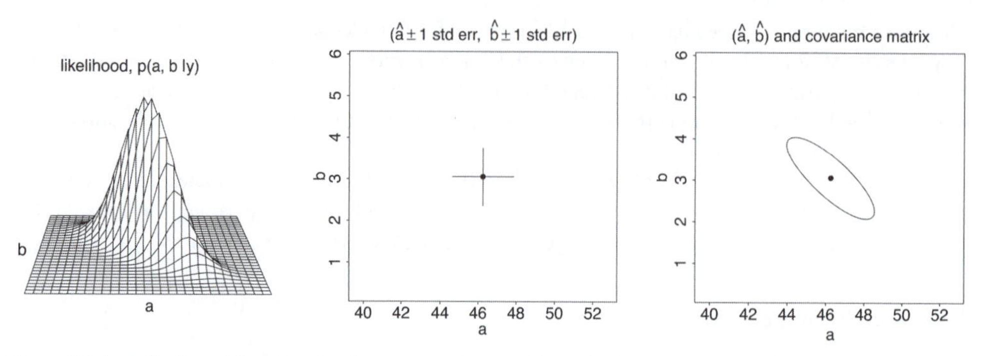


Figure 8.1 *(a) Likelihood function for the parameters a and b in the linear regression $y = a + bx$ + error, of election outcomes, $y_i$, on economic growth, $x_i$. (b) Mode of the likelihood function (that is, the maximum likelihood estimate $(\hat{a}, \hat{b})$) with ± 1 standard error bars shown for each parameter. (c) Mode of the likelihood function with an ellipse summarizing the inverse-second-derivative-matrix of the log likelihood at the mode.*

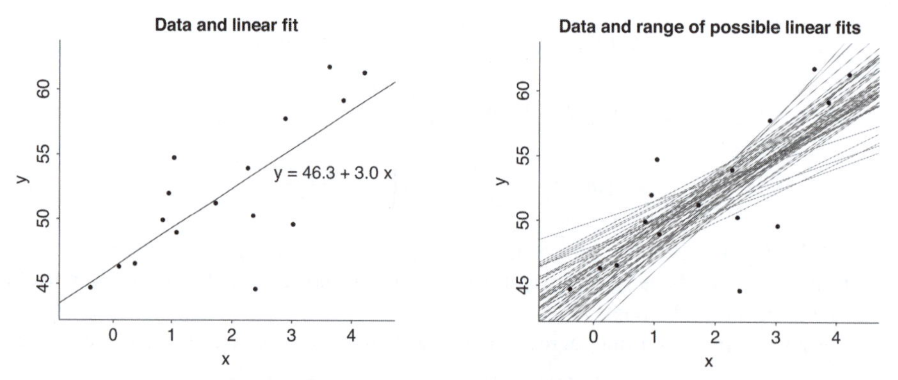


Figure 8.2 *(a) Election data with the linear fit, $y = 46.3 + 3.0x$, repeated from Figure 7.2b. (b) Several lines that are are roughly consistent with the data. Where the slope is higher, the intercept (the value of the line when $x = 0$) is lower; hence there is a negative correlation between a and b in the likelihood.*

the lines representing 50 draws from the Bayesian posterior distribution (see below). Lines of higher slope (for which $b$ is higher) intersect the $y$-axis at lower values (and thus have lower values of $a$), and vice versa, hence the negative correlation in Figure 8.1c.

## Bayesian inference

Least squares or maximum likelihood finds the parameters that best fit the data (according to some pre-specified criterion), but without otherwise constraining or guiding the fit. But, as discussed in Section 9.3 and elsewhere, we typically have prior information about the parameters of the model. Bayesian inference produces a compromise between prior information and data, doing this by multiplying the likelihood with a *prior distribution* that probabilistically encodes external information about the parameters. The product of the likelihood (in the above example, it is $p(y|a, b, \sigma)$ in (8.6), considered as a function of $a$, $b$, and $\sigma$) and the prior distribution is called the *posterior distribution* and it summarizes our knowledge of the parameter, *after* seeing the data. ("Posterior" is Latin for "later.")

In Chapter 9 we consider Bayesian inference from many angles. In the present section we focus on the posterior distribution as a modification of the likelihood function. The generalization of maximum likelihood estimation is maximum *penalized* likelihood estimation, in which the prior distribution is considered as a "penalty function" that downgrades the likelihood for less-favored values of the parameter, again, thus giving an estimate that is typically somewhere in between the prior and what would be obtained by the data alone. Because of the anchoring of the prior, the maximum penalized likelihood estimate can be more stable than the raw maximum likelihood or least squares estimate.

In addition to adding prior information, Bayesian inference is also distinctive in that it expresses uncertainty using probability. When we fit a model using `stan_glm`, we obtain a set of simulation draws that represent the posterior distribution, and which we typically summarize using medians, median absolute deviations, and uncertainty intervals based on these simulations, as we shall discuss in Section 9.1.

We prefer to use Bayesian methods because of the flexibility of propagation of uncertainty using probability and simulation, and because inclusion of prior information can make inferences more stable. That said, it can be helpful to see the connections to least squares and maximum likelihood, as these simpler methods can be easier to understand and are overwhelmingly the most popular ways to fit regression in current practice. So in the following section we show how to fit models in both the classical and Bayesian frameworks.

### Point estimate, mode-based approximation, and posterior simulations

The least squares solution is a *point estimate* that represents the vector of coefficients that provides the best overall fit to data. For a Bayesian model, the corresponding point estimate is the *posterior mode*, which provides the best overall fit to data and prior distribution. The least squares or maximum likelihood estimate is the posterior mode corresponding to the model with a uniform or flat prior distribution.

But we do not just want an estimate; we also want uncertainty. For a model with just one parameter, the uncertainty can be represented by the estimate $\pm$ standard error; more generally we use a bell-shaped probability distribution representing the multivariate uncertainty, as illustrated in Figure 8.1.

As discussed more fully in Chapter 9, it is convenient to summarize uncertainty using simulations from this mode-based approximation or, more generally, from the posterior distribution of the model parameters. By default, when we fit models using `stan_glm`, we get posterior simulations summarized by median and mad sd; see Section 5.3.

## 8.2 Influence of individual points in a fitted regression

From expressions (8.3) and (8.4), we can see that the least squares estimated regression coefficients $\hat{a}$ and $\hat{b}$ are linear functions of the data, $y$. We can use these linear expressions to understand the *influence* of each data point by looking at how much a change in each $y_i$ would change $\hat{b}$. We could also work out the influence on the intercept—the predicted value when $x = 0$—or any other prediction under the model, but typically it is the slope that is most of interest.

From equation (8.3), we see that an increase of 1 in $y_i$ corresponds to a change in $\hat{b}$ that is proportional to $(x_i - \bar{x})$:

- If $x_i = \bar{x}$, the influence of point $i$ on the regression slope is 0. This makes sense: taking a point in the center and moving it up or down will affect the height of the fitted line but not its slope.
- If $x_i > \bar{x}$, the influence of point $i$ is positive, with greater influence the further $x_i$ is from the mean.
- If $x_i < \bar{x}$, the influence of point $i$ is negative, with greater absolute influence the further $x_i$ is from the mean.

One way to understand influence is to consider the fitted regression line as a rod attached to the data by rubber bands; then imagine how the position and orientation of the rod changes as individual data points are moved up and down. Figure 8.3 illustrates.

Influence can also be computed for multiple regression, using the matrix expression (equation

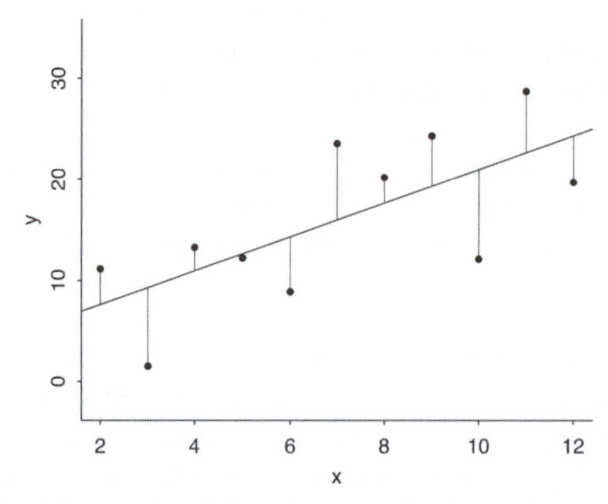


Figure 8.3 *Understanding the influence of individual data points on the fitted regression line. Picture the vertical lines as rubber bands connecting each data point to the least squares line. Take one of the points on the left side of the graph and move it up, and the slope of the line will* decrease. *Take one of the points on the right side and move it up, and the slope will* increase. *Moving the point in the center of the graph up or down will not change the slope of the fitted line.*

(8.2)), which reveals how the estimated vector of regression coefficients $\hat{\beta}$ is a linear function of the data vector $y$, and for generalized linear models, by re-fitting the regression after altering data points one at a time.

## 8.3 Least squares slope as a weighted average of slopes of pairs

In Section 7.3, we discussed that, when a regression $y = a + bx + \text{error}$ is fit with just an indicator variable (that is, where $x$ just takes on the values 0 and 1), the least squares estimate of its coefficient $b$ is simply the average difference in the outcome between the two groups; that is, $\bar{y}_1 - \bar{y}_0$.

There is a similar identity when the predictor $x$ is continuous; in this case, we can express the estimated slope $\hat{b}$ from (8.3) as a weighted average of slopes.

The basic idea goes as follows. With $n$ data points $(x, y)$ there are $n^2$ pairs (including the possibility of taking the same data point twice). For each pair $i, j$ we can compute the slope of the line connecting them:

$$\text{slope}_{ij} = \frac{y_j - y_i}{x_j - x_i}.$$

This expression is not defined when the two points have the same value of the predictor—that is, when $x_i = x_j$—but don't worry about that now; it will turn out that these cases drop out of our equation.

We would like to define the best-fit regression slope as an average of the individual slopes, but it makes sense to use a weighted average, in which $\text{slope}_{ij}$ counts more if the two points are further apart in $x$. We might, then, weight each slope by the difference between the two values, $|x_j - x_i|$. For mathematical reasons that we do not discuss here but which relate to the use of the normal distribution for errors (which is in turn motivated by the Central Limit Theorem, as discussed in Section 3.5), it makes sense to weight each slope by the squared separation, $(x_j - x_i)^2$.

We can then compute the weighted average:

$$\begin{aligned}\text{weighted average of slopes} &= \frac{\sum_{i,j}(x_j - x_i)^2 \frac{y_j - y_i}{x_j - x_i}}{\sum_{i,j}(x_j - x_i)^2} \\ &= \frac{\sum_{i,j}(x_j - x_i)(y_j - y_i)}{\sum_{i,j}(x_j - x_i)^2}. \qquad (8.8)\end{aligned}$$

If you collect the terms carefully, you can show that this expression is the same as $\hat{b}$ in (8.3), so we can interpret the estimated coefficient $\hat{b}$ as the weighted average slope in the data, and we can interpret the underlying parameter $b$ as the weighted average slope in the population.

## 8.4 Comparing two fitting functions: `lm` and `stan_glm`

The standard routine for fitting linear regressions in R is `lm`, which performs classical least squares regression and returns estimates and standard errors; indeed, that is what we used in our earlier book on regression, and there are comparable functions in all other statistical software. In this book, though, we use `stan_glm`, a program that performs Bayesian inference and returns estimates, standard errors, and posterior simulations.

We switch to `stan_glm` for two reasons. First, the simulations automatically computed by the new program represent uncertainty and can be used to obtain standard errors and predictive distributions for any function of current data, future data, and parameters. Second, the new program performs Bayesian inference, which can be used to get more stable estimates and predictions incorporating prior information. We discuss both these aspects of Bayesian inference—the probabilistic expression of uncertainty, and the inclusion of prior information—in Chapter 9, along with examples of how to perform probabilistic predictions and express prior information using regressions in `stan_glm`.

For many simple problems, classical and Bayesian inferences are essentially the same. For example, in Chapters 6 and 7, we fit regressions of earnings on incomes, daughters' height on mothers' heights, and elections on economic growth, in each case using `stan_glm` with its default settings. Switching to `lm` would give essentially identical results. The default prior is weak, hence it yields similar inferences to those obtained under maximum likelihood in these examples. The main role of the default prior in this sort of regression is to keep the coefficient estimates stable in some pathological cases such as near-collinearity; otherwise, it does not make much difference.

Bayesian inference makes more of a difference with weak data or strong priors. In addition, for advanced and hierarchical models, there are more differences between classical and Bayesian inferences: this is because more complex models can have latent or weakly identified parameters, so that a Bayesian inferential structure can provide smoother estimates more generally. In addition, advanced and multilevel models tend to increase in complexity as sample sizes increases, and so Bayesian inference can make a difference even with large datasets.

### Reproducing maximum likelihood using `stan_glm` with flat priors and optimization

Here we demonstrate the commands for bridging between the Bayesian and classical estimates. Our default regression fit to some data frame `mydata` is

```
stan_glm(y ~ x, data=mydata)
```

When `stan_glm` is called without setting the prior arguments, it by default uses weak priors that partially pool the coefficients toward zero, as discussed in Section 9.5.

If we want to get a closer match to classical inference, we can use a flat prior, so that the posterior distribution is the same as the likelihood. Here is the function call:

```
stan_glm(y ~ x, data=mydata, prior_intercept=NULL, prior=NULL, prior_aux=NULL)
```

The three different `NULL`s set flat priors for the intercept, the other coefficients in the model, and $\sigma$, respectively.

Again, we explain all this in the next chapter; at this point we just want to demonstrate the function calls.

To move even closer to standard regression, we can tell Stan to perform optimization instead of sampling. This yields the maximum penalized likelihood estimate, which in the case of a flat prior is simply the maximum likelihood estimate:

```
stan_glm(y ~ x, data=mydata, prior_intercept=NULL, prior=NULL, prior_aux=NULL,
  algorithm="optimizing")
```

### Running lm

As discussed in Section 1.6, if you only want to do maximum likelihood, and you have no interest in probabilistic predictions, you could instead fit regressions using the lm function in R:

```
lm(y ~ x, data=mydata)
```

In general, though, we prefer to use stan_glm even for maximum likelihood estimation because doing so allows us to propagate uncertainty in functions of parameters and data, as we shall explain in Chapter 9.

### Confidence intervals, uncertainty intervals, compatibility intervals

In Section 4.2 we discussed the general idea of confidence intervals, which are a way of conveying inferential uncertainty. The starting point, based on the assumption of an unbiased, normally distributed estimate, is that an estimate $\pm 1$ standard error has an approximate 68% chance of including the true value of the quantity being estimated, and the estimate $\pm 2$ standard errors has an approximately 95% chance of covering the true value.

For linear regression, the residual standard deviation $\sigma$ is itself estimated with error. If a regression with $k$ coefficients is fit to $n$ data points, there are said to be $n - k$ degrees of freedom (see page 53), and confidence intervals for regression coefficients are constructed using the $t_{n-k}$ distribution. For example, the model $y = a + bx +$ error has 2 coefficients; if it is fit to 10 data points, then we can type qt(0.975, 8) in R, yielding the value 2.31, and so the 95% confidence interval for either of these coefficients is the estimate $\pm 2.31$ standard errors.

When fitting models using stan_glm, we can get approximate 68% and 95% intervals using median $\pm 1$ or 2 mad sd, or we can construct intervals directly from the simulations. Again, we shall discuss this further in Chapter 9, but, very briefly, the fit from stan_glm yields a matrix of simulations, each column of which corresponds to one of the parameters in the model. Here is a simple example:

```
x <- 1:10
y <- c(1,1,2,3,5,8,13,21,34,55)
fake <- data.frame(x, y)
fit <- stan_glm(y ~ x, data=fake)
print(fit)
```

which yields,

```
            Median MAD_SD
(Intercept) -13.8    6.7
x             5.1    1.1

Auxiliary parameter(s):
      Median MAD_SD
sigma 10.0    2.5
```

We then extract the simulations:

```
sims <- as.matrix(fit)
```

yielding a matrix sims with three columns, containing simulations of the intercept, the coefficient for $x$, and the residual standard deviation $\sigma$. To extract a 95% interval for the coefficient for $x$, we type quantile(sims[,2], c(0.025, 0.975)) or quantile(sims[,"x"], c(0.025, 0.975)), either of which returns the interval [2.6, 7.5], which is not far from the approximation $[5.1 \pm 2 * 1.1]$ obtained using the median and mad sd.

All the confidence intervals discussed above can be called "uncertainty intervals," in that they represent uncertainty about the quantity being estimated, or "compatibility intervals," in that they give a range of parameter values that are most compatible with the observed data.

## 8.5 Bibliographic note

Various textbooks, for example Neter et al. (1996), derive least squares and maximum likelihood estimates and explain the relevant matrix algebra. Gelman and Greenland (2019) and Greenland (2019) discuss the ideas of "uncertainty intervals" and "compatibility intervals" as replacements for confidence intervals. Gelman et al. (2013) provide an introduction to Hamiltonian Monte Carlo and convergence diagnostics. See Stan Development Team (2020) for details of the algorithms implemented in `stan_glm` and R Core Team (2019) for background on `lm`.

## 8.6 Exercises

8.1 *Least squares*: The folder `ElectionsEconomy` contains the data for the example in Section 7.1. Load these data, type in the R function `rss()` from page 104, and evaluate it at several different values of $(a, b)$. Make two graphs: a plot of the sum of squares of residuals as a function of $a$, with $b$ fixed at its least squares estimate given in Section 7.1, and a plot of the sum of squares of residuals as a function of $b$, with $a$ fixed at its least squares estimate. Confirm that the residual sum of squares is indeed minimized at the least squares estimate.

8.2 *Maximum likelihood*: Repeat the previous exercise but this time write a function, similar to `rss()` on page 104, that computes the logarithm of the likelihood (8.6) as a function of the data and the parameters $a, b, \sigma$. Evaluate this function as several values of these parameters, and make a plot demonstrating that it is maximized at the values computed from the formulas in the text (with $\sigma$ computed using $\frac{1}{n}$, not $\frac{1}{n-2}$; see page 104).

8.3 *Least absolute deviation*: Repeat 8.1, but instead of calculating and minimizing the sum of squares of residuals, do this for the sum of absolute values of residuals. Find the $(a, b)$ that minimizes the sum of absolute values of residuals, and plot the sum of absolute values of residuals as a function of $a$ and of $b$. Compare the least squares and least absolute deviation estimates of $(a, b)$.

8.4 *Least squares and least absolute deviation*: Construct a set of data $(x, y)_i$, $i = 1, \ldots, n$, for which the least squares and least absolute deviation (see Exercise 8.3) estimates of $(a, b)$ in the fit, $y = a + bx$, are much different. What did you have to do to make this happen?

8.5 *Influence of individual data points*: A linear regression is fit to the data below. Which point has the most influence (see Section 8.2) on the slope?

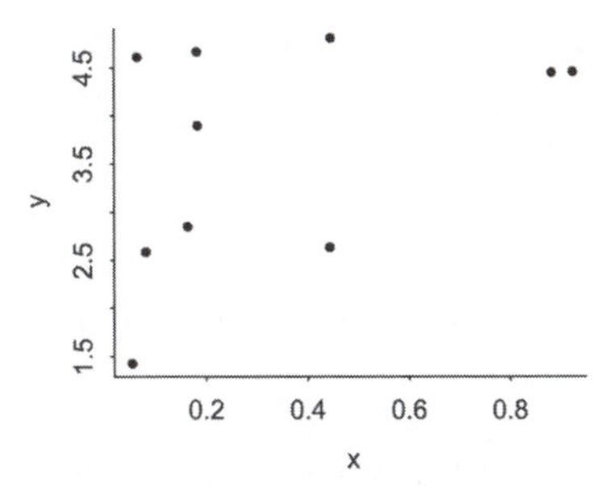


8.6 *Influence of individual data points*:

(a) Using expression (8.3), compute the influence of each of the data points in the election forecasting example on the fitted slope of the model. Make a graph plotting influence of point $i$ vs. $x_i$.

(b) Re-fit the model $n$ times, for each data point $i$ adding 1 to $y_i$. Save $\hat{b}$ from each of these altered regressions, compare to the $\hat{b}$ from the original data, and check that the influence is approximately the same as computed above using the formula. (The two calculations will not give identical results because `stan_glm` uses a prior distribution and so it does not exactly yield the least squares estimate.)

8.7 *Least squares slope as a weighted average of individual slopes*:

(a) Prove that the weighted average slope defined in equation (8.8) is equivalent to the least squares regression slope in equation (8.3).

(b) Demonstrate how this works in a simple case with three data points, $(x, y) = (0,0),\ (4,1),\ (5,5)$.

8.8 *Comparing* `lm` *and* `stan_glm`: Use simulated data to compare least squares estimation to default Bayesian regression:

(a) Simulate 100 data points from the model, $y = 2 + 3x + \text{error}$, with predictors $x$ drawn from a uniform distribution from 0 to 20, and with independent errors drawn from the normal distribution with mean 0 and standard deviation 5. Fit the regression of $y$ on $x$ data using `lm` and `stan_glm` (using its default settings) and check that the two programs give nearly identical results.

(b) Plot the simulated data and the two fitted regression lines.

(c) Repeat the two steps above, but try to create conditions for your simulation so that `lm` and `stan_glm` give much different results.

8.9 *Leave-one-out cross validation*: As discussed in the context of (8.5), the root mean square of residuals, $\sqrt{\frac{1}{n}\sum_{i=1}^{n}(y_i - (\hat{a} + \hat{b}x_i))^2}$, is an underestimate of the error standard deviation $\sigma$ of the regression model, because of overfitting, that the parameters $a$ and $b$ are estimated from the same $n$ data points as are being used to compute the residuals.

*Cross validation*, which we discuss in detail in Section 11.8, is an alternative approach to assessing predictive error that avoids some of the problems of overfitting. The simplest version of cross validation is the leave-one-out approach, in which the model is fit $n$ times, in each case excluding one data point, fitting the model to the remaining $n-1$ data points, and using this fitted model to predict the held-out observation:

- For $i = 1, \dots, n$:
  - Fit the model $y = a + bx + \text{error}$ to the $n-1$ data points $(x, y)_j,\ j \neq i$. Label the estimated regression coefficients as $\hat{a}_{-i}, \hat{b}_{-i}$.
  - Compute the cross-validated residual, $r_i^{\text{CV}} = y_i - (\hat{a}_{-i} + \hat{b}_{-i}x_i)$.
- Compute the estimate $\hat{\sigma}^{\text{CV}} = \sqrt{\frac{1}{n}\sum_{i=1}^{n} r_i^2}$.

(a) Perform the above steps for the elections model from Section 7.1. Compare three estimates of $\sigma$: (i) the estimate produced by `stan_glm`, (ii) formula (8.5), and (iii) $\hat{\sigma}^{\text{CV}}$ as defined above.

(b) Discuss any differences between the three estimates.

8.10 *Leave-one-out cross validation*: Create a fake dataset $(x, y)_i,\ i = 1, \dots, n$, in such a way that there is a big difference between $\hat{\sigma}^{\text{CV}}$ as defined in the previous exercise, and the estimated residual standard deviation from (8.5). Explain what you did to create this discrepancy.

8.11 *Working through your own example*: Continuing the example from the final exercises of the earlier chapters, take your fitted regression line from Exercise 7.10 and compute the influence of each data point on the estimated slope. Where could you add a new data point so it would have zero influence on the estimated slope? Where could you add a point so it would have a large influence?

# Chapter 9

# Prediction and Bayesian inference

Bayesian inference involves three steps that go beyond classical estimation. First, the data and model are combined to form a *posterior distribution*, which we typically summarize by a set of simulations of the *parameters* in the model. Second, we can propagate uncertainty in this distribution—that is, we can get simulation-based *predictions* for unobserved or future outcomes that accounts for uncertainty in the model parameters. Third, we can include additional information into the model using a *prior distribution*. The present chapter describes all three of these steps in the context of examples capturing challenges of prediction and inference.

## 9.1 Propagating uncertainty in inference using posterior simulations

Later in this chapter we discuss the role of the prior distribution. But we start by considering Bayesian simulation simply as a way of expressing uncertainty in inferences and predictions, using the function stan_glm to fit a model and then using simulations in R to form probabilistic predictions.

Example: Elections and the economy

For example, here is the fit for the election forecasting model from Section 7.1:[1]

```
            Median MAD_SD
(Intercept) 46.3    1.7
growth       3.0    0.7

Auxiliary parameter(s):
      Median MAD_SD
sigma 3.9    0.7
```

These numbers are summaries of a matrix of simulations representing different possible values of the parameter vector $(a, b, \sigma)$: the intercept, the slope, and the residual standard deviation of the regression. We have a set of *posterior simulations* rather than a single point estimate because we have *uncertainty* about these parameters.

We can access the simulations by extracting the fitted model as a matrix in R:

```
sims <- as.matrix(M1)
```

And then, to check that we understand how the summaries are defined, we can compute them directly from the simulations:

```
Median <- apply(sims, 2, median)
MAD_SD <- apply(sims, 2, mad)
print(cbind(Median, MAD_SD))
```

The mad sd is the scaled median absolute deviation (see Section 5.3) of the posterior simulations, which we use to summarize uncertainty, as this can be more stable than the standard error or posterior standard deviation. For convenience we sometimes call the posterior mad sd the "standard error," as this is the term typically used by practitioners to represent inferential uncertainty.

[1]Data and code for this example are in the folder ElectionsEconomy.

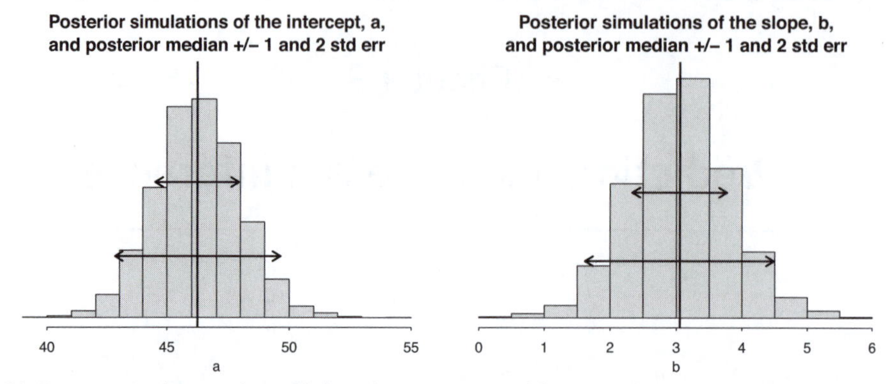


Figure 9.1 *Posterior simulations of a and b from the regression model, $y = a + bx + error$. For each parameter, we display the simulation draws produced by* `stan_glm` *and also the posterior median* $\pm$ *1 and 2 mad sd's.*

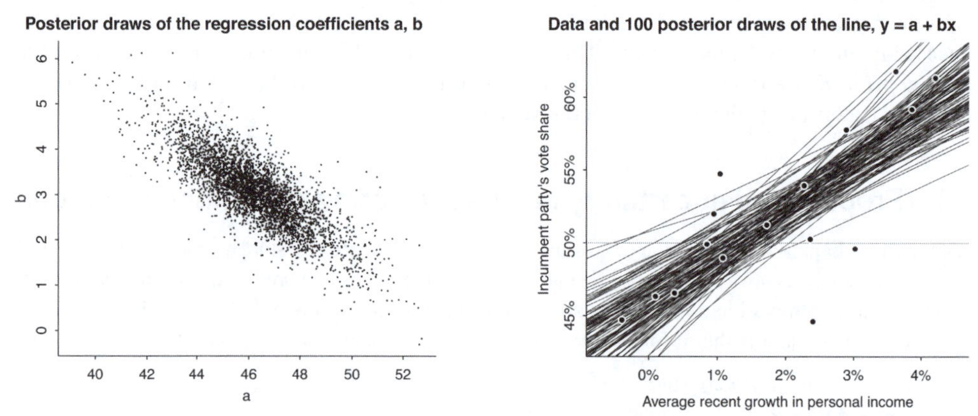


Figure 9.2 *(a) 4000 posterior simulations of $(a, b)$ from the election forecasting model, $y = a + bx + error$, as fit using* `stan_glm`*. (b) The lines $y = a + bx$ corresponding to 100 of these simulations of $(a, b)$.*

### Uncertainty in the regression coefficients and implied uncertainty in the regression line

To understand the Bayesian idea of an uncertainty distribution, it can be helpful to plot the simulation draws of the coefficients. Figure 9.1 shows histograms of the posterior simulations for the intercept and slope of the fitted election forecasting model. We have overlaid the estimate $\pm$ 1 and 2 mad sd's for each parameter to show the relationship between the numerical summaries and the simulations.

The intercept and slope go together; Figure 9.2a plots the scatterplot indicating the posterior distribution. Each of these dots $(a, b)$ can be mapped to a line, $y = a + bx$, and Figure 9.2b displays 100 of these lines. This plot shows the inferential uncertainty in the fitted regression. As the sample size increases, the standard errors of the coefficients decreases, and the plot corresponding to Figure 9.2b approaches a graph with many data points and just a single narrow band indicating high precision in estimation; see Exercise 7.5.

### Using the matrix of posterior simulations to express uncertainty about a parameter estimate or function of parameter estimates

The real advantage of summarizing inference by simulations is that we can directly use these to propagate uncertainty. For a simple example, suppose we want an estimate and standard error for some combination of parameters, for example $a/b$ (not that there would be any good reason to do this for this problem, just to demonstrate it as an example). We can work with the simulations directly:

```
a <- sims[,1]
b <- sims[,2]
z <- a/b
print(c(median(z), mad(z)))
```

## 9.2 Prediction and uncertainty: predict, posterior_linpred, and posterior_predict

After fitting a regression, $y = a + bx$ + error, we can use it to predict a new data point, or a set of new data points, with predictors $x^{\text{new}}$. We can make three sorts of predictions, corresponding to increasing levels of uncertainty:

- The *point prediction*, $\hat{a} + \hat{b}x^{\text{new}}$: Based on the fitted model, this is the best point estimate of the average value of $y$ for new data points with this new value of $x$. We use $\hat{a}$ and $\hat{b}$ here because the point prediction ignores uncertainty.
- The *linear predictor with uncertainty*, $a + bx^{\text{new}}$, propagating the inferential uncertainty in $(a, b)$: This represents the distribution of uncertainty about the expected or average value of $y$ for new data points with predictors $x^{\text{new}}$.
- The *predictive distribution for a new observation*, $a + bx^{\text{new}}$ + error: This represents uncertainty about a new observation $y$ with predictors $x^{\text{new}}$.

For example, consider a study in which blood pressure, $y$, is predicted from the dose, $x$, of a drug. For any given $x^{\text{new}}$, the point prediction is the best estimate of the average blood pressure in the population, conditional on dose $x^{\text{new}}$; the linear predictor is the modeled average blood pressure of people with dose $x^{\text{new}}$ in the population, with uncertainty corresponding to inferential uncertainty in the coefficients $a$ and $b$; and the predictive distribution represents the blood pressure of a single person drawn at random drawn from this population, under the model conditional on the specified value of $x^{\text{new}}$.

As sample size approaches infinity, the coefficients $a$ and $b$ are estimated more and more precisely, and the uncertainty in the linear predictor approaches zero, but the uncertainty in the predictive distribution for a new observation does not approach zero; it approaches the residual standard deviation $\sigma$.

### Point prediction using predict

We now show how to construct all three predictions, using the example from Section 7.1 of forecasting elections from the economy.[2] We shall predict the incumbent party's vote share, conditional on economic growth of 2.0%. The fitted model, M1, appears on page 113. We first create a new dataset with the hypothetical value of $x$:

```
new <- data.frame(growth=2.0)
```

We can then compute the point prediction:

```
y_point_pred <- predict(M1, newdata=new)
```

Equivalently, we could perform this computation "by hand":

```
a_hat <- coef(M1)[1]
b_hat <- coef(M1)[2]
y_point_pred <- a_hat + b_hat*new
```

Either way, the point prediction is a single number, $46.3 + 3.0 * 2 = 52.3$.

[2] Data and code for this example are in the folder ElectionsEconomy.

### Linear predictor with uncertainty using posterior_linpred or posterior_epred

We can use posterior_linpred to get uncertainty in the value of the fitted regression line:

```
y_linpred <- posterior_linpred(M1, newdata=new)
```

This returns a vector of posterior simulations from a distribution whose mean equals the point prediction obtained above and whose standard deviation represents uncertainty in the fitted model.

Equivalently we can use the function posterior_epred, which returns the expected prediction for a new data point. For linear regression, the expected value, $\mathrm{E}(y|x^{\mathrm{new}})$, is the same as the linear predictor, $X^{\mathrm{new}}\beta$, but as we discuss in Chapter 13, these two quantities differ for nonlinear models.

One way to understand this vector of simulations is to compute it "by hand" from the posterior simulations of the regression coefficients:

```
sims <- as.matrix(M1)
a <- sims[,1]
b <- sims[,2]
y_linpred <- a + b*new
```

### Predictive distribution for a new observation using posterior_predict

Finally, we can construct a vector representing predictive uncertainty in a single election:

```
y_pred <- posterior_predict(M1, newdata=new)
```

To obtain this result "by hand" using the simulations, we just add an error term to the computation from before:

```
n_sims <- nrow(sims)
sigma <- sims[,3]
y_pred <- as.numeric(a + b*new) + rnorm(n_sims, 0, sigma)
```

In any case, we can summarize the vector of simulations graphically:

```
hist(y_pred)
```

and numerically:

```
y_pred_median <- median(y_pred)
y_pred_mad <- mad(y_pred)
win_prob <- mean(y_pred > 50)
cat("Predicted Clinton percentage of 2-party vote: ", round(y_pred_median,1),
  ", with s.e. ", round(y_pred_mad, 1), "\nPr (Clinton win) = ", round(win_prob, 2),
  sep="")
```

Here is the result:

```
Predicted Clinton percentage of 2-party vote: 52.3, with s.e. 3.9
Pr (Clinton win) = 0.72
```

### Prediction given a range of input values

We can also use predict, posterior_linpred, and posterior_predict to generate a range of predicted values, for example predicting the election outcome for a grid of possible values of economic growth from −2% to +4%:

```
new_grid <- data.frame(growth=seq(-2.0, 4.0, 0.5))
y_point_pred_grid <- predict(M1, newdata=new_grid)
y_linpred_grid <- posterior_linpred(M1, newdata=new_grid)
y_pred_grid <- posterior_predict(M1, newdata=new_grid)
```

The result is a vector of length 13 (y_point_pred_grid) and two n_sims × 13 matrices (y_linpred_grid and y_pred_grid) corresponding to predictions for each of the 13 values specified for growth.

### Propagating uncertainty

In the above calculations we expressed uncertainty in the election outcome conditional on various pre-set values of economic growth. But growth is estimated only approximately in the lead-up to the election, and later the figures are often revised by the government. Hence it makes sense when applying our model to account for uncertainty in this predictor.

Let us say that, in advance of the election, our best estimate of economic growth was 2.0% but with some uncertainty that we shall express as a normal distribution with standard deviation 0.3%. We can then *propagate the uncertainty* in this predictor to obtain a forecast distribution that more completely expresses our uncertainty.

We just need to add a line to our R code to simulate the distribution of the prediction:

```
x_new <- rnorm(n_sims, 2.0, 0.3)
y_pred <- rnorm(n_sims, a + b*x_new, sigma)
```

Following this, we summarize as before, obtaining this result:

```
Predicted Clinton percentage of 2-party vote: 52.3, with s.e. 4.1
Pr (Clinton win) = 0.71.
```

The point prediction is unchanged at 52.3% of the two-party vote, but the standard deviation has increased slightly to reflect this extra uncertainty.

### Simulating uncertainty for the linear predictor and new observations

Example: Height and weight

We go over these ideas one more time using a model predicting weight (in pounds) from height (in inches), from a survey whose responses have been cleaned and saved in a data frame called earnings:[3]

```
fit_1 <- stan_glm(weight ~ height, data=earnings)
print(fit_1)
```

which yields,

```
            Median MAD_SD
(Intercept) -171.7   11.7
height         4.9    0.2

Auxiliary parameter(s):
      Median MAD_SD
sigma 29.1   0.5
```

The intercept is difficult to interpret—it represents the predicted weight of someone who is zero inches tall—and so we transform:

```
earnings$c_height <- earnings$height - 66
fit_2 <- stan_glm(weight ~ c_height, data=earnings)
print(fit_2)
```

The predictor in this regression is height relative to 66 inches. This approximate centering of the predictor enables more stable and interpretable inferences for the regression coefficients:

```
            Median MAD_SD
(Intercept) 153.2    0.6
c_height      4.9    0.2

Auxiliary parameter(s):
      Median MAD_SD
sigma 29.1  0.5
```

[3]Data and code for this example are in the folder Earnings.

Inferences for the slope and error standard deviation have not changed, but the intercept is different; it now corresponds to the predicted weight of a 66-inch-tall person. From the fitted model we can estimate the average weight of 66-inchers in the *population* as 153.2 pounds with an uncertainty of 0.6. But if we wanted to predict the weight of any *particular* 66-inch-tall person chosen at random from the population, we would need to include the predictive uncertainty, whose standard deviation is estimated to be 29.1.

In general, when applying a fitted regression to a new data point $x^{\text{new}}$, we can make inferences about the *linear predictor* $a + bx^{\text{new}}$ or about the *predicted value* $y^{\text{new}} = a + bx^{\text{new}} + \epsilon$.

For example, let's predict the weight of a person who is 70 inches tall, so that `c_height` = `height` $- 66 = 4$:

$$\begin{aligned}\text{linear predictor: } & a + 4.0\, b,\\ \text{predicted value: } & a + 4.0\, b + \epsilon.\end{aligned}$$

Sometimes we are interested in the predicted expectation or linear predictor, as it represents the predicted average weight for everyone of this height in the population; in other settings we want to predict the weight of an individual person. It depends on the context, and we need to be able to assess the uncertainty for each.

The above model gives a point prediction for a 70-inch-tall person to weigh $153.2+4.0*4.9 = 172.8$ pounds. If this equation represented the true model, rather than an estimated model, then we could use $\hat{\sigma} = 29.1$ as an estimate of the standard deviation for the predicted value. Actually, though, the estimated error standard deviation is slightly higher than $\hat{\sigma}$, because of uncertainty in the estimate of the regression parameters—a complication that gives rise to those special prediction standard deviation formulas seen in some regression texts. For example, in the regression model $y = a + bx + \text{error}$, the standard deviation for the linear predictor $a + bx$ is

$$\hat{\sigma}_{\text{linpred}} = \hat{\sigma}\sqrt{\frac{1}{n} + \frac{(x^{\text{new}} - \bar{x})^2}{\sum_{i=1}^{n}(x_i - \bar{x})^2}}, \tag{9.1}$$

and the standard deviation for the predicted value $a + bx + \epsilon$ is

$$\hat{\sigma}_{\text{prediction}} = \hat{\sigma}\sqrt{1 + \frac{1}{n} + \frac{(x^{\text{new}} - \bar{x})^2}{\sum_{i=1}^{n}(x_i - \bar{x})^2}}. \tag{9.2}$$

Rather than trying to use such formulas (which do not even work once we move to more complicated models and nonlinear predictions), we simply compute all predictive uncertainties using simulations in R.

For a linear model, the two sorts of predictive distributions have the same mean, the point prediction $\hat{a} + \hat{b}x$.

The difference comes in the uncertainties: as the sample size $n$ in the fitted regression increases, the standard deviation of the linear predictor goes to zero (it scales like $1/\sqrt{n}$), while the standard deviation of the predicted value approaches the nonzero $\sigma$, as illustrated in formula (9.2). Even if the parameters of the fitted model are perfectly known, there is uncertainty in predicting a new data point.

After fitting a model with `stan_glm`, we can obtain posterior simulations of the two kinds of prediction using `posterior_linpred` and `posterior_predict`.

We demonstrate with the height and weight example, using the Bayesian regression result `fit_2` from above, then defining a data frame for the new person who is 70 inches tall (so that the centered height is 4.0 inches):

```
new <- data.frame(c_height=4.0)
```

We can compute the point prediction:

```
y_point_pred_2 <- predict(fit_2, newdata=new)
```

and simulations of the linear predictor:

```
y_linpred_2 <- posterior_linpred(fit_2, newdata=new)
```

This yields a vector representing possible values of $a + 4.0\,b$, with variation coming from posterior uncertainty in the coefficients.

And we can compute posterior predictive simulations for a single new person of height 70 inches:

```
y_postpred_2 <- posterior_predict(fit_2, newdata=new)
```

yielding a vector of simulations representing possible values of $a + 4.0\,b + \epsilon$.

In either case, we can summarize by the median, mad sd, histogram, and so forth.

## 9.3 Prior information and Bayesian synthesis

Classical statistical methods produce summaries and inferences based on a single dataset. *Bayesian methods* combine a model of the data with *prior information* with the goal of obtaining inferences that are consistent with both sources of information.

Full understanding of statistical inference, Bayesian or otherwise, requires probability theory at a level beyond what is covered in this book. However, users of statistics can go far without fully digesting the mathematics of probability, and a key tenet of this book is to move straight to the methods and how to understand them, with the hope that readers who wish to go deeper can later study the theory and derivations; see the bibliographic note at the end of this chapter for some pointers on where to go next.

### Expressing data and prior information on the same scale

We begin with the formulas for Bayesian inference for the normal distribution, with $\theta$ being the continuous parameter we want to estimate. The goal is to combine a *prior estimate* $\hat{\theta}_{\text{prior}}$ (with *prior standard error* $\text{se}_{\text{prior}}$) and a *data estimate* $\hat{\theta}_{\text{data}}$ (with standard error $\text{se}_{\text{data}}$).

The resulting Bayesian estimate is $\hat{\theta}_{\text{Bayes}}$, with standard error $\text{se}_{\text{Bayes}}$, where:

$$\hat{\theta}_{\text{Bayes}} = \left(\frac{1}{\text{se}^2_{\text{prior}}}\hat{\theta}_{\text{prior}} + \frac{1}{\text{se}^2_{\text{data}}}\hat{\theta}_{\text{data}}\right) \Bigg/ \left(\frac{1}{\text{se}^2_{\text{prior}}} + \frac{1}{\text{se}^2_{\text{data}}}\right), \tag{9.3}$$

$$\text{se}_{\text{Bayes}} = 1 \Bigg/ \sqrt{\frac{1}{\text{se}^2_{\text{prior}}} + \frac{1}{\text{se}^2_{\text{data}}}}\,. \tag{9.4}$$

These formulas are somewhat awkward, but when studied carefully they reveal the way in which Bayesian inference in this simple setting is a compromise between prior and data. In particular, expression (9.3) has the form of a weighted average of prior and data estimates, where each estimate gets a weight proportional to the inverse square of its standard error. Expression (9.3) can be understood by considering $1/\text{se}^2$ as the *precision* and seeing that the posterior precision equals the precision from the prior plus the precision from the data.

We have shared the above formulas because this weighted-average idea can be a helpful way to understand Bayesian reasoning in general. The details of the compromise depend on the prior and data standard errors: if the two standard errors are equal, the Bayes estimate is midway between the prior and data estimates; otherwise, the Bayes estimate is closer to whichever source of information has the lower standard error.

Equation (9.4) is less intuitive; it can be understood by saying that the inverse variances from prior and data add; as a result, $\text{se}_{\text{Bayes}}$ is necessarily lower than either of $\text{se}_{\text{prior}}$ and $\text{se}_{\text{data}}$: combining information increases precision.

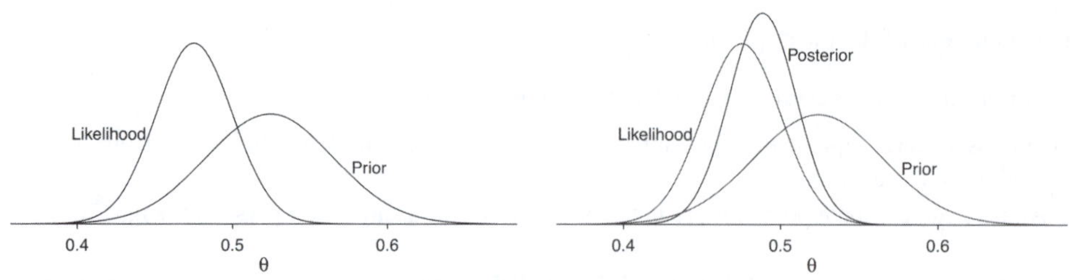


Figure 9.3 *(a) Likelihood (distribution representing data estimate and standard error) and prior distribution for the example combining estimate from a poll (the data) with the forecast from a fitted model (the prior). In this case, the data estimate is 0.475 with standard error 0.025, and the prior estimate is 0.524 with prior standard error 0.041. (b) Bayesian inference yields the posterior distribution, which is a compromise between the likelihood and the prior. In this example, the data are more informative than the prior, and so the posterior distribution is closer to the likelihood than the prior distribution.*

## Bayesian information aggregation

We use a simple example to demonstrate the use of the formulas (9.3)–(9.4) for the posterior estimate and standard error, combining prior information and data.

Example: Combining information from a forecast and a poll

Suppose an election is coming up, and a previously fitted model using economic and political conditions gives a forecast that the Democratic candidate will receive 52.4% of the two-party vote, with a predictive uncertainty of 4.1%. Using the notation above, $\hat{\theta}_{\text{prior}} = 0.524$ and $\text{se}_{\text{prior}} = 0.041$.

We now conduct a survey of 400 people, of whom 190 say they will vote for the Democratic candidate and 210 support the Republican. If the survey was a simple random sample of voters with no nonresponse, with voters who will not change their minds between the time of the survey and the election, then the data estimate is $\hat{\theta}_{\text{data}} = 190/400 = 0.475$ with standard error $\text{se}_{\text{data}} = \sqrt{0.475 * (1 - 0.475)/400} = 0.025$. A more realistic analysis, accounting for possible nonsampling error, would increase the standard error and might change the point estimate as well, but we do not discuss that here.

Figure 9.3a shows the prior distribution and likelihood (the information from the data) described in the preceding two paragraphs.

We now combine the prior and data estimates, making the key assumption, which is completely reasonable in this example, that the prior and data represent two different sources of information. More precisely, the assumption is that the uncertainties in the two estimates are statistically independent. Here, the prior distribution comes from a forecast based on past elections, along with a measure of current economic performance, with forecast uncertainty coming from nonzero residuals in the fitted model. The data come from a survey of potential voters in the new election, and the uncertainty in the data estimate comes from sampling variability, along with whatever nonsampling error has been accounted for in the analysis.

In this example, the prior standard error is 4.1% and the data standard error is 2.5%, so the data are more informative than the prior, and the Bayes estimate will be closer to the data.

We can easily do the calculation:[4]

```
theta_hat_prior <- 0.524
se_prior <- 0.041
n <- 400
y <- 190
theta_hat_data <- y/n
se_data <- sqrt((y/n)*(1-y/n)/n)
theta_hat_bayes <- (theta_hat_prior/se_prior^2 + theta_hat_data/se_data^2) /
  (1/se_prior^2 + 1/se_data^2)
se_bayes <- sqrt(1/(1/se_prior^2 + 1/se_data^2))
```

[4]Data and code for this example are in the folder `ElectionsEconomy`.

The result is $\hat{\theta}_{\text{Bayes}} = 0.488$ and $\text{se}_{\text{Bayes}} = 0.021$. The estimate is indeed between the prior and data estimates, closer to the data, as shown in Figure 9.3b.

Now consider the same situation but with much more uncertain data. Just take the same point estimate and increase the data standard error $\text{se}_{\text{data}}$ from 0.025 to, say, 0.075. The Bayes estimate then becomes 0.512; it is now closer to the prior.

### Different ways of assigning prior distributions and performing Bayesian calculations

One thing that can be confusing in statistics is that similar analyses can be performed in different ways. For example, in Section 7.3 we discussed how a comparison between two groups can also be expressed as a regression on an indicator variable, and the corresponding standard error can be calculated using formula (4.1) or using the algebra of least squares regression.

Similarly, Bayesian inference, at least in simple examples with one source of prior information and another source of data, can be performed directly using equations (9.3) and (9.4) or using the linear algebra of regression analysis as implemented in `stan_glm`.

Another complication is that Bayesian inference can be applied to any uncertain quantity. In the above example, the uncertain quantity is an observable outcome, the incumbent party's vote share in the next election. In Section 9.4, the uncertain quantity is the difference in the probability of a baby being a girl, comparing two different categories of parents. In Section 9.5 we discuss the more general problem where the uncertain quantity is a regression coefficient or set of regression coefficients.

The prior distribution will be set up differently in these different situations. For the voting example, a regression model fit on past data gives predictive inference for a single observable outcome that can be used as prior information when combining with a poll on the current election. In the births example, we need prior information on a population difference. In the general example of regression modeling, we must specify prior information on all the coefficients, which in practice often entails weak priors on coefficients for which we have little knowledge or about which we do not want to make any strong assumptions.

## 9.4 Example of Bayesian inference: beauty and sex ratio

xample:
eauty and
ex ratio

We can use prior information to refine estimates from noisy studies. For example, several years ago a researcher analyzed data from a survey of 3000 Americans and observed a correlation between attractiveness of parents and the sex of their children. The survey coded adults into five attractiveness categories, and it turned out that 56% of the children of parents in the highest attractiveness category were girls, compared to 48% of the children of parents in the other categories. The observed difference of 8 percentage points had a standard error (based on the usual formula for the difference in proportions) of 3 percentage points.[5]

The observed difference is more than two standard errors from zero, meeting the usual standard of statistical significance, and indeed the claim that beautiful parents have more daughters was published in a scientific journal and received wide publicity.

### Prior information

More information is available, however. It is well known that the variation in the human sex ratio occurs in a very narrow range. For example, a recent count in the United States reported 48.7% girls among whites and 49.2% among blacks. Similar differences of half of a percentage point or less have been found when comparing based on factors such as birth order, maternal age, or season of birth. Given that attractiveness is itself only subjectively measured, we would find it hard to believe that any difference between more and less attractive parents could be as large as 0.5%.

[5]Data and code for this example are in the folder `SexRatio`.

We now perform Bayesian inference using the template above. The parameter of interest here, $\theta$, is the probability of girl births among beautiful parents, minus the probability among other parents, all in the general population of Americans. As is often the case, we are interested in the comparison of $\theta$ to zero: Is there strong evidence that $\theta > 0$, which would imply that more attractive parents are more likely to have girls?

### Prior estimate and standard error

We can express our scientific knowledge as a prior distribution on $\theta$ with mean 0% and standard deviation 0.25%. The prior mean of zero says that before seeing the data, we would have no reason to expect beautiful parents to have an elevated or depressed rate of girl births. The prior standard deviation of 0.25% says that we find it highly implausible that the true value of $\theta$ is higher than 0.5% or lower than −0.5%.

For convenience we are expressing our estimates and uncertainties on a percentage scale to avoid the awkwardness of working with expressions such as 0.0025 and possibly dropping a zero somewhere.

### Data estimate and standard error

On the percentage scale the survey gives us the estimate $\hat{\theta}_{\text{data}} = 8\%$ with standard error $\text{se}_{\text{data}} = 3\%$, and we can now see that the prior is much more informative than the data: the data standard error is more than 10 times the prior uncertainty.

### Bayes estimate

Using (9.3)–(9.4) to combine the prior and data, we get $\hat{\theta}_{\text{Bayes}} = 0.06$ and $\text{se}_{\text{Bayes}} = 0.25$: the Bayes estimate of the difference between the proportion of girl births from beautiful and non-beautiful parents is 0.06% (that is, less than one-tenth of a percentage point) with a standard error of 0.25%. The magnitude of the estimate is tiny and is quite a bit lower than the standard error, indicating that the data from this survey tell us essentially nothing about variation in the sex ratio.

### Understanding the Bayes estimate

How could this be? A sample size of 3000 sounds pretty good, so how is it that it provides almost no information here? The reason is that we need more precision here than in the usual survey setting.

Here are some quick calculations to make the point:

- An estimated proportion from a simple random sample of size 3000 has standard error $\sqrt{p(1-p)/3000}$, which, if $p$ is near 0.5, is approximately $\sqrt{0.5 * 0.5/3000} = 0.009$. A survey of 3000 people lets us estimate a proportion to within a precision of about 1% (setting aside usual concerns about nonsampling error).
- Now consider a comparison of proportions, comparing two groups, each of size 1500. The standard error of the difference between these proportions is $\sqrt{p_1(1-p_1)/1500 + p_2(1-p_2)/1500}$, which, if $p_1$ and $p_2$ are close to 0.5, is approximately $\sqrt{2 * 0.5 * 0.5/1500} = 0.018$. Thus, under optimal conditions we can estimate the difference in proportions to within about 2 percentage points.
- The actual standard error from the survey was a bit higher, at 3.3 percentage points, because there were not equal numbers of people in the two comparison groups; only about 10% of respondents were labeled as "very attractive." For example, $\sqrt{0.5 * 0.5/300 + 0.5 * 0.5/2700} = 0.03$.
- Estimating a comparison to an accuracy of 2 or 3 percentage points is pretty good—for many purposes. But for studying differences in the sex ratio, we need much higher precision, more on the scale of 0.1 percentage point.

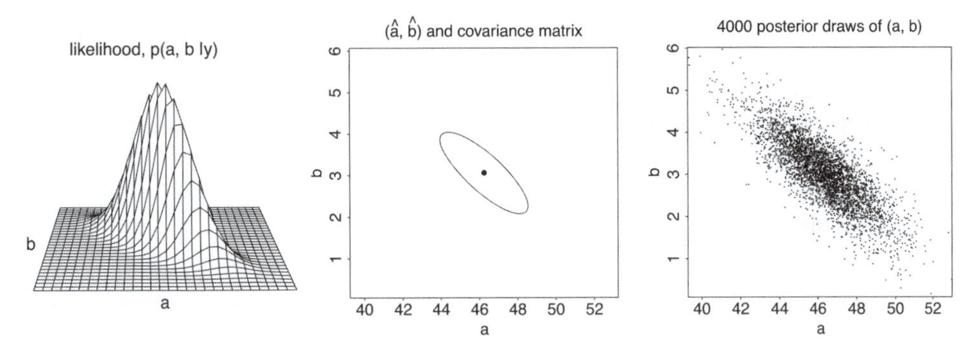


Figure 9.4 *(a) Reproduced from Figure 8.1a: Likelihood function for the parameters a and b in the linear regression $y = a + bx$ + error, of election outcomes, $y_i$, on economic growth, $x_i$. (b) Reproduced from Figure 8.1c: Mode of the likelihood function with an ellipse summarizing the inverse-second-derivative-matrix of the log likelihood at the mode. (c) 4000 random simulation draws $(a, b)$ from the posterior distribution corresponding to a flat prior.*

## 9.5 Uniform, weakly informative, and informative priors in regression

In Sections 9.1 and 9.2, we performed default Bayesian analyses using simulation to capture uncertainty in estimation and prediction. We were not particularly concerned with prior distributions, as in these examples the available data were strong relative to other readily available information.

When data are less informative, it can make sense to think more carefully about what prior information to include in the analysis. We gave some examples in Sections 9.3 and 9.4, and here we consider the topic more formally.

In Bayesian inference the likelihood is multiplied by a *prior distribution* to yield a *posterior distribution*, which we take as a summary of our inference given model and data. In this section we consider uniform, weakly informative, and informative prior distributions and how to work with them using `stan_glm`.

The `stan_glm` function includes prior distributions for each of the regression coefficients $a, b$ (or, more generally, the vector of coefficients in a regression with multiple predictors) and also the scale parameter, $\sigma$.

### Uniform prior distribution

First consider the case when the prior distribution is flat or uniform, which is sometimes called *noninformative*. With a flat prior, the posterior is simply the likelihood function multiplied by a constant, and the maximum likelihood estimate is also the mode of the posterior distribution.

As mentioned in Section 8.4, we can run `stan_glm` with uniform prior densities for the coefficients and scale parameter by using the `NULL` option:

```
M3 <- stan_glm(vote ~ growth, data=hibbs,
  prior_intercept=NULL, prior=NULL, prior_aux=NULL)
```

To understand the connection to the likelihood, we extract the posterior simulations of the coefficients and graph them:

```
sims <- as.data.frame(M3)
a <- sims[,1]
b <- sims[,2]
plot(a, b)
```

Figure 9.4c shows the result, with the graph formatted to be more readable using various R plotting options not shown here. Figures 9.4a and 9.4b repeat Figures 8.1a and 8.1c, showing the likelihood surface, the maximum likelihood estimate, and the uncertainty ellipse. The third plot displays the posterior simulations of $(a, b)$, which for this simple model are centered at the point estimate and distributed with a shape corresponding to the ellipse. The point of this example is to give a sense of inferential uncertainty in a simple example by displaying the distribution of this uncertainty in three different ways.

### Default prior distribution

The next step beyond a flat prior is a weak prior distribution that pulls the parameter estimates in a model by some small amount—this is sometimes called a *soft constraint*—enough to keep inferences in a reasonable range when data are sparse.

By default, `stan_glm` uses a weakly informative family of prior distributions. For a model of the form, $y = a + b_1 x_1 + b_2 x_2 + \cdots + b_K x_K + \text{error}$, each coefficient $b_k$ is given a normal prior distribution with mean 0 and standard deviation $2.5\,\text{sd}(y)/\text{sd}(x_k)$. The intercept $a$ is not given a prior distribution directly; instead we assign the prior to the expected value of $y$ with the predictors $x$ set to their mean value in the data, that is, $a + b_1\bar{x}_1 + b_2\bar{x}_2 + \cdots + b_K\bar{x}_K$. This centered intercept is given a normal prior distribution with mean $\mu_y$ and standard deviation $2.5\,\text{sd}(y)$. Finally, the residual standard deviation is given an exponential prior distribution with rate $1/\text{sd}(y)$. We have not yet discussed the exponential distribution in this book; here it is enough to say that it is a weak prior that will typically have almost no effect on inferences except when sample size is very low.

The default priors in `stan_glm` are not fully Bayesian, in that they are scaled based on the data, whereas a true "prior distribution" must be defined before any data have been seen. Technically, we can think of the above default priors as approximations to fully Bayesian priors, where we would like to scale based on the mean and standard deviation of $x$ and $y$ in the population, and we are using the sample averages as placeholders for these population values.

To display the priors used in any model fit with `stan_glm`, use the `prior_summary()` function. Type `vignette("priors", package="rstanarm")` in R to see further details on how priors work in `rstanarm`.

To understand the default prior, recall that a regression coefficient corresponds to the expected difference in the outcome, $y$, comparing two people who differ by 1 unit in a predictor, $x$. If $x$ and $y$ have both been standardized, then the coefficient is the expected difference in standard deviations in $y$ corresponding to a change of 1 standard deviation in $x$. We typically expect such a difference to be less than 1 in absolute value, hence a normal prior with mean 0 and scale 2.5 will partially pool noisy coefficient estimates toward that range. The particular choice of 2.5 as a scaling factor is somewhat arbitrary, chosen to provide some stability in estimation while having little influence on the coefficient estimate when data are even moderately informative.

The default prior distribution can be specified implicitly:

```
M1 <- stan_glm(vote ~ growth, data=hibbs)
```

or explicitly:

```
sd_x <- sd(hibbs$growth)
sd_y <- sd(hibbs$vote)
mean_y <- mean(hibbs$vote)
M1a <- stan_glm(vote ~ growth, data=hibbs, prior=normal(0, 2.5*sd_y/sd_x),
  prior_intercept=normal(mean_y, 2.5*sd_y), prior_aux=exponential(1/sd_y))
```

These two models are the same, with the only difference being that for `M1a` we specified the priors directly.

### Weakly informative prior distribution based on subject-matter knowledge

The default prior is intended to be enough to keep inferences stable. In many problems we can do better by including prior information specific to the problem at hand.

Consider the election forecasting example. What can be said, without reference to the data, regarding the coefficients in the model, vote = $a + b *$ growth + error?

First, the intercept. The prior_intercept argument in stan_glm is defined with the predictors set to their average values—that is, $a + b\bar{x}$—which in this case corresponds to the average incumbent's share of the two-party vote when economic growth is at its historical average. Without pointing to the data at hand, we know that vote percentage $y$ must be between 0 and 100, and likely to be near 50, so we might consider a normal prior distribution with mean 50 and standard deviation 10 for $a + b\bar{x}$.

What about the coefficient of economic growth on the incumbent party's vote percentage? It would seem to make sense to expect $b$ to be positive, but it would be a mistake to be too sure on prior grounds alone: we want to retain the capacity of being surprised. How large might $b$ be in magnitude? Economic growth in the past century has typically been in the range 0–4%, and the variable growth is coded in percentage terms. Could a difference of 1% in economic growth correspond to a predicted difference of 10% of the vote? Maybe, but it's hard to imagine the coefficient being much larger. We will assign a normal prior distribution for $b$ with mean 5 and standard deviation 5: this implies that $b$ is most likely positive and is most likely less than 10.

The above is not the only reasonable choice of prior distribution, but it indicates how we can use some knowledge of the structure of the problem to set a weakly informative prior specifying a general range of what we might expect to see.

We can now fit the model including this information:

```
M4 <- stan_glm(vote ~ growth, data=hibbs,
  prior=normal(5, 5), prior_intercept=normal(50, 10))
```

And here is the result of print(M4):

```
            Median MAD_SD
(Intercept) 46.2    1.7
growth       3.1    0.7

Auxiliary parameter(s):
      Median MAD_SD
sigma 3.9    0.7
```

These inferences are essentially identical to the earlier results with the flat prior and the default prior, which makes sense in this particular example, because here our prior contains very little information compared to the data (as can be seen because the posterior standard error (mad sd) of $b$ is 0.7, is much smaller than the prior standard deviation of 5).

### Example where an informative prior makes a difference: Beauty and sex ratio

In the election forecasting example, the informative prior doesn't do much—it's fine to include it but there's really no need. In other settings, though, data are noisy enough that prior information can be valuable in keeping estimates reasonable.

xample: eauty and ex ratio

As an example, we return to the beauty-and-sex ratio example from Section 9.4 and place it in a regression context.

The relevant data came from a study of American adolescents whose attractiveness on a five-point scale was assessed by interviewers in a face-to-face survey. Years later, many of these survey respondents had children, and Figure 9.5a shows the percentage of girl births among parents in each attractiveness category. The original analysis of these data compared the sex ratio of children of the "very attractive" parents to all others (that is the source of the estimate of 8 percentage points discussed in Section 9.4), but here we consider beauty as a continuous predictor and fit a least squares regression using lm, yielding the following result:[6]

[6]Data and code for this example are in the folder SexRatio.

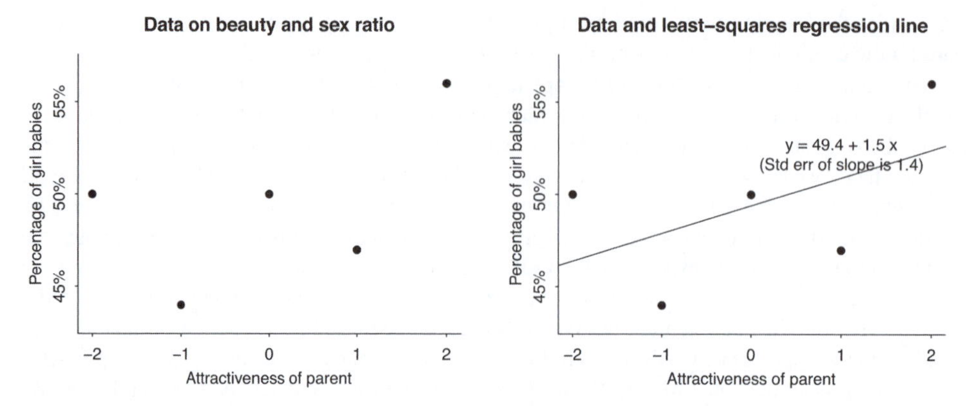


Figure 9.5 *(a) Data from a survey showing the percentage of girl births among parents of five different attractiveness categories; (b) data with fitted regression line. Figure 9.6 shows different expressions of uncertainty in the fit.*

```
            coef.est coef.se
(Intercept) 49.4     1.9
x            1.5     1.4
```

Figure 9.5b displays the least squares line. The slope is highly uncertain, as can be seen from its standard error, indicating that the data alone tell us little about the relation between $x$ and $y$ in this example.

Next we perform Bayesian inference with a default prior:

```
fit_default <- stan_glm(y ~ x, data=sexratio)
print(fit_default)
```

which yields,

```
            Median MAD_SD
(Intercept) 49.4    2.0
x            1.4    1.4

Auxiliary parameter(s):
      Median MAD_SD
sigma 4.6    1.7
```

This is nearly identical to the least squares estimate, which makes sense given that the default prior is, by design, weak.

We next consider an informative prior. As discussed in Section 9.4, the percentage of girl births is remarkably stable at roughly 48.5% to 49% girls. We can use this information to set weakly informative priors for the regression coefficients:

- *The intercept, a.* Again, `prior_intercept` in `stan_glm` is set corresponding to predictors set to the average in the data; thus we need a prior for the percentage of girls for parents of average beauty. For this unknown quantity we choose a normal distribution centered at 48.8 with standard deviation 0.5, indicating that the expected value of $y$ at the average of attractiveness is roughly in the range $48.8 \pm 0.5$.
- *The slope, b.* We choose a normal distribution centered at 0 with standard deviation 0.2, indicating that we have no prior expectation that girl births will be positively or negatively correlated with parents' beauty, and that we expect the coefficient to be roughly between $-0.2$ and $0.2$. The predictor $x$ in the data has a range of 4, so this is equivalent to saying that we expect any differences in percentage of girl births in the population to be no more than 0.8, comparing the highest and lowest category of attractiveness.

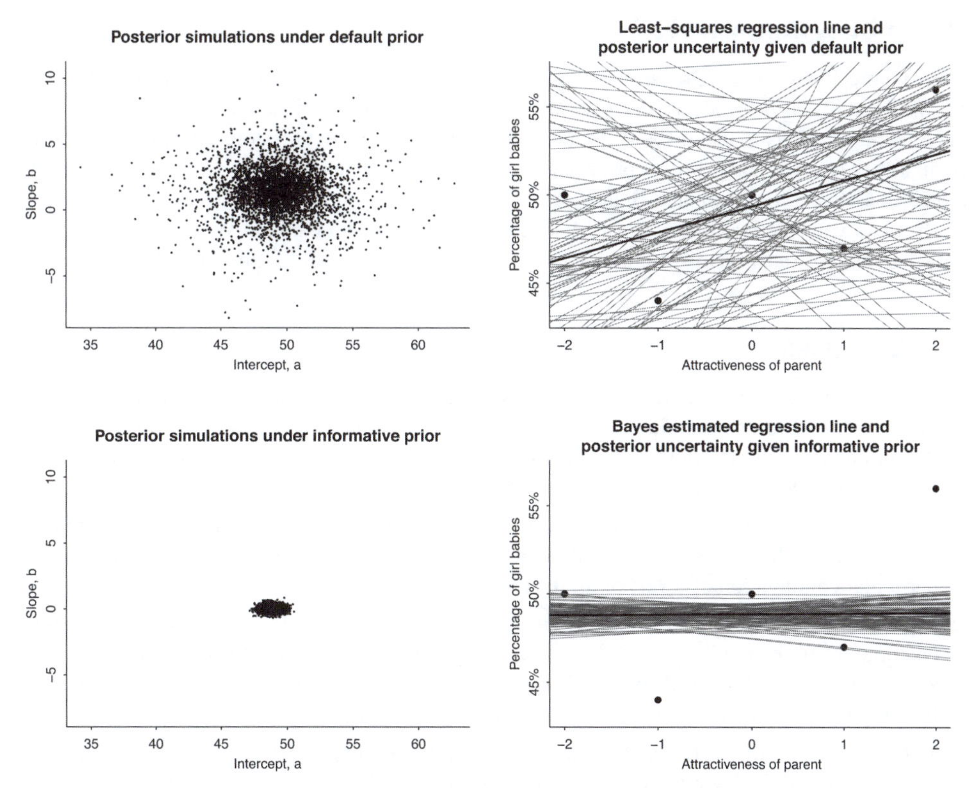


Figure 9.6 *100 posterior simulations of coefficients and regression lines, $y = a + bx$ for the beauty-and-sex ratio data from Figure 9.5, fitted given two different priors. The top row shows the wide range of parameter values and corresponding regression lines consistent with the data and default prior. The bottom row shows results for the model fit with informative prior, which greatly narrows the range of possibilities. Given this prior, the data supply almost no evidence about a and b.*

We then fit an informative-prior regression:

```
fit_post <- stan_glm(y ~ x, data=sexratio,
  prior=normal(0, 0.2), prior_intercept=normal(48.8, 0.5))
print(fit_post)
```

And here is the summary of the fitted model:

```
            Median MAD_SD
(Intercept) 48.8    0.5
x            0.0    0.2

Auxiliary parameter(s):
      Median MAD_SD
sigma 4.3    1.3
```

The estimated coefficient of $x$ is not exactly zero; if we were interested we could display the results to additional significant digits, but in any case it is clear that any estimate is overwhelmed by uncertainty. The data supply so little information in this case that the posterior is essentially the same as the prior.

Figure 9.6 shows posterior simulations of $(a, b)$ and the corresponding regression lines, $y = a+bx$, for the model fitted with the default prior (top row of graphs) and the informative prior (bottom row). In this case, the prior supplies much more information than the data.

## 9.6 Bibliographic note

To learn more about Bayesian ideas, methods, and applications, we recommend Gelman et al. (2013) and McElreath (2020). Modern Bayesian inference is typically performed using simulation rather than formulas such as (9.3) and (9.4), but we find these analytic expressions helpful in providing insight into the relative weighting of prior information and local data. For more on weakly informative priors, see Gelman, Simpson, and Betancourt (2017).

The beauty and sex ratio example is discussed in detail by Gelman and Weakliem (2009), in a paper that demonstrates Bayesian and non-Bayesian uses of prior information in statistical analysis.

## 9.7 Exercises

9.1 *Prediction for a comparison*: A linear regression is fit on high school students modeling grade point average given household income. Write R code to compute the 90% predictive interval for the difference in grade point average comparing two students, one with household incomes of \$40 000 and one with household income of \$80 000.

9.2 *Predictive simulation for linear regression*: Using data of interest to you, fit a linear regression. Use the output from this model to simulate a predictive distribution for observations with a particular combination of levels of all the predictors in the regression.

9.3 *Uncertainty in the predicted expectation and the forecast*: Consider the economy and voting example from Section 7.1. Fit the linear regression model to the data through 2012; these are available in the folder `ElectionsEconomy`. Make a forecast for the incumbent party's share of the two-party vote in a future election where economic growth is 2%.

   (a) Compute the point forecast, the standard deviation of the predicted expectation from (9.1), and the standard deviation of the predicted value from (9.2).

   (b) Now compute these using the relevant prediction functions discussed in Section 9.2. Check that you get the same values as in part (a) of this problem.

9.4 *Partial pooling*: Consider the example in Section 9.3 of combining prior information and survey data to forecast an election. Suppose your prior forecast for a given candidate's share of the two-party vote is 42% with a forecast standard deviation of 5 percentage points, and you have survey data in which this candidate has 54% support. Assume that the survey is a simple random sample of voters with no nonresponse, and that voters will not change their mind between the time of the survey and the election.

   (a) When you do the calculation, it turns out that your posterior mean estimate of the candidate's vote share is 49%. What must the sample size of the survey be?

   (b) Given this information, what is the posterior probability that the candidate wins the election?

9.5 *Combining prior information and data*: A new job training program is being tested. Based on the successes and failures of previously proposed innovations, your prior distribution on the effect size on log(income) is normal with a mean of −0.02 and a standard deviation of 0.05. You then conduct an experiment which gives an unbiased estimate of the treatment effect of 0.16 with a standard deviation of 0.08. What is the posterior mean and standard deviation of the treatment effect?

9.6 *Bayesian inference with a zero-centered informative prior on the log scale*: Perform the Bayesian analysis for the Jamaica experiment described on page 15. We shall work on the logarithmic scale:

   (a) Do Exercise 3.8 to get the estimate and standard error of the log of the multiplicative treatment effect.

   (b) Combine these with a normal prior distribution with mean 0 and standard deviation 0.10 to get the posterior distribution of the log of the multiplicative treatment effect.

(c) Exponentiate the mean of this distribution and check that it comes to 1.09, that is, an estimated treatment effect of +9%.

(d) Compute the 95% interval on the log scale and exponentiate it; check that this comes to [0.92, 1.28], that is, a range from −8% to +28% on the original scale.

9.7 *Uniform, weakly informative, and informative priors*: Follow the steps of Section 9.5 for a different example, a regression of earnings on height using the data from the folder `Earnings`. You will need to think what could be an informative prior distribution in this setting.

9.8 *Simulation for decision analysis*: An experiment is performed to measure the efficacy of a television advertising program. The result is an estimate that each minute spent on a national advertising program will increase sales by $500 000, and this estimate has a standard error of $200 000. Assume the uncertainty in the treatment effect can be approximated by a normal distribution. Suppose ads cost $300 000 per minute. What is the expected net gain for purchasing 20 minutes of ads? What is the probability that the net gain is negative?

9.9 *Prior distributions*: Consider a regression predicting final exam score $y$ from midterm exam score $x$. Suppose that both exams are on the scale of 0–100, that typical scores range from 50–100, that the correlation of midterm and final is somewhere in the range 0.5–0.8, and that the average score on the final exam might be up to 20 points higher or lower than the average score on the midterm.

Given the above information, set up reasonable priors for the slope and the intercept after centering.

9.10 *Prior distribution and likelihood*: Consider the model, $y = a + bx + \text{error}$, with predictors $x$ uniformly sampled from the range $(-1, 1)$, independent prior distributions for the coefficients $a$ and $b$, and the default prior distribution for the residual standard deviation $\sigma$. For $a$, assume a normal prior distribution with mean 0 and standard deviation 1; for $b$, assume a normal prior prior distribution with mean 0 and standard deviation 0.2.

(a) Simulate $n = 100$ data points from this model, assuming the true parameter values are $a = 1, b = 0.1, \sigma = 0.5$. Compute the least squares estimate of $a, b$ and compare to the Bayesian estimate obtained from `stan_glm` using the above priors.

(b) Repeat the simulations with different values of $n$. Graph the Bayes estimates for $a$ and $b$ as a function of $n$. What will these values be in the limit of $n = 0$? $n = \infty$?

(c) For what value of $n$ is the Bayes estimate for $a$ halfway between the prior mean and the least squares estimate? For what value of $n$ is the Bayes estimate for $b$ halfway between the prior mean and the least squares estimate?

9.11 *Working through your own example*: Continuing the example from the final exercises of the earlier chapters, take your regression from Exercise 7.10 and re-fit it using informative prior for the regression slope. Explain how you have chosen the parameters for your prior distribution.

# Chapter 10

# Linear regression with multiple predictors

As we move from the simple model, $y = a + bx + \text{error}$ to the more general $y = \beta_0 + \beta_1 x_1 + \beta_2 x_2 + \cdots + \text{error}$, complexities arise, involving choices of what predictors $x$ to include in the model, interpretations of the coefficients and how they interact, and construction of new predictors from existing variables to capture discreteness and nonlinearity. We need to learn how to build and understand models as new predictors are added. We discuss these challenges through a series of examples illustrated with R code and graphs of data and fitted models.

## 10.1 Adding predictors to a model

mple:
ldren's
tests

Regression coefficients are typically more complicated to interpret with multiple predictors because the interpretation for any given coefficient is, in part, contingent on the other variables in the model. The coefficient $\beta_k$ is the average or expected difference in outcome $y_k$, comparing two people who differ by one unit in the predictor $x_k$ while being equal in all the other predictors. This is sometimes stated in shorthand as comparing two people (or, more generally, two observational units) that differ in $x_k$ with all the other predictors held constant. We illustrate with an example, starting with single predictors and then putting them together. We fit a series of regressions predicting cognitive test scores of preschoolers given characteristics of their mothers, using data from a survey of adult American women and their children (a subsample from the National Longitudinal Survey of Youth).

### Starting with a binary predictor

We start by modeling the children's test scores given an indicator for whether the mother graduated from high school (coded as 1) or not (coded as 0). We fit the model in R as `stan_glm(kid_score ~ mom_hs, data=kidiq)`, and the result is,[1]

$$\text{kid\_score} = 78 + 12 * \text{mom\_hs} + \text{error}. \tag{10.1}$$

This model summarizes the difference in average test scores between the children of mothers who completed high school and those with mothers who did not. Figure 10.1 displays how the regression line runs through the mean of each subpopulation.

The intercept, 78, is the average (or predicted) score for children whose mothers did not complete high school. To see this algebraically, consider that to obtain predicted scores for these children we would just set `mom_hs` to 0. To get average test scores for children (or the predicted score for a single child) whose mothers were high school graduates, we would just plug in 1 to obtain $78 + 12 * 1 = 90$.

The difference between these two subpopulation means is equal to the coefficient on `mom_hs`, and it tells us that children of mothers who have completed high school score 12 points higher on average than children of mothers who have not completed high school.

[1] Data and code for this example are in the folder `KidIQ`.

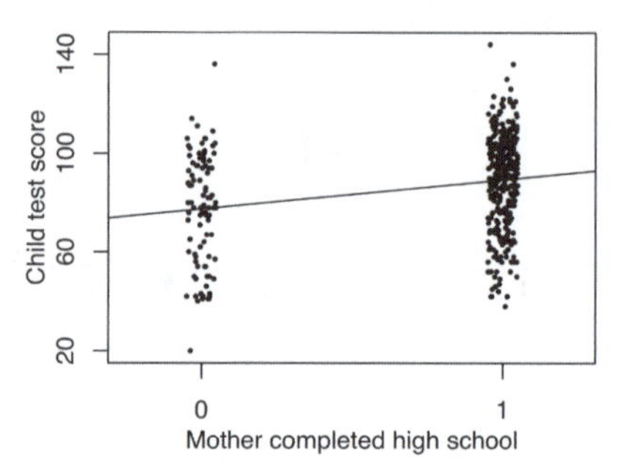


Figure 10.1 *Child's test score plotted versus an indicator for whether mother completed high school. Superimposed is the regression line, which runs through the average of each subpopulation defined by maternal education level. The indicator variable for high school completion has been* jittered; *that is, a random number has been added to each x-value so that the points do not lie on top of each other.*

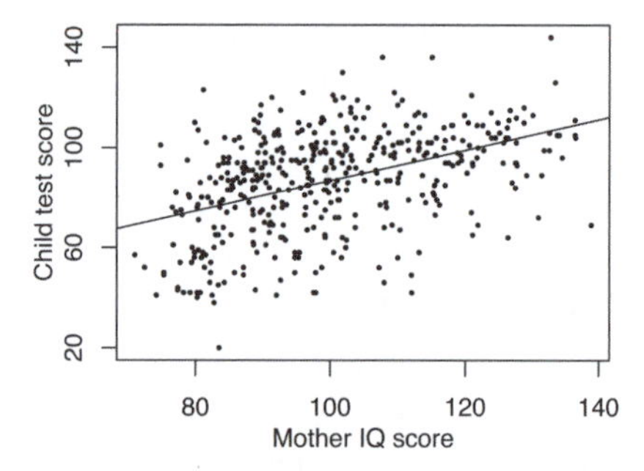


Figure 10.2 *Child's test score plotted versus maternal IQ with regression line superimposed. Each point on the line can be conceived of either as a predicted child test score for children with mothers who have the corresponding IQ, or as the average score for a subpopulation of children with mothers with that IQ.*

## A single continuous predictor

If we instead try a continuous predictor, mother's score on an IQ test, the fitted model is

$$\text{kid\_score} = 26 + 0.6 * \text{mom\_iq} + \text{error} \tag{10.2}$$

and is shown in Figure 10.2. We can think of the line as representing predicted test scores for children at each of several maternal IQ levels, or average test scores for subpopulations defined by these scores.

If we compare average child test scores for subpopulations that differ in maternal IQ by 1 point, we expect to see that the group with higher maternal IQ achieves 0.6 points more on average. Perhaps a more interesting comparison would be between groups of children whose mothers' IQ differed by 10 points—these children would be expected to have scores that differed by 6 points on average.

To understand the constant term in the regression, we must consider a case with zero values of all the other predictors. In this example, the intercept of 26 reflects the predicted test scores for children whose mothers have IQ scores of zero. This is not the most helpful quantity—we don't observe any women with zero IQ. We will discuss a simple transformation in the next section that gives the intercept a more useful interpretation.

## Including both predictors

Now consider a linear regression of child test scores on two predictors: the maternal high school indicator and maternal IQ. In R, we fit and display a regression with two predictors like this:

```
fit_3 <- stan_glm(kid_score ~ mom_hs + mom_iq, data=kidiq)
print(fit_3)
```

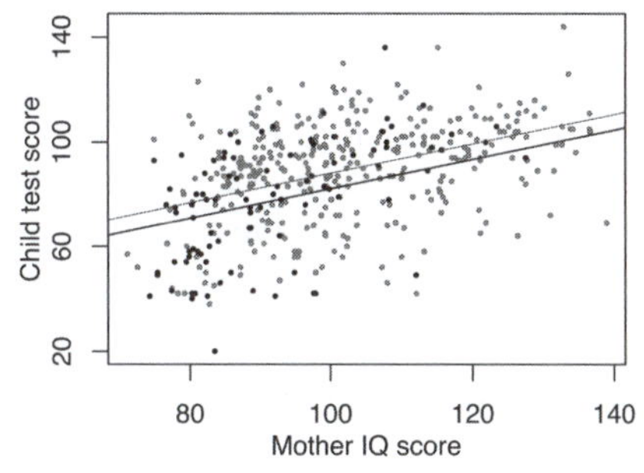


Figure 10.3 *Child's test score plotted versus maternal IQ. Light dots represent children whose mothers graduated from high school and dark dots represent children whose mothers did not graduate from high school. Superimposed are the lines from the regression of child's test score on maternal IQ and maternal high school indicator (the darker line for children whose mothers did not complete high school, the lighter line for children whose mothers did complete high school).*

And here is the result:

```
            Median MAD_SD
(Intercept) 25.7    5.9
mom_hs       6.0    2.4
mom_iq       0.6    0.1

Auxiliary parameter(s):
      Median MAD_SD
sigma 18.2   0.6
```

### Understanding the fitted model

The fitted line from the above regression is shown in Figure 10.3 and has the form,

$$\text{kid\_score} = 26 + 6 * \text{mom\_hs} + 0.6 * \text{mom\_iq} + \text{error} \tag{10.3}$$

This model forces the slope of the regression of child's test score on mother's IQ score to be the same for each maternal education subgroup. In Section 10.3 we consider an *interaction* model in which the slopes of the two lines differ. First, however, we interpret the coefficients in model (10.3):

- *The intercept.* If a child had a mother with an IQ of 0 and who did not complete high school (thus, `mom_hs` = 0), then we would predict this child's test score to be 26. This is not a useful prediction, since no mothers have IQs of 0. In Sections 12.1–12.2 we discuss ways to make the intercept more interpretable.
- *The coefficient of maternal high school completion.* Comparing children whose mothers have the same IQ, but who differed in whether they completed high school, the model predicts an expected difference of 6 in their test scores.
- *The coefficient of maternal IQ.* Comparing children with the same value of `mom_hs`, but whose mothers differ by 1 point in IQ, we would expect to see a difference of 0.6 points in the child's test score (equivalently, a difference of 10 in mothers' IQs corresponds to a difference of 6 points for their children).

## 10.2 Interpreting regression coefficients

### It's not always possible to change one predictor while holding all others constant

We interpret regression slopes as comparisons of individuals that differ in one predictor while being *at the same levels of the other predictors.* In some settings, one can also imagine manipulating the predictors to change some or hold others constant—but such an interpretation is not necessary. This becomes clearer when we consider situations in which it is logically impossible to change the value of one predictor while keeping the value of another constant. For example, if a model includes both IQ and $IQ^2$ as predictors, it does not make sense to consider changes in IQ with $IQ^2$ held constant. Or, as we discuss in the next section, if a model includes `mom_hs`, `mom_IQ`, and their interaction, `mom_hs:mom_IQ`, it is not meaningful to consider any of these three with the other two held constant.

### Counterfactual and predictive interpretations

In the more general context of multiple linear regression, it pays to be more explicit about how we interpret coefficients in general. We distinguish between two interpretations of regression coefficients.

- The *predictive interpretation* considers how the outcome variable differs, on average, when comparing two groups of items that differ by 1 in the relevant predictor while being identical in all the other predictors. Under the linear model, the coefficient is the expected difference in $y$ between these two items. This is the sort of interpretation we have described thus far.
- The *counterfactual interpretation* is expressed in terms of changes within individuals, rather than comparisons between individuals. Here, the coefficient is the expected change in $y$ caused by adding 1 to the relevant predictor, while leaving all the other predictors in the model unchanged. For example, "changing maternal IQ from 100 to 101 would lead to an expected increase of 0.6 in child's test score." This sort of interpretation arises in causal inference.

Introductory statistics and regression texts sometimes warn against the latter interpretation but then allow for similar phrasings such as "a change of 10 in maternal IQ is *associated* with a change of 6 points in child's score." The latter expression is not necessarily correct either. From the data alone, a regression only tells us about *comparisons between units*, not about *changes within units*.

Thus, the most careful interpretation of regression coefficients is in terms of comparisons, for example, "When comparing two children whose mothers have the same level of education, the child whose mother is $x$ IQ points higher is predicted to have a test score that is $6x$ higher, on average." Or, "Comparing two items $i$ and $j$ that differ by an amount $x$ on predictor $k$ but are identical on all other predictors, the predicted difference $y_i - y_j$ is $\beta_k x$, on average." This is an awkward way to put things, which helps explain why people often prefer simpler formulations such as "a change of 1 in $x_k$ causes, or is associated with, a change of $\beta$ in $y$"—but those sorts of expressions can be terribly misleading. You just have to accept that regression, while a powerful data-analytic tool, can be difficult to interpret. We return in Chapters 18–21 to conditions under which regressions can be interpreted causally.

## 10.3 Interactions

In model (10.3), the slope of the regression of child's test score on mother's IQ was forced to be equal across subgroups defined by mother's high school completion, but inspection of the data in Figure 10.3 suggests that the slopes differ substantially. A remedy for this is to include an *interaction* between `mom_hs` and `mom_iq`—that is, a new predictor defined as the product of these two variables. This allows the slope to vary across subgroups. In, R, we fit and display the model with,

```
fit_4 <- stan_glm(kid_score ~ mom_hs + mom_iq + mom_hs:mom_iq, data=kidiq)
print(fit_4)
```

thus including the *main effects* and their interaction, `mom_hs:mom_iq`. The fitted model,

$$\text{kid\_score} = -11 + 51 * \text{mom\_hs} + 1.1 * \text{mom\_iq} - 0.5 * \text{mom\_hs} * \text{mom\_iq} + \text{error},$$

is displayed in Figure 10.4a, with separate lines for each subgroup defined by maternal education.

Figure 10.4b shows the regression lines on a scale with the $x$-axis extended to zero to display the

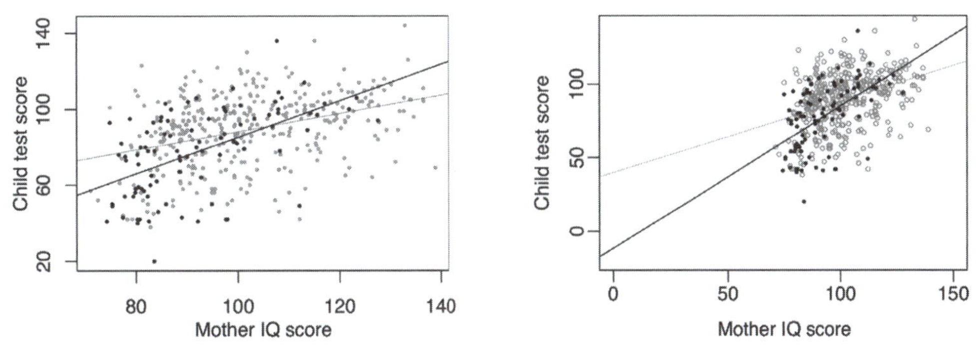


Figure 10.4 *(a) Regression lines of child's test score on mother's IQ with different symbols for children of mothers who completed high school (light circles) and those whose mothers did not complete high school (dark dots). The interaction allows for a different slope in each group, with light and dark lines corresponding to the light and dark points. (b) Same plot but with horizontal and vertical axes extended to zero to reveal the intercepts.*

intercepts—the points on the $y$-axis where the lines cross zero. This highlights that the intercept has no direct interpretation when the range of possible values of the predictor is far from zero.

Care must be taken in interpreting the coefficients in this model. We derive meaning from the fitted model by examining average or predicted test scores within and across specific subgroups. Some coefficients are interpretable only for certain subgroups.

- *The intercept* represents the predicted test scores for children whose mothers did not complete high school and had IQs of 0—not a meaningful scenario. As we discuss in Sections 12.1–12.2, intercepts can be more interpretable if input variables are centered before including them as regression predictors.
- *The coefficient of* `mom_hs` can be conceived as the difference between the predicted test scores for children whose mothers did not complete high school and had IQs of 0, and children whose mothers did complete high school and had IQs of 0. You can see this by just plugging in the appropriate numbers and comparing the equations. Since it is implausible to imagine mothers with IQs of 0, this coefficient is not easily interpretable.
- *The coefficient of* `mom_iq` can be thought of as the comparison of mean test scores across children whose mothers did not complete high school, but whose mothers differ by 1 point in IQ. This is the slope of the dark line in Figure 10.4.
- *The coefficient on the interaction term* represents the *difference* in the slope for `mom_iq`, comparing children with mothers who did and did not complete high school: that is, the difference between the slopes of the light and dark lines in Figure 10.4.

An equivalent way to understand the model is to look at the separate regression lines for children of mothers who completed high school and those whose mothers did not:

$$\begin{aligned}
\text{mom_hs} = 0: \quad \text{kid_score} &= -11 + 51 * 0 + 1.1 * \text{mom_iq} - 0.5 * 0 * \text{mom_iq} \\
&= -11 + 1.1 * \text{mom_iq} \\
\text{mom_hs} = 1: \quad \text{kid_score} &= -11 + 51 * 1 + 1.1 * \text{mom_iq} - 0.5 * 1 * \text{mom_iq} \\
&= 40 + 0.6 * \text{mom_iq}.
\end{aligned}$$

The estimated slopes of 1.1 for children whose mothers did not complete high school and 0.6 for children of mothers who did are directly interpretable. The intercepts still suffer from the problem of only being interpretable at mother's IQs of 0.

### When should we look for interactions?

Example: Radon, smoking, and lung cancer

Interactions can be important, and the first place we typically look for them is with predictors that have large coefficients when not interacted. For a familiar example, smoking is strongly associated with cancer. In epidemiological studies of other carcinogens, it is crucial to adjust for smoking both as an uninteracted predictor and as an interaction, because the strength of association between other risk factors and cancer can depend on whether the individual is a smoker. We illustrated this interaction in Figure 1.7 with the example of home radon exposure: high levels of radon are associated with greater likelihood of cancer—but this difference is much greater for smokers than for nonsmokers.

Including interactions is a way to allow a model to be fit differently to different subsets of data. These two approaches—fitting models separately within distinct subgroups versus fitting an interacted model to the full sample—are related, as we discuss later in the context of multilevel models.

### Interpreting regression coefficients in the presence of interactions

Models with interactions can often be more easily interpreted if we preprocess the data by centering each input variable about its mean or some other convenient reference point. We discuss this in Section 12.2 in the context of linear transformations.

## 10.4 Indicator variables

Example: Height and weight

In Section 7.3 we discussed how to express comparisons using regression with indicator ("dummy") variables. We further explore this idea here, fitting models to predict height from weight and other variables based on survey data. Here are data from the first rows of the data frame `earnings`:[2]

```
  height weight male  earn ethnicity education walk exercise smokenow tense angry age
1     74    210    1 50000     White        16    3        3        2     0     0  45
2     66    125    0 60000     White        16    6        5        1     0     0  58
3     64    126    0 30000     White        16    8        1        2     1     1  29
4     65    200    0 25000     White        17    8        1        2     0     0  57
```

We start by predicting weight (in pounds) from height (in inches):

```
fit_1 <- stan_glm(weight ~ height, data=earnings)
print(fit_1)
```

which yields,

```
            Median MAD_SD
(Intercept) -172.9   11.6
height         4.9    0.2

Auxiliary parameter(s):
      Median MAD_SD
sigma 29.1    0.5
```

The fitted regression line is weight $= -172.9 + 4.9 *$ height:

- Comparing two people who differ in height by one inch, their expected difference in weight is 4.9 pounds.
- The predicted weight for a 0-inch-tall person is $-172.9$ pounds . . . hmmm, this doesn't mean very much. The average height of American adults is about 66 inches, so let's state it this way: the predicted weight for a 66-inch-tall person is $-172.9 + 4.9 * 66$ pounds:

```
coefs_1 <- coef(fit_1)
predicted_1 <- coefs_1[1] + coefs_1[2]*66
```

[2]Data and code for this example are in the folder `Earnings`.

Or we can simply use posterior_predict:

```
new <- data.frame(height=66)
pred <- posterior_predict(fit_1, newdata=new)
```

which returns a vector of length n_sims, which we can summarize by its mean and standard deviation:

```
cat("Predicted weight for a 66-inch-tall person is", round(mean(pred)),
  "pounds with a sd of", round(sd(pred)), "\n")
```

### Centering a predictor

To improve interpretation of the fitted models, we use a centered version of height as a predictor:

```
earnings$c_height <- earnings$height - 66
fit_2 <- stan_glm(weight ~ c_height, data=earnings)
```

yielding,

```
            Median MAD_SD
(Intercept) 153.4    0.6
c_height      4.9    0.2

Auxiliary parameter(s):
      Median MAD_SD
sigma 29.1   0.5
```

### Including a binary variable in a regression

Next we expand the model by including an indicator variable for sex:

```
fit_3 <- stan_glm(weight ~ c_height + male, data=earnings)
```

which yields,

```
            Median MAD_SD
(Intercept) 149.6    1.0
c_height      3.9    0.3
male         12.0    2.0

Auxiliary parameter(s):
      Median MAD_SD
sigma 28.8    0.5
```

The coefficient of 12.0 on male tells us that, in these data, comparing a man to a woman of the same height, the man will be predicted to be 12 pounds heavier.

To compute the predicted weight for a 70-inch-tall woman, say:

```
coefs_3 <- coef(fit_3)
predicted <- coefs_3[1] + coefs_3[2]*4.0 + coefs_3[3]*0
```

Or,

```
new <- data.frame(c_height=4.0, male=0)
pred <- posterior_predict(fit_3, newdata=new)
cat("Predicted weight for a 70-inch-tall woman is", round(mean(pred)),
  "pounds with a sd of", round(sd(pred)), "\n")
```

Either way, the result is 165 pounds. The corresponding point prediction for a 70-inch-tall man is mean(posterior_predict(fit_3, newdata=data.frame(c_height=4.0, male=1))), which comes to 177 pounds, which is indeed 12 pounds higher than the prediction for the woman.

### Using indicator variables for multiple levels of a categorical predictor

Next we shall include ethnicity in the regression. In our data this variable takes on four levels, as we can see by typing table(earnings$ethnicity) in the R console:

```
Black Hispanic    Other    White
  177      103       37     1473
```

We can include ethnicity in our regression as a *factor*:

```
fit_4 <- stan_glm(weight ~ c_height + male + factor(ethnicity), data=earnings)
print(fit_4)
```

which yields,

```
                           Median MAD_SD
(Intercept)                154.1    2.2
c_height                     3.8    0.3
male                        12.2    2.0
factor(ethnicity)Hispanic   -5.9    3.6
factor(ethnicity)Other     -12.6    5.2
factor(ethnicity)White      -5.0    2.3

Auxiliary parameter(s):
      Median MAD_SD
sigma 28.7    0.5
```

Ethnicity has four levels in our data, but looking carefully at the output, we see only three coefficients for ethnicity, for Hispanics, Others, and Whites. The missing group is Blacks. In computing the regression, R took Black to be the *baseline* category against which all other groups are measured.

Thus, the above coefficient of –5.9 implies that, when comparing a Hispanic person and a Black person with the same height and sex, the fitted model predicts the Hispanic person to be 5.9 pounds lighter, on average. Similarly, the model predicts an Other person to be 12.6 pounds lighter and a White person to be 5.0 pounds lighter than a Black person of the same height and sex.

### Changing the baseline factor level

When including a factor variable in a regression, any of the levels can be used as the baseline. By default, R orders the factors in alphabetical order, hence in this case Black is the first category and is used as the baseline.

We can change the baseline category by directly setting the levels when constructing the factor:

```
earnings$eth <- factor(earnings$ethnicity,
  levels=c("White", "Black", "Hispanic", "Other"))
fit_5 <- stan_glm(weight ~ c_height + male + eth, data=earnings)
print(fit_5)
```

which yields,

```
            Median MAD_SD
(Intercept) 149.1    1.0
c_height      3.8    0.2
male         12.2    2.0
ethBlack      5.0    2.2
ethHispanic  -0.9    2.9
ethOther     -7.6    4.6

Auxiliary parameter(s):
      Median MAD_SD
sigma 28.3    0.4
```

This model uses White as the baseline category because we listed it first when setting up the factor variable eth. Going through the coefficients:

- In the earlier fit, the intercept of 154.1 was the predicted weight for a person with c_height = 0, male = 0, and ethnicity = Black. In the new version, the intercept of 149.1 is the predicted value with c_height = 0, male = 0, and ethnicity = White, the new baseline category. The change of 5.0 corresponds to negative of the coefficient of White in the original regression.
- The coefficients for height and male do not change.
- The coefficient for Black, which earlier was 0 by implication, as Black was the baseline category and thus not included in the regression, is now 5.0, which is the difference relative to the new baseline category of White.
- The coefficient for Hispanic has increased from −5.9 to −0.9, a change of 5.0 corresponding to the shift in the baseline from Black to White.
- The coefficient for Other also increases by this same 5.0.
- The coefficient for White has increased from −5.6 to the implicit value of 0.

An alternative approach is to create indicators for the four ethnic groups directly:

```
earnings$eth_White <- ifelse(earnings$ethnicity=="White", 1, 0)
earnings$eth_black <- ifelse(earnings$ethnicity=="Black", 1, 0)
earnings$eth_hispanic <- ifelse(earnings$ethnicity=="Hispanic", 1, 0)
earnings$eth_other <- ifelse(earnings$ethnicity=="Other", 1, 0)
```

It is not necessary to name the new variables in this way, but this sort of naming can sometimes make it easier to keep track. In any case, once we have created these numerical variables we can include them in the usual way, for example:

```
fit_6 <- stan_glm(weight ~ height + male + eth_Black + eth_Hispanic +
  eth_Other, data=earnings)
```

### Using an index variable to access a group-level predictor

Sometimes we are fitting a regression at the individual level but with predictors at the group level. For example, we might be predicting students' test scores given student background variables and also the avarege income level of parents at the students' schools. Suppose we have data from 1000 students in 20 schools, and we have an index variable called school which is a vector of length 1000 that takes on values from 1 through 20, telling which school each student is in. Further suppose we have a vector of length 20 called income which is some estimate of average parent income within each school. We can then create a student-level predictor, a vector of length 1000:

```
school_income <- income[school]
```

And then we can include it in the regression, for example,

```
stan_glm(score ~ pretest + age + male + school_income, data=students)
```

## 10.5 Formulating paired or blocked designs as a regression problem

We have repeatedly discussed how regression coefficients can be interpreted as comparisons. Conversely, it can often be helpful to express comparisons as regressions.

### Completely randomized experiment

Consider a simple experiment in which $n$ people are randomly assigned to treatment and control groups, with $n/2$ in each group. The straightforward estimate of the treatment effect is then $\bar{y}_T - \bar{y}_C$, with standard error $\sqrt{\mathrm{sd}_T^2/(n/2) + \mathrm{sd}_C^2/(n/2)}$.

As discussed in Section 7.3, we can express this inference in a regression framework by using the group indicator as a predictor: the resulting least-squares estimate of the coefficient is simply the difference, $\bar{y}_T - \bar{y}_C$, and the standard error will be close to that obtained from the difference-in-means formula, with potentially a minor difference when going from unpooled to pooled variance estimate.

For the simple case of no pre-treatment predictors, the difference and regression inferences are essentially the same; regression has the advantage of generalizing to more complicated settings.

### Paired design

Next consider a more complicated example in which the $n$ people are first paired, with the two people in each pair being randomly assigned to treatment and control. The standard recommendation for analyzing such data is to compute the difference within each pair, labeling these as $z_i$, for $i = 1, \dots, n/2$, and then estimate the treatment effect and standard error as $\bar{z}$ and $\mathrm{sd}(z)/\sqrt{n/2}$.

Alternatively, the data from the paired design can be analyzed using regression, in this case by fitting a model on all $n$ data points and including a treatment indicator and indicators for the pairs. We first create an index variable `pairs` taking on values between 1 and $n/2$; then we fit the regression,

```
fit <- stan_glm(y ~ treatment + factor(pairs), data=expt)
```

using `factor` to create the corresponding $n/2$ indicator variables for the regression. As discussed in Section 10.4, R will avoid collinearity in this model by taking one of the pairs as the baseline and including indicators for all the others, so that the regression has $(n/2) + 1$ predictors: a constant term, a treatment indicator, and $(n/2) - 1$ remaining group indicators. The treatment effect is the coefficient of the treatment indicator, and its estimate will be approximately the same as the estimate $\bar{z}$ with standard error $\mathrm{sd}(z)/\sqrt{n/2}$ obtained by working with the average of the paired differences. Again, the regression framework has the advantage of being easily extended to the setting with pre-treatment information, which can be included as additional predictors in the model.

### Block design

The general principle is to include any pre-treatment information that could be relevant to predicting the outcome measurements. For example, consider a block design in which the $n$ people are in $J$ groups—they could be students in classrooms, for example—with random assignment to treatment and control within each group. Then we can fit a regression predicting the outcome variable on the treatment indicator and $J - 1$ group indicators. For the purpose of estimating the treatment effect, it does not matter which group is taken as a baseline. Again, if other pre-treatment variables are available, they can be included as additional predictors in the regression.

## 10.6 Example: uncertainty in predicting congressional elections

Example: Predicting congressional elections

We illustrate simulation-based predictions in the context of a model of elections for the U.S. Congress.[3] We first construct a model to predict the 1988 election from the 1986 election, then apply the model to predict 1990 from 1988, and then check these predictions against the actual outcomes in 1990.

### Background

The United States is divided into 435 congressional districts, and we define the outcome $y_i$, for $i = 1, \dots, n = 435$ to be the Democratic party's share of the two-party vote (that is, excluding the votes for parties other than the Democrats and the Republicans) in district $i$ in 1988. Figure 10.5 shows a histogram of the data $y$.

[3] Data and code for this example are in the folder `Congress`.

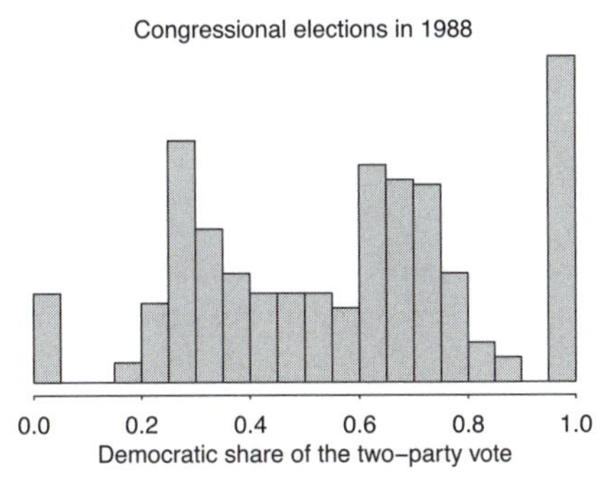


Figure 10.5 *Histogram of congressional election data from 1988. The spikes at the left and right ends represent uncontested Republicans and Democrats, respectively.*

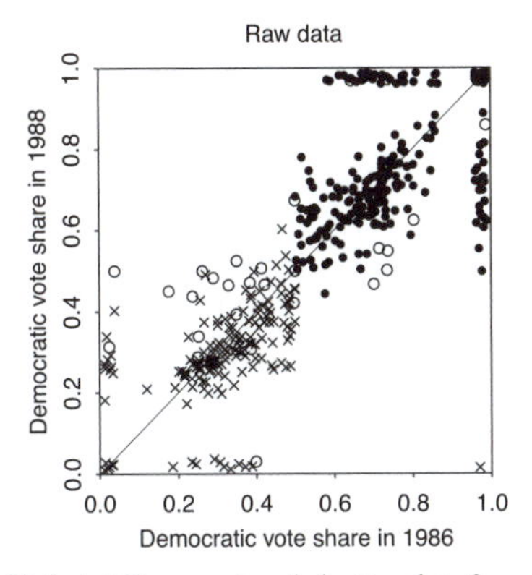


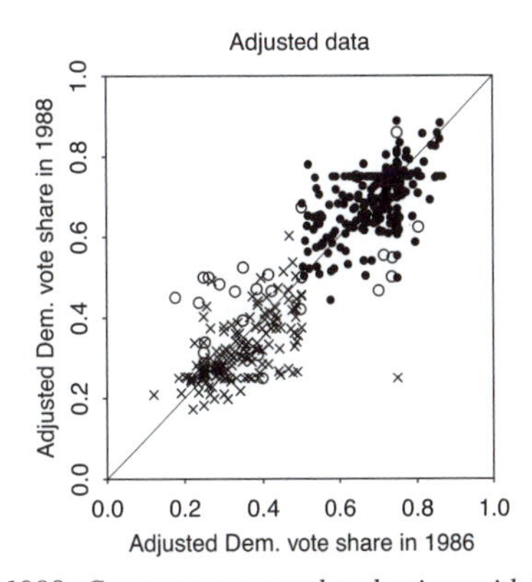


Figure 10.6 *(a) Congressional election data from 1986 and 1988. Crosses correspond to elections with Republican incumbents running in 1988, dots correspond to Democratic incumbents, and open circles correspond to open seats. The "incumbency" predictor in the regression model equals 0 for the circles, +1 for the dots, and −1 for the crosses. Uncontested election outcomes (at 0 and 1) have been jittered slightly. (b) Data for the regression analysis, with uncontested 1988 elections removed and uncontested 1986 election values replaced by 0.25 and 0.75. The $y = x$ line is included as a comparison on both plots.*

How can the variation in the data be understood? What information would be relevant in predicting the outcome of a congressional election? First of all, it is useful to know whether both parties are contesting the election; the spikes at the two ends of the histogram reveal that many of the elections were uncontested. After that, it would seem to make sense to use the outcome in each district of the most recent previous election, which was in 1986. In addition, we use the knowledge of whether the *incumbent*—the current occupant of the congressional seat—is running for reelection.

Our regression model has the following predictors:

- A constant term.
- The Democratic share of the two-party vote in district $i$ in the previous election.
- Incumbency: an indicator that equals +1 if district $i$ was occupied in 1988 by a Democrat who is running for reelection, −1 if a Republican was running for reelection, and 0 if the election was *open*—that is, if neither of the two candidates was occupying the seat at that time.

Because the incumbency predictor is categorical, we can display the data in a single scatterplot using different symbols for Republican incumbents, Democratic incumbents, and open seats; see Figure 10.6a.

We shall fit a linear regression. The data—the number of votes for each candidate—are discrete, but the number of votes within each district is large enough that essentially nothing is lost by fitting a continuous model.

### Data issues

We are predicting district-by-district results given the outcome in the previous election and the knowledge of whether the incumbent is running for reelection. Primary elections are typically in September, and so it is reasonable to expect to have this information about two months before the November general election.

Many elections were uncontested in one year or another, so that the vote share is 0 or 1 exactly. It would be possible to simply include these in the model as is; however, instead we impute the value 0.25 for uncontested Republicans and 0.75 for uncontested Democrats, which have been chosen to approximate the proportion of votes received by the Democratic candidate had the election actually been contested. More generally, we can impute random values from the distribution of contested election outcomes preceding an uncontested race, but for our purposes here the simple imputation is sufficient. The adjusted dataset is displayed in Figure 10.6b.

### Fitting the model

We set up a regression to predict `vote` (the Democratic share of the two-party vote in each district), given `past_vote` (the Democrats' share in the previous election) and `incumbency` (the $-1/0/1$ variable just defined). First we predict 1988 from 1986, which requires pulling out the appropriate variables from the dataset:

```
data88 <- data.frame(vote=congress$v88_adj, past_vote=congress$v86_adj,
  inc=congress$inc88)
fit88 <- stan_glm(vote ~ past_vote + inc, data=data88)
```

which yields,

```
            Median MAD_SD
(Intercept) 0.24   0.02
past_vote   0.52   0.03
inc         0.10   0.01

Auxiliary parameter(s):
      Median MAD_SD
sigma 0.07   0.00
```

This model has some problems, as can be seen by careful examination of the before-after plot in Figure 10.6b: the open-seat elections (shown by open circles on that plot) are mostly off the regression line, suggesting a possible interaction—that is, different slopes for incumbents and open seats. In addition, the jump between the average $y$-values just below and just above $x = 0.5$ is not completely fit by the `incumbency_88` predictor. And we are ignoring any uncertainty in the potential vote shares for the uncontested elections, which were simply imputed as 0.25 and 0.75. Better models can be fit to these data (see Exercise 19.10), but the simple regression fit here is sufficient to demonstrate the principles of simulation-based predictive inference.

### Simulation for inferences and predictions of new data points

Running `stan_glm` produces a set of simulation draws expressing uncertainty in the parameters in the fitted model. We can access these simulations by extracting them from the fitted model object:

```
sims88 <- as.matrix(fit88)
```

| sim | $\sigma$ | $\beta_0$ | $\beta_1$ | $\beta_2$ | $\tilde{y}_1$ | $\tilde{y}_2$ | $\cdots$ | $\tilde{y}_{435}$ | $\sum_i I(\tilde{y}_i > 0.5)$ |
|---|---|---|---|---|---|---|---|---|---|
| 1 | .070 | .26 | .48 | .10 | .73 | .59 | $\cdots$ | .68 | 269 |
| 2 | .069 | .26 | .47 | .11 | .85 | .56 | $\cdots$ | .69 | 260 |
| $\vdots$ | $\vdots$ | $\vdots$ | $\vdots$ | $\vdots$ | $\vdots$ | $\vdots$ | $\ddots$ | $\vdots$ | $\vdots$ |
| 4000 | .068 | .22 | .58 | .08 | .80 | .74 | $\cdots$ | .68 | 271 |
| median | .067 | .24 | .52 | .10 | .74 | .67 | $\cdots$ | .73 | 260 |
| mean | .090 | .24 | .52 | .10 | .74 | .67 | $\cdots$ | .73 | 262.1 |
| sd | .002 | .02 | .03 | .01 | .07 | .07 | $\cdots$ | .07 | 7.1 |

Figure 10.7 *Simulation results for the congressional election forecasting model. The predicted values $\tilde{y}_i$ correspond to the 1990 election.*

The first five columns of Figure 10.7 show the simulated parameter values and summaries of their posterior distributions. We use the simulations, along with the data from 1988 and incumbency information in 1990, to predict the district-by-district election outcome in 1990. We start by creating a new matrix of predictors, $\tilde{X}$:

```
data90 <- data.frame(past_vote=congress$v88_adj, inc=congress$inc90)
```

We then simulate predictive simulations of the vector of new outcomes:

```
pred90 <- posterior_predict(fit88, newdata=data90)
```

This is a matrix with `n_sims` rows and one column for each predicted congressional district, thus, the subset of the matrix in Figure 10.7 labeled $\tilde{y}_1$ through $\tilde{y}_{435}$.

### Predictive simulation for a nonlinear function of new data

For the congressional elections example, to perform inference on the summary, $\sum_{i=1}^{\tilde{n}} I(\tilde{y}_i > 0.5)$, the number of elections won by the Democrats in 1990, we sum over the rows in the matrix:

```
dems_pred <- rowSums(pred90 > 0.5)
```

The last column of Figure 10.7 shows the results. Each row represents the outcome of a different random simulation. We could also calculate that sum in a loop:

```
dems_pred <- rep(NA, n_sims)
for (s in 1:n_sims) {
  dems_pred[s] <- sum(pred90[s,] > 0.5)
}
```

The lower lines of Figure 10.7 show the median, mean, and standard deviation for each parameter and simulated outcome. The means and medians of the parameters $\sigma$ and $\beta$ are nearly identical to the point estimates. The differences are due to variation because there are only 4000 simulation draws. The 4000 simulations show some posterior variation, inferential uncertainty that is reflected in the standard errors (the mad sd's for the parameters reported in the regression output), along with posterior correlations that are not displayed in the output but are captured in the likelihood and the simulations.

The next columns in the table show the future election outcomes, which in each district has a predictive uncertainty of about 0.07, which makes sense since the estimated standard deviation from the regression is $\hat{\sigma} = 0.07$. The predictive uncertainties are slightly higher than $\hat{\sigma}$ but by only a very small amount since the number of data points in the original regression is large, and the $x$-values for the predictions are all within the range of the original data; recall equation (9.2).

Finally, the entries in the lower-right corner of Figure 10.7 give a predictive mean of 260.1 and standard deviation of 7.1 for the number of districts to be won by the Democrats. This estimate and standard deviation *could not* simply be calculated from the estimates and uncertainties for the individual districts. Simulation is the only practical method of assessing the predictive uncertainty for this nonlinear function of the predicted outcomes.

The actual number of seats won by the Democrats in the 1990 election was 267, which happens to be within the range of the predictions, but this is in some sense a bit of luck, as the model as constructed not allow for national partisan swings of the sort that happen from election to election. To put it another way, the intercept of the model predicting 1988 from 1986 is not the same as the intercept for predicting 1990 from 1988, and it would be a mistake to directly apply the first model to make predictions from the second dataset.

### Combining simulation and analytic calculations

In some settings it is helpful to supplement simulation-based inference with mathematical analysis. For example we return to the problem in Section 3.6 of estimating the probability that the election in a particular district will be tied, or within one vote of being exactly tied. This calculation is relevant in estimating the probability that an individual vote will be decisive, and comparison of these probabilities across districts or states can be relevant for decisions of allocating campaign resources.

Example: Probability of a decisive vote

In Section 3.6 we estimated the probability of a tied election given a forecast by using the normal distribution. Here we discuss how to compute this probability using predictive simulations. Consider a district, $i$, with $n_i$ voters. We have approximated the distribution of $\tilde{y}$ as continuous—which is perfectly reasonable given that the $n_i$'s are in the tens or hundreds of thousands—and so a tie is equivalent to $\tilde{y}_i$ being in the range $[\frac{1}{2} - \frac{1}{2n_i}, \frac{1}{2} + \frac{1}{2n_i}]$.

How can we estimate this probability by simulation? The most direct way is to perform many predictive simulations and count the proportion for which $\tilde{y}_i$ falls in the range $0.5 \pm 1/(2n_i)$. Unfortunately, for realistic $n_i$'s, this range is so tiny that thousands or millions of simulations could be required to estimate this probability accurately. For example, it would not be very helpful to learn that 0 out of 1000 simulations fell within the interval.

A better approach is to combine simulation and analytical results: first compute 1000 simulations of $\tilde{y}$, as shown, then for each district compute the proportion of simulations that fall between 0.49 and 0.51, say, and divide by $0.02\,n_i$ (that is, the number of intervals of width $1/n_i$ that fit between 0.49 and 0.51). Or compute the proportion falling between 0.45 and 0.55, and divide by $0.1n_i$. For some districts, the probability will still be estimated at zero after 1000 simulation draws, but the estimated zero will be much more precise.

Estimated probabilities for extremely rare events can be computed in this example using the fact that predictive distributions from a linear regression follow the $t$ distribution with $n-k$ degrees of freedom. We can use 1000 simulations to compute a predictive mean and standard deviation for each $\tilde{y}_i$, then use tail probabilities of the $t_{n-3}$ distribution (in this example, the model has three coefficients) to compute the probability of falling in the range $0.5 \pm 1/(2n_i)$.

## 10.7 Mathematical notation and statistical inference

When illustrating specific examples, it helps to use descriptive variable names. In order to discuss more general theory and data manipulations, however, we shall adopt generic mathematical notation. This section introduces this notation and discusses the stochastic aspect of the model as well.

### Predictors

Example: Children's IQ tests

We use the term *predictors* for the columns in the $X$ matrix (other than the constant term), and we also sometimes use the term when we want to emphasize the information that goes into the predictors. For example, consider the model that includes the interaction of maternal education and maternal IQ:

$$\text{kid\_score} = 58 + 16 * \text{mom\_hs} + 0.5 * \text{mom\_iq} - 0.2 * \text{mom\_hs} * \text{mom\_iq} + \text{error}.$$

| y | X | | | | | | |
|---|---|---|---|---|---|---|---|
| 1.4 | 1 | 0.69 | −1 | −0.69 | 0.5 | 2.6 | 0.31 |
| 1.8 | 1 | 1.85 | 1 | 1.85 | 1.94 | 2.71 | 3.18 |
| 0.3 | 1 | 3.83 | 1 | 3.83 | 2.23 | 2.53 | 3.81 |
| 1.5 | 1 | 0.5 | −1 | −0.5 | 1.85 | 2.5 | 1.73 |
| 2.0 | 1 | 2.29 | −1 | −2.29 | 2.99 | 3.26 | 2.51 |
| 2.3 | 1 | 1.62 | 1 | 1.62 | 0.51 | 0.77 | 1.01 |
| 0.2 | 1 | 2.29 | −1 | −2.29 | 1.57 | 1.8 | 2.44 |
| 0.9 | 1 | 1.8 | 1 | [illegible] | 3.72 | 1.1 | 1.32 |
| 1.8 | 1 | 1.22 | 1 | 1.22 | 1.13 | 1.05 | 2.66 |
| 1.8 | 1 | 0.92 | −1 | −0.92 | 2.29 | 2.2 | 2.95 |
| 0.2 | 1 | 1.7 | 1 | 1.7 | 0.12 | 0.17 | 2.86 |
| 2.3 | 1 | 1.46 | −1 | −1.46 | 2.28 | 2.4 | 2.04 |
| −0.3 | 1 | 4.3 | 1 | 4.3 | 2.3 | 1.87 | 0.48 |
| 0.4 | 1 | 3.64 | −1 | −3.64 | 1.9 | 1.13 | 0.51 |
| 1.5 | 1 | 2.27 | 1 | 2.27 | 0.47 | 3.04 | 3.12 |
| $\tilde{y}$ | $\tilde{X}$ | | | | | | |
| ? | 1 | 1.63 | −1 | −1.63 | 0.84 | 2.35 | 1.25 |
| ? | 1 | 0.65 | −1 | −0.65 | 2.08 | 1.26 | 2.3 |
| ? | 1 | 1.83 | −1 | [illegible] | 1.84 | 1.58 | 2.99 |
| ? | 1 | 2.58 | 1 | 2.58 | 2.03 | 1.8 | 1.39 |
| ? | 1 | 0.07 | −1 | −0.07 | 2.1 | 2.32 | 1.27 |

Figure 10.8 *Notation for regression modeling. The model is fit to the observed outcomes $y$ given predictors $X$. As described in the text, the model can then be applied to predict unobserved outcomes $\tilde{y}$ (indicated by small question marks), given predictors on new data $\tilde{X}$.*

This regression has three *predictors*: maternal high school, maternal IQ, and maternal high school * IQ. Depending on context, the constant term is also sometimes called a predictor.

## Regression in vector-matrix notation

We follow the usual notation and label the outcome for the $i^{\text{th}}$ individual as $y_i$ and the deterministic prediction as $X_i\beta = \beta_1 X_{i1} + \cdots + \beta_k X_{ik}$, indexing the people in the data as $i = 1, \ldots, n$. In our most recent example, $y_i$ is the $i^{\text{th}}$ child's test score, and there are $n = 1378$ data points and $k = 4$ items in the vector $X_i$ (the $i^{\text{th}}$ row of the matrix $X$): $X_{i1}$, a *constant term* that is defined to equal 1 for all people; $X_{i2}$, the mother's high school completion status (coded as 0 or 1); $X_{i3}$, the mother's test score; and $X_{i4}$, the interaction between mother's test score and high school completion status. The vector $\beta$ of coefficients has length $k = 4$ as well.

The deviations of the outcomes from the model, called *errors*, are labeled as $\epsilon_i$ and assumed to follow a normal distribution with mean 0 and standard deviation $\sigma$, which we write as $\mathrm{N}(0, \sigma^2)$. The term *residual* is used for the differences between the outcomes and predictions from the estimated model. Thus, $y - X\beta$ and $y - X\hat{\beta}$ are the vectors of errors and residuals, respectively. We use the notation $\tilde{y}$ for predictions from the model, given new data $\tilde{X}$; see Figure 10.8.

Conventions vary across disciplines regarding what terms to use for the variables we refer to as predictors and outcomes (or responses). Some use the terms "independent variable" for predictors and "dependent variable" for the outcome. These terms came from a time when regression models were used to model outcomes from experiments where the manipulation of the input variables might have led to predictors that were independent of each other. This is rarely the case in social science, however, so we avoid these terms. Other times the predictors and outcome are called the "left-hand side" and "right-hand side" variables.

### Two ways of writing the model

The classical linear regression model can then be written mathematically as

$$y_i = \beta_1 X_{i1} + \cdots + \beta_k X_{ik} + \epsilon_i, \quad \text{for } i = 1, \ldots, n, \tag{10.4}$$

where the errors $\epsilon_i$ have independent normal distributions with mean 0 and standard deviation $\sigma$.

An equivalent representation is,

$$y_i = X_i \beta + \epsilon_i, \quad \text{for } i = 1, \ldots, n,$$

where $X$ is an $n$ by $k$ matrix with $i^{\text{th}}$ row $X_i$, or, using multivariate notation,

$$y_i \sim \mathrm{N}(X_i \beta, \sigma^2), \text{ for } i = 1, \ldots, n.$$

For even more compact notation we can use,

$$y \sim \mathrm{N}(X\beta, \sigma^2 I),$$

where $y$ is a vector of length $n$, $X$ is a $n \times k$ matrix of predictors, $\beta$ is a column vector of length $k$, and $I$ is the $n \times n$ identity matrix. Fitting the model (in any of its forms) using least squares yields estimates $\hat{\beta}$ and $\hat{\sigma}$, as we demonstrated in Section 8.1 for simple regression with just one predictor and a constant term.

### Least squares, maximum likelihood, and Bayesian inference

The steps of estimation and statistical inference in linear regression with multiple predictors are the same as with one predictor, as described in Sections 8.1 and 9.5. The starting point is the least squares estimate, that is, the vector $\hat{\beta}$ that minimizes the sum of the squared residuals, $\mathrm{RSS} = \sum_{i=1}^{n}(y_i - X\hat{\beta})^2$. For the standard linear regression model with predictors that are measured accurately and errors that are independent, of equal variance, and normally distributed, the least squares solution is also the maximum likelihood estimate. The only slight difference from Section 8.1 is that the standard estimate of the residual standard deviation is

$$\hat{\sigma} = \sqrt{\mathrm{RSS}/(n-k)}, \tag{10.5}$$

where $k$ is the number of regression coefficients. This reduces to formula (8.5) on page 104 for a regression with one predictor, in which case $k = 2$.

### Nonidentified parameters, collinearity, and the likelihood function

In maximum likelihood, parameters are nonidentified if they can be changed without altering the likelihood. Continuing with the "hill" analogy from Section 8.1, nonidentifiability corresponds to a "ridge" in the likelihood—a direction in parameter space in which the likelihood is flat.

To put it another way, a model is said to be *nonidentifiable* if it contains parameters that cannot be estimated uniquely—or, to put it another way, that have standard errors of infinity. The offending parameters are called *nonidentified.* The most familiar and important example of nonidentifiability arises from *collinearity* (also called multicollinearity) of regression predictors. A set of predictors is collinear if there is a linear combination of them that equals 0 for all the data. We discuss this problem in the context of indicator variables at the end of Section 12.5.

A simple example of collinearity is a model predicting family outcomes given the number of boys in the family, the number of girls, and the total number of children. Labeling these predictors as $x_{2i}$, $x_{3i}$, and $x_{4i}$, respectively, for family $i$ (with $x_{1i} = 1$ being reserved for the constant term in the regression), we have $x_{4i} = x_{2i} + x_{3i}$, thus $-x_{2i} - x_{3i} + x_{4i} = 0$ and the set of predictors is collinear.

There can also be problems with *near*-collinearity, which leads to poor identification. For example, suppose you try to predict people's heights from the lengths of their left and right feet. These two predictors are nearly (but not exactly) collinear, and as a result it will be difficult to untangle the two coefficients, and we can characterize the fitted regression as unstable, in the sense that if it were re-fitted with new data, a new sample from the same population, we could see much different results. This instability should be reflected in large standard errors for the estimated coefficients.

### Hypothesis testing: why we do not like *t* tests and *F* tests

One thing that we do *not* recommend is traditional null hypothesis significance tests. For reference, we review here the two most common such procedures.

The $t$ test is used to demonstrate that a regression coefficient is statistically significantly different from zero, and it is formally a test of the null hypothesis that a particular coefficient $\beta_j$ equals zero. Under certain assumptions, the standardized regression coefficient $\hat{\beta}_j/\text{s.e.}_j$ will approximately follow a $t_{n-k}$ distribution under the null hypothesis, and so the null hypothesis can be rejected at a specified significance level if $|\hat{\beta}_j/\text{s.e.}_j|$ exceeds the corresponding quantile of the $t_{n-k}$ distribution.

The $F$ test is used to demonstrate that there is evidence that an entire regression model—not just any particular coefficient—adds predictive power, and it is formally a test of the null hypothesis that *all* the coefficients in the model, except the constant term, equal zero. Under certain assumptions, the ratio of total to residual sum of squares, suitably scaled, follows something called the $F$ distribution, and so the null hypothesis can be rejected if the ratio of sums of squares exceeds some level that depends on both the sample size $n$ and the number of predictors $k$.

We have essentially no interest in using hypothesis tests for regression because we almost never encounter problems where it would make sense to think of coefficients as being exactly zero. Thus, rejection of null hypotheses is irrelevant, since this just amounts to rejecting something we never took seriously in the first place. In the real world, with enough data, any hypothesis can be rejected.

That said, uncertainty in estimation is real, and we do respect the deeper issue being addressed by hypothesis testing, which is assessing when an estimate is overwhelmed by noise, so that some particular coefficient or set of coefficients could just as well be zero, as far as the data are concerned. We recommend addressing such issues by looking at standard errors as well as parameter estimates, and by using Bayesian inference when estimates are noisy (see Section 9.5), as the use of prior information should stabilize estimates and predictions. When there is the goal of seeing whether good predictions can be made without including some variables in the model, we recommend comparing models using cross validation, as discussed in Section 11.8.

## 10.8 Weighted regression

As discussed in Section 8.1, least squares regression is equivalent to maximum likelihood estimation, under the model in which errors are independent and normally distributed with equal variance. In the sum of squared residuals, each term gets equal weight.

But in some settings it makes sense to weight some data points more than others when fitting the model, and one can perform *weighted least squares*, where the estimate $\hat{\beta}_{\text{wls}}$ is that which minimizes $\sum_{i=1}^{n} w_i(y_i - X_i\beta)^2$, for some specified $w = (w_1, \dots, w_n)$ of nonnegative weights. Points with higher weights count for more in this formula, so the regression line is constrained to be closer to them.

As with the ordinary least squares estimate (8.2), one can use matrix algebra to derive the weighted least squares estimate, which comes to,

$$\hat{\beta} = (X^t W^{-1} X)^{-1} X^t W^{-1} y, \tag{10.6}$$

where $W$ is the matrix of weights, $W = \text{Diag}(w)$.

### Three models leading to weighted regression

Weighted least squares can be derived from three different models:

1. *Using observed data to represent a larger population.* This is the most common way that regression weights are used in practice. A weighted regression is fit to sample data in order to estimate the (unweighted) linear model that would be obtained if it could be fit to the entire population. For example, suppose our data come from a survey that oversamples older white women, and we are interested in estimating the population regression. Then we would assign to survey respondent a weight that is proportional to the number of people of that type in the population represented by that person in the sample. In this example, men, younger people, and members of ethnic minorities would have higher weights. Including these weights in the regression is a way to approximately minimize the sum of squared errors with respect to the population rather than the sample.
2. *Duplicate observations.* More directly, suppose each data point can represent one or more actual observations, so that $i$ represents a collection of $w_i$ data points, all of which happen to have $x_i$ as their vector of predictors, and where $y_i$ is the average of the corresponding $w_i$ outcome variables. Then weighted regression on the compressed dataset, $(x, y, w)$, is equivalent to unweighted regression on the original data.
3. *Unequal variances.* From a completely different direction, weighted least squares is the maximum likelihood estimate for the regression model with independent normally distributed errors with unequal variances, where $\text{sd}(\epsilon_i)$ is proportional to $1/\sqrt{w_i}$. That is, measurements with higher variance get lower weight when fitting the model. As discussed further in Section 11.1, unequal variances are not typically a major issue for the goal of estimating regression coefficients, but they become more important when making predictions about individual cases.

These three models all result in the same point estimate but imply different standard errors and different predictive distributions. For the most usual scenario in which the weights are used to adjust for differences between sample and population, once the weights have been constructed and made available to us, we first renormalize the vector of weights to have mean 1 (in R, we set `w <- w/mean(w)`), and then we can include them as an argument in the regression (for example, `stan_glm(y ~ x, data=data, weights=w)`).

### Using a matrix of weights to account for correlated errors

Formula (10.6) can also be applied using a weight matrix $W$, with this yielding the maximum likelihood estimate for the model with normally distributed errors with covariance matrix $W^{-1}$. This is sometimes called generalized least squares. Models with correlations appear in the analysis of time series, spatial statistics, cluster samples, and other settings with structured data.

## 10.9 Fitting the same model to many datasets

It is common to fit a regression model repeatedly, either for different datasets or to subsets of an existing dataset. For example, one could estimate the relation between height and earnings using surveys from several years, or from several countries, or within different regions or states within the United States.

Beyond the scope of this book is *multilevel modeling*, a way to estimate a regression repeatedly, partially pooling information from the different fits. Here we consider the more informal procedure of estimating the regression separately—with no pooling between years or groups—and then displaying all these estimates together, which can be considered as an informal precursor to multilevel modeling. This is a form of replication that can be easily applied in many settings.

The method of repeated modeling, followed by time-series plots of estimates, is sometimes called the "secret weapon" because it is so easy and powerful but yet is rarely used as a data-analytic tool. We suspect that one reason for its rarity of use is that, once one acknowledges the time-series structure

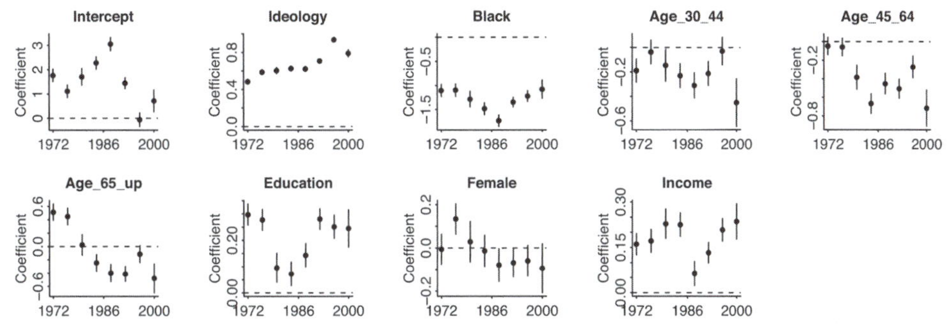


Figure 10.9 *Estimated coefficients* $\pm 0.67$ *standard errors (thus, 50% intervals) for the regression of party identification on political ideology, ethnicity, and other predictors, as fit separately to poll data from each presidential election campaign from 1976 through 2000. The plots are on different scales, with the predictors ordered roughly in declining order of the magnitudes of their coefficients. The set of plots illustrates the display of inferences from a series of regressions.*

of a dataset, it is natural to want to take the next step and model that directly. In practice, however, there is a broad range of problems for which a cross-sectional analysis is informative, and for which a time-series display is appropriate to give a sense of trends.

### Predicting party identification

xample: 'redicting arty identi- cation

We illustrate the secret weapon with a series of cross-sectional regressions modeling party identification given political ideology and demographic variables.[4]

Political scientists have long been interested in party identification and its changes over time. Here we study these trends using the National Election Study, which asks about party identification on a 1–7 scale (1 = strong Democrat, 2 = Democrat, 3 = weak Democrat, 4 = independent, . . . , 7 = strong Republican), which we treat as a continuous variable. We include the following predictors: political ideology (1 = strong liberal, 2 = liberal, . . . , 7 = strong conservative), ethnicity (0 = white, 1 = black, 0.5 = other), age (as categories: 18–29, 30–44, 45–64, and 65+ years, with the lowest age category as a baseline), education (1 = no high school, 2 = high school graduate, 3 = some college, 4 = college graduate), sex (0 = male, 1 = female), and income (1 = 0–16$^{\text{th}}$ percentile, 2 = 17–33$^{rd}$ percentile, 3 = 34–67$^{\text{th}}$ percentile, 4 = 68–95$^{\text{th}}$ percentile, 5 = 96–100$^{\text{th}}$ percentile).

Figure 10.9 shows the estimated coefficients tracked over time. Ideology and ethnicity are the most important, and their coefficients have been changing over time. Ideology is on a seven-point scale, so that its coefficients must be multiplied by 4 to get the expected change when comparing a liberal (ideology = 2) to a conservative (ideology = 6). The predictive differences for age and sex change fairly dramatically during the 30-year period. More generally, Figure 10.9 demonstrates the power of displaying multiple model fits next to each other in a graph, thus revealing average patterns (in comparison to the zero lines shown on each plot) and trends.

## 10.10 Bibliographic note

Linear regression has been used for centuries in applications in the social and physical sciences; see Stigler (1986). There are many strong textbooks on regression, including Neter et al. (1996); Ramsey and Schafer (2001), which focuses on issues such as model understanding, graphical display, and experimental design; and Woolridge (2001), which presents regression modeling from an econometric

[4]Data and code for this example are in the folder NES.

perspective. Fox (2002) teaches R in the context of applied regression. Carlin and Forbes (2004) provide an excellent conceptual introduction to the concepts of linear modeling and regression.

In the mothers' and children's test scores example, we used data from the Armed Forces Qualification Test, which we have normed in the same way IQ scores are normed, to have a mean of 100 and standard deviation of 15. For more on children's test scores and maternal employment, see Hill et al. (2005). The height and foot length example comes from McElreath (2020).

## 10.11 Exercises

10.1 *Regression with interactions*: Simulate 100 data points from the model, $y = b_0 + b_1 x + b_2 z + b_3 xz + \text{error}$, with a continuous predictor $x$ and a binary predictor $z$, coefficients $b = c(1, 2, -1, -2)$, and errors drawn independently from a normal distribution with mean 0 and standard deviation 3, as follows. For each data point $i$, first draw $z_i$, equally likely to take on the values 0 and 1. Then draw $x_i$ from a normal distribution with mean $z_i$ and standard deviation 1. Then draw the error from its normal distribution and compute $y_i$.

(a) Display your simulated data as a graph of $y$ vs. $x$, using dots and circles for the points with $z = 0$ and 1, respectively.

(b) Fit a regression predicting $y$ from $x$ and $z$ with no interaction. Make a graph with the data and two parallel lines showing the fitted model.

(c) Fit a regression predicting $y$ from $x$, $z$, and their interaction. Make a graph with the data and two lines showing the fitted model.

10.2 *Regression with interactions*: Here is the output from a fitted linear regression of outcome $y$ on pre-treatment predictor $x$, treatment indicator $z$, and their interaction:

```
            Median MAD_SD
(Intercept) 1.2    0.2
x           1.6    0.4
z           2.7    0.3
x:z         0.7    0.5

Auxiliary parameter(s):
      Median MAD_SD
sigma 0.5    0.0
```

(a) Write the equation of the estimated regression line of $y$ on $x$ for the treatment group and the control group, and the equation of the estimated regression line of $y$ on $x$ for the control group.

(b) Graph with pen on paper the two regression lines, assuming the values of $x$ fall in the range (0, 10). On this graph also include a scatterplot of data (using open circles for treated units and dots for controls) that are consistent with the fitted model.

10.3 *Checking statistical significance*: In this exercise and the next, you will simulate two variables that are statistically independent of each other to see what happens when we run a regression to predict one from the other. Generate 1000 data points from a normal distribution with mean 0 and standard deviation 1 by typing `var1 <- rnorm(1000,0,1)` in R. Generate another variable in the same way (call it `var2`). Run a regression of one variable on the other. Is the slope coefficient "statistically significant"? We do not recommend summarizing regressions in this way, but it can be useful to understand how this works, given that others will do so.

10.4 *Simulation study of statistical significance*: Continuing the previous exercise, run a simulation repeating this process 100 times. This can be done using a loop. From each simulation, save the $z$-score (the estimated coefficient of `var1` divided by its standard error). If the absolute value of the $z$-score exceeds 2, the estimate is "statistically significant."

To perform this computation, we start by creating an empty vector of $z$-scores filled with missing values (NAs). Another approach is to start with `z_scores <- numeric(length=100)`, which would set up a vector of zeroes. In general, however, we prefer to initialize with NAs, because then when there is a bug in the code, it sometimes shows up as NAs in the final results, alerting us to the problem. Here is code to perform the simulation:

```
z_scores <- rep(NA, 100)
for (k in 1:100) {
  var1 <- rnorm(1000, 0, 1)
  var2 <- rnorm(1000, 0, 1)
  fake <- data.frame(var1, var2)
  fit <- stan_glm(var2 ~ var1, data=fake)
  z_scores[k] <- coef(fit)[2] / se(fit)[2]
}
```

How many of these 100 $z$-scores exceed 2 in absolute value, thus achieving the conventional level of statistical significance?

10.5 *Regression modeling and prediction*: The folder `KidIQ` contains a subset of the children and mother data discussed earlier in the chapter. You have access to children's test scores at age 3, mother's education, and the mother's age at the time she gave birth for a sample of 400 children.

(a) Fit a regression of child test scores on mother's age, display the data and fitted model, check assumptions, and interpret the slope coefficient. Based on this analysis, when do you recommend mothers should give birth? What are you assuming in making this recommendation?

(b) Repeat this for a regression that further includes mother's education, interpreting both slope coefficients in this model. Have your conclusions about the timing of birth changed?

(c) Now create an indicator variable reflecting whether the mother has completed high school or not. Consider interactions between high school completion and mother's age. Also create a plot that shows the separate regression lines for each high school completion status group.

(d) Finally, fit a regression of child test scores on mother's age and education level for the first 200 children and use this model to predict test scores for the next 200. Graphically display comparisons of the predicted and actual scores for the final 200 children.

ample:
auty and
aching
aluations

10.6 *Regression models with interactions*: The folder `Beauty` contains data (use file `beauty.csv`) from Hamermesh and Parker (2005) on student evaluations of instructors' beauty and teaching quality for several courses at the University of Texas. The teaching evaluations were conducted at the end of the semester, and the beauty judgments were made later, by six students who had not attended the classes and were not aware of the course evaluations.

(a) Run a regression using beauty (the variable `beauty`) to predict course evaluations (`eval`), adjusting for various other predictors. Graph the data and fitted model, and explain the meaning of each of the coefficients along with the residual standard deviation. Plot the residuals versus fitted values.

(b) Fit some other models, including beauty and also other predictors. Consider at least one model with interactions. For each model, explain the meaning of each of its estimated coefficients.

See also Felton, Mitchell, and Stinson (2003) for more on this topic.

10.7 *Predictive simulation for linear regression*: Take one of the models from the previous exercise.

(a) Instructor A is a 50-year-old woman who is a native English speaker and has a beauty score of $-1$. Instructor B is a 60-year-old man who is a native English speaker and has a beauty score of $-0.5$. Simulate 1000 random draws of the course evaluation rating of these two instructors. In your simulation, use `posterior_predict` to account for the uncertainty in the regression parameters as well as predictive uncertainty.

(b) Make a histogram of the difference between the course evaluations for A and B. What is the probability that A will have a higher evaluation?

10.8 *How many simulation draws*: Take the model from Exercise 10.6 that predicts course evaluations from beauty and other predictors.

(a) Display and discuss the fitted model. Focus on the estimate and standard error for the coefficient of beauty.

(b) Compute the median and mad sd of the posterior simulations of the coefficient of beauty, and check that these are the same as the output from printing the fit.

(c) Fit again, this time setting `iter` = 1000 in your `stan_glm` call. Do this a few times in order to get a sense of the simulation variability.

(d) Repeat the previous step, setting `iter` = 100 and then `iter` = 10.

(e) How many simulations were needed to give a good approximation to the mean and standard error for the coefficient of beauty?

10.9 *Collinearity*: Consider the elections and economy example from Chapter 7.

(a) Create a variable that is collinear to the economic growth predictor in the model. Graph the two predictors to confirm that they are collinear.

(b) Add this new predictor to the model, in addition to economic growth. Fit the regression and report what happens.

(c) Create a variable that is *nearly* collinear to the economic growth predictor, with a 0.9 correlation between the two variables. Fit the regression predicting election outcome given economic growth and this new predictor you have created, and again describe and explain what happens.

10.10 *Regression with few data points and many predictors*: Re-fit the elections and economy example from Chapter 7, adding several economic predictors such as unemployment rate in the year of the election, inflation rate in the year of the election, etc. Discuss the difficulties in interpretation of the fitted model.

10.11 *Working through your own example*: Continuing the example from the final exercises of the earlier chapters, fit a linear regression with multiple predictors and interpret the estimated parameters and their uncertainties. Your regression should include at least one interaction term.

# Chapter 11

# Assumptions, diagnostics, and model evaluation

In this chapter we turn to the assumptions of the regression model, along with diagnostics that can be used to assess whether some of these assumptions are reasonable. Some of the most important assumptions rely on the researcher's knowledge of the subject area and may not be directly testable from the available data alone. Hence it is good to understand the ideas underlying the model, while recognizing that there is no substitute for engagement with data and the purposes for which they are being used. We show different sorts of plots of data, fitted models, and residuals, developing these methods in the context of real and simulated-data examples. We consider diagnostics based on predictive simulation from the fitted model, along with numerical summaries of fit, including residual error, explained variance, external validation, and cross validation. The goal is to develop a set of tools that you can use in constructing, interpreting, and evaluating regression models with multiple predictors.

## 11.1 Assumptions of regression analysis

We list the assumptions of the regression model in decreasing order of importance.

1. *Validity.* Most important is that the data you are analyzing should map to the research question you are trying to answer. This sounds obvious but is often overlooked or ignored because it can be inconvenient. Optimally, this means that the outcome measure should accurately reflect the phenomenon of interest, the model should include all relevant predictors, and the model should generalize to the cases to which it will be applied.
   For example, with regard to the outcome variable, a model of incomes will not necessarily tell you about patterns of total assets. A model of test scores will not necessarily tell you about child intelligence or cognitive development.
   Choosing inputs to a regression is often the most challenging step in the analysis. We are generally encouraged to include all relevant predictors, but in practice it can be difficult to determine which are necessary and how to interpret coefficients with large standard errors. Chapter 19 discusses the choice of inputs for regressions used in causal inference.
   A sample that is representative of all mothers and children may not be the most appropriate for making inferences about mothers and children who participate in the Temporary Assistance for Needy Families program. However, a carefully selected subsample may reflect the distribution of this population well. Similarly, results regarding diet and exercise obtained from a study performed on patients at risk for heart disease may not be generally applicable to generally healthy individuals. In this case assumptions would have to be made about how results for the at-risk population might relate to those for the healthy population.
   Data used in empirical research rarely meet all (if any) of these criteria precisely. However, keeping these goals in mind can help you be precise about the types of questions you can and cannot answer reliably.

2. *Representativeness.* A regression model is fit to data and is used to make inferences about a larger population, hence the implicit assumption in interpreting regression coefficients is that the sample is representative of the population.
   To be more precise, the key assumption is that the data are representative of the distribution of the outcome $y$ given the predictors $x_1, x_2, \ldots$, that are included in the model. For example, in a regression of earnings on height and sex, it would be acceptable for women and tall people to be overrepresented in the sample, compared to the general population, but problems would arise if the sample includes too many rich people. Selection on $x$ does not interfere with inferences from the regression model, but selection on $y$ does. This is one motivation to include more predictors in our regressions, to allow the assumption of representativeness, conditional on $X$, to be more reasonable.
   Representativeness is a concern even with data that would not conventionally be considered as a sample. For example, the forecasting model in Section 7.1 contains data from 16 consecutive elections, and these are not a sample from anything, but one purpose of the model is to predict future elections. Using the regression fit to past data to predict the next election is mathematically equivalent to considering the observed data and the new outcome as a random sample from a hypothetical superpopulation. Or, to be more precise, it is like treating the errors as a random sample from the normal error distribution. For another example, if a regression model is fit to data from the 50 states, we are not interested in making predictions for a hypothetical $51^{\text{st}}$ state, but you may well be interested in the hypothetical outcome in the 50 states in a future year. As long as some generalization is involved, ideas of statistical sampling arise, and we need to think about representativeness of the sample to the implicit or explicit population about which inferences will be drawn. This is related to the idea of generative modeling, as discussed in Section 4.1.
3. *Additivity and linearity.* The most important mathematical assumption of the linear regression model is that its deterministic component is a linear function of the separate predictors: $y = \beta_0 + \beta_1 x_1 + \beta_2 x_2 + \cdots$.
   If additivity is violated, it might make sense to transform the data (for example, if $y = abc$, then $\log y = \log a + \log b + \log c$) or to add interactions. If linearity is violated, perhaps a predictor should be put in as $1/x$ or $\log(x)$ instead of simply linearly. Or a more complicated relationship could be expressed using a nonlinear function such as a spline or Gaussian process, which we discuss briefly in Chapter 22.
   For example, in medical and public health examples, a nonlinear function can allow a health measure to decline with higher ages, with the rate of decline becoming steeper as age increases. In political examples, including non-linear function allows the possibility of increasing slopes with age and also U-shaped patterns if, for example, the young and old favor taxes more than the middle-aged.
   Just about every problem we ever study will be nonlinear, but linear regression can still be useful in estimating an average relationship, for example as discussed in Section 8.3. As we get more data, and for problems where nonlinearity is of interest, it will make more sense to put in the effort to model nonlinearity. And for some problems a linear model would not make sense at all; for example, see Exercise 1.9.
4. *Independence of errors.* The simple regression model assumes that the errors from the prediction line are independent, an assumption that is violated in time series, spatial, and multilevel settings.
5. *Equal variance of errors.* Unequal error variance (also called heteroscedasticity, in contrast to equal variances, or homoscedasticity) can be an issue when a regression is used for probabilistic prediction, but it does not affect what is typically the most important aspect of a regression model, which is the information that goes into the predictors and how they are combined. If the variance of the regression errors are unequal, estimation is more efficiently performed by accounting for this in the model, as with weighted least squares discussed in Section 10.8. In most cases, however, this issue is minor.
6. *Normality of errors.* The distribution of the error term is relevant when predicting individual data

points. For the purpose of estimating the regression line (as compared to predicting individual data points), the assumption of normality is typically barely important at all. Thus we do *not* recommend diagnostics of the normality of regression residuals. For example, many textbooks recommend quantile-quantile (Q-Q) plots, in which the ordered residuals are plotted vs. the corresponding expected values of ordered draws from a normal distribution, with departures of this plot from linearity indicating nonnormality of the error term. There is nothing wrong with making such a plot, and it can be relevant when evaluating the use of the model for predicting individual data points, but we are typically more concerned with the assumptions of validity, representativeness, additivity, linearity, and so on, listed above.

If the distribution of errors is of interest, perhaps because of predictive goals, this should be distinguished from the distribution of the data, $y$. For example, consider a regression on a single discrete predictor, $x$, which takes on the values 0, 1, and 2, with one-third of the population in each category. Suppose the true regression line is $y = 0.2 + 0.5x$ with normally-distributed errors with standard deviation 0.1. Then a graph of the data $y$ will show three fairly sharp modes centered at 0.2, 0.7, and 1.2. Other examples of such mixture distributions arise in economics, when including both employed and unemployed people, or the study of elections, when comparing districts with incumbent legislators of different parties.

The regression model does *not* assume or require that predictors be normally distributed. In addition, the normal distribution on the outcome refers to the regression errors, not to the raw data. Depending on the structure of the predictors, it is possible for data $y$ to be far from normally distributed even when coming from a linear regression model.

### Failures of the assumptions

What do we do when the assumptions break down?

Most directly, one can extend the model, for example, with measurement error models to address problems with validity, selection models to handle nonrepresentative data, nonadditive and nonlinear models, correlated errors or latent variables to capture violations of the independence assumption, and models for varying variances and nonnormal errors.

Other times, it is simpler to change the data or model so the assumptions are more reasonable: these steps could include obtaining cleaner data, adding predictors to make representativeness assumptions more reasonable, adding interactions to capture nonlinearity, and transforming predictors and outcomes so that an additive model can make more sense.

Alternatively, one can change or restrict the questions to align them closer to the data, making conclusions that are more descriptive and less causal or extrapolative, defining the population to match the sample, and predicting averages rather than individual cases. In practice, we typically meet in the middle, applying some mix of model expansion, data processing, and care in extrapolation beyond the data.

### Causal inference

Further assumptions are necessary if a regression coefficient is to be given a causal interpretation, as we discuss in Part 4 of this book. From a regression context, causal inference can be considered as a form of prediction in which we are interested in what would happen if various predictors were set to particular values. This relates to the common, but often mistaken, interpretation of a regression coefficient as "the effect of a variable with all else held constant." To see the fundamental error in automatically giving a causal interpretation to a regression coefficient, consider the model predicting earnings from height, and imagine interpreting the slope as the effect of an additional inch of height. The problem here is the regression is fit to data from different people, but the causal question addresses what would happen to a single person. Strictly speaking, the slope of this regression represents the average difference in earnings, comparing two people who differ by an inch in height. (This interpretation is not quite correct either, as it relies on the assumption of linearity, but set that aside for now.) Tall people make more money on average than short people, but that does not necessarily imply that making individual people taller would increase their earnings, even on average. Indeed, even to post this question makes it clear that it is not well defined, as we would need to consider how this (hypothetical) increase in individual heights was to be done, before considering its average effect on earnings. Similar issues arise in other fields: for example, if blood pressure correlates with some negative health outcome, this does not necessarily imply that a decrease in blood pressure will have beneficial effects. It can all depend on *how* blood pressure would be reduced—in causal terms, what is the treatment?—and, even when that is clear, assumptions need to be made for a model fit to existing data to directly apply to predictions of what would happen under an intervention.

## 11.2 Plotting the data and fitted model

Graphics are helpful for visualizing data, understanding models, and revealing patterns in the data not explained by fitted models. We first discuss general principles of data display in the context of linear regression and then in Section 11.3 consider plots specifically designed to reveal model misfit.

### Displaying a regression line as a function of one input variable

Example: Children's IQ tests

We return to the children's test score example to demonstrate some of the details of displaying fitted regressions in R.[1] We displayed some aspects of the data in Figures 10.1–10.3. We can make a plot such as Figure 10.2 as follows:

```
fit_2 <- stan_glm(kid_score ~ mom_iq, data=kidiq)
plot(kidiq$mom_iq, kidiq$kid_score, xlab="Mother IQ score", ylab="Child test score")
abline(coef(fit_2)[1], coef(fit_2)[2])
```

The function `plot` creates the scatterplot of observations, and `abline` superimposes the line $y = \hat{a} + \hat{b}x$ using the estimated coefficients from the fitted regression.

### Displaying two fitted regression lines

**Model with no interaction.** For the model with two predictors, we can create a graph with two sets of points and two regression lines, as in Figure 10.3:

```
fit_3 <- stan_glm(kid_score ~ mom_hs + mom_iq, data=kidiq)
colors <- ifelse(kidiq$mom_hs==1, "black", "gray")
plot(kidiq$mom_iq, kidiq$kid_score,
  xlab="Mother IQ score", ylab="Child test score", col=colors, pch=20)
b_hat <- coef(fit_3)
abline(b_hat[1] + b_hat[2], b_hat[3], col="black")
abline(b_hat[1], b_hat[3], col="gray")
```

Setting `pch=20` tells the `plot` function to display the data using small dots, and the `col` option sets the colors of the points, which we have assigned to black or gray according to the value of `mom_hs`. Finally, the calls to `abline` superimpose the fitted regression lines for the two groups defined by maternal high school completion.

If `mom_hs` were a continuous predictor here, we could display the fitted model by plotting `kid_score` vs. `mom_iq` for two different values of `mom_hs` that are in the range of the data.

**Model with interaction.** We can set up the same sort of plot for the model with interactions, with the only difference being that the two lines have different slopes:

[1] Data and code for this example are in the folder `KidIQ`.

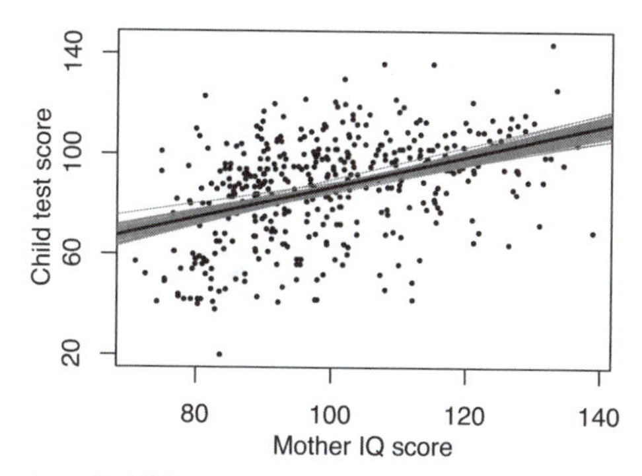


Figure 11.1 *Data and regression of child's test score on maternal IQ, with the solid line showing the fitted regression model and light lines indicating uncertainty in the fitted regression.*

```
fit_4 <- stan_glm(kid_score ~ mom_hs + mom_iq + mom_hs:mom_iq, data=kidiq)
colors <- ifelse(kidiq$mom_hs==1, "black", "gray")
plot(kidiq$mom_iq, kidiq$kid_score,
  xlab="Mother IQ score", ylab="Child test score", col=colors, pch=20)
b_hat <- coef(fit_4)
abline(b_hat[1] + b_hat[2], b_hat[3] + b_hat[4], col="black")
abline(b_hat[1], b_hat[3], col="gray")
```

The result is shown in Figure 10.4.

### Displaying uncertainty in the fitted regression

In Section 9.1 we discussed how posterior simulations represent our uncertainty in the estimated regression coefficients. Here we briefly describe how to use these simulations to display this inferential uncertainty graphically. Consider this simple model:

```
fit_2 <- stan_glm(kid_score ~ mom_iq, data=kidiq)
```

The following code creates Figure 11.1, which shows the fitted regression line along with 10 simulations representing uncertainty about the line:

```
sims_2 <- as.matrix(fit_2)
n_sims_2 <- nrow(sims_2)
beta_hat_2 <- apply(sims_2, 2, median)
plot(kidiq$mom_iq, kidiq$kid_score, xlab="Mother IQ score", ylab="Child test score")
sims_display <- sample(n_sims_2, 10)
for (i in sims_display){
  abline(sims_2[i,1], sims_2[i,2], col="gray")
}
abline(coef(fit_2)[1], coef(fit_2)[2], col="black")
```

We use the loop to display 10 different simulation draws.

### Displaying using one plot for each input variable

Now consider the regression including the indicator for maternal high school completion:

```
fit_3 <- stan_glm(kid_score ~ mom_hs + mom_iq, data=kidiq)
sims_3 <- as.matrix(fit_3)
n_sims_3 <- nrow(sims_3)
```

We display this model in Figure 11.2 as two plots, one for each of the two input variables with the other held at its average value:

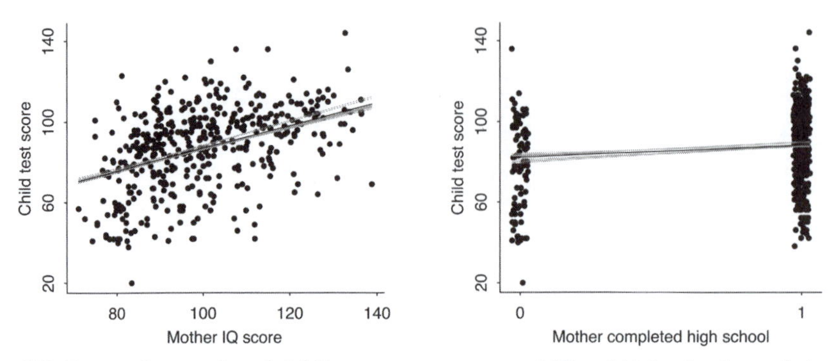


Figure 11.2 *Data and regression of child's test score on maternal IQ and high school completion, shown as a function of each of the two input variables with the other held at its average value. Light lines indicate uncertainty in the regressions. Values for mother's high school completion have been jittered to make the points more distinct.*

```
par(mfrow=c(1,2))
plot(kidiq$mom_iq, kidiq$kid_score, xlab="Mother IQ score", ylab="Child test score")
mom_hs_bar <- mean(kidiq$mom_hs)
sims_display <- sample(n_sims_3, 10)
for (i in sims_display){
  curve(cbind(1, mom_hs_bar, x) %*% sims_3[i,1:3], lwd=0.5, col="gray", add=TRUE)
}
curve(cbind(1, mom_hs_bar, x) %*% coef(fit_3), col="black", add=TRUE)

plot(kidiq$mom_hs, kidiq$kid_score, xlab="Mother completed high school",
  ylab="Child test score")
mom_iq_bar <- mean(kidiq$mom_iq)
for (i in sims_display){
  curve(cbind(1, x, mom_iq_bar) %*% sims_3[i,1:3], lwd=0.5, col="gray", add=TRUE)
}
curve(cbind(1, x, mom_iq_bar) %*% coef(fit_3), col="black", add=TRUE)
```

This example demonstrates how we can display a fitted model using multiple plots, one for each predictor.

## Plotting the outcome vs. a continuous predictor

Consider the following common situation: a continuous outcome $y$ is modeled given a treatment indicator $z$ and a continuous pre-treatment predictor $x$:

$$y = a + bx + \theta z + \text{error}.$$

Example: Simulated regression plots

We simulate some fake data:[2]

```
N <- 100
x <- runif(N, 0, 1)
z <- sample(c(0, 1), N, replace=TRUE)
a <- 1
b <- 2
theta <- 5
sigma <- 2
y <- a + b*x + theta*z +  rnorm(N, 0, sigma)
fake <- data.frame(x=x, y=y, z=z)
```

[2]Code for this example is in the folder Residuals.

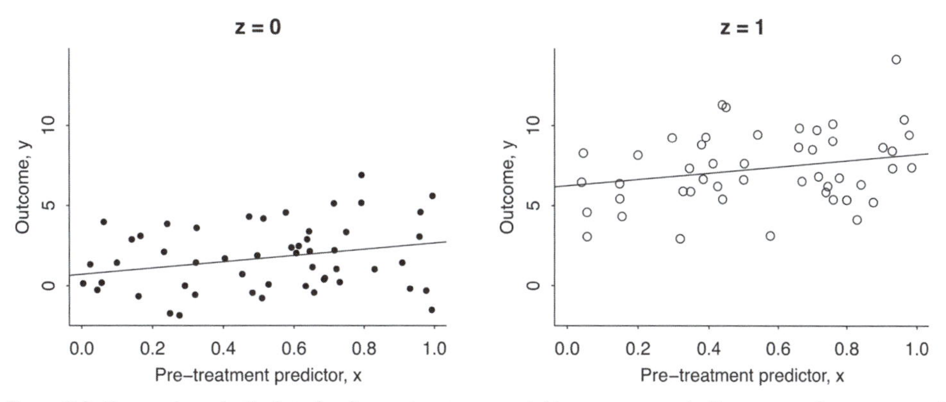


Figure 11.3 *From a hypothetical study of a pre-treatment variable x, treatment indicator z, and outcome y: data and fitted regression line plotted separately for control and treatment groups.*

We then fit a regression and plot the data and fitted model:

```
fit <- stan_glm(y ~ x + z, data=fake)
par(mfrow=c(1,2))
for (i in 0:1){
  plot(range(x), range(y), type="n", main=paste("z =", i))
  points(x[z==i], y[z==i], pch=20+i)
  abline(coef(fit)["(Intercept)"] + coef(fit)["z"]*i, coef(fit)["x"])
}
```

Figure 11.3 shows the result. The key here is to recognize that $x$ is continuous and $z$ is discrete, hence it makes sense to make graphs showing $y$ vs. $x$ for different values of $z$.

## Forming a linear predictor from a multiple regression

Now extend the previous example so that the treatment indicator $z$ is accompanied by $k$ pre-treatment predictors $x_k$, $k = 1, \dots, K$:

$$y = b_0 + b_1 x_1 + \cdots + b_K x_K + \theta z + \text{error}.$$

How should we graph these data?

One option is to graph $y$ vs. $x_k$, for each $k$, holding each of the other pre-treatment predictors at its average, again making two graphs for $z = 0$ and $z = 1$ as in Figure 11.2. For each $k$, then, the data $(x_{ki}, y_i)$ would be plotted along with the line, $y = \hat{\mu} + \hat{b}_k x_k$ (corresponding to $z = 0$) or $y = (\hat{\mu} + \hat{\theta}) + \hat{b}_k x_k$ (for $z = 1$), where $\hat{\mu} = (\hat{b}_0 + \hat{b}_1 \bar{x}_1 + \dots \hat{b}_K \bar{x}_K) - \hat{b}_k \bar{x}_k$, and in total this would yield a $K \times 2$ grid of plots, with each row corresponding to a different predictor $x_k$.

Another approach is to plot the outcome, $y$, against the *linear predictor*, $\hat{y} = \hat{b}_0 + \hat{b}_1 x_1 + \cdots + \hat{b}_K x_K + \hat{\theta} z$, separately for the control ($z = 0$) and treatment ($z = 1$) groups. In either case, the linear predictor can be viewed as a unidimensional summary of the pre-treatment information.

Here is an example. First we simulate some fake data from a regression with $K = 10$ predictors and a constant term:

```
N <- 100
K <- 10
X <- array(runif(N*K, 0, 1), c(N, K))
z <- sample(c(0, 1), N, replace=TRUE)
a <- 1
b <- 1:K
theta <- 5
```

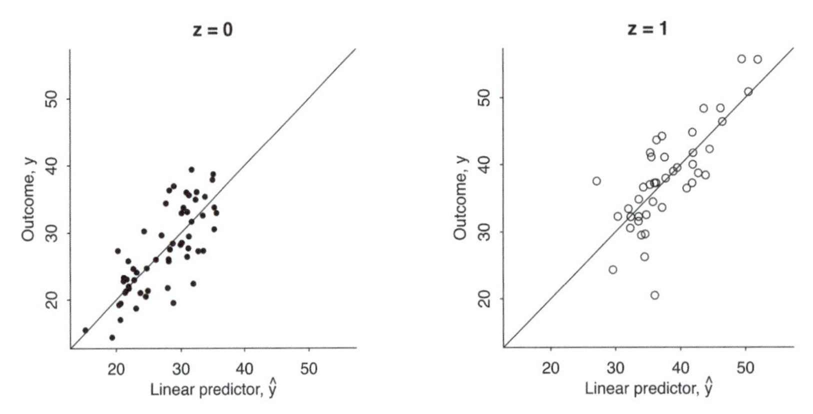


Figure 11.4 *From a hypothetical study of 10 pre-treatment variables X, treatment indicator z, and outcome $y$: Outcome plotted vs. fitted linear predictor $\hat{y}$ (so that the fitted regression line is by definition $y = \hat{y}$), plotted separately for control and treatment groups.*

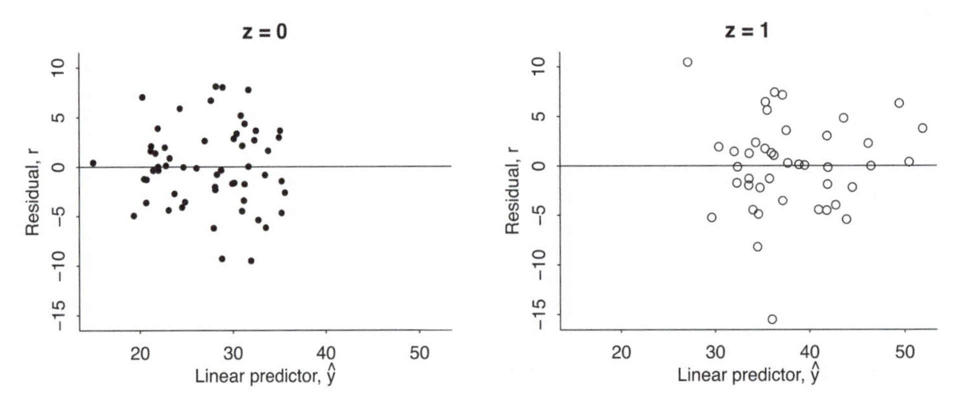


Figure 11.5 *Continuing Figure 11.4, the residual, $r = y - \hat{y}$, vs. the fitted linear predictor $\hat{y}$, plotted separately for control and treatment groups.*

```
sigma <- 2
y <- a + X %*% b + theta*z +  rnorm(N, 0, sigma)
fake <- data.frame(X=X, y=y, z=z)
fit <- stan_glm(y ~ X + z, data=fake)
```

Now we compute the linear predictor based on the point estimate of the model:

```
y_hat <- predict(fit)
```

And now we are ready to plot the data vs. the linear predictor for each value of $z$:

```
par(mfrow=c(1,2))
for (i in 0:1){
  plot(range(y_hat,y), range(y_hat,y), type="n", main=paste("z =", i))
  points(y_hat[z==i], y[z==i], pch=20+i)
  abline(0, 1)
}
```

Figure 11.4 shows the results. Because $u$ is defined as the linear predictor, $E(y)$ (the expected or predicted average value of the data given the estimated coefficients and all the predictors in the model), by construction the line tracing the expectation of $y$ given $u$ is the 45° line on each graph.

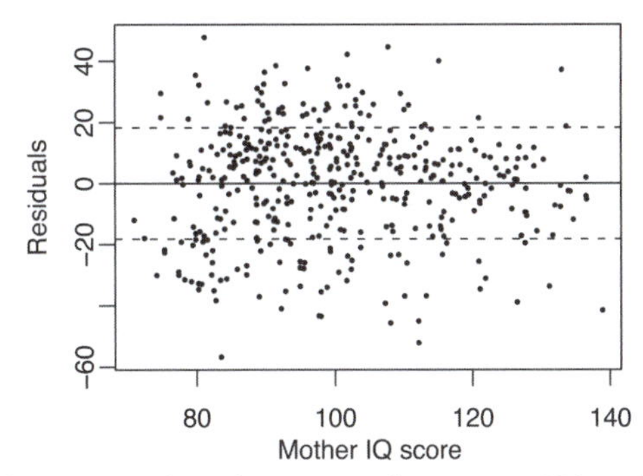


Figure 11.6 *Residual plot for child test score data when regressed on maternal IQ, with dotted lines showing ±1 standard-deviation bounds. The residuals show no striking patterns.*

## 11.3 Residual plots

Once we have plotted the data and fitted regression lines, we can evaluate this fit by looking at the difference between data and their expectations: the *residuals*,

$$r_i = y_i - X_i\hat{\beta}.$$

In the example concluding the previous section, this is $r_i = y_i - (\hat{b}_0 + \hat{b}_1 x_{i1} + \cdots + \hat{b}_K x_{iK} + \hat{\theta} z_i)$.

One advantage of plotting residuals is that, if the model is correct, they should look roughly randomly scattered in comparison to a horizontal line, which can be visually more clear than comparing to a fitted line. Figure 11.5 illustrates.

Figure 11.6 shows a residual plot for the test scores example where child's test score is regressed simply on mother's IQ.[3] The plot looks fine in that there do not appear to be any strong patterns. In other settings, residual plots can reveal systematic problems with model fit, as is illustrated, for example, in Chapter 15. Plots of residuals and binned residuals can be seen as visual comparisons to the hypothesis that the errors from a model are independent with zero mean.

### Using fake-data simulation to understand residual plots

Simulation of fake data can be used to validate statistical algorithms and to check the properties of estimation procedures. We illustrated in Section 7.2 with a simple regression, where we simulated fake data from the model, $y = X\beta + \epsilon$, re-fit the model to the simulated data, and checked the coverage of the 68% and 95% intervals for the coefficients $\beta$.

For another illustration of the power of fake data, we simulate from a regression model to get insight into residual plots, in particular, to understand why we plot residuals versus fitted values rather than versus observed values.

### A confusing choice: plot residuals vs. predicted values, or residuals vs. observed values?

Example: Midterm and final exams

We illustrate with a simple model fit to real data, predicting final exam scores from midterms in an introductory statistics class:[4]

```
fit_1 <- stan_glm(final ~ midterm, data=introclass)
print(fit_1)
```

yielding,

[3]Data and code for this example are in the folder KidIQ.
[4]Data and code for this example are in the folder Introclass.

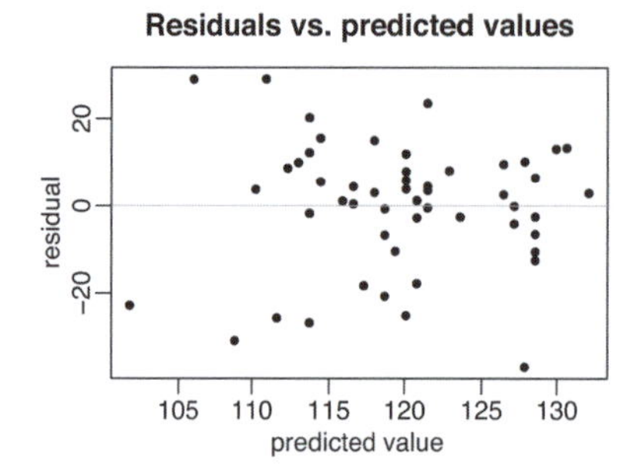


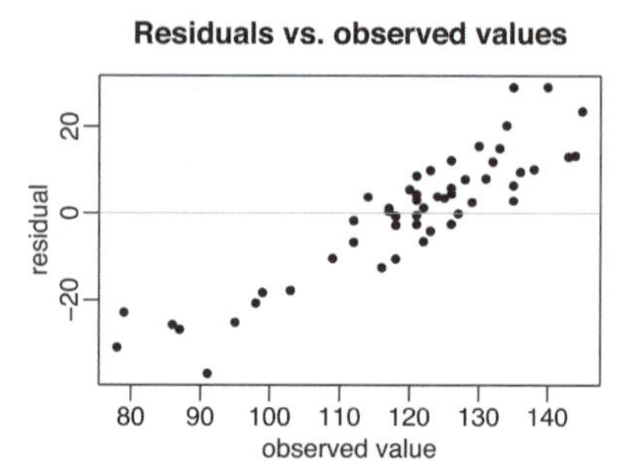


Figure 11.7 *From a model predicting final exam grades from midterms: plots of regression residuals versus predicted and versus observed values. The left plot looks reasonable but the right plot shows strong patterns. How to understand these? An exploration using fake data (see Figure 11.8) shows that, even if the model were correct, we would expect the right plot to show strong patterns. The plot of residuals versus observed thus does not indicate a problem with the model.*

```
            Median MAD_SD
(Intercept) 64.4   17.2
midterm      0.7    0.2

Auxiliary parameter(s):
      Median MAD_SD
sigma 14.9   1.5
```

We extract simulated coefficients and compute mean predictions $y_i^{\text{pred}}$ and residuals $y_i - y_i^{\text{pred}}$:

```
sims <- as.matrix(fit_1)
predicted <- predict(fit_1)
resid <- introclass$final - predicted
```

Figure 11.7 shows the residuals from this model, plotted in two different ways: (a) residuals versus fitted values, and (b) residuals versus observed values. Figure 11.7a looks reasonable: the residuals are centered around zero for all fitted values. But Figure 11.7b looks troubling.

It turns out that the first plot is what we should be looking at, and the second plot is misleading. This can be understood using probability theory (from the regression model, the errors $\epsilon_i$ should be independent of the predictors $x_i$, not the data $y_i$) but a perhaps more convincing demonstration uses fake data, as we now illustrate.

## Understanding the choice using fake-data simulation

For this example, we set the coefficients and residual standard deviation to reasonable values given the model estimates, and then we simulate fake final exam data using the real midterm as a predictor:

```
a <- 64.5
b <- 0.7
sigma <- 14.8
n <- nrow(introclass)
introclass$final_fake <- a + b*introclass$midterm + rnorm(n, 0, sigma)
```

Next we fit the regression model to the fake data and compute fitted values and residuals:

```
fit_fake <- stan_glm(final_fake ~ midterm, data=introclass)
sims <- as.matrix(fit_fake)
predicted_fake <- colMeans(sims[,1] + sims[,2] %*% t(introclass$midterm))
```

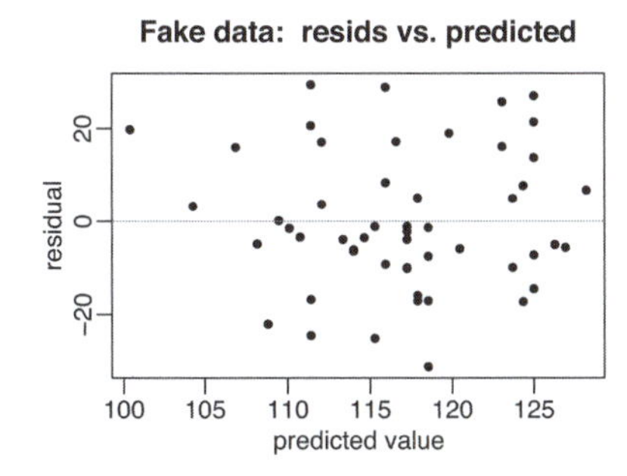


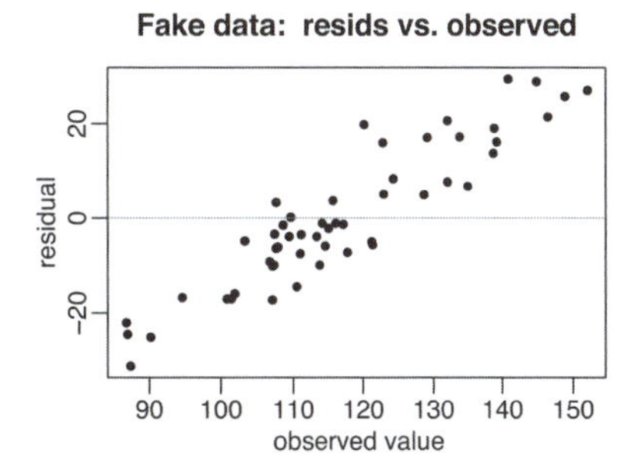


Figure 11.8 *From fake data: plots of regression residuals versus (a) predicted or (b) observed values. Data were simulated from the fitted family of regression models, and so we know that the pattern on the right does not represent any sort of model failure. This is an illustration of the use of fake data to evaluate diagnostic plots. Compare to the corresponding plots of real data in Figure 11.7.*

Figure 11.8 shows the plots of `resid_fake` versus `predicted_fake` and `final_fake`. These are the sorts of residual plots we would see *if the model were correct.* This simulation shows why we prefer to plot residuals versus predicted rather than observed values.

## 11.4 Comparing data to replications from a fitted model

So far we have considered several uses of simulation: exploring the implications of hypothesized probability models (Section 5.1); exploring the implications of statistical models that were fit to data (Section 10.6); and studying the properties of statistical procedures by comparing estimates to known true values of parameters (Sections 7.2 and 11.3). Here we introduce *posterior predictive checking*: simulating replicated datasets under the fitted model and then comparing these to the observed data.

### Example: simulation-based checking of a fitted normal distribution

The most fundamental way to check model fit is to display replicated datasets and compare them to the actual data. Here we illustrate with a simple case from a famous historical dataset that did not fit the normal distribution. The goal of this example is to demonstrate how the lack of fit can be seen using predictive replications.

Example: Speed of light

Figure 11.9 shows the data, a set of measurements taken by Simon Newcomb in 1882 as part of an experiment to estimate the speed of light.[5] We (inappropriately) fit a normal distribution to these data, which in the regression context can be done by fitting a linear regression with no predictors. Our implicit model for linear regression is normally-distributed errors, and, as discussed in Section 7.3, estimating the mean is equivalent to regressing on a constant term:

```
fit <- stan_glm(y ~ 1, data=newcomb)
```

The next step is to simulate replications from the parameters in the fitted model (in this case, simply the constant term $\beta_0$ and the residual standard deviation $\sigma$):

```
sims <- as.matrix(fit)
n_sims <- nrow(sims)
```

We can then use these simulations to create `n_sims` fake datasets of 66 observations each:

```
n <- length(newcomb$y)
y_rep <- array(NA, c(n_sims, n))
for (s in 1:n_sims) {
```

[5]Data and code for this example are in the folder `Newcomb`.

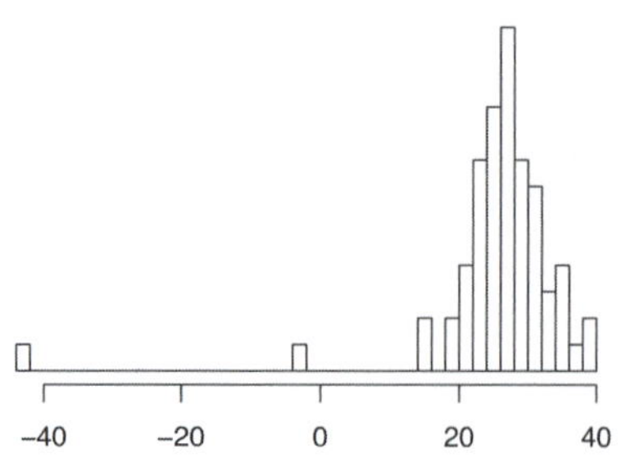


Figure 11.9 *Histogram of Simon Newcomb's measurements for estimating the speed of light, from Stigler (1977). The data represent the amount of time required for light to travel a distance of 7442 meters and are recorded as deviations from* 24 800 *nanoseconds.*

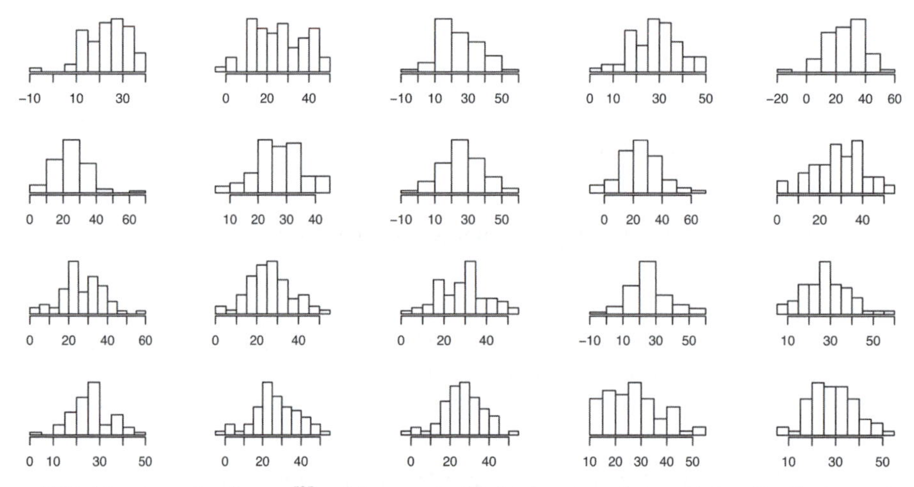

Figure 11.10 *Twenty replications, $y^{\text{rep}}$, of the speed-of-light data from the predictive distribution under the normal model; compare to observed data, $y$, in Figure 11.9. Each histogram displays the result of drawing 66 independent values $y_i^{\text{rep}}$ from a common normal distribution with mean and standard deviation $(\mu, \sigma)$ simulated from the fitted model.*

```
  y_rep[s,] <- rnorm(n, sims[s,1], sims[s,2])
}
```

More simply we can just use the built-in function for posterior predictive replications:

```
y_rep <- posterior_predict(fit)
```

**Visual comparison of actual and replicated datasets.** Figure 11.10 shows a plot of 20 randomly sampled datasets, produced as follows:

```
par(mfrow=c(5,4))
for (s in sample(n_sims, 20)) {
  hist(y_rep[s,])
}
```

The systematic differences between data and replications are clear. Figure 11.11 shows an alternative visualization using overlaid density estimates of the data and the replicates. In more complicated problems, more effort may be needed to effectively display the data and replications for useful comparisons, but the same general idea holds.

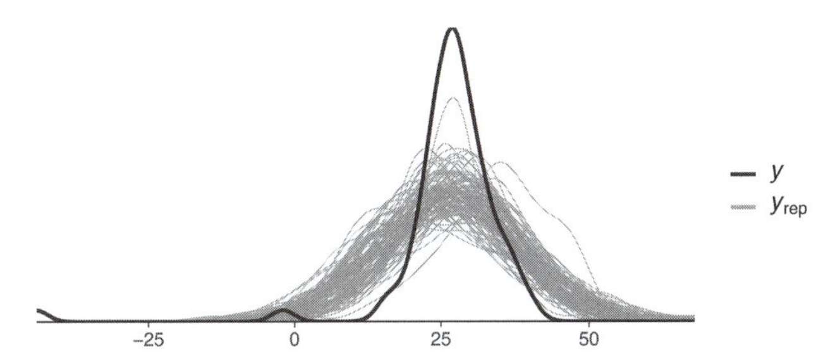


Figure 11.11 *Density estimates of the speed-of-light data $y$ and 100 replications, $y^{\text{rep}}$, from the predictive distribution under the normal model. Each density estimate displays the result of original data or drawing 66 independent values $y_i^{\text{rep}}$ from a common normal distribution with mean and standard deviation $(\mu, \sigma)$ simulated from the fitted model.*

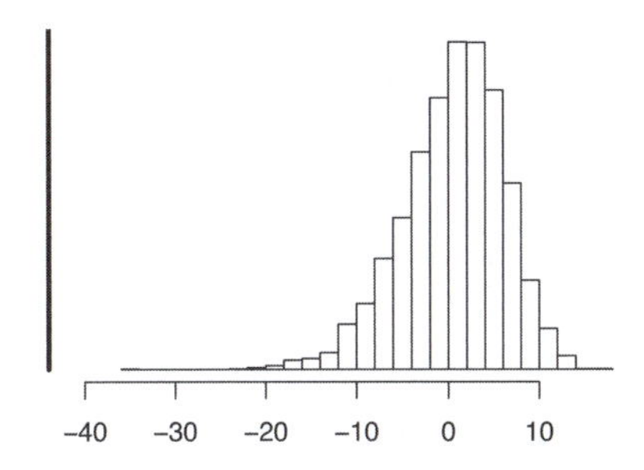


Figure 11.12 *Smallest observation of Newcomb's speed-of-light data (the vertical line at the left of the graph), compared to the smallest observations from each of 20 posterior predictive simulated datasets displayed in Figure 11.10.*

**Checking model fit using a numerical data summary.** Data displays can suggest more focused test statistics with which to check model fit. Here we demonstrate a simple example with the speed-of-light measurements. The graphical check in Figures 11.9 and 11.10 shows that the data have some extremely low values that do not appear in the replications. We can formalize this check by defining a *test statistic*, $T(y)$, equal to the minimum value of the data, and then calculating $T(y^{\text{rep}})$ for each of the replicated datasets:

```
test <- function(y) {
  min(y)
}
test_rep <- apply(y_rep, 1, test)
```

We then plot a histogram of the minima of the replicated datasets, with a vertical line indicating the minimum of the observed data:

```
hist(test_rep, xlim=range(test(y), test_rep))
lines(rep(test(y),2), c(0,n))
```

Figure 11.12 shows the result: the smallest observations in each of the hypothetical replications are all much larger than Newcomb's smallest observation, which is indicated by a vertical line on the graph. The normal model clearly does not capture the variation that Newcomb observed. A revised model might use an asymmetric contaminated normal distribution or a symmetric long-tailed distribution in place of the normal measurement model.

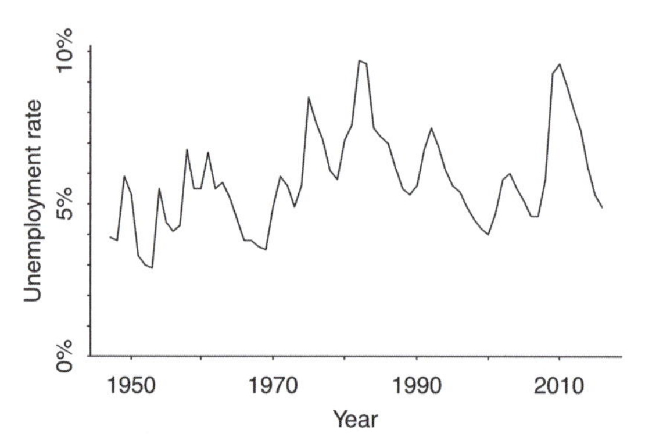


Figure 11.13 *Time series of U.S. annual unemployment rates from 1947 to 2016. We fit a first-order autoregression to these data and then simulate several datasets, shown in Figure 11.14, from the fitted model.*

## 11.5 Example: predictive simulation to check the fit of a time-series model

Predictive simulation is more complicated in time-series models, which are typically set up so that the distribution for each point depends on the earlier data. We illustrate with a simple autoregressive model.

### Fitting a first-order autoregression to the unemployment series

Example: Unemployment time series

Figure 11.13 shows the time series of annual unemployment rates in the United States from 1947 to 2004.[6] We would like to see how well these data are fit by a first-order autoregression, that is, a regression on last year's unemployment rate. Such a model is easy to set up and fit:

```
n <- nrow(unemp)
unemp$y_lag <- c(NA, unemp$y[1:(n-1)])
fit_lag <- stan_glm(y ~ y_lag, data=unemp)
```

yielding the following fit:

```
            Median MAD_SD
(Intercept) 1.35   0.44
y_lag       0.77   0.07

Auxiliary parameter(s):
      Median MAD_SD
sigma 1.03   0.09
```

This is potentially informative but does not give a sense of the fit of model to data. To examine the fit, we will simulate replicated data from the fitted model.

### Simulating replicated datasets

We can extract the simulations from the fitted model:

```
sims <- as.matrix(fit_lag)
n_sims <- nrow(sims)
```

Then we create a container for replicated datasets and fill it in with simulated time series:

[6]Data and code for this example are in the folder `Unemployment`.

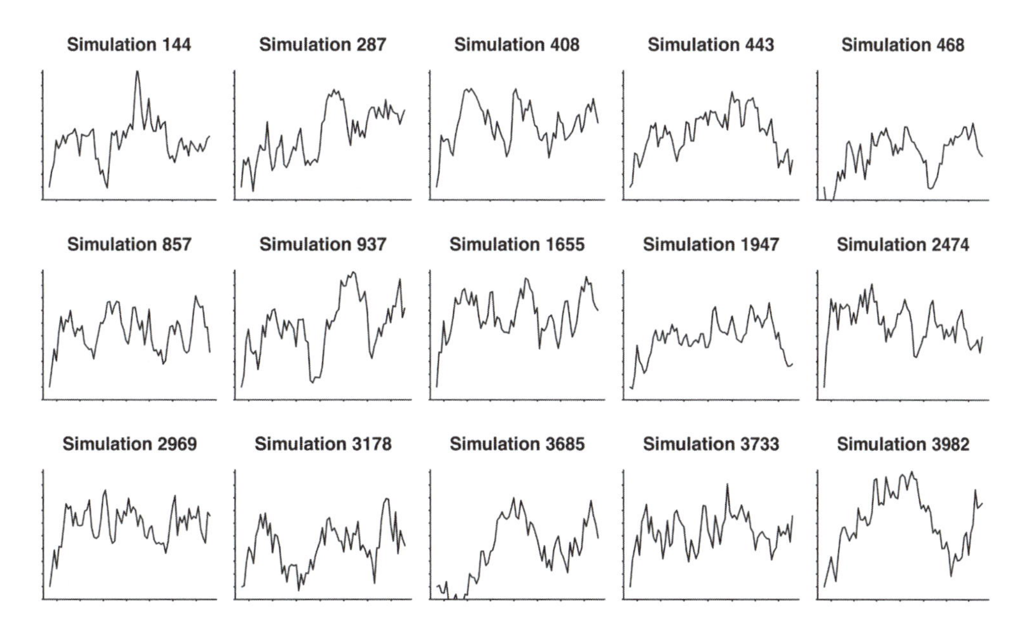


Figure 11.14 *Graphing a random sample of the simulated replications of the unemployment series from the fitted autoregressive model. The replications capture many of the features of the actual data in Figure 11.13 but show slightly more short-term variation.*

```
y_rep <- array(NA, c(n_sims, n))
for (s in 1:n_sims){
  y_rep[s,1] <- y[1]
  for (t in 2:n){
    y_rep[s,t] <- sims[s,"(Intercept)"] + sims[s,"y_lag"] * y_rep[s,t-1] +
                  rnorm(1, 0, sims[s,"sigma"])
  }
}
```

We could not simply create these simulations using `posterior_predict` because with this time-series model we need to simulate each year conditional on the last.

## Visual and numerical comparisons of replicated to actual data

Our first step in model checking is to plot some simulated datasets, which we do in Figure 11.14, and compare them visually to the actual data in Figure 11.13. The 15 simulations show different patterns, with many of them capturing the broad features of the data—its range, lack of overall trend, and irregular rises and falls. This autoregressive model clearly can represent many different sorts of time-series patterns.

Looking carefully at Figure 11.14, we see one pattern in all these replicated data that was not in the original data in Figure 11.13, and that is a jaggedness, a level of short-term ups and downs that contrasts to the smoother appearance of the actual time series.

To quantify this discrepancy, we define a test statistic that is the frequency of "switches"—the number of years in which an increase in unemployment is immediately followed by a decrease, or vice versa:

```
test <- function(y){
  n <- length(y)
  y_lag <- c(NA, y[1:(n-1)])
```

```
  y_lag_2 <- c(NA, NA, y[1:(n-2)])
  sum(sign(y-y_lag) != sign(y_lag-y_lag_2), na.rm=TRUE)
}
```

As with the examples in the previous section, we compute this test for the data and for the replicated datasets:

```
test_y <- test(unemp$y)
test_rep <- apply(y_rep, 1, test)
```

The actual unemployment series featured 26 switches. In this case, 99% of the n_sims = 4000 replications had more than 26 switches, with 80% in the range [31, 41], implying that this aspect of the data was not captured well by the model. The point of this test is not to "reject" the autoregression—no model is perfect, after all—but rather to see that this particular aspect of the data, its smoothness, is not well captured by the fitted model. The real time series switched direction much less frequently than would be expected from the model. The test is a confirmation of our visual impression of a systematic difference between the observed and replicated datasets.

## 11.6 Residual standard deviation $\sigma$ and explained variance $R^2$

The residual standard deviation, $\sigma$, summarizes the scale of the residuals $r_i = y_i - X_i\hat{\beta}$. For example, in the children's test scores example, $\hat{\sigma} = 18$, which tells us that the linear model can predict scores to about an accuracy of 18 points. Said another way, we can think of this standard deviation as a measure of the average distance each observation falls from its prediction from the model.

The magnitude of this standard deviation is more salient when compared to the total variation in the outcome variable. The fit of the model can be summarized by $\sigma$ (the smaller the residual standard deviation, the better the fit) and by $R^2$ (also sometimes called the coefficient of determination), the fraction of variance "explained" by the model. The "unexplained" variance is $\sigma^2$, and if we label $s_y$ as the standard deviation of the data, then, using a classic definition,

$$R^2 = 1 - \left(\hat{\sigma}^2/s_y^2\right). \tag{11.1}$$

When the model is fit using least squares, the expression above is equivalent to an alternative formula that uses explained variance directly:

$$R^2 = V_{i=1}^n\, \hat{y}_i / s_y^2, \tag{11.2}$$

where $\hat{y}_i = X_i\hat{\beta}$ and we are using the notation $V$ for the sample variance:

$$V_{i=1}^n\, z_i = \frac{1}{n-1}\sum_{i=1}^n (z_i - \bar{z})^2, \text{ for any vector } z \text{ of length } n. \tag{11.3}$$

In the model fit to test scores in Section 10.1, $R^2$ is a perhaps disappointing 22%. However, in a deeper sense, it is presumably a good thing that this regression has a low $R^2$—that is, that a child's achievement cannot be accurately predicted given only these maternal characteristics.

To understand $R^2$, consider two special cases in the one-predictor scenario, so that $\hat{y} = \hat{a} + \hat{b}x$.

1. First, suppose that the fitted line is nearly the same as a horizontal line at $\bar{y}$, so that $\hat{a} + \hat{b}x_i \approx \bar{y}$ for all $i$. In this case, each residual $y_i - (\hat{a} + \hat{b}x_i)$ is nearly identical to $y_i - \bar{y}$, so that $\sigma \approx s_y$. Hence, $R^2 \approx 0$. Thus, $R^2 \approx 0$ indicates that the regression line explains practically none of the variation in $y$. In this case, the estimated intercept $\hat{a}$ is approximately $\bar{y}$ and the estimated slope $\hat{b}$ is approximately 0.
2. At the other extreme, suppose that the fitted line passes through all the points nearly perfectly, so that each residual is almost 0. In this case, $\sigma$ will be close to 0 and much smaller than $s_y$, so that $R^2 \approx 1$. Thus, $R^2 \approx 1$ indicates that the regression line explains practically all of the variation in $y$. In general, the closer $R^2$ is to 1, the happier we are with the model fit.

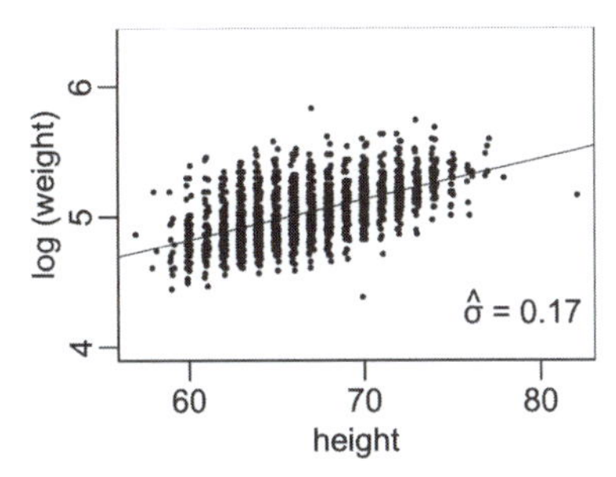


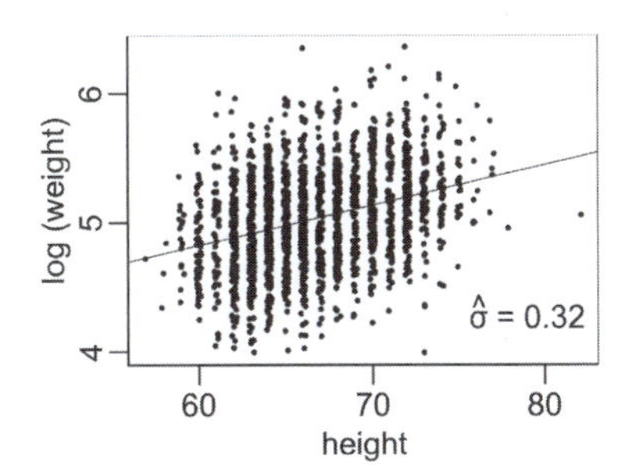


Figure 11.15 *Two hypothetical datasets with the same fitted regression line, $\hat{a} + \hat{b}x$, but different values of the residual standard deviation, $\sigma$. (a) Actual data from a survey of adults; (b) data with noise added to $y$.*

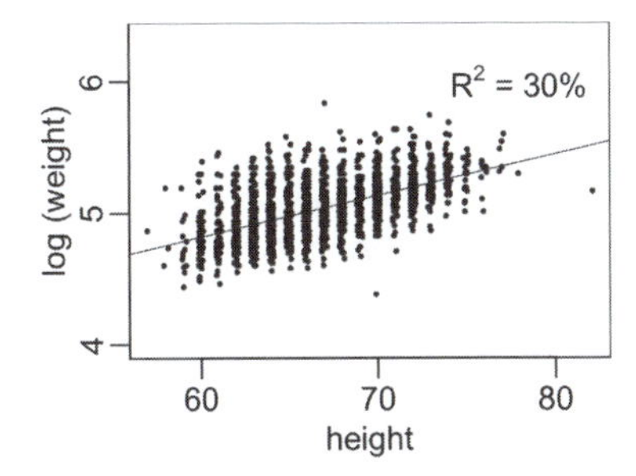


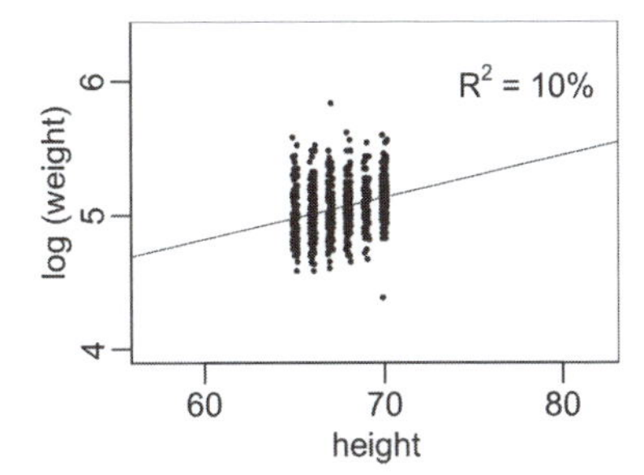


Figure 11.16 *Two hypothetical datasets with the same regression line, $\hat{a} + \hat{b}x$, and residual standard deviation, $\sigma$, but different values of the explained variance, $R^2$. (a) Actual data from a survey of adults; (b) data restricted to heights between 65 and 70 inches.*

There are a few things about $R^2$ that you should know. First, it does not change if you multiply $x$ or $y$ in the regression by a constant. So, if you want to change the units of the predictors or response to aid interpretation, you won't change the summary of the fit of the model to the data. Second, in a least squares regression with one predictor, one can show that $R^2$ equals the square of the correlation between $x$ and $y$; see Exercise 11.7. There is no such interpretation for regressions with more than one predictor.

The quantity $n - k$, the number of data points minus the number of estimated coefficients, is called the *degrees of freedom* for estimating the residual errors. In classical regression, $k$ must be less than $n$—otherwise, the data could be fit perfectly, and it would not be possible to estimate the regression errors at all.

## Difficulties in interpreting residual standard deviation and explained variance

As we make clear throughout the book, we are generally more interested in the deterministic part of the model, $X\beta$, than in the variation, $\epsilon$. However, when we do look at the residual standard deviation, $\sigma$, we are typically interested in it for its own sake—as a measure of the unexplained variation in the data—or because of its relevance to the precision of inferences about the regression coefficients $\beta$. Figure 11.15 illustrates two regressions with the same deterministic model, $\hat{a} + \hat{b}x$, but different values of $\sigma$.

Interpreting the proportion of explained variance, $R^2$, can be tricky because its numerator and denominator can be changed in different ways. Figure 11.16 illustrates with an example where the regression model is identical, but $R^2$ decreases because the model is estimated on a subset of the data. Going from the left to right plots in the figure, the residual standard deviation $\sigma$ is unchanged but the standard deviation of the raw data, $s_y$, decreases when we restrict to this subset; thus, $R^2 = 1 - \hat{\sigma}^2/s_y^2$

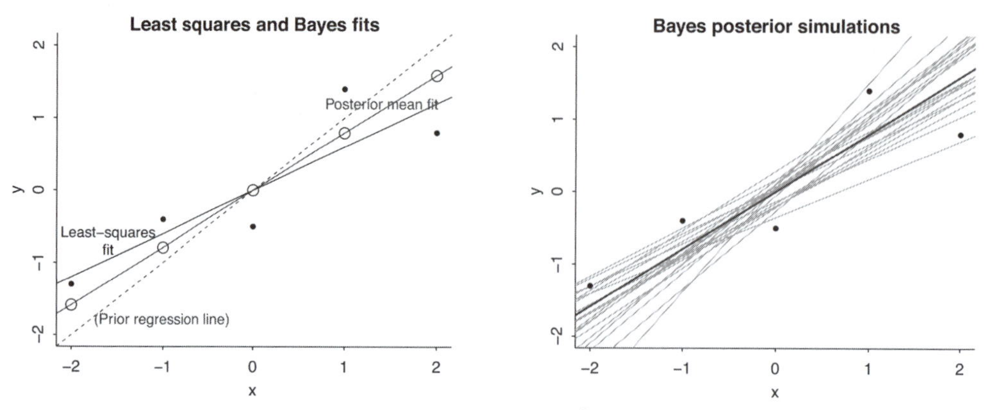


Figure 11.17 *Simple example showing the challenge of defining $R^2$ for a fitted Bayesian model. (a) Data, least squares regression line, and fitted Bayes line, which is a compromise between the prior and the least squares fit. The standard deviation of the fitted values from the Bayes model (the open circles on the line) is greater than the standard deviation of the data, so the usual definition of $R^2$ will not work. (b) Posterior mean fitted regression line along with 20 draws of the line from the posterior distribution. We compute Bayesian $R^2$ by evaluating formula (11.5) for each posterior simulation draw and then taking the median.*

declines. Even though $R^2$ is much lower in Figure 11.16a, the model fits the data just as well as in Figure 11.16b.

## Bayesian $R^2$

$R^2$ is usually defined based on the point estimate of the fitted model. In Bayesian inference, however, we are also concerned with uncertainty; indeed, as discussed in Chapter 9, one of the main motivations for using Bayesian inference in regression is to be able to propagate uncertainty in predictions.

Our first thought for Bayesian $R^2$ is simply to use the posterior mean estimate of $\beta$ to create the vector of Bayesian predictions $X\hat{\beta}$, use this to compute the residual standard deviation, and then plug this into formula (11.1) or compute the standard deviation of mean predictions and then plug this into formula (11.2). This approach has two problems: first, it dismisses uncertainty to use a point estimate in Bayesian computation; and, second, the two equations can be in disagreement, and (11.2) can give $R^2$ greater than 1. When $\beta$ is estimated using least squares, and assuming the regression model includes a constant term, the numerator of (11.2) is less than or equal to the denominator by definition; for general estimates, though, there is no requirement that this be the case, and it would be awkward to say that a fitted model explains more than 100% of the variance. On the other hand, in formula (11.1) it is possible to have a case where the estimated residual variance is higher than the data variance, and then estimated $R^2$ would be less than 0.

Example: Simulation illustrating Bayesian $R^2$

To see an example where the simple $R^2$ would be inappropriate, consider the model $y = \beta_0 + \beta_1 x + \text{error}$ with a strong prior on $\beta$ and only a few data points.[7] Figure 11.17a shows data and the least squares regression line, with $R^2$ of 0.77. We then do a Bayes fit with strong priors: normal with mean 0 and standard deviation 0.2 for $\beta_0$ and normal with mean 1 and standard deviation 0.2 for $\beta_1$. The standard deviation of the fitted values from the Bayes model is 1.3, while the standard deviation of the data is only 1.08, so the square of this ratio—$R^2$ as defined in (11.2)—is greater than 1. Figure 11.17b shows the posterior mean fitted regression line along with 20 draws of the line $\beta_0 + \beta_1 x$ from the fitted posterior distribution of $\beta$.

Instead of working with (11.1) or (11.2) directly, we define $R^2$ explicitly based on the predictive

[7]Code for this example is in the folder Rsquared.

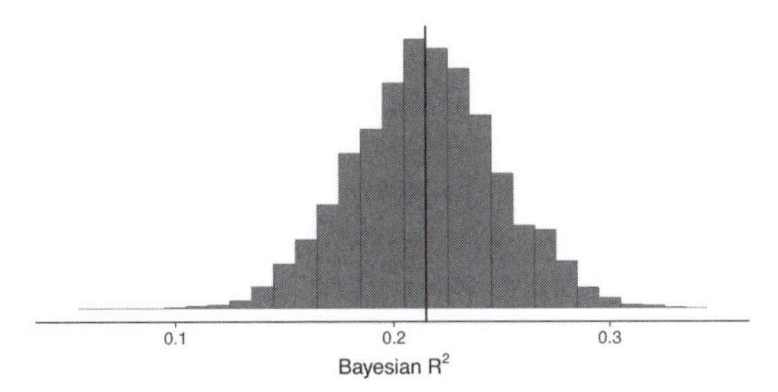


Figure 11.18 *The posterior distribution of Bayesian $R^2$ for a model predicting child's test score given mother's score on an IQ test and mother's high school status.*

distribution of data, using the following variance decomposition for the denominator:

$$\text{alternative } R^2 = \frac{\text{Explained variance}}{\text{Explained variance} + \text{Residual variance}} = \frac{\text{var}_{\text{fit}}}{\text{var}_{\text{fit}} + \text{var}_{\text{res}}}, \tag{11.4}$$

where $\text{var}_{\text{fit}}$ is the variance of the modeled predictive means, $V_{i=1}^n \hat{y}_i$, and $\text{var}_{\text{res}}$ is the modeled residual variance $\sigma^2$. The first of these quantities is the variance among the expectations of the new data; the second term is the expected variance for new residuals, in both cases assuming the same predictors $X$ as in the observed data.

In Bayesian inference we work with a set of posterior simulation draws, $s = 1, \dots, S$. For each simulation draw $s$, we can compute the vector of predicted values, $\hat{y}_i^s = X_i \beta^s$, the residual variance $\sigma_s^2$, and the proportion of variance explained,

$$\text{Bayesian } R_s^2 = \frac{V_{i=1}^n \hat{y}_i^s}{V_{i=1}^n \hat{y}_i^s + \sigma_s^2}. \tag{11.5}$$

This expression is always between 0 and 1 by construction, no matter what procedure is used to create the predictions $\hat{y}$. In this Bayesian version, we define predicted values based on the simulations of $\beta$ rather than on the point estimate $\hat{\beta}$.

We can compute this Bayesian $R^2$ from a regression that has been fit in rstanarm. For example if the fitted model from `stan_glm` is called `fit`, then `bayes_R2(fit)`, returns a vector representing posterior uncertainty. If a point summary is desired, one can compute `median(bayes_R2(fit))`. For the example in Figure 11.17, the Bayesian $R^2$ from (11.5) has posterior median 0.75, mean 0.70, and standard deviation 0.17. In the model fit to test scores in Section 10.1, Bayesian $R^2$ is 0.21, but it is useful to see the related uncertainty as shown in Figure 11.18.

## 11.7 External validation: checking fitted model on new data

The most fundamental way to test a model is to use it to make predictions and then compare to actual data. Figure 11.19 illustrates with the children's test score model, which was fit to data collected from children who were born before 1987. Having fit the model using `stan_glm`, we then use `posterior_predict` to obtain simulations representing the predictive distribution for new cases.

We apply the model to predict the outcomes for children born in 1987 or later. This is not an ideal example for prediction, because we would not necessarily expect the model for the older children to be appropriate for the younger children, even though tests for all children were taken at age 3 or 4. However, we can use it to demonstrate the methods for computing and evaluating predictions.

The new data, $y^{\text{new}}$, are the outcomes for the 336 new children predicted from `mom_iq` and `mom_hs`, using the model fit using the data from the older children. Figure 11.19a plots actual values $y_i^{\text{new}}$ versus predicted values $X_i^{\text{new}} \hat{\beta}$, and Figure 11.19b plots residuals versus predicted values with dotted lines at $\pm \hat{\sigma}$ (approximate 68% error bounds; see Section 3.5). The error plot shows no obvious

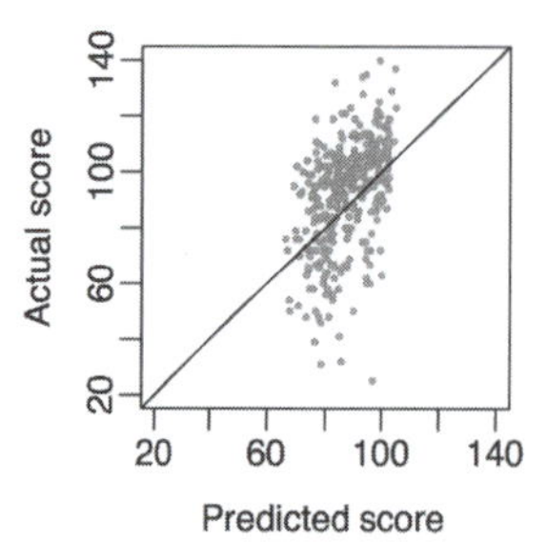


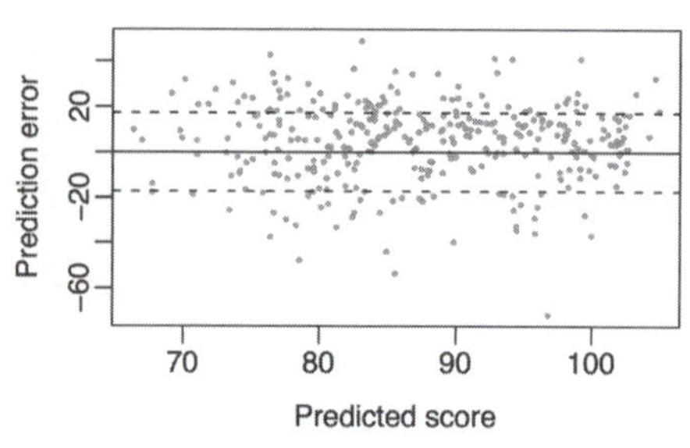


Figure 11.19 *Plots assessing how well the model fit to older children works in making predictions for younger children: (a) comparing predictions for younger children from a model against their actual values, (b) comparing residuals from these predictions against the predicted values.*

problems with applying the older-child model to the younger children, though from the scale we detect that the predictions have wide variability. In this example, what we are plotting as the "predicted score" for each child is simply the mean of the `n_sims` predictive simulation draws. We could also check coverage of the predictions by looking, for example, at how many of the actual data points fell within the 50% predictive intervals computed from the posterior predictive simulations.

Even if we had detected clear problems with these predictions, this would not mean necessarily that there is anything wrong with the model as fit to the original dataset. However, we would need to understand it further before generalizing to other children.

## 11.8 Cross validation

Often we would like to evaluate and compare models without waiting for new data. One can simply evaluate predictions on the observed data. But since these data have already been used to fit the model parameters, these predictions are optimistic for assessing generalization.

In cross validation, part of the data is used to fit the model and the rest of the data—the *hold-out set*—is used as a proxy for future data. When there is no natural prediction task for future data, we can think of cross validation as a way to assess generalization from one part of the data to another part.

Different data partitions can be used, depending on the modeling task. We can hold out individual observations (*leave-one-out (LOO)* cross validation) or groups of observations (*leave-one-group-out*), or use past data to predict future observations (*leave-future-out*). Leave-one-group-out and leave-future-out cross validation can be useful in multilevel and time series settings, which are beyond the scope of this book. In any form of cross validation, the model is re-fit leaving out one part of the data and then the prediction for the held-out part is evaluated. Cross validation removes the overfitting problem arising from using the same data for estimation and evaluation, but at the cost of requiring the model to be fit as many times as the number of partitions.

### Leave-one-out cross validation

The naive implementation of LOO would require the model to be fit $n$ times, once for each held-out data point. The good news is that we have built an R package, `loo`, containing a function, `loo`, with a computational shortcut that uses probability calculations to approximate the leave-one-out evaluation, while only fitting the model once, and also accounting for posterior uncertainty in the fit.

We first illustrate LOO cross validation with a small simulated dataset:[8]

```
n <- 20
x <- 1:n
a <- 0.2
```

[8]Code for this example is in the folder `CrossValidation`.

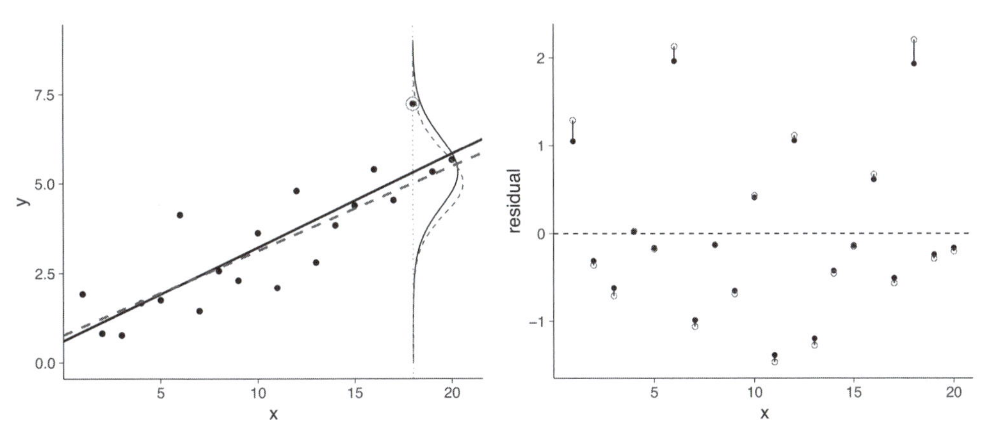


Figure 11.20 *(a) Simulated data, along with the posterior mean and predictive distribution for the* 18th *data point from fitted linear regressions. The solid curve shows the posterior predictive distribution for* $y_{18}$ *given the line fit to all the data, and the dashed line shows the posterior predictive distribution given the line fit to all the data except the* 18th *observation. (b) Residuals from the fitted models. Solid circles show residuals from the model fit to all the data; open circles show leave-one-out residuals, where each observation is compared to its expected value from the model holding that data point out of the fit.*

```
b <- 0.3
sigma <- 1
set.seed(2141)
y <- a + b*x + sigma*rnorm(n)
fake <- data.frame(x, y)
```

We use the function `set.seed` to start the random number generator at a fixed value. The particular value chosen is arbitrary, but now we can inspect the data and choose a point with a large residual to see where cross validation can make a difference.

For this particular simulated dataset we notice that the 18th data point is far from the line, and we demonstrate the principle of cross validation by fitting two regression models, one fit to all the data and one fit to the data excluding the 18th observation:

```
fit_all <- stan_glm(y ~ x, data = fake)
fit_minus_18 <- stan_glm(y ~ x, data = fake[-18,])
```

We next evaluate the prediction for the 18th data point. The posterior predictive distribution is obtained by averaging over all possible parameter values. Using posterior simulation draws, the posterior predictive distribution for a new data point $y^{\text{new}}$ with predictors $x^{\text{new}}$, conditional on the observed data $y$ and $X$, can be approximated in probability notation as,

$$p(y^{\text{new}}|x^{\text{new}}) \approx \frac{1}{S}\sum_{s=1}^{S} p(y^{\text{new}}|x^{\text{new}}, \beta^s, \sigma^s). \tag{11.6}$$

Figure 11.20a shows our simulated data, along with the posterior mean and predictive distribution for $y_{18}$ from two linear regressions, one fit to all the data and the other fit to the data excluding the held-out data point. Figure 11.20a also shows the fitted regression lines, revealing that the fit when excluding the data point is different from the fit from all the data. When a data point is held out, the posterior mean and predictive distribution move away from that observation.

In LOO cross validation, the process of leaving out one observation is repeated for all the data points. Figure 11.20b shows posterior and LOO residuals computed as the observation minus either the posterior mean predicted value (that is, the result from `predict`) applied to the fitted model, or the predicted mean from the fitted LOO model excluding the held-out data point (that is, the result from loo_predict). In this example, the LOO residuals all have larger magnitudes, reflecting that it is more difficult to predict something that was not used in model fitting.

To further compare posterior residuals to LOO residuals, we can examine the standard deviation of the residuals. In our simulation, the residual standard deviation was set to 1, the LOO residual standard deviation is 1.01, and the posterior residual standard deviation is 0.92; this is a typical example in which the direct residuals show some overfitting.

The connection of residual standard deviation to explained variance $R^2$ was discussed in Section 11.6. Analogously to the explained proportion of variance, $R^2$, we can calculate LOO $R^2$, where the residual standard deviation is calculated from the LOO residuals. For the above simulation example, $R^2$ is 0.74 and LOO $R^2$ is 0.68.

### Fast leave-one-out cross validation

LOO cross validation could be computed by fitting the model $n$ times, once with each data point excluded. But this can be time consuming for large $n$. The loo function uses a shortcut that makes use of the mathematics of Bayesian inference, where the posterior distribution can be written as the prior distribution multiplied by the likelihood. If the observations are conditionally independent (as is typically the case in regression models), the likelihood is the product of $n$ factors, one for each data point: in Bayesian notation, the posterior distribution is $p(\theta|y) \propto p(\theta) \prod_{i=1}^{n} p(y_i|\theta)$, where $\theta$ represents all the parameters in the model and conditioning on the predictors $X$ is implicit.

In leave-one-out cross validation, we want to perform inferences excluding each data point $i$, one at a time. In the above expression, excluding that one data point is equivalent to multiplying the posterior distribution by the factor $1/p(y_i|\theta)$. The LOO posterior excluding point $i$ is written as $p(\theta|y_{-i}) = p(\theta|y)/p(y_i|\theta)$, and the LOO distribution is computed by taking the posterior simulations for $\theta$ obtained from stan_glm and giving each simulation a weight of $1/p(y_i|\theta)$. This set of weighted simulations is used to approximate the predictive distribution of $y_i$, the held-out data point. Using the raw weights can be noisy and so the loo function smooths them before doing the computation.

### Summarizing prediction error using the log score and deviance

There are various ways to assess the accuracy of a model's predictions. Residual standard deviation or $R^2$ are interpretable summaries for linear regression but do not always make sense for logistic and other discrete-data models. In addition, residual standard deviation and $R^2$ measure only the prediction error relative to the mean of the predictive distribution, ignoring the uncertainty represented by the predictive distribution. We should be able to do better using the entire predictive distribution.

A summary that applies to probability models more generally is the *log score*, which is defined as the logarithm of the probability or density of outcome $y$ given predictors $x$. For linear regression with normally-distributed errors, the density function, $p(y_i|\beta, \sigma)$, is $\frac{1}{\sqrt{2\pi}\sigma} \exp\left(-\frac{1}{2\sigma^2}(y_i - X_i\beta)^2\right)$ (see equation (8.7)), and so the log score is $-\frac{1}{2}\log\sigma - \frac{1}{2\sigma^2}(y_i - X_i\beta)^2$. Strictly speaking, this last expression should also include a term of $-\frac{1}{2}\log(2\pi)$, but typically we only look at differences in the log score, comparing models, so any constant shift can be ignored. For the purpose of this book, you will not need this formula, as the log score is computed automatically by stan_glm, but we include it here to connect practice to theory.

Figure 11.21 shows log posterior predictive densities and log LOO predictive densities for our simulated linear regression example. To summarize the predictive performance the usual practice is to add these log densities; the result is called the *expected log predictive density* (elpd).

The within-sample log score multiplied by $-2$ is called the *deviance*—for historical reasons, this has been a commonly used scale when using maximum likelihood inference—and it is reported when displaying a glm fit in R. Better fits to the data corresponds to higher values of the log score and lower values of the deviance for the fitted model.

More generally, the log score is also useful for different non-normal error distributions and can be

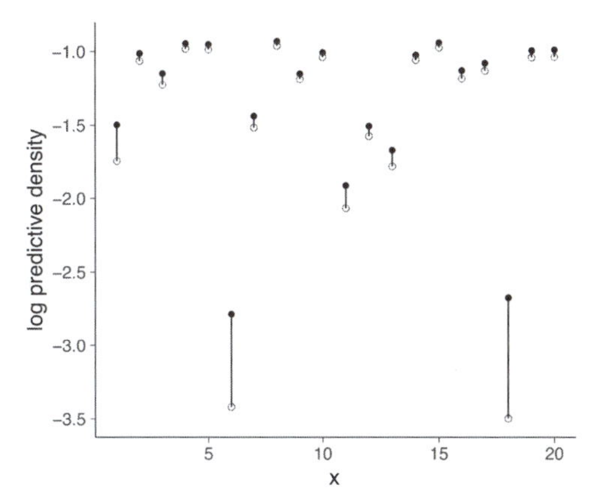


Figure 11.21 *Log posterior predictive densities and log LOO predictive densities from a regression model fitted to the data shown in Figure 11.20a. The solid circles represent log predictive densities relative to the model fit to all the data, and the open circles correspond to leave-one-out fits. The open circles are always lower than the solid circles, because removing an observation from the likelihood causes the model to fit less well to that data point.*

justified when there is interest in the accuracy of the whole predictive distribution instead of just some point estimates. We discuss the log score further in Section 13.5 in the context of logistic regression.

## Overfitting and AIC

As discussed above, cross validation avoids the overfitting problem that arises from simple calculations of within-sample predictive error that use the same data to fit and evaluate the model. A simple alternative way to correct for overfitting is to adjust the log score to account for the number of parameters fitted in the model. For maximum likelihood models, this corresponds to the *Akaike information criterion* (AIC) and is defined as AIC = deviance + $2k$, where $k$ is the number of coefficients fit in the model. AIC is an estimate of the deviance that would be expected if the fitted model is used to predict new data. Deviance is −2 times the log score, so AIC is equivalent to subtracting $k$ from the log score to account for overfitting. The correction in AIC for number of parameters has a mathematical connection to the degrees-of-freedom adjustment in which $n - k$ rather than $n$ is used in the denominator of formula (10.5) for $\hat{\sigma}$. We prefer to use cross validation as it is more general and directly interpretable as an estimate of prediction error, but for simple models the two methods should give similar answers.

## Interpreting differences in log scores

The log score is rarely given a direct interpretation; almost always we look at differences in the log score, comparing two or more models. The reference point is that adding a linear predictor that is pure noise and with a weak prior on the coefficient should increase the log score of the fitted by 0.5, in expectation, but should result in an expected *decrease* of 0.5 in the LOO log score. This difference is additive if more predictors are included in the model. For example, adding 10 noise predictors should result in an increase of approximately 5 in the raw log score and a decrease of approximately 5 in the LOO log score. The log score relative to the full model increases, but this is an illusion due to overfitting; the LOO log score decreases because the inclusion of noise parameters really does, on average, degrade the fit.

Changes in the LOO log score can thus be calibrated in terms of the decrease in predictive accuracy arising from adding noise predictors to the model. That said, these log scores are themselves directly computed from data and can be noisy, hence we should also be aware of our uncertainty in these calculations, as we illustrate next with an example.

### Demonstration of adding pure noise predictors to a model

Example: Regression with noise predictors

When predictors are added to a model—even if the predictors are pure noise—the fit to data and within-sample predictive measures will generally improve. The ability of model to fit to the data is desired in modeling, but when the model is fit using finite data the model will fit partly also to random noise. We want to be able to separate which parts of the data allow us to generalize to new data or from one part of the data to the rest of the data and which part of the data is random noise which we should not be able to predict.

We demonstrate with an example. In Chapter 10 we introduced a model predicting children's test scores given maternal high school indicator and maternal IQ. Here is the fitted model:

```
            Median MAD_SD
(Intercept) 25.8    5.8
mom_hs       5.9    2.2
mom_iq       0.6    0.1

Auxiliary parameter(s):
      Median MAD_SD
sigma 18.1   0.6
```

We then add five pure noise predictors:

```
n <- nrow(kidiq)
kidiqr <- kidiq
kidiqr$noise <- array(rnorm(5*n), c(n,5))
fit_3n <- stan_glm(kid_score ~ mom_hs + mom_iq + noise, data=kidiqr)
```

And here is the new result:

```
            Median MAD_SD
(Intercept) 25.4    5.8
mom_hs       5.9    2.3
mom_iq       0.6    0.1
noise1      -0.1    0.9
noise2      -1.2    0.9
noise3       0.1    1.0
noise4       0.2    0.9
noise5      -0.5    0.9

Auxiliary parameter(s):
      Median MAD_SD
sigma 18.1   0.6
```

Unsurprisingly, the estimated coefficients for these new predictors are statistically indistinguishable from zero and residual $\sigma$ estimate has slightly increased. But the fit to data has improved: $R^2$ has gone from 0.21 to 0.22, an increase of 0.01, and the *posterior predictive log score* has gone from $-1872.0$ to $-1871.1$, an increase of 0.9. This pattern, that fit to data improves even with noise predictors, is an example of overfitting, and it is one reason we have to be careful when throwing predictors into a regression. Using LOO cross validation we see that the predictive performance for new data or generalization capability has not improved: LOO $R^2$ has gone from 0.2 to 0.19, a decrease of 0.01, and the LOO predictive log score has gone from $-1876.1$ to $-1879.9$, a decrease of 3.7.

In this simple example, the estimate of `elpd_loo`, $-1876.1$, is nearly the same as what is obtained by subtracting the number of additional parameters from the log score: $-1872 - 4 = -1876$. For the model with five additional pure noise predictors, the estimate of `elpd_loo` is $-1879.9$, which is close to this: log score $- k = -1871.1 - 9 = -1880$. For more complicated problems, though, the two estimates can differ, and when that occurs, we prefer the LOO estimate, as in general it better accounts for uncertainty.

In this example, a default weak prior was used which makes it more likely that overfitting can happen when pure noise predictors are added. It is possible to reduce the overfitting by using more informative priors, but this is beyond the scope of this book.

Having fit the model, we use the `loo` to compute the leave-one-out cross validation log score:

```
loo_3 <- loo(fit_3)
print(loo_3)
```

which yields,

```
Computed from 4000 by 434 log-likelihood matrix
          Estimate   SE
elpd_loo   -1876.1 14.2
p_loo          4.1  0.4
looic       3752.3 28.4
------
Monte Carlo SE of elpd_loo is 0.0.
All Pareto k estimates are good (k < 0.5).
See help('pareto-k-diagnostic') for details.
```

Here is how we interpret all this:

- The "log-likelihood matrix" is $\log p(y_i|x_i, \beta, \sigma)$ at $n = 434$ data points $(x_i, y_i)$ using 4000 draws of $\beta$ and $\sigma$ from the posterior distribution.
- `elpd_loo` is the estimated log score along with a standard error representing uncertainty due to using only 434 data points.
- `p_loo` is the estimated "effective number of parameters" in the model. The above model has 4 parameters, so it makes sense that `p_loo` is close to 4 here, but in the case of models with stronger prior information, weaker data, or model misspecification, this direct identification will not always work.
- `looic` is the LOO information criterion, $-2$ `elpd_loo`, which we compute for comparability to deviance.
- The Monte Carlo SE of `elpd_loo` gives the uncertainty of the estimate based on the posterior simulations. If the number of simulations is large enough, this Monte Carlo SE will approach zero, in contrast to the standard errors of `epld_loo`, `p_loo`, and `looic`, whose uncertainties drive from the finite number of data points in the regression.
- The two lines at the end of the display give a warning if the Pareto $k$ estimates are unstable, in which case an alternative computation is recommended, as we discuss on page 178.

We continue the example by performing a model comparison. Let's take the model above (without pure noise predictors) and compare it to the model using only the maternal high school indicator:

```
fit_1 <- stan_glm(kid_score ~ mom_hs, data=kidiq)
loo_1 <- loo(fit_1)
print(loo_1)
```

Here is what we get from the LOO computation:

```
Computed from 4000 by 434 log-likelihood matrix
          Estimate   SE
elpd_loo   -1914.9 13.8
p_loo          3.2  0.3
looic       3829.8 27.6
------
Monte Carlo SE of elpd_loo is 0.0.
All Pareto k estimates are good (k < 0.5).
See help('pareto-k-diagnostic') for details.
```

The simpler model performs worse in cross validation: $\text{elpd}_{\text{loo}}$ has decreased from $-1876.1$ to $-1914.9$, a decline of 38.8.

To get a sense of the uncertainty in this improvement for future data, we need to compare the two models directly for each data point, which can be done using the loo_compare function:

```
loo_compare(loo_3, loo_1)
```

which yields,

```
Model comparison:
(negative 'elpd_diff' favors 1st model, positive favors 2nd)
elpd_diff        se
    -38.8       8.3
```

The difference of 38.8 has a standard error of only 8.3. When the difference (elpd_diff) is larger than 4, the number of observations is larger than 100, and the model is not badly misspecified then the standard error (se_diff) is a reliable measure of the uncertainty in the difference. Differences smaller than 4 are hard to distinguish from noise.

We can continue with other models. For example:

```
fit_4 <- stan_glm(kid_score ~ mom_hs + mom_iq + mom_hs:mom_iq, data=kidiq)
loo_4 <- loo(fit_4)
loo_compare(loo_3, loo_4)
```

which yields,

```
Model comparison:
(negative 'elpd_diff' favors 1st model, positive favors 2nd)
elpd_diff        se
      3.6       2.6
```

There is no clear predictive advantage in adding an interaction between maternal high school indicator and maternal IQ.

To take into account the accuracy of the whole predictive distribution, we recommend the log score for model comparison in general. With care, log densities and log probabilities can be interpreted when transformed to densities and probabilities, but it can also be useful to compute more familiar measures of predictive performance such as $R^2$ to get an understanding of the fit of a model. For example, we can compute and compute LOO $R^2$ for the above models. For model fit_1, the LOO $R^2$ is 0.05, a decrease of 0.16 with standard error 0.04 when compared to model fit_3. For model fit_4, the LOO $R^2$ is 0.22, an increase of 0.01 with standard error 0.05 when compared to model fit_3.

### *K*-fold cross validation

As discussed earlier, leave-one-out cross validation in theory requires re-fitting the model $n$ times, once with each data point left out, which can take too long for a dataset of even moderate size. The loo function has a computationally-efficient approximation that requires the model to be fit only once, but in some cases it can be unstable. Difficulties arise when the probability densities $p(y_i|\theta)$ take on very low values—if this density is near zero, then the resulting weight, $1/p(y_i|\theta)$, becomes very high. A data point for which $p(y_i|\theta)$ near zero can be thought of as an outlier in probabilistic sense, in that it is a data point from an area that has low probability under the fitted model.

When leave-one-out cross validation is unstable—which would be indicated by a warning message after running loo—you can switch to *K-fold cross validation* in which the data are randomly partitioned into $K$ subsets, and the fit of each subset is evaluated based on a model fit to the rest of the data. This requires only $K$ new model fits. It is conventional to use $K = 10$, which represents a tradeoff between $K$ being too low (in which case the estimate has larger bias) or too high (and then the procedure becomes too computationally expensive).

### Demonstration of $K$-fold cross validation using simulated data

mple: ulated a for ss dation

We illustrate 10-fold cross validation with another fake-data example. We start by simulating a $60 \times 30$ matrix representing 30 predictors that are random but not independent; rather, we draw them from a multivariate normal distribution with correlations 0.8, using the `mvrnorm` function, which is in the `MASS` package in R:[9]

```
library("MASS")
k <- 30
rho <- 0.8
Sigma <- rho*array(1, c(k,k)) + (1-rho)*diag(k)
X <- mvrnorm(n, rep(0,k), Sigma)
```

We simulate $y$ from the model $X\beta$ + error with the three first coefficients $\beta_1, \beta_2, \beta_3$ set somewhat arbitrarily to $(-1, 1, 2)$ and the rest set to zero, and with independent errors with mean 0 and standard deviation 2:

```
b <- c(c(-1,1,2), rep(0, k-3))
y <- X %*% b + 2*rnorm(n)
fake <- data.frame(X, y)
```

We then fit linear regression using a weak prior distribution:

```
fit_1 <- stan_glm(y ~ ., prior=normal(0, 10), data=fake)
loo_1 <- loo(fit_1)
```

Now `loo(fit_1)` returns the following warning:

```
1: Found 14 observations with a pareto_k > 0.7.
With this many problematic observations we recommend calling 'kfold' with argument
'K=10' to perform 10-fold cross validation rather than LOO.
```

So we follow that advice:

```
kfold_1 <- kfold(fit_1, K=10)
print(kfold_1)
```

This randomly partitions the data into 10 roughly equal-sized pieces, then fits the model 10 times, each time to 9/10 of the data, and returns:

```
            Estimate  SE
elpd_kfold   -155.4 4.9
```

This is the estimated log score of the data under cross validation, and it is on the same scale as LOO (although we cannot use the result of `loo` for this example given the instability in the approximation).

We next compare the fit to that obtained from an alternative model using a more informative prior, the "regularized horseshoe," which is weakly informative, stating that it is likely that only a small number of predictors are relevant, but we don't know which ones. This prior is often useful when the number of predictors in the model is relatively large compared to the number of observations. Here is the computation:

```
fit_2 <- update(fit_2, prior=hs())
kfold_2 <- kfold(fit_2, K=10)
loo_compare(kfold_1, kfold_2)
```

Here is the result of comparing the two cross validation estimates of the log score:

```
Model comparison:
(negative 'elpd_diff' favors 1st model, positive favors 2nd)
elpd_diff        se
     13.3       5.5
```

Model 2 outperforms model 1 on this measure—an increase of 13 in the estimated log score—so for predictive purposes it seems better here to use the regularized horseshoe prior.

[9]Code for this example is in the folder FakeKCV.

### Concerns about model selection

Given several candidate models, we can use cross validation to compare their predictive performance. But it does not always make sense to choose the model that optimizes estimated predicted performance, for several reasons. First, if the only goal is prediction on data similar to what was observed, it can be better to average across models rather than to choose just one. Second, if the goal is prediction on a new set of data, we should reweight the predictive errors accordingly to account for differences between sample and population, as with the poststratification discussed in Section 17.1. Third, rather than choosing among or averaging over an existing set of models, a better choice can be to perform continuous model expansion, constructing a larger model that encompasses the earlier models as special cases. For example, instead of selecting a subset of variables in a regression, or averaging over subsets, one can include all the candidate predictors, regularizing the inference by partially pooling the coefficient estimates toward zero. Finally, in causal inference it can become difficult to interpret regression coefficients when intermediate outcomes are included in the model (see Section 19.6); hence typically we recommend not including post-treatment variables in such models even when they increase predictive power. If post-treatment predictors are included in a causal regression, then additional effort is required to extract estimated treatment effects; it is not enough to simply look at the coefficient on the treatment variable.

## 11.9 Bibliographic note

Berk (2004) considers some of the assumptions implicit in regression analysis and Harrell (2001) discusses regression diagnostics.

Simulation-based posterior predictive checking was introduced by Rubin (1984) and developed further by Gelman, Meng, and Stern (1996) and Gelman et al. (2013). Gabry et al. (2019) discuss visualizations and prior, posterior, and cross validation predictive checks as part of Bayesian workflow.

Influential ideas in predictive model comparison related to the Akaike (1973) information criterion (AIC) include $C_p$ (Mallows, 1973), cross validation (Stone, 1974, 1977, Geisser and Eddy, 1979), and the deviance information criterion (DIC; Spiegelhalter et al., 2002). See also Fox (2002) for an applied overview, and Vehtari and Ojanen (2012), and Gelman, Hwang, and Vehtari (2014) for a Bayesian perspective. Vehtari, Gelman, and Gabry (2017) summarize some of our thinking on cross validation, information criteria, and predictive model evaluation and discuss the calculation of leave-one-out cross validation in R and Stan. Yao et al. (2018) discuss how to use LOO for averaging over a set of predictive models. Buerkner and Vehtari (2019, 2020) discuss how to use cross validation for time series and certain non-factorized models. Magnusson, Andersen, Jonasson and Vehtari (2020) discuss how leave-one-out cross validation can be further sped up using subsampling.

Piironen and Vehtari (2017a) discuss and demonstrate the overfitting if cross validation or information criteria are used to select among large number of models. Bayesian $R^2$ is discussed by Gelman, Goodrich, et al. (2019). The regularized horseshoe prior was introduced by Piironen and Vehtari (2017b).

## 11.10 Exercises

11.1 *Assumptions of the regression model*: For the model in Section 7.1 predicting presidential vote share from the economy, discuss each of the assumptions in the numbered list in Section 11.1. For each assumption, state where it is made (implicitly or explicitly) in the model, whether it seems reasonable, and how you might address violations of the assumptions.

11.2 *Descriptive and causal inference*:

(a) For the model in Section 7.1 predicting presidential vote share from the economy, describe the coefficient for economic growth in purely descriptive, non-causal terms.

(b) Explain the difficulties of interpreting that coefficient as the effect of economic growth on the incumbent party's vote share.

11.3 *Coverage of confidence intervals*: Consider the following procedure:

- Set $n = 100$ and draw $n$ continuous values $x_i$ uniformly distributed between 0 and 10. Then simulate data from the model $y_i = a + bx_i + \text{error}_i$, for $i = 1, \dots, n$, with $a = 2$, $b = 3$, and independent errors from a normal distribution.
- Regress $y$ on $x$. Look at the median and mad sd of $b$. Check to see if the interval formed by the median $\pm$ 2 mad sd includes the true value, $b = 3$.
- Repeat the above two steps 1000 times.

(a) True or false: the interval should contain the true value approximately 950 times. Explain your answer.

(b) Same as above, except the error distribution is bimodal, not normal. True or false: the interval should contain the true value approximately 950 times. Explain your answer.

11.4 *Interpreting residual plots*: Anna takes continuous data $x_1$ and binary data $x_2$, creates fake data $y$ from the model, $y = a + b_1x_1 + b_2x_2 + b_3x_1x_2 + \text{error}$, and gives these data to Barb, who, not knowing how the data were constructed, fits a linear regression predicting $y$ from $x_1$ and $x_2$ but without the interaction. In these data, Barb makes a residual plot of $y$ vs. $x_1$, using dots and circles to display points with $x_2 = 0$ and $x_2 = 1$, respectively. The residual plot indicates to Barb that she should fit the interaction model. Sketch with pen on paper a residual plot that Barb could have seen after fitting the regression without interaction.

11.5 *Residuals and predictions*: The folder `Pyth` contains outcome $y$ and predictors $x_1, x_2$ for 40 data points, with a further 20 points with the predictors but no observed outcome. Save the file to your working directory, then read it into R using `read.table()`.

(a) Use R to fit a linear regression model predicting $y$ from $x_1, x_2$, using the first 40 data points in the file. Summarize the inferences and check the fit of your model.

(b) Display the estimated model graphically as in Figure 10.2.

(c) Make a residual plot for this model. Do the assumptions appear to be met?

(d) Make predictions for the remaining 20 data points in the file. How confident do you feel about these predictions?

After doing this exercise, take a look at Gelman and Nolan (2017, section 10.4) to see where these data came from.

11.6 *Fitting a wrong model*: Suppose you have 100 data points that arose from the following model: $y = 3 + 0.1\,x_1 + 0.5\,x_2 + \text{error}$, with independent errors drawn from a $t$ distribution with mean 0, scale 5, and 4 degrees of freedom. We shall explore the implications of fitting a standard linear regression to these data.

(a) Simulate data from this model. For simplicity, suppose the values of $x_1$ are simply the integers from 1 to 100, and that the values of $x_2$ are random and equally likely to be 0 or 1. In R, you can define `x_1 <- 1:100`, simulate `x_2` using `rbinom`, then create the linear predictor, and finally simulate the random errors in `y` using the `rt` function. Fit a linear regression (with normal errors) to these data and see if the 68% confidence intervals for the regression coefficients (for each, the estimates $\pm 1$ standard error) cover the true values.

(b) Put the above step in a loop and repeat 1000 times. Calculate the confidence coverage for the 68% intervals for each of the three coefficients in the model.

11.7 *Correlation and explained variance*: In a least squares regression with one predictor, show that $R^2$ equals the square of the correlation between $x$ and $y$.

11.8 *Using simulation to check the fit of a time-series model*: Find time-series data and fit a first-order autoregression model to it. Then use predictive simulation to check the fit of this model as in Section 11.5.

11.9 *Leave-one-out cross validation*: Use LOO to compare different models fit to the beauty and teaching evaluations example from Exercise 10.6:

(a) Discuss the LOO results for the different models and what this implies, or should imply, for model choice in this example.

(b) Compare predictive errors pointwise. Are there some data points that have high predictive errors for all the fitted models?

11.10 *K-fold cross validation*: Repeat part (a) of the previous example, but using 5-fold cross validation:

(a) Randomly partition the data into five parts using the `sample` function in R.

(b) For each part, re-fitting the model excluding that part, then use each fitted model to predict the outcomes for the left-out part, and compute the sum of squared errors for the prediction.

(c) For each model, add up the sum of squared errors for the five steps in (b). Compare the different models based on this fit.

11.11 *Working through your own example*: Continuing the example from the final exercises of the earlier chapters, graph and check the fit of the model you fit in Exercise 10.11 and discuss the assumptions needed to use it to make real-world inferences.

# Chapter 12

# Transformations and regression

It is not always best to fit a regression using data in their raw form. In this chapter we start by discussing linear transformations for standardizing predictors and outcomes in a regression, which connects to "regression to the mean," earlier discussed in Chapter 6, and how it relates to linear transformations and correlation. We then discuss logarithmic and other transformations with a series of examples in which input and outcome variables are transformed and combined in various ways in order to get more understandable models and better predictions. This leads us to more general thoughts about building and comparing regression models in applications, which we develop in the context of an additional example.

## 12.1 Linear transformations

### Scaling of predictors and regression coefficients

xample: arnings nd height

The coefficient $\beta_j$ represents the average difference in $y$, comparing items that differ by 1 unit on the $j^{\text{th}}$ predictor and are otherwise identical. In some cases, though, a difference of 1 unit in $x$ is not the most relevant comparison. Consider, from page 12, a model fit to data we downloaded from a survey of adult Americans in 1990 that predicts their earnings (in dollars) given their height (in inches):

$$\text{earnings} = -85\,000 + 1600 * \text{height} + \text{error}, \quad (12.1)$$

with a residual standard deviation of 22 000.

A linear model is not really appropriate for this problem, as we shall discuss soon, but we'll stick with the simple example for introducing the concept of linear transformations.

Figure 12.1a shows the regression line and uncertainty along with the data, and Figure 12.1b extends the $x$-axis to zero to display the intercept—the point on the $y$-axis where the line crosses zero. The estimated intercept of −85 000 has little meaning since it corresponds to the predicted earnings for a person of zero height!

Now consider the following alternative forms of the model:

$$\text{earnings} = -85\,000 + 63 * \text{height (in millimeters)} + \text{error},$$
$$\text{earnings} = -85\,000 + 101\,000\,000 * \text{height (in miles)} + \text{error}.$$

How important is height? While \$63 does not seem to matter much, \$101 000 000 is a lot. Yet, both these equations reflect the same underlying information. To understand these coefficients better, we need some sense of the variation in height in the population to which we plan to apply the model. One approach is to scale by the standard deviation of heights in the data, which is 3.8 inches (or 97 millimeters, or 0.000 060 miles). The expected difference in earnings corresponding to a 3.8-inch difference in height is $\$1600 * 3.8 = \$63 * 97 = \$101\,000\,000 * 0.000\,060 = \$610$, which is reasonably large but much smaller than the residual standard deviation of \$22 000 unexplained by the regression.

Linear transformations of the predictors $X$ or the outcome $y$ do not affect the fit of a classical regression model, and they do not affect predictions; the changes in the inputs and the coefficients

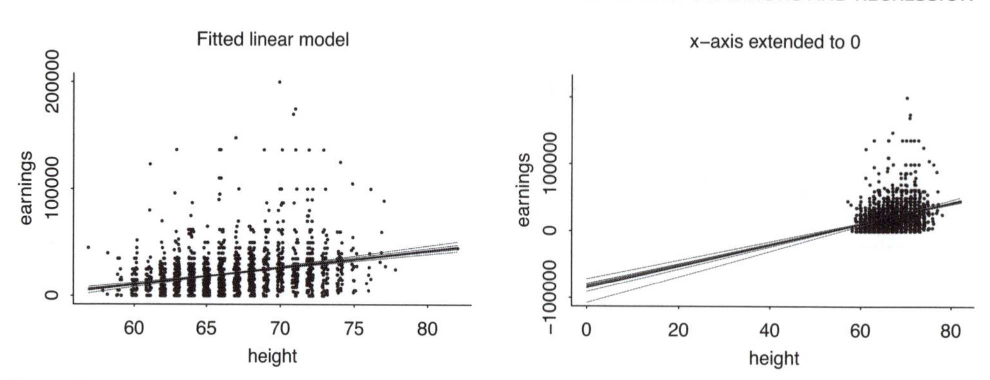


Figure 12.1 *(a) Regression of earnings on height, earnings* = −85 000 + 1600 ∗ *height, with lines indicating uncertainty in the fitted regression. (b) Extending the x-scale to zero reveals the estimate and uncertainty for the intercept of the regression line. To improve resolution, a data point at earnings of* \$400 000 *has been excluded from the graphs.*

cancel in forming the predicted value $X\beta$. However, well-chosen linear transformations can improve interpretability of coefficients and make a fitted model easier to understand. We saw in Chapters 4 and 10 how linear transformations can help with the interpretation of the intercept; this section and the next provide examples involving the interpretation of the other coefficients in the model.

## Standardization using z-scores

Another way to scale the coefficients is to *standardize* the predictor by subtracting the mean and dividing by the standard deviation to yield a "$z$-score." For these `height` would be replaced by `z_height` = (`height` − 66.6)/3.8, and the coefficient for `z_height` becomes 6100. Then coefficients are interpreted in units of standard deviations with respect to the corresponding predictor just as they were, after the fact, in the previous example. This is helpful because standard deviations can be seen as a measure of practical significance; in this case, a difference in one standard deviation on the input scale is a meaningful difference in that it roughly reflects a typical difference between the mean and a randomly drawn observation. In addition, standardizing predictors using $z$-scores will change our interpretation of the intercept to the mean of $y$ when all predictor values are at their mean values.

It can often be preferable, however, to divide by 2 standard deviations to allow inferences to be more consistent with those for binary inputs, as we discuss in Section 12.2.

Standardization using the sample mean and standard deviation of the predictors uses raw estimates from the data and thus should be used only when the the number of observations is big enough that these estimates are stable. When sample size is small, we recommend standardizing using an externally specified population distribution or other externally specified reasonable scales.

## Standardization using an externally specified population distribution

A related approach is to rescale based on some standard set outside the data. For example, in analyses of test scores it is common to express estimates on the scale of standard deviations of test scores across all students in a grade. A test might be on a 0–100 scale, with fourth graders having a national mean score of 55 and standard deviation of 18. Then if the analysis is done on the scale of points on the exam, all coefficient estimates and standard errors from analyses of fourth graders are divided by 18 so that they are on this universal scale. Equivalently, one could first define $z = (y - 55)/18$ for all fourth graders and then run all regressions on $z$. The virtue of using a fixed scaling, rather than standardizing each dataset separately, is that estimates are all directly comparable.

### Standardization using reasonable scales

Sometimes it is useful to keep inputs on familiar scales such as inches, dollars, or years, but make convenient rescalings to aid in the interpretability of coefficients. For example, we might work with income/$10 000 or age in decades.

For another example, in Section 10.9 we analyzed party identification, a variable on a 1–7 scale: 1 = strong Democrat, 2 = Democrat, 3 = weak Democrat, 4 = independent, 5 = weak Republican, 6 = Republican, 7 = strong Republican. Rescaling to (pid − 4)/4 gives us a variable that equals −0.5 for Democrats, 0 for moderates, and +0.5 for Republicans, and so the coefficient on this variable is directly interpretable, with a change of 1 comparing a Democrat to a Republican.

## 12.2 Centering and standardizing for models with interactions

xample: hildren's ꞯ tests

Figure 12.1b illustrates the difficulty of interpreting the intercept term in a regression in a setting where it does not make sense to consider predictors set to zero. More generally, similar challenges arise in interpreting coefficients in models with interactions, as we saw in Section 10.3 with the following model:[1]

```
              Median MAD_SD
(Intercept)    -8.0   13.2
mom_hs         47.0   14.5
mom_iq          0.9    0.1
mom_hs:mom_iq -0.5    0.2

Auxiliary parameter(s):
      Median MAD_SD
sigma 18.0   0.6
```

The coefficient on mom_hs is 47.0—does this mean that children with mothers who graduated from high school perform, on average, 47.0 points better on their tests? No. The model includes an interaction, and 47.0 is the predicted difference for kids that differ in mom_hs, *among those with* mom_iq = 0. Since mom_iq is never even close to zero (see Figure 10.4), the comparison at zero, and thus the coefficient of 47.0, is essentially meaningless.

Similarly, the coefficient of 0.9 for the "main effect" of mom_iq is the slope for this variable, among those children for whom mom_hs = 0. This is less of a stretch, as mom_hs actually does equal zero for many of the cases in the data (see Figure 10.1) but still can be somewhat misleading since mom_hs = 0 is at the edge of the data so that this coefficient cannot be interpreted as an average over the general population.

### Centering by subtracting the mean of the data

We can simplify the interpretation of the regression model by first subtracting the mean of each input variable:

```
kidiq$c_mom_hs <- kidiq$mom_hs - mean(kidiq$mom_hs)
kidiq$c_mom_iq <- kidiq$mom_iq - mean(kidiq$mom_iq)
```

Each main effect now corresponds to a predictive difference with the other input at its average value:

```
                  Median MAD_SD
(Intercept)       87.6    0.9
c_mom_hs           2.9    2.4
c_mom_iq           0.6    0.1
c_mom_hs:c_mom_iq -0.5    0.2
```

[1]Data and code for this example are in the folder KidIQ.

```
Auxiliary parameter(s):
      Median MAD_SD
sigma 18.0   0.6
```

The residual standard deviation does not change—linear transformation of the predictors does not affect the fit of the model—and the coefficient and standard error of the interaction did not change, but the main effects and the intercept change a lot and are now interpretable based on comparison to the mean of the data.

### Using a conventional centering point

Another option is to center based on an understandable reference point, for example, the midpoint of the range for mom_hs and the population average IQ:

```
kidiq$c2_mom_hs <- kidiq$mom_hs - 0.5
kidiq$c2_mom_iq <- kidiq$mom_iq - 100
```

In this parameterization, the coefficient of c2_mom_hs is the average predictive difference between a child with mom_hs = 1 and a child with mom_hs = 0, among those children with mom_iq = 100. Similarly, the coefficient of c2_mom_iq corresponds to a comparison under the condition mom_hs = 0.5, which includes no actual data but represents a midpoint of the range.

```
                    Median MAD_SD
(Intercept)         86.8    1.2
c2_mom_hs            2.9    2.3
c2_mom_iq            0.7    0.1
c2_mom_hs:c2_mom_iq -0.5    0.2

Auxiliary parameter(s):
      Median MAD_SD
sigma 18.0   0.6
```

Once again, the residual standard deviation and coefficient for the interaction have not changed. The intercept and main effect have changed very little, because the points 0.5 and 100 happen to be close to the mean of mom_hs and mom_iq in the data.

### Standardizing by subtracting the mean and dividing by 2 standard deviations

Centering helped us interpret the main effects in the regression, but it still leaves us with a scaling problem. The coefficient of mom_hs is much larger than that of mom_iq, but this is misleading, considering that we are comparing the complete change in one variable (mother completed high school or not) to a mere 1-point change in mother's IQ, which is not much at all; see Figure 10.4.

A natural step is to scale the predictors by dividing by 2 standard deviations—we shall explain shortly why we use 2 rather than 1—so that a 1-unit change in the rescaled predictor corresponds to a change from 1 standard deviation below the mean, to 1 standard deviation above. Here are the rescaled predictors in the child testing example:

```
kidiq$z_mom_hs <- (kidiq$mom_hs - mean(kidiq$mom_hs))/(2*sd(kidiq$mom_hs))
kidiq$z_mom_iq <- (kidiq$mom_iq - mean(kidiq$mom_iq))/(2*sd(kidiq$mom_iq))
```

We can now interpret all the coefficients on a roughly common scale (except for the intercept, which now corresponds to the average predicted outcome with all inputs at their mean):

```
                  Median MAD_SD
(Intercept)        87.6    0.9
z_mom_hs            2.3    2.1
z_mom_iq           17.7    1.8
z_mom_hs:z_mom_iq -11.9    4.0
```

```
Auxiliary parameter(s):
      Median MAD_SD
sigma 18.0   0.6
```

### Why scale by 2 standard deviations?

We divide by 2 standard deviations rather than 1 because this is consistent with what we do with binary input variables. To see this, consider the simplest binary $x$ variable, which takes on the values 0 and 1, each with probability 0.5. The standard deviation of $x$ is then $\sqrt{0.5 * 0.5} = 0.5$, and so the standardized variable, $(x - \mu_x)/(2\sigma_x)$, takes on the values $\pm 0.5$, and its coefficient reflects comparisons between $x = 0$ and $x = 1$. In contrast, if we had divided by 1 standard deviation, the rescaled variable takes on the values $\pm 1$, and its coefficient corresponds to half the difference between the two possible values of $x$. This identity is close to precise for binary inputs even when the frequencies are not exactly equal, since $\sqrt{p(1-p)} \approx 0.5$ when $p$ is not too far from 0.5.

In a complicated regression with many predictors, it can make sense to leave binary inputs as is and linearly transform continuous inputs, possibly by scaling using the standard deviation. In this case, dividing by 2 standard deviations ensures a rough comparability in the coefficients. In our children's testing example, the predictive difference corresponding to 2 standard deviations of mother's IQ is clearly much higher than the comparison of mothers with and without a high school education.

### Multiplying each regression coefficient by 2 standard deviations of its predictor

For models with no interactions, a procedure that is equivalent to centering and rescaling is to leave the regression predictors as is, and then create rescaled regression coefficients by multiplying each $\beta$ by two times the standard deviation of its corresponding $x$. This gives a sense of the importance of each variable, adjusting for all the others in the linear model. As noted, scaling by 2 (rather than 1) standard deviations allows these scaled coefficients to be comparable to unscaled coefficients for binary predictors.

## 12.3 Correlation and "regression to the mean"

Consider a regression with a constant term and one predictor; thus, $y = a + bx + \text{error}$. If both of the variables $x$ and $y$ are standardized—that is, if they are defined as `x <- (x-mean(x))/sd(x)` and `y <- (y-mean(y))/sd(y)`—then the regression intercept is zero, and the slope is simply the correlation between $x$ and $y$. Thus, the slope of a regression of two standardized variables must always be between $-1$ and 1, or, to put it another way, if a regression slope is more than 1 in absolute value, then the variance of $y$ must exceed that of $x$. In general, the slope of a regression with one predictor is $b = \rho\sigma_y/\sigma_x$, where $\rho$ is the correlation between the two variables and $\sigma_x$ and $\sigma_y$ are the standard deviations of $x$ and $y$.

### The principal component line and the regression line

Some of the confusing aspects of regression can be understood in the simple case of standardized variables. Figure 12.2 shows a simulated-data example of standardized variables with correlation (and thus regression slope) 0.5. Figure 12.2a shows the *principal component line*, which goes closest through the cloud of points, in the sense of minimizing the sum of squared distances between the points and the line. The principal component line in this case is simply $y = x$.

Figure 12.2b shows the *regression line*, which minimizes the sum of the squares of the *vertical* distances between the points and the line—it is the familiar least squares line, $y = \hat{a} + \hat{b}x$, with $\hat{a}, \hat{b}$

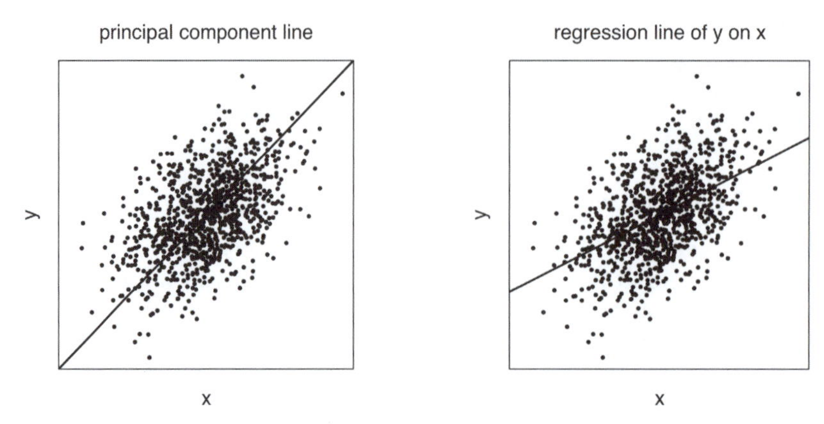


Figure 12.2 *Data simulated from a bivariate normal distribution with correlation 0.5. (a) The principal component line goes closest through the cloud of points. (b) The regression line, which represents the best prediction of $y$ given $x$, has half the slope of the principal component line.*

chosen to minimize $\sum_{i=1}^{n}(y_i - (\hat{a} + \hat{b}x_i))^2$. In this case, $\hat{a} = 0$ and $\hat{b} = 0.5$; the regression line thus has slope 0.5.

When given this sort of scatterplot (without any lines superimposed) and asked to draw the regression line of $y$ on $x$, students tend to draw the principal component line, which is shown in Figure 12.2a. However, for the goal of predicting $y$ from $x$, or for estimating the average of $y$ for any given value of $x$, the regression line is in fact better—even if it does not appear so at first.

The superiority of the regression line for estimating the average of $y$ given $x$ can be seen from a careful study of Figure 12.2. For example, consider the points at the extreme left of either graph. They all lie above the principal component line but are roughly half below and half above the regression line. Thus, the principal component line underpredicts $y$ for low values of $x$. Similarly, a careful study of the right side of each graph shows that the principal component line overpredicts $y$ for high values of $x$. In contrast, the regression line again gives unbiased predictions, in the sense of going through the average of $y$ given $x$.

## Regression to the mean

This all connects to our earlier discussion of "regression to the mean" in Section 6.5. When $x$ and $y$ are standardized (that is, placed on a common scale, as in Figure 12.2), the regression line always has slope less than 1. Thus, when $x$ is 1 standard deviation above the mean, the predicted value of $y$ is somewhere between 0 and 1 standard deviations above the mean. This phenomenon in linear models—that $y$ is predicted to be closer to the mean (in standard-deviation units) than $x$—is called *regression to the mean* and occurs in many vivid contexts.

For example, if a woman is 10 inches taller than the average for her sex, and the correlation of mothers' and adult daughters' heights is 0.5, then her daughter's predicted height is 5 inches taller than the average. She is expected to be taller than average, but not so much taller—thus a "regression" (in the nonstatistical sense) to the average.

A similar calculation can be performed for any variables that are not perfectly correlated. For example, let $x_i$ and $y_i$ be the number of games won by football team $i$ in two successive seasons. They will not be correlated 100%; thus, we expect the teams that did the best in season 1 (that is, with highest values of $x$) to do not as well in season 2 (that is, we expect their values of $y$ to be closer to the average for all the teams). Similarly, we expect teams with poor records in season 1 to improve on average in season 2, relative to the other teams.

A naive interpretation of regression to the mean is that heights, or football records, or other variable phenomena become more and more "average" over time. As discussed in Section 6.5, this view is mistaken because it ignores the error in the regression predicting $y$ from $x$. For any data point $x_i$, the point prediction for its $y_i$ will be regressed toward the mean, but the actual observed $y_i$ will not be exactly where it is predicted. Some points end up falling closer to the mean and some fall further. This can be seen in Figure 12.2b.

## 12.4 Logarithmic transformations

When additivity and linearity are not reasonable assumptions (see Section 11.1), a nonlinear transformation can sometimes remedy the situation. It commonly makes sense to take the logarithm of outcomes that are all-positive. For outcome variables, this becomes clear when we think about making predictions on the original scale. The regression model imposes no constraints that would force these predictions to be positive as well. However, if we take the logarithm of the variable, run the model, make predictions on the log scale, and then transform back by exponentiating, the resulting predictions are necessarily positive because for any real $a$, $\exp(a) > 0$.

Perhaps more important, a linear model on the logarithmic scale corresponds to a multiplicative model on the original scale. Consider the linear regression model,

$$\log y_i = b_0 + b_1 X_{i1} + b_2 X_{i2} + \cdots + \epsilon_i.$$

Exponentiating both sides yields

$$\begin{aligned} y_i &= e^{b_0 + b_1 X_{i1} + b_2 X_{i2} + \cdots + \epsilon_i} \\ &= B_0 B_1^{X_{i1}} B_2^{X_{i2}} \cdots E_i, \end{aligned}$$

where $B_0 = e^{b_0}$, $B_1 = e^{b_1}$, $B_2 = e^{b_2}$, ... are exponentiated regression coefficients (and thus are positive), and $E_i = e^{\epsilon_i}$ is the exponentiated error term (also positive). On the scale of the original data $y_i$, the predictors $X_{i1}, X_{i2}, \ldots$ come in multiplicatively.

In Section 3.4, we discussed the connections between logarithmic transformations and exponential and power-law relationships; here we consider these in the context of regression.

### Earnings and height example

ample: rnings d height

We illustrate logarithmic regression by considering models predicting earnings from height.[2] Expression (12.1) shows a linear regression of earnings on height. However, it really makes more sense to model earnings on the logarithmic scale, as long as we exclude those people who reported zero earnings. We can fit a regression to log earnings and then take the exponential to get predictions on the original scale.

**Direct interpretation of small coefficients on the log scale.** We take the logarithm of earnings and regress on height,

```
logmodel_1 <- stan_glm(log(earn) ~ height, data=earnings, subset=earn>0)
print(logmodel_1)
```

yielding the following estimate:

```
            Median MAD_SD
(Intercept) 5.91   0.38
height      0.06   0.01

Auxiliary parameter(s):
      Median MAD_SD
sigma 0.88   0.02
```

[2]Data and code for this example are in the folder Earnings.

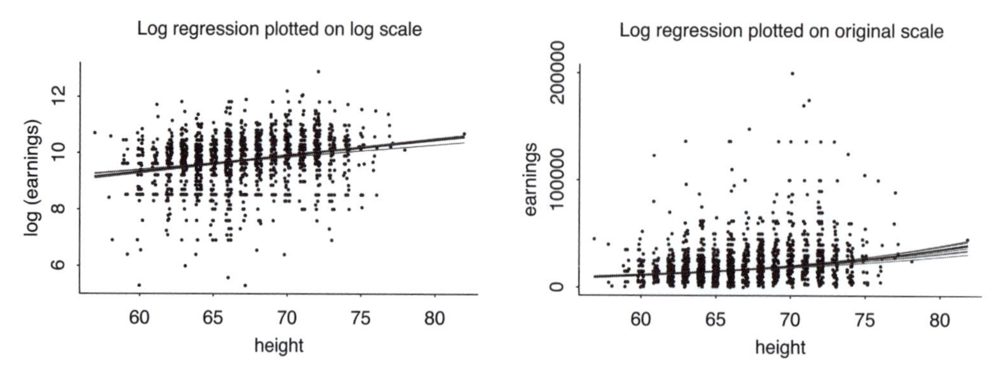


Figure 12.3 *Regression of earnings on log(height), with curves showing uncertainty the model, log(earnings) = $a + b * $height, fit to data with positive earnings. The data and fit are plotted on the logarithmic and original scales. Compare to the linear model, shown in Figure 12.1a. To improve resolution, a data point at earnings of* \$400 000 *has been excluded from the original-scale graph.*

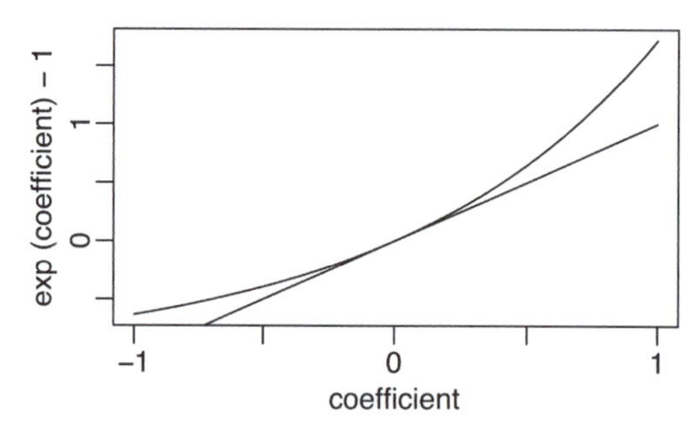


Figure 12.4 *Interpretation of exponentiated coefficients in a logarithmic regression model as relative difference (curved upper line), and the approximation exp($x$) = $1 + x$, which is valid for small coefficients $x$ (straight line).*

Figure 12.3 shows the data and fitted regression on the log and linear scales.

The estimate $\hat{\beta}_1 = 0.06$ implies that a difference of 1 inch in height corresponds to an expected difference of 0.06 in log(earnings), so that earnings are multiplied by exp(0.06). But exp(0.06) $\approx$ 1.06 (more precisely, 1.062). Thus, a difference of 1 in the predictor corresponds to an expected positive difference of about 5% in the outcome variable. Similarly, if $\beta_1$ were $-0.06$, then a positive difference of 1 inch of height would correspond to an expected *negative* difference of about 6% in earnings.

This correspondence becomes more nonlinear as the magnitude of the coefficient increases. Figure 12.4 displays the deterioration of the correspondence as the coefficient size increases. The plot is restricted to coefficients in the range $(-1, 1)$ because, on the log scale, regression coefficients are typically (though not always) less than 1. A coefficient of 1 on the log scale implies that a change of one unit in the predictor is associated with a change of exp(1) = 2.7 in the outcome, and, if predictors are parameterized in a reasonable way, it is unusual to see effects of this magnitude.

**Predictive checking.** One way to get a sense of fit is to simulate replicated datasets from the fitted model and compare them to the observed data. We demonstrate for the height and earnings regression.

First we simulate new data:

```
yrep_1 <- posterior_predict(fit_1)
```

The above code returns a matrix in which each row is a replicated dataset from the posterior distribution of the fitted regression of earnings on height. We then plot the density of the observed earnings data, along with 100 draws of the distribution of replicated data:

```
n_sims <- nrow(yrep_1)
```

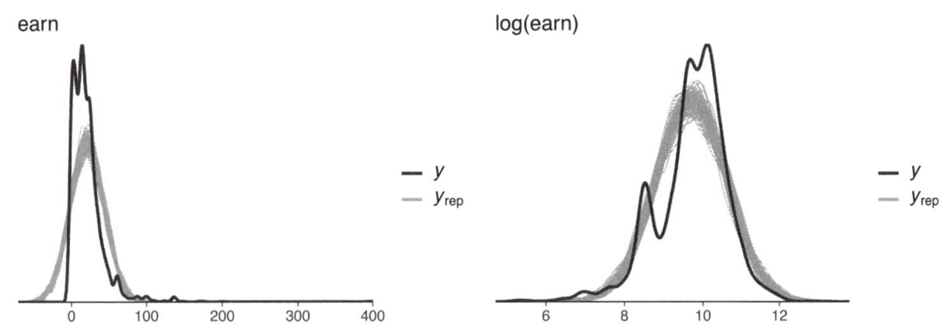


Figure 12.5 *Posterior predictive checks comparing the density plot of earnings data (dark line) to 100 predictive replications (gray lines) of replicated data from fitted models (a) on the original scale, and (b) on the log scale. Both models show some lack of fit. The problem is particularly obvious with the linear-scale regression, as the observed earnings are all positive (with the density function including a small negative tail just as an artifact of the smoothing procedure) and the replicated data include many negative values. The non-smoothed aspects of the observed data arise from discreteness in the survey responses.*

```
subset <- sample(n_sims, 100)
library("bayesplot")
ppc_dens_overlay(earnings$earn, yrep_1[subset,])
```

The result is shown in Figure 12.5a. Unsurprisingly, the fit on the untransformed scale is poor: observed earnings in these data are always positive, while the predictive replications contain many negative values.

We can then do the same predictive checking procedure for the model fit on the log scale, first simulating the predictions:

```
yrep_log_1 <- posterior_predict(logmodel_1)
```

Then we plot 100 simulations along with the observed data:

```
n_sims <- nrow(yrep_log_1)
subset <- sample(n_sims, 100)
ppc_dens_overlay(log(earnings$earn[earnings$earn>0]), yrep_log_1[subset,])
```

The resulting fit on the log scale is not perfect (see Figure 12.5b), which could be of interest, depending on one's goal in fitting the model. The point of this example is that, as a model is altered, we can perform predictive checks to assess different aspects of fit. Here we looked at the marginal distribution of the data, but more generally one can look at other graphical summaries.

## Why we use natural log rather than log base 10

We prefer natural logs (that is, logarithms base $e$) because, as described above, coefficients on the natural-log scale are directly interpretable as approximate proportional differences: with a coefficient of 0.05, a difference of 1 in $x$ corresponds to an approximate 5% difference in $y$, and so forth. Natural log is sometimes written as "ln," but we simply write "log" since this is our default.

Another approach is to take logarithms base 10, which we write as $\log_{10}$. The connection between the two different scales is that $\log_{10}(x) = \log(x)/\log(10) = \log(x)/2.30$. The advantage of $\log_{10}$ is that the predicted values themselves are easier to interpret; for example, when considering the earnings regressions, $\log_{10}(10\,000) = 4$ and $\log_{10}(100\,000) = 5$, and with some experience we can also quickly read off intermediate values—for example, if $\log_{10}(\text{earnings}) = 4.5$, then earnings $\approx 30\,000$.

The disadvantage of $\log_{10}$ is that the resulting coefficients are harder to interpret. For example, if we fit the earnings regression on the $\log_{10}$ scale,

```
logmodel_1a <- stan_glm(log10(earn) ~ height, data=earnings, subset=earn>0)
```

we get,

```
            Median MAD_SD
(Intercept) 2.57   0.16
height      0.02   0.00

Auxiliary parameter(s):
      Median MAD_SD
sigma 0.38   0.01
```

The coefficient of 0.02 tells us that a difference of 1 inch in height corresponds to a difference of 0.02 in $\log_{10}$(earnings), that is, a multiplicative difference of $10^{0.02} = 1.06$ (after fixing roundoff error). This is the same 6% change as before, but it cannot be seen by simply looking at the coefficient as could be done on the natural-log scale.

## Building a regression model on the log scale

**Adding another predictor.** A difference of an inch of height corresponds to a difference of 6% in earnings—that seems like a lot! But men are mostly taller than women and also tend to have higher earnings. Perhaps the 6% predictive difference can be understood by differences between the sexes. Do taller people earn more, on average, than shorter people of the same sex?

```
logmodel_2 <- stan_glm(log(earn) ~ height + male, data=earnings, subset=earn>0)
```

```
            Median MAD_SD
(Intercept) 7.97   0.51
height      0.02   0.01
male        0.37   0.06

Auxiliary parameter(s):
      Median MAD_SD
sigma 0.87   0.02
```

After adjusting for sex, each inch of height corresponds to an estimated predictive difference of 2%: under this model, two people of the same sex but differing by 1 inch in height will differ, on average, by 2% in earnings. The predictive comparison of sex, however, is huge: comparing a man and a woman of the same height, the man's earnings are exp(0.37) = 1.45 times the woman's, that is, 45% more. (We cannot simply convert the 0.37 to 45% because this coefficient is not so close to zero; see Figure 12.4.) This coefficient also is not easily interpretable. Does it mean that "being a man" *causes* one to earn nearly 50% more than a woman? We will explore this sort of troubling question in the causal inference chapters in Part 5 of the book.

**Naming inputs.** Incidentally, we named this new input variable `male` so that it could be immediately interpreted. Had we named it `sex`, for example, we would always have to go back to the coding to check whether 0 and 1 referred to men and women, or vice versa. Another approach would be to consider `sex` as a factor with two named levels, `male` and `female`; see page 198. Our point here is that, if the variable is coded numerically, it is convenient to give it the name `male` corresponding to the coding of 1.

**Residual standard deviation and $R^2$.** Finally, the regression model has a residual standard deviation $\sigma$ of 0.87, implying that approximately 68% of log earnings will be within 0.87 of the predicted value. On the original scale, approximately 68% of earnings will be within a factor of exp(0.87) = 2.4 of the prediction. For example, a 70-inch person has predicted earnings of $7.97 + 0.02 * 70 = 9.37$, with a predictive standard deviation of approximately 0.87. Thus, there is an approximate 68% chance that this person has log earnings in the range $[9.37 \pm 0.87] = [8.50, 10.24]$, which corresponds to earnings in the range [exp(8.50), exp(10.24)]] = [5000, 28 000]. This wide range tells us that the regression model does not predict earnings well—it is not very impressive to have a prediction that can be wrong by a factor of 2.4—and this is also reflected in $R^2$, which is only 0.08, indicating that only 8% of the variance in the log transformed data is explained by the regression model. This low $R^2$ manifests itself graphically in Figure 12.3, where the range of the regression predictions is clearly much narrower than the range of the data.

**Including an interaction.** We now consider a model with an interaction between height and sex, so that the predictive comparison for height can differ for men and women:

```
logmodel_3 <- stan_glm(log(earn) ~ height + male + height:male,
  data=earnings, subset=earn>0)
```

which yields,

```
            Median MAD_SD
(Intercept)  8.48   0.66
height       0.02   0.01
male        -0.76   0.94
height:male  0.02   0.01

Auxiliary parameter(s):
      Median MAD_SD
sigma 0.87   0.02
```

That is,

$$\log(\text{earnings}) = 8.48 + 0.02 * \text{height} - 0.76 * \text{male} + 0.02 * \text{height} * \text{male}. \quad (12.2)$$

We shall try to interpret each of the four coefficients in this model.

- The *intercept* is the predicted log earnings if `height` and `male` both equal zero. Because heights are never close to zero, the intercept has no direct interpretation.
- The coefficient for `height`, 0.02, is the predicted difference in log earnings corresponding to a 1-inch difference in height, if `male` equals zero. Thus, the estimated predictive difference per inch of height is 2% for women, with some uncertainty as indicated by the standard error of 0.01.
- The coefficient for `male` is the predicted difference in log earnings between women and men, if `height` equals 0. Heights are never close to zero, and so the coefficient for `male` has *no direct interpretation* in this model. If you want to interpret it, you can move to a more relevant value for `height`; as discussed in Section 12.2, it makes sense to use a centered parameterization.
- The coefficient for `height:male` is the difference in slopes of the lines predicting log earnings on height, comparing men to women. Thus, a difference of an inch of height corresponds to 2% more of a difference in earnings among men than among women, and the estimated predictive difference per inch of height among men is 2% + 2% = 4%.

The interaction coefficient has a large standard error, which tells us that the point estimate is uncertain and could change in sign and magnitude if additional data were fed into the analysis.

**Linear transformation to make coefficients more interpretable.** We can make the parameters in the interaction model more easily interpretable by rescaling the height predictor to have a mean of 0 and standard deviation 1:

```
earnings$z_height <- (earnings$height - mean(earnings$height))/sd(earnings$height)
```

The mean and standard deviation of heights in these data are 66.6 inches and 3.8 inches, respectively. Fitting the model to `z_height`, `male`, and their interaction yields,

```
             Median MAD_SD
(Intercept)  9.55   0.04
z_height     0.06   0.04
```

```
male           0.35   0.06
z_height:male  0.08   0.06

Auxiliary parameter(s):
      Median MAD_SD
sigma 0.87   0.02
```

We can now interpret all four of the coefficients:

- The *intercept* is the predicted log earnings if z_height and male both equal zero. Thus, a 66.6-inch-tall woman is predicted to have log earnings of 9.55, or earnings of exp(9.55) = 14 000.
- The coefficient for z_height is the predicted difference in log earnings corresponding to a 1 standard deviation difference in height, if male equals zero. Thus, the estimated predictive difference for a 3.8-inch increase in height is 6% for women (but with a standard error indicating much uncertainty in this coefficient).
- The coefficient for male is the predicted difference in log earnings between women and men, if z_height equals 0. Thus, a 66.6-inch-tall man is predicted to have log earnings that are 0.35 higher than that of a 66.6-inch-tall woman. This corresponds to a ratio of exp(0.35) = 1.42, so the man is predicted to have 42% higher earnings than the woman.
- The coefficient for z_height:male is the difference in slopes between the predictive differences for height among women and men. Thus, comparing two men who differ by 3.8 inches in height, the model predicts a difference of $0.06 + 0.08 = 0.14$ in log earnings, thus a ratio of exp(0.14) = 1.15, a difference of 15%.

One might also consider centering the predictor for sex, but here it is easy enough to interpret male = 0, which corresponds to the baseline category (in this case, women).

### Further difficulties in interpretation

For a glimpse into yet another challenge in interpreting regression coefficients, consider the simpler log earnings regression without the interaction term. The predictive interpretation of the height coefficient is simple enough: comparing two adults of the same sex, the taller person will be expected to earn 2% more per inch of height; see the model on page 192. This seems to be a reasonable comparison.

To interpret the coefficient for male, we would say that comparing two adults of the same height but different sex, the man will be expected to earn 45% more on average. But how clear is it for us to interpret this comparison? For example, if we are comparing a 66-inch woman to a 66-inch man, then we are comparing a tall woman to a short man. So, in some sense, they do not differ only in sex. Perhaps a more reasonable or relevant comparison would be of an "average woman" to an "average man."

The ultimate solution to this sort of problem must depend on why the model is being fit in the first place. For now we shall focus on the technical issues of fitting reasonable models to data. We discuss causal interpretations in Chapters 18–21.

### Log-log model: transforming the input and outcome variables

If the log transformation is applied to an input variable as well as the outcome, the coefficient can be interpreted as the expected proportional difference in $y$ per proportional difference in $x$. For example:

```
earnings$log_height <- log(earnings$height)
logmodel_5 <- stan_glm(log(earn) ~ log_height + male, data=earnings, subset=earn>0)
```

yields,

```
             Median MAD_SD
(Intercept) 2.76   2.19
log_height  1.62   0.53
male        0.37   0.06

Auxiliary parameter(s):
      Median MAD_SD
sigma 0.87   0.01
```

For each 1% difference in height, the predicted difference in earnings is 1.62%. The other input, male, is categorical so it does not make sense to take its logarithm.

In economics, the coefficient in a log-log model is sometimes called an "elasticity"; see Exercise 12.11 for an example.

### Taking logarithms even when not necessary

If a variable has a narrow dynamic range (that is, if the ratio between the high and low values is close to 1), then it will not make much of a difference in fit if the regression is on the logarithmic or the original scale. For example, the standard deviation of log_height in our survey data is 0.06, meaning that heights in the data vary by only approximately a factor of 6%.

In such a situation, it might seem to make sense to stay on the original scale for reasons of simplicity. However, the logarithmic transformation can make sense even here, because coefficients are often more easily understood on the log scale. The choice of scale comes down to interpretability: whether it is easier to understand the model as proportional increase in earnings per inch, or per proportional increase in height.

For an input with a larger amount of relative variation (for example, heights of children, or weights of animals), it would make sense to work with its logarithm immediately, both as an aid in interpretation and likely as an improvement in fit too.

## 12.5 Other transformations

### Square root transformations

The square root is sometimes useful for compressing high values more mildly than is done by the logarithm. Consider again our height and earnings example.

Fitting a linear model on the raw, untransformed scale seemed inappropriate. Expressed in a different way than before, we would expect the differences between people earning nothing versus those earning \$10 000 to be far greater than the differences between people earning, say, \$80 000 versus \$90 000. But under the linear model, these are all equal increments as in model (12.1), where an extra inch is worth \$1300 more in earnings at all levels.

On the other hand, the log transformation seems too severe with these data. With logarithms, the differences between populations earning \$5000 versus \$10 000 is equivalent to the differences between those earning \$40 000 and those earning \$80 000. On the square root scale, however, the predicted differences between the \$0 earnings and \$10 000 earnings groups are the same as comparisons between \$10 000 and \$40 000 or between \$40 000 and \$90 000, in each case stepping up by 100 in square root of earnings. See Chapter 17 for more on this example.

Unfortunately, models on the square root scale lack the clean interpretation of the original-scale and log-transformed models. For one thing, large negative predictions on this scale get squared and become large positive values on the original scale, thus introducing a nonmonotonicity in the model. We are more likely to use the square root model for prediction than within models whose coefficients we want to understand.

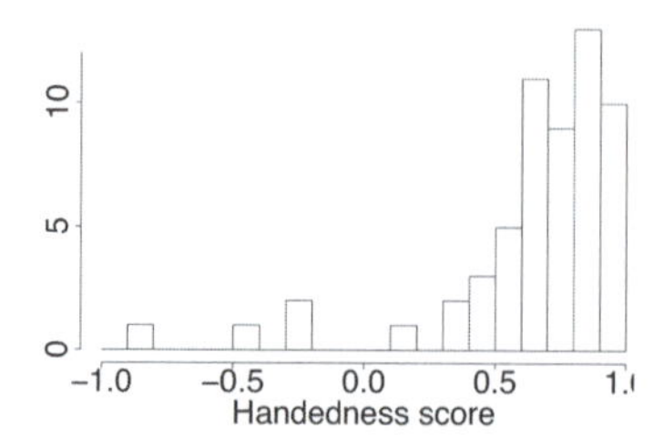


Figure 12.6 *Histogram of handedness scores of a sample of students. Scores range from −1 (completely left-handed) to +1 (completely right-handed) and are based on the responses to 10 questions such as "Which hand do you write with?" and "Which hand do you use to hold a spoon?" The continuous range of responses shows the limitations of treating handedness as a dichotomous variable.*

### Idiosyncratic transformations

Sometimes it is useful to develop transformations tailored for specific problems. For example, with the original height-earnings data, it would have not been possible to simply take the logarithm of earnings, as many observations had zero values. Instead, a model can be constructed in two steps: first model the probability that earnings exceed zero (for example, using a logistic regression; see Chapter 13); then fit a linear regression, conditional on earnings being positive, which is what we did in the example above. One could also model total income, but economists are often interested in modeling earnings alone, excluding so-called unearned income.

In any case, plots and simulations should definitely be used to summarize inferences, since the coefficients of the two parts of the model combine nonlinearly in their joint prediction of earnings. We discuss this sort of model further in Section 15.8.

What sort of transformed scale would be appropriate for a variable such as "assets" that can be negative, positive, or zero? One possibility is a discrete coding that compresses the high range, for example, 0 for assets between −\$100 and \$100, 1 for assets between \$100 and \$1000, 2 for assets between \$1000 and \$10 000, −1 for assets between −\$100 and −\$1000, and so forth. Such a mapping could be expressed more fully as a continuous transformation, but for explanatory purposes it can be convenient to use a discrete scale.

### Using continuous rather than discrete predictors

Many variables that appear binary or discrete can usefully be viewed as continuous. For example, rather than define "handedness" as −1 for left-handers and +1 for right-handers, one can use a standard 10-question handedness scale that gives an essentially continuous scale from −1 to 1 (see Figure 12.6).

We avoid discretizing continuous variables (except as a way of simplifying a complicated transformation, as described previously, or to model nonlinearity, as described later). A common mistake is to take a numerical measure and replace it with a binary "pass/fail" score. For example, suppose we tried to predict election winners, rather than continuous votes. Such a model would not work as well, as it would discard much of the information in the data (for example, the distinction between a candidate receiving 51% or 65% of the vote). Even if our only goal is to predict the winners, we are better off predicting continuous vote shares and then transforming them into predictions about winners, as in our example with congressional elections in Section 10.6.

### Using discrete rather than continuous predictors

In some cases, however, it is convenient to discretize a continuous variable if a simple parametric relation does not seem appropriate. For example, in modeling political preferences, it can make sense to include age with four indicator variables: 18–29, 30–44, 45–64, and 65+, to allow for different sorts of generational patterns. This kind of discretization is convenient, since, conditional on the discretization, the model remains linear. We briefly mention more elaborate nonlinear models for continuous predictors in Section 22.7.

xample: hildren's ) tests

We demonstrate inference with discrete predictors using an example from Chapter 10 of models for children's test scores given information about their mothers. Another input variable that can be used in these models is maternal employment, which is defined on a four-point ordered scale:

- mom_work = 1: mother did not work in first three years of child's life
- mom_work = 2: mother worked in second or third year of child's life
- mom_work = 3: mother worked part-time in first year of child's life
- mom_work = 4: mother worked full-time in first year of child's life.

Fitting a simple model using discrete predictors yields,

```
                      Median MAD_SD
(Intercept)           82.0    2.2
as.factor(mom_work)2  3.8     3.0
as.factor(mom_work)3 11.4     3.5
as.factor(mom_work)4  5.1     2.7

Auxiliary parameter(s):
      Median MAD_SD
sigma 20.2   0.7
```

This parameterization of the model allows for different averages for the children of mothers corresponding to each category of maternal employment. The "baseline" category (mom_work = 1) corresponds to children whose mothers do not go back to work at all in the first three years after the child is born; the average test score for these children is estimated by the intercept, 82.0. The average test scores for the children in the other categories is found by adding the corresponding coefficient to this baseline average. This parameterization allows us to see that the children of mothers who work part-time in the first year after the child is born achieve the highest average test scores, $82.0 + 11.4$. These families also tend to be the most advantaged in terms of many other sociodemographic characteristics as well, so a causal interpretation is not warranted unless these variables are included in the model.

### Index and indicator variables

*Index variables* divide a population into categories. For example:

- male = 1 for males and 0 for females
- age = 1 for ages 18–29, 2 for ages 30–44, 3 for ages 45–64, 4 for ages 65+
- state = 1 for Alabama, ..., 50 for Wyoming
- county indexes for the 3082 counties in the United States.

*Indicator variables* are 0/1 predictors based on index variables, as discussed in Section 10.4. For example:

- sex_1 = 1 for females and 0 otherwise
  sex_2 = 1 for males and 0 otherwise
- age_1 = 1 for ages 18–29 and 0 otherwise
  age_2 = 1 for ages 30–44 and 0 otherwise
  age_3 = 1 for ages 45–64 and 0 otherwise
  age_4 = 1 for ages 65+ and 0 otherwise
- 50 indicators for state
- 3082 indicators for county.

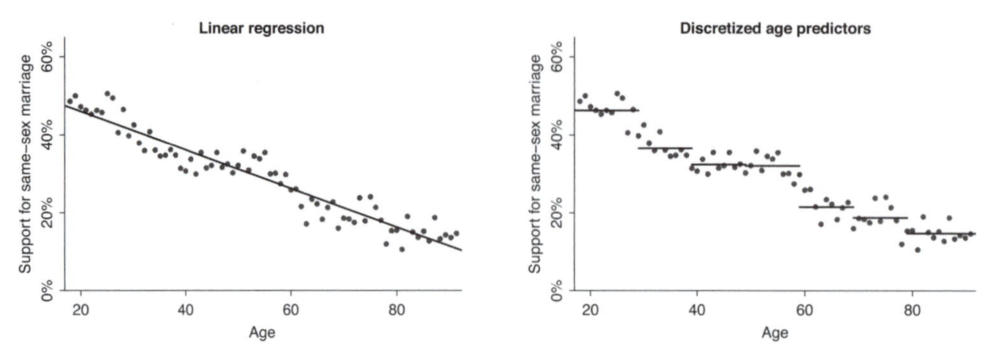


Figure 12.7 *Support for same-sex marriage as a function of age, from a national survey taken in 2004. Fits are shown from two linear regression: (a) using age as a predictor, (b) using indicators for age, discretized into categories.*

Including these variables as regression predictors allows for different means for the populations corresponding to each of the categories delineated by the variable.

When an input has only two levels, we prefer to code it with a single variable and name it appropriately; for example, as discussed earlier with the earnings example, the name `male` is more descriptive than `sex_1` and `sex_2`.

R also allows variables to be included as *factors* with named *levels*; for example, `sex` could have the levels `male` and `female`.

Figure 12.7 demonstrates with a simple example showing support for same-sex marriage as a function of age (and with the few respondents reporting ages greater than 90 all assigned the age of 91 for the purpose of this analysis). Here is the result of a linear regression using an indicator for each decade of age, with age under 30 as the reference category:

```
                              Median MAD_SD
(Intercept)                    0.46   0.01
factor(age_discrete)(29,39]   -0.10   0.01
factor(age_discrete)(39,49]   -0.14   0.01
factor(age_discrete)(49,59]   -0.14   0.01
factor(age_discrete)(59,69]   -0.25   0.01
factor(age_discrete)(69,79]   -0.28   0.01
factor(age_discrete)(79,100]  -0.32   0.01

Auxiliary parameter(s):
      Median MAD_SD
sigma 0.03   0.00
```

Figure 12.7a shows the fitted linear regression on age, and Figure 12.7b shows the fit from the linear regression using age indicators: the first bar is at $y = 0.46$, the second is at $0.46 - 0.10$, the third is at $0.46 - 0.14$, and so on. Neither of the two fits in Figure 12.7 is perfect; indeed Figure 12.7b gives a somewhat misleading picture, with the eye being drawn too strongly to the horizontal lines. One reason why we show both graphs is to give two perspectives on the data. For example, the dots in Figure 12.7a show a steady downward trend between the ages of 25 and 40, but in Figure 12.7b, that pattern is obscured by the fitted lines.

## Indicator variables, identifiability, and the baseline condition

As discussed in Section 10.7, a regression model is nonidentifiable if its predictors are collinear, that is, if there is a linear combination of them that equals 0 for all the data. This can arise with indicator variables. If a factor takes on $J$ levels, then there are $J$ associated indicator variables. A classical regression can include only $J-1$ of any set of indicators—if all $J$ were included, they would be collinear with the constant term. You could include a full set of $J$ indicators by excluding the constant term, but then the same problem would arise if you wanted to include a new set of indicators. For example, you could not include both of the sex categories and all four of the age categories. It is simpler just to keep the constant term and all but one of each set of indicators.

For each index variable, the indicator that is excluded from the regression is known as the default, reference, or baseline condition because it is the implied category if all the $J-1$ indicators are set to zero. As discussed in Section 10.4, the default in R is to set the alphabetically first level of a factor as the reference condition; other options include using the last level as baseline, selecting the baseline, and constraining the coefficients to sum to zero. An option that we often prefer is to embed the varying coefficients in a multilevel model, but this goes beyond the scope of this book.

## 12.6 Building and comparing regression models for prediction

A model must be created before it can be fit and checked, and yet we put "model building" near the end of this chapter. Why? It is best to have a theoretical model laid out before any data analyses begin. But in practical data analysis it is usually easiest to start with a simple model and then build in additional complexity, taking care to check for problems along the way.

There are typically many reasonable ways in which a model can be constructed. Models may differ depending on the inferential goals or the way the data were collected. Key choices include how the input variables should be combined or transformed in creating predictors, and which predictors should be included in the model. In classical regression, these are huge issues, because if you include too many predictors in a model, the parameter estimates become so variable as to be useless. Some of these issues are less important in regularized regression (as we discuss in our follow-up book on advanced regression and multilevel models) but they certainly do not disappear completely.

This section focuses on the problem of building models for prediction. Building models that can yield causal inferences is a related but separate topic that is addressed in Chapters 18–21.

### General principles

Our general principles for building regression models for prediction are as follows:

1. Include all input variables that, for substantive reasons, might be expected to be important in predicting the outcome.
2. It is not always necessary to include these inputs as separate predictors—for example, sometimes several inputs can be averaged or summed to create a "total score" that can be used as a single predictor in the model, and that can result in more stable predictions when coefficients are estimated using maximum likelihood or least squares.
3. For inputs that have large effects, consider including their interactions as well.
4. Use standard errors to get a sense of uncertainties in parameter estimates. Recognize that if new data are added to the model, the estimate can change.
5. Make decisions about including or excluding predictors based on a combination of contextual understanding (prior knowledge), data, and the uses to which the regression will be put:
   (a) If the coefficient of a predictor is estimated precisely (that is, if it has a small standard error), it generally makes sense to keep it in the model as it should improve predictions.
   (b) If the standard error of a coefficient is large and there seems to be no good substantive reason for the variable to be included, it can make sense to remove it, as this can allow the other coefficients in the model to be estimated more stably and can even reduce prediction errors.
   (c) If a predictor is important for the problem at hand (for example, indicators for groups that we are interested in comparing or adjusting for), then we generally recommend keeping it in, even

if the estimate has a large standard error and is not "statistically significant." In such settings one must acknowledge the resulting uncertainty and perhaps try to reduce it, either by gathering more data points for the regression or by adding a Bayesian prior (see Section 9.5).

(d) If a coefficient seems not to make sense (for example, a negative coefficient for years of education in an income regression), try to understand how this could happen. If the standard error is large, the estimate could be explainable from random variation. If the standard error is small, it can make sense to put more effort into understanding the coefficient. In the education and income example, for example, the data could be coming from a subpopulation in which the more educated people are younger and have been in their jobs for a shorter period of time and have lower average incomes.

These strategies do not completely solve our problems, but they help keep us from making mistakes such as discarding important information. They are predicated on having thought hard about these relationships *before* fitting the model. It's always easier to justify a coefficient's sign once we have seen it than to think hard ahead of time about what we expect. On the other hand, an explanation that is determined after running the model can still be valid. We should be able to adjust our theories in light of new information.

It is important to record and describe the choices made in modeling, as these choices represent degrees of freedom that, if not understood, can lead to a garden of forking paths and overconfident conclusions. Model performance estimates such as LOO log score can alleviate the problem if there are not too many models.

### Example: predicting the yields of mesquite bushes

Example: Mesquite bushes

We illustrate some ideas of model checking with a real-data example that is nonetheless somewhat artificial in being presented in isolation from its applied context. Partly because this example is not a clear success story and our results are inconclusive, it represents the sort of analysis one might perform in exploring a new dataset.

Data were collected in order to develop a method of estimating the total production (biomass) of mesquite leaves using easily measured parameters of the plant, before actual harvesting takes place.[3] Two separate sets of measurements were taken, one on a group of 26 mesquite bushes and the other on a different group of 20 mesquite bushes measured at a different time of year. All the data were obtained in the same geographical location (ranch), but neither constituted a strictly random sample.

The outcome variable is the total weight (in grams) of photosynthetic material as derived from actual harvesting of the bush. The input variables are:

| | |
|---|---|
| `diam1`: | diameter of the canopy (the leafy area of the bush) in meters, measured along the longer axis of the bush |
| `diam2`: | canopy diameter measured along the shorter axis |
| `canopy_height`: | height of the canopy |
| `total_height`: | total height of the bush |
| `density`: | plant unit density (# of primary stems per plant unit) |
| `group`: | group of measurements (0 for the first group, 1 for the second) |

It is reasonable to predict the leaf weight using some sort of regression model. Many formulations are possible. The simplest approach is to regress `weight` on all of the predictors, yielding the estimates,

```
fit_1 <- stan_glm(formula = weight ~ diam1 + diam2 + canopy_height +
                  total_height + density + group, data=mesquite)

              Median  MAD_SD
(Intercept)  -1092.1   176.0
diam1          195.7   118.6
```

[3] Data and code for this example are in the folder `Mesquite`.

```
diam2           369.0   130.0
canopy_height   349.2   211.1
total_height    -99.3   186.6
density         130.8    34.7
group           360.1   102.7

Auxiliary parameter(s):
      Median MAD_SD
sigma 274.7  32.8
```

We evaluate this model using leave-one-out (LOO) cross validation (see Section 11.8):

```
(loo_1 <- loo(fit_1))
Computed from 4000 by 46 log-likelihood matrix
         Estimate   SE
elpd_loo   -336.1 13.8
p_loo        17.8  9.9
looic       672.2 27.6
------
Monte Carlo SE of elpd_loo is NA.
Pareto k diagnostic values:
                         Count Pct.    Min. n_eff
(-Inf, 0.5]   (good)     41    89.1%   801
 (0.5, 0.7]   (ok)        2     4.3%   433
   (0.7, 1]   (bad)       2     4.3%   21
   (1, Inf)   (very bad)  1     2.2%   2
```

Diagnostics indicate that the approximate LOO computation is unstable, and K-fold-CV is recommended instead, as discussed in Section 11.8. Furthermore, a p_loo value larger than the total number of parameters (in this case, 9) also indicates that the normal distribution is not a good fit for the residuals. We could make a posterior predictive check, but here continue with model comparison and run K-fold-CV to get a more accurate estimate of out-of-sample prediction error:

```
(kfold_1 <- kfold(fit_1, K=10))
 10-fold cross-validation
           Estimate   SE
elpd_kfold   -354.0 30.4
```

The loo diagnostic was correctly indicating problems, as in this case the kfold estimate is much more pessimistic.

To get a sense of the importance of each predictor, it is useful to know the range of each variable:

```
              min   q25 median  q75    max   IQR
diam1         0.8   1.4    2.0  2.5    5.2   1.1
diam2         0.4   1.0    1.5  1.9    4.0   0.9
canopy_height 0.5   0.9    1.1  1.3    2.5   0.4
total_height  0.6   1.2    1.5  1.7    3.0   0.5
density       1.0   1.0    1.0  2.0    9.0   1.0
group         0.0   0.0    0.0  1.0    1.0   1.0

weight         60   220    360  690   4050   470
```

"IQR" in the last column refers to the *interquartile range*—the difference between the 75$^{th}$ and 25$^{th}$ percentile points of each variable.

But perhaps it is more reasonable to fit the model on the logarithmic scale, so that effects are multiplicative rather than additive:

```
fit_2 <- stan_glm(formula = log(weight) ~ log(diam1) + log(diam2) + log(canopy_height) +
                   log(total_height) + log(density) + group, data=mesquite)
```

```
                    Median MAD_SD
(Intercept)         4.8    0.2
log(diam1)          0.4    0.3
log(diam2)          1.1    0.2
log(canopy_height) 0.4    0.3
log(total_height)  0.4    0.3
log(density)        0.1    0.1
group               0.6    0.1

Auxiliary parameter(s):
      Median MAD_SD
sigma 0.3    0.0
```

Instead of, "each meter difference in canopy height is associated with an additional 349 grams of leaf weight," we have, "a difference of $x\%$ in canopy height is associated with an (approximate) positive difference of $0.4x\%$ in leaf weight" (evaluated at the same levels of all other variables across comparisons).

For predictive model comparison, we perform leave-one-out cross validation:

```
(loo_2 <- loo(fit_2))
Computed from 4000 by 46 log-likelihood matrix
         Estimate   SE
elpd_loo    -19.6  5.4
p_loo         7.7  1.6
looic        39.2 10.7
------
Monte Carlo SE of elpd_loo is 0.1.
Pareto k diagnostic values:
                         Count Pct.    Min. n_eff
(-Inf, 0.5]   (good)     42    91.3%   939
 (0.5, 0.7]   (ok)        4     8.7%   241
   (0.7, 1]   (bad)       0     0.0%   <NA>
   (1, Inf)   (very bad) 0     0.0%   <NA>
All Pareto k estimates are ok (k < 0.7).
```

Diagnostics indicate that the approximate LOO computation is stable.

## Using the Jacobian to adjust the predictive comparison after a transformation

The `elpd` values, $-354$ and $-19.6$, from the two fitted models are not directly comparable. When we compare models with transformed continuous outcomes, we must take into account how the nonlinear transformation warps the continuous variable. To reduce the possibility of confusion, the `loo_compare` function we introduced in Chapter 11 refuses to make the comparison in this case. We have to adjust LOO by adding the log of the Jacobian, a mathematical correction for the nonlinear transformation, which we do not explain in this book. For this particular example, the transformation is $\log y$, which has Jacobian $1/y$, leading to the adjustment,

```
loo_2_with_jacobian <- loo_2
loo_2_with_jacobian$pointwise[,1] <- loo_2_with_jacobian$pointwise[,1] -
                                         log(mesquite$weight)
```

We can then compute adjusted LOO and use it to compare the two models:

```
sum(loo_2_with_jacobian$pointwise[,1])
loo_compare(kfold_1, loo_2_with_jacobian)
```

The resulting output tells us that the Jacobian-corrected `elpd_loo` for the second model is $-292$ and the `loo_compare` function gives the difference of 62 with standard error 28. The positive difference

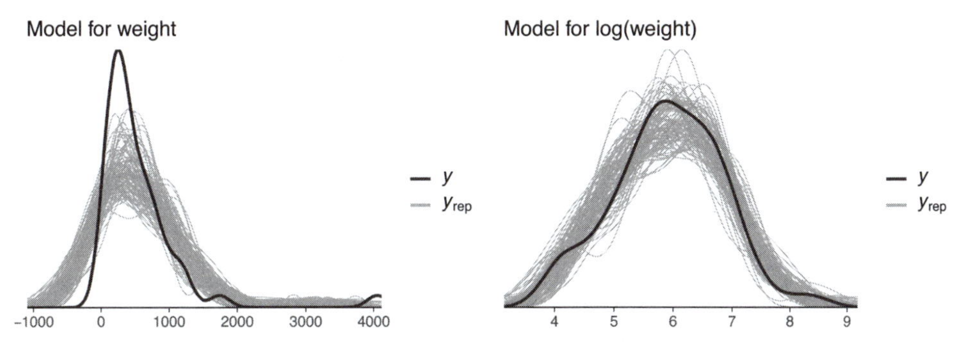


Figure 12.8 *Posterior predictive checks comparing the distribution of leaf weights in the mesquite data (dark line) to 100 predictive replications (gray lines) of replicated data from fitted models (a) on the original scale, and (b) on the log scale. The model on the original scale shows clear lack of fit. The model on the log scale shows a good fit to the data distribution.*

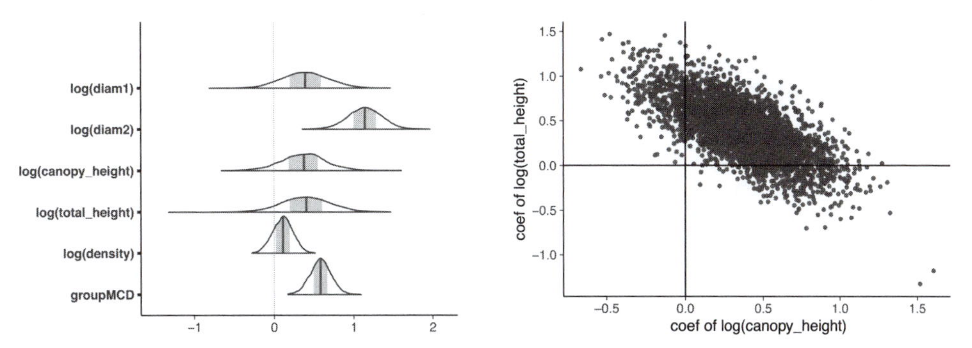


Figure 12.9 *Marginal posterior density estimates for the second mesquite model and a scatterplot of joint posterior simulations for the coefficients of log canopy height and log total height in the fitted regression.*

in predictive density implies an improvement in predictions, averaging over the data at hand, so we continue with this new model.

We can also simulate replicated datasets from the fitted models and compare them to the observed data to get a better understanding of how the log transformation is useful:

```
yrep_1 <- posterior_predict(fit_1)
n_sims <- nrow(yrep_1)
subset <- sample(n_sims, 100)
ppc_dens_overlay(mesquite$weight, yrep_1[subset,])
yrep_2 <- posterior_predict(fit_2)
ppc_dens_overlay(log(mesquite$weight), yrep_2[subset,])
```

The result is shown in Figure 12.8. Observed weights in these data are always positive, but regression on the original scale did not take this into account, and negative weights were predicted. The fit on the log scale is much better, as reflected also in a cross validation comparison.

Figure 12.9a shows the marginal posterior density estimates (with a small negative tail that we can ignore, as it is an artifact of the simple automatic smoothing procedure we have used). This kind of plot can be misleading when predictors are strongly correlated, as is the case for canopy height and total height in this dataset. A clue here is that the coefficients of these predictors have high uncertainties. If we look at the joint density of posterior simulations in Figure 12.9b, we can see that although the univariate marginal densities overlap with zero, the joint distribution is clearly separated from zero. It is fine to keep both predictors in the model, understanding that with the given data it is difficult to untangle their contributions to the linear predictor.

### Constructing a simpler model

So far, we have been throwing all the predictors directly into the model. A more minimalist approach is to try to come up with a simple model that makes sense and which has a good predictive performance estimated by cross validation. Thinking geometrically, we can predict leaf weight from the volume of the leaf canopy, which we shall roughly approximate as

$$\text{canopy\_volume} = \text{diam1} * \text{diam2} * \text{canopy\_height}.$$

This model oversimplifies: the leaves are mostly on the surface of a bush, not in its interior, and so some measure of surface area is perhaps more appropriate. We shall return to this point shortly.

It still makes sense to work on the logarithmic scale:

```
fit_3 <- stan_glm(log(weight) ~ log(canopy_volume), data=mesquite)

                   Median MAD_SD
(Intercept)        5.2    0.1
log(canopy_volume) 0.7    0.1

Auxiliary parameter(s):
      Median MAD_SD
sigma 0.4    0.0
```

Thus, leaf weight is approximately proportional to `canopy_volume` to the 0.7 power. It is perhaps surprising that this power is not closer to 1. The usual explanation is that there is variation in `canopy_volume` that is unrelated to the weight of the leaves, and this tends to *attenuate* the regression coefficient—that is, to decrease its absolute value from the "natural" value of 1 to something lower. Similarly, regressions of "after" versus "before" typically have slopes of less than 1. For another example, Section 10.6 has an example of forecasting congressional elections in which the vote in the previous election has a coefficient of only 0.58.

Let's compare the model to the previous model by computing the LOO log score (we do not need to adjust for the Jacobian now, as both models have the same log weight target):

```
loo_3 <- loo(fit_3)
loo_compare(loo_2, loo_3)
Model comparison:
(negative 'elpd_diff' favors 1st model, positive favors 2nd)
elpd_diff        se
     -7.0       5.0
```

and LOO $R^2$:

```
median(loo_R2(fit_2))
[1] 0.84
median(loo_R2(fit_3))
[1] 0.78
```

The regression with only `canopy_volume` is satisfyingly simple, with an impressive LOO $R^2$ of 78%. However, the predictions are still worse than the model with all the predictors. Perhaps we should go back and put that other information back in. We shall define

$$\text{canopy\_area} = \text{diam1} * \text{diam2},$$
$$\text{canopy\_shape} = \text{diam1}/\text{diam2}.$$

The set (`canopy_volume`, `canopy_area`, `canopy_shape`) is just a different parameterization of the three canopy dimensions. Including them all in the model yields,

```
fit_4 <- stan_glm(formula = log(weight) ~ log(canopy_volume) + log(canopy_area) +
    log(canopy_shape) + log(total_height) + log(density) + group, data=mesquite)
```

```
                   Median MAD_SD
(Intercept)         4.8    0.2
log(canopy_volume)  0.4    0.3
log(canopy_area)    0.4    0.3
log(canopy_shape)  -0.4    0.2
log(total_height)   0.4    0.3
log(density)        0.1    0.1
group               0.6    0.1

Auxiliary parameter(s):
      Median MAD_SD
sigma 0.3    0.0
```

And we get this predictive comparison:

```
loo_compare(loo_2, loo_4)
Model comparison:
(negative 'elpd_diff' favors 1st model, positive favors 2nd)
elpd_diff        se
      0.2       0.1
```

This fit is identical to that of the earlier log-scale model (just a linear transformation of the predictors), but to us these coefficient estimates are more directly interpretable:

- Canopy volume and area are both positively associated with leaf weight. This makes sense: (a) a larger-volume canopy should have more leaves, and (b) conditional on volume, a canopy with larger cross-sectional area should have more exposure to the sun. However, both coefficients have large standard errors in relation to their estimates, which is likely due to near-collinearity of these predictors (unsurprising, given that log(`canopy_volume`) = log(`diam1`) + log(`diam2`) + log(`height`), and log(`canopy_area`) = log(`diam1`) + log(`diam2`)), and we should examine their joint posterior to assess their relevance. The near-collinearity implies that dropping one of these predictors from the model should not reduce the performance much.
- The negative coefficient of `canopy_shape` implies that bushes that are more circular in cross section have more leaf weight (after adjusting for volume and area). Again, the standard error is high, implying that this could be arising from collinearity with other predictors. As log(`canopy_shape`) = log(`diam1`) − log(`diam2`), it should have low correlation with `canopy_volume` and `canopy_area`.
- Total height is positively associated with weight, which could make sense if the bushes are planted close together—taller bushes get more sun. The standard error is large, which is likely due to the near-collinearity.
- It is not clear how to interpret the coefficient for `density`. We have no prior information suggesting this should be an important predictor and its coefficient is highly uncertain, so we are inclined to exclude it from the model.
- The coefficient for `group` is large and precisely estimated, so we should keep it in, even though we had no particular reason ahead of time to think it would be important. It would be a good idea to understand how the two groups differ so that a more relevant measurement could be included for which `group` is a proxy.

We want to include both volume and shape in the model, since for geometrical reasons we expect both to be predictive of leaf volume. This leaves us with a model such as,

```
fit_5 <- stan_glm(log(weight) ~ log(canopy_volume) + log(canopy_shape) + group,
    data=mesquite)
```

```
                    Median MAD_SD
(Intercept)          4.9    0.1
log(canopy_volume)   0.8    0.1
log(canopy_shape)   -0.4    0.2
group                0.6    0.1

Auxiliary parameter(s):
      Median MAD_SD
sigma 0.3    0.0
```

Comparing to the earlier model yields,

```
loo_compare(loo_4, loo_5)
Model comparison:
(negative 'elpd_diff' favors 1st model, positive favors 2nd)
elpd_diff        se
      1.3       1.5
```

This fit is in practice identical to the previous model but with only three predictors. After dropping nearly collinear predictors, the marginal uncertainties for the remaining predictors are also smaller. Leaf weight is approximately proportional to canopy volume to the 0.8 power, whereas in the previous model with near-collinearity the posterior mean was only 0.4, as the predictive value of this information was shared among other predictors.

One reason we get as good or better performance and understanding from a simpler model is that we are fitting all these regressions with weak prior distributions. With weak priors, the data drive the inferences, and when sample sizes are small or predictors are highly correlated, the coefficients are poorly identified, and so their estimates are noisy. Excluding or combining predictors is a way to get more stable estimates, which can be easier to interpret without losing much in predictive power. An alternative approach is to keep all the predictors in the model but with strong priors on their coefficients to stabilize the estimates. Such a prior can be constructed based on previous analyses or using a multilevel model.

Finally, it would seem like a good idea to include interactions of `group` with the other predictors. Unfortunately, without strong priors, it turns out to be impossible to estimate these interactions accurately from these 46 data points: when we include these interactions when fitting the model using least squares or using `stan_glm` with default priors, their coefficients of these interactions all end up with large standard errors. More could be learned with data from more bushes, and it would also make sense to look at some residual plots to look for any patterns in the data beyond what has been fitted by the model.

To conclude this example: we had some success in transforming the outcome and input variables to obtain a reasonable predictive model. For this problem, the log model was clearly better. In general, cross validation can be used to choose among multiple models fit to the same data, but if too many models are compared, the selection process itself can lead to undesired overfitting. The log score can be used to choose between the linear and log-transformed target models, and in the continuous case we needed to take care because of mathematical subtlety of the Jacobian of the nonlinear transformation, as discussed on page 202.

## 12.7 Models for regression coefficients

Example: Predicting student grades

In this section we consider regression with more than a handful of predictors. We demonstrate the usefulness of standardizing predictors and including models for regression coefficients.

We work with an example of modeling grades from a sample of high school students in Portugal.[4]

[4]Data and code for this example are in the folder `Student`.

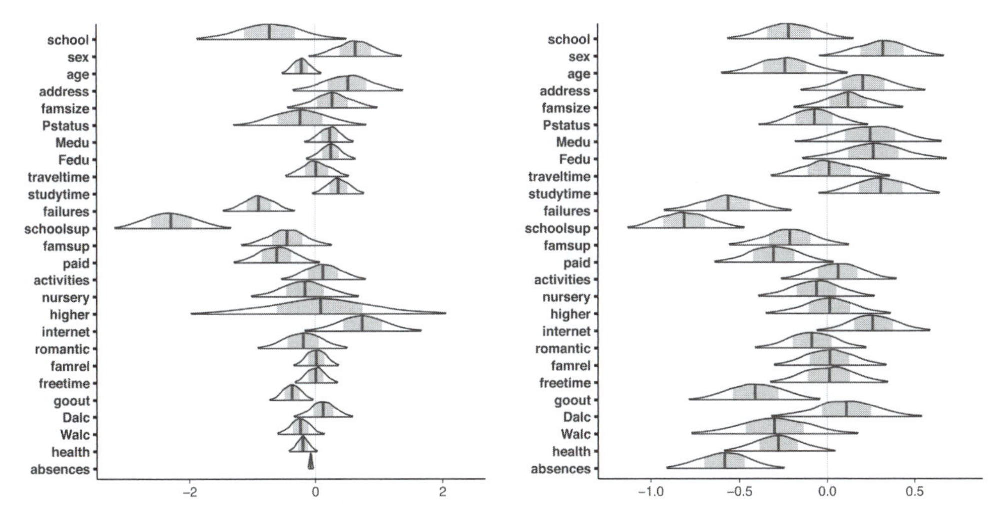


Figure 12.10 *Marginal posterior density estimates for a linear regression predicting student grades from many predictors using regression with default weak priors. Models are fit and displayed (a) without and (b) with standardization of the predictors.*

We predict the final-year mathematics grade for 343 students given a large number of potentially relevant predictors: student's school, student's sex, student's age, student's home address type, family size, parents' cohabitation status, mother's education, father's education, home-to-school travel time, weekly study time, number of past class failures, extra educational support, extra paid classes within the course subject, extra-curricular activities, whether the student attended nursery school, whether the student wants to take higher education, internet access at home, whether the student has a romantic relationship, quality of family relationships, free time after school, going out with friends, weekday alcohol consumption, weekend alcohol consumption, current health status, and number of school absences.

We first fit a regression model without standardizing the predictors:

```
predictors <- c("school","sex","age","address","famsize","Pstatus","Medu","Fedu",
  "traveltime","studytime","failures","schoolsup","famsup","paid","activities",
  "nursery", "higher", "internet", "romantic","famrel","freetime","goout","Dalc",
  "Walc","health","absences")
data_G3mat <- subset(data, subset=G3mat>0, select=c("G3mat",predictors))
fit0 <- stan_glm(G3mat ~ ., data=data_G3mat)
```

Figure 12.10a shows that without standardization of predictors, it looks like there is a different amount of uncertainty on the relevance of the predictors. For example, it looks like absences has really small relevance and high certainty.

We standardize all the predictors (for simplicity including also binary and ordinal) and re-fit:

```
datastd_G3mat <- data_G3mat
datastd_G3mat[,predictors] <-scale(data_G3mat[,predictors])
fit1 <- stan_glm(G3mat ~ ., data=datastd_G3mat)
```

Figure 12.10b shows that after all predictors have been standardized to have equal standard deviation, the uncertainties on the relevances are similar. For example, it is now easier to see that absences has relatively high relevance compared to other predictors in the model.

As we have many potential predictors, we may worry that maybe some of them add more noise than predictive power to the model. As discussed in Section 11.8 we can compare Bayesian $R^2$ and LOO $R^2$ and look at the effective number of parameters obtained from the LOO log score. In this case, median LOO $R^2$, at 0.17, is much lower than the median Bayesian $R^2$ of 0.31, indicating

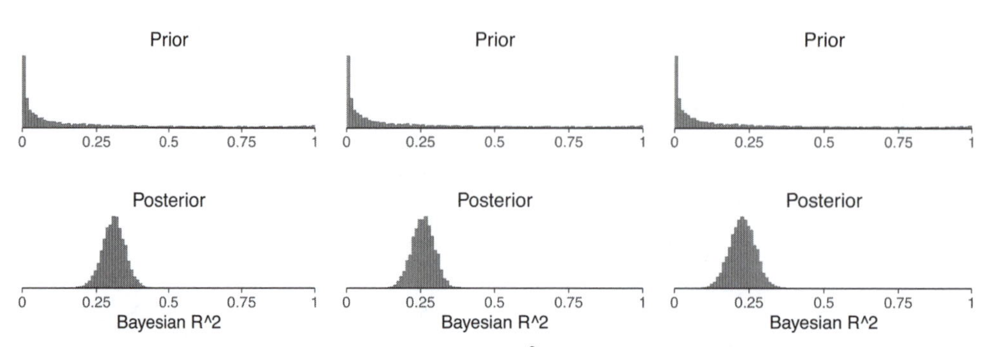


Figure 12.11 *Prior and posterior distribution of Bayesian $R^2$ for the regression predicting student grades from many predictors, using three different priors for the coefficients: (a) default weak prior, (b) normal prior scaled with the number of predictors, and (c) regularized horseshoe prior.*

overfitting. In addition, the model has 26 coefficients and the effective number of parameters from LOO is approximately 26, which indicates that the model is fitting to all predictors. In `stan_glm` at its default setting, the coefficients are given independent priors, each with mean 0 and standard deviation 2.5 if the data have been standardized as above; see Section 9.5. With many predictors, it can be useful to think more carefully about the information we have, so we can construct better models for regression coefficients.

If the predictors have been standardized to have standard deviation 1 and we give the regression coefficients independent normal priors with mean 0 and standard deviation 2.5, this implies that the prior standard deviation of the modeled predictive means is $2.5\sqrt{26} = 12.7$. The default prior for $\sigma$ is an exponential distribution, scaled to have mean equal to data standard deviation which in this case is approximately 3.3 which is much less than 12.7. We can simulate from these prior distributions and examine what is the corresponding prior distribution for explained variance $R^2$.

Figure 12.11a shows that with the default prior on regression coefficients and $\sigma$, the implied prior distribution for $R^2$ is strongly favoring larger values and thus is favoring overfitted models. The priors often considered as weakly informative for regression coefficients turn out to be, in the multiple predictor case, highly informative for the explained variance.

What would be more sensible prior models for regression coefficients when we have many predictors? It is unlikely that all predictors would be strongly related to the outcome. We may assume either many predictors having small relevance each or only some of the predictors having high relevance and the rest of the predictors having negligible relevance. We present models for both cases and then discuss other alternatives on page 210.

If we assume that many predictors may each have small relevance, we can scale the independent priors so that the sum of the prior variance stays reasonable. In this case we have 26 predictors and could have a prior guess that proportion of the explained variance is near 0.3. Then a simple approach would be to assign independent priors to regression coefficients with mean 0 and standard deviation $\sqrt{0.3/26}\,\mathrm{sd}(y)$, and an exponential prior for $\sigma$ with mean $\sqrt{0.7}\,\mathrm{sd}(y)$. Then the expected prior predictive variance is approximately the same as the data variance and the implied prior on explained variance $R^2$ is more evenly distributed, as shown in Figure 12.11b. We can see that this prior is not strongly concentrated near our prior guess for $R^2$, which makes it weakly informative. This simple approach still seems to have a bit too much mass on values very near 0 and 1, but this is not a problem if the data is not supporting those extreme values. The implied prior on $R^2$ could be further improved by using a joint prior on $\beta$ and $\sigma$.

We fit a regression with standardized predictors, using a prior scaled by the number of predictors:

```
fit2 <- stan_glm(G3mat ~ ., data=datastd_G3mat,
                 prior=normal(scale=sd(datastd_G3mat$G3mat)/sqrt(0.3*26)))
```

Figure 12.12a shows the posterior distributions of coefficients, which are slightly more concentrated

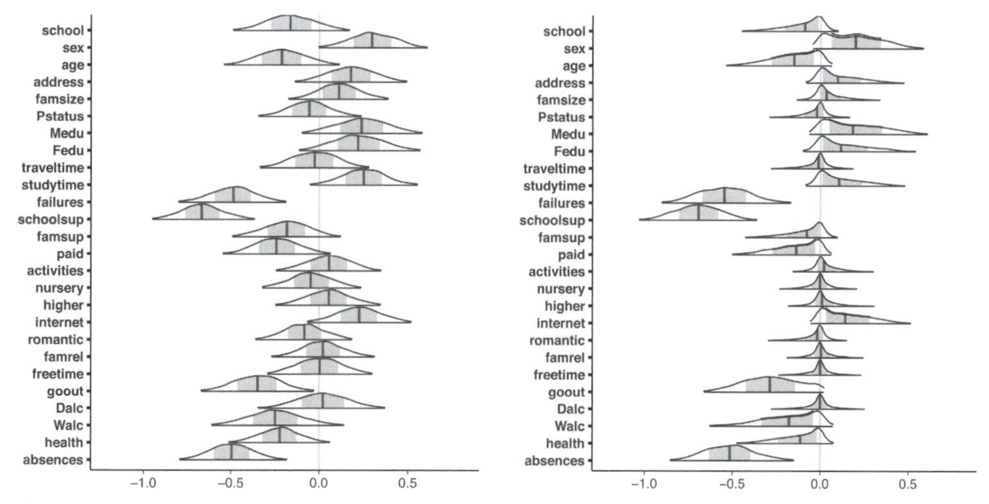


Figure 12.12 *Posterior distributions for coefficients in the regression predicting student grades from many predictors, using two different priors: (a) normal prior scaled with the number of predictors, and (b) regularized horseshoe prior.*

than for the previous model shown in Figure 12.10b. The medians of Bayesian $R^2$ and LOO $R^2$ are 0.26 and 0.19, respectively, indicating less overfitting than with the default prior. In addition, comparison of the LOO log score reveals that the new model has better leave-one-out prediction.

Alternatively we may fit a model assuming that only some of the predictors have high relevance and the rest of the predictors have negligible relevance. One possibility for modeling this assumption is the *regularized horseshoe prior*, with the name coming from the implied U-shaped distribution on the amount of partial pooling for coefficients, with the shrinkage factor likely being close to 0 (no shrinkage) or 1 (complete shrinkage to zero). The prior starts with independent normal prior distributions for the coefficients $\beta_j$ with means 0 and standard deviations $\tau\lambda_j$, which has in addition of common global scale $\tau$, a local scale $\lambda_j$ for each coefficient $\beta_j$. By setting different priors for $\lambda_j$, different assumptions about the distribution of relevances can be presented. The horseshoe prior fits a half-Cauchy model to local scales $\lambda_j$. For $\tau$ we set a half-Cauchy prior with scale $\frac{p_0}{D-p_0}\frac{\sigma}{\sqrt{n}}$, where $p_0$ is the expected number of relevant predictors, and the factor of $\frac{\sigma}{\sqrt{n}}$ works with the prior on the residual variance. The regularized horseshoe prior has an additional regularization term called "slab scale," as the prior can be considered as a so-called spike-and-slab model, with the horseshoe representing the spike for non-relevant predictors and the normal distribution representing the slab for the relevant predictors.

We assume in this example that the expected number of relevant predictors is near $p_0 = 6$, and the prior scale for the relevant predictors is chosen as in the previous model but using $p_0$ for scaling. We can then simulate from this prior and examine the corresponding prior for $R^2$. Figure 12.11c shows that the regularized horseshoe prior with sensible parameters implies a more cautious prior on explained variance $R^2$ than is implicitly assumed by the default wide prior. The horseshoe prior favors simpler models, but is quite flat around most $R^2$ values.

We fit a regression model with standardized predictors and using the regularized horseshoe prior (`prior=hs`):

```
p <- length(predictors)
n <- nrow(datastd_G3mat)
p0 <- 6
slab_scale <- sqrt(0.3/p0)*sd(datastd_G3mat$G3mat)
# global scale without sigma, as the scaling by sigma is done inside stan_glm
global_scale <- (p0/(p - p0))/sqrt(n)
```

```
fit3 <- stan_glm(G3mat ~ ., data=datastd_G3mat,
                 prior=hs(global_scale=global_scale, slab_scale=slab_scale))
```

Figure 12.12b shows that the regularized horseshoe prior has the benefit of shrinking the posterior for many regression coefficients more tightly towards 0, making it easier to see the most relevant predictors. Failures, school support, going out, and the number of absences appear to be the most relevant predictors.

The medians of Bayesian $R^2$ and LOO $R^2$ for `fit3` are 0.23 and 0.19 respectively, indicating even less overfitting than with the previous prior, but the difference is small. When we compare models using the LOO log score, the new model is better than the default prior model, but there is no difference compared to the model with a normal prior scaled with the predictors. It is common that the data do not have strong information about how many predictors are relevant, and different types of priors can produce similar predictive accuracies.

We can test how good predictions we could get if we only use failures, school support, going out, and the number of absences as the predictors.

```
fit4 <- stan_glm(G3mat ~ failures + schoolsup + goout + absences, data=datastd_G3mat)
```

For this new model, the medians of Bayesian $R^2$ and LOO $R^2$ are 0.20 and 0.17 respectively; that is, the difference is small and there is less overfit than when using all predictors with wide prior. LOO $R^2$ is just slightly smaller than for models with all predictors and better priors. In this case, the prediction performance can not improved much by adding more predictors. By observing more students it might be possible to learn regression coefficients for other predictors with sufficiently small uncertainty that predictions for new students could be improved.

### Other models for regression coefficients

There are several different regularization methods for regression coefficients that are governed by a "tuning parameter" $\tau$ that governs the amount of partial pooling of the coefficients toward zero. These methods correspond to priors that are independent conditional on $\tau$: the conditional prior for each coefficient $\beta_j$ is normal with mean 0 and standard deviation $\tau\lambda_j$. Different methods correspond to different priors on the $\lambda_j$'s, and for certain priors on $\lambda_j$ the corresponding prior on the explained variance $R^2$ (see Section 11.6) can be derived analytically, which makes it easy to define the desired prior on $R^2$. This analysis assumes the predictors are independent but can also be useful in the case of moderate collinearity. To take into account the collinearity, the function `stan_lm` has a multivariate normal version of the prior for coefficients which has an analytic form for the implied prior on $R^2$.

The lasso procedure for predictor selection mixes regularization and point estimate inference, often producing useful results. The success of lasso regularization has inspired the use of the double-exponential (Laplace) distribution as a prior model for regression coefficients. In settings where there is an expectation that only a few predictors will be relevant, we recommend the horseshoe class of priors instead.

We can also use these sorts of tools as guidance in considering models to pursue further, for example choosing predictors for a smaller model by inspecting coefficient inferences from the regularized horseshoe prior. If the predictors would be completely independent without collinearity, we could consider assessing the relevance from the marginals, but in case of collinearity we may miss relevant predictors as illustrated in Figure 12.9b. The prior model for regression coefficients is useful even without predictor selection, and ideally the selection and averaging of models should be made using formal decision-theoretic cost-benefit analysis, which is beyond the scope of this book.

## 12.8 Bibliographic note

For additional reading on transformations, see Atkinson (1985), Mosteller and Tukey (1977), Box and Cox (1964), and Carroll and Ruppert (1981). Bring (1994) has a thorough discussion on standardizing regression coefficients; see also Blalock (1961), Greenland, Schlessman, and Criqui (1986), and Greenland et al. (1991). Gelman (2007) discusses scaling inputs by dividing by two standard deviations. Harrell (2001) discusses strategies for regression modeling.

For more on the earnings and height example, see Persico, Postlewaite, and Silverman (2004) and Gelman and Nolan (2017). For more on the handedness example, see Gelman and Nolan (2017, sections 2.5 and 3.3.2). We further explore the same-sex marriage example in Section 22.7. The mesquite bushes example in Section 12.6 comes from an exam problem from the 1980s. The student grades data come from Cortez and Silva (2008).

The lasso and horseshoe priors were introduced by Tibshirani (1996) and Carvalho, Polson, and Scott (2010), respectively. Selection of horseshoe prior parameters is discussed by Piironen and Vehtari (2017b, c), and the regularized horseshoe prior was introduced by Piironen and Vehtari (2017c). A decision-theoretic approach for variable selection after creating a good predictive model is discussed by Piironen and Vehtari (2017a) and Piironen, Paasiniemi, and Vehtari (2020).

## 12.9 Exercises

12.1 *Plotting linear and quadratic regressions*: The folder `Earnings` has data on weight (in pounds), age (in years), and other information from a sample of American adults. We create a new variable, `age10` = age/10, and and fit the following regression predicting weight:

```
            Median MAD_SD
(Intercept) 148.7    2.2
age10         1.8    0.5

Auxiliary parameter(s):
      Median MAD_SD
sigma 34.5    0.6
```

(a) With pen on paper, sketch a scatterplot of weights versus age (that is, weight on $y$-axis, age on $x$-axis) that is consistent with the above information, also drawing the fitted regression line. Do this just given the information here and your general knowledge about adult heights and weights; do not download the data.

(b) Next, we define `age10_sq` = $(\text{age}/10)^2$ and predict weight as a quadratic function of age:

```
            Median MAD_SD
(Intercept) 108.0    5.7
age10        21.3    2.6
age10sq      -2.0    0.3

Auxiliary parameter(s):
      Median MAD_SD
sigma 33.9    0.6
```

Draw this fitted curve on the graph you already sketched above.

12.2 *Plotting regression with a continuous variable broken into categories*: Continuing Exercise 12.1, we divide age into 4 categories and create corresponding indicator variables, `age18_29`, `age30_44`, `age45_64`, and `age65_up`. We then fit the following regression:

```
stan_glm(weight ~ age30_44 + age45_64 + age65_up, data=earnings)
```

```
             Median MAD_SD
(Intercept)  147.8    1.6
age30_44TRUE   9.6    2.1
age45_64TRUE  16.6    2.3
age65_upTRUE   7.5    2.7
```

```
Auxiliary parameter(s):
      Median MAD_SD
sigma 34.1   0.6
```

(a) Why did we not include an indicator for the youngest group, age18_29?

(b) Using the same axes and scale as in your graph for Exercise 12.1, sketch with pen on paper the scatterplot, along with the above regression function, which will be discontinuous.

12.3 *Scale of regression coefficients*: A regression was fit to data from different countries, predicting the rate of civil conflicts given a set of geographic and political predictors. Here are the estimated coefficients and their $z$-scores (coefficient divided by standard error), given to three decimal places:

| | Estimate | $z$-score |
|---|---|---|
| Intercept | −3.814 | −20.178 |
| Conflict before 2000 | 0.020 | 1.861 |
| Distance to border | 0.000 | 0.450 |
| Distance to capital | 0.000 | 1.629 |
| Population | 0.000 | 2.482 |
| % mountainous | 1.641 | 8.518 |
| % irrigated | −0.027 | −1.663 |
| GDP per capita | −0.000 | −3.589 |

Why are the coefficients for distance to border, distance to capital population, and GDP per capita so small?

12.4 *Coding a predictor as both categorical and continuous*: A linear regression is fit on a group of employed adults, predicting their physical flexibility given age. Flexibility is defined on a 0–30 scale based on measurements from a series of stretching tasks. Your model includes age in categories (under 30, 30–44, 45–59, 60+) and also age as a linear predictor. Sketch a graph of flexibility vs. age, showing what the fitted regression might look like.

12.5 *Logarithmic transformation and regression*: Consider the following regression:

$$\log(\text{weight}) = -3.8 + 2.1 \log(\text{height}) + \text{error},$$

with errors that have standard deviation 0.25. Weights are in pounds and heights are in inches.

(a) Fill in the blanks: Approximately 68% of the people will have weights within a factor of ___ and ___ of their predicted values from the regression.

(b) Using pen and paper, sketch the regression line and scatterplot of log(weight) versus log(height) that make sense and are consistent with the fitted model. Be sure to label the axes of your graph.

12.6 *Logarithmic transformations*: The folder `Pollution` contains mortality rates and various environmental factors from 60 U.S. metropolitan areas (see McDonald and Schwing, 1973). For this exercise we shall model mortality rate given nitric oxides, sulfur dioxide, and hydrocarbons as inputs. This model is an extreme oversimplification, as it combines all sources of mortality and does not adjust for crucial factors such as age and smoking. We use it to illustrate log transformations in regression.

(a) Create a scatterplot of mortality rate versus level of nitric oxides. Do you think linear regression will fit these data well? Fit the regression and evaluate a residual plot from the regression.

(b) Find an appropriate transformation that will result in data more appropriate for linear regression. Fit a regression to the transformed data and evaluate the new residual plot.

(c) Interpret the slope coefficient from the model you chose in (b).

(d) Now fit a model predicting mortality rate using levels of nitric oxides, sulfur dioxide, and hydrocarbons as inputs. Use appropriate transformations when helpful. Plot the fitted regression model and interpret the coefficients.

(e) Cross validate: fit the model you chose above to the first half of the data and then predict for the second half. You used all the data to construct the model in (d), so this is not really cross validation, but it gives a sense of how the steps of cross validation can be implemented.

12.7 *Cross validation comparison of models with different transformations of outcomes:* When we compare models with transformed continuous outcomes, we must take into account how the nonlinear transformation warps the continuous variable. Follow the procedure used to compare models for the mesquite bushes example on page 202.

(a) Compare models for earnings and for log(earnings) given height and sex as shown on pages 84 and 192. Use `earnk` and `log(earnk)` as outcomes.

(b) Compare models from other exercises in this chapter.

12.8 *Log-log transformations*: Suppose that, for a certain population of animals, we can predict log weight from log height as follows:

- An animal that is 50 centimeters tall is predicted to weigh 10 kg.
- Every increase of 1% in height corresponds to a predicted increase of 2% in weight.
- The weights of approximately 95% of the animals fall within a factor of 1.1 of predicted values.

(a) Give the equation of the regression line and the residual standard deviation of the regression.

(b) Suppose the standard deviation of log weights is 20% in this population. What, then, is the $R^2$ of the regression model described here?

12.9 *Linear and logarithmic transformations*: For a study of congressional elections, you would like a measure of the relative amount of money raised by each of the two major-party candidates in each district. Suppose that you know the amount of money raised by each candidate; label these dollar values $D_i$ and $R_i$. You would like to combine these into a single variable that can be included as an input variable into a model predicting vote share for the Democrats.
Discuss the advantages and disadvantages of the following measures:

(a) The simple difference, $D_i - R_i$

(b) The ratio, $D_i/R_i$

(c) The difference on the logarithmic scale, $\log D_i - \log R_i$

(d) The relative proportion, $D_i/(D_i + R_i)$.

12.10 *Special-purpose transformations*: For the congressional elections example in the previous exercise, propose an idiosyncratic transformation as in the example on page 196 and discuss the advantages and disadvantages of using it as a regression input.

12.11 *Elasticity*: An economist runs a regression examining the relations between the average price of cigarettes, $P$, and the quantity purchased, $Q$, across a large sample of counties in the United States, assuming the functional form, $\log Q = \alpha + \beta \log P$. Suppose the estimate for $\beta$ is 0.3. Interpret this coefficient.

12.12 *Sequence of regressions*: Find a regression problem that is of interest to you and can be performed repeatedly (for example, data from several years, or for several countries). Perform a separate analysis for each year, or country, and display the estimates in a plot as in Figure 10.9.

12.13 *Building regression models*: Return to the teaching evaluations data from Exercise 10.6. Fit regression models predicting evaluations given many of the inputs in the dataset. Consider interactions, combinations of predictors, and transformations, as appropriate. Consider several models, discuss in detail the final model that you choose, and also explain why you chose it rather than the others you had considered.

12.14 *Prediction from a fitted regression*: Consider one of the fitted models for mesquite leaves, for example `fit_4`, in Section 12.6. Suppose you wish to use this model to make inferences about the average mesquite yield in a new set of trees whose predictors are in data frame called `new_trees`. Give R code to obtain an estimate and standard error for this population average. You do not need to make the prediction; just give the code.

12.15 *Models for regression coefficients*: Using the Portuguese student data from the `Student` folder, repeat the analyses in Section 12.7 with the same predictors, but using as outcome the Portuguese language grade rather than the mathematics grade.

12.16 *Applying ideas of regression*: Read a published article that uses regression modeling and is on a topic of interest to you. Write a page or two evaluating and criticizing the article, addressing issues discussed in Chapters 1–12, such as measurement, data visualization, modeling, inference, simulation, regression, assumptions, model checking, interactions, and transformations. The point of this exercise is not to come up with a comprehensive critique of the article but rather to review the key points of this book so far in the context of a live example.

12.17 *Working through your own example*: Continuing the example from the final exercises of the earlier chapters, fit a linear regression that includes transformations and interpret the estimated parameters and their uncertainties. Compare your results to the model you fit in Exercise 10.11.

# Part 3: Generalized linear models

# Chapter 13

# Logistic regression

Linear regression is an additive model, which does not work for binary outcomes—that is, data $y$ that take on the values 0 or 1. To model binary data, we need to add two features to the base model $y = a + bx$: a nonlinear transformation that bounds the output between 0 and 1 (unlike $a + bx$, which is unbounded), and a model that treats the resulting numbers as probabilities and maps them into random binary outcomes. This chapter and the next describe one such model—logistic regression—and then in Chapter 15 we discuss *generalized linear models*, a larger class that includes linear and logistic regression as special cases. In the present chapter we introduce the mathematics of logistic regression and also its latent-data formulation, in which the binary outcome $y$ is a discretized version of an unobserved or latent continuous measurement $z$. As with the linear model, we show how to fit logistic regression, interpret its coefficients, and plot data and fitted curves. The nonlinearity of the model increases the challenges of interpretation and model-building, as we discuss in the context of several examples.

## 13.1 Logistic regression with a single predictor

The *logistic* function,

$$\text{logit}(x) = \log\left(\frac{x}{1-x}\right),$$

maps the range (0, 1) to $(-\infty, \infty)$ and is useful for modeling probabilities. Its inverse function, graphed in Figure 13.1a, maps back to the unit range:

$$\text{logit}^{-1}(x) = \frac{e^x}{1+e^x}. \tag{13.1}$$

In R, we can access these functions using the logistic distribution:

```
logit <- qlogis
invlogit <- plogis
```

We next discuss the use of the logistic transformation in modeling binary data.

### Example: modeling political preference given income

Example: Income and voting

Conservative parties have traditionally received more support among voters with higher incomes. We illustrate classical logistic regression with a simple analysis of this pattern from the National Election Study in 1992. For each respondent $i$ in this poll, we label $y_i = 1$ if he or she preferred George Bush (the Republican candidate for president) or 0 if he or she preferred Bill Clinton (the Democratic candidate), excluding respondents who preferred other candidates or had no opinion.[1]

We predict preferences given the respondent's income level, which is characterized on a five-point scale as described on page 149. An alternative would be to use dollar income as a continuous

[1]Data and code for this example are in the folder NES.

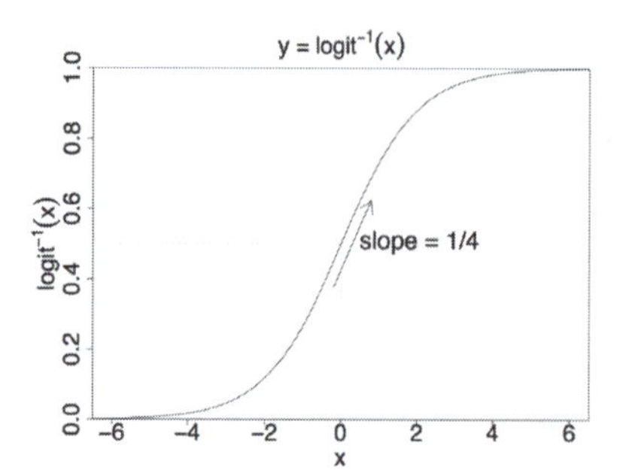


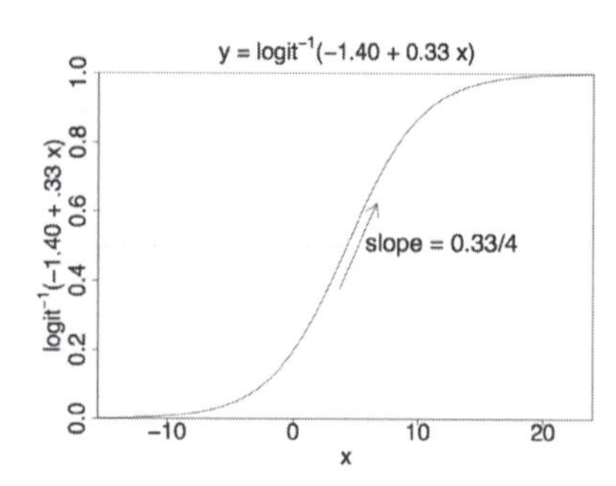


Figure 13.1 *(a) Inverse logit function* $logit^{-1}(x)$ *from (13.1): the transformation from linear predictors to probabilities that is used in logistic regression. (b) An example of the predicted probabilities from a logistic regression model:* $y = logit^{-1}(-1.40 + 0.33x)$*. The shape of the curve is the same, but its location and scale have changed; compare the x-axes on the two graphs. On each graph, the vertical dotted line shows where the predicted probability is 0.5: in graph (a), this is at* $logit(0.5) = 0$*; in graph (b), the halfway point is where* $-1.40 + 0.33x = 0$*, which is* $x = 1.40/0.33 = 4.2$*. As discussed in Section 13.2, the slope of the curve at the halfway point is the logistic regression coefficient divided by 4, thus 1/4 for* $y = logit^{-1}(x)$ *and* $0.33/4$ *for* $y = logit^{-1}(-1.40 + 0.33x)$*. The slope of the logistic regression curve is steepest at this halfway point.*

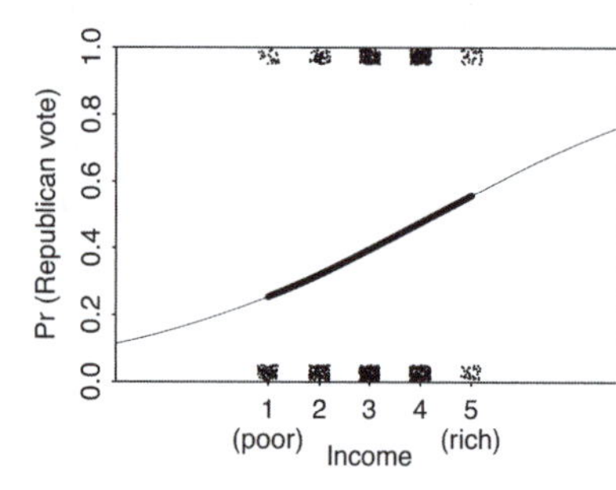


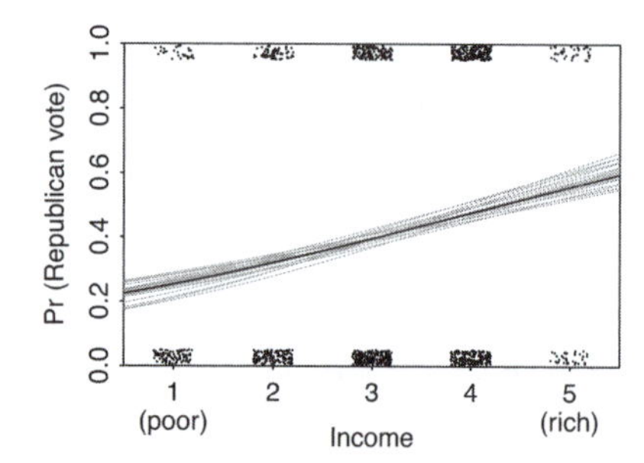


Figure 13.2 *Logistic regression estimating the probability of supporting George Bush in the 1992 presidential election, as a function of discretized income level. Survey data are indicated by jittered dots. In this example, little is revealed by these jittered points, but we want to emphasize here that the data and fitted model can be put on a common scale. (a) Fitted logistic regression: the thick line indicates the curve in the range of the data; the thinner lines at the ends show how the logistic curve approaches 0 and 1 in the limits. (b) In the range of the data, the solid line shows the best-fit logistic regression, and the light lines show uncertainty in the fit.*

predictor, but for predicting partisanship it seems to make more sense to use quantiles, as this captures relative income, which seems more relevant when studying political attitudes. In this case, five income categories were available from the survey response.

The data are shown as (jittered) dots in Figure 13.2, along with the fitted *logistic regression* line, a curve that is constrained to lie between 0 and 1. We interpret the line as the probability that $y = 1$ given $x$—in mathematical notation, $\Pr(y = 1|x)$.

We fit and display the model using `stan_glm`, specifying logistic regression as follows:

```
fit_1 <- stan_glm(rvote ~ income, family=binomial(link="logit"), data=nes92)
print(fit_1)
```

to yield,

```
            Median MAD_SD
(Intercept) -1.4    0.2
income       0.3    0.1
```

We write the fitted model as $\Pr(y_i = 1) = \text{logit}^{-1}(-1.4 + 0.3 * \text{income})$. We shall define this model mathematically and then return to discuss its interpretation.

Unlike with linear regression, there is no sigma in the output. Logistic regression has no separate variance term; its uncertainty comes from its probabilistic prediction of binary outcomes.

Figure 13.2a shows a graph of the data and fitted line, which can be made as follows:

```
plot(nes92$income, nes92$rvote)
curve(invlogit(coef(fit_1)[1] + coef(fit_1)[2]*x), add=TRUE)
```

The R code we actually use to make the figure has more steps so as to display axis labels, jitter the points, adjust line thickness, and so forth.

## The logistic regression model

We would run into problems if we tried to fit a linear regression model, $X\beta$ + error, to data $y$ that take on the values 0 and 1. The coefficients in such a model could be interpreted as differences in probabilities, but we could have difficulties using the model for prediction—it would be possible to have predicted probabilities below 0 or above 1—and information would be lost by modeling discrete outcomes as if they were continuous.

Instead, we model the probability that $y = 1$,

$$\Pr(y_i = 1) = \text{logit}^{-1}(X_i\beta), \tag{13.2}$$

under the assumption that the outcomes $y_i$ are independent given these probabilities. We refer to $X\beta$ as the *linear predictor*. We illustrate logistic regression curves schematically in Figure 13.1 and as fitted to the election polling example in Figure 13.2.

Equivalently, model (13.2) can be written,

$$\begin{aligned} \Pr(y_i = 1) &= p_i \\ \text{logit}(p_i) &= X_i\beta, \end{aligned} \tag{13.3}$$

We prefer to work with $\text{logit}^{-1}$ as in (13.2) because we find it more natural to focus on the mapping from the linear predictor to the probabilities, rather than the reverse. However, you will need to understand formulation (13.3) to follow the literature.

The inverse logistic function is curved, and so the expected difference in $y$ corresponding to a fixed difference in $x$ is not a constant. As can be seen in Figure 13.1, the steepest change occurs at the middle of the curve. For example:

- $\text{logit}(0.5) = 0$, and $\text{logit}(0.6) = 0.4$. Here, adding 0.4 on the logit scale corresponds to a change from 50% to 60% on the probability scale.
- $\text{logit}(0.9) = 2.2$, and $\text{logit}(0.93) = 2.6$. Here, adding 0.4 on the logit scale corresponds to a change from 90% to 93% on the probability scale.

Similarly, adding 0.4 at the low end of the scale moves a probability from 7% to 10%. In general, any particular change on the logit scale is compressed at the ends of the probability scale, which is needed to keep probabilities bounded between 0 and 1.

## Fitting the model using stan_glm and displaying uncertainty in the fitted model

As with linear regression, we can use the stan_glm function simulations to work with predictive uncertainty. We can extract random draws of the parameter simulations to display uncertainty in the coefficients by adding the following to the plotting commands from page 219. Here we sample 20 random draws from the simulations of the fitted model, because if we plotted all 4000 the graph would be impossible to read:

```
sims_1 <- as.matrix(fit_1)
n_sims <- nrow(sims_1)
for (j in sample(n_sims, 20)){
  curve(invlogit(sims_1[j,1] + sims_1[j,2]*x), col="gray", lwd=0.5, add=TRUE)
}
```

## 13.2 Interpreting logistic regression coefficients and the divide-by-4 rule

Coefficients in logistic regression can be challenging to interpret because of the nonlinearity just noted. We shall try to generalize the procedure for understanding coefficients one at a time, as was done for linear regression in Chapter 10. We illustrate with the model, Pr(Bush support) = $\text{logit}^{-1}(-1.40 + 0.33 * \text{income})$. We present some simple approaches here and return in Section 14.4 to more comprehensive numerical summaries.

### Evaluation at and near the mean of the data

The nonlinearity of the logistic function requires us to choose where to evaluate changes, if we want to interpret on the probability scale. The average of the predictors in the data can be a useful start.

- As with linear regression, the *intercept* can only be interpreted assuming zero values for the other predictors. When zero is not interesting or not even in the model (as in the voting example, where income is on a 1–5 scale), the intercept must be evaluated at some other point. For example, we can evaluate Pr(Bush support) at the highest income category and get $\text{logit}^{-1}(-1.40+0.33*5) = 0.56$. Or we can evaluate Pr(Bush support) at the mean of respondents' incomes: $\text{logit}^{-1}(-1.40+0.33\,\bar{x})$. In R we can compute,

```
invlogit(-1.40 + 0.33*mean(nes92$income))
```

  or, more generally,

```
invlogit(coef(fit_1)[1] + coef(fit_1)[2]*mean(nes92$income))
```

  with `invlogit = plogis` as defined on page 217. For this dataset, $\bar{x} = 3.1$, yielding Pr(Bush support) = $\text{logit}^{-1}(-1.40 + 0.33 * 3.1) = 0.41$ at this central point.

- A difference of 1 in income (on this 1–5 scale) corresponds to a positive difference of 0.33 in the logit probability of supporting Bush. We can evaluate how the probability differs with a unit difference in $x$ near the central value. Since $\bar{x} = 3.1$ in this example, we can evaluate the logistic regression function at $x = 3$ and $x = 2$; the difference in $\Pr(y = 1)$ corresponding to adding 1 to $x$ is $\text{logit}^{-1}(-1.40 + 0.33 * 3) - \text{logit}^{-1}(-1.40 + 0.33 * 2) = 0.08$.

### The divide-by-4 rule

The logistic curve is steepest at its center, at which point $\alpha + \beta x = 0$ so that $\text{logit}^{-1}(\alpha + \beta x) = 0.5$; see Figure 13.1. The slope of the curve—the derivative of the logistic function—is maximized at this point and attains the value $\beta e^0/(1 + e^0)^2 = \beta/4$. Thus, $\beta/4$ is the maximum difference in $\Pr(y = 1)$ corresponding to a unit difference in $x$.

As a rule of convenience, we can take logistic regression coefficients (other than the constant term) and divide them by 4 to get an upper bound of the predictive difference corresponding to a unit difference in $x$. This upper bound is a reasonable approximation near the midpoint of the logistic curve, where probabilities are close to 0.5.

For example, in the model Pr(Bush support) = $\text{logit}^{-1}(-1.40 + 0.33 * \text{income})$, we can divide 0.33 by 4 to get 0.08: a difference of 1 in income category corresponds to no more than an 8% positive difference in the probability of supporting Bush. In this case the probabilities are not far from 50% (see Figure 13.2), and so the divide-by-4 approximation works well.

### Interpretation of coefficients as odds ratios

Another way to interpret logistic regression coefficients is in terms of *odds ratios*. If two outcomes have the probabilities $(p, 1-p)$, then $\frac{p}{1-p}$ is called the *odds*. An odds of 1 is equivalent to a probability of 0.5—that is, equally likely outcomes. Odds of 0.5 or 2.0 represent probabilities of $(\frac{1}{3}, \frac{2}{3})$. Dividing

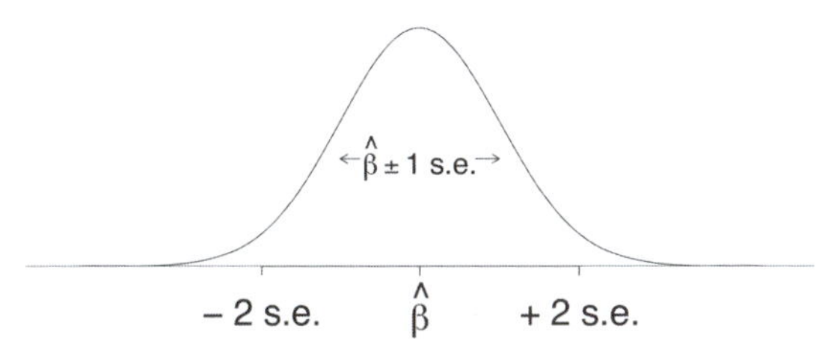


Figure 13.3 *Distribution representing uncertainty in an estimated regression coefficient (repeated from Figure 4.1). The range of this distribution corresponds to the possible values of $\beta$ that are consistent with the data. When using this as an uncertainty distribution, we assign an approximate 68% chance that $\beta$ will lie within 1 standard error of the point estimate, $\hat{\beta}$, and an approximate 95% chance that $\beta$ will lie within 2 standard errors. Assuming the regression model is correct, it should happen only about 5% of the time that the estimate, $\hat{\beta}$, falls more than 2 standard errors away from the true $\beta$.*

two odds, $\frac{p_1}{1-p_1}/\frac{p_2}{1-p_2}$, gives an odds ratio. Thus, an odds ratio of 2 corresponds to a change from $p = 0.33$ to $p = 0.5$, or a change from $p = 0.5$ to $p = 0.67$.

An advantage of working with odds ratios (instead of probabilities) is that it is possible to keep scaling up odds ratios indefinitely without running into the boundary points of 0 and 1. For example, going from an odds of 2 to an odds of 4 increases the probability from $\frac{2}{3}$ to $\frac{4}{5}$; doubling the odds again increases the probability to $\frac{8}{9}$, and so forth.

Exponentiated logistic regression coefficients can be interpreted as odds ratios. For simplicity, we illustrate with a model with one predictor, so that

$$\log\left(\frac{\Pr(y=1|x)}{\Pr(y=0|x)}\right) = \alpha + \beta x. \tag{13.4}$$

Adding 1 to $x$ (that is, changing $x$ to $x + 1$ in the above expression) has the effect of adding $\beta$ to both sides of the equation. Exponentiating both sides, the odds are then multiplied by $e^{\beta}$. For example, if $\beta = 0.2$, then a unit difference in $x$ corresponds to a multiplicative change of $e^{0.2} = 1.22$ in the odds (for example, changing the odds from 1 to 1.22, or changing $p$ from 0.5 to 0.55).

We find that the concept of odds can be somewhat difficult to understand, and odds ratios are even more obscure. Therefore we prefer to interpret coefficients on the original scale of the data when possible, for example, saying that adding 0.2 on the logit scale corresponds to a change in probability from $\text{logit}^{-1}(0)$ to $\text{logit}^{-1}(0.2)$.

### Coefficient estimates and standard errors

The coefficients in classical logistic regression are estimated using maximum likelihood, a procedure that can often work well for models with few predictors fit to reasonably large samples (but see Section 14.6 for a potential problem).

As with the linear model, the standard errors represent estimation uncertainty. We can roughly say that coefficient estimates within 2 standard errors of $\hat{\beta}$ are consistent with the data. Figure 13.3 shows the normal distribution that approximately represents the range of possible values of $\beta$. For the voting example, the coefficient of income has an estimate $\hat{\beta}$ of 0.33 and a standard error of 0.06; thus the data are roughly consistent with values of $\beta$ in the range $[0.33 \pm 2 * 0.06] = [0.21, 0.45]$.

### Statistical significance

As with linear regression, an estimate is conventionally labeled as statistically significant if it is at least 2 standard errors away from zero or, more formally, if its 95% confidence interval excludes zero.

For reasons discussed in earlier chapters, we do *not* recommend using statistical significance to

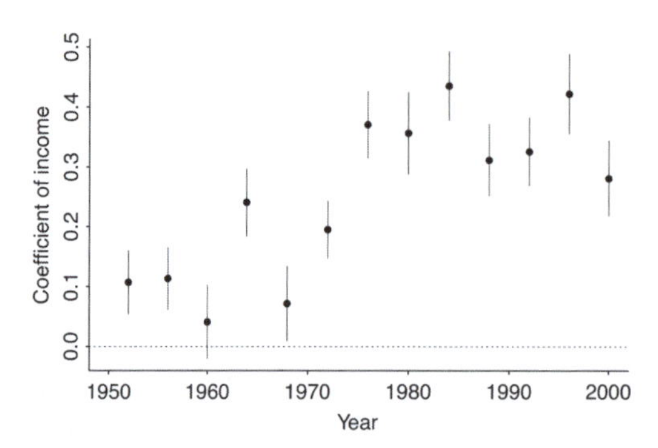


Figure 13.4 *Coefficient of income (on a 1–5 scale) with ±1 standard-error bounds in logistic regressions predicting Republican preference for president, as estimated separately from surveys in the second half of the twentieth century. The pattern of richer voters supporting Republicans has increased since 1970. The data used in the estimate for 1992 appear in Figure 13.2.*

make decisions in logistic regression about whether to include predictors or whether to interpret effects as real. When fitting many models, or when exploring data and deciding on models to fit, it is easy to find patterns—including patterns that are "statistically significant"—arising from chance alone. Conversely, if an estimate is *not* labeled as statistically significant—that is, if it is less than 2 standard errors from zero—there could still be an underlying effect or relationship in the larger population. Thus, we recognize that standard errors convey uncertainty in estimation without trying to make a one-to-one mapping between statistically significant coefficients and real effects.

When considering multiple inputs, we follow the same principles as with linear regression when deciding when and how to include and combine inputs in a model, as discussed in Section 12.6.

### Displaying the results of several logistic regressions

Example: Income and voting

We can display estimates from a series of logistic regressions in a single graph, just as was done in Section 10.9 for linear regression coefficients. Figure 13.4 illustrates with the estimate ±1 standard error for the coefficient for income on presidential preference, fit to National Election Studies pre-election polls from 1952 through 2000. Higher income has consistently been predictive of Republican support, but the connection became stronger toward the end of the century.

## 13.3 Predictions and comparisons

Logistic regression predictions are probabilistic, so for each unobserved future data point $y_i^{\text{new}}$, there is a predictive probability,

$$p_i^{\text{new}} = \Pr(y_i^{\text{new}} = 1) = \text{logit}^{-1}(X_i^{\text{new}}\beta),$$

rather than a point prediction.

We discussed predictions for linear regression in Section 9.2; for logistic regression and other generalized linear models we can similarly get point predictions for the expected value of a new observation using `predict` and then use the functions `posterior_linpred`, `posterior_epred`, and `posterior_predict` to get posterior simulations representing uncertainty about the linear predictor $X^{\text{new}}\beta$, the predicted probability $\text{logit}^{-1}(X^{\text{new}}\beta)$, and the random binary outcome $y^{\text{new}}$, respectively. We demonstrate with a simple example.

### Point prediction using predict

Earlier we fit a logistic regression to predict the probability that respondents in a survey from 1992 would support George Bush for president. The fitted model is Pr(Bush support) = $\text{logit}^{-1}(-1.40 + 0.33 * \text{income})$, and it is saved as the R object fit1, which includes posterior simulations representing uncertainty in the fitted coefficients.

Now suppose we want to predict the vote preference of a person with income level 5 (recall the five-point scale in Figure 13.2). We can use the predict function to compute the estimated probability that he or she supports Bush, $\text{E}(y^{\text{new}}|x^{\text{new}}) = \text{Pr}(y^{\text{new}} = 1|x^{\text{new}})$:

```
new <- data.frame(income=5)
pred <- predict(fit_1, type="response", newdata=new)
```

The result is 0.56, a predicted 56% chance of supporting Bush. Alternatively we could directly compute the prediction using the point estimate, invlogit(-1.40 + 0.33*5). Because the logistic transformation is nonlinear, the result from performing invlogit on the fitted linear predictor is not quite identical to what is obtained from predict, but in this case the difference is tiny: the coefficients are estimated with enough precision that there is very little uncertainty in the linear predictor, and the logistic transformation is approximately linear over any narrow range.

In the above call to predict, we specified type="response" to get the prediction on the probability scale. Alternatively, we could set type="link" to get the estimated linear predictor, the prediction on the logit scale, which in this case is $0.25 = -1.40 + 0.33 * 5$. The command invlogit(predict(fit_1, type="link", newdata=new)) gives the same result as predict(fit_1, type="response", newdata=new).

### Linear predictor with uncertainty using posterior_linpred

As with linear regression, the function posterior_linpred yields simulation draws for the linear predictor, $X^{\text{new}}\beta$:

```
linpred <- posterior_linpred(fit_1, newdata=new)
```

The result in this case is a vector representing the posterior distribution of $a + b * 5$. Typically, though, we are interested in the predicted value on the probability scale.

### Expected outcome with uncertainty using posterior_epred

We can compute the uncertainty in the expected value or predicted probability, $\text{E}(y^{\text{new}}) = \text{Pr}(y^{\text{new}} = 1) = \text{logit}^{-1}(X^{\text{new}}\beta)$ using the posterior_epred function:

```
epred <- posterior_epred(fit_1, newdata=new)
```

The result is a vector of length equal to the number of posterior simulation draws; we can summarize, for example, using the mean and standard deviation:

```
print(c(mean(epred), sd(epred)))
```

The result is 0.56 (the same as computed using predict) and 0.03. According to our fitted model, the percentage of Bush supporters in the population at that time, among people with income level 5, was probably in the range 56% ± 3%. This inference implicitly assumes that the survey respondents within each income category are a random sample from the population of interest, and that the logistic regression model accurately describes the association between income and vote preference.

The command invlogit(posterior_linpred(fit_1, newdata=new)) gives the same result as posterior_epred(fit_1, newdata=new), just converting from the linear predictor to the probability scale.

### Predictive distribution for a new observation using posterior_predict

To get the uncertainty for a single voter corresponding to some individual data point $x^{\text{new}}$, we use the posterior prediction:

```
postpred <- posterior_predict(fit_1, newdata=new)
```

which returns a vector of length n_sims, taking on values 0, 1, 1, 0, 1, ..., corresponding to possible values of vote among people with income=5. Taking the average of these simulations, mean(postpred), gives 0.56, as this is equivalent to computing the point prediction, $\text{E}(y^{\text{new}}|x^{\text{new}})$, for this new data point.

### Prediction given a range of input values

Above we used the prediction functions for a single new data point. We can also use them to make predictions for a vector of new observations. For example, suppose we want to make predictions for five new people whose incomes take on the values 1 through 5:

```
new <- data.frame(income=1:5)
pred <- predict(fit_1, type="response", newdata=new)
linpred <- posterior_linpred(fit_1, newdata=new)
epred <- posterior_epred(fit_1, newdata=new)
postpred <- posterior_predict(fit_1, newdata=new)
```

The first two objects above are vectors of length 5; the last three are n_sims $\times$ 5 matrices.

We can use epred to make statements about the population. For example, this will compute the posterior probability, according to the fitted model, that Bush was more popular among people with income level 5 than among people with income level 4:

```
mean(epred[,5] > epred[,4])
```

And this will compute a 95% posterior distribution for the difference in support for Bush, comparing people in the richest to the second-richest category:

```
quantile(epred[,5] - epred[,4], c(0.025, 0.975))
```

We can use postpred to make statements about individual people. For example, this will compute the posterior simulations of the number of these new survey respondents who support Bush:

```
total <- apply(postpred, 1, sum)
```

And here is how to compute the probability that at least three of them support Bush:

```
mean(total >= 3)
```

### Logistic regression with just an intercept

As discussed in Section 7.3, linear regression with just an intercept is the same as estimating an average, and linear regression with a single binary predictor is the same as estimating a difference in averages. Similarly, logistic regression with just an intercept is equivalent to the estimate of a proportion.

Here is an example. A random sample of 50 people are tested, and 10 have a particular disease. The proportion is 0.20 with standard error $\sqrt{0.2 * 0.8/50} = 0.06$. Alternatively we can set this up as logistic regression:

```
y <- rep(c(0, 1), c(40, 10))
simple <- data.frame(y)
fit <- stan_glm(y ~ 1, family=binomial(link="logit"), data=simple)
```

yielding the result,

```
            Median MAD_SD
(Intercept) -1.38   0.35
```

We can transform the prediction to the probability scale and get $\text{logit}^{-1}(-1.38) = 0.20$, as with the direct calculation, and then obtain approximate $\pm 1$ standard error bounds using $\text{logit}^{-1}(-1.38 \pm 0.35) = (\text{logit}^{-1}(1.73), \text{logit}^{-1}(1.03)) = (0.15, 0.26)$, which is essentially the same as the classical estimate with uncertainty of $0.20 \pm 0.06$.

Alternatively, we can get the inference on the probability scale using `posterior_epred`:

```
new <- data.frame(x=0)
epred <- posterior_epred(fit, newdata=new)
print(c(mean(epred), sd(epred)))
```

This again returns 0.20 and 0.06. If you go to more decimal places, the classical and logistic regression estimates differ, in part because `stan_glm` uses a prior distribution (see Section 13.5) and in part because the classical standard error is only an approximation to the inferential uncertainty arising from discrete data.

**Data on the boundary.** What happens when $y = 0$ or $n$? As discussed on page 52, the classical standard error formula gives 0 at these extremes, so instead we use the 95% interval $\hat{p} \pm \sqrt{\hat{p}(1 - \hat{p})/(n + 4)}$, where $\hat{p} = (y + 2)/(n + 4)$. Let's compare this to logistic regression, using the example of a test with zero positive results out of 50:

```
y <- rep(c(0, 1), c(50, 0))
simple <- data.frame(y)
fit <- stan_glm(y ~ 1, family=binomial(link="logit"), data=simple)
```

Here is the result:

```
            Median MAD_SD
(Intercept) -4.58   1.14
```

The approximate 95% interval is $\text{logit}^{-1}(-4.58 \pm 2 * 1.14) = (0, 0.09)$. The lower endpoint is not exactly zero but we bring it down to zero in any case, as the data are obviously consistent with an underlying rate of zero. In this case, the classical $\hat{p} \pm \sqrt{\hat{p}(1 - \hat{p})/(n + 4)}$ interval is also $(0, 0.09)$ to two decimal places. Again, the intervals based on the simple formula and logistic regression will not be identical, as the two approaches make different assumptions.

## Logistic regression with a single binary predictor

Logistic regression on an indicator variable is equivalent to a comparison of proportions. For a simple example, consider tests for a disease on samples from two different populations, where 10 out of 50 from population A test positive, as compared to 20 out of 60 from population B. The classical estimate is 0.13 with standard error 0.08. Here is the setup as logistic regression:

```
x <- rep(c(0, 1), c(50, 60))
y <- rep(c(0, 1, 0, 1), c(40, 10, 40, 20))
simple <- data.frame(x, y)
fit <- stan_glm(y ~ x, family=binomial(link="logit"), data=simple)
```

which yields:

```
            Median MAD_SD
(Intercept) -1.39   0.35
x            0.69   0.44
```

To get inference for the difference in probabilities, we compare predictions on the probability scale for $x = 0$ and $x = 1$:

```
new <- data.frame(x=c(0, 1))
epred <- posterior_epred(fit, newdata=new)
diff <- epred[,2] - epred[,1]
print(c(mean(diff), sd(diff)))
```

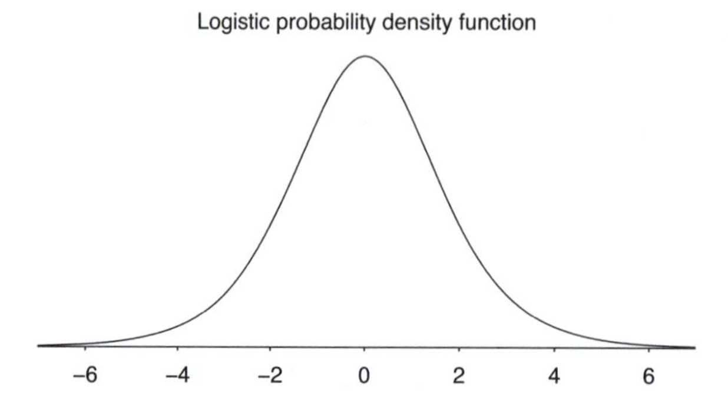


Figure 13.5 *The probability density function of the logistic distribution, which is used for the error term in the latent-data formulation (13.5) of logistic regression. The logistic curve in Figure 13.1a is the cumulative distribution function of this density. The maximum of the density is 0.25, which corresponds to the maximum slope of 0.25 in the inverse logit function of Figure 13.1a.*

This returns 0.13 and 0.08, the same estimate and standard error as computed using the simple formula. The point of this example is not that you should use logistic regression to do simple comparisons of proportions; rather, we are demonstrating the underlying unity of comparisons and logistic regression for binary data, in the same way that Section 7.3 shows how comparisons of continuous data is a special case of linear regression.

## 13.4 Latent-data formulation

We can interpret logistic regression directly—as a nonlinear model for the probability of a "success" or "yes" response given some predictors—and also indirectly, using what are called unobserved or *latent* variables.

In this formulation, each discrete outcome $y_i$ is associated with a continuous, unobserved outcome $z_i$, defined as follows:

$$\begin{aligned} y_i &= \begin{cases} 1 & \text{if } z_i > 0 \\ 0 & \text{if } z_i < 0 \end{cases} \\ z_i &= X_i\beta + \epsilon_i, \end{aligned} \tag{13.5}$$

with independent errors $\epsilon_i$ that have the *logistic* probability distribution. The logistic distribution is shown in Figure 13.5 and is defined so that

$$\Pr(\epsilon_i < x) = \text{logit}^{-1}(x) \text{ for all } x.$$

Thus, $\Pr(y_i = 1) = \Pr(z_i > 0) = \Pr(\epsilon_i > -X_i\beta) = \text{logit}^{-1}(X_i\beta)$, and so models (13.2) and (13.5) are equivalent.

Figure 13.6 illustrates for an observation $i$ with income level $x_i = 1$ (that is, a person in the lowest income category), whose linear predictor, $X_i\beta$, thus has the value $-1.40 + 0.33 * 1 = -1.07$. The curve illustrates the distribution of the latent variable $z_i$, and the shaded area corresponds to the probability that $z_i > 0$, so that $y_i = 1$. In this example, $\Pr(y_i = 1) = \text{logit}^{-1}(-1.07) = 0.26$.

### Interpretation of the latent variables

Latent variables are a computational trick but they can also be interpreted substantively. For example, in the pre-election survey, $y_i = 1$ for Bush supporters and 0 for Clinton supporters. The unobserved

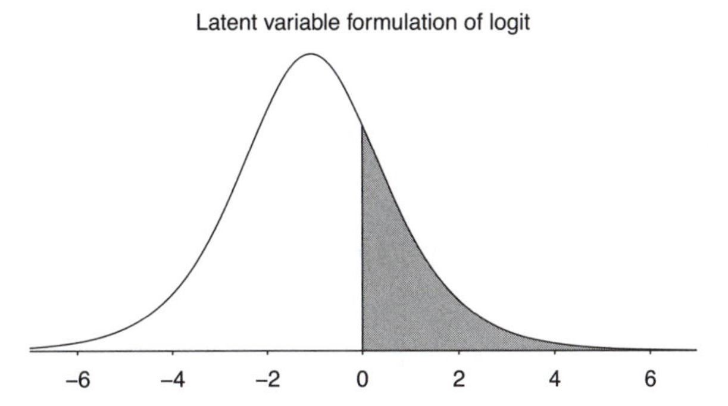


Figure 13.6 *The probability density function of the latent variable $z_i$ in model (13.5) if the linear predictor, $X_i\beta$, has the value −1.07. The shaded area indicates the probability that $z_i > 0$, so that $y_i = 1$ in the logistic regression.*

continuous $z_i$ can be interpreted as the respondent's "utility" or preference for Bush, compared to Clinton: the sign of the utility tells us which candidate is preferred, and its magnitude reveals the strength of the preference.

Only the sign of $z_i$, not its magnitude, can be determined directly from binary data. However, we can learn more about the $z_i$'s given the logistic regression predictors. In addition, in some settings direct information is available about the $z_i$'s; for example, a survey can ask "feeling thermometer" questions such as, "Rate your feelings about Barack Obama on a 1–10 scale, with 1 being the most negative and 10 being the most positive."

### Nonidentifiability of the latent scale parameter

In Section 14.6 we discuss challenges of identification, situations where the data do not supply precise information about parameters in a model, a concept introduced earlier on page 146 in Section 10.7. A particular identification issue arises with the scaling of the latent model in logistic regression.

The logistic probability density function in Figure 13.5 appears bell-shaped, much like the normal density that is used for errors in linear regression. In fact, the logistic distribution is very close to the normal distribution with mean 0 and standard deviation 1.6—an identity that we discuss further on page 272 in the context of "probit regression." For now, we merely note that the logistic model (13.5) for the latent variable $z$ is closely approximated by the normal regression model,

$$z_i = X_i\beta + \epsilon_i, \quad \epsilon_i \sim \mathrm{N}(0, \sigma^2), \tag{13.6}$$

with $\sigma = 1.6$. This then raises the question, Why not estimate $\sigma$?

We cannot estimate the parameter $\sigma$ in model (13.6) because it is not identified when considered jointly with$\beta$. If all the elements of $\beta$ are multiplied by a positive constant and $\sigma$ is also multiplied by that constant, then the model does not change. For example, suppose we fit the model

$$z_i = -1.40 + 0.33x_i + \epsilon_i, \quad \epsilon_i \sim \mathrm{N}(0, 1.6^2).$$

This is equivalent to the model

$$z_i = -14.0 + 3.3x_i + \epsilon_i, \quad \epsilon_i \sim \mathrm{N}(0, 16^2),$$

or

$$z_i = -140 + 33x_i + \epsilon_i, \quad \epsilon_i \sim \mathrm{N}(0, 160^2).$$

As we move from each of these models to the next, $z$ is multiplied by 10, but the *sign* of $z$ does not change. Thus all the models have the same implications for the observed data $y$: for each model,

$\Pr(y_i = 1) \approx \text{logit}^{-1}(-1.40 + 0.33x_i)$. This is only approximate because the logistic distribution is not exactly normal.

Thus, model (13.6) has an essential indeterminacy when fit to binary data, and it is standard to resolve this by setting the variance parameter $\sigma$ to a fixed value such as 1.6, chosen because the normal distribution with standard deviation 1.6 is close to the unit logistic distribution.

## 13.5 Maximum likelihood and Bayesian inference for logistic regression

Classical linear regression can be motivated in a purely algorithmic fashion (as least squares) or as maximum likelihood inference under a normal model; see Section 10.7. With logistic regression (and other generalized linear models, as described in the next chapter), the usual practice is to start with maximum likelihood, followed by Bayesian inference and other regularizing techniques.

### Maximum likelihood using iteratively weighted least squares

For binary logistic regression with data $y_i = 0$ or 1, the likelihood is

$$p(y|\beta, X) = \prod_{i=1}^{n} \begin{cases} \text{logit}^{-1}(X_i\beta) & \text{if } y_i = 1 \\ 1 - \text{logit}^{-1}(X_i\beta) & \text{if } y_i = 0, \end{cases}$$

which can be written more compactly, but equivalently, as

$$p(y|\beta, X) = \prod_{i=1}^{n} \left(\text{logit}^{-1}(X_i\beta)\right)^{y_i} \left(1 - \text{logit}^{-1}(X_i\beta)\right)^{1-y_i}. \qquad (13.7)$$

To find the $\beta$ that maximizes this expression, one can compute the derivative $dp(y|\beta, X)/d\beta$ of the likelihood (or, more conveniently, the derivative of the logarithm of the likelihood), set this derivative equal to 0, and solve for $\beta$. There is no closed-form solution, but the maximum likelihood estimate can be found using an iterative optimization algorithm that converges to a point of zero derivative and thus the vector of coefficients $\beta$ that maximizes (13.7), when such a maximum exists. We shall not discuss this further here, as these computations are all now done invisibly by the computer. We mention it only to give a sense of how likelihood functions are used in classical estimation. Section 14.6 discusses collinearity and separation, which are two conditions under which no maximum likelihood estimate exists.

### Bayesian inference with a uniform prior distribution

If the prior distribution on the parameters is uniform, then the posterior density is proportional to the likelihood function, and the posterior mode—the vector of coefficients $\beta$ that maximizes the posterior density—is the same as the maximum likelihood estimate.

As with linear regression, the benefit of Bayesian inference with a noninformative prior is that we can use simulations from the entire posterior distribution—not just a maximum or any other point estimate—to summarize uncertainty, and we can also use these simulations to make probabilistic predictions. In `stan_glm` we can declare a uniform prior by setting `prior=NULL, prior_intercept=NULL`. We do not generally recommend that choice, as this can result in having no estimate: as we discuss in the Section 14.6, there are real-world logistic regressions in which the coefficients are not identified from data alone, and prior information is required to obtain any estimates at all.

### Default prior in `stan_glm`

By default, `stan_glm` uses a weakly informative family of prior distributions. For a logistic regression of the form, $\Pr(y = 1) = \text{logit}^{-1}(a + b_1x_1 + b_2x_2 + \cdots + b_Kx_K)$, each coefficient $b_k$ is given a normal prior distribution with mean 0 and standard deviation $2.5/\mathrm{sd}(x_k)$. The intercept $a$ is not given a prior distribution directly; instead we assign the prior to the linear predictor with each $x$ set to its mean value in the data, that is, $a + b_1\bar{x}_1 + b_2\bar{x}_2 + \cdots + b_K\bar{x}_K$. This centered intercept is given a normal prior distribution with mean 0 and standard deviation 2.5. This is similar to the default priors for linear regression discussed on page 124, except that for logistic regression there is no need to scale based on the outcome variable.

### Bayesian inference with some prior information

When prior information is available, it can be used to increase the precision of estimates and predictions. We demonstrate with a simple hypothetical example of a logistic regression with one predictor, $\Pr(y_i = 1) = \mathrm{logit}^{-1}(a + bx)$, in some area of application for which we expect the slope $b$ to fall between 0 and 1. We assume this is a "soft constraint," not a "hard constraint"; that is, we expect the value of $b$ for any given problem in this domain to be between 0 and 1, but with it being possible on occasion for $b$ to fall outside that range. We shall summarize this soft constraint as a normal prior distribution on $b$ with mean 0.5 and standard deviation 0.5, which implies that, for any given problem, $b$ has a 68% chance of being between 0 and 1 and a 95% chance of being between −0.5 and 1.5. We also include by default a weak prior for the intercept.

### Comparing maximum likelihood and Bayesian inference using a simulation study

mple:
r and
a for
stic
ession

To understand the role of the prior distribution, we consider several data scenarios, each time assuming that the true parameter values are $a = -2$ and $b = 0.8$ and that data are gathered from an experiment where the values of $x$ are drawn from a uniform distribution between −1 and 1. These choices are entirely arbitrary, just to have a specific model for our simulation study. Here is the R code:[2]

```
library("arm", "rstanarm")
bayes_sim <- function(n, a=-2, b=0.8){
  x <- runif(n, -1, 1)
  z <- rlogis(n, a + b*x, 1)
  y <- ifelse(z>0, 1, 0)
  fake <- data.frame(x, y, z)
  glm_fit <- glm(y ~ x, family=binomial(link="logit"), data=fake)
  stan_glm_fit <- stan_glm(y ~ x, family=binomial(link="logit"), data=fake,
    prior=normal(0.5, 0.5))
  display(glm_fit, digits=1)
  print(stan_glm_fit, digits=1)
}
```

We set `digits=1` in the displays to avoid being distracted by details.

We then simulate for a range of sample sizes, each time focusing on inference about $b$:

- $n = 0$. With no new data, the above function does not run (as it would have to work with empty data vectors), but in any case the posterior is the same as the prior, centered at 0.5 with standard deviation 0.5.
- $n = 10$. With only 10 observations, the maximum likelihood estimate is noisy, and we expect the Bayesian posterior to be close to the prior. Here's an example of what happens after typing `bayes_sim(10)` in the R console:

```
glm:         coef.est coef.se
(Intercept) -4.4      2.9
x            5.3      4.4
---
stan_glm:   Median MAD_SD
```

[2]Code for this example is in the folder `LogisticPriors`.

```
(Intercept) -1.8    0.8
x            0.6    0.5
```

We shall focus on the coefficient for $x$, which represents the parameter of interest in this hypothetical study. In the above simulation, glm gives a maximum likelihood estimate of 5.3, which is far from the specified belief that $b$ is likely to be in the range (0, 1)—but that estimate also has a large standard error, indicating that the likelihood provides little information in this $n = 10$ setting. In contrast, the inference from stan_glm relies heavily on the prior distribution: the Bayes estimate of 0.6 is close to the prior mean of 0.5, being pulled away from the data only slightly.

- $n = 100$. Next we try bayes_sim(100), which yields,

```
glm:         coef.est coef.se
(Intercept) -2.6       0.5
x            1.4       0.7
---
stan_glm:    Median MAD_SD
(Intercept) -2.4    0.4
x            0.8    0.4
```

The maximum likelihood estimate is again extreme, but less so than before, and the Bayes estimate is again pulled toward the prior mean of 0.5, but less so than before. This is just one realization of the process, and another random simulation will give different results—try it yourself!—but it illustrates the general pattern of the Bayesian posterior estimate being a compromise between data and prior.

- $n = 1000$. Finally, we simulate an experiment with 1000 data points. Here the sample size is large enough that we expect the likelihood to dominate. Let's look at a simulation:

```
glm:         coef.est coef.se
(Intercept) -2.0       0.1
x            1.1       0.2
---
stan_glm:    Median MAD_SD
(Intercept) -2.0    0.1
x            1.0    0.2
```

The Bayes estimate is only very slightly pooled toward the prior mean. In this particular example, once $n$ is as large as 1000, the prior distribution doesn't really make a difference.

## 13.6 Cross validation and log score for logistic regression

When evaluating predictive and generalization performance, we could compare binary observations to binary predictions and compute, for example, the percentage of correct predictions, but that completely ignores the additional information in the predicted probabilities. Probability forecasts are sometimes evaluated using the Brier score, which is defined as $\frac{1}{n}\sum_{i=1}^{n}(p_i - y_i)^2$; this is equivalent to the mean of squared residuals and is proportional to the posterior residual standard deviation discussed in Section 11.6. When computing Bayesian $R^2$ for linear regression, the posterior residual standard deviation is replaced with the modeled $\sigma$, and correspondingly for logistic regression we can use the model-based approximation, $\frac{1}{n}\sum_{i=1}^{n} p_i(1 - p_i)$. However, $R^2$ for logistic regression has shortcomings, as the residuals are not additive and thus it is not as interpretable as for linear regression. In Section 11.8 we discussed summarizing prediction errors by the log score, which is also a natural choice when the predictions are probabilities, as with logistic regression. Sometimes *deviance*, which is $-2$ times the log score, is reported.

As discussed in Section 11.8 in the context of linear regression, we can use cross validation to avoid the overfitting associated with using the same data for model fitting and evaluation. Thus, rather than evaluating log predictive probabilities directly, we can compute leave-one-out (LOO) log prediction probabilities, which can be used to compute the LOO log score (or, if desired, the LOO Brier score, LOO $R^2$, or any other measure of fit).

### Understanding the log score for discrete predictions

We develop intuition about the logarithms of the modeled probability of the observed outcomes using a simple example.

Consider the following scenario: We fit a logistic regression model with estimated parameters $\hat{\beta}$ and apply it to $n^{\text{new}}$ new data points with predictor matrix $X^{\text{new}}$, thus yielding a vector of predictive probabilities $p_i^{\text{new}} = \text{logit}^{-1}(X_i^{\text{new}}\hat{\beta})$. We then find out the outcomes $y^{\text{new}}$ for the new cases and evaluate the model fit using

$$\text{out-of-sample log score} = \sum_{i=1}^{n^{\text{new}}} \begin{cases} \log p_i^{\text{new}} & \text{if } y_i^{\text{new}} = 1 \\ \log(1 - p_i^{\text{new}}) & \text{if } y_i^{\text{new}} = 0. \end{cases}$$

The same idea applies more generally to any statistical model: the predictive log score is the sum of the logarithms of the predicted probabilities of each data point.

Probabilities are between 0 and 1, and the logarithm of any number less than 1 is negative, with $\log(1) = 0$. Thus the log score is necessarily negative (except in the trivial case of a model that predicts events with perfect probabilities of 1). In general, higher values of log score are better. For example, if two models are used to predict the same data, a log score of $-150$ is better than a log score of $-160$.

How large should we expect a log score to be? We can calibrate by considering a model with no information that assigns 50% probability to every outcome. When $p_i^{\text{new}} = 0.5$, the log score equals $\log(0.5) = -0.693$ for every data point $y_i^{\text{new}}$, whether 0 or 1, and thus the log score for $n^{\text{new}}$ new data points is simply $-0.693\, n^{\text{new}}$.

What about something better? Consider a model that predicts $p_i = 0.8$ for half the cases and $p_i = 0.6$ for the other half. (This could be obtained, for example, from the model, $\Pr(y_i = 1) = \text{logit}^{-1}(0.4 + x_i)$, for a dataset in which half the cases have $x_i = 1$ and half have $x_i = 0$.) Further suppose that this model has been fit to data that are representative of the general population, so that when we apply it to new cases it will be *calibrated*: when $p_i^{\text{new}} = 0.8$, the new observation $y_i^{\text{new}}$ will equal 1 in 80% of the cases; and when $p_i^{\text{new}} = 0.6$, the new observation $y_i^{\text{new}}$ will equal 1 in 60% of the cases. In that case, the expected value of the log score for a single new observation chosen at random will be $0.5 * (0.8 \log 0.8 + 0.2 \log 0.2) + 0.5 * (0.6 \log 0.6 + 0.4 \log 0.4) = -0.587$, and so we would expect the log score for a new dataset to be approximately $-0.587\, n^{\text{new}}$.

In this simple example, when we move from the simple coin-flip model to the model with probabilities 0.8 and 0.6, the log score improves by $0.693\, n^{\text{new}} - 0.587\, n^{\text{new}} = 0.106\, n^{\text{new}}$, an improvement of about 1 in the log score per 10 data points.

### Log score for logistic regression

ample:
come and
ting

We explore the log score using the example of the model in Section 13.1 of polling during the 1992 U.S. presidential election campaign. The survey we are analyzing had 702 respondents who expressed support for Bill Clinton and 477 who preferred George Bush. We compute the log score for a sequence of models using increasing amounts of information:

- The null model gives a 0.5 probability of each outcome; the log score for our 1179 survey respondents is is $1179 \log(0.5) = -817$.
- Just using the fact that 59.5% of survey respondents supported Clinton and 40.5% supported Bush, we can assign 40.5% to $\Pr(y_i = 1)$ for each respondent, which improves the log score to $477 \log(0.405) + 702 \log(-.595) = -796$. Equivalently, we can compute this by fitting a logistic regression with only the constant term:

```
fit_1a <- stan_glm(rvote ~ 1, family=binomial(link="logit"), data=nes92)
```

and then summing the logarithms of the predicted probabilities:

```
predp_1a <- predict(fit_1a, type="response")
y <- nes92$rvote
logscore_1a <- sum(y*log(predp_1a) + (1-y)*log(1 - predp_1a))
```

The regression yields an estimate of the constant term of −0.383, which is the logit of the predicted probability, 0.405, and the log score is −796, as above.

- The fit further improves when we include income as a predictor:

```
fit_1 <- stan_glm(rvote ~ income, family=binomial(link="logit"), data=nes92)
predp_1 <- predict(fit_1, type="response")
logscore_1 <- sum(y*log(predp_1) + (1-y)*log(1 - predp_1))
```

The new log score is −778.

Dividing these log scores by the sample size, we can summarize as follows: using a constant probability estimated from the data improves the log score per observation by $(817-796)/1179 = 0.018$ compared to the null ($p = 0.5$) model, and including income as a predictor increases the log score per observation by a further $(796 - 778)/1179 = 0.015$.

In Section 11.7 we discussed how within-sample validation is optimistic and how cross validation can be used to approximate external validation. Running leave-one-out cross validation,

```
loo(fit_1)
```

gives output,

```
         Estimate   SE
elpd_loo   -780.4  8.6
p_loo         2.0  0.1
looic      1560.9 17.2
```

The LOO estimated log score (`elpd_loo`) of −780 is 2 lower than the within-sample log score of −778 computed above; this difference is about what we would expect, given that the fitted model has 2 parameters or degrees of freedom.

## 13.7 Building a logistic regression model: wells in Bangladesh

Example: Arsenic in Bangladesh

We illustrate the steps of building, understanding, and checking the fit of a logistic regression model using an example from economics (or perhaps it is psychology or public health): modeling the decisions of households in Bangladesh about whether to change their source of drinking water.[3]

### Background

Many of the wells used for drinking water in Bangladesh and other South Asian countries are contaminated with natural arsenic, affecting an estimated 100 million people. Arsenic is a cumulative poison, and exposure increases the risk of cancer and other diseases, with risks estimated to be proportional to exposure.

Any locality can include wells with a range of arsenic levels, as can be seen from the map in Figure 13.7 of all the wells in a collection of villages in a small area of Bangladesh. The bad news is that even if your neighbor's well is safe, it does not mean that yours is also safe. However, the corresponding good news is that, if your well has a high arsenic level, you can probably find a safe well nearby to get your water from—if you are willing to walk the distance and your neighbor is willing to share. (The amount of water needed for drinking is low enough that adding users to a

[3]Data and code for this example are in the folder `Arsenic`.

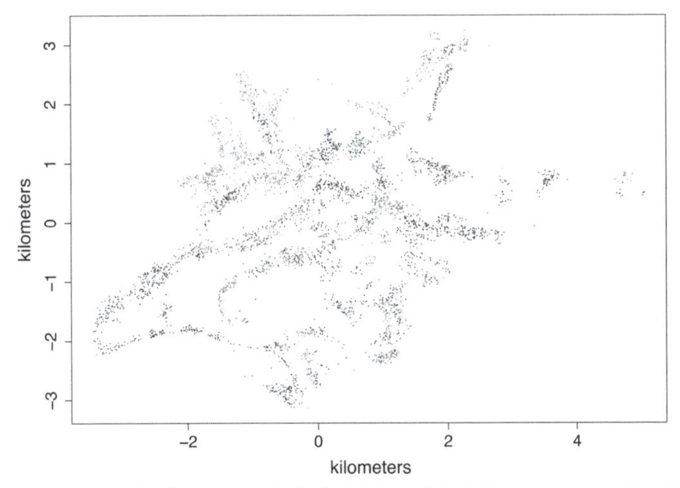


Figure 13.7 *Wells in an area of Araihazar, Bangladesh. Light and dark dots represent wells with arsenic greater than and less than the safety standard of 0.5 (in units of hundreds of micrograms per liter). The wells are located where people live. The empty areas between the wells are mostly cropland. Safe and unsafe wells are intermingled in most of the area, which suggests that users of unsafe wells can switch to nearby safe wells.*

well would not exhaust its capacity, and the surface water in this area is subject to contamination by microbes, hence the desire to use water from deep wells.)

In the area shown in Figure 13.7, a research team from the United States and Bangladesh measured all the wells and labeled them with their arsenic level as well as a characterization as "safe" (below 0.5 in units of hundreds of micrograms per liter, the Bangladesh standard for arsenic in drinking water) or "unsafe" (above 0.5). People with unsafe wells were encouraged to switch to nearby private or community wells or to new wells of their own construction.

A few years later, the researchers returned to find out who had switched wells. We shall perform a logistic regression analysis to understand the factors predictive of well switching among the users of unsafe wells. In the notation of the previous section, our outcome variable is

$$y_i = \begin{cases} 1 & \text{if household } i \text{ switched to a new well} \\ 0 & \text{if household } i \text{ continued using its own well.} \end{cases}$$

We consider the following inputs:

- The distance (in meters) to the closest known safe well.
- The arsenic level of respondent's well.
- Whether any members of the household are active in community organizations.
- The education level of the head of household.

We shall first fit the model just using distance to the nearest well and then put in arsenic concentration, organizational membership, and education.

## Logistic regression with just one predictor

We fit the logistic regression in R:

```
fit_1 <- stan_glm(switch ~ dist, family=binomial(link="logit"), data=wells)
```

Displaying this yields,

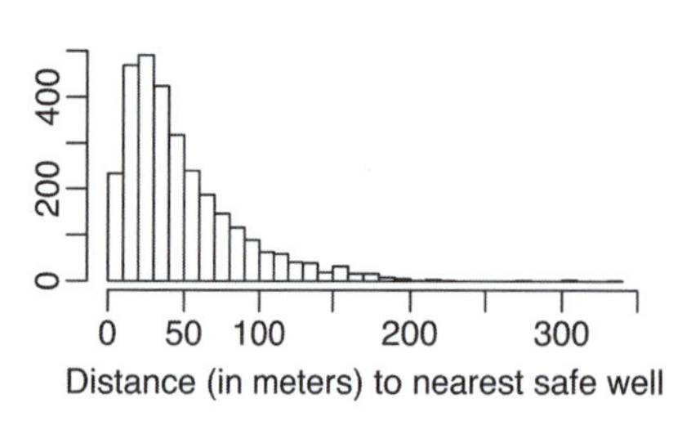


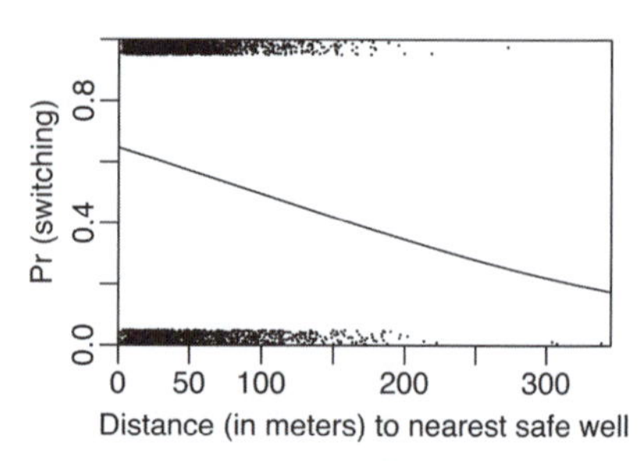


Figure 13.8 *(a) Histogram of distance to the nearest safe well, for each of the unsafe wells in our data from Araihazar, Bangladesh (see Figure 13.7). (b) Graphical expression of the fitted logistic regression,* $Pr(\text{switching wells}) = logit^{-1}(0.61 - 0.62 * \texttt{dist100})$*, with (jittered) data overlaid. The predictor* `dist100` *is* `dist`*/100: distance to the nearest safe well in 100-meter units.*

```
            Median MAD_SD
(Intercept)  0.606  0.059
dist        -0.006  0.001
```

The coefficient for `dist` is −0.006, which seems low, but this is misleading since distance is measured in meters, so this coefficient corresponds to the difference between, say, a house that is 90 meters away from the nearest safe well and a house that is 91 meters away.

Figure 13.8a shows the distribution of `dist` in the data. It seems more reasonable to rescale distance in 100-meter units:

```
wells$dist100 <- wells$dist/100
```

and re-fitting the logistic regression yields,

```
            Median MAD_SD
(Intercept)  0.6    0.1
dist100     -0.6    0.1
```

## Graphing the fitted model

In preparing to plot the data, we first create a function to jitter the binary outcome while keeping the points between 0 and 1:

```
jitter_binary <- function(a, jitt=0.05){
  ifelse(a==0, runif(length(a), 0, jitt), runif(length(a), 1 - jitt, 1))
}
```

We can then graph the data and fitted model:

```
wells$switch_jitter <- jitter_binary(wells$switch)
plot(wells$dist, wells$switch_jitter)
curve(invlogit(coef(fit_1)[1] + coef(fit_1)[2]*x), add=TRUE)
```

The result is displayed in Figure 13.8b. The probability of switching is about 60% for people who live near a safe well, declining to about 20% for people who live more than 300 meters from any safe well. This makes sense: the probability of switching is higher for people who live closer to a safe well. Another display option, which would more clearly show the differences between households that did and did not switch, would be to overlay separate histograms of `dist` for the switchers and nonswitchers.

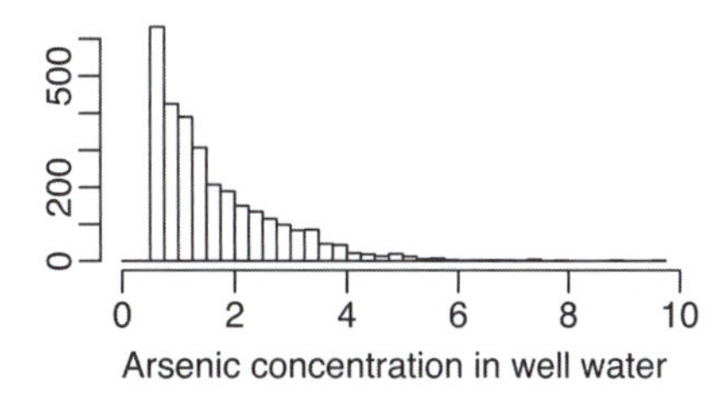


Figure 13.9 *Histogram of arsenic levels in unsafe wells (those exceeding 0.5) in the measured area of Araihazar, Bangladesh (see Figure 13.7).*

## Interpreting the logistic regression coefficients

We can interpret the coefficient estimates using evaluations of the inverse logit function and its derivative, as in the example of Section 13.1. Our model here is,

$$\Pr(\text{switch}) = \text{logit}^{-1}(0.61 - 0.62 * \text{dist100}).$$

1. The constant term can be interpreted when `dist100` = 0, in which case the probability of switching is $\text{logit}^{-1}(0.61) = 0.65$. Thus, the model estimates a 65% probability of switching if you live right next to an existing safe well.
2. We can evaluate the predictive difference with respect to `dist100` by computing the derivative at the average value of `dist100` in the dataset, which is 0.48 (that is, 48 meters; see Figure 13.8a). The value of the linear predictor here is $0.61 - 0.62 * 0.48 = 0.31$, and so the slope of the curve at this point is $-0.62\, e^{0.31}/(1 + e^{0.31})^2 = -0.15$. Thus, adding 1 to `dist100`—that is, adding 100 meters to the distance to the nearest safe well—corresponds to a negative difference in the probability of switching of about 15%.
3. More quickly, the divide-by-4 rule gives us $-0.62/4 = -0.15$. This comes out the same, to two decimal places, as the result from calculating the derivative, because in this case the curve happens to pass through the 50% point right in the middle of the data; see Figure 13.8b.

In addition to interpreting its magnitude, we can look at the uncertainty of the coefficient for distance. The slope is estimated well, with a standard error of only 0.10, which is tiny compared to the coefficient estimate of −0.62. The approximate 95% interval is [−0.73, −0.49]. Overall probability of switching a well is 57% and the corresponding log score is −2059. The LOO log score for the model with distance is −2040; distance supplies some predictive information.

## Adding a second input variable

We now extend the well-switching example by adding the arsenic level of the existing well as a regression input. At the levels present in the Bangladesh drinking water, the health risks from arsenic are roughly proportional to exposure, and so we would expect switching to be more likely from wells with high arsenic levels. Figure 13.9 shows the arsenic levels of the unsafe wells before switching, and here is the logistic regression predicting switching given arsenic level:

```
fit_3 <- stan_glm(switch ~ dist100 + arsenic, family=binomial(link="logit"), data=wells)
```

which, when displayed, yields,

```
            Median MAD_SD
(Intercept)  0.00   0.08
dist100     -0.90   0.11
arsenic      0.46   0.04
```

Thus, comparing two wells with the same arsenic level, every 100 meters in distance to the nearest safe well corresponds to a *negative* difference of 0.90 in the logit probability of switching. Similarly, a difference of 1 in arsenic concentration corresponds to a 0.46 *positive* difference in the logit probability of switching. Both coefficients are large compared to their standard errors. And both their signs make sense: switching is easier if there is a nearby safe well, and if a household's existing well has a high arsenic level, there should be more motivation to switch.

For a quick interpretation, we divide each of the coefficients by 4: thus, 100 meters more in distance corresponds to an approximately 22% lower probability of switching, and 1 unit more in arsenic concentration corresponds to an approximately 11% positive difference in switching probability.

Comparing these two coefficients, it would at first seem that distance is a more important factor than arsenic level in determining the probability of switching. Such a statement is misleading, however, because in our data `dist100` shows less variation than `arsenic`: the standard deviation of distances to the nearest well is 0.38 (in units of 100 meters), whereas arsenic levels have a standard deviation of 1.10 on the scale used here. Thus, the logistic regression coefficients corresponding to 1-standard-deviation differences are $-0.90 * 0.38 = -0.34$ for distance and $0.46 * 1.10 = 0.51$ for arsenic level. Dividing by 4 yields the quick summary estimate that a difference of 1 standard deviation in distance corresponds to an expected $-0.34/4$ or approximately 8% negative difference in Pr(switch), and a difference of 1 standard deviation in arsenic level corresponds to an expected $0.51/4$ or approximately 13% positive difference in Pr(switch).

The LOO log score for the new model is −1968.5, and comparison to the model with distance as the only predictor yields,

```
Model comparison:
(negative 'elpd_diff' favors 1st model, positive favors 2nd)
elpd_diff        se
     71.6      12.2
```

Including arsenic level in the model clearly improves the predictive accuracy.

The log score is good for model comparison as it takes properly into account the increase (or decrease) in the predictive information. It can be also useful to consider other summaries that are more directly interpretable. The average improvement in LOO predictive probabilities describes how much more accurately on average we are predicting the probabilities, although with the drawback that that probabilities close to zero can't improve much in absolute value even if they can improve a lot relatively.

### Comparing the coefficient estimates when adding a predictor

The coefficient for `dist100` changes from −0.62 in the original model to −0.90 when arsenic level is added to the model. This change occurs because wells that are far from the nearest safe well are also likely to be particularly high in arsenic.

### Graphing the fitted model with two predictors

The most natural way to graph the regression of $y$ on two predictors might be as a three-dimensional surface, with the vertical axis showing Pr(y=1) as a function of predictors plotted on two horizontal axes.

However, we find such graphs hard to read, so instead we make separate plots as a function of each of the two variables; see Figure 13.10. As with the lines in Figure 10.4, we can plot the focus input variable on the $x$-axis and use multiple lines to show the fit for different values of the other input. To produce Figure 13.10a, we first plot the (jittered) data points, forcing zero to be included in the $x$-range of the plot because it is a natural baseline comparison for distance:

```
plot(wells$dist, wells$switch_jitter, xlim=c(0,max(wells$dist)))
```

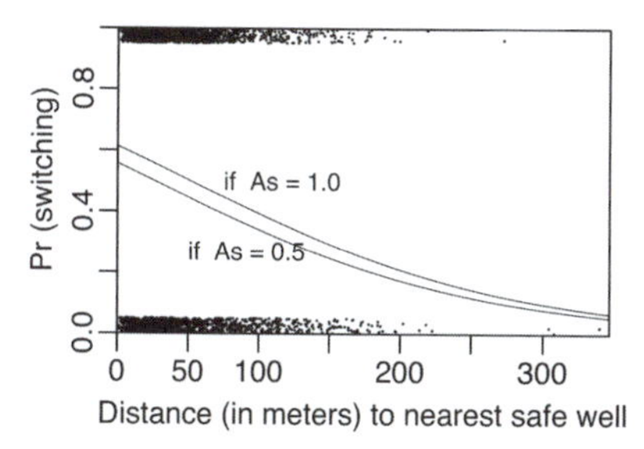


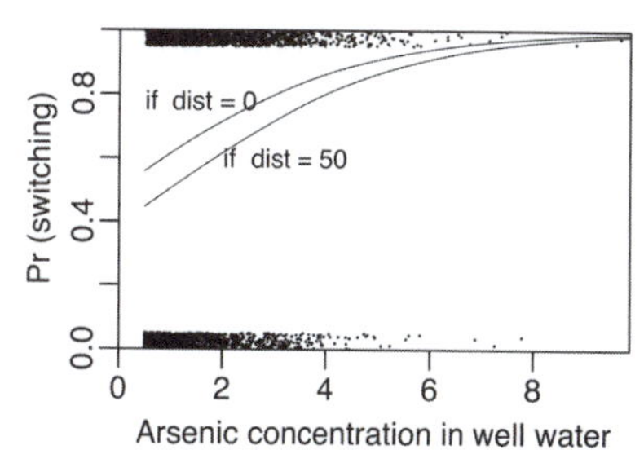


Figure 13.10 *Fitted logistic regression of probability of switching from an unsafe well as a function of two variables, plotted (a) as a function of distance to nearest safe well and (b) as a function of arsenic level of existing well. For each plot, the other input variable is held constant at different representative values.*

We next add the fitted curves:

```
curve(invlogit(cbind(1, x/100, 0.5) %*% coef(fit_3)), add=TRUE)
curve(invlogit(cbind(1, x/100, 1.0) %*% coef(fit_3)), add=TRUE)
```

We need to divide $x$ by 100 here because the plot is in the scale of meters but the model is defined in terms of `dist100` = `dist/100`.

The object created by `cbind(1, x/100, 0.5)` is an $n \times 3$ matrix constructed from a column of 1's, the vector `x` (used internally by the `curve` function), and a vector of 0.5's. In constructing the matrix, R automatically expands the scalars 1 and 0.5 to the length of the vector `x`. For the two lines, we pick arsenic levels of 0.5 and 1.0 because 0.5 is the minimum value of arsenic concentration (since we are only studying users of unsafe wells), and a difference of 0.5 represents a reasonable comparison, given the distribution of arsenic levels in the data (see Figure 13.9).

Similar commands are used to make Figure 13.10b, showing the probability of switching as a function of arsenic concentration with distance held constant:

```
plot(wells$arsenic, wells$switch_jitter, xlim=c(0,max(wells$arsenic)))
curve(invlogit(cbind(1, 0, x) %*% coef(fit_3)), add=TRUE)
curve(invlogit(cbind(1,.5, x) %*% coef(fit_3)), add=TRUE)
```

## 13.8 Bibliographic note

According to Cramer (2003, chapter 9), logistic regression was introduced for binary data in the mid-twentieth century and has become increasingly popular as computational improvements have allowed it to become a routine data-analytic tool. McCullagh and Nelder (1989) discuss logistic regression in the context of generalized linear models, a topic we consider further in Chapter 15.

For more on income and voting in presidential elections, see Gelman, Park, et al. (2009). The example of drinking water in Bangladesh is described further by van Geen et al. (2003) and Gelman, Trevisani, et al. (2004).

## 13.9 Exercises

13.1 *Fitting logistic regression to data*: The folder `NES` contains the survey data of presidential preference and income for the 1992 election analyzed in Section 13.1, along with other variables including sex, ethnicity, education, party identification, and political ideology.

(a) Fit a logistic regression predicting support for Bush given all these inputs. Consider how to include these as regression predictors and also consider possible interactions.

(b) Evaluate and compare the different models you have fit.

(c) For your chosen model, discuss and compare the importance of each input variable in the prediction.

13.2 *Sketching the logistic curve*: Sketch the following logistic regression curves with pen on paper:

(a) $\Pr(y = 1) = \text{logit}^{-1}(x)$

(b) $\Pr(y = 1) = \text{logit}^{-1}(2 + x)$

(c) $\Pr(y = 1) = \text{logit}^{-1}(2x)$

(d) $\Pr(y = 1) = \text{logit}^{-1}(2 + 2x)$

(e) $\Pr(y = 1) = \text{logit}^{-1}(-2x)$

13.3 *Understanding logistic regression coefficients*: In Chapter 7 we fit a model predicting incumbent party's two-party vote percentage given economic growth: vote $= 46.2 + 3.1 *$ growth $+$ error, where `growth` ranges from $-0.5$ to $4.5$ in the data, and errors are approximately normally distributed with mean 0 and standard deviation 3.8. Suppose instead we were to fit a logistic regression, $\Pr(\text{vote} > 50) = \text{logit}^{-1}(a + b * \text{growth})$. Approximately what are the estimates of $(a, b)$?

Figure this out in four steps: (i) use the fitted linear regression model to estimate $\Pr(\text{vote} > 50)$ for different values of `growth`; (ii) second, plot these probabilities and draw a logistic curve through them; (iii) use the divide-by-4 rule to estimate the slope of the logistic regression model; (iv) use the point where the probability goes through 0.5 to deduce the intercept. Do all this using the above information, without downloading the data and fitting the model.

13.4 *Logistic regression with two predictors*: The following logistic regression has been fit:

```
            Median MAD_SD
(Intercept) -1.9    0.6
x            0.7    0.8
z            0.7    0.5
```

Here, $x$ is a continuous predictor ranging from 0 to 10, and $z$ is a binary predictor taking on the values 0 and 1. Display the fitted model as two curves on a graph of $\Pr(y = 1)$ vs. $x$.

13.5 *Interpreting logistic regression coefficients*: Here is a fitted model from the Bangladesh analysis predicting whether a person with high-arsenic drinking water will switch wells, given the arsenic level in their existing well and the distance to the nearest safe well:

```
stan_glm(formula = switch ~ dist100 + arsenic, family=binomial(link="logit"),
  data=wells)
              Median MAD_SD
(Intercept)    0.00   0.08
dist100       -0.90   0.10
arsenic        0.46   0.04
```

Compare two people who live the same distance from the nearest well but whose arsenic levels differ, with one person having an arsenic level of 0.5 and the other person having a level of 1.0. You will estimate how much more likely this second person is to switch wells. Give an approximate estimate, standard error, 50% interval, and 95% interval, using two different methods:

(a) Use the divide-by-4 rule, based on the information from this regression output.

(b) Use predictive simulation from the fitted model in R, under the assumption that these two people each live 50 meters from the nearest safe well.

13.6 *Interpreting logistic regression coefficient uncertainties*: In Section 13.7, there were two models, `fit_4` and `fit_5`, with distance and arsenic levels as predictors along with an interaction term. The model `fit_5` differed by using centered predictors. Compare the reported uncertainty estimates (mad sd) for the coefficients, and use for example the `mcmc_pairs` function in the `bayesplot` package to examine the pairwise joint posterior distributions. Explain why the mad sd values are different for `fit_4` and `fit_5`.

13.7 *Graphing a fitted logistic regression*: We downloaded data with weight (in pounds) and age (in years) from a random sample of American adults. We then defined a new variable:

```
heavy <- weight > 200
```

and fit a logistic regression, predicting heavy from height (in inches):

```
stan_glm(formula = heavy ~ height, family=binomial(link="logit"), data=health)
              Median MAD_SD
(Intercept)  -21.51   1.60
height         0.28   0.02
```

(a) Graph the logistic regression curve (the probability that someone is heavy) over the approximate range of the data. Be clear where the line goes through the 50% probability point.

(b) Fill in the blank: near the 50% point, comparing two people who differ by one inch in height, you'll expect a difference of ____ in the probability of being heavy.

13.8 *Linear transformations*: In the regression from the previous exercise, suppose you replaced height in inches by height in centimeters. What would then be the intercept and slope?

13.9 *The algebra of logistic regression with one predictor*: You are interested in how well the combined earnings of the parents in a child's family predicts high school graduation. You are told that the probability a child graduates from high school is 27% for children whose parents earn no income and is 88% for children whose parents earn \$60 000. Determine the logistic regression model that is consistent with this information. For simplicity, you may want to assume that income is measured in units of \$10 000.

13.10 *Expressing a comparison of proportions as a logistic regression*: A randomized experiment is performed within a survey, and 1000 people are contacted. Half the people contacted are promised a \$5 incentive to participate, and half are not promised an incentive. The result is a 50% response rate among the treated group and 40% response rate among the control group.

(a) Set up these results as data in R. From these data, fit a logistic regression of response on the treatment indicator.

(b) Compare to the results from Exercise 4.1.

13.11 *Building a logistic regression model*: The folder `Rodents` contains data on rodents in a sample of New York City apartments.

(a) Build a logistic regression model to predict the presence of rodents (the variable `rodent2` in the dataset) given indicators for the ethnic groups (`race`). Combine categories as appropriate. Discuss the estimated coefficients in the model.

(b) Add to your model some other potentially relevant predictors describing the apartment, building, and community district. Build your model using the general principles explained in Section 12.6. Discuss the coefficients for the ethnicity indicators in your model.

13.12 *Fake-data simulation to evaluate a statistical procedure*: When can we get away with fitting linear regression to binary data? You will explore this question by simulating data from a logistic regression, then fitting a linear regression, then looping this procedure to compute the coverage of the estimates.

(a) You will be simulating independent binary data, $y_i, i = 1, \dots, n$, from the model, $\Pr(y_i = 1) = \text{logit}^{-1}(a + bx_i + \theta z_i)$, where the $x_i$'s are drawn uniformly from the range (0, 100) and the $z_i$'s are randomly set to 0 or 1. The "cover story" here is that $y$ represents passing or failing an exam, $x$ is the score on a pre-test, and $z$ is a treatment.

To do this simulation, you will need to set true values of $a$, $b$, and $\theta$. Choose $a$ and $b$ so that 60% of the students in the control group will pass the exam, with the probability of passing being 80% for students in the control group who scored 100 on the midterm.

Choose $\theta$ so that the average probability of passing increases by 10 percentage points under the treatment.

Report your values for $a, b, \theta$ and explain your reasoning (including simulation code). It's not enough just to guess.

(b) Simulate $n = 50$ data points from your model, and then fit a linear regression of $y$ on $x$ and $z$. Look at the estimate and standard error for the coefficient of $z$. Does the true average treatment effect fall inside this interval?

(c) Repeat your simulation in (b) 10 000 times. Compute the coverage of the normal-theory 50% and 95% intervals (that is, the estimates $\pm\, 0.67$ and 1.96 standard errors).

13.13 *Working through your own example*: Continuing the example from the final exercises of the earlier chapters, fit a logistic regression, graph the data and fitted model, and interpret the estimated parameters and their uncertainties.

# Chapter 14

# Working with logistic regression

With logistic, as with linear regression, fitting is only part of the story. In this chapter, we develop more advanced graphics to visualize data and fitted logistic regressions with one or more predictors. We discuss the challenges of interpreting coefficients in the presence of interactions and the use of linear transformations to aid understanding. We show how to make probabilistic predictions and how to average these predictions to obtain summaries—average predictive comparisons—that can be more interpretable than logistic regression coefficients. We discuss the evaluation of fitted models using binned residual plots and predictive errors, and we present all these tools in the context of a worked example. The chapter concludes with a discussion of the use of Bayesian inference and prior distributions to resolve a challenge of inference that arises with sparse discrete data, which again we illustrate with an applied example.

## 14.1 Graphing logistic regression and binary data

As can be seen from examples such as Figure 13.8b, it can be a challenge to display discrete data: even with jittering, the data points overlap, and a scatterplot of $y$ vs. $x$ is not nearly as informative as a corresponding scatterplot of continuous data.

ample: mulated ta for gistic gression

That said, a simple scatterplot can be helpful in clarifying how logistic regression works.[1] Figure 14.1a shows 50 data points simulated independently from the logistic model, $\Pr(y = 1) = \text{logit}^{-1}(2 + 3x)$, and then plotted along with the curve, $\text{logit}^{-1}(\hat{a} + \hat{b}x)$, using the point estimate $(\hat{a}, \hat{b})$ from the model fitted using `stan_glm`. The graph shows the pattern of logistic regression that the data fall into three zones: on the left side of the plot the outcomes $y$ are all 0, on the right side they are all 1, and there is a band of overlap in the middle.

For visualizing patterns in discrete data, it can be helpful to plot binned averages. Figure 14.1b illustrates for the same simulated data. For real data, such a binned average plot can show more complex patterns than could be seen from plotting individual 0's and 1's; examples appear in Figures 22.1 and 22.5.

When two continuous predictors are available, we can plot discrete data by using different symbols for different values of $y$. Figure 14.2 illustrates with 100 data points simulated independently from the logistic model, $\Pr(y = 1) = \text{logit}^{-1}(2 + 3x_1 + 4x_2)$, plotted as open circles for $y = 1$ and solid circles for $y = 0$. Also shown on the graph are lines representing equal-probability lines from a fitted logistic regression.

The steps go as follows: We run `stan_glm` and obtain the fitted model, which for the data in Figure 14.2 comes to $\Pr(y = 1) = \text{logit}^{-1}(2.0 + 4.4x_1 + 4.1x_2)$. From this model, the $\Pr(y = 1) = 0.5$ when $2.0 + 4.4x_1 + 4.1x_2 = 0$, which corresponds to the line, $x_2 = -2.0/4.1 - (4.0/4.1)x_1$. This is plotted as the solid line in Figure 14.2, and indeed this appears to be the line that best splits the data into open and solid circles.

Figure 14.2 also shows dotted lines corresponding to 10% and 90% probabilities, which we compute in similar ways. Under the fitted model, $\Pr(y = 1) = 0.9$ when $2.0 + 4.4x_1 + 4.1x_2 =$

[1]Code for this example is in the folder `LogitGraphs`.

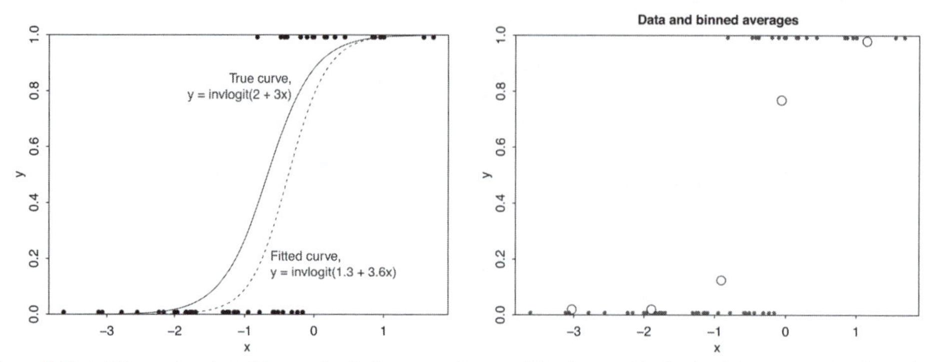


Figure 14.1 *(a) Data simulated from a logistic regression model, along with the logistic regression fit to these data; (b) Binned averages, plotting $\bar{y}$ vs. $\bar{x}$ for the data divided into five bins based on the values of x.*

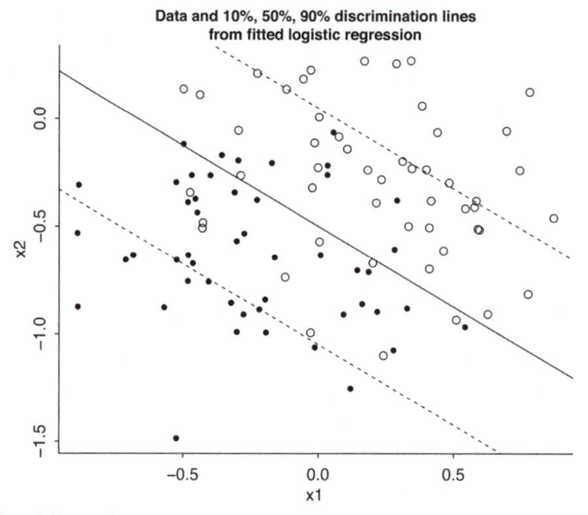


Figure 14.2 *Data simulated from a logistic regression model with two predictors, along with lines corresponding to Pr($y$ = 1) = 0.1, 0.5, 0.9, based on the fitted model.*

logit(0.9) = 2.2, which corresponds to the line, $x_2 = 0.2/4.1 - (4.0/4.1)x_1$, and $\Pr(y = 1) = 0.9$ when $2.0 + 4.4x_1 + 4.1x_2 = \text{logit}(0.1) = -2.2$, thus $x_2 = -4.2/4.1 - (4.0/4.1)x_1$.

These dotted lines in Figure 14.2 do *not* represent uncertainty in the line; rather, they convey the variation inherent in the logistic regression model. The solid line corresponds to a predicted probability of 0.5; thus for data with predictors $(x_1, x_2)$ are close to this line, we should expect about half 0's and half 1's, and indeed that is what we see in the plot. For data near the dotted line on the upper right, we see about 90% 1's, and for data near the dotted line on the lower left, the frequency of 1's is about 10%. Moving further away from the center line, the data become more homogeneous.

The patterns in Figure 14.2 are particularly clear because the data are simulated from the model. In settings where the model does not fit so well, problems could appear on the graph.

## 14.2 Logistic regression with interactions

Example: Arsenic in Bangladesh

We continue the well-switching example from the previous chapter by adding the interaction between the two predictors:[2]

[2]Data and code for this example are in the folder `Arsenic`.

```
fit_4 <- stan_glm(switch ~ dist100 + arsenic + dist100:arsenic,
  family=binomial(link="logit"), data=wells)
print(fit_4)
```

which yields,

```
                Median MAD_SD
(Intercept)     -0.15   0.12
dist100         -0.58   0.21
arsenic          0.56   0.07
dist100:arsenic -0.18   0.10
```

To understand the numbers in the table, we use the following tricks:

- Evaluating predictions and interactions at the mean of the data, which have average values of 0.48 for `dist100` and 1.66 for `arsenic` (that is, a mean distance of 48 meters to the nearest safe well, and a mean arsenic level of 1.66 among the unsafe wells).
- Dividing by 4 to get approximate predictive differences on the probability scale.

We now interpret each regression coefficient in turn.

- *Constant term*: $\text{logit}^{-1}(-0.15) = 0.47$ is the estimated probability of switching, if the distance to the nearest safe well is 0 and the arsenic level of the current well is 0. This is an impossible condition (since arsenic levels all exceed 0.5 in our set of unsafe wells), so we do not try to interpret the constant term. Instead, we can evaluate the prediction at the average values of `dist100` = 0.48 and `arsenic` = 1.66, where the probability of switching is $\text{logit}^{-1}(-0.15 - 0.58 * 0.48 + 0.56 * 1.66 - 0.18 * 0.48 * 1.66) = 0.59$.
- *Coefficient for distance*: this corresponds to comparing two wells that differ by 1 in `dist100`, if the arsenic level is 0 for both wells. Once again, we should not try to interpret this.
  Instead, we can look at the average value, `arsenic` = 1.66, where distance has a coefficient of $-0.58 - 0.18 * 1.66 = -0.88$ on the logit scale. To quickly interpret this on the probability scale, we divide by 4: $-0.88/4 = -0.22$. Thus, at the mean level of arsenic in the data, each 100 meters of distance corresponds to an approximate 22% *negative* difference in probability of switching.
- *Coefficient for arsenic*: this corresponds to comparing two wells that differ by 1 in `arsenic`, if the distance to the nearest safe well is 0 for both. Again, this is not so interpretable, so instead we evaluate the comparison at the average value for distance, `dist100` = 0.48, where arsenic has a coefficient of $0.56 - 0.18 * 0.48 = 0.47$ on the logit scale. To quickly interpret this on the probability scale, we divide by 4: $0.47/4 = 0.12$. Thus, at the mean level of distance in the data, each additional unit of arsenic corresponds to an approximate 12% *positive* difference in probability of switching.
- *Coefficient for the interaction term*: this can be interpreted in two ways. Looking from one direction, for each additional unit of arsenic, the value −0.18 is added to the coefficient for distance. We have already seen that the coefficient for distance is −0.88 at the average level of arsenic, and so we can understand the interaction as saying that the importance of distance as a predictor increases for households with higher existing arsenic levels.
  Looking at it the other way, for each additional 100 meters of distance to the nearest well, the value −0.18 is added to the coefficient for arsenic. We have already seen that the coefficient for distance is 0.47 at the average distance to nearest safe well, and so we can understand the interaction as saying that the importance of arsenic as a predictor decreases for households that are farther from existing safe wells.

## Centering the input variables

As we discussed in the context of linear regression, before fitting interactions it makes sense to center the input variables so that we can more easily interpret the coefficients. The centered inputs are,

```
wells$c_dist100 <- wells$dist100 - mean(wells$dist100)
wells$c_arsenic <- wells$arsenic - mean(wells$arsenic)
```

We do not fully standardize these—that is, we do not scale by their standard deviations—because it is convenient to be able to consider known differences on the scales of the data: arsenic-concentration units and 100-meter distances.

### Re-fitting the interaction model using the centered inputs

We can re-fit the model using the centered input variables, which will make the coefficients much easier to interpret:

```
fit_5 <- stan_glm(switch ~ c_dist100 + c_arsenic + c_dist100:c_arsenic,
  family=binomial(link="logit"), data=wells)
```

We center the *inputs*, not the *predictors*. Hence, we do not center the interaction (`dist100*arsenic`); rather, we include the interaction of the two centered input variables. Displaying `fit_5` yields,

```
                    Median MAD_SD
(Intercept)          0.35   0.04
c_dist100           -0.88   0.10
c_arsenic            0.47   0.04
c_dist100:c_arsenic -0.18   0.10
```

Interpreting the inferences on this new scale:

- *Constant term*: $\text{logit}^{-1}(0.35) = 0.59$ is the estimated probability of switching, if `c_dist100` = `c_arsenic` = 0, that is, if distance to nearest safe well and arsenic level are at their averages in the data. We obtained this same calculation, but with more effort, from our earlier model with uncentered inputs.
- *Coefficient for distance*: this is the coefficient for distance (on the logit scale) if arsenic level is at its average value. To quickly interpret this on the probability scale, we divide by 4: $-0.88/4 = -0.22$. Thus, at the mean level of arsenic in the data, each 100 meters of distance corresponds to an approximate 22% *negative* difference in probability of switching.
- *Coefficient for arsenic*: this is the coefficient for arsenic level if distance to nearest safe well is at its average value. To quickly interpret this on the probability scale, we divide by 4: $0.47/4 = 0.12$. Thus, at the mean level of distance in the data, each additional unit of arsenic corresponds to an approximate 12% *positive* difference in probability of switching.
- *Coefficient for the interaction term*: this is unchanged by centering and has the same interpretation as before.

The predictions for new observations are unchanged. Linear centering preserved the eventual substantive interpretation of the prediction based on distance and arsenic level, but the value and meaning of the corresponding coefficients are different in the context of the reparameterized model.

### Statistical significance of the interaction

As can be seen from the above regression table, `c_dist100:c_arsenic` has an estimated coefficient of $-0.18$ with a standard error of 0.10. The estimate is not quite two standard errors away from zero and so is not quite "statistically significant." However, as discussed before, this does not mean that we should treat the underlying coefficient as zero. Rather we must accept our uncertainty in the context that it is plausible that arsenic level becomes a less important predictor for households that are farther from the nearest safe well, and the magnitude of the association is also plausible. The LOO log score for the model with the interaction is $-1967.9$, and comparison to the model without the interaction yields,

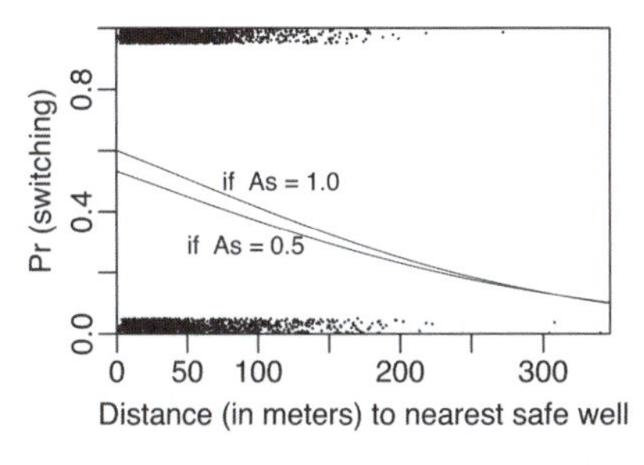


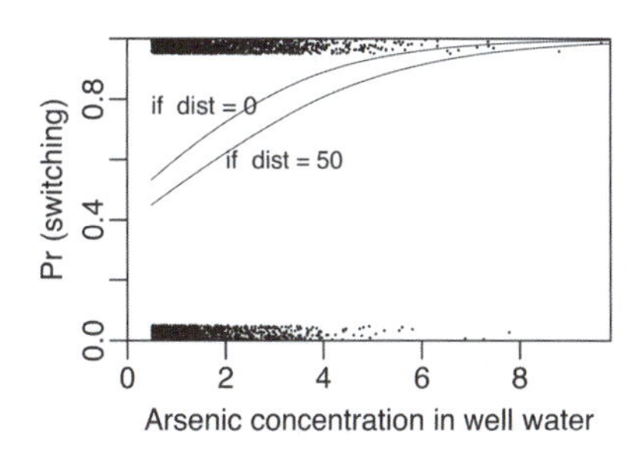


Figure 14.3 *Fitted logistic regression of probability of switching from an unsafe well as a function of distance to nearest safe well and arsenic level of existing well, for the model with interactions. Compare to the no-interaction model in Figure 13.10.*

```
Model comparison:
(negative 'elpd_diff' favors 1st model, positive favors 2nd)
elpd_diff        se
      0.6       1.9
```

Adding the interaction doesn't change the predictive performance, and there is no need to keep it in the model for the predictive purposes (unless new information can be obtained).

### Graphing the model with interactions

The clearest way to visualize the interaction model is to plot the regression curves as a function for each picture. The result is shown in Figure 14.3, the first graph of which we make in R as follows (with similar commands for the other graph):

```
plot(wells$dist, wells$y_jitter, xlim=c(0,max(wells$dist)))
curve(invlogit(cbind(1, x/100, 0.5, 0.5*x/100) %*% coef(fit_4)), add=TRUE)
curve(invlogit(cbind(1, x/100, 1.0, 1.0*x/100) %*% coef(fit_4)), add=TRUE)
```

As Figure 14.3 makes clear, the interaction is small in the range of most of the data. The largest pattern shows up in Figure 14.3a, where the two lines intersect at around 300 meters. This graph shows evidence that the differences in switching associated with differences in arsenic level are large if you are close to a safe well, but with a diminishing effect if you are far from any safe well. This interaction makes some sense; however, there is some uncertainty in the size of the interaction (from the earlier regression table, an estimate of $-0.18$ with a standard error of 0.10), and as Figure 14.3a shows, there are only a few data points in the area where the interaction makes much of a difference.

The interaction also appears in Figure 14.3b, this time in a plot of probability of switching as a function of arsenic concentration, at two different levels of distance.

### Adding social predictors

Are well users more likely to switch if they have community connections or more education? To see, we add two inputs:

- educ = years of education of the well user
- assoc = 1 if a household member is in any community organization.

We actually work with educ4 = educ/4 for the usual reasons of making the regression coefficient more interpretable—it now represents the predictive difference corresponding to four years of education, for example comparing a high school graduate to an elementary school graduate or to a college graduate. The levels of education among the 3000 respondents vary from 0 to 17 years, with nearly a third having zero.

```
             Median MAD_SD
(Intercept) -0.16   0.10
dist100     -0.90   0.10
arsenic      0.47   0.04
educ4        0.17   0.04
assoc       -0.12   0.08
```

Respondents with higher education are more likely to say they would switch wells: the crude estimated difference is $0.17/4 = 0.04$, or a 4% positive difference in switching probability when comparing households that differ by 4 years of education. We repeated our analysis with a discrete recoding of the education variable (0 = 0 years, 1 = 1–8 years, 2 = 9–12 years, 3 = 12+ years), and our results were essentially unchanged. The coefficient for education makes sense and is estimated fairly precisely—its standard error is much lower than the coefficient estimate.

Throughout this example, we refer to "coefficients" and "differences," rather than to "effects" and "changes," because the observational nature of the data makes it difficult to directly interpret the regression model causally. We continue causal inference more carefully in Chapters 18–21, briefly discussing the arsenic problem at the end of Section 19.7.

Belonging to a community association, perhaps surprisingly, is associated in our data with a *lower* probability of switching, after adjusting for the other factors in the model. However, this coefficient is not estimated precisely, and so for clarity and stability we remove it from the model, leaving us with the following fit:

```
             Median MAD_SD
(Intercept) -0.22   0.09
dist100     -0.90   0.11
arsenic      0.47   0.04
educ4        0.17   0.04
```

Comparing this to the model `fit_4` without social predictors gives

```
elpd_diff        se
      8.5       4.9
```

Adding education improves predictive log score, but there is considerable uncertainty.

### Adding further interactions

When inputs have large main effects, it is our general practice to include their interactions as well. We first create a centered education variable:

```
wells$c_educ4 <- wells$educ4 - mean(wells$educ4)
```

and then fit a new model interacting it with distance to nearest safe well and arsenic level of the existing well:

```
                    Median MAD_SD
(Intercept)          0.12   0.06
c_dist100           -0.92   0.10
c_arsenic            0.49   0.04
educ4                0.19   0.04
c_dist100:c_educ4   0.33    0.11
c_arsenic:c_educ4   0.08    0.04
```

We can interpret these new interactions by understanding how education modifies the predictive difference corresponding to distance and arsenic.

- *Interaction of distance and education*: a difference of 4 years of education corresponds to a difference of 0.33 in the coefficient for `dist100`. As we have already seen, `dist100` has a negative coefficient on average; thus positive changes in education *reduce* distance's negative association.

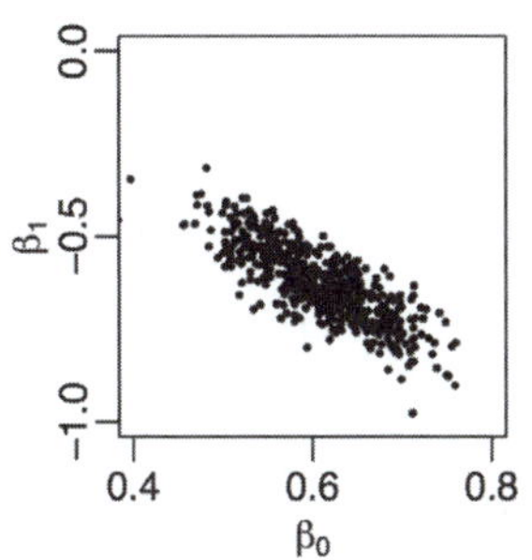


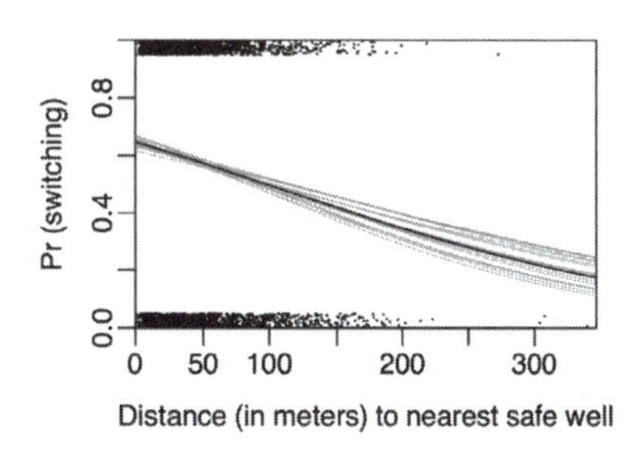


Figure 14.4 *(a) Uncertainty in the estimated coefficients $\beta_0$ and $\beta_1$ in the logistic regression, Pr(switching wells) = logit*$^{-1}$*($\beta_0 - \beta_1 *$ `dist100`). (b) Graphical expression of the best-fit model, Pr(switching wells) = logit*$^{-1}$*(0.61 − 0.62 ∗* `dist100`*), with (jittered) data overlaid. Light lines represent estimation uncertainty in the logistic regression coefficients, corresponding to the distribution of $\beta$ shown to the left. Compare to Figure 13.8b.*

This makes sense: people with more education probably have other resources so that walking an extra distance to get water is not such a burden.

- *Interaction of arsenic and education*: a difference of 4 years of education corresponds to a difference of 0.08 in the coefficient for `arsenic`. As we have already seen, `arsenic` has a positive coefficient on average; thus increasing education *increases* arsenic's positive association. This makes sense: people with more education could be more informed about the risks of arsenic and thus more sensitive to increasing arsenic levels (or, conversely, less in a hurry to switch from wells with arsenic levels that are relatively low).

As before, centering allows us to interpret the main effects as coefficients when other inputs are held at their average values in the data.

Adding interactions with `educ4` further improved the LOO log score, and the comparison to a model without `educ4` gives,

```
elpd_diff          se
     15.8         6.3
```

### Standardizing predictors

We should think seriously about standardizing all predictors as a default option when fitting models with interactions. The struggles with `dist100` and `educ4` in this example suggest that standardization—by subtracting the mean from each of the continuous input variables and dividing by 2 standard deviations, as suggested near the end of Section 12.2—might be the simplest approach.

## 14.3 Predictive simulation

As discussed in Section 13.3, we can use the inferences from `stan_glm` to obtain simulations that we can then use to make probabilistic predictions. We demonstrate the general principles using the model of the probability of switching wells given the distance from the nearest safe well.

### Simulating the uncertainty in the estimated coefficients

Figure 14.4a shows the uncertainty in the regression coefficients, computed as follows:

```
fit <- stan_glm(switch ~ dist100, family=binomial(link="logit"), data=wells)
sims <- as.matrix(fit)
n_sims <- nrow(sims)
```

| sim | $\beta_0$ | $\beta_1$ | $y_1^{\text{new}}$ | $y_2^{\text{new}}$ | $\cdots$ | $y_{10}^{\text{new}}$ |
|---|---|---|---|---|---|---|
| 1 | 0.68 | −0.007 | 1 | 0 | $\cdots$ | 1 |
| 2 | 0.61 | −0.005 | 0 | 0 | $\cdots$ | 1 |
| $\vdots$ | $\vdots$ | $\vdots$ | $\vdots$ | $\vdots$ | $\ddots$ | $\vdots$ |
| 4000 | 0.69 | −0.006 | 1 | 1 | $\cdots$ | 1 |
| mean | 0.61 | −0.006 | 0.60 | 0.59 | $\cdots$ | 0.52 |

Figure 14.5 *Simulation results for 10 hypothetical new households in the well-switching example, predicting based only on distance to the nearest well. The inferences for $(\beta_0, \beta_1)$ are displayed as a scatterplot in Figure 14.4a. Each of the first 4000 rows of the table corresponds to one plausible set of model parameters, along with one corresponding simulation of data for 10 new households, given their predictors x, which are not shown in the table. The bottom row—the mean of the simulated values of $y_i^{\text{new}}$ for each household i—gives the estimated probabilities of switching.*

```
plot(sims[,1], sims[,2], xlab=expression(beta[0]), ylab=expression(beta[1]))
```

and Figure 14.4b shows the corresponding uncertainty in the logistic regression curve, displayed as follows:

```
plot(wells$dist100, wells$y)
for(s in 1:20){
  curve(invlogit(sims[s,1] + sims[s,2]*x), col="gray", lwd=0.5, add=TRUE)
}
curve(invlogit(mean(sims[,1]) + mean(sims[,2])*x), add=TRUE)
```

Fitting the model produced `n_sims` simulations (by default `stan_glm` saves 1000 iterations for each of 4 chains, thus `n_sims` = 4000) but in this plot we only display 20 curves, each using thinner lines (setting the line width argument, `lwd`, to 0.5) so as to preserve readability.

## Predictive simulation using the binomial distribution

Now suppose, for example, that we would like to predict the switching behavior for $n^{\text{new}}$ new households, given a predictor matrix $X^{\text{new}}$ (which will have $n^{\text{new}}$ rows and, in this example, two columns, corresponding to the constant term and the distance to the nearest safe well). As with linear regression, we can use simulation to account for the predictive uncertainty. In this case, we use the binomial distribution to simulate the prediction errors:

```
n_new <- nrow(X_new)
y_new <- array(NA, c(n_sims, n_new))
for (s in 1:n_sims){
  p_new <- invlogit(X_new %*% sims[s,])
  y_new[s,] <- rbinom(n_new, 1, p_new)
}
```

Figure 14.5 shows an example set of `n_sims` = 4000 simulations corresponding to a particular set of `n_new` = 10 new households.

## Predictive simulation using the latent logistic distribution

An alternative way to simulate logistic regression predictions uses the latent-data formulation (see Section 13.4). We obtain simulations for the latent data $z^{\text{new}}$ by adding independent errors $\epsilon^{\text{new}}$ to the linear predictor, and then convert to binary data by setting $y_i^{\text{new}} = 1$ if $z_i^{\text{new}} > 0$ for each new household $i$:

```
y_new <- array(NA, c(n_sims, n_new))
for (s in 1:n_sims){
```

```
    epsilon_new <- logit(runif(n_new, 0, 1))
    z_new <- X_new %*% t(sims[s,]) + epsilon_new
    y_new[s,] <- ifelse(z_new > 0, 1, 0)
}
```

## 14.4 Average predictive comparisons on the probability scale

As illustrated for example by Figure 13.10 on page 237, logistic regressions are nonlinear on the probability scale and linear on the logit scale. This is because logistic regression is linear in the parameters but nonlinear in the relation of inputs to outcome. A specified difference in one of the $x$ variables does *not* correspond to a constant difference in $\Pr(y=1)$. As a result, logistic regression coefficients cannot directly be interpreted on the scale of the data. Logistic regressions are inherently more difficult than linear regressions to interpret.

Graphs such as Figure 13.10 are useful, but for models with many predictors, or where graphing is inconvenient, it is helpful to have a summary, comparable to the linear regression coefficient, which gives the expected, or average, difference in $\Pr(y=1)$ corresponding to a unit difference in each of the input variables. In this section we describe a way to construct such a summary, and this approach can be used not just for logistic regression but also for predictive models more generally.

Assume we have fit a probability model predicting a numerical outcome $y$ based on inputs $x$ and parameters $\theta$. Consider the scalar inputs one at a time, using the notation,

$$u: \text{ the input of interest,}$$
$$v: \text{ all the other inputs.}$$

Thus, $x = (u, v)$.

Suppose we are considering comparisons of $u = u^{\text{hi}}$ to $u = u^{\text{lo}}$ with all other inputs held constant (for example, we might compare the expected behaviors of households that are 0 meters or 100 meters from the nearest safe well). The *predictive difference* in probabilities between two cases, differing only in $u$, is

$$\text{predictive difference} = \mathrm{E}(y|u^{\text{hi}}, v, \theta) - \mathrm{E}(y|u^{\text{lo}}, v, \theta),$$

where the vertical bar in these expressions is read, "conditional on."

Averaging over $u^{\text{lo}}$, $u^{\text{hi}}$, and $v$ in the data corresponds to counting all pairs of transitions of $(u^{\text{lo}}, v)$ to $(u^{\text{hi}}, v)$—that is, differences in $u$ with $v$ held constant. Figure 14.6 illustrates.

More generally we can compute the expected predictive difference in $y$ per unit difference in the input of interest, $u$, again comparing two cases that do not differ in $v$:

$$\text{predictive comparison} = \frac{\mathrm{E}(y|u^{\text{hi}}, v, \theta) - \mathrm{E}(y|u^{\text{lo}}, v, \theta)}{u^{\text{hi}} - u^{\text{lo}}}. \tag{14.1}$$

Except in the simple case of a linear model with no interactions, these expressions depend on $u$, $v$, and $\theta$, so to get a single summary for the average predictive comparison for a fitted model and input of interest $u$, it will be necessary to fix values of the data $x = (u, v)$ and parameters $\theta$ or to average over some distribution of them when evaluating (14.1).

### Problems with evaluating predictive comparisons at a central value

A rough approach that is sometimes used is to evaluate $\mathrm{E}(y|u, v, \theta)$ at a point estimate of $\theta$, fixed comparison points $(u^{\text{hi}}, u^{\text{lo}})$, and a central value of the other predictors $v$—perhaps using the mean or the median of these predictors in the data—and then estimate predictive comparisons by holding $v$ constant at this value. Evaluating differences about a central value can work well in practice, but one can run into problems when the space of inputs is very spread out (in which case no single central value can be representative) or if many of the inputs are binary or bimodal (in which case the concept

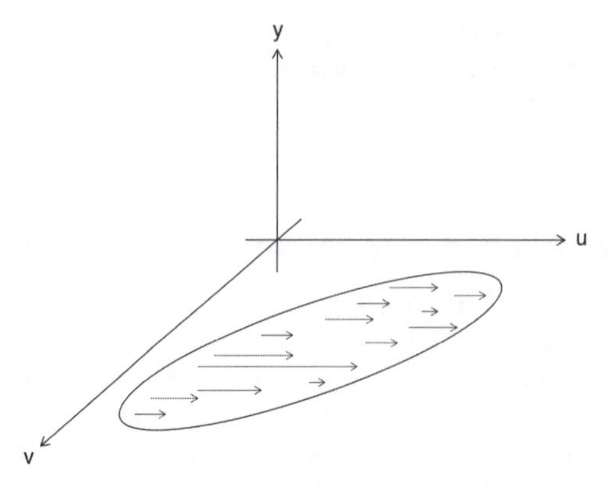


Figure 14.6 *Diagram illustrating differences in the input of interest u, with the other inputs v held constant. The ellipse in (u, v)-space represents the joint distribution p(u, v), and as the arrows indicate, we wish to consider differences in u in the region of support of this distribution.*

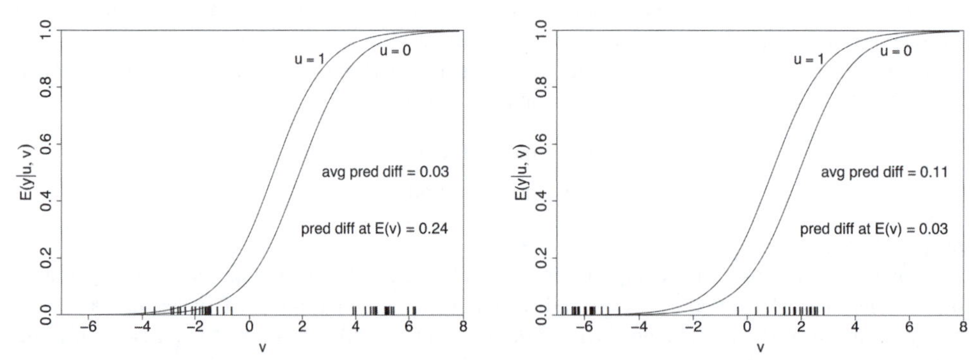


Figure 14.7 *Hypothetical example illustrating the need for averaging over the distribution of v (rather than simply working with a central value) in summarizing a predictive comparison. Each graph shows a logistic regression given a binary input of interest, u, and a continuous input variable, v. The vertical lines on the x-axis indicate the values of v in the data. (a) In the left plot, data $v_i$ are concentrated near the ends of the predictive range. Hence the predictive comparison averaged over the data is small. In contrast, the average of v is near the center of the range; hence the predictive comparison at this average value is large, even though this is not appropriate for any of the data points individually. (b) Conversely, in the right plot, the predictive comparison averaged over the data is reasonably large, but this would not be seen if the predictive comparison were evaluated at the average value of v.*

of a "central value" is less meaningful). In addition, this approach is hard to automate since it requires choices about how to set up the range for each input variable. In fact, our research in this area was motivated by practical difficulties that can arise in trying to implement this central-value approach.

We illustrate some challenges in defining predictive comparisons with a simple hypothetical example of a logistic regression of data $y$ on a binary input of interest, $u$ and a continuous input variable, $v$. The curves in each plot of Figure 14.7 show the assumed predictive relationship. In this example, $u$ has a constant effect on the logit scale but, on the scale of $\mathrm{E}(y)$, the predictive comparison (as defined in (14.1), with $u^{\mathrm{lo}} = 0$ and $u^{\mathrm{hi}} = 1$) is high for $v$ in the middle of the plotted range and low at the extremes. As a result, any average of (14.1) will depend strongly on the distribution of $v$.

The plots in Figure 14.7 show two settings in which the average predictive comparison differs from the predictive comparison evaluated at a central value. In the first plot, the data are at the extremes, so the average predictive comparison is small—but the predictive comparison evaluated at $\mathrm{E}(v)$ is misleadingly large. The predictive comparison in $y$ corresponding to a difference in $u$ is small, because switching $u$ from 0 to 1 typically has the effect of switching $\mathrm{E}(y|u, v)$ from, say, 0.02 to 0.05 or from 0.96 to 0.99. In contrast, the predictive comparison if evaluated at the mean of the data is large, where switching $u$ from 0 to 1 switches $\mathrm{E}(y|u, v)$ from 0.36 to 0.60.

The second plot in Figure 14.7 shows a similar example, but where the centrally computed predictive comparison is too low compared to the population-average predictive comparison. Here, the centrally located value of $v$ is already near the edge of the curve, at which point a difference in $u$ corresponds to only a small difference in $\mathrm{E}(y|u, v)$ of only 0.03. In comparison, the average predictive comparison, averaging over the data locations, has the larger value of 0.11, which appropriately reflects that many of the sample data are in the range where a difference in $u$ can correspond to a large difference in $\mathrm{E}(y)$.

## Demonstration with the well-switching example

mple:
enic in
gladesh

For a model with nonlinearity or interactions, or both, the average predictive comparison depends on the values of the input variables, as we shall illustrate with the well-switching example. To keep the presentation clean at this point, we shall work with a simple no-interaction model fit to the well-switching data:

```
fit_7 <- stan_glm(switch ~ dist100 + arsenic + educ4, family=binomial(link="logit"),
  data=wells)
```

which yields,

```
            Median MAD_SD
(Intercept) -0.22   0.09
dist100     -0.90   0.10
arsenic      0.47   0.04
educ4        0.17   0.04
```

giving the probability of switching as a function of distance to the nearest well (in 100-meter units), arsenic level, and education (in 4-year units).

**Average predictive difference in probability of switching.** Let us compare two households—one with `dist100` = 0 and one with `dist100` = 1—but identical in the other input variables, `arsenic` and `educ4`. The *predictive difference* in probability of switching between these two households is,

$$\begin{aligned}\delta(\text{arsenic}, \text{educ4}) = \ & \text{logit}^{-1}(-0.22 - 0.90 * 1 + 0.47 * \text{arsenic} + 0.17 * \text{educ4}) - \\ & \text{logit}^{-1}(-0.22 - 0.90 * 0 + 0.47 * \text{arsenic} + 0.17 * \text{educ4}). \quad (14.2)\end{aligned}$$

We write $\delta$ as a function of `arsenic` and `educ4` to emphasize that it depends on the levels of these other variables.

We average the predictive differences over the $n$ households in the data to obtain:

$$\text{average predictive difference: } \frac{1}{n}\sum_{i=1}^{n} \delta(\text{arsenic}_i, \text{educ4}_i). \quad (14.3)$$

In R:

```
b <- coef(fit_7)
hi <- 1
lo <- 0
delta <- invlogit(b[1] + b[2]*hi + b[3]*wells$arsenic + b[4]*wells$educ4) -
         invlogit(b[1] + b[2]*lo + b[3]*wells$arsenic + b[4]*wells$educ4)
round(mean(delta), 2)
```

The result is −0.21, implying that, on average in the data, households that are 100 meters from the nearest safe well are 21% less likely to switch, compared to households that are right next to the nearest safe well, at the same arsenic and education levels.

**Comparing probabilities of switching for households differing in arsenic levels.** We can similarly compute the predictive difference, and average predictive difference, comparing households at two different arsenic levels, assuming equality in distance to nearest safe well and education levels. We choose arsenic = 0.5 and 1.0 as comparison points because 0.5 is the lowest unsafe level, 1.0 is twice that, and this comparison captures much of the range of the data; see Figure 13.9. Here is the computation:

```
hi <- 1.0
lo <- 0.5
delta <- invlogit(b[1] + b[2]*wells$dist100 + b[3]*hi + b[4]*wells$educ4) -
         invlogit(b[1] + b[2]*wells$dist100 + b[3]*lo + b[4]*wells$educ4)
round(mean(delta), 2)
```

The result is 0.06—so this comparison corresponds to a 6% difference in probability of switching.

**Average predictive difference in probability of switching, comparing householders with 0 and 12 years of education.** Similarly, we can compute an average predictive difference of the probability of switching for householders with 0 compared to 12 years of education (that is, comparing educ4 = 0 to educ4 = 3):

```
hi <- 3
lo <- 0
delta <- invlogit(b[1] + b[2]*wells$dist100 + b[3]*wells$arsenic + b[4]*hi) -
         invlogit(b[1] + b[2]*wells$dist100 + b[3]*wells$arsenic + b[4]*lo)
round(mean(delta), 2)
```

which comes to 0.12, a difference of 12 percentage points.

### Average predictive comparisons in the presence of interactions

We can perform similar calculations for models with interactions. For example, consider the average predictive difference, comparing dist = 0 to dist = 100, for the model that includes interactions:

```
fit_8 <- stan_glm(switch ~ c_dist100 + c_arsenic + c_educ4 + c_dist100:c_educ4 +
  c_arsenic:c_educ4, family=binomial(link="logit"), data=wells)
```

which, when displayed, yields,

```
                  Median MAD_SD
(Intercept)        0.35   0.04
c_dist100         -0.92   0.10
c_arsenic          0.49   0.04
c_educ4            0.19   0.04
c_dist100:c_educ4  0.33   0.11
c_arsenic:c_educ4  0.08   0.04
```

Here is R code for computing the average predictive difference comparing dist1 = 1 to dist1 = 0:

```
b <- coef(fit_8)
hi <- 1
lo <- 0
delta <- invlogit(b[1] + b[2]*hi + b[3]*wells$c_arsenic + b[4]*wells$c_educ4 +
           b[5]*hi*wells$c_educ4 + b[6]*wells$c_arsenic*wells$c_educ4) -
         invlogit(b[1] + b[2]*lo + b[3]*wells$c_arsenic + b[4]*wells$c_educ4 +
           b[5]*lo*wells$c_educ4 + b[6]*wells$c_arsenic*wells$c_educ4)
round(mean(delta), 2)
```

which comes to −0.21.

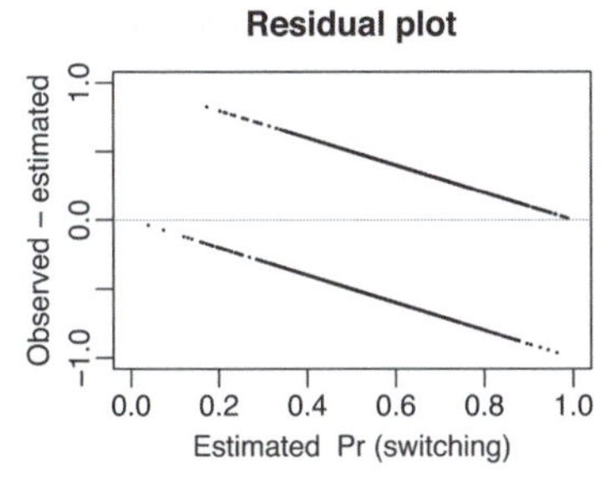


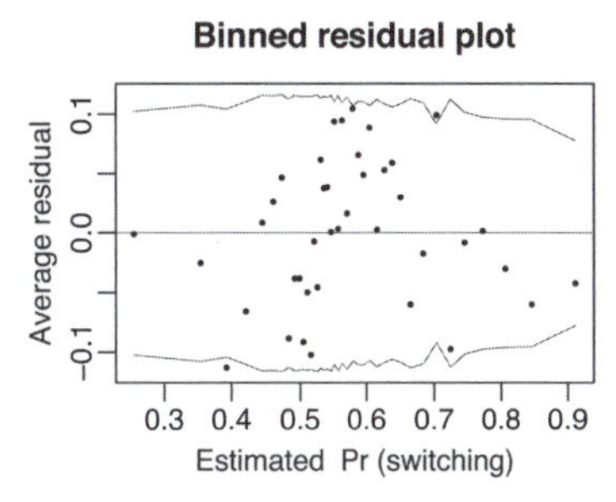


Figure 14.8 *(a) Residual plot and (b) binned residual plot for the well-switching model shown on page 246. The strong patterns in the raw residual plot arise from the discreteness of the data and inspire us to use the binned residual plot instead. The bins are not equally spaced; rather, each bin has an equal number of data points. The light lines in the binned residual plot indicate theoretical 95% error bounds.*

## 14.5 Residuals for discrete-data regression

### Residuals and binned residuals

We can define residuals for logistic regression, as with linear regression, as observed minus expected values:

$$\text{residual}_i = y_i - \mathrm{E}(y_i|X_i) = y_i - \text{logit}^{-1}(X_i\beta).$$

The data $y_i$ are discrete and so are the residuals. For example, if $\text{logit}^{-1}(X_i\beta) = 0.7$, then $\text{residual}_i = -0.7$ or $+0.3$, depending on whether $y_i = 0$ or 1. As a result, plots of raw residuals from logistic regression are generally not useful. For example, Figure 14.8a plots residuals versus fitted values for the well-switching regression.

Instead, we plot *binned residuals* by dividing the data into categories (bins) based on their fitted values, and then plotting the average residual versus the average fitted value for each bin. The result appears in Figure 14.8b; here we divided the data into 40 bins of equal size. There is typically some arbitrariness in choosing the number of bins: we want each bin to contain enough points so that the averaged residuals are not too noisy, but it helps to have many bins to see local patterns in the residuals. For this example, 40 bins seemed to give sufficient resolution, while still having enough points per bin. The dotted lines (computed as $2\sqrt{p_j(1-p_j)/n_j}$, where $n_j$ is the number of points in bin $j$, $3020/40 \approx 75$ in this case) indicate $\pm 2$ standard errors, within which one would expect about 95% of the binned residuals to fall, if the model were true. One of the 40 binned residuals in Figure 14.8b falls outside the bounds, which is not a surprise, and no dramatic pattern appears.

### Plotting binned residuals versus inputs of interest

We can also look at residuals in a more structured way by binning and plotting them with respect to individual input variables or combinations of inputs. For example, in the well-switching example, Figure 14.9a displays the average residual in each bin as defined by distance to the nearest safe well, and Figure 14.9b shows average residuals, binned by arsenic levels.

This latter plot shows a disturbing pattern, with an extreme negative residual in the first three bins: people with wells in the lowest bin, which turns out to correspond to arsenic levels between 0.51 and 0.53, are about 20% less likely to switch than is predicted by the model: the average predicted probability of switching for these users is 49%, but actually only 32% of them switched. There is also a slight pattern in the residuals as a whole, with positive residuals (on average) in the middle of the range of arsenic and negative residuals at the high end.

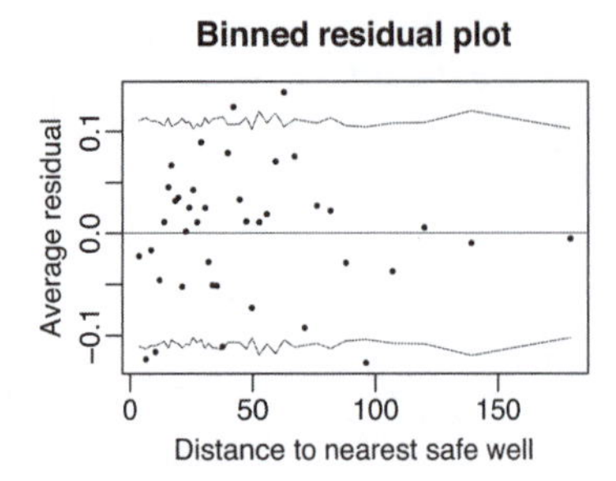


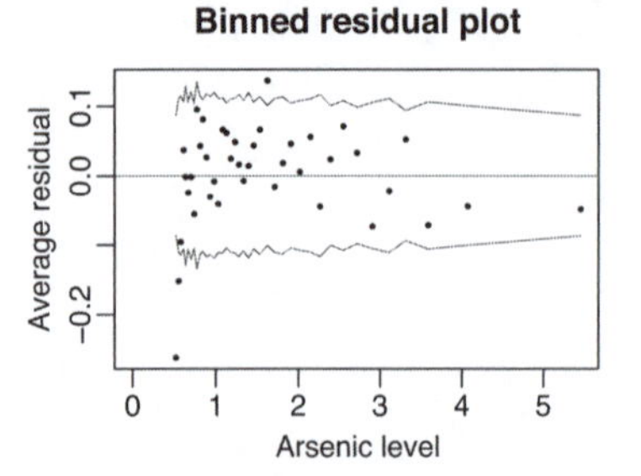


Figure 14.9 *Plots of residuals for the well-switching model, binned and plotted versus (a) distance to nearest safe well and (b) arsenic level. The dotted lines in the binned residual plot indicate theoretical 95% error bounds that would be appropriate if the model were true. The second plot shows a problem with the model in the lowest bins of arsenic levels.*

## Improving a model by transformation

To experienced regression modelers, a rising and then falling pattern of residuals such as in Figure 14.9b is a signal to consider taking the logarithm of the predictor on the $x$-axis—in this case, arsenic level. Another option would be to add a quadratic term to the regression; however, since arsenic is an all-positive variable, it makes sense to consider its logarithm. We do not, however, perform a log transform on distance to the nearest well, since the residual plot, as shown in Figure 14.9a, indicates a good fit of the linear model.

We define,

```
wells$log_arsenic <- log(wells$arsenic)
wells$c_log_arsenic <- wells$log_arsenic - mean(wells$log_arsenic)
```

and then fit a model using `log_arsenic` in place of `arsenic`:

```
                     Median MAD_SD
(Intercept)           0.34   0.04
c_dist100            -1.01   0.11
c_log_arsenic         0.91   0.07
c_educ4               0.18   0.04
c_dist100:c_educ4     0.35   0.11
c_log_arsenic:c_educ4 0.06   0.07
```

This is qualitatively similar to the model on the original scale: the interactions have the same sign as before, and the signs of the main effects are also unchanged. The LOO log score comparison shows that log transformation increases the predictive performance:

```
elpd_diff        se
     14.6       4.3
```

Figure 14.10a shows the predicted probability of switching as a function of arsenic level. Compared to the model in which arsenic was included as a linear predictor (see Figure 13.10b), the curves are compressed at the left and stretched at the right.

Figure 14.10b displays the residuals for the log model, again binned by arsenic level. Compared to the earlier model, the residuals look better, but a problem remains at the very low end. Users of wells with arsenic levels just above 0.50 are less likely to switch than predicted by the model. At this point, we do not know if this can be explained psychologically (values just over the threshold do not seem so bad), by measurement error (perhaps some of the wells we have recorded as 0.51 or 0.52 were measured before or after and found to have arsenic levels below 0.5), or by some other reason.

Compared to the model, the data show an unexpectedly low rate of switching from wells that were just barely over the dangerous level for arsenic, possibly suggesting that people were moderating their decisions when in this ambiguous zone, or that there was other information not included in the model that could explain these decisions.

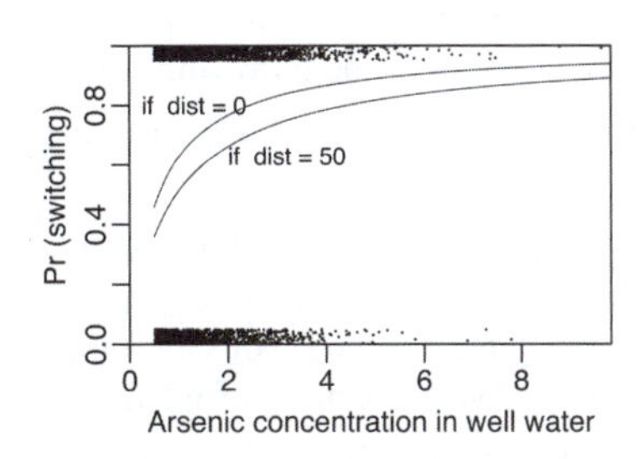


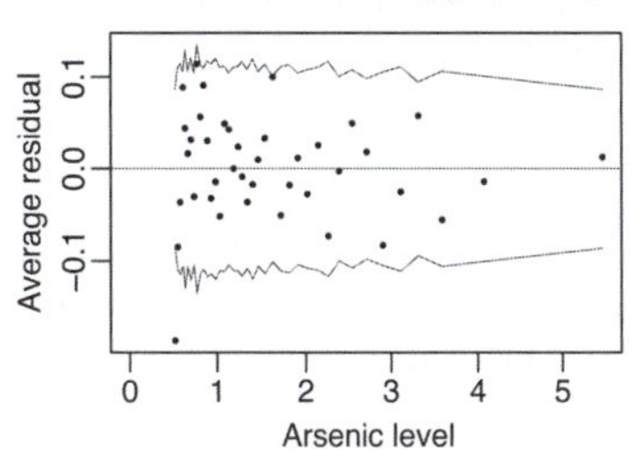


Figure 14.10 *(a) Probability of switching as a function of arsenic level (at two different values of distance and with education at its average value in the data), for the model that includes arsenic on the logarithmic scale. Compared to Figure 13.10b (the corresponding plot with arsenic level included as a linear predictor), the model looks similar, but with a steeper slope at the low end of the curve and a more gradual slope at the high end. (b) Average residuals for this model, binned by arsenic level. Compared to Figure 14.9b, the residual plot from the model that used unlogged arsenic level as a predictor, this new residual plot looks cleaner, although it still reveals a problem at the lowest arsenic levels.*

## Error rate and comparison to the null model

The *error rate* is defined as the proportion of cases for which the deterministic prediction—guessing $y_i = 1$ if $\text{logit}^{-1}(X_i\beta) > 0.5$ and guessing $y_i = 0$ if $\text{logit}^{-1}(X_i\beta) < 0.5$—is wrong. Here is the code to compute it in R:

```
error_rate <- mean((predicted>0.5 & y==0) | (predicted<0.5 & y==1))
```

The error rate should always be less than 1/2 (otherwise, we could simply set all the $\beta$'s to 0 and get a better-fitting model), but in many cases we would expect it to be much lower. We can compare it to the error rate of the *null model*, which is simply to assign the same probability to each $y_i$. This is simply logistic regression with only a constant term, and the estimated probability will simply be the proportion of 1's in the data, or $p = \sum_{i=1}^{n} y_i/n$ (recalling that each $y_i = 0$ or 1). The error rate of the null model is then $p$ or $1-p$, whichever is lower.

For example, in the well-switching example, the null model has an error rate of 42%; that is, 58% of the respondents are switchers and 42% are not, thus the model with no predictors gives each person a 58% chance of switching, which corresponds to a point prediction of switching for each person, and that guess will be wrong 42% of the time. Our final logistic regression model has an error rate of 36%. If used to make deterministic predictions, this model correctly predicts the behavior of 64% of the respondents in the data.

Error rate is not a perfect summary of model misfit, because it does not distinguish between predictions of 0.6 and 0.9, for example. Still, it can provide a simple starting point for evaluating a model. For example, an error rate of 36%, is only 6 percentage points better than the 42% error rate obtainable in this dataset by simply guessing that all people will switch.

This high error rate does not mean the model is useless—as the plots showed, the fitted model is highly predictive of the probability of switching. But most of the data are close to the mean level of the inputs (distances of less than 100 meters to the nearest safe well, and arsenic levels between 0.5 and 1.0), and so for most of the data, the simple mean prediction, Pr(switch) = 0.58, works well. The model is informative near the extremes, but relatively few data points are out there and so the overall predictive accuracy of the model is not high.

## Where error rate can mislead

The error rate can be a useful summary, but it does not tell the whole story, especially in settings where predicted probabilities are close to zero.

Example: Hypothetical rare disease

We illustrate with a simple example in which models are fit to predict the occurrence or non-occurrence of a rare disease that has 1% prevalence in the dataset and also in the population. Each model will then be used to make predictions out of sample. Consider the following two models:

1. Logistic regression with only a constant term. The predicted probability of disease is then 0.01 for each person.
2. Logistic regression with several useful predictors (for example, age, sex, and some measures of existing symptoms). Suppose that, under this model, predictive probabilities are very low (for example, in the range of 0.001 or 0.0001) for most people, but for some small subgroup, the model will identify them as high risk and give them a 0.40 predictive probability of having the disease.

Model 2, assuming it fits the data well and that the sample is representative of the population, is much better than model 1, as it successfully classifies some people as being of high disease risk, while reducing worry about the vast majority of the population.

But now consider comparing the two models using error rate:

1. The simple model, which assigns probability 0.01 to everyone, will give "no disease" as its point prediction for every person. We have stipulated that 1% of people actually have the disease, so the error rate is 1%.
2. The more complex model still happens to return predictive probability of disease at less then 50% for all people, so it still gives a point prediction of "no disease" for every person, and the error rate is still 1%.

Model 2, despite being much more accurate, and identifying some people to have as high as a 40% chance of having the disease, still gives no improvement in error rate. This does not mean that Model 2 is no better than Model 1; it just tells us that error rate is an incomplete summary.

## 14.6 Identification and separation

There are two reasons that a logistic regression can be nonidentified (that is, have parameters that cannot be estimated from the available data and model, as discussed in Section 10.7 in the context of linear regression):

- As with linear regression, if predictors are collinear, then estimation of the linear predictor, $X\beta$, does not allow separate estimation of the individual parameters $\beta$. We can handle this kind of nonidentifiability in the same way that we would proceed for linear regression, as described in Section 10.7.
- A completely separate identifiability problem, called *separation*, can arise from the discreteness of the data.
  - If a predictor $x_j$ is completely aligned with the outcome, so that $y = 1$ for all the cases where $x_j$ exceeds some threshold $T$, and $y = 0$ for all cases where $x_j < T$, then the best estimate for the coefficient $\beta_j$ is $\infty$. Figure 14.11 shows an example.
  - Conversely, if $y = 1$ for all cases where $x_j < T$, and $y = 0$ for all cases where $x_j > T$, then $\hat{\beta}_j$ will be $-\infty$.
  - More generally, this problem will occur if any linear combination of predictors is perfectly aligned with the outcome. For example, suppose that $7x_1 + x_2 - 3x_3$ is completely positively aligned with the data, with $y = 1$ if and only if this linear combination of predictors exceeds some threshold. Then the linear combination $7\hat{\beta}_1 + \hat{\beta}_2 - 3\hat{\beta}_3$ will be estimated at $\infty$, which will cause at least one of the three coefficients $\beta_1, \beta_2, \beta_3$ to be estimated at $\infty$ or $-\infty$.

One way to handle separation is using a Bayesian approach that provides some amount of information on all the regression coefficients, including those that are not identified from the data alone.

When a fitted model "separates" (that is, perfectly predicts some subset of) discrete data, the maximum likelihood estimate can be undefined or unreasonable, and we can get a more reasonable

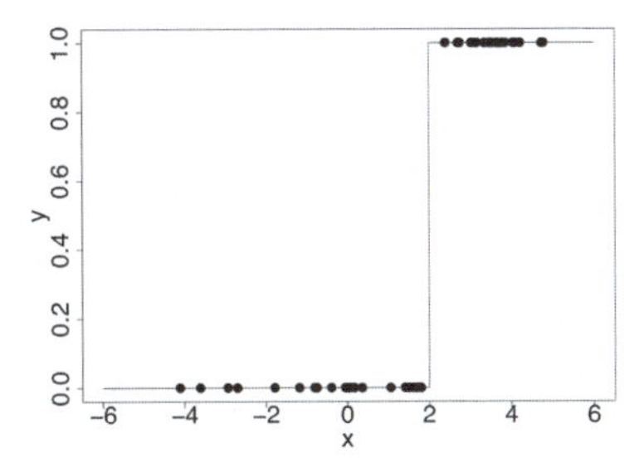


Figure 14.11 *Example of data for which a logistic regression model is nonidentifiable. The outcome $y$ equals 0 for all data below $x = 2$ and 1 for all data above $x = 2$; hence the best-fit logistic regression line is $y = \text{logit}^{-1}(\infty(x-2))$, which has an infinite slope at $x = 2$.*

```
1960
              coef.est coef.se
(Intercept)   -0.14     0.23
female         0.24     0.14
black         -1.03     0.36
income         0.03     0.06
  n = 875, k = 4
```

```
1964
              coef.est coef.se
(Intercept)   -1.15     0.22
female        -0.08     0.14
black        -16.83   420.40
income         0.19     0.06
  n = 1058, k = 4
```

```
1968
              coef.est coef.se
(Intercept)    0.26     0.24
female        -0.03     0.14
black         -3.53     0.59
income         0.00     0.07
  n = 876, k = 4
```

```
1972
              coef.est coef.se
(Intercept)    0.67     0.18
female        -0.25     0.12
black         -2.63     0.27
income         0.09     0.05
  n = 1518, k = 4
```

Figure 14.12 *Estimates and standard errors from logistic regressions (maximum likelihood estimates, which roughly correspond to Bayesian estimates with uniform prior distributions) predicting Republican vote intention in pre-election polls, fit separately to survey data from four presidential elections from 1960 through 1972. The estimates are reasonable except in 1964, where there is complete separation (with none of black respondents supporting the Republican candidate, Barry Goldwater), which leads to an essentially infinite estimate of the coefficient for* `black`.

answer using a Bayesian prior information to effectively constrain the inference. We illustrate with an example that arose in one of our routine analyses, where we were fitting logistic regressions to a series of datasets.

xample:
ncome and
oting

Figure 14.12 shows the estimated coefficients in a model predicting probability of Republican vote for president given sex, ethnicity, and income (coded as $-2, -1, 0, 1, 2$), fit separately to pre-election polls for a series of elections.[3] The estimates look fine except in 1964, where there is complete separation: of the 87 African Americans in the survey that year, none reported a preference for the Republican candidate. We fit the model in R, which actually yielded a finite estimate for the coefficient of `black` even in 1964, but that number and its standard error are essentially meaningless, being a function of how long the iterative fitting procedure goes before giving up. The maximum likelihood estimate for the coefficient of `black` in that year is $-\infty$.

Figure 14.13 displays the *profile likelihood* for the coefficient of `black` in the voting example, that is, the maximum value of the likelihood function conditional on this coefficient. The maximum of the profile likelihood is at $-\infty$, a value that makes no sense in this application. As we shall see, an informative prior distribution will bring the estimate down to a reasonable value.

[3]Data and code for this example are in the folder `NES`.

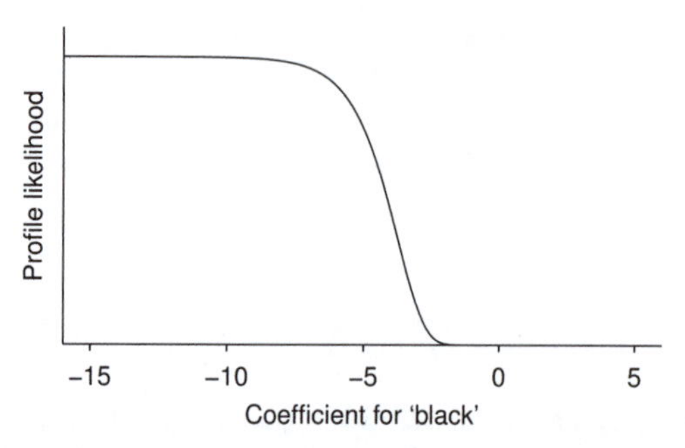


Figure 14.13 *Profile likelihood (in this case, essentially the posterior density given a uniform prior distribution) of the coefficient of* `black` *from the logistic regression of Republican vote in 1964 (displayed in the lower left of Figure 14.12), conditional on point estimates of the other coefficients in the model. The maximum occurs as* $\beta \to -\infty$*, indicating that the best fit to the data would occur at this unreasonable limit. The* $y$*-axis starts at 0; we do not otherwise label it because all that matters are the relative values of this curve, not its absolute level.*

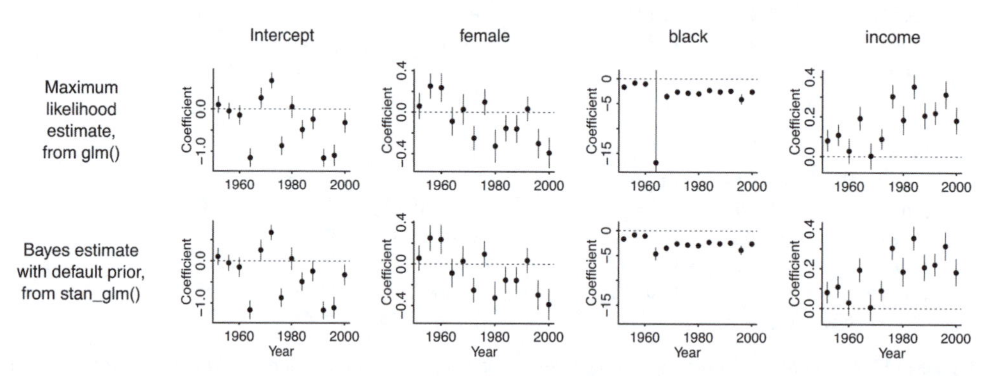


Figure 14.14 *The top row shows the estimated coefficients (±1 standard error) for a logistic regression predicting probability of Republican vote for president, given sex, race, and income, as fit separately to data from the National Election Study for each election 1952 through 2000. The complete separation in 1964 led to a coefficient estimate of* $-\infty$ *that year. (The particular finite values of the estimate and standard error are determined by the number of iterations used by the* `glm` *function in R before stopping.) The bottom row shows the posterior estimates and standard deviations for the same model fit each year using Bayesian inference with default priors from* `stan_glm`*. The Bayesian procedure does a reasonable job at stabilizing the estimates for 1964, while leaving the estimates for other years essentially unchanged.*

We apply our default prior distribution to the pre-election polls discussed earlier in this section, in which we could easily fit a logistic regression with flat priors for every election year except 1964, where the coefficient for `black` blew up due to complete separation in the data.

The top row of Figure 14.14 shows the time series of estimated coefficients and error bounds for the four coefficients of the logistic regression fit using `glm` separately to poll data for each election year from 1952 through 2000. All the estimates are reasonable except for the intercept and the coefficient for `black` in 1964, when the maximum likelihood estimates are infinite. As discussed already, we do not believe that these coefficient estimates for 1964 are reasonable: in the population as a whole, we do *not* believe that the probability was zero that an African American in the population would vote Republican. It is, however, completely predictable that with moderate sample sizes there will occasionally be separation, yielding infinite estimates. (The estimates from `glm` shown here for 1964 are finite only because the generalized linear model fitting routine in R stops after a finite number of iterations.)

The bottom row of Figure 14.14 shows the coefficient estimates fit using `stan_glm` on its default setting. The estimated coefficient for `black` in 1964 has been stabilized, with the other coefficients being essentially unchanged. This example illustrates how a weakly informative prior distribution can work in routine practice.

### Weakly informative default prior compared to actual prior information

The default prior distribution used by `stan_glm` does not represent our prior knowledge about the coefficient for `black` in the logistic regression for 1964 or any other year. Even before seeing data from this particular series of surveys, we knew that blacks have been less likely than whites to vote for Republican candidates; thus our prior belief was that the coefficient was negative. Furthermore, our prior for any given year would be informed by data from other years. For example, given the series of estimates in Figure 14.14, we would guess that the coefficient for `black` in 2004 is probably between $-2$ and $-5$. Finally, we are not using specific prior knowledge about these elections. The Republican candidate in 1964 was particularly conservative on racial issues and opposed the Civil Rights Act; on those grounds we would expect him to poll particularly poorly among African Americans (as indeed he did). In sum, we feel comfortable using a default model that excludes much potentially useful information, recognizing that we could add such information if it were judged to be worth the trouble (for example, instead of performing separate estimates for each election year, setting up a hierarchical model allowing the coefficients to gradually vary over time and including election-level predictors including information such as the candidates' positions on economic and racial issues).

## 14.7 Bibliographic note

Binned residual plots and related tools for checking the fit of logistic regressions are discussed by Landwehr, Pregibon, and Shoemaker (1984); Gelman, Goegebeur, et al. (2000); and Pardoe and Cook (2002).

See Gelman and Pardoe (2007) for more on average predictive comparisons and Gelman, Goegebeur et al. (2000) for more on binned residuals.

Nonidentifiability of logistic regression and separation in discrete data are discussed by Albert and Anderson (1984), Lesaffre and Albert (1989), Heinze and Schemper (2003), as well as in the book by Agresti (2002). Firth (1993), Zorn (2005), and Gelman et al. (2008) present Bayesian solutions.

## 14.8 Exercises

14.1 *Graphing binary data and logistic regression*: Reproduce Figure 14.1 with the model, $\Pr(y = 1) = \text{logit}^{-1}(0.4 - 0.3x)$, with 50 data points $x$ sampled uniformly in the range $[A, B]$. (In Figure 14.1 the $x$'s were drawn from a normal distribution.) Choose the values $A$ and $B$ so that the plot includes a zone where values of $y$ are all 1, a zone where they are all 0, and a band of overlap in the middle.

14.2 *Logistic regression and discrimination lines*: Reproduce Figure 14.2 with the model, $\Pr(y = 1) = \text{logit}^{-1}(0.4-0.3x_1+0.2x_2)$, with $(x_1, x_2)$ sampled uniformly from the rectangle $[A_1, B_1]\times[A_2, B_2]$. Choose the values $A_1, B_1, A_2, B_2$ so that the plot includes a zone where values of $y$ are all 1, a zone where they are all 0, and a band of overlap in the middle, and with the three lines corresponding to $\Pr(y = 1) = 0.1, 0.5$, and 0.9 are all visible.

14.3 *Graphing logistic regressions*: The well-switching data described in Section 13.7 are in the folder `Arsenic`.

(a) Fit a logistic regression for the probability of switching using log (distance to nearest safe well) as a predictor.

(b) Make a graph similar to Figure 13.8b displaying Pr(switch) as a function of distance to nearest safe well, along with the data.

(c) Make a residual plot and binned residual plot as in Figure 14.8.

(d) Compute the error rate of the fitted model and compare to the error rate of the null model.

(e) Create indicator variables corresponding to `dist` $< 100$; `dist` between 100 and 200; and `dist` $> 200$. Fit a logistic regression for Pr(switch) using these indicators. With this new model, repeat the computations and graphs for part (a) of this exercise.

14.4 *Working with logistic regression*: Perform a logistic regression for a problem of interest to you. This can be from a research project, a previous class, or data you download. Choose one variable of interest to be the outcome, which will take on the values 0 and 1 (since you are doing logistic regression).

(a) Analyze the data in R.

(b) Fit several different versions of your model. Try including different predictors, interactions, and transformations of the inputs.

(c) Choose one particular formulation of the model and do the following:

i. Describe how each input affects $\Pr(y = 1)$ in the fitted model. You must consider the estimated coefficient, the range of the input values, and the nonlinear inverse logit function.

ii. What is the error rate of the fitted model? What is the error rate of the null model?

iii. Compare the fitted and null models using leave-one-out cross validation (see Section 11.8). Does the improvement in fit seem to be real?

iv. Use the model to make predictions for some test cases of interest.

v. Use the simulations from the `stan_glm` output to create a 50% interval for some nonlinear function of parameters (for example, $\beta_1/\beta_2$).

14.5 *Working with logistic regression*: In a class of 50 students, a logistic regression is performed of course grade (pass or fail) on midterm exam score (continuous values with mean 60 and standard deviation 15). The fitted model is $\Pr(\text{pass}) = \text{logit}^{-1}(-24 + 0.4x)$.

(a) Graph the fitted model. Also on this graph put a scatterplot of hypothetical data consistent with the information given.

(b) Suppose the midterm scores were transformed to have a mean of 0 and standard deviation of 1. What would be the equation of the logistic regression using these transformed scores as a predictor?

(c) Create a new predictor that is pure noise; for example, in R you can create `newpred <- rnorm(n,0,1)`. Add it to your model. How much does the leave-one-out cross validation score decrease?

14.6 *Limitations of logistic regression*: Consider a dataset with $n = 20$ points, a single predictor $x$ that takes on the values $1, \dots, 20$, and binary data $y$. Construct data values $y_1, \dots, y_{20}$ that are inconsistent with any logistic regression on $x$. Fit a logistic regression to these data, plot the data and fitted curve, and explain why you can say that the model does not fit the data.

14.7 *Model building and comparison*: Continue with the well-switching data described in the previous exercise.

(a) Fit a logistic regression for the probability of switching using, as predictors, distance, log(arsenic), and their interaction. Interpret the estimated coefficients and their standard errors.

(b) Make graphs as in Figure 14.3 to show the relation between probability of switching, distance, and arsenic level.

(c) Following the procedure described in Section 14.4, compute the average predictive differences corresponding to:

i. A comparison of `dist` $= 0$ to `dist` $= 100$, with `arsenic` held constant.

ii. A comparison of `dist` $= 100$ to `dist` $= 200$, with `arsenic` held constant.

iii. A comparison of `arsenic` = 0.5 to `arsenic` = 1.0, with `dist` held constant.
iv. A comparison of `arsenic` = 1.0 to `arsenic` = 2.0, with `dist` held constant.

Discuss these results.

14.8 *Learning from social science data*: The General Social Survey (GSS) has been conducted in the United States every two years since 1972.

(a) Go to the GSS website and download the data. Consider a question of interest that was asked in many rounds of the survey and convert it to a binary outcome, if it is not binary already. Decide how you will handle nonresponse in your analysis.

(b) Make a graph of the average response of this binary variable over time, each year giving $\pm 1$ standard error bounds as in Figure 4.3.

(c) Set up a logistic regression of this outcome variable given predictors for age, sex, education, and ethnicity. Fit the model separately for each year that the question was asked, and make a grid of plots with the time series of coefficient estimates $\pm$ standard errors over time.

(d) Discuss the results and how you might want to expand your model to answer some social science question of interest.

14.9 *Linear or logistic regression for discrete data*: Simulate continuous data from the regression model, $z = a + bx +$ error. Set the parameters so that the outcomes $z$ are positive about half the time and negative about half the time.

(a) Create a binary variable $y$ that equals 1 if $z$ is positive or 0 if $z$ is negative. Fit a logistic regression predicting $y$ from $x$.

(b) Fit a linear regression predicting $y$ from $x$: you can do this, even though the data $y$ are discrete.

(c) Estimate the average predictive comparison—the expected difference in $y$, corresponding to a unit difference in $x$—based on the fitted logistic regression in (a). Compare this average predictive comparison to the linear regression coefficient in (b).

14.10 *Linear or logistic regression for discrete data*: In the setup of the previous exercise:

(a) Set the parameters of your simulation so that the coefficient estimate in (b) and the average predictive comparison in (c) are close.

(b) Set the parameters of your simulation so that the coefficient estimate in (b) and the average predictive comparison in (c) are much different.

(c) In general, when will it work reasonably well to fit a linear model to predict a binary outcome?

See also Exercise 13.12.

14.11 *Working through your own example*: Continuing the example from the final exercises of the earlier chapters, take the logistic regression you fit in Exercise 13.13 and compute average predictive comparisons.

# Chapter 15

# Other generalized linear models

We can apply the principle of logistic regression—taking a linear "link function" $y = a + bx$ and extending it through a nonlinear transformation and a probability model—to allow it to predict bounded or discrete data of different forms. This chapter presents this generalized linear modeling framework and goes through several important special cases, including Poisson or negative binomial regression for count data, the logistic-binomial and probit models, ordered logistic regression, robust regression, and some extensions. As always, we explain these models with a variety of examples, with graphs of data and fitted models along with associated R code, with the goal that you should be able to build, fit, understand, and evaluate these models on new problems.

## 15.1 Definition and notation

*Generalized linear modeling* is a framework for statistical analysis that includes linear and logistic regression as special cases. Linear regression directly predicts continuous data $y$ from a *linear predictor* $X\beta = \beta_0 + X_1\beta_1 + \cdots + X_k\beta_k$. Logistic regression predicts $\Pr(y = 1)$ for binary data from a linear predictor with an inverse logit transformation. A generalized linear model involves:

1. A vector of outcome data $y = (y_1, \dots, y_n)$.
2. A matrix of predictors $X$ and vector of coefficients $\beta$, forming a linear predictor vector $X\beta$.
3. A *link function* $g$, yielding a vector of transformed data $\hat{y} = g^{-1}(X\beta)$ that are used to model the data.
4. A data distribution, $p(y|\hat{y})$.
5. Possibly other parameters, such as variances, overdispersions, and cutpoints, involved in the predictors, link function, and data distribution.

The options in a generalized linear model are the transformation $g$ and the distribution $p$.

- In *linear regression*, the transformation is the identity (that is, $g(u) \equiv u$) and the data distribution is normal, with standard deviation $\sigma$ estimated from data.
- In *logistic regression*, the transformation is the inverse logit, $g^{-1}(u) = \text{logit}^{-1}(u)$ (see Figure 13.1a) and the data distribution is defined by the probability for binary data: $\Pr(y=1) = \hat{y}$.

This chapter discusses several other generalized linear models, which we list here for convenience:

- The *Poisson* and *negative binomial* models (Section 15.2) are used for count data, that is, where each data point $y_i$ can equal 0, 1, 2, ..., as discussed on page 44. The usual transformation $g$ used here is the logarithmic, so that $g(u) = \exp(u)$ transforms a continuous linear predictor $X_i\beta$ to a positive $\hat{y}_i$. The negative binomial distribution includes an extra parameter to capture *overdispersion*, that is, variation in the data beyond what would be predicted from the Poisson distribution alone.
- The *logistic-binomial* model (Section 15.3) is used in settings where each data point $y_i$ represents the number of successes in some number $n_i$ of tries. (This $n_i$, the number of tries for data point $i$,

is not the same as $n$, the number of data points.) In this model, the transformation $g$ is the inverse logit and the data distribution is binomial.
As with Poisson regression, the binomial model is typically improved by the inclusion of an overdispersion parameter, which can be done using the *beta-binomial* distribution.

- The *probit* model (Section 15.4) is the same as logistic regression but with the logit function replaced by the normal cumulative distribution, or equivalently with the normal distribution instead of the logistic in the latent-data errors.
- *Multinomial* logit and probit models (Section 15.5) are extensions of logistic and probit regressions for categorical data with more than two options, for example survey responses such as Strongly Agree, Agree, Indifferent, Disagree, Strongly Disagree. These models use the logit or probit transformation and the multinomial distribution and require additional parameters to model the multiple possibilities of the data.
Multinomial models are further classified as *ordered* (for example, Strongly Agree, ..., Strongly Disagree) or *unordered* (for example, Vanilla, Chocolate, Strawberry).
- *Robust* regression models (Section 15.6) replace the usual normal or logistic models by other distributions (usually the so-called Student-$t$ family of models) that allow occasional extreme values.
In the statistical literature, generalized linear models have been defined using the exponential-family class of data distributions, which includes the normal, binomial, Poisson, and several others but excludes, for example, the $t$ distribution. For our purposes, however, we use the term *generalized linear model* to apply to any model with a linear predictor, link function, and data distribution, not restricting to exponential-family models.

This chapter briefly covers several of these models, with an example of count-data regression in Section 15.2 and an ordered logistic example in Section 15.5. In Section 15.7 we discuss the connections between generalized linear models and behavioral models of choice that are used in psychology and economics, using as an example the logistic regression for well switching in Bangladesh. The chapter is not intended to be a comprehensive overview of generalized linear models; rather, we want to give a sense of the variety of regression models that can be appropriate for different data structures that we have seen in applications.

### Fitting generalized linear models in R

Because of the variety of options involved, generalized linear modeling can be more complicated than linear and logistic regression. The most commonly used generalized limear models have been programmed in Stan and can be accessed using the `rstanarm` and `brms` packages, as we discuss below.

## 15.2 Poisson and negative binomial regression

In count-data regressions, each unit $i$ corresponds to a setting (typically a spatial location or a time interval) in which $y_i$ events are observed. For example, $i$ could index street intersections in a city, and $y_i$ could be the number of traffic accidents at intersection $i$ in a given year.

As with linear and logistic regression, the variation in $y$ can be explained with linear predictors $X$. In the traffic accidents example, these predictors could include: a constant term, a measure of the average speed of traffic near the intersection, and an indicator for whether the intersection has a traffic signal.

### Poisson model

The simplest regression model for count data is,

$$y_i \sim \text{Poisson}(e^{X_i\beta}), \tag{15.1}$$

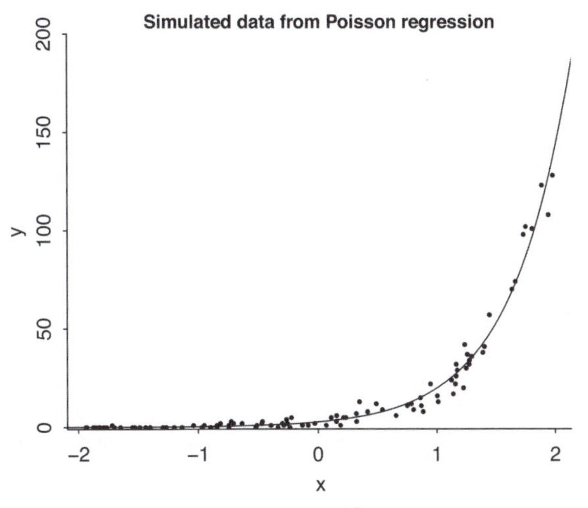


Figure 15.1 *Simulated data from the model, $y_i \sim \text{Poisson}(e^{a+bx_i})$ along with the fitted curve, $y = e^{\hat{a}+\hat{b}x}$. The vertical deviations of the points from the line are consistent with the property of the Poisson distribution that $sd(y) = \sqrt{E(y)}$.*

so that the linear predictor $X_i\beta$ is the logarithm of the expected value of measurement $y_i$. Under the Poisson model, $\text{sd}(y_i) = \sqrt{\text{E}(y_i)}$; thus if the model accurately describes the data, we also have a sense of how much variation we would expect from the fitted curve.

xample: imulated ata for oisson egression

We demonstrate with a simple example.[1] First we simulate a predictor $x$ and then simulate fake data $y$ from a Poisson regression:

```
n <- 50
x <- runif(n, -2, 2)
a <- 1
b <- 2
linpred <- a + b*x
y <- rpois(n, exp(linpred))
fake <- data.frame(x=x, y=y)
```

Then we fit the model to our simulated data:

```
fit_fake <- stan_glm(y ~ x, family=poisson(link="log"), data=fake)
print(fit_fake)
```

Then we plot the data along with the fitted curve:

```
plot(x, y)
curve(exp(coef(fit_fake)[1] + coef(fit_fake)[2]*x), add=TRUE)
```

Here is the fitted model:

```
            Median MAD_SD
(Intercept) 1.0    0.1
x           2.0    0.1
```

Figure 15.1 shows the simulated data and fitted line. The Poisson distribution has its own internal scale of variation: unlike with the normal distribution, there is no `sigma` parameter to be fit. From the Poisson distribution, we would expect variation on the order of $\sqrt{\text{E}(y)}$: for example, where expected number of counts is 10, prediction errors should be mostly in the range $\pm 3$, where expected number of counts is 100, prediction errors should be mostly in the range $\pm 10$, and so on.

[1]Code for this example is in the folder PoissonExample.

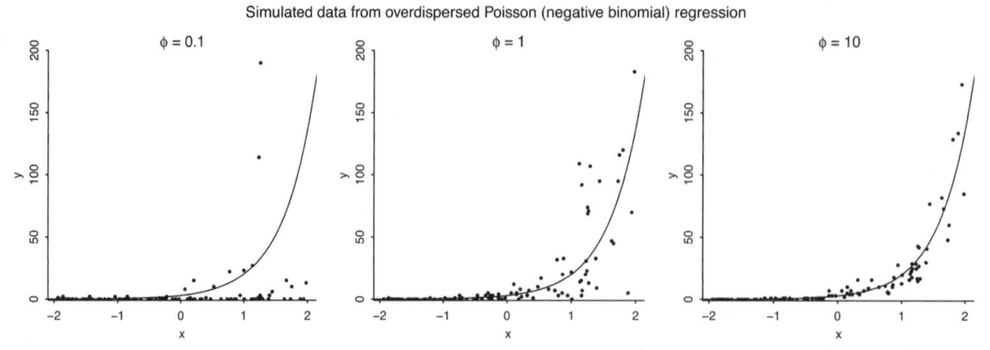


Figure 15.2 *Simulated data from the model, $y_i \sim$ negative binomial$(e^{a+bx_i}, \phi)$, for three different values of the reciprocal dispersion parameter $\phi$. For each, we also plot the fitted curve, $y = e^{\hat{a}+\hat{b}x}$. As can be seen from this example, the lower the parameter $\phi$, the greater the vertical deviations of the points from the line.*

## Overdispersion and underdispersion

Overdispersion and underdispersion refer to data that show more or less variation than expected based on a fitted probability model. We discuss overdispersion for binomial models in Section 15.3 and for Poisson models here.

To continue our earlier story, suppose you look at the data for traffic accidents, compare them to the fitted regression, and see much more variation than would be expected from the Poisson model: that would be overdispersion. Count data where the variation of the data from the fitted line is much less than the square root of the predicted value would indicate underdispersion, but this is rare in actual data.

## Negative binomial model for overdispersion

To generalize the Poisson model to allow for overdispersion, we use the *negative binomial*—a probability distribution whose name comes from a sampling procedure that we do not discuss here—which includes an additional "reciprocal dispersion" parameter $\phi$ so that $\text{sd}(y|x) = \sqrt{\text{E}(y|x) + \text{E}(y|x)^2/\phi}$. In this parameterization, the parameter $\phi$ is restricted to be positive, with lower values corresponding to more overdispersion, and the limit $\phi \to \infty$ representing the Poisson model (that is, zero overdispersion).

Example: Simulated data for negative binomial regression

To get a sense of what this model looks like, we simulate three datasets, using the coefficients and the same values of the predictor $x$ as in Figure 15.1, but replacing the Poisson with the negative binomial distribution with parameter $\phi$ set to 0.1, 1, or 10. First we set up containers for the data and fitted models:[2]

```
phi_grid <- c(0.1, 1, 10)
K <- length(phi_grid)
y_nb <- as.list(rep(NA, K))
fake_nb <- as.list(rep(NA, K))
fit_nb <- as.list(rep(NA, K))
```

Then, for each of the three values of $\phi$, we simulate the data and fit and display a negative binomial regression:

```
library("MASS")
for (k in 1:K){
  y_nb[[k]] <- rnegbin(n, exp(linpred), phi_grid[k])
  fake_nb[[k]] <- data.frame(x=x, y=y_nb[[k]])
```

[2]Code for this example is in the folder PoissonExample.

```
    fit_nb[[k]] <- stan_glm(y ~ x, family=neg_binomial_2(link="log"), data=fake)
    print(fit_nb[[k]])
  }
```

Finally, we plot each dataset along with the fitted curve:

```
  for (k in 1:K) {
    plot(x, y_nb[[k]])
    curve(exp(coef(fit_nb[[k]])[1] + coef(fit_nb[[k]])[2]*x), add=TRUE)
  }
```

Figure 15.2 shows the results. The curve of $\mathrm{E}(y|x)$ is the same for all three models, but the variation of $y$ is much higher when $\phi$ is near zero.

### Interpreting Poisson or negative binomial regression coefficients

The coefficients $\beta$ in a logarithmic regression model can be exponentiated and treated as multiplicative effects. For example, suppose the traffic accident model is

$$y_i \sim \text{negative binomial}\left(e^{2.8+0.012X_{i1}-0.20X_{i2}}, \phi\right),$$

where $X_{i1}$ is average speed in miles per hour (mph) on the nearby streets and $X_{i2} = 1$ if the intersection has a traffic signal or 0 otherwise. We can then interpret each coefficient as follows:

- The constant term gives the intercept of the regression, that is, the prediction if $X_{i1} = 0$ and $X_{i2} = 0$. Since this is not possible (no street will have an average traffic speed of 0), we will not try to interpret the constant term.
- The coefficient of $X_{i1}$ is the expected difference in $y$ (on the logarithmic scale) for each additional mph of traffic speed. Thus, the expected multiplicative increase is $e^{0.012} = 1.012$, or a 1.2% positive difference in the rate of traffic accidents per mph. Since traffic speeds vary by tens of mph, it would make sense to define $X_{i1}$ as speed in tens of mph, in which case its coefficient would be 0.12, corresponding to a 12% increase (more precisely, $e^{0.12} = 1.127$: a 12.7% increase) in accident rate per ten mph.
- The coefficient of $X_{i2}$ tells us that the predictive difference of having a traffic signal can be found by multiplying the accident rate by $e^{-0.20} = 0.82$ yielding a reduction of 18%.

As usual, each regression coefficient is interpreted as a comparison in which one predictor differs by one unit while all the other predictors remain at the same level, which is not necessarily the most appropriate assumption when extending the model to new settings. For example, installing traffic signals in all the intersections in the city would *not* necessarily be expected to reduce accidents by 18%.

### Exposure

In most applications of count-data regression, there is a baseline or *exposure*, some value such as the average flow of vehicles that travel through the intersection in the traffic accidents example. We can model $y_i$ as the number of cases in a process with rate $\theta_i$ and exposure $u_i$.

$$y_i \sim \text{negative binomial}(u_i\theta_i, \phi), \tag{15.2}$$

where, as before, $\theta_i = e^{X_i\beta}$. Expression (15.2) includes Poisson regression as the special case of $\phi \to \infty$.

The logarithm of the exposure, $\log(u_i)$, is called the *offset* in generalized linear model terminology. The regression coefficients $\beta$ now reflect the associations between the predictors and $\theta_i$ (in our example, the rate of traffic accidents per vehicle).

### Including log(exposure) as a predictor in a Poisson or negative binomial regression

Putting the logarithm of the exposure into the model as an offset, as in model (15.2), is equivalent to including it as a regression predictor, but with its coefficient fixed to the value 1; see Exercise 15.2. Another option is to include it as a predictor and let its coefficient be estimated from the data. In some settings, this makes sense in that it can allow the data to be fit better; in other settings, it is simpler to just keep it as an offset so that the estimated rate $\theta$ has a more direct interpretation.

### Differences between the binomial and Poisson or negative binomial models

The Poisson or negative binomial model is similar to the binomial model for count data (see Section 15.3) but they are applied in slightly different situations:

- If each data point $y_i$ can be interpreted as the number of successes out of $n_i$ trials, then it is standard to use the binomial/logistic model, as described in Section 15.3, or its overdispersed generalization.
- If each data point $y_i$ does not have a natural limit—it is not based on a number of independent trials—then it is standard to use the Poisson or negative binomial model with logarithmic link, as described just above.
- If each data point $y_i$ has a natural limit that is much higher than the expected number of counts, then Poisson/logarithmic regression can be used to approximate the binomial model. For example, if a county has a population of 100 000, then the Poisson distribution has nonzero probability for counts larger than 100 000, but if the expected count is 4.52, then the probability of exceeding the limit is practically zero, and the Poisson distribution can be used to approximate the binomial.

### Example: zeroes in count data

Example: Roaches

For a more complicated example, we consider a study of the effect of integrated pest management on reducing cockroach levels in urban apartments.[3] In this experiment, the treatment and control were applied to 158 and 104 apartments, respectively, and the outcome measurement $y_i$ in each apartment $i$ was the number of roaches caught in a set of traps. Different apartments had traps for different numbers of days, and we label as $u_i$ the number of trap-days. The natural model for the roach counts is then $y_i \sim$ negative binomial$(u_i \exp(X_i\beta), \phi)$, where $X$ represents the regression predictors (in this case, a pre-treatment roach level, a treatment indicator, and an indicator for whether the apartment is in a "senior" building restricted to the elderly, and the constant term). The logarithm of the exposure, $\log(u_i)$, plays the role of the offset in the logarithmic regression; see model (15.2).

Fitting the model,

```
roaches$roach100 <- roaches$roach1/100
fit_1 <- stan_glm(y ~ roach100 + treatment + senior, family=neg_binomial_2,
  offset=log(exposure), data=roaches)
print(fit_1, digits=2)
```

yields,

```
                     Median MAD_SD
(Intercept)           2.84   0.24
roach100              1.31   0.25
treatment            -0.78   0.25
senior               -0.34   0.27

Auxiliary parameter(s):
                      Median MAD_SD
reciprocal_dispersion 0.27   0.03
```

[3] Data and code for this example are in the folder Roaches.

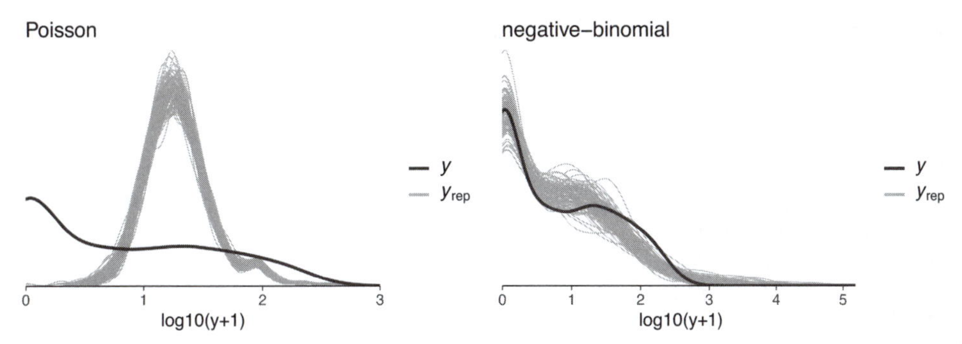


Figure 15.3 *Posterior predictive checks comparing density plot of roaches data (dark line) to 100 predictive replications (gray lines) of replicated data from (a) a fitted Poisson model, and (b) a fitted negative binomial model. The x axis is on the* $\log 10(y+1)$ *scale to better show the whole range of the data. The Poisson model shows strong lack of fit, and the negative binomial model shows some lack of fit.*

The first four rows of the output give the coefficient estimates and standard errors from the fitted model, and the last row shows the estimate and standard error of the reciprocal of the overdispersion parameter. So in this case the estimated overdispersion is $1/0.27 = 3.7$. As discussed in Section 15.2, the coefficient estimates are the same as before, but the standard errors are much larger, reflecting the variation in the fitted overdispersion.

**Checking model fit by comparing the data, $y$, to replicated datasets, $y^{\text{rep}}$.** How well does this model fit the data? We explore by simulating replicated datasets $y^{\text{rep}}$ that might be seen if the model were true and the study were performed again. The following function call produces an array of n_sims ×262 values—that is, n_sims posterior predictive simulations of the vector $y^{\text{rep}}$ of length $n = 262$:

```
y_rep_1 <- posterior_predict(fit_1)
```

We can compare the replicated datasets $y^{\text{rep}}$ to the original data $y$ in various ways. Figure 15.3b shows a density estimate of $\log_{10}(\text{count} + 1)$ and the corresponding posterior predictive distribution for 100 replications. The following code was used to make the plot:

```
n_sims <- nrow(yrep_1)
subset <- sample(n_sims, 100)
ppc_dens_overlay(log10(roaches$y+1), log10(yrep_1[subset,]+1))
```

One might be concerned that the number of zeroes in the data is not what would be predicted by the model. To check this formally, we can define a test statistic and compute it for each of the replications separately:

```
test <- function (y){
  mean(y==0)
}
test_rep_1 <- apply(yrep_1, 1, test)
```

The proportion of zeroes in the replicated datasets varies from 0.21 to 0.47, with 80% of the simulations falling in the range [0.29, 0.39], so the observed proportion of zeroes of 0.36 fits right in. However, other aspects of the data might not be so well fit, as could be discovered by looking at the posterior predictive distribution of other data summaries. Figure 15.3b shows a density estimate of $\log_{10}(\text{count} + 1)$ and the corresponding posterior predictive distribution for 100 replications. The model predictions are overall much closer than with Poisson model (as discussed below), but there is still clear lack of fit and especially the model is predicting much larger counts. The maximum roach count in the data is 357, but the maximum count in the negative binomial model replications is close to one million, which would make sense only in a horror movie. We will introduce another improvement later in this chapter.

**What if we had used Poisson regression?** The above fit reveals that the data show overdispersion. We can compare to the result of (mistakenly) fitting Poisson regression:

```
fit_2 <- stan_glm(y ~ roach100 + treatment + senior, family=poisson,
  offset=log(exposure2), data=roaches)
```

which yields,

```
            Median MAD_SD
(Intercept)  3.09   0.02
roach100     0.70   0.01
treatment   -0.52   0.02
senior      -0.38   0.03
```

The treatment appears to be effective in reducing roach counts. For now, we are simply interested in evaluating the model as a description of the data, without worrying about causal issues or the interpretation of the coefficients.

**Checking the fit of the non-overdispersed Poisson regression.** As before, we can check the fit of the model by simulating a matrix of posterior predictive replicated datasets:

```
yrep_2 <- posterior_predict(fit_2)
```

The true data have more smaller counts than what the model predicts. We can also use quantitative checks, and illustrate with a simple test of the number of zeroes in the data:

```
print(mean(roaches$y==0))
print(mean(yrep_2==0))
```

which reveals that 36% of the observed data points, but a very small fraction of the replicated data points, equal zero. This suggests a potential problem with the model: in reality, many apartments have zero roaches, but this would not be happening if the model were true, at least to judge from one simulation.

We can formally perform the posterior predictive check:

```
test_rep_2 <- apply(y_rep_2, 1, test)
```

The values of `test_rep_1` vary from 0 to 0.008—all of which are much lower than the observed test statistic of 0.36. Thus the Poisson regression model would not have replicated the frequency of zeroes in the data.

## 15.3 Logistic-binomial model

Chapter 13 discussed logistic regression for binary (Yes/No or 0/1) data. The logistic model can also be used for count data, using the binomial distribution (see page 43) to model the number of "successes" out of a specified number of possibilities, with the probability of success being fit to a logistic regression.

We demonstrate with a simple model of basketball shooting. We first simulate $N = 100$ players each shooting $n = 20$ shots, where the probability of a successful shot is a linear function of height (30% for a 5'9" player, 40% for a 6' tall player, and so forth):

```
N <- 100
height <- rnorm(N, 72, 3)
p <- 0.4 + 0.1*(height - 72)/3
n <- rep(20, N)
y <- rbinom(N, n, p)
data <- data.frame(n=n, y=y, height=height)
```

We can then fit and display a logistic regression predicting success probability given height:

```
fit_1a <- stan_glm(cbind(y, n-y) ~ height, family=binomial(link="logit"),
  data=data)
print(fit_1a)
```

### The binomial model for count data, applied to death sentences

Example: Death sentences

We illustrate binomial logistic regression in the context of a study of the proportion of death penalty verdicts that were overturned, in each of 34 states in the 23 years, 1973–1995. The units of this analysis are the $34 * 23 = 784$ state-years (actually, we only have $n = 450$ state-years in our analysis, since different states have restarted the death penalty at different times since 1973). For each state-year $i$, we label $n_i$ as the number of death sentences in that state in that year and $y_i$ as the number of these verdicts that were later overturned by higher courts. Our model has the form,

$$\begin{aligned} y_i &\sim \text{Binomial}(n_i, p_i), \\ p_i &= \text{logit}^{-1}(X_i \beta), \end{aligned} \tag{15.3}$$

where $X$ is a matrix of predictors. To start, we include

- A constant term.
- 33 indicators for states.
- A time trend for years (that is, a variable that equals 1 for 1973, 2 for 1974, 3 for 1975, and so on).

This model could also be written as,

$$\begin{aligned} y_{st} &\sim \text{Binomial}(n_{st}, p_{st}), \\ p_{st} &= \text{logit}^{-1}(\mu + \alpha_s + \beta_t), \end{aligned}$$

with subscripts $s$ for state and $t$ for time (that is, year relative to 1972). We prefer the form (15.3) because of its greater generality. But it is useful to be able to go back and forth between the two formulations; the probabilities $p$ are the same in the two models and are just indexed differently.

In either version of the model, we only include indicators for 33 of the 34 states, with the left-out state representing the default case: if we were to include a constant term and also indicators for all 34 states, the model would be collinear. In this sort of problem we prefer to fit a multilevel model with a varying intercept for state, but that is beyond the scope of this book.

### Overdispersion

When logistic regression is applied to count data, it is possible—in fact, usual—for the data to have more variation than is explained by the model. This overdispersion problem arises because the logistic regression model does not have a variance parameter $\sigma$.

More specifically, if data $y$ have a binomial distribution with parameters $n$ and $p$, then the mean of $y$ is $np$ and the standard deviation of $y$ is $\sqrt{np(1-p)}$. We define the standardized residual for each data point $i$ as,

$$\begin{aligned} z_i &= \frac{y_i - \hat{y}_i}{\text{sd}(\hat{y}_i)} \\ &= \frac{y_i - n_i \hat{p}_i}{\sqrt{n_i \hat{p}_i (1 - \hat{p}_i)}}, \end{aligned} \tag{15.4}$$

where $p_i = \text{logit}^{-1}(X_i \hat{\beta})$. If the binomial model is true, then the $z_i$'s should be approximately independent, each with mean 0 and standard deviation 1.

As with the Poisson model, we can then compute the estimated overdispersion $\frac{1}{N-k} \sum_{i=1}^N z_i^2$ and formally test for overdispersion by comparing $\sum_{i=1}^N z_i^2$ to a $\chi^2_{N-k}$ distribution. The $N$ here represents the number of data points and is unrelated to the notation $n_i$ in models (15.3) and (15.4) referring to the number of cases in state-year $i$.

In practice, overdispersion happens almost all the time that logistic regression (or Poisson regression, as discussed in Section 15.2) is applied to count data. In the more general family of distributions known as overdispersed models, the standard deviation can have the form $\sqrt{\omega np(1-p)}$, where $\omega > 1$ is known as the *overdispersion parameter*. The overdispersed model reduces to binomial logistic regression when $\omega = 1$.

Overdispersed binomial regressions can be fit using the `brm` function with a custom beta-binomial family or by writing the model directly in Stan.

### Binary-data model as a special case of the count-data model

Logistic regression for binary data (see Chapters 13 and 14) is a special case of the binomial form (15.3) with $n_i \equiv 1$ for all $i$. Overdispersion at the level of the individual data points cannot occur in the binary model, which is why we did not introduce overdispersed models in those earlier chapters.

### Count-data model as a special case of the binary-data model

Conversely, the binomial model (15.3) can be expressed in the binary-data form (13.2) by considering each of the $n_i$ cases as a separate data point. The sample size of this expanded regression is $\sum_i n_i$, and the data points are 0's and 1's: each unit $i$ corresponds to $y_i$ ones and $n_i - y_i$ zeroes. Finally, the $X$ matrix is expanded to have $\sum_i n_i$ rows, where the $i^{\text{th}}$ row of the original $X$ matrix becomes $n_i$ identical rows in the expanded matrix.

## 15.4 Probit regression: normally distributed latent data

The *probit* model is the same as the logit, except it replaces the logistic distribution (recall Figure 13.5) by the normal. We can write the model directly as

$$\Pr(y_i = 1) = \Phi(X_i\beta),$$

where $\Phi$ is the normal cumulative distribution function. In the latent-data formulation,

$$\begin{aligned} y_i &= \begin{cases} 1 & \text{if } z_i > 0 \\ 0 & \text{if } z_i < 0, \end{cases} \\ z_i &= X_i\beta + \epsilon_i, \\ \epsilon_i &\sim \mathrm{N}(0,1), \end{aligned} \tag{15.5}$$

that is, a normal distribution for the latent errors with mean 0 and standard deviation 1.

More generally, the model can have an error variance, so that the last line of (15.5) is replaced by

$$\epsilon_i \sim \mathrm{N}(0, \sigma^2),$$

but then $\sigma$ is nonidentified, because the model is unchanged if we multiply $\sigma$ by some constant $c$ and then multiply the vector $\beta$ by $c$ also. Hence we need some restriction on the parameters, and the standard approach is to fix $\sigma = 1$ as in (15.5).

### Probit or logit?

The logit and probit models have the same structure, with the only difference being the latent error distribution, which is either unit logistic or unit normal. It turns out that the unit logistic distribution is close to the normal with standard deviation 1.6; compare Figure 13.5 to Figure 15.4. The two models are so close that, except in some unusual situations involving points in the far tails of the distributions, we can go back and forth between logit and probit just by a simple scaling. Coefficients in a probit regression are typically close to logistic regression coefficients divided by 1.6.

For example, here is the probit version of the logistic regression model on page 234 for well switching:

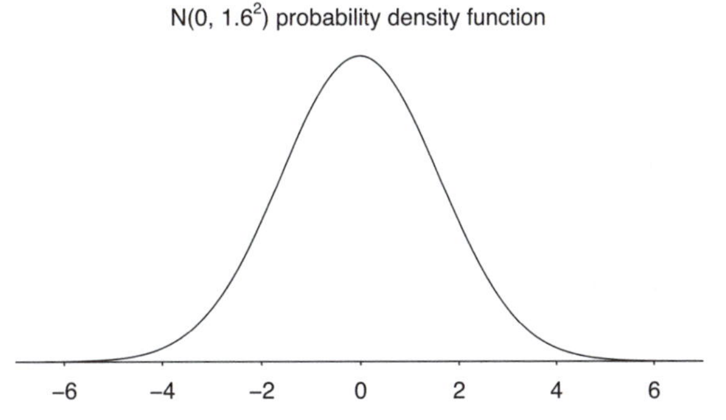


Figure 15.4 *Normal density function with mean 0 and standard deviation 1.6. For most practical purposes, this is indistinguishable from the logistic density shown in Figure 13.5. Thus we can interpret coefficients in probit models as logistic regression coefficients divided by 1.6.*

```
fit_probit <- stan_glm(y ~ dist100, family=binomial(link="probit"), data=wells)
print(fit_probit)
```

which yields,

```
            Median MAD_SD
(Intercept)  0.38   0.04
dist100     -0.39   0.06
```

For the examples we have seen, the choice of logit or probit model is a matter of taste or convenience, for example, in interpreting the latent normal errors of probit models. When we see probit regression coefficients, we can simply multiply them by 1.6 to obtain the equivalent logistic coefficients. For example, the model we have just fit, $\Pr(y = 1) = \Phi(0.38 - 0.39x)$, is essentially equivalent to the logistic model $\Pr(y = 1) = \text{logit}^{-1}(1.6 * (0.38 - 0.39x)) = \text{logit}^{-1}(0.61 - 0.62x)$, which indeed is the logit model estimated on page 234.

## 15.5 Ordered and unordered categorical regression

Logistic and probit regression can be extended to multiple categories, which can be ordered or unordered. Examples of ordered categorical outcomes include Democrat, Independent, Republican; Yes, Maybe, No; Always, Frequently, Often, Rarely, Never. Examples of unordered categorical outcomes include Liberal, Labour, Conservative; Football, Basketball, Baseball, Hockey; Train, Bus, Automobile, Walk; White, Black, Hispanic, Asian, Other. We discuss ordered categories first, including an extended example, and then briefly discuss regression models for unordered categorical variables.

### The ordered multinomial logit model

Consider a categorical outcome $y$ that can take on the values $1, 2, \dots, K$. The ordered logistic model can be written in two equivalent ways. First we express it as a series of logistic regressions:

$$\begin{aligned}
\Pr(y > 1) &= \text{logit}^{-1}(X\beta), \\
\Pr(y > 2) &= \text{logit}^{-1}(X\beta - c_2), \\
\Pr(y > 3) &= \text{logit}^{-1}(X\beta - c_3), \\
&\dots \\
\Pr(y > K-1) &= \text{logit}^{-1}(X\beta - c_{K-1}).
\end{aligned} \tag{15.6}$$

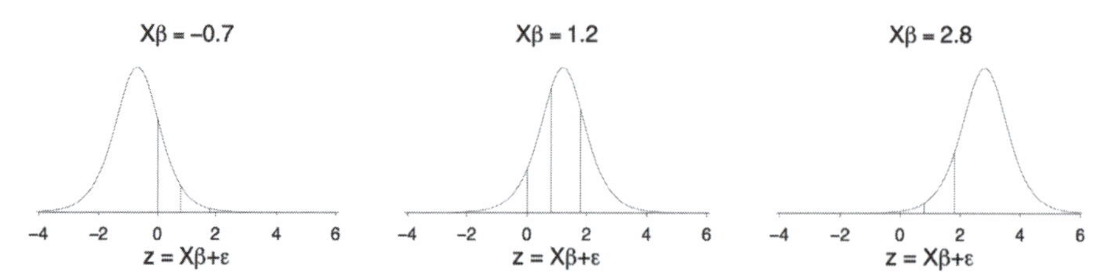


Figure 15.5 *Illustration of cutpoints in an ordered categorical logistic model. In this example, there are $K = 4$ categories and the cutpoints are $c_1 = 0, c_2 = 0.8, c_3 = 1.8$. The three graphs illustrate the distribution of the latent outcome $z$ corresponding to three different values of the linear predictor, $X\beta$. For each, the cutpoints show where the outcome $y$ will equal 1, 2, 3, or 4.*

The parameters $c_k$ (which are called thresholds or *cutpoints*, for reasons that we shall explain shortly) are constrained to increase: $0 = c_1 < c_2 < \cdots < c_{K-1}$, because the probabilities in (15.6) are strictly decreasing (assuming that all $K$ outcomes have nonzero probabilities of occurring). Since $c_1$ is defined to be 0, the model with $K$ categories has $K-2$ free parameters $c_k$ in addition to $\beta$. This makes sense since $K=2$ for the usual logistic regression, for which only $\beta$ needs to be estimated.

The cutpoints $c_2, \ldots, c_{K-1}$ can be estimated using Bayesian or likelihood approaches, simultaneously with the coefficients $\beta$. For some datasets, the parameters can be nonidentified, as with logistic regression for binary data (see Section 14.6), in which case prior distributions can be important.

The expressions in (15.6) can be subtracted to get the probabilities of individual outcomes:

$$\begin{aligned}\Pr(y = k) &= \Pr(y > k-1) \;-\; \Pr(y > k)\\ &= \text{logit}^{-1}(X\beta - c_{k-1}) \;-\; \text{logit}^{-1}(X\beta - c_k).\end{aligned}$$

## Latent variable interpretation with cutpoints

The ordered categorical model is easiest to understand by generalizing the latent variable formulation (13.5) to $K$ categories:

$$\begin{aligned}y_i &= \begin{cases} 1 & \text{if } z_i < 0\\ 2 & \text{if } z_i \in (0, c_2)\\ 3 & \text{if } z_i \in (c_2, c_3)\\ & \ldots\\ K-1 & \text{if } z_i \in (c_{K-2}, c_{K-1})\\ K & \text{if } z_i > c_{K-1}, \end{cases}\\ z_i &= X_i\beta + \epsilon_i, \end{aligned} \qquad (15.7)$$

with independent errors $\epsilon_i$ that have the logistic distribution, as in (13.5).

Figure 15.5 illustrates the latent variable model and shows how the distance between any two adjacent cutpoints $c_{k-1}, c_k$ affects the probability that $y = k$. The lowest and highest categories in (15.7) are unbounded, so if the linear predictor $X\beta$ is high enough, $y$ will almost certainly take on the highest possible value, and if $X\beta$ is low enough, $y$ will almost certainly equal the lowest possible value.

## Example of ordered categorical regression

Example: Storable votes

We illustrate ordered categorical data analysis with a study from experimental economics, on the topic of storable voting.[4] This somewhat complicated example illustrates both the use and potential

[4]Data and code for this example are in the folder `Storable`.

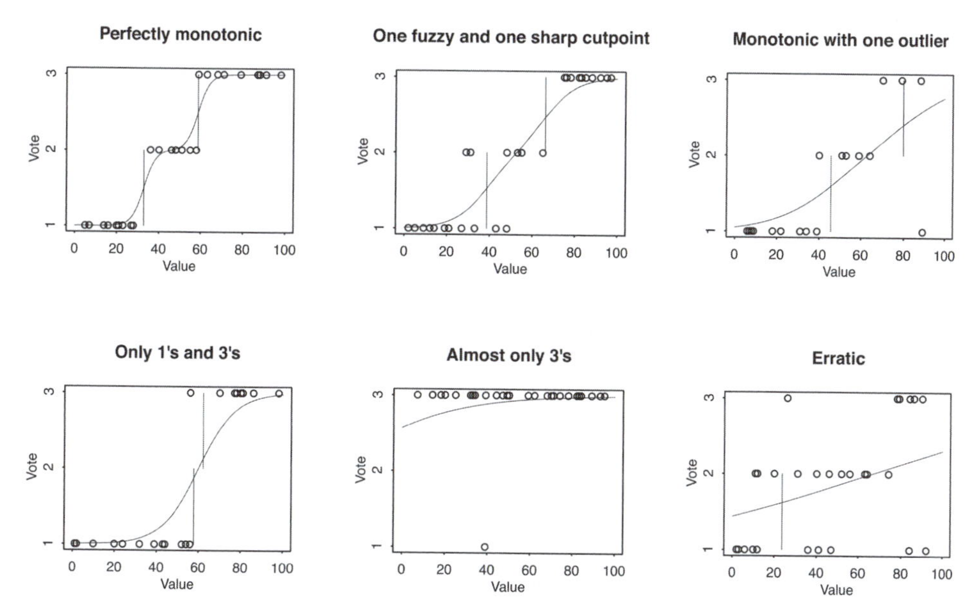


Figure 15.6 *Data from some example individuals in the storable votes study. Vertical lines show estimated cutpoints, and curves show expected responses as estimated using ordered logistic regressions. The two left graphs show data that fit the model reasonably well; the others fit the model in some ways but not perfectly.*

limitations of the standard ordered logistic model. In the experiment under study, college students were recruited to play a series of voting games. In each game, a set of players are brought together to vote on two issues, with the twist being that each player is given a total of 4 votes. On the first issue, a player has the choice of casting 1, 2, or 3 votes, with the remaining votes cast on the second issue. The winning side of each issue is decided by majority vote, at which point the players on the winning side each get positive payoffs, which are drawn from a uniform distribution on the interval $[1, 100]$.

To increase their expected payoffs, players should follow a strategy of spending more of their votes for issues where their potential payoffs are higher. The way this experiment is conducted, the players are told the distribution of possible payoffs, and they are told their potential payoff for each issue just before the vote. Thus, in making the choice of how many votes to cast in the first issue, each player knows his or her potential payoff for that vote only. Then, the players are told their potential payoffs for the second vote, but no choice is involved at this point since they will automatically spend all their remaining votes. Players' strategies can thus be summarized as their choices of initial votes, $y = 1, 2$, or $3$, given their potential payoff, $x$.

Figure 15.6 graphs the responses from six out of the hundred or so students in the experiment, with those six chosen to represent several different patterns of data. We were not surprised to see that responses were generally monotonic—students tend to spend more votes when their potential payoff is higher—but it was interesting to see the variety of approximately monotonic strategies that were chosen.

Most individuals' behaviors can be summarized by three parameters—the cutpoint between votes of 1 and 2, the cutpoint between 2 and 3, and the fuzziness of these divisions. The two cutpoints characterize the chosen monotone strategy, and the sharpness of the divisions indicates the consistency with which the strategy is followed.

**Three parameterizations of the ordered logistic model.** It is convenient to model the responses using an ordered logit, using a parameterization slightly different from that of model (15.7) to match up with our understanding of the monotone strategies. The model is,

$$y_i = \begin{cases} 1 & \text{if } z_i < c_{1.5} \\ 2 & \text{if } z_i \in (c_{1.5}, c_{2.5}) \\ 3 & \text{if } z_i > c_{2.5}, \end{cases}$$
$$z_i \sim \text{logistic}(x_i, \sigma^2). \tag{15.8}$$

In this parameterization, the cutpoints $c_{1.5}$ and $c_{2.5}$ are on the 1–100 scale of the data $x$, and the scale $\sigma$ of the errors $\epsilon$ corresponds to the fuzziness of the cutpoints.

This model has the same number of parameters as the conventional parameterization (15.7)—two regression coefficients have disappeared, while one additional free cutpoint and an error variance have been added. Here is model (15.7) with $K = 3$ categories and one predictor $x$,

$$y_i = \begin{cases} 1 & \text{if } z_i < 0 \\ 2 & \text{if } z_i \in (0, c_2) \\ 3 & \text{if } z_i > c_2, \end{cases}$$
$$z_i = \alpha + \beta x_i + \epsilon_i, \tag{15.9}$$

with independent errors $\epsilon_i \sim \text{logistic}(0, 1)$.

Yet another version of the model keeps the two distinct cutpoints but removes the constant term, $\alpha$; thus,

$$y_i = \begin{cases} 1 & \text{if } z_i < c_{1|2} \\ 2 & \text{if } z_i \in (c_{1|2}, c_{2|3}) \\ 3 & \text{if } z_i > c_{2|3}, \end{cases}$$
$$z_i = \beta x_i + \epsilon_i, \tag{15.10}$$

with independent errors $\epsilon_i \sim \text{logistic}(0, 1)$.

The three models are equivalent, with $z_i/\beta$ in (15.10) and $(z_i - \alpha)/\beta$ in (15.9) corresponding to $z_i$ in (15.8) and the parameters matching up as follows:

| Model (15.8) | Model (15.9) | Model (15.10) |
|---|---|---|
| $c_{1.5}$ | $-\alpha/\beta$ | $-c_{1\|2}/\beta$ |
| $c_{2.5}$ | $(c_2 - \alpha)/\beta$ | $-c_{2\|3}/\beta$ |
| $\sigma$ | $1/\beta$ | $1/\beta$ |

We prefer parameterization (15.8) because we can directly interpret $c_{1.5}$ and $c_{2.5}$ as thresholds on the scale of the input $x$, and $\sigma$ corresponds to the gradualness of the transitions from 1's to 2's and from 2's to 3's. It is sometimes convenient, however, to fit the model using the standard parameterizations (15.9) and (15.10), and so it is helpful to be able to go back and forth between the models.

**Fitting the model in R.** We can fit ordered logit (or probit) models using the `stan_polr` ("proportional odds logistic regression") function in R. We illustrate with data from one of the people in the storable votes study:

```
fit_1 <- stan_polr(factor(vote) ~ value, data=data_401, prior=R2(0.3, "mean"))
print(fit_1)
```

which yields,

```
      Median MAD_SD
value 0.09   0.03

Cutpoints:
    Median MAD_SD
1|2 2.72   1.46
2|3 5.92   2.18
```

This is the fit of the model of the form (15.10), with estimates $\hat{\beta} = 0.09$, $\hat{c}_{1|2} = 2.72$ and $\hat{c}_{2|3} = 5.92$.

If we want to transform to the alternative expression (15.8), we need to linearly rescale by dividing by $\beta$, which yields estimates of $\hat{c}_{1.5} = 2.72/0.09 = 30.2$, $\hat{c}_{2.5} = 5.92/0.09 = 65.8$, and $\hat{\sigma} = 1/0.09 = 11.1$.

**Displaying the fitted model.** Figure 15.6 shows the cutpoints $c_{1.5}, c_{2.5}$ and expected votes $\mathrm{E}(y)$ as a function of $x$, as estimated from the data from each of several students. From the model (15.8), the expected votes can be written as

$$\begin{aligned}\mathrm{E}(y|x) &= 1 * \Pr(y = 1|x) + 2 * \Pr(y = 2|x) + 3 * \Pr(y = 3|x) \\ &= 1 * \left(1 - \operatorname{logit}^{-1}\left(\frac{x - c_{1.5}}{\sigma}\right)\right) + \\ &\quad + 2 * \left(\operatorname{logit}^{-1}\left(\frac{x - c_{1.5}}{\sigma}\right) - \operatorname{logit}^{-1}\left(\frac{x - c_{2.5}}{\sigma}\right)\right) + \\ &\quad + 3 * \operatorname{logit}^{-1}\left(\frac{x - c_{2.5}}{\sigma}\right),\end{aligned} \tag{15.11}$$

where $\operatorname{logit}^{-1}(x) = e^x/(1 + e^x)$ is the logistic curve displayed in Figure 13.1a. Expression (15.11) looks complicated but is easy to program as a function in R:

```
expected <- function (x, c1.5, c2.5, sigma){
  p1.5 <- invlogit((x-c1.5)/sigma)
  p2.5 <- invlogit((x-c2.5)/sigma)
  return((1*(1-p1.5) + 2*(p1.5-p2.5) + 3*p2.5))
}
```

The data, cutpoints, and curves in Figure 15.6 can then be plotted as follows:

```
plot(x, y, xlim=c(0,100), ylim=c(1,3), xlab="Value", ylab="Vote")
lines(rep(c1.5, 2), c(1,2))
lines(rep(c2.5, 2), c(2,3))
curve(expected(x, c1.5, c2.5, sigma), add=TRUE)
```

Having displayed these estimates for individuals, the next step is to study the distribution of the parameters in the population, to understand the range of strategies applied by the students. In this context, the data have a multilevel structure—30 observations for each of several students—and we pursue this example further in our book on multilevel models.

## Alternative approaches to modeling ordered categorical data

Ordered categorical data can be modeled in several ways, including:

- Ordered logit model with $K-1$ cutpoint parameters, as we have just illustrated.
- The same model in probit form.
- Simple linear regression (possibly preceded by a simple transformation of the outcome values). This can be a good idea if the number of categories is large and if they can be considered equally spaced. This presupposes that a reasonable range of the categories is actually used. For example, if ratings are potentially on a 1 to 10 scale but in practice always equal 9 or 10, then a linear model probably will not work well.
- Nested logistic regressions—for example, a logistic regression model for $y = 1$ versus $y = 2, \dots, K$; then, if $y \geq 2$, a logistic regression for $y = 2$ versus $y = 3, \dots, K$; and so on up to a model, if $y \geq K - 1$ for $y = K-1$ versus $y = K$. Separate logistic (or probit) regressions have the advantage of more flexibility in fitting data but the disadvantage of losing the simple latent-variable interpretation of the cutpoint model we have described.
- Finally, robit regression, which we discuss in Section 15.6, is a competitor to logistic regression that accounts for occasional aberrant data such as the outlier in the upper-right plot of Figure 15.6.

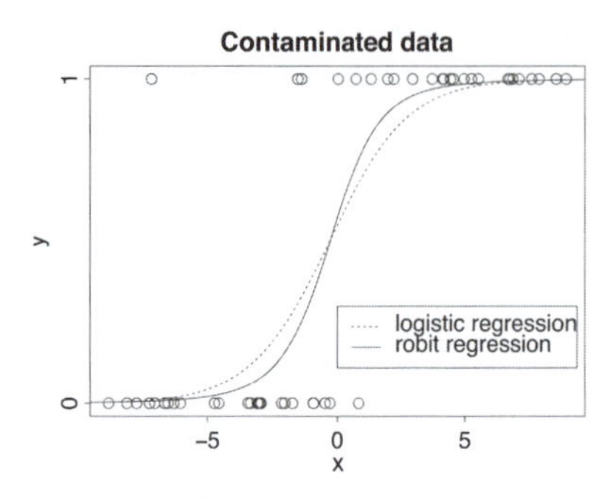


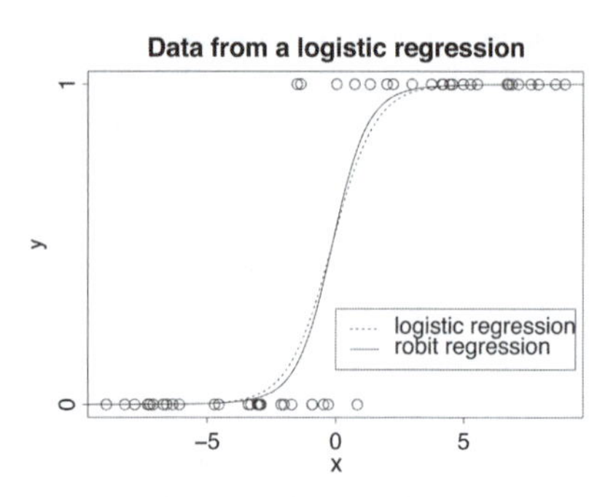


Figure 15.7 *Hypothetical data to be fitted using logistic regression: (a) a dataset with an "outlier" (the unexpected $y = 1$ value near the upper left); (b) data simulated from a logistic regression model, with no outliers. In each plot, the dotted and solid lines show the fitted logit and robit regressions, respectively. In each case, the robit line is steeper—especially for the contaminated data—because it effectively downweights the influence of points that do not appear to fit the model.*

### Unordered categorical regression

As discussed at the beginning of Section 15.5, it is sometimes appropriate to model discrete outcomes as unordered. An example that arose in our research was the well-switching problem. As described in Section 13.7, households with unsafe wells had the option to switch to safer wells. But the actual alternatives are more complicated and can be summarized as: (0) do nothing, (1) switch to an existing private well, (2) switch to an existing community well, (3) install a new well yourself. If these are coded as 0, 1, 2, 3, then we can model $\Pr(y \geq 1), \Pr(y \geq 2 \mid y \geq 1), \Pr(y = 3 \mid y \geq 2)$. Although the four options could be considered to be ordered in some way, it does not make sense to apply the ordered multinomial logit or probit model, since different factors likely influence the three different decisions. Rather, it makes more sense to fit separate logit (or probit) models to each of the three components of the decision: (a) Do you switch or do nothing? (b) If you switch, do you switch to an existing well or build a new well yourself? (c) If you switch to an existing well, is it a private or community well? More about this important category of model can be found in the references at the end of this chapter.

## 15.6 Robust regression using the *t* model

### The *t* distribution instead of the normal

When a regression model can have occasional very large errors, it is generally more appropriate to use a $t$ distribution rather than a normal distribution for the errors. The basic form of the regression is unchanged—$y = X\beta + \epsilon$—but with a different distribution for the $\epsilon$'s and thus a slightly different method for estimating $\beta$ (see the discussion of maximum likelihood estimation in Chapter 7) and a different distribution for predictions. Regressions estimated using the $t$ model are said to be *robust* in that the coefficient estimates are less influenced by individual outlying data points. Regressions with $t$ errors can be fit using `brm` with the `student` family in the `brms` package in R.

### Robit instead of logit or probit

Example: Simulated discrete data with an outlier

Logistic regression (and the essentially equivalent probit regression) are flexible and convenient for modeling binary data, but they can run into problems with outliers. Outliers are usually thought of as extreme observations, but in the context of discrete data, an "outlier" is more of an *unexpected* observation. Figure 15.7a illustrates with data simulated from a logistic regression, with an extreme point switched from 0 to 1. In the context of the logistic model, an observation of $y = 1$ for this value of $x$ would be extremely unlikely, but in real data we can encounter this sort of "misclassification." Hence this graph represents the sort of data to which we might fit a logistic regression, even though this model is not quite appropriate.

For another illustration of a logistic regression with an aberrant data point, see the upper-right plot in Figure 15.6. That is an example with three outcomes; for simplicity, we restrict our attention here to binary outcomes.

Logistic regression can be conveniently "robustified" by generalizing the latent-data formulation (13.5):

$$y_i = \begin{cases} 1 & \text{if } z_i > 0 \\ 0 & \text{if } z_i < 0, \end{cases}$$
$$z_i = X_i\beta + \epsilon_i,$$

to give the latent errors $\epsilon$ a $t$ distribution:

$$\epsilon_i \sim t_\nu\left(0, \frac{\nu - 2}{\nu}\right), \tag{15.12}$$

with the degrees-of-freedom parameter $\nu > 2$ estimated from the data and the $t$ distribution scaled so that its standard deviation equals 1.

The $t$ model for the $\epsilon_i$'s allows the occasional unexpected prediction—a positive value of $z$ for a highly negative value of the linear predictor $X\beta$, or vice versa. Figure 15.7a illustrates with the simulated "contaminated" dataset: the solid line shows $\Pr(y=1)$ as a function of the $x$ for the fitted robit regression, and it is quite a bit steeper than the fitted logistic model. The $t$ distribution effectively downweights the discordant data point so that the model better fits the main part of the data.

Figure 15.7b shows what happens with data that actually come from a logistic model: here, the robit model is close to the logit, which makes sense since it does not find discrepancies.

Mathematically, the robit model can be considered as a generalization of probit and an approximate generalization of logit. Probit corresponds to the degrees of freedom $\nu = \infty$, and logit is close to the robit model with $\nu = 7$.

## 15.7 Constructive choice models

So far, we have considered regression modeling as a descriptive tool for studying how an outcome can be predicted given some input variables. A completely different approach is to model a decision outcome as a balancing of goals or utilities.

xample: rsenic in angladesh

We demonstrate this idea using the example of well switching in Bangladesh (see Section 13.7). How can we understand the relation between distance, arsenic level, and the decision to switch? It makes sense that people with higher arsenic levels would be more likely to switch, but what coefficient values should we expect? Should the relation be on the log or linear scale? The actual health risk is believed to be linear in arsenic concentration; does that mean that a logarithmic model is inappropriate? Such questions can be addressed using a model for individual decisions.

To set up a *choice model*, we must specify a *value function*, which represents the strength of preference for one decision over the other—in this case, the preference for switching as compared to not switching. The value function is scaled so that zero represents indifference, positive values correspond to a preference for switching, and negative values result in not switching. This model is thus similar to the latent-data interpretation of logistic regression (see page 226); and that model is a special case, as we shall see here.

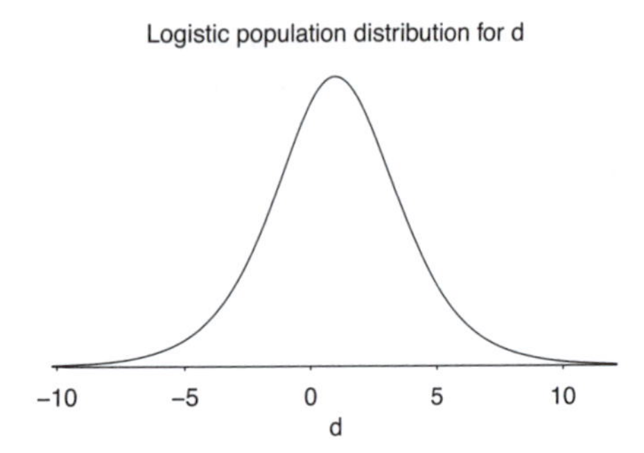


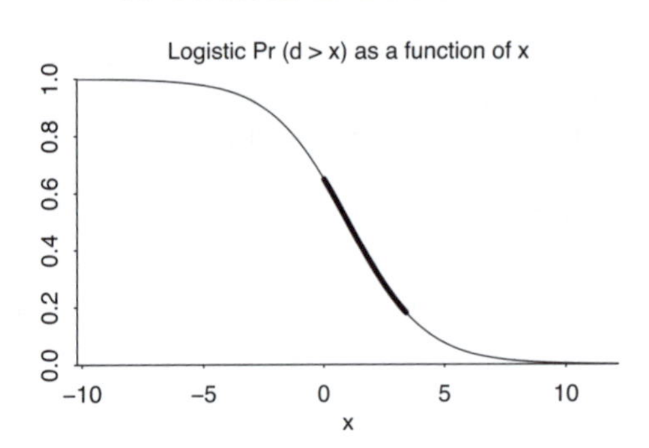


Figure 15.8 *(a) Hypothesized logistic distribution of $d_i = (a_i - b_i)/c_i$ in the population and (b) corresponding logistic regression curve of the probability of switching given distance. These both correspond to the model, $Pr(y_i = 1) = Pr(d_i > x_i) = logit^{-1}(0.61 - 0.62x_i)$. The dark part of the curve in (b) corresponds to the range of x (distance in 100-meter units) in the well-switching data; see Figure 13.8b.*

## Logistic or probit regression as a choice model in one dimension

There are simple one-dimensional choice models that reduce to probit or logit regression with a single predictor, as we illustrate with the model of switching given distance to nearest well. From page 234, the logistic regression is,

```
             Median MAD_SD
(Intercept)  0.61   0.06
dist100     -0.62   0.10
```

Now let us think about switching from first principles as a decision problem. For household $i$, define

- $a_i$: the benefit of switching from an unsafe to a safe well,
- $b_i + c_i x_i$: the cost of switching to a new well a distance $x_i$ away.

We are assuming a utility theory in which the benefit (in reduced risk of disease) can be expressed on the same scale as the cost (the inconvenience of no longer using one's own well, plus the additional effort—proportional to distance—required to carry the water).

**Logit model.** Under the utility model, household $i$ will switch if $a_i > b_i + c_i x_i$. However, we do not have direct measurements of the $a_i$'s, $b_i$'s, and $c_i$'s. All we can learn from the data is the probability of switching as a function of $x_i$; that is,

$$\text{Pr(switch)} = \text{Pr}(y_i = 1) = \text{Pr}(a_i > b_i + c_i x_i), \tag{15.13}$$

treating $a_i, b_i, c_i$ as random variables whose distribution is determined by the (unknown) values of these parameters in the population.

Expression (15.13) can be written as,

$$\text{Pr}(y_i = 1) = \text{Pr}\left(\frac{a_i - b_i}{c_i} > x_i\right),$$

a re-expression that is useful in that it puts all the random variables in the same place and reveals that the population relation between $y$ and $x$ depends on the distribution of $(a - b)/c$ in the population.

For convenience, label $d_i = (a_i - b_i)/c_i$: the net benefit of switching to a neighboring well, divided by the cost per distance traveled to a new well. If $d_i$ has a logistic distribution in the population, and if $d$ is independent of $x$, then $\text{Pr}(y = 1)$ will have the form of a logistic regression on $x$, as we shall show here.

If $d_i$ has a logistic distribution with center $\mu$ and scale $\sigma$, then $d_i = \mu + \sigma\epsilon_i$, where $\epsilon_i$ has the

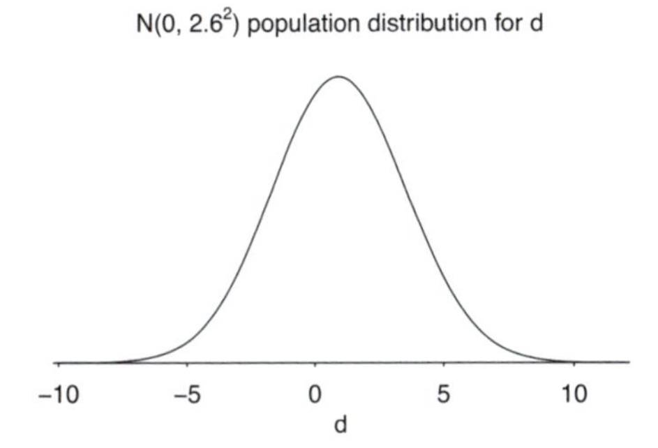


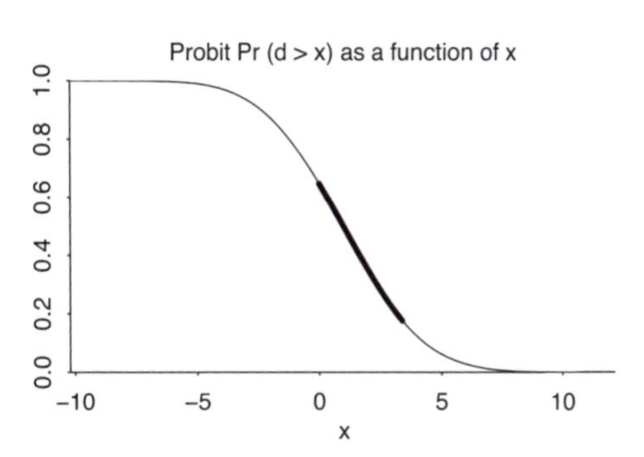


Figure 15.9 *(a) Hypothesized normal distribution of $d_i = (a_i - b_i)/c_i$ with mean 0.98 and standard deviation 2.6 and (b) corresponding probit regression curve of the probability of switching given distance. These both correspond to the model, $Pr(y_i = 1) = Pr(d_i > x_i) = \Phi(0.38 - 0.39x_i)$. Compare to Figure 15.8.*

unit logistic density; see Figure 13.1. Then

$$\begin{aligned}\text{Pr(switch)} &= \text{Pr}(d_i > x) = \text{Pr}\left(\frac{d_i - \mu}{\sigma} > \frac{x - \mu}{\sigma}\right) \\ &= \text{logit}^{-1}\left(\frac{\mu - x}{\sigma}\right) = \text{logit}^{-1}\left(\frac{\mu}{\sigma} - \frac{1}{\sigma}x\right),\end{aligned}$$

which is simply a logistic regression with coefficients $\mu/\sigma$ and $-1/\sigma$. We can then fit the logistic regression and solve for $\mu$ and $\sigma$. For example, the well-switching model, $\text{Pr}(y = 1) = \text{logit}^{-1}(0.61 - 0.62x)$, corresponds to $\mu/\sigma = 0.61$ and $-1/\sigma = -0.62$; thus $\sigma = 1/0.62 = 1.6$ and $\mu = 0.61/0.62 = 0.98$. Figure 15.8 shows the distribution of $d$, along with the curve of $\text{Pr}(d > x)$ as a function of $x$.

**Probit model.** A similar model is obtained by starting with a normal distribution for the utility parameter: $d \sim \text{N}(\mu, \sigma^2)$. In this case,

$$\begin{aligned}\text{Pr(switch)} &= \text{Pr}(d_i > x) = \text{Pr}\left(\frac{d_i - \mu}{\sigma} > \frac{x - \mu}{\sigma}\right) \\ &= \Phi\left(\frac{\mu - x}{\sigma}\right) = \Phi\left(\frac{\mu}{\sigma} - \frac{1}{\sigma}x\right),\end{aligned}$$

which is simply a probit regression. The model $\text{Pr}(y = 1) = \Phi(0.38 - 0.39x)$ corresponds to $\mu/\sigma = 0.38$ and $-1/\sigma = -0.39$; thus $\sigma = 1/0.39 = 2.6$ and $\mu = 0.38/0.39 = 0.98$. Figure 15.9 shows this model, which is nearly identical to the logistic model shown in Figure 15.8.

## Choice models, discrete data regressions, and latent data

Logistic regression and generalized linear models are usually set up as methods for estimating the probabilities of different outcomes $y$ given predictors $x$. A fitted model represents an entire population, with the "error" in the model coming in through probabilities that are not simply 0 or 1 (hence, the gap between data points and fitted curves in graphs such as Figure 13.8b.

In contrast, choice models are defined at the level of the individual, as we can see in the well-switching example, where each household $i$ has, along with its own data $X_i$, $y_i$, its own parameters $a_i, b_i, c_i$ that determine its utility function and thus its decision of whether to switch.

## Logistic or probit regression as a choice model in multiple dimensions

We can extend the well-switching model to multiple dimensions by considering the arsenic level of the current well as a factor in the decision.

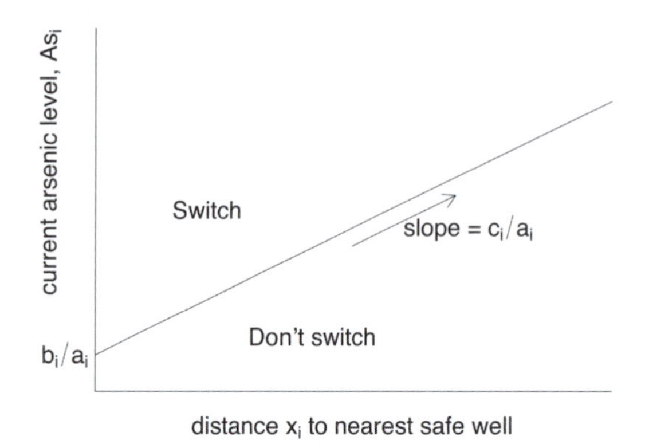


Figure 15.10 *Decision options for well switching given arsenic level of current well and distance to the nearest safe well, based on the decision rule: switch if* $a_i * (\text{As})_i > b_i + cx_i$.

- $a_i * (\text{As})_i$ = the benefit of switching from an unsafe well with arsenic level $\text{As}_i$ to a safe well. (It makes sense for the benefit to be proportional to the current arsenic level, because risk is believed to be essentially proportional to cumulative exposure to arsenic.)
- $b_i + c_i x_i$ = the cost of switching to a new well a distance $x_i$ away.

Household $i$ should then switch if $a_i * (\text{As})_i > b_i + cx_i$—the decision thus depends on the household's arsenic level $(\text{As})_i$, its distance $x_i$ to the nearest well, and its utility parameters $a_i, b_i, c_i$.

Figure 15.10 shows the decision space for an individual household, depending on its arsenic level and distance to the nearest safe well. Given $a_i, b_i, c_i$, the decision under this model is deterministic. However, $a_i, b_i, c_i$ are not directly observable—all we see are the decisions ($y_i = 0$ or 1) for households, given their arsenic levels $\text{As}_i$ and distances $x_i$ to the nearest safe well.

Certain distributions of $(a, b, c)$ in the population reduce to the fitted logistic regression, for example, if $a_i$ and $c_i$ are constants and $b_i/a_i$ has a logistic distribution that is independent of $(\text{As})_i$ and $x_i$. More generally, choice models reduce to logistic regressions if the factors come in additively, with coefficients that do not vary in the population, and if there is a fixed cost ($b_i$ in this example) that has a logistic distribution in the population.

Other distributions of $(a, b, c)$ are possible and could be fit using a Bayesian model, using techniques that go beyond the scope of this book.

### Insights from decision models

A choice model can give us some insight even if we do not formally fit it. For example, in fitting logistic regressions, we found that distance worked well as a linear predictor, whereas arsenic level fit better on the logarithmic scale. A simple utility analysis would suggest that both these factors should come in linearly, and the transformation for arsenic suggests that people are (incorrectly) perceiving the risks on a logarithmic scale—seeing the difference between 4 to 8, say, as no worse than the difference between 1 and 2. (In addition, our residual plot showed the complication that people seem to underestimate risks from arsenic levels very close to 0.5. And behind this is the simplifying assumption that all wells with arsenic levels below 0.5 are "safe.")

We can also use the utility model to interpret the coefficient for education in the model—more educated people are more likely to switch, indicating that their costs of switching are lower, or their perceived benefits from reducing arsenic exposure are higher. Interactions correspond to dependence among the latent utility parameters in the population.

The model could also be elaborated to consider the full range of individual options, which include doing nothing, switching to an existing private well, switching to an existing community well, or digging a new private well. The decision depends on the cost of walking, perception of health risks, financial resources, and future plans.

## 15.8 Going beyond generalized linear models

The models we have considered so far can handle many regression problems in practice. For continuous data, we start with linear regression with normal errors, consider appropriate transformations and interactions as discussed in Chapter 12, and switch to a $t$ error model for data with occasional large errors. For binary data we can use logit, probit, or perhaps robit, building and checking the model as discussed in Chapters 13 and 14. For count data, our starting points are the beta-binomial (overdispersed binomial) and negative binomial (overdispersed Poisson) models, and for discrete outcomes with more than two categories we can fit ordered or unordered multinomial logit or probit regression. Here we briefly describe some situations where it is helpful to consider other options.

### Extending existing models

The usual linear regression model, $y_i \sim \mathrm{N}(a + bx_i, \sigma^2)$, assumes a constant error standard deviation, $\sigma$; this assumption is called homoscedasticity and is mentioned in Section 11.1. We can also allow the residual standard deviation to vary (heteroscedasticity) and build a model for the error variance, for example, as $y_i \sim \mathrm{N}(a + bx_i, e^{c+dx_i})$. Such a model can be easily fit using `brm`. For example, for the earnings data we could fit the model, `bf(log(earn) ~ height + male, sigma ~ male)` which allows different error variances for women and men. Similarly we can extend other regression models. For example, for negative binomial regression, unequal overdispersion can be modeled as $y_i \sim \text{negative binomial}(e^{a+bx_i}, e^{c+dx_i})$.

Any standard set of models is constrained by its functional form. In Chapter 12, we discussed how linear regression can be extended by transforming predictor and outcome variables in different ways. For discrete regression models, it generally does not make sense to transform the outcome variable; instead, the predictive model can be changed. There are many such examples; we illustrate here with a class of models that generalizes logistic regression.

In logistic regression, $\Pr(y = 1)$ asymptotes to 0 and 1 at the extremes of the linear predictor, $X\beta$. But suppose we want other asymptotes? One way this can arise is in multiple-choice testing. If a question has four options, then a student should be able to have a 1/4 chance of success just by guessing. This suggests the following model: $\Pr(y = 1) = 0.25 + 0.75\,\mathrm{logit}^{-1}(X\beta)$.

Another time we might want limiting probabilities other than 0 or 1 is to allow the occasionally completely unexpected prediction. The new model, $\Pr(y = 1) = \epsilon_1 + (1 - (\epsilon_0 + \epsilon_1))\,\mathrm{logit}^{-1}(X\beta)$, generalizes logistic regression by allowing "contamination" of the data by 0's and 1's completely at random at rates $\epsilon_0$ and $\epsilon_1$, which might be set to low values such as 0.01. Fitting this sort of model allows the logistic probability function to be fit to the mass of the data without being unduly influenced by these aberrant data points, which might arise in survey responses, for example.

These are not generalized linear models and cannot be fit with `stan_glm` but can be programmed directly without much difficulty in Stan, although this is beyond the scope of this book.

### Mixed discrete/continuous data

Example: Earnings and height

Earnings is an example of an outcome variable with both discrete and continuous aspects. In our earnings and height regressions in Chapter 12, we preprocessed the data by removing all respondents with zero earnings. In general, however, it can be appropriate to model a variable such as earnings in two steps: first a logistic regression for $\Pr(y > 0)$ fit to all the data, then a linear regression on $\log(y)$, fit just to the subset of the data for which $y > 0$. Predictions for such a model then must be done in two steps, most conveniently by first simulating the yes/no outcome of whether $y$ is positive, and then simulating $y$ for the positive cases.

When modeling an outcome in several steps, programming effort is sometimes required to convert inferences onto the original scale of the data. We demonstrate with the example of predicting earnings from height and sex.[5] We first fit a logistic regression to predict whether earnings are positive:

```
fit_1a <- stan_glm((earn > 0) ~ height + male, family=binomial(link="logit"),
  data=earnings)
```

yielding,

```
            Median MAD_SD
(Intercept) -2.97   1.95
height       0.07   0.03
male         1.66   0.31
```

We then fit a log regression model to the respondents with positive earnings:

```
fit_1b <- stan_glm(log(earn) ~ height + male, data=earnings, subset=earn>0)
```

yielding the fit,

```
            Median MAD_SD
(Intercept) 7.97   0.51
height      0.02   0.01
male        0.37   0.06

Auxiliary parameter(s):
      Median MAD_SD
sigma 0.87   0.02
```

Thus, for example, a 66-inch-tall woman has an estimated probability $\text{logit}^{-1}(-2.97 + 0.07 * 66 + 1.66 * 0) = 0.84$, or an 84% chance, of having positive earnings. If her earnings are positive, their predicted value is $\exp(7.97 + 0.02 * 66 + 0.37 * 0) = 10\,800$. Combining these gives a mixture of a spike at 0 and a lognormal distribution, which is most easily manipulated using simulations, as we discuss in Sections 15.8 and 17.5. This kind of model can also be fit using `brm` using the `hurdle_lognormal` family.

### Latent-data models

Another way to model mixed data is through latent data, for example, positing an "underlying" income level $z_i$—the income that person $i$ would have if he or she were employed—that is observed only if $y_i > 0$. *Tobit regression* is one such model that is popular in econometrics.

### Cockroaches and the zero-inflated negative binomial model

The binomial and Poisson models, and their overdispersed generalizations, all can be expressed in terms of an underlying continuous probability or rate of occurrence of an event. Sometimes, however, the underlying rate itself has discrete aspects.

Example: Roaches

For example, in the study of cockroach infestation in city apartments discussed on page 268, each apartment $i$ was set up with traps for several days. We label $u_i$ as the number of trap-days and $y_i$ as the number of cockroaches trapped. With a goal of predicting cockroach infestation given predictors $X$ (including income and ethnicity of the apartment dwellers, indicators for neighborhood, and measures of quality of the apartment), we would start with the model,

$$y_i \sim \text{negative-binomial}\,(u_i e^{X_i \beta}, \phi). \tag{15.14}$$

It is possible, however, for the data to have more zeroes (that is, apartments $i$ with cockroach counts $y_i = 0$) than predicted by this model. A natural explanation would then be that some apartments have

[5]Data and code for this example are in the folder `Earnings`.

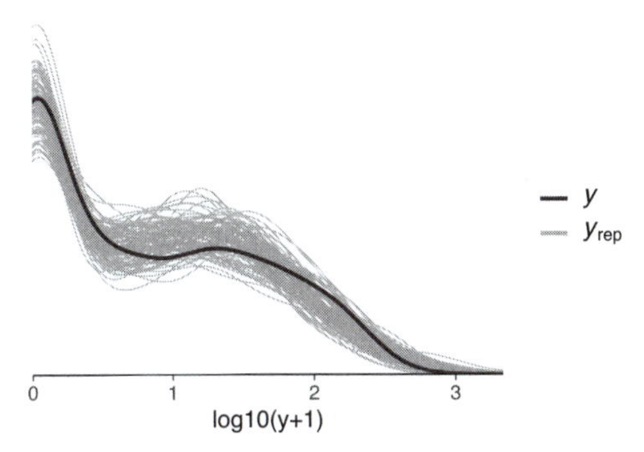


Figure 15.11 *Posterior predictive checks comparing the density of the roaches data (dark line) to 100 predictive replications (gray lines) of replicated data from (a fitted zero-inflated negative binomial model. The x axis is on the scale of* $\log 10(y+1)$ *to better show the whole scale of the data. The model shows better fit than the Poisson and negative binomial models in Figure 15.3.*

truly a zero (or very near-zero) rate of cockroaches, whereas others simply have zero counts from the discreteness of the data. The *zero-inflated* regression places (15.14) into a *mixture model*:

$$y_i \begin{cases} = 0, \text{ if } S_i = 0, \\ \sim \text{negative-binomial}\,(u_i e^{X_i\beta}, \phi), \text{ if } S_i = 1. \end{cases}$$

Here, $S_i$ is an indicator of whether apartment $i$ has any cockroaches at all, and it could be modeled using logistic regression:

$$\Pr(S_i = 1) = \text{logit}^{-1}(X_i\gamma),$$

where $\gamma$ is a new set of regression coefficients for this part of the model. This two-stage model can be fit in one line of code using `brm` using the `zero_inflated_negbinomial` family or by writing the model directly in Stan. Figure 15.11 shows a density estimate of $\log_{10}(\text{count}+1)$ and the corresponding model-based posterior predictive distribution for 100 replications. The zero-inflated negative binomial model predictions are visually better than with negative binomial model shown in Figure 15.3b. The LOO log score comparison also supports better predictive performance for the zero-inflated negative binomial model.[6]

## Prediction and Bayesian inference

The function `rnegbin` in the `MASS` package samples from the negative binomial distribution. Another option is to sample first from the gamma, then the Poisson. For the overdispersed binomial model, simulations from the beta-binomial distribution can be obtained by drawing first from the beta distribution, then the binomial.

Simulation is the easiest way of summarizing inferences from more complex models. For example, as discussed in Section 15.8, we can model earnings given height in two steps:

$$\Pr(\text{earnings} > 0) = \text{logit}^{-1}(-3.97 + 0.08 * \text{height} + 1.70 * \text{male}),$$
$$\text{If earnings} > 0, \text{ then earnings} = e^{8.15 + 0.02*\text{height} + 0.43*\text{male} + \epsilon}$$

with the error term $\epsilon$ having a normal distribution with mean 0 and standard deviation 0.88.

To demonstrate the predictive distribution, we simulate the earnings of a randomly chosen 68-inch tall woman using the models `fit_1a` and `fit_1b` from page 284:

[6]Data and code for this example are in the folder `Roaches`.

```
new <- data.frame(height=68, male=0)
pred_1a <- posterior_predict(fit_1a, newdata=new)
pred_1b <- posterior_predict(fit_1b, newdata=new)
pred <- ifelse(pred_1a==1, exp(pred_1b), 0)
```

The result is a vector of length `n_sims` representing the predictive distribution, including some zeroes and some positive values. To understand this process further, you can perform the probabilistic prediction "manually" by simulating from the binomial and normal distributions.

## 15.9 Bibliographic note

The concept of the generalized linear model was introduced by Nelder and Wedderburn (1972) and developed further, with many examples, by McCullagh and Nelder (1989). Dobson (2001) is an accessible introductory text. For more on overdispersion, see Anderson (1988) and Liang and McCullagh (1993). Fienberg (1977) and Agresti (2002) are other useful references.

The police stops example comes from Spitzer (1999) and Gelman, Fagan, and Kiss (2007). The death penalty example comes from Gelman, Liebman, et al. (2004). Models for traffic accidents are discussed by Chapman (1973) and Hauer, Ng, and Lovell (1988).

Maddala (1983) presents discrete-data regressions and choice models from an econometric perspective, and McCullagh (1980) considers general forms for latent-parameter models for ordered data. Amemiya (1981) discusses the factor of 1.6 for converting from logit to probit coefficients.

Walker and Duncan (1967) introduce the ordered logistic regression model, and Imai and van Dyk (2003) discuss the models underlying multinomial logit and probit regression. The storable votes example comes from Casella, Gelman, and Palfrey (2006). See Agresti (2002) and Imai and van Dyk (2003) for more on categorical regression models, ordered and unordered.

Robust regression using the $t$ distribution is discussed by Zellner (1976) and Lange, Little, and Taylor (1989), and the robit model is introduced by Liu (2004). See Stigler (1977) and Mosteller and Tukey (1977) for further discussions of robust inference from an applied perspective. Wiens (1999) and Newton et al. (2001) discuss the gamma and lognormal models for positive continuous data. For generalized additive models and other nonparametric methods, see Hastie and Tibshirani (1990) and Hastie, Tibshirani, and Friedman (2009).

Connections between logit/probit regressions and choice models have been studied in psychology, economics, and political science; some important references are Thurstone (1927a, b), Wallis and Friedman (1942), Mosteller (1951), Bradley and Terry (1952), and McFadden (1973). Tobit models are named after Tobin (1958) and are covered in econometrics texts such as Woolridge (2001).

## 15.10 Exercises

15.1 *Poisson and negative binomial regression*: The folder `RiskyBehavior` contains data from a randomized trial targeting couples at high risk of HIV infection. The intervention provided counseling sessions regarding practices that could reduce their likelihood of contracting HIV. Couples were randomized either to a control group, a group in which just the woman participated, or a group in which both members of the couple participated. One of the outcomes examined after three months was "number of unprotected sex acts."

(a) Model this outcome as a function of treatment assignment using a Poisson regression. Does the model fit well? Is there evidence of overdispersion?

(b) Next extend the model to include pre-treatment measures of the outcome and the additional pre-treatment variables included in the dataset. Does the model fit well? Is there evidence of overdispersion?

(c) Fit a negative binomial (overdispersed Poisson) model. What do you conclude regarding effectiveness of the intervention?

(d) These data include responses from both men and women from the participating couples. Does this give you any concern with regard to our modeling assumptions?

15.2 *Offset in a Poisson or negative binomial regression*: Explain why putting the logarithm of the exposure into a Poisson or negative binomial model as an offset, is equivalent to including it as a regression predictor, but with its coefficient fixed to the value 1.

15.3 *Binomial regression*: Redo the basketball shooting example on page 270, making some changes:

(a) Instead of having each player shoot 20 times, let the number of shots per player vary, drawn from the uniform distribution between 10 and 30.

(b) Instead of having the true probability of success be linear, have the true probability be a logistic function, set so that Pr(success) = 0.3 for a player who is 5'9" and 0.4 for a 6' tall player.

15.4 *Multinomial logit*: Using the individual-level survey data from the 2000 National Election Study (data in folder NES), predict party identification (which is on a five-point scale) using ideology and demographics with an ordered multinomial logit model.

(a) Summarize the parameter estimates numerically and also graphically.

(b) Explain the results from the fitted model.

(c) Use a binned residual plot to assess the fit of the model.

15.5 *Comparing logit and probit*: Take one of the data examples from Chapter 13 or 14. Fit these data using both logit and probit models. Check that the results are essentially the same after scaling by factor of 1.6 (see Figure 15.4).

15.6 *Comparing logit and probit*: Construct a dataset where the logit and probit models give clearly *different* estimates, not just differing by a factor of 1.6.

15.7 *Tobit model for mixed discrete/continuous data*: Experimental data from the National Supported Work example are in the folder Lalonde. Use the treatment indicator and pre-treatment variables to predict post-treatment (1978) earnings using a Tobit model. Interpret the model coefficients.

15.8 *Robust linear regression using the t model*: The folder Congress has the votes for the Democratic and Republican candidates in each U.S. congressional district in 1988, along with the parties' vote proportions in 1986 and an indicator for whether the incumbent was running for reelection in 1988. For your analysis, just use the elections that were contested by both parties in both years.

(a) Fit a linear regression using stan_glm with the usual normal-distribution model for the errors predicting 1988 Democratic vote share from the other variables and assess model fit.

(b) Fit the same sort of model using the brms package with a $t$ distribution, using the brm function with the student family. Again assess model fit.

(c) Which model do you prefer?

15.9 *Robust regression for binary data using the robit model*: Use the same data as the previous example with the goal instead of predicting for each district whether it was won by the Democratic or Republican candidate.

(a) Fit a standard logistic or probit regression and assess model fit.

(b) Fit a robit regression and assess model fit.

(c) Which model do you prefer?

15.10 *Logistic regression and choice models*: Using the individual-level survey data from the election example described in Section 10.9 (data in the folder NES), fit a logistic regression model for the choice of supporting Democrats or Republicans. Then interpret the output from this regression in terms of a utility/choice model.

15.11 *Multinomial logistic regression and choice models*: Repeat the previous exercise but now with three options: Democrat, no opinion, Republican. That is, fit an ordered logit model and then express it as a utility/choice model.

15.12 *Spatial voting models*: Suppose that competing political candidates $A$ and $B$ have positions that can be located spatially in a one-dimensional space (that is, on a line). Suppose that voters have "ideal points" with regard to these positions that are normally distributed in this space, defined so that voters will prefer candidates whose positions are closest to their ideal points. Further suppose that voters' ideal points can be modeled as a linear regression given inputs such as party identification, ideology, and demographics.

(a) Write this model in terms of utilities.

(b) Express the probability that a voter supports candidate $S$ as a probit regression on the voter-level inputs.

See Erikson and Romero (1990) and Clinton, Jackman, and Rivers (2004) for background.

15.13 *Multinomial choice models*: Pardoe and Simonton (2008) fit a discrete choice model to predict winners of the Academy Awards. Their data are in the folder `AcademyAwards`.

(a) Fit your own model to these data.

(b) Display the fitted model on a plot that also shows the data.

(c) Make a plot displaying the uncertainty in inferences from the fitted model.

15.14 *Model checking for count data*: The folder `RiskyBehavior` contains data from a study of behavior of couples at risk for HIV; see Exercise 15.1.

(a) Fit a Poisson regression predicting number of unprotected sex acts from baseline HIV status. Perform predictive simulation to generate 1000 datasets and record the percentage of observations that are equal to 0 and the percentage that are greater than 10 (the third quartile in the observed data) for each. Compare these to the observed value in the original data.

(b) Repeat (a) using a negative binomial (overdispersed Poisson) regression.

(c) Repeat (b), also including ethnicity and baseline number of unprotected sex acts as inputs.

15.15 *Summarizing inferences and predictions using simulation*: Exercise 15.7 used a Tobit model to fit a regression with an outcome that had mixed discrete and continuous data. In this exercise you will revisit these data and build a two-step model: (1) logistic regression for zero earnings versus positive earnings, and (2) linear regression for level of earnings given earnings are positive. Compare predictions that result from each of these models with each other.

15.16 *Average predictive comparisons*:

(a) Take the roach model of well switching from Section 15.2 and estimate the average predictive comparisons for logarithm of roach count, for each of the input variables in the model.

(b) Do the same thing but estimating average predictive comparisons for roach count itself.

15.17 *Learning from social science data*: The General Social Survey (GSS) has been conducted in the United States every two years since 1972.

(a) Go to the GSS website and download the data. Consider a question of interest that was asked in many rounds of the survey and which is a count variable. Decide how you will handle nonresponse in your analysis.

(b) Make a graph of the average response of this count variable over time, each year giving $\pm 1$ standard error bounds, as in Figure 4.3, computing the standard error for each year as $\mathrm{sd}(y)/\sqrt{n}$ using the data from that year.

(c) Set up a negative binomial regression for this outcome given predictors for age, sex, education, and ethnicity. Fit the model separately for each year that the question was asked, and make a grid of plots with the time series of coefficient estimates $\pm$ standard errors over time.

(d) Discuss the results and how you might want to expand your model to answer some social science question of interest.

15.18 *Working through your own example*: Continuing the example from the final exercises of the earlier chapters, fit a generalized linear model that is not a logistic regression, graph the data and fitted model, and interpret the estimated parameters and their uncertainties.

# Part 4: Before and after fitting a regression

# Chapter 16

# Design and sample size decisions

This chapter is a departure from the rest of the book, which focuses on data analysis: building, fitting, understanding, and evaluating models fit to existing data. In the present chapter, we consider the design of studies, in particular asking the question of what sample size is required to estimate a quantity of interest to some desired precision. We focus on the paradigmatic inferential tasks of estimating population averages, proportions, and comparisons in sample surveys, or estimating treatment effects in experiments and observational studies. However, the general principles apply for other inferential goals such as prediction and data reduction. We present the relevant algebra and formulas for sample size decisions and demonstrating with a range of examples, but we also criticize the standard design framework of "statistical power," which when studied naively yields unrealistic expectations of success and can lead to the design of ineffective, noisy studies. As we frame it, the goal of design is not to attain statistical significance with some high probability, but rather to have a sense—before and after data have been collected—about what can realistically be learned from statistical analysis of an empirical study.

## 16.1 The problem with statistical power

Statistical *power* is defined as the probability, before a study is performed, that a particular comparison will achieve "statistical significance" at some predetermined level (typically a $p$-value below 0.05), given some assumed true effect size. A power analysis is performed by first hypothesizing an effect size, then making some assumptions about the variation in the data and the sample size of the study to be conducted, and finally using probability calculations to determine the chance of the $p$-value being below the threshold.

The conventional view is that you should avoid low-power studies because they are unlikely to succeed. This, for example, comes from an influential paper in criminology:

> Statistical power provides the most direct measure of whether a study has been designed to allow a fair test of its research hypothesis. When a study is underpowered it is unlikely to yield a statistically significant result even when a relatively large program or intervention effect is found.

This statement is correct but too simply presents statistical significance as a goal.

To see the problem with aiming for statistical significance, suppose that a study is low power but can be performed for free, or for a cost that it is very low compared to the potential benefits that would arise from a research success. Then a researcher might conclude that a lower-power study is still worth doing, that it is a gamble worth undertaking.

The traditional power threshold is 80%; funding agencies are reluctant to approve studies that are not deemed to have at least an 80% chance of obtaining a statistically significant result. But under a simple cost-benefit calculation, there would be cases where 50% power, or even 10% power, would suffice, for simple studies such as psychology experiments where human and dollar costs are low. Hence, when costs are low, researchers are often inclined to roll the dice, on the belief that a successful finding could potentially bring large benefits (to society as well as to the researcher's career). But this is not necessarily a good idea, as we discuss next.

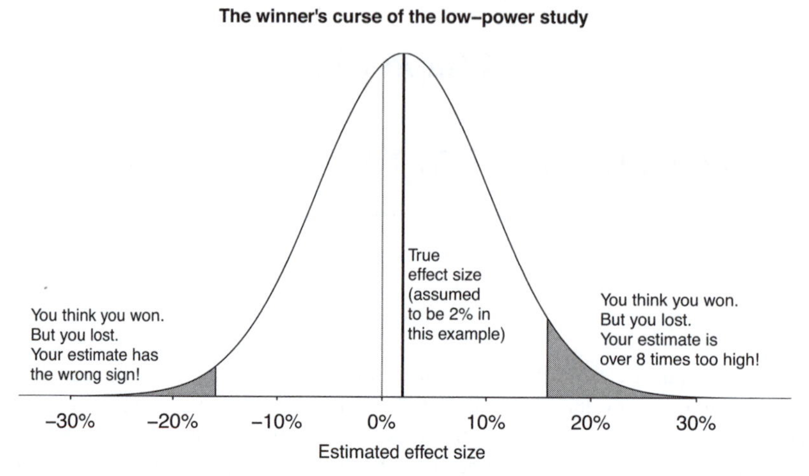


Figure 16.1 *When the effect size is small compared to the standard error, statistical power is low. In this diagram, the bell-shaped curve represents the distribution of possible estimates, and the gray shaded zones correspond to estimates that are "statistically significant" (at least two standard errors away from zero). In this example, statistical significance is unlikely to be achieved, but in the rare cases where it does happen, it is highly misleading: there is a large chance the estimate has the wrong sign (a type S error) and, in any case, the magnitude of the effect size will be vastly overstated (a type M error) if it happens to be statistically significant. Thus, what would naively appear to be a "win" or a lucky draw—a statistically significant result from a low-power study—is, in the larger sense, a loss to science and to policy evaluation.*

## The winner's curse in low-power studies

The problem with the conventional reasoning is that, in a low-power study, the seeming "win" of statistical significance can actually be a trap. Economists speak of a "winner's curse" in which the highest bidder in an auction will, on average, be overpaying. Research studies—even randomized experiments—suffer from a similar winner's curse, that by focusing on comparisons that are statistically significant, we (the scholarly community as well as individual researchers) get a systematically biased and over-optimistic picture of the world.

Put simply, when signal is low and noise is high, statistically significant patterns in data are likely to be wrong, in the sense that the results are unlikely to replicate.

To put it in technical terms, statistically significant results are subject to type M and type S errors, as described in Section 4.4. Figure 16.1 illustrates for a study where the true effect could not realistically be more than 2 percentage points and is estimated with a standard error of 8.1 percentage points. We can examine the statistical properties of the estimate using the normal distribution: conditional on it being statistically significant (that is, at least two standard errors from zero), the estimate has at least a 24% probability of being in the wrong direction and is, by necessity, over 8 times larger than the true effect.

A study with these characteristics has essentially no chance of providing useful information, and we can say this even before the data have been collected. Given the numbers above for standard error and possible effect size, the study has a power of at most 6% (see Exercise 16.4), but it would be misleading to say it has even a 6% chance of success. From the perspective of scientific learning, the real failures are the 6% of the time that the study appears to succeed, in that these correspond to ridiculous overestimates of treatment effects that are likely to be in the wrong direction as well. In such an experiment, to win is to lose.

Thus, a key risk for a low-power study is not so much that it has a small chance of succeeding, but rather that an apparent success merely masks a larger failure. Publication of noisy findings in turn can contribute to the replication crisis when these fragile claims collapse under more careful analysis or do not show up in attempted replications, as discussed in Section 4.5.

### Hypothesizing an effect size

The other challenge is that any power analysis or sample size calculations is conditional on an assumed effect size, and this is something that is the target of the study and is thus never known ahead of time.

There are different ways to choose an effect size for performing an analysis of a planned study design. One strategy, which we demonstrate in Section 16.5, is to try a range of values consistent with the previous literature on the topic. Another approach is to decide what magnitude of effect would be of practical interest: for example, in a social intervention we might feel that we are only interested in pursuing a particular treatment if it increases some outcome by at least 10%; we could then perform a design analysis to see what sample size would be needed to reliably detect an effect of that size.

One common practice that we do *not* recommend is to make design decisions based on the estimate from a single noisy study. Section 16.3 gives an example of how one can use a patchwork of information from earlier studies to make informed judgments about statistical power and sample size.

## 16.2 General principles of design, as illustrated by estimates of proportions

### Effect sizes and sample sizes

In designing a study, it is generally better, if possible, to double the effect size $\theta$ than to double the sample size $n$, since standard errors of estimation decrease with the square root of the sample size. This is one reason, for example, why potential toxins are tested on animals at many times their exposure levels in humans; see Exercise 16.8.

Studies are designed in several ways to maximize effect size:

- In drug studies, setting doses as low as ethically possible in the control group and as high as ethically possible in the experimental group.
- To the extent possible, choosing individuals that are likely to respond strongly to the treatment. For example, an educational intervention in schools might be performed on poorly performing classes in each grade, for which there will be more room for improvement.

In practice, this advice cannot be followed completely. Sometimes it can be difficult to find an intervention with *any* noticeable positive effect, let alone to design one where the effect would be doubled. Also, when treatments in an experiment are set to extreme values, generalizations to more realistic levels can be suspect. Further, treatment effects discovered on a sensitive subgroup may not generalize to the entire population. But, on the whole, conclusive effects on a subgroup are generally preferred to inconclusive but more generalizable results, and so conditions are usually set up to make effects as large as possible.

### Published results tend to be overestimates

There are various reasons why we would typically expect future effects to be smaller than published estimates. First, as noted just above, interventions are often tested on people and in scenarios where they will be most effective—indeed, this is good design advice—and effects will be smaller in the general population "in the wild." Second, results are more likely to be reported and more likely to be published when they are "statistically significant," which leads to overestimation: type M errors, as discussed in Section 4.4. Some understanding of the big picture is helpful when considering how to interpret the results of published studies, even beyond the uncertainty captured in the standard error.

## Design calculations

Before data are collected, it can be useful to estimate the precision of inferences that one expects to achieve with a given sample size, or to estimate the sample size required to attain a certain precision. This goal is typically set in one of two ways:

- Specifying the standard error of a parameter or quantity to be estimated, or
- specifying the probability that a particular estimate will be "statistically significant," which typically is equivalent to ensuring that its 95% confidence interval will exclude the null value.

In either case, the sample size calculation requires assumptions that typically cannot really be tested until the data have been collected. Sample size calculations are thus inherently hypothetical.

## Sample size to achieve a specified standard error

To understand these two kinds of calculations, consider the simple example of estimating the proportion of the population who support the death penalty (under a particular question wording). Suppose we suspect the population proportion is around 60%. First, consider the goal of estimating the true proportion $p$ to an accuracy (that is, standard error) of no worse than 0.05, or 5 percentage points, from a simple random sample of size $n$. The standard error of the mean is $\sqrt{p(1-p)/n}$. Substituting the guessed value of 0.6 for $p$ yields a standard error of $\sqrt{0.6 * 0.4/n} = 0.49/\sqrt{n}$, and so we need $0.49/\sqrt{n} \leq 0.05$, or $n \geq 96$. More generally, we do not know $p$, so we would use a conservative standard error of $\sqrt{0.5 * 0.5/n} = 0.5/\sqrt{n}$, so that $0.5/\sqrt{n} \leq 0.05$, or $n \geq 100$.

## Sample size to achieve a specified probability of obtaining statistical significance

Second, suppose we have the goal of demonstrating that more than half the population supports the death penalty—that is, that $p > 1/2$—based on the estimate $\hat{p} = y/n$ from a sample of size $n$. As above, we shall evaluate this under the hypothesis that the true proportion is $p = 0.60$, using the conservative standard error for $\hat{p}$ of $\sqrt{0.5 * 0.5/n} = 0.5/\sqrt{n}$. The 95% confidence interval for $p$ is $[\hat{p} \pm 1.96 * 0.5/\sqrt{n}]$, and classically we would say we have demonstrated that $p > 1/2$ if the interval lies entirely above 1/2; that is, if $\hat{p} > 0.5 + 1.96 * 0.5/\sqrt{n}$. The estimate must be at least 1.96 standard errors away from the comparison point of 0.5.

A simple, but not quite correct, calculation, would set $\hat{p}$ to the hypothesized value of 0.6, so that the requirement is $0.6 > 0.5 + 1.96 * 0.5/\sqrt{n}$, or $n > (1.96 * 0.5/0.1)^2 = 96$. This is mistaken, however, because it confuses the assumption that $p = 0.6$ with the claim that $\hat{p} > 0.6$. In fact, if $p = 0.6$, then $\hat{p}$ depends on the sample, and it has an approximate normal distribution with mean 0.6 and standard deviation $\sqrt{0.6 * 0.4/n} = 0.49/\sqrt{n}$; see the top half of Figure 16.2.

To determine the appropriate sample size, we must specify the desired *power*—that is, the probability that a 95% interval will be entirely above the comparison point of 0.5. Under the assumption that $p = 0.6$, choosing $n = 96$ yields 50% power: there is a 50% chance that $\hat{p}$ will be more than 1.96 standard deviations away from 0.5, and thus a 50% chance that the 95% interval will be entirely greater than 0.5.

The conventional level of power in sample size calculations is 80%: the goal is to choose $n$ such that 80% of the possible 95% confidence intervals will not include 0.5. When $n$ is increased, the estimate becomes closer (on average) to the true value, and the width of the confidence interval decreases. Both these effects (decreasing variability of the estimator and narrowing of the confidence interval) can be seen in going from the top half to the bottom half of Figure 16.2.

To find the value of $n$ such that exactly 80% of the estimates will be at least 1.96 standard errors from 0.5, we need

$$0.5 + 1.96 * \text{s.e.} = 0.6 - 0.84 * \text{s.e.}$$

Some algebra then yields $(1.96 + 0.84) * \text{s.e.} = 0.1$. We can then substitute $\text{s.e.} = 0.5/\sqrt{n}$ and solve for $n$, as we discuss next.

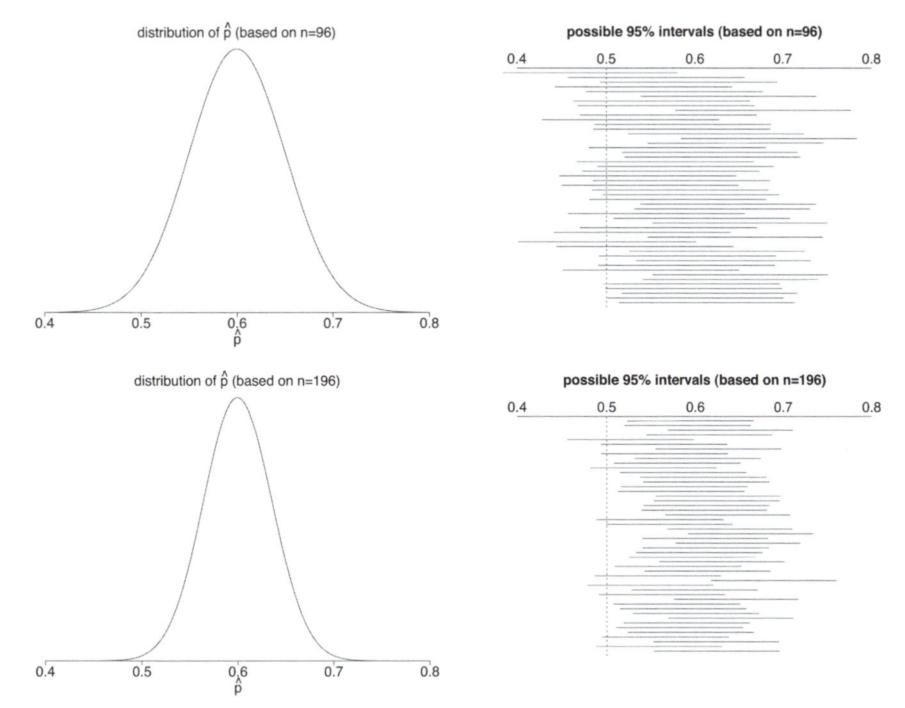


Figure 16.2 *Illustration of simple sample size calculations.* Top row: *(left) distribution of the sample proportion $\hat{p}$ if the true population proportion is $p = 0.6$, based on a sample size of 96; (right) several possible 95% intervals for p based on a sample size of 96. The power is 50%—that is, the probability is 50% that a randomly generated interval will be entirely to the right of the comparison point of 0.5.* Bottom row: *corresponding graphs for a sample size of 196. Here the power is 80%.*

In summary, *to have 80% power, the true value of the parameter must be 2.8 standard errors away from the comparison point*: the value 2.8 is 1.96 from the 95% interval, plus 0.84 to reach the $80^{\text{th}}$ percentile of the normal distribution. The bottom row of Figures 16.2 and 16.3 illustrate: with $n = (2.8 * 0.49/0.1)^2 = 196$, and if the true population proportion is $p = 0.6$, there is an 80% chance that the 95% confidence interval will be entirely greater than 0.5, thus conclusively demonstrating that more than half the people support the death penalty.

These calculations are only as good as their assumptions; in particular, one would generally not know the true value of $p$ before doing the study. Nonetheless, design analyses can be useful in giving a sense of the size of effects that one could reasonably expect to demonstrate with a study of given size. For example, a survey of size 196 has 80% power to demonstrate that $p > 0.5$ if the true value is 0.6, and it would easily detect the difference if the true value were 0.7; but if the true $p$ were equal to 0.56, say, then the difference would be only $0.06/(0.5/\sqrt{196}) = 1.6$ standard errors away from zero, and it would be likely that the 95% interval for $p$ would include 0.5, even in the presence of this true effect. Thus, if the goal of the survey is to conclusively detect a difference from 0.5, it would probably not be wise to use a sample of only $n = 196$ unless we suspect the true $p$ is at least 0.6. Such a small survey would not have the power to reliably detect differences of less than 0.1.

## Estimates of hypothesized proportions

The standard error of a proportion $p$, if it is estimated from a simple random sample of size $n$, is $\sqrt{p(1-p)/n}$, which has an upper bound of $0.5/\sqrt{n}$. This upper bound is very close to the actual standard error for a wide range of probabilities $p$ near 1/2: for example, if the probability is 0.5, then the standard error is $\sqrt{0.5 * 0.5/n} = 0.5/\sqrt{n}$ exactly; if probabilities are 60/40, then we get

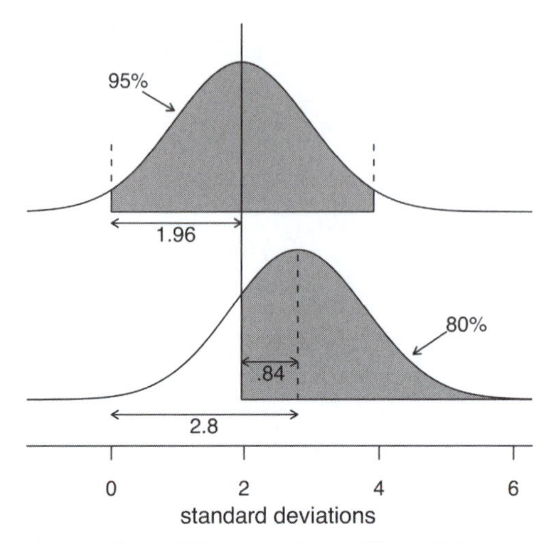


Figure 16.3 *Sketch illustrating that, to obtain 80% power for a 95% confidence interval, the true effect size must be at least 2.8 standard errors from zero (assuming a normal distribution for estimation error). The top curve shows that the estimate must be at least 1.96 standard errors from zero for the 95% interval to be entirely positive. The bottom curve shows the distribution of the parameter estimates that might occur, if the true effect size is 2.8. Under this assumption, there is an 80% probability that the estimate will exceed 1.96. The two curves together show that the lower curve must be centered all the way at 2.8 to get an 80% probability that the 95% interval will be entirely positive.*

$\sqrt{0.6 * 0.4/n} = 0.49/\sqrt{n}$; and if probabilities are 70/30, then we get $\sqrt{0.7 * 0.3/n} = 0.46/\sqrt{n}$, which is still not far from $0.5/\sqrt{n}$.

If the goal is a specified standard error, then the required sample size is determined conservatively by s.e. $= 0.5/\sqrt{n}$, so that $n = (0.5/\text{s.e.})^2$ or, more precisely, $n = p(1-p)/(\text{s.e.})^2$. If the goal is 80% power to distinguish $p$ from a specified value $p_0$, then a conservative required sample size is that needed for the true parameter value to be 2.8 standard errors from zero; solving for this standard error yields $n = (2.8 * 0.5/(p-p_0))^2$ or, more precisely, $n = p(1-p)(2.8/(p-p_0))^2$.

## Simple comparisons of proportions: equal sample sizes

The standard error of a difference between two proportions is, by a simple probability calculation, $\sqrt{p_1(1-p_1)/n_1 + p_2(1-p_2)/n_2}$, which has an upper bound of $0.5\sqrt{1/n_1 + 1/n_2}$. If we assume $n_1 = n_2 = n/2$ (equal sample sizes in the two groups), the upper bound on the standard error becomes simply $1/\sqrt{n}$. A specified standard error can then be attained with a sample size of $n = 1/(\text{s.e.})^2$. If the goal is 80% power to distinguish between hypothesized proportions $p_1$ and $p_2$ with a study of size $n$, equally divided between the two groups, a conservative sample size is $n = ((2.8/(p_1-p_2))^2$ or, more precisely, $n = 2(p_1(1-p_1) + p_2(1-p_2))(2.8/(p_1-p_2))^2$.

For example, suppose we suspect that the death penalty is 10% more popular in the United States than in Canada, and we plan to conduct surveys in both countries on the topic. If the surveys are of equal sample size, $n/2$, how large must $n$ be so that there is an 80% chance of achieving statistical significance, if the true difference in proportions is 10%? The standard error of $\hat{p}_1 - \hat{p}_2$ is approximately $1/\sqrt{n}$, so for 10% to be 2.8 standard errors from zero, we must have $n > (2.8/0.10)^2 = 784$, or a survey of 392 people in each country.

## Simple comparisons of proportions: unequal sample sizes

In epidemiology, it is common to have unequal sample sizes in comparison groups. For example, consider a study in which 20% of units are exposed and 80% are controls.

First, consider the goal of estimating the difference between the exposed and control groups, to some specified precision. The standard error of the difference is $\sqrt{p_1(1-p_1)/(0.2n)+p_2(1-p_2)/(0.8n)}$, and this expression has an upper bound of $0.5\sqrt{1/(0.2n)+1/(0.8n)} = 0.5\sqrt{1/(0.2)+1/(0.8)}/\sqrt{n} = 1.25/\sqrt{n}$. A specified standard error can then be attained with a sample size of $n = (1.25/\text{s.e.})^2$.

Second, suppose we want sufficient total sample size $n$ to achieve 80% power to detect a difference of 10%, again with 20% of the sample size in one group and 80% in the other. Again, the standard error of $\hat{p}_1 - \hat{p}_2$ is bounded by $1.25/\sqrt{n}$, so for 10% to be 2.8 standard errors from zero, we must have $n > (2.8 * 1.25/0.10)^2 = 1225$, or 245 cases and 980 controls.

## 16.3 Sample size and design calculations for continuous outcomes

ample: ic experi- ents

Sample size calculations proceed much the same way with continuous outcomes, with the added difficulty that the population standard deviation must also be specified along with the hypothesized effect size. We shall illustrate with a proposed experiment adding zinc to the diet of HIV-positive children in South Africa. In various other populations, zinc and other micronutrients have been found to reduce the occurrence of diarrhea, which is associated with immune system problems, as well as to slow the progress of HIV. We first consider the one-sample problem—how large a sample size we would expect to need to measure various outcomes to a specified precision—and then move to two-sample problems comparing treatment to control groups.

### Estimates of means

Suppose we are trying to estimate a population mean value $\theta$ from data $y_1, \dots, y_n$, a random sample of size $n$. The quick estimate of $\theta$ is the sample mean, $\bar{y}$, which has a standard error of $\sigma/\sqrt{n}$, where $\sigma$ is the standard deviation of $y$ in the population. So if the goal is to achieve a specified s.e. for $\bar{y}$, then the sample size must be at least $n = (\sigma/\text{s.e.})^2$. If the goal is 80% power to distinguish $\theta$ from a specified value $\theta_0$, then a conservative required sample size is $n = (2.8\,\sigma/(\theta - \theta_0))^2$.

### The t distribution and uncertainty in standard deviations

In this section, we perform all design analyses using the normal distribution, which is appropriate for linear regression when the residual standard deviation $\sigma$ is known. For very small studies, though, degrees of freedom are low, the residual standard deviation is not estimated precisely from data, and inferential uncertainties (confidence intervals or posterior intervals) follow the $t$ distribution. In that case, the value 2.8 needs to be replaced with a larger number to capture this additional source of uncertainty. For example, when designing a study comparing two groups of 6 patients each, the degrees of freedom are 10 (calculated as 12 data points minus two coefficients being estimated; see the beginning of Section 4.4), and the normal distributions in the power calculations are replaced by $t_{10}$. In R, `qnorm(0.8) + qnorm(0.975)` yields the value 2.8, while `qt(0.8,10) + qt(0.975,10)` yields the value 3.1, so we would replace 2.8 by 3.1 in the calculations for 80% power. We usually don't worry about the $t$ correction because it is minor except when sample sizes are very small.

### Simple comparisons of means

The standard error of $\bar{y}_1 - \bar{y}_2$ is $\sqrt{\sigma_1^2/n_1 + \sigma_2^2/n_2}$. If we make the restriction $n_1 = n_2 = n/2$ (equal sample sizes in the two groups), the standard error becomes simply s.e. $= \sqrt{2(\sigma_1^2+\sigma_2^2)/n}$. A specified standard error can then be attained with a sample size of $n = 2(\sigma_1^2+\sigma_2^2)/(\text{s.e.})^2$. If we further suppose that the variation is the same within each of the groups ($\sigma_1 = \sigma_2 = \sigma$), then s.e. $= 2\sigma/\sqrt{n}$, and the required sample size is $n = (2\sigma/\text{s.e.})^2$.

If the goal is 80% power to detect a difference of $\Delta$, with a study of size $n$, equally divided

Rosado et al. (1997), Mexico

| Treatment | Sample size | Avg. # episodes in a year ± s.e. |
|---|---|---|
| placebo | 56 | 1.1 ± 0.2 |
| iron | 54 | 1.4 ± 0.2 |
| zinc | 54 | 0.7 ± 0.1 |
| zinc + iron | 55 | 0.8 ± 0.1 |

Ruel et al. (1997), Guatemala

| Treatment | Sample size | Avg. # episodes per 100 days [95% c.i.] |
|---|---|---|
| placebo | 44 | 8.1 [5.8, 10.2] |
| zinc | 45 | 6.3 [4.2, 8.9] |

Lira et al. (1998), Brazil

| Treatment | Sample size | % days with diarrhea | Prevalence ratio [95% c.i.] |
|---|---|---|---|
| placebo | 66 | 5% | 1 |
| 1 mg zinc | 68 | 5% | 1.00 [0.72, 1.40] |
| 5 mg zinc | 71 | 3% | 0.68 [0.49, 0.95] |

Muller et al. (2001), West Africa

| Treatment | Sample size | # days with diarrhea/ total # days |
|---|---|---|
| placebo | 329 | 997/49 021 = 0.020 |
| zinc | 332 | 869/49 086 = 0.018 |

Figure 16.4 *Results from various experiments studying zinc supplements for children with diarrhea. We use this information to hypothesize the effect size $\Delta$ and within-group standard deviation $\sigma$ for our planned experiment.*

between the two groups, then the required sample size is $n = 2(\sigma_1^2 + \sigma_2^2)(2.8/\Delta)^2$. If $\sigma_1 = \sigma_2 = \sigma$, this simplifies to $(5.6\,\sigma/\Delta)^2$.

For example, consider the effect of zinc supplements on young children's growth. Results of published studies suggest that zinc can improve growth by approximately 0.5 standard deviations. That is, $\Delta = 0.5\sigma$ in the our notation. To have 80% power to detect an effect size, it would be sufficient to have a total sample size of $n = (5.6/0.5)^2 = 126$, or $n/2 = 63$ in each group.

## Estimating standard deviations using results from previous studies

Sample size calculations for continuous outcomes are based on estimated effect sizes and standard deviations in the population—that is, $\Delta$ and $\sigma$. Guesses for these parameters can be estimated or deduced from previous studies. We illustrate with the design of a study to estimate the effects of zinc on diarrhea in children. Various experiments have been performed on this topic—Figure 16.4 summarizes the results, which we shall use to get a sense of the sample size required for our study.

We consider the studies reported in Figure 16.4 in order. For Rosado et al. (1997), we estimate the effect of zinc by averaging over the iron and no-iron cases, thus an estimated $\Delta$ of $\frac{1}{2}(1.1+1.4)-\frac{1}{2}(0.7+0.8) = 0.5$ episodes in a year, with a standard error of $\sqrt{\frac{1}{4}(0.2^2+0.2^2)+\frac{1}{4}(0.1^2+0.1^2)} = 0.15$. From this study, we estimate that zinc reduces diarrhea in that population by an average of about 0.3 to 0.7 episodes per year. Next, we deduce the within-group standard deviations $\sigma$ using the formula s.e.$= \sigma/\sqrt{n}$; thus the standard deviations are $0.2 * \sqrt{56} = 1.5$ for the placebo group and are for 1.5, 0.7, and 0.7 for the other three groups. The number of episodes is bounded below by zero, so it makes sense that when the mean level goes down, the standard deviation decreases also.

Assuming an effect size of $\Delta = 0.5$ episodes per year and within-group standard deviations of 1.5 and 0.7 for the control and treatment groups, we can evaluate the power of a future study with $n/2$ children in each group. The estimated difference would have a standard error of $\sqrt{1.5^2/(n/2)+0.7^2/(n/2)} = 2.4/\sqrt{n}$, and so for the effect size to be at least 2.8 standard errors away from zero (and thus to have 80% power to attain statistical significance), $n$ would have to be at least $(2.8 * 2.4/0.5)^2 = 180$ people in the two groups.

Now turning to the Ruel et al. (1997) study, we first see that rates of diarrhea—for control and treated children both—are much higher than in the previous study: 8 episodes per hundred days, which corresponds to 30 episodes per year, more than 20 times the rate in the earlier group. We are dealing with very different populations here. In any case, we can divide the uncertainty interval widths by 4 to get standard errors—thus, 1.1 for the placebo group and 1.2 for the treated group—yielding an estimated treatment effect of 1.8 with standard error 1.6, which is consistent with a treatment effect of nearly zero or as high as about 4 episodes per 100 days. When compared to the average observed rate in the control group, the estimated treatment effect from this study is about half that of the Rosado et al. (1997) experiment: $1.8/8.1 = 0.22$, compared to $0.5/1.15 = 0.43$, which suggests a higher sample size might be required. However, the wide uncertainty bounds of the Ruel et al. (1997) study make it consistent with the larger effect size.

Next, Lira et al. (1998) report the average percentage of days with diarrhea of children in the control and two treatment groups corresponding to a low (1 mg) or high (5 mg) dose of zinc. We shall consider only the 5 mg condition, as this is closer to the treatment for our experiment. The estimated effect of the treatment is to multiply the number of days with diarrhea by 68%—that is, a reduction of 32%, which again is consistent with the approximate 40% decrease found in the first study. To make a power calculation, we first convert the uncertainty interval $[0.49, 0.95]$ for this multiplicative effect to the logarithmic scale—thus, an additive effect of $[-0.71, -0.05]$ on the logarithm—then divide by 4 to get an estimated standard error of 0.16 on this scale. The estimated effect of 0.68 is $-0.38$ on the log scale, thus 2.4 standard errors away from zero. For this effect size to be 2.8 standard errors from zero, we would need to increase the sample size by a factor of $(2.8/2.4)^2 = 1.4$, thus moving from approximately 70 children to approximately 100 in each of the two groups.

Finally, Muller et al. (2001) compare the proportion of days with diarrhea, which declined from 2.03% in the controls to 1.77% among children who received zinc. Unfortunately, no standard error is reported for this 13% decrease, and it is not possible to compute it from the information in the article. However, the estimates of within-group variation $\sigma$ from the other studies would lead us to conclude that we would need a very large sample size to be likely to reach statistical significance, if the true effect size were only 10%. For example, from the Lira et al. (1998) study, we estimate a sample size of 100 in each group is needed to detect an effect of 32%; thus, to detect a true effect of 13%, we would need a sample size of $100 * (0.32/0.13)^2 = 600$.

These calculations are necessarily speculative; for example, to detect an effect of 10% (instead of 13%), the required sample size would be $100 * (0.32/0.10)^2 = 1000$ per group, a huge change considering the very small change in hypothesized treatment effects. Thus, it is misleading to think of these as required sample sizes. Rather, these calculations tell us how large the effects are that we could expect to have a good chance of discovering, given any specified sample size.

The first two studies in Figure 16.4 report the frequency of episodes, and the last two give the proportion of days with diarrhea, which is proportional to the frequency of episodes multiplied by the average duration of each episode. Other data (not shown here) show no effect of zinc on average duration, and so we treat all four studies as estimating the effects on frequency of episodes.

In conclusion, a sample size of about 100 per treatment group should give adequate power to detect an effect of zinc on diarrhea, if its true effect is to reduce the frequency, on average, by 30%–50% compared to no treatment. A sample size of 200 per group would have the same power to detect effects a factor $\sqrt{2}$ smaller, that is, effects in the 20%–35% range.

### Including more regression predictors

Now suppose we are comparing treatment and control groups with additional pre-treatment data on the children (for example, age, height, weight, and health status at the start of the experiment). These can be included in a regression. For simplicity, consider a model with no interactions—that is, with coefficients for the treatment indicator and the other inputs—in which case, the treatment coefficient represents the causal effect, the comparison after adjusting for pre-treatment differences.

Sample size calculations for this new study are exactly as before, except that the within-group standard deviation $\sigma$ is replaced by the residual standard deviation of the regression. This can be hypothesized in its own right or in terms of the added predictive power of the pre-treatment data. For example, if we hypothesize a within-group standard deviation of 0.2, then a residual standard deviation of 0.14 would imply that half the variance within any group is explained by the regression model, which would actually be pretty good.

Adding relevant predictors should decrease the residual standard deviation and thus reduce the required sample size for any specified level of precision or power.

### Estimation of regression coefficients more generally

More generally, sample sizes for regression coefficients and other estimands can be calculated using the rule that standard errors are proportional to $1/\sqrt{n}$; thus, if inferences exist under a current sample size, effect sizes can be estimated and standard errors extrapolated for other hypothetical samples.

We illustrate with the example of the survey earnings and height discussed in Chapter 4. The coefficient for the sex-earnings interaction in model (12.2) is plausible (a positive interaction, implying that an extra inch of height is worth 0.7% more for men than for women), but it is not statistically significant—the standard error is 1.9%, yielding a 95% interval of $[-3.1, 4.5]$, which contains zero.

How large a sample size would be needed for the coefficient on the interaction to be statistically significant? A simple calculation uses the fact that standard errors are proportional to $1/\sqrt{n}$. For a point estimate of 0.7% to achieve statistical significance, it would need a standard error of 0.35%, which would require the sample size to be increased by a factor of $(1.9\%/0.35\%)^2 = 29$. The original survey had a sample of 1192; this implies a required sample size of $29 * 1192 = 35\,000$.

To extend this to a power calculation, we suppose that the true $\beta$ for the interaction is equal to 0.7% and that the standard error is as we have just calculated. With a standard error of 0.35%, the estimate from the regression would then be statistically significant only if $\hat{\beta} > 0.7\%$ (or, strictly speaking, if $\hat{\beta} < -0.7\%$, but that latter possibility is highly unlikely given our assumptions). If the true coefficient is $\beta$, we would expect the estimate from the regression to possibly take on values in the range $\beta \pm 0.35\%$ (that is what is meant by "a standard error of 0.35%"), and thus if $\beta$ truly equals 0.7%, we would expect $\hat{\beta}$ to exceed 0.7%, and thus achieve statistical significance, with a probability of 1/2—that is, 50% power. To get 80% power, we need the true $\beta$ to be 2.8 standard errors from zero, so that there is an 80% probability that $\hat{\beta}$ is at least 2 standard errors from zero. If $\beta = 0.7\%$, then its standard error would have to be no greater than $0.7\%/2.8 = 0.25\%$, so that the survey would need a sample size of $(1.9\%/0.25\%)^2 * 1192 = 70\,000$.

This design calculation is close to meaningless, however, because it makes the very strong assumption that the true value of $\beta$ is 0.7%, the estimate that we happened to obtain from our survey. But the estimate from the regression is $0.7\% \pm 1.9\%$, which implies that these data are consistent with a low, zero, or even negative value of the true $\beta$ (or, in the other direction, a true value that is greater than the point estimate of 0.7%). If the true $\beta$ is actually less than 0.7%, then even a sample size of 70 000 would be insufficient for 80% power.

This is not to say the design analysis is useless but just to point out that, even when done correctly, it is based on an assumption that is inherently untestable from the available data (hence the need for a larger study). So we should not necessarily expect statistical significance from a proposed study, even if the sample size has been calculated correctly. To put it another way, the value of the above calculations is *not* to tell us the power of the study that was just performed, or to choose a sample size of a new study, but rather to develop our intuitions of the relation between inferential uncertainty, standard error, and sample size.

### Sample size, design, and interactions

Sample size is never large enough. As $n$ increases, we can estimate more interactions, which typically are smaller and have relatively larger standard errors than main effects; for example, see the fitted regression on page 193 of log earnings on sex, standardized height, and their interaction. Estimating interactions is similar to comparing coefficients estimated from subsets of the data (for example, the coefficient for height among men, compared to the coefficient among women), thus reducing power because the sample size for each subset is halved, and also the differences themselves may be small. As more data are included in an analysis, it becomes possible to estimate these interactions (or, using multilevel modeling, to include them and partially pool them as appropriate), so this is not a problem. We are just emphasizing that, just as you never have enough money, because perceived needs increase with resources, your inferential needs will increase with your sample size.

## 16.4 Interactions are harder to estimate than main effects

In causal inference, it is often important to study varying effects: for example, a treatment could be more effective for men than for women, or for healthy than for unhealthy patients. We are often interested in interactions in predictive models as well.

### You need 4 times the sample size to estimate an interaction that is the same size as the main effect

Suppose a study is designed to have 80% power to detect a main effect at a 95% confidence level. As discussed earlier in this chapter, that implies that the true effect size is 2.8 standard errors from zero. That is, the $z$-score has a mean of 2.8 and standard deviation of 1, and there's an 80% chance that the $z$-score exceeds 1.96 (in R, `pnorm(2.8,1.96,1)` = 0.8).

Further suppose that an interaction of interest is the same size as the main effect. For example, if the average treatment effect on the entire population is $\theta$, with an effect of $0.5\,\theta$ among women and $1.5\,\theta$ among men, then the interaction—the difference in treatment effect comparing men to women—is the same size as the main effect.

The standard error of an interaction is roughly *twice* the standard error of the main effect, as we can see from some simple algebra:

- The estimate of the main effect is $\bar{y}_T - \bar{y}_C$, and this has standard error $\sqrt{\sigma^2/(n/2) + \sigma^2/(n/2)} = 2\sigma/\sqrt{n}$; for simplicity we are assuming a constant variance within groups, which will typically be a good approximation for binary data, for example.
- The estimate of the interaction is $(\bar{y}_{T,\text{men}} - \bar{y}_{C,\text{men}}) - (\bar{y}_{T,\text{women}} - \bar{y}_{C,\text{women}})$, which has standard error $\sqrt{\sigma^2/(n/4) + \sigma^2/(n/4) + \sigma^2/(n/4) + \sigma^2/(n/4)} = 4\sigma/\sqrt{n}$. By using the same $\sigma$ here as in the earlier calculation, we are assuming that the residual standard deviation is unchanged (or essentially unchanged) after including the interaction in the model; that is, we are assuming that inclusion of the interaction does not change $R^2$ much.

To put it another way, to be able to estimate the interaction to the same level of accuracy as the main effect, we would need four times the sample size.

What is the power of the estimate of the interaction, as estimated from the original experiment of size $n$? The probability of seeing a difference that is "statistically significant" at the 5% level is the probability that the $z$-score exceeds 1.96; that is, `pnorm(1.4,1.96,1)` = 0.29. And, if you do perform the analysis and report it if the 95% interval excludes zero, you will overestimate the size of the interaction by a lot, as we can see by simulating a million runs of the experiment:

```
raw <- rnorm(1e6, 1.4, 1)
significant <- raw > 1.96
mean(raw[significant])
```

The result is 2.6, implying that, on average, a statistically significant result will overestimate the size of the interaction by a factor of 2.6.

This implies a big problem with the common plan of designing a study with a focus on the main effect and then looking to see what shows up in the interactions. Or, even worse, designing a study, not finding the anticipated main effect, and then using the interactions to bail you out. The problem is not just that this sort of analysis is "exploratory"; it's that these data are a lot noisier than you realize, so what you think of as interesting exploratory findings could be just a bunch of noise.

### You need 16 times the sample size to estimate an interaction that is half the size as the main effect

As demonstrated above, if an interaction is the same size as the main effect—for example, a treatment effect of 0.5 among women, 1.5 among men, and 1.0 overall—then it will require four times the sample size to estimate with the same accuracy from a balanced experiment.

There are cases where main effects are small and interactions are large. Indeed, in general, these labels have some arbitrariness to them; for example, when studying U.S. congressional elections, recode the outcome from Democratic or Republican vote share to incumbent party vote share, and interactions with incumbent party become main effects, and main effects become interactions. So the above analysis is in the context of main effects that are modified by interactions; there's the implicit assumption that if the main effect is positive, then it will be positive in the subgroups we look at, just maybe a bit larger or smaller.

It makes sense, where possible, to code variables in a regression so that the larger comparisons appear as main effects and the smaller comparisons appear as interactions. The very nature of a "main effect" is that it is supposed to tell as much of the story as possible. When interactions are important, they are important as modifications of some main effect. This is not always the case—for example, you could have a treatment that flat-out hurts men while helping women—but in such examples it's not clear that the main-effects-plus-interaction framework is the best way of looking at things.

When a large number of interactions are being considered, we would expect most interactions to be smaller than the main effect. Consider a treatment that could interact with many possible individual characteristics, including age, sex, education, health status, and so forth. We would not expect all or most of the interactions of treatment effect with these variables to be large. Thus, when considering the challenge of estimating interactions that are not chosen ahead of time, it could be more realistic to suppose something like half the size of main effects. In that case—for example, a treatment effect of 0.75 in one group and 1.25 in the other—one would need 16 times the sample size to estimate the interaction with the same relative precision as is needed to estimate the main effect.

The message we take from this analysis is *not* that interactions are too difficult to estimate and should be ignored. Rather, interactions can be important; we just need to accept that in many settings we won't be able to attain anything like near-certainty regarding the magnitude or even direction of particular interactions. It is typically not appropriate to aim for "statistical significance" or 95% intervals that exclude zero, and it often will be appropriate to use prior information to get more stable and reasonable estimates, and to accept uncertainty, not acting as if interactions of interest are zero just because their estimate is not statistically significant.

### Understanding the problem by simulating regressions in R

We can play around in R to get a sense of how standard errors for main effects and interactions depend on parameterization. For simplicity, all our simulations assume that the true (underlying) coefficients are 0. In this case, the true values are irrelevant for our goal of computing the standard error.

Example: Simulation of main effects and interactions

We start with a basic model in which we simulate 1000 data points with two predictors, each taking on the value −0.5 or 0.5. This is the same as the model above: the estimated main effects are simple differences, and the estimated interaction is a difference in differences. We also have assumed the two predictors are independent, which is what would happen in a randomized experiment where, on average, the treatment and control groups would each be expected to be evenly divided between men and women. Here is the simulation:[1]

[1] Code for this example is in the folder `SampleSize`.

```
n <- 1000
sigma <- 10
y <- rnorm(n, 0, sigma)
x1 <- sample(c(-0.5,0.5), n, replace=TRUE)
x2 <- sample(c(-0.5,0.5), n, replace=TRUE)
fake <- data.frame(c(y,x1,x2))
fit_1 <- stan_glm(y ~ x1, data=fake)
fit_2 <- stan_glm(y ~ x1 + x2 + x1:x2, data=fake)
print(fit_1)
print(fit_2)
```

And here is the result:

```
            Median MAD_SD
(Intercept) -0.1    0.3
x1           0.7    0.6
x2           0.8    0.6
x1:x2        1.2    1.3

Auxiliary parameter(s):
      Median MAD_SD
sigma 10.0    0.2
```

Ignore the estimates; they're pure noise. Just look at the standard errors. They go just as in the above formulas: $2\sigma/\sqrt{n} = 2 * 10/\sqrt{1000} = 0.6$, and $4\sigma/\sqrt{n} = 1.3$.

Now let's do the exact same thing but make the predictors take on the values 0 and 1 rather than −0.5 and 0.5:

```
fake$x1 <- sample(c(0,1), n, replace=TRUE)
fake$x2 <- sample(c(0,1), n, replace=TRUE)
fit_1 <- stan_glm(y ~ x1, data=fake)
fit_2 <- stan_glm(y ~ x1 + x2 + x1:x2, data=fake)
print(fit_1)
print(fit_2)
```

And this is what happens:

```
            Median MAD_SD
(Intercept) -0.1    0.6
x1           1.0    0.9
x2           0.1    0.9
x1:x2       -1.9    1.3

Auxiliary parameter(s):
      Median MAD_SD
sigma 10.0    0.2
```

Again, just look at the standard errors. The standard error for the interaction is still 1.3, but the standard errors for the main effects went up to 0.9. What happened?

What happened was that the main effects are now estimated at the edge of the data: the estimated coefficient of $x_1$ is now the difference in $y$, comparing the two values of $x_1$, just at $x_2 = 0$. So its standard error is $\sqrt{\sigma^2/(n/4) + \sigma^2/(n/4)} = 2\sqrt{2}\sigma/\sqrt{n}$. Under this parameterization, the coefficient of $x_1$ is estimated just from the half of the data for which $x_2 = 0$, so the standard error is $\sqrt{2}$ times as big as before. Similarly for $x_2$.

But these aren't really "main effects"; in the context of the above problem, the main effect of the treatment is the average over men and women. If we put the problem in a regression framework, we should be coding the predictors not as 0, 1 but as −0.5, 0.5, so that the main effect for each predictor corresponds to the other predictor set to its average level.

But here's another possibility: What about coding each predictor as −1, 1? Let's take a look:

```
fake$x1 <- sample(c(-1,1), n, replace=TRUE)
fake$x2 <- sample(c(-1,1), n, replace=TRUE)
fit_1 <- stan_glm(y ~ x1, data=fake)
fit_2 <- stan_glm(y ~ x1 + x2 + x1:x2, data=fake)
print(fit_1)
print(fit_2)
```

This yields:

```
            Median MAD_SD
(Intercept) -0.4    0.3
x1          -0.5    0.3
x2           0.0    0.3
x1:x2        0.7    0.3

Auxiliary parameter(s):
      Median MAD_SD
sigma 9.9    0.2
```

Again, ignore the coefficient estimates and look at the standard errors. Compared to the fitted model with the −0.5, 0.5 coding on page 303, the standard errors for the main effects are smaller by a factor of 2, and now the standard error for the interaction has been divided by 4. What happened in this simulation?

The factor of 2 for the main effect is clear enough: If you multiply $x$ by 2, and $\beta * x$ doesn't change, then you have to divide $\beta$ by 2 to compensate, and its standard error gets divided by 2 as well. But what happened to the interaction? That's clear too: we've multiplied $x_1$ and $x_2$ each by 2, so $x_1 x_2$ is multiplied by 4.

So to make sense of all these standard errors, you have to have a feel for the appropriate scale for the coefficients.

## 16.5 Design calculations after the data have been collected

We return to the beauty and sex ratio example, introduced in Sections 9.4 and 9.5 to demonstrate Bayesian inference. Here we attack the problem in a slightly different way using design analysis. Either way, the message is that we can use available prior information to interpret results from particular data.

Example: Beauty and sex ratio

As a result of the intrinsic interest of the topic and the availability of data from birth records, there have been many studies of factors affecting the probability of male and female births. Most have found little or no evidence of any effects, but the study described in Section 9.4 appeared to be an exception, reporting data from a survey in which attractive parents were more likely to have daughters, a finding that was then given an explanation in terms of evolutionary biology, on the grounds that physical attractiveness enhances the reproductive success of women more than that of men.

For our discussion here we shall work with the simple analysis from Section 9.4, comparing the "very attractive" parents in the survey (56% of their children were girls) to the other parents (only 44% of their children were girls). The difference was 8% with a standard error of 3%. The classical 95% interval is $[8\% \pm 2 * 3\%] = [2\%, 14\%]$, which tells us that effects as low as 2 percentage points or as high as 14 percentage are roughly consistent with the data.

The challenge is to interpret this finding in light of our knowledge from the scientific literature that any difference in sex ratios between two such groups in the population is probably much less than 0.5% (for example, the probability of a girl birth shifting from 48.5% to 49.0%).

How, then, do we account for the fact that the 95% interval for the difference is [2%, 14%], which excludes the range of plausible differences in the population? One answer is that unusual things happen: 5% events do occur 5% of the time. A longer answer is that researchers typically have many choices or "degrees of freedom" in their analysis. For example, in this particular example, survey respondents were placed in five attractiveness categories, and the published comparison was category

5 compared to categories 1–4, pooled; see Figure 9.5. But the researcher could just as well have compared categories 4–5 to categories 1–3, or compared 3–5 to 1–2, or compared 4–5 to 1–2, and so forth. Looked at this way, it's no surprise that a determined data analyst was able to find a comparison somewhere in the data for which the 95% interval was far from zero.

What, then, can be learned from the published estimate of 8%? For the present example, the standard error of 3% means that statistical significance would only happen with an estimate of at least 6% in either direction: more than 12 times larger than any true effect that could reasonably be expected based on previous research. Thus, even if the inference of an association between parental beauty and child's sex is valid for the general population, the magnitude of the estimate from a study of this size is likely to be much larger than the true effect. This is an example of a type M (magnitude) error, as defined in Section 4.4. We can also consider the possibility of type S (sign) errors, in which a statistically significant estimate is in the opposite direction of the true effect.

We may get a sense of the probabilities of these errors by considering three scenarios of studies with standard errors of 3 percentage points:

1. *True difference of zero.* If there is no correlation between parental beauty and sex ratio of children, then a statistically significant estimate will occur 5% of the time, and it will always be misleading—a type 1 error.
2. *True difference of 0.2%.* If the probability of girl births is actually 0.2 percentage points higher among beautiful than among other parents, then what might happen with an estimate whose standard error is 3%? We can do the calculation in R: the probability of the estimate being at least 6% (two standard errors away from zero, thus "statistically significant") is `1 - pnorm(6,0.2,3)`, or 0.027, and the probability of it being at least 6% in the *negative* direction is `pnorm(-6,0.2,3)`, which comes to 0.019. The type S error rate is 0.019/(0.019 + 0.027) = 42%.
   Thus, before the data were collected, we could say that if the true population difference were 0.2%, that this study has a 3% probability of being statistically significant and positive—and a 2% chance of being statistically significant negative result. If the estimate is statistically significant, it must be at least 6 percentage points, thus at least 30 times higher than the true effect, and with a 40% chance of going in the wrong direction.
3. *True difference of 0.5%.* If the probability of girl births is actually 0.5 percentage point higher among beautiful than among other parents—which, based on the literature, is well beyond the high end of possible effect sizes—then there is a 0.033 chance of a statistically significant positive result, and a 0.015 chance of a statistically significant result in the wrong direction. The type S error rate is 0.015/(0.015 + 0.033) = 31%.
   So, if the true difference is 0.5%, any statistically significant estimated effect will be at least 12 times the magnitude of the true effect and with a 30% chance of having the wrong sign. Thus, again, the experiment gives little information about the sign or the magnitude of the true effect.

A sample of this size is just not useful for estimating variation on the order of half a percentage points or less, which is why most studies of the human sex ratio use much larger samples, typically from demographic databases. The example shows that if the sample is too small relative to the expected size of any differences, it is not possible to draw strong conclusions even when estimates are seemingly statistically significant.

Indeed, with this level of noise, *only* very large estimated effects could make it through the statistical significance filter. The result is almost a machine for producing exaggerated claims, which become only more exaggerated when they hit the news media with the seal of scientific approval.

It is well known that with a large enough sample size, even a very small estimate can be statistically significantly different from zero. Many textbooks contain warnings about mistaking statistical significance in a large sample for practical importance. It is also well known that it is difficult to obtain statistically significant results in a small sample. Consequently, when results are significant despite the handicap of a small sample, it is natural to think that they are real and important. The above example shows then this is not necessarily the case.

If the estimated effects in the sample are much larger than those that might reasonably be expected in the population, even seemingly statistically significant results provide only weak evidence of any effect. Yet one cannot simply ask researchers to avoid using small samples. There are cases in which it is difficult or impossible to obtain more data, and researchers must make do with what is available.

In such settings, researchers should determine plausible effect sizes based on previous research or theory, and carry out design calculations based on the observed test statistics. Conventional significance levels tell us how often the observed test statistic would be obtained if there were no effect, but one should also ask how often the observed test statistic would be obtained under a reasonable assumption about the size of effects. Estimates that are much larger than expected might reflect population effects that are much larger than previously imagined. Often, however, large estimates will merely reflect the influence of random variation. It may be disappointing to researchers to learn that even estimates that are both "statistically" and "practically" significant do not necessarily provide strong evidence. Accurately identifying findings that are suggestive rather than definitive, however, should benefit both the scientific community and the general public.

## 16.6 Design analysis using fake-data simulation

Example: Fake-data simulation for experimental design

The most general and often the clearest method for studying the statistical properties of a proposed design is to simulate the data that might be collected along with the analyses that could be performed. We demonstrate with an artificial example of a randomized experiment on 100 students designed to test an intervention for improving final exam scores.[2]

### Simulating a randomized experiment

We start by assigning the potential outcomes, the final exam scores that would be observed for each student if he or she gets the control or the treatment:

```
n <- 100
y_if_control <- rnorm(n, 60, 20)
y_if_treated <- y_if_control + 5
```

In this very simple model, the intervention would add 5 points to each student's score.

We then assign treatments ($z = 0$ for control or 1 for treatment), which then determine which outcome is observed for each person:

```
z <- sample(rep(c(0,1), n/2))
y <- ifelse(z==1, y_if_treated, y_if_control)
fake <- data.frame(y, z)
```

Having simulated the data, we can now compare treated to control outcomes and compute the standard error for the difference:

```
diff <- mean(y[z==1]) - mean(y[z==0])
se_diff <- sqrt(sd(y[z==0])^2/sum(z==0) + sd(y[z==1])^2/sum(z==1))
```

Equivalently (see Section 7.3), we can run the regression:

```
fit_1a <- stan_glm(y ~ z, data=fake)
```

which yields,

```
            Median MAD_SD
(Intercept) 66.0    2.7
z           -2.8    4.3

Auxiliary parameter(s):
      Median MAD_SD
sigma 21.3    1.5
```

[2]Code for this example is in the folder FakeMidtermFinal.

The parameter of interest here is the coefficient of $z$, and its standard error is 4.3, suggesting that, under these conditions, a sample size of 100 would not be enough to get a good estimate of a treatment effect of 5 points. The standard error of 4.3 is fairly precisely estimated, as we can tell because the uncertainty in sigma is low compared to its estimate.

When looking at the above simulation result to assess this design choice, we should focus on the standard error of the parameter of interest (in this case, 4.0) and compare it to the assumed parameter value (in this case, 5), *not* to the noisy point estimate from the simulation (in this case −2.8).

To give a sense of why it would be a mistake to focus on the point estimate, we repeat the above steps, simulating for a new batch of 100 students simulated from the model. Here is the result:

```
(Intercept) 59.7    2.9
z           11.8    4.0

Auxiliary parameter(s):
      Median MAD_SD
sigma 20.1    1.4
```

A naive read of this table would be that the design with 100 students is just fine, as the estimate is well over two standard errors away from zero. But that conclusion would be a mistake, as the coefficient estimate here is too noisy to be useful.

The above simulation indicates that, under the given assumptions, the randomized design with 100 students gives an estimate of the treatment effect with standard error of approximately 4 points. If that is acceptable, fine. If not, one approach would be to increase the sample size. Standard error decreases with the square root of sample size, so if, for example, we wanted to reduce the standard error to 2 points, we would need a sample size of approximately 400.

## Including a pre-treatment predictor

Another approach to increase efficiency is to consider a pre-test. Suppose pre-test scores $x$ have the same distribution as post-test scores $y$ but with a slightly lower average:

```
fake$x <- rnorm(n, 50, 20)
```

We can then adjust for pre-test in our regression:

```
fit_1b <- stan_glm(y ~ z + x, data=fake)

            Median MAD_SD
(Intercept) 51.3    5.9
z           10.9    4.5
x            0.2    0.1

Auxiliary parameter(s):
      Median MAD_SD
sigma 21.1    1.5
```

Again, the coefficient of $z$ estimates the treatment effect, and it still has a standard error of about 4, which might seem surprising: shouldn't the inclusion of a pre-treatment predictor increase the precision of our estimate? The answer is that, the way we constructed the pre-test variable, it wasn't much of a pre-treatment predictor at all, as we simulated it independently of the potential outcomes for the final test score.

To perform a realistic simulation, we must simulate both test scores in a correlated way, which we do here by borrowing a trick from the example of simulated midterm and final exams in Section 6.5:

1. Each student is assumed to have a true ability drawn from a distribution with mean 50 and standard deviation 16.

2. Each student's score on the pre-test, $x$, is the sum of two components: the student's true ability, and a random component with mean 0 and standard deviation 12, reflecting that performance on any given test will be unpredictable.
3. Each student's score on the post-test, $y$, is his or her true ability, plus another, independent, random component, plus an additional 10 points if a student receives the control or 15 points if he or she receives the treatment.

These are the same conditions as in Section 6.5, except that (i) we have increased the standard deviations of each component of the model so that the standard deviation of the final scores, $\sqrt{16^2 + 12^2} = 20$, is consistent with the distribution assumed for $y$ in our simulations above, and (ii) we have increased the average score level on the post-test along with a treatment effect.

Here is the code to create the artificial world:

```
n <- 100
true_ability <- rnorm(n, 50, 16)
x <- true_ability + rnorm(n, 0, 12)
y_if_control <- true_ability + rnorm(n, 0, 12) + 10
y_if_treated <- y_if_control + 5
```

As above, we assign treatments, construct the observed outcome, and put the data into a frame:

```
z <- sample(rep(c(0,1), n/2))
y <- ifelse(z==1, y_if_treated, y_if_control)
fake_2 <- data.frame(x, y, z)
```

The simple comparison is equivalent to a regression on the treatment indicator:

```
fit_2a <- stan_glm(y ~ z, data=fake_2)

            Median MAD_SD
(Intercept) 59.2    3.0
z            9.5    4.3

Auxiliary parameter(s):
      Median MAD_SD
sigma 21.6    1.5
```

And the estimate adjusting for pre-test:

```
fit_2b <- stan_glm(y ~ z + x, data=fake_2)

            Median MAD_SD
(Intercept) 27.4    4.3
z            6.1    3.3
x            0.7    0.1

Auxiliary parameter(s):
      Median MAD_SD
sigma 16.2    1.2
```

In this case, with the strong dependence between pre-test and post-test, this adjustment has reduced the residual standard deviation by about a third.

### Simulating an experiment with selection bias

With data coming from a randomized experiment, all the regressions considered above give unbiased estimates of the treatment effect. But suppose we are concerned about bias in the treatment assignment. We can simulate that too.

For example, suppose that school administrators, out of kindness, are more likely to give the

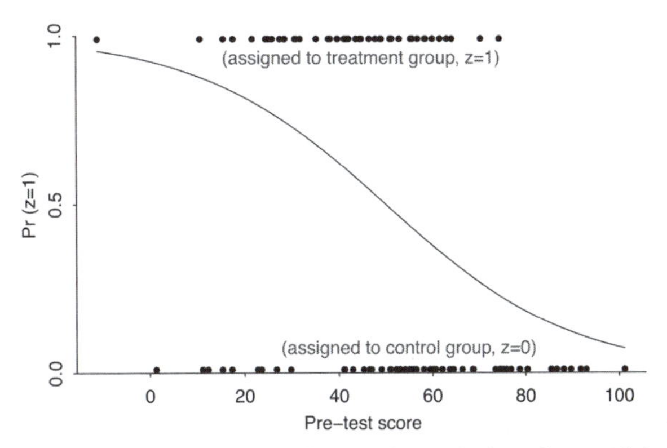


Figure 16.5 *Simulated treatment assignments based on a rule in which students with lower pre-test scores are more likely to get the treatment. We use this to demonstrate how a simulation study can be used to assess the bias in a design and estimation procedure.*

treatment to students who are performing poorly. We could simulate this behavior with an unequal-probability assignment rule such as $\Pr(z_i = 1) = \text{logit}^{-1}(-(x_i - 50)/20)$, where we have chosen the logistic curve for convenience and set its parameters so that the probability averages to approximately 0.5, with a bit of variation from one end of the data to the other. Figure 16.5 shows the assumed logistic curve and the simulated treatment assignments for the 100 students in this example, as produced by the following code:

```
z <- rbinom(n, 1, invlogit(-(x-50)/20))
```

We then record the observed post-test and save as a data frame:

```
y <- ifelse(z==1, y_if_treated, y_if_control)
fake_3 <- data.frame(x, y, z)
```

By construction, the true treatment effect is 5 points, as before, but a simple comparison yields a biased estimate, while the linear regression adjusting for pre-test is better.

To see this, we should not just perform one simulation; as discussed earlier in this section, not much can be learned from the estimate obtained from any single simulation. Instead we first write a function to simulate the fake data, assign the treatments, and perform the simple comparison and the regression adjusting for pre-test:

```
experiment <- function(n) {
  true_ability <- rnorm(n, 50, 16)
  x <- true_ability + rnorm(n, 0, 12)
  y_if_control <- true_ability + rnorm(n, 0, 12) + 10
  y_if_treated <- y_if_control + 5
  z <- rbinom(n, 1, invlogit(-(x-50)/20))
  y <- ifelse(z==1, y_if_treated, y_if_control)
  fake_3 <- data.frame(x, y, z)
  fit_3a <- stan_glm(y ~ z, data=fake_3, refresh=0)
  fit_3b <- stan_glm(y ~ z + x, data=fake_3, refresh=0)
  rbind(c(coef(fit_3a)["z"], se(fit_3a)["z"]), c(coef(fit_3b)["z"], se(fit_3b)["z"]))
}
```

We then loop this simulation 50 times:

```
n <- 100
n_loop <- 50
results <- array(NA, c(n_loop, 2, 2),
  dimnames=list(1:n_loop, c("Simple", "Adjusted"), c("Estimate", "SE")))
  for (loop in 1:n_loop){
```

```
    results[loop,,] <- experiment(n)
  }
```

The above steps produce a $50 \times 2 \times 2$ matrix which we then average over to compute a $2 \times 2$ matrix of average estimate and average standard error for the two procedures:

```
results_avg <- apply(results, c(2,3), mean)
```

Here is the result:

```
         Estimate  SE
Simple       -6.4 3.9
Adjusted      4.6 3.4
```

The true parameter value here is 5.0, so in this case the simple comparison is horribly biased—no surprise if you reflect upon the big differences between treatment and control groups from the simulation shown in Figure 16.5. In contrast, the bias of the adjusted estimate is low. In other settings, for example if the underlying relation between pre-test and post-test is nonlinear, or if there is selection on an unobserved or unmodeled variable, the regression-adjusted estimate can have a large bias too. We discuss these topics further in Chapters 18–21; our point here is that you can assess such biases using simulation, conditional on a model for data, measurement, and treatment assignment.

## 16.7 Bibliographic note

The quote at the beginning of Section 16.1 is from Weisburd, Petrosino, and Mason (1993); see also Gelman, Skardhamar, and Aaltonen (2017). The problems of statistical power are discussed by Button et al. (2013) and Gelman (2019a). Figure 16.1 comes from Gelman (2015d).

Cochran (1977) and Lohr (2009) are standard and useful references for classical models in survey sampling. Groves et al. (2009) and Heeringa, West, and Berglund (2017) go over practical aspects of survey design and analysis. Yates (1967), Montgomery (1986), and Box, Hunter, and Hunter (2005) review the statistical aspects of experimental design.

Hoenig and Heisey (2001), Lenth (2001), and Gelman and Carlin (2014) provide some general warnings and advice on sample size and power calculations. Assmann et al. (2000) discuss the general difficulty of estimating interactions. The design calculations for the sex ratio example in Section 16.5 are taken from Gelman and Weakliem (2009).

## 16.8 Exercises

16.1 *Sample size calculations for estimating a proportion*:

(a) How large a sample survey would be required to estimate, to within a standard error of $\pm$ 3%, the proportion of the U.S. population who support the death penalty?

(b) About 14% of the U.S. population is Latino. How large would a national sample of Americans have to be in order to estimate, to within a standard error of $\pm$3%, the proportion of Latinos in the United States who support the death penalty?

(c) How large would a national sample of Americans have to be in order to estimate, to within a standard error of $\pm$ 1%, the proportion who are Latino?

16.2 *Sample size calculation for estimating a difference*: Consider an election with two major candidates, A and B, and a minor candidate, C, who are believed to have support of approximately 45%, 35%, and 20% in the population. A poll is to be conducted with the goal of estimating the difference in support between candidates $A$ and $B$. How large a sample would you estimate is needed to estimate this difference to within a standard error of 5 percentage points? (Hint: consider an outcome variable that is coded as $+1$, $-1$, and 0 for supporters of A, B, and C, respectively.)

16.3 *Power*: Following Figure 16.3, determine the power (the probability of getting an estimate that is "statistically significantly" different from zero at the 5% level) of a study where the true effect size is $X$ standard errors from zero. Answer for the following values of $X$: 0, 1, 2, and 3.

16.4 *Power, type M error, and type S error*: Consider the experiment shown in Figure 16.1 where the true effect could not realistically be more than 2 percentage points and it is estimated with a standard error of 8.1 percentage points.

(a) Assuming the estimate is unbiased and normally distributed and the true effect size is 2 percentage points, use simulation to answer the following questions: What is the power of this study? What is the type M error rate? What is the type S error rate?

(b) Assuming the estimate is unbiased and normally distributed and the true effect size is *no more than* 2 percentage points in absolute value, what can you say about the power, type M error rate, and type S error rate?

16.5 *Design analysis for an experiment*: You conduct an experiment in which half the people get a special get-out-the-vote message and others do not. Then you follow up after the election with a random sample of 500 people to see if they voted.

(a) What will be the standard error of your estimate of effect size? Figure this out making reasonable assumptions about voter turnout and the true effect size.

(b) Check how sensitive your standard error calculation is to your assumptions.

(c) For a range of plausible effect sizes, consider conclusions from this study, in light of the statistical significance filter. As a researcher, how can you avoid this problem?

16.6 *Design analysis with pre-treatment information*: A new teaching method is hoped to increase scores by 5 points on a certain standardized test. An experiment is performed on $n$ students, where half get this intervention and half get the control. Suppose that the standard deviation of test scores in the population is 20 points. Further suppose that a pre-test is available which has a correlation of 0.8 with the post-test under the control condition. What will be the standard error of the estimated treatment effect based on a fitted regression, assuming that the treatment effect is constant and independent of the value of the pre-test?

16.7 *Decline effect*: After a study is published on the effect of some treatment or intervention, it is common for the estimated effect in future studies to be lower. Give five reasons why you might expect this to happen.

16.8 *Effect size and sample size*: Consider a toxin that can be tested on animals at different doses. Suppose a typical exposure level for humans is 1 (in some units), and at this level the toxin is hypothesized to introduce a risk of 0.01% of death per person.

(a) Consider different animal studies, each time assuming a linear dose-response relation (that is, 0.01% risk of death per animal per unit of the toxin), with doses of 1, 100, and 10 000. At each of these exposure levels, what sample size is needed to have 80% power of detecting the effect?

(b) This time assume that response is a logged function of dose and redo the calculations in (a).

16.9 *Cluster sampling with equal-sized clusters*: A survey is being planned with the goal of interviewing $n$ people in some number $J$ of clusters. For simplicity, assume simple random sampling of clusters and a simple random sample of size $n/J$ (appropriately rounded) within each sampled cluster.

Consider inferences for the proportion of Yes responses in the population for some question of interest. The estimate will be simply the average response for the $n$ people in the sample. Suppose that the true proportion of Yes responses is not too far from 0.5 and that the standard deviation among the mean responses of clusters is 0.1.

(a) Suppose the total sample size is $n = 1000$. What is the standard error for the sample average if $J = 1000$? What if $J = 100, 10, 1$?

(b) Suppose the cost of the survey is \$50 per interview, plus \$500 per cluster. Further suppose that the goal is to estimate the proportion of Yes responses in the population with a standard error of no more than 2%. What values of $n$ and $J$ will achieve this at the lowest cost?

16.10 *Simulation for design analysis*: The folder `ElectricCompany` contains data from the Electric Company experiment analyzed in Chapter 19. Suppose you wanted to perform a new experiment under similar conditions, but for simplicity just for second graders, with the goal of having 80% power to find a statistically significant result (at the 5% level) in grade 2.

(a) State clearly the assumptions you are making for your design calculations. (Hint: you can set the numerical values for these assumptions based on the analysis of the existing Electric Company data.)

(b) Suppose that the new data will be analyzed by simply comparing the average scores for the treated classrooms to the average scores for the controls. How many classrooms would be needed for 80% power?

(c) Repeat (b), but supposing that the new data will be analyzed by comparing the average gain scores for the treated classrooms to the average gain scores of the controls.

(d) Repeat, but supposing that the new data will be analyzed by regression, adjusting for pre-test scores as well as the treatment indicator.

16.11 *Optimal design*:

(a) Suppose that the zinc study described in Section 16.3 would cost \$150 for each treated child and \$100 for each control. Under the assumptions given in that section, determine the number of control and treated children needed to attain 80% power at minimal total cost. You will need to set up a loop of simulations as illustrated for the example in the text. Assume that the number of measurements per child is fixed at $K = 7$ (that is, measuring every two months for a year).

(b) Make a generalization of Figure 16.1 with several lines corresponding to different values of the design parameter $K$, the number of measurements for each child.

16.12 *Experiment with pre-treatment information:* An intervention is hoped to increase voter turnout in a local election from 20% to 25%.

(a) In a simple randomized experiment, how large a sample size would be needed so that the standard error of the estimated treatment effect is less than 2 percentage points?

(b) Now suppose that previous voter turnout was known for all participants in the experiment. Make a reasonable assumption about the correlation between turnout in two successive elections. Under this assumption, how much would the standard error decrease if previous voter turnout was included as a pre-treatment predictor in a regression to estimate the treatment effect?

16.13 *Sample size calculations for main effects and interactions*: In causal inference, it is often important to study varying treatment effects: for example, a treatment could be more effective for men than for women, or for healthy than for unhealthy patients. Suppose a study is designed to have 80% power to detect a main effect at a 95% confidence level. Further suppose that interactions of interest are half the size of main effects.

(a) What is its power for detecting an interaction, comparing men to women (say) in a study that is half men and half women?

(b) Suppose 1000 studies of this size are performed. How many of the studies would you expect to report a "statistically significant" interaction? Of these, what is the expectation of the ratio of estimated effect size to actual effect size?

16.14 *Working through your own example*: Continuing the example from the final exercises of the earlier chapters, think of a new data survey, experiment, or observational study that could be relevant and perform a design analysis for it, addressing issues of measurement, precision, and sample size. Simulate fake data for this study and analyze the simulated data.

# Chapter 17

# Poststratification and missing-data imputation

When fitting a model to data, parameter estimates are often just the beginning. Indeed, it is unusual for regression coefficients to directly map to questions of substantive interest, and it is also unusual for us to be completely sure of what model we want to fit. In practice, much of our effort goes into post-processing regression results and mapping them to questions of interest. All these steps require programming, and it will be important for us to manipulate the output of fitted models and not just look at default displays.

The present chapter considers two sorts of operations that are done as adjuncts to fitting a regression. In poststratification, the output from a fitted model are combined to make predictions about a new population that can differ systematically from the data. The model allows us to adjust for differences between sample and population—as long as the relevant adjustment variables are included as predictors in the regression, and as long as their distribution is known in the target population.

Poststratification is a form of post-processing of inferences that is important in survey research and also arises in causal inference for varying treatment effects, as discussed in subsequent chapters. In contrast, missing-data analysis is a pre-processing step in which data are cleaned or imputed in some ways so as to allow them to be used more easily in a statistical analysis. This chapter introduces the basic ideas of poststratification and missing-data imputation using a mix of real and simulated-data examples.

## 17.1 Poststratification: using regression to generalize to a new population

A typical assumption, often unstated, in statistical analysis is that the inference obtained from the data at hand can be applied to other scenarios. This assumption makes sense if data are a random sample from some well-defined population, but we should be careful if our data are a convenience sample or even a survey but with real-world imperfections.

Regression can reduce some of the concerns that arise in generalizing an inference to new data. The idea is that even if existing observations are not a representative sample of the population of interest, it may be possible to construct a regression to allow prediction of new cases, and then *poststratify* to average the fitted model's predictions over the population of interest. We demonstrate with two examples from political polling.

### Example: adjusting pre-election polls for party identification

Opinion polls during election campaigns are highly volatile, even though direct evidence from panel surveys suggests that only a small percentage of voters actually change their vote intentions. One explanation for poll-to-poll variation is that surveys taken at different times reach different groups of people. It is difficult for survey organizations to reach people on the phone, and, when reached, most potential respondents decline to be interviewed. Pollsters deal with nonresponse by adjusting their samples to match the population.

### Adjusting for a single factor

Example: Survey adjustment

Before getting to the general procedure, we demonstrate using a simplified example of adjustment on a single factor. The CBS News poll conducted from October 12–16, 2016, reported that, among likely voters who preferred one of the two major-party candidates, 45% intended to vote for Donald Trump and 55% for Hillary Clinton. Of these respondents, 33% reported Republican party identification, 40% affiliated themselves with the Democrats, and 27% did not declare a major-party affiliation.[1]

The party identification numbers reveal a sample that is more Democratic than voters in general, which are, let us say, 33% Republican, 36% Democratic, and 31% other. Estimating partisanship in the voting population is itself a challenge, involving the analysis of current opinion polls as well as extrapolating from exit polls from previous elections. Here for simplicity we treat these population proportions as known.

The CBS News survey was itself adjusted to match the demographics of the voting population, but we shall ignore that here, for simplicity treating it as a sample of voters that just happens to be not quite representative in party identification. More precisely, we make the assumption that the probability of a survey respondent supporting Trump (among those who prefer either Trump or Clinton), conditional on party identification, is the same as the corresponding conditional probability among the population of voters. We do not have the raw data from the survey; for the purpose of this exercise we construct a set of fake data consistent with the reported summaries.

We shall use poststratification to adjust the sample to match the population, performing the calculations in two ways, first using a simple weighted average that makes the idea transparent and then using regression prediction, a method that better generalizes when using multiple predictors.

**Poststratification using a weighted average.** Among the survey respondents who supported a major-party candidate, Trump had the support of 91% of self-declared Republicans, 5% of Democrats, and 49% of others. Weight these numbers by the proportion of Republicans, Democrats, and others in the population, and you get

$$\text{population-weighted average} = \frac{N_1}{N}\bar{y}_1 + \frac{N_2}{N}\bar{y}_2 + \frac{N_3}{N}\bar{y}_3, \tag{17.1}$$

where $N_j$ is the size of each group $j = 1, 2, 3$ in the population. In this case, the estimate is $0.33 * 0.91 + 0.36 * 0.05 + 0.31 * 0.49 = 0.47$. By comparison, the raw estimate from the survey was 0.45, which was too low because the sample did not include enough Republicans.

**What if we had weighted using the distribution of party identification in the sample?** Suppose instead we were to take Trump's support in each of the three poststratification cells and weight, not by the *population* proportion of Republicans, Democrats, and others, but by the proportion of each group in the *sample*? We would just get the simple average of the data: $0.33*0.91+0.40*0.05+0.27*0.49 = 0.45$.

This is simple algebra. Label $\bar{y}_j$ as the proportion of $n_j$ respondents in cell $j$ who support Trump, where $j = 1, \dots, J$, and the $J = 3$ *poststratification cells* correspond to Republicans, Democrats, and others. Further label the total sample size (again, just considering respondents who support Clinton or Trump) as $n = n_1 + n_2 + n_3$. Then the data-weighted average of the stratum proportions is

$$\begin{aligned} \text{data-weighted average} &= \frac{n_1}{n}\bar{y}_1 + \frac{n_2}{n}\bar{y}_2 + \frac{n_3}{n}\bar{y}_3 \\ &= \frac{n_1\bar{y}_1 + n_2\bar{y}_2 + n_3\bar{y}_3}{n} \\ &= \bar{y}, \end{aligned} \tag{17.2}$$

with this last equality holding because the numerator of the second line above, $n_1\bar{y}_1 + n_2\bar{y}_2 + n_3\bar{y}_3$, is simply the number of Trump supporters in the sample. We want to use formula (17.1), not (17.2), because the former adjusts for the sizes of the three groups in the population.

[1] Code for this example is in the folder `Poststrat`.

| $i$ | Vote | Party |
|---|---|---|
| 1 | 1 | "Independent" |
| 2 | 0 | "Democrat" |
| 3 | 0 | "Independent" |
| 4 | 1 | "Republican" |
| 5 | 0 | "Democrat" |
| | | ... |
| 778 | 0 | "Republican" |

| $j$ | Party | $N_j/N$ |
|---|---|---|
| 1 | "Republican" | 0.33 |
| 2 | "Democrat" | 0.36 |
| 3 | "Independent" | 0.31 |

Figure 17.1 *(a) Some of the (reconstructed) data from an opinion poll showing the outcome (vote preference) and a single input variable (political party identification). (b) The complete poststratification table for this example, with three cells. For clarity, we label survey respondents as $i = 1, \ldots, n$, and poststratification cells as $j = 1, \ldots, J$. The stratum weights $N_j/N$ in the poststratification table are proportional to the number of people in each cell in the population.*

**Poststratification using regression prediction.** Another way to estimate Trump's support among all the voters is to first fit a regression model to the data, and then use that fitted model to make predictions on the population. In this case, the model is simple because we are adjusting for just one factor, party identification:

```
fit_1 <- stan_glm(vote ~ factor(pid), data=poll)
```

A subset of the data are shown in Figure 17.1a, and here is the fitted regression:

```
                          Median MAD_SD
(Intercept)                0.05   0.02
factor(pid)Independent 0.43   0.03
factor(pid)Republican  0.85   0.03

Auxiliary parameter(s):
      Median MAD_SD
sigma 0.34   0.01
```

We set up the poststratification matrix as the data frame shown in Figure 17.1b and use the fitted model to make predictions for the cells:

```
poststrat_data <- data.frame(pid = c("Republican", "Democrat", "Independent"),
  N = c(0.33, 0.36, 0.31))
epred_1 <- posterior_epred(fit_1, newdata=poststrat_data)
```

We use `posterior_epred` because we are interested in the expected value—the population average—within each cell. Here we are doing linear regression, so `posterior_linpred` would give the same result, but for nonlinear models we would have to use `posterior_epred`; see Section 13.3. The result is a `n_sims` × 3 matrix representing posterior uncertainty in the proportion of Trump supporters in each of the three strata. We sum these to get the population average:

```
poststrat_est_1 <- epred_1 %*% poststrat_data$N/sum(poststrat_data$N)
print(c(mean(poststrat_est_1), mad(poststrat_est_1)), digits=2)
```

The poststratified estimate is 0.47 with standard error 0.013.

One might find this estimate to be uncomfortably precise—can we *really* predict a candidate's support so accurately from a single poll? The problem here is that the survey may be nonrepresentative (even after adjusting for party identification), and also vote preferences can change. The estimate and standard error represent our inference about the population based on the sample-as-snapshot, and the model can be extended to allow for other sources of error.

Suppose, for example, we would like to say that political polls can easily be off by 2 percentage points, merely due to nonresponse and changes of opinion. We could then add this uncertainty to our prediction:

| $i$ | Vote | Party | Male |
|---|---|---|---|
| 1 | 1 | "Independent" | 1 |
| 2 | 0 | "Democrat" | 1 |
| 3 | 0 | "Independent" | 0 |
| 4 | 1 | "Republican" | 0 |
| 5 | 0 | "Democrat" | 0 |
| | | ... | |
| 778 | 0 | "Republican" | 1 |

| $j$ | Party | Male | $N_j/N$ |
|---|---|---|---|
| 1 | "Republican" | 0 | 0.16 |
| 2 | "Republican" | 1 | 0.17 |
| 3 | "Democrat" | 0 | 0.22 |
| 4 | "Democrat" | 1 | 0.14 |
| 5 | "Independent" | 0 | 0.14 |
| 6 | "Independent" | 1 | 0.17 |

Figure 17.2 *(a) Some of the (reconstructed) data from an opinion poll showing the outcome (vote preference) and two input variables (sex and political party identification). (b) The complete poststratification table for this example, with six cells.*

```
n_sims <- nrow(epred_1)
poststrat_est_2 <- poststrat_est_1 + rnorm(n_sims, 0, 0.02)
print(c(mean(poststrat_est_2), mad(poststrat_est_2)), digits=2)
```

This new estimate has the same mean but a larger standard error.

### Adjusting for two factors

We can apply the same tools to poststratify on multiple factors, but the coding gets more involved and more modeling choices arise. We illustrate with a two-dimensional example, adjusting our survey to party identification and sex. To perform this analysis, we need the two datasets that are shown in Figure 17.2: the survey responses including vote intention, party, and sex; and the poststratification table, which in this case has 6 rows, corresponding to the $3 \times 2$ cells defined by party and sex, and which we can create as follows:

```
poststrat_data_2 <- data.frame(male = rep(c(0,1), 3),
  pid = rep(c("Republican", "Democrat", "Independent"), c(2, 2, 2)),
  N = c(0.16, 0.17, 0.19, 0.17, 0.16, 0.15))
```

Again, for our purposes here we treat the poststratification table as known, even though it also is estimated from fallible data.

We then fit a model to the data using our general principles of regression modeling. The simplest approach is regression with no interactions:

```
fit_2 <- stan_glm(vote ~ factor(pid) + male, data=poll)
```

Or we could include an interaction:

```
fit_2 <- stan_glm(vote ~ factor(pid) + male + factor(pid):male, data=poll)
```

And the fit could include prior information. In any case, we would just continue by applying the fitted model to an updated `poststrat` matrix:

```
epred_2 <- posterior_epred(fit_2, newdata=poststrat_data_2)
```

The remaining steps look the same. The only difficulty was setting up the poststratification matrix, which required looping through the values of each factor and obtaining the population sizes $N_j$.

### Adjusting the 2012 Xbox survey for many political and demographic variables

Example: Xbox survey

We next discuss regression and poststratification in a more realistic survey example. For decades, public opinion polls in the United States have mostly been using probability-based sampling methods such as random-digit dialing. Though historically effective, such traditional methods are often slow and expensive, and with declining response rates, even their accuracy has come into question. At the same time, non-representative polls, such as opt-in online surveys, have become increasingly fast and cheap. Such nonprobability-based sampling has generally been dismissed by the polling community,

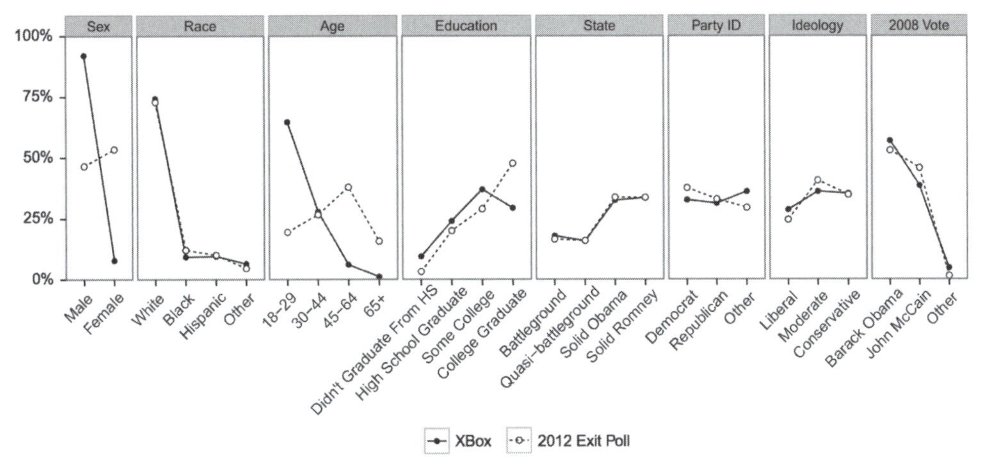


Figure 17.3 *A comparison of the demographic composition of participants in the Xbox polls and the 2012 electorate as measured by adjusted exit polls. Compared to the general population of voters, Xbox respondents are younger, have less education, more likely to be men, and more likely to have no affiliation with either party.*

and not without reason. It is not obvious how to extract meaningful signal from a collection of participants whose opinions are often far from representative of the population at large.

Here we demonstrate how with proper statistical adjustment, even highly non-representative polls can yield accurate estimates of population-level attitudes. Our approach is to use regression and poststratification based on a partition of the data and population into many demographic cells: first, we estimate attitudes at the cell level using a regression model; second, we aggregate cell-level estimates in accordance with the target population's demographic composition.

This procedure can also be used to adjust for confounding (systematic differences between treatment and control groups) in causal inference: first, the regression should include interactions between treatment and pre-treatment variables of interest; this model will yield estimated treatment effects that vary across the population. The second step is then to include these variables into the poststratification to estimate average treatment effects. We discuss this idea further in Section 19.4.

We now discuss the example at hand. For the last 45 days before the 2012 election, we operated an opt-in polling application on the Xbox gaming platform; questions were changed daily and users could respond up to once per day. There was limited coverage for this survey in that the only way to answer the polling questions was via the Xbox Live platform. And the survey was truly opt-in: there was no invitation or permanent link to the poll, so respondents had to locate it daily on the Xbox Live home page. Each daily poll had three to five questions, always including the question, "If the election were held today, who would you vote for?" Before answering the poll the first time, each respondent was asked to provide a variety of demographic information, including sex, ethnicity, race, age, education, state of residence, party identification, political ideology, and 2008 U.S. presidential vote. We introduced this example on page 6 in the opening chapter of this book, and here we explain a bit about how the analysis worked.

Figure 17.3 shows the distributions of the demographic and political background variables for Xbox respondents compared to the general population of voters as measured by adjusted exit polls from the 2012 election. Xbox respondents differ from the electorate in many ways: this is nothing like a representative sample. And, indeed, the raw estimates of national opinion from the Xbox sample are terrible. Figure 17.4 shows Barack Obama's share of the two-party support on each day from the Xbox polls, and from these data alone, Obama would look like a near-certain loser to his Republican opponent, Mitt Romney. The graph also shows a series of aggregated results from national telephone polls, which showed a close race throughout.

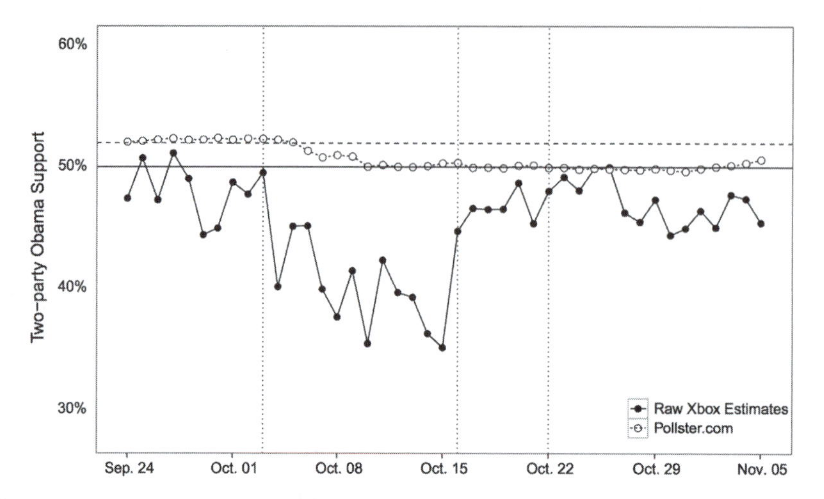


Figure 17.4 *Daily unadjusted Xbox and Pollster estimates of Barack Obama's share of the two-party support during the 45 days leading up to the 2012 presidential election. The Xbox series (black dots) appears to show a highly volatile electorate concluding with a landslide loss for Obama. The open circles, from the daily aggregated polling results at Pollster.com, show a much closer race. The horizontal dashed line at 52% indicates the actual two-party vote share obtained by Obama, and the vertical dotted lines give the dates of the three presidential debates.*

## Using regression to make inferences about the larger population

We can do better, though. Figure 17.3 doesn't just show how the Xbox sample is off; it also gives us a clue of how to adjust the sample to match the population. We shall now show how to do this, but demonstrating with a simplified model and skipping many of the details. On page 319 we briefly discuss the complications we have avoided in this description.

The first step of our simplified procedure is to fit a regression predicting vote from demographics:

```
fit <- stan_glm(y ~ female + factor(ethnicity) + factor(age) + factor(education) +
  factor(state_type) + factor(party_id) + factor(ideology) + factor(previous_vote),
  data=poll)
```

Here, $y$ is a variable that equals 1 if the respondent supports Obama for president and 0 if he or she supports Romney, with undecided voters and supporters of other candidates excluded from the analysis. We include all the input variables as discrete factors (with levels shown in Figure 17.3), even age, because of the nonlinear relation between age and vote preference.

Having fit the model, we can use it to make predictions about the general population. We do this by constructing a matrix of hypothetical voters. This matrix has 8 named columns (corresponding to the predictors in the above regression) and $2 \times 4 \times 4 \times 4 \times 4 \times 3 \times 3 \times 3 = 13\,824$ rows (corresponding to all the possible combinations of levels of all these factors). We call this matrix X_pop, and each of its rows represents a different segment of the population of voters. Our poststratification information is encoded as a vector, N_pop, that represents the number of voters in each of these categories (not known, but treated as known, extrapolated based on results of the previous presidential election).

We take the fitted model and use it to make predictions for each of the 13 824 sorts of voters:

```
epred_fit <- posterior_epred(fit, newdata=X_pop)
```

This yields a n_sims $\times$ 13 824 matrix representing posterior simulations of the model's predictions for each cell. We can then sum these over the population to get a vector of n_sims simulations representing uncertainty in Obama's national vote share, adjusting for differences between sample and population:

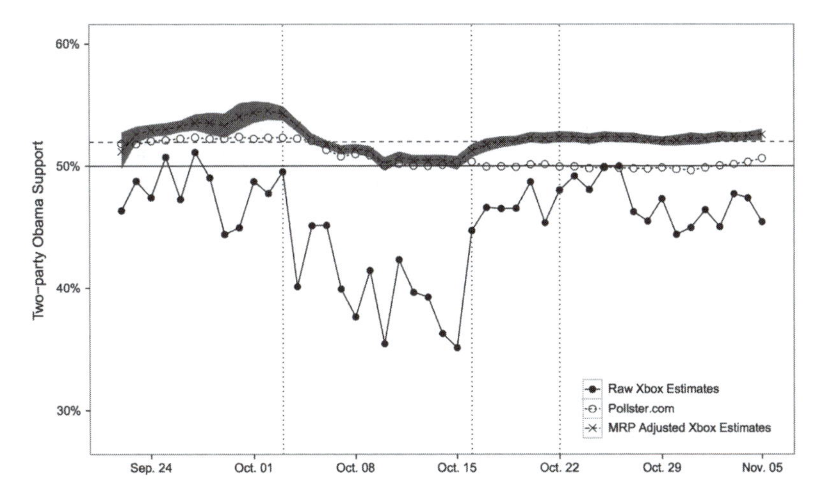


Figure 17.5 *Daily adjusted estimate from Xbox survey after regression and poststratification (with 95% uncertainty bounds), along with unadjusted Xbox and Pollster estimates of Barack Obama's share of the two-party support during the 45 days leading up to the 2012 presidential election.*

```
avg_pop <- epred_fit %*% N_pop / sum(N_pop)
```

And it really works. Figure 17.5 shows the adjusted estimates from the Xbox survey using regression and poststratification. Compared with the raw responses, the adjusted estimates of voter intent is much more reasonable, and voter intent in the last few days is close to the actual outcome.

This is not to say that online polling is perfect. A similar adjustment of online polls in the 2016 election estimated Hillary Clinton at 52% of the two-party vote on election eve. She actually received 51%—a small error, but enough to lose her the Electoral College. The polls were off by several percentage points in some key swing states, and the differences between sample and population in those states were not captured by the demographic and political variables used in the adjustment.

So our point is not that poststratification will necessarily get to the right answer. Poststratification adjusts for specific discrepancies between sample and population and not others. In 2012 and 2016, however, suitably adjusted internet polls performed as well as conventional telephone polls, in a way that would not have been possible without adjustment.

## Some real-world complications

In the above description, we simplified many of the steps of the Xbox example so we could focus on the key idea of using regression to generalize to a new population. Our actual analysis was more complicated, in several ways.

First, the outcome is binary and so we actually fit logistic, not linear, regression, but above we presented the simpler linear model just to keep the focus on the poststratification, which is a general process that can be done after fitting any regression model.

Second, the numbers $N_j$ in the poststratification cells are themselves estimates, and in this and other examples a certain amount of statistical modeling is required to come up with them. In some examples it can make sense to propagate the uncertainty into later inferences. Even if our model contains continuous predictors, we typically poststratify into discrete cells, in which case we use our regression to make some sort of average prediction for each cell.

Third, we obtained separate estimates from each day's survey responses. But estimating the above regression from each separate day's data would be too noisy, so we actually fit a single model to the data from all 45 days, allowing some of the coefficients to vary over time.

Fourth, we did not estimate the simple regression shown above. Instead, we included many interactions, and we fit the model using *multilevel regression*, an application of Bayesian inference that allows us to fit model with many varying coefficients in a way that is more stable than would be possible using least squares or maximum likelihood. This leads to multilevel regression and poststratification or, more generally, regularized prediction and poststratification.

In Exercise 17.2, you will have a chance to do regression and poststratification in a simplified survey setting, but we thought it would be valuable to see a real-world example here, even if it is somewhat stylized.

## 17.2 Fake-data simulation for regression and poststratification

We can get further insight into poststratification adjustments, as well as regression inferences more generally, using a simulated example. Here we demonstrate by setting up a scenario in which a Yes/No survey response is fit given sex, age (18–29, 30–44, 45–64, 65+), and ethnicity (non-Hispanic white, black, Hispanic, other): thus $2 \times 4 \times 4$ categories.[2]

### Creating the artificial world

Example: Simulated poststratification

First we construct the poststratification table. The right way to do this is to download data from the U.S. Census, which collects demographic information on all residents of the United States every 10 years, with supplemental surveys in between. To demonstrate how this can all be done from scratch, though, we shall make up the numbers based on vague guesses. It's fine to use made-up numbers in a simulation; you can always go back later and insert more accurate values.

The first step is to create the empty poststratification table, with rows for each of the $2 \times 4 \times 4$ sorts of people.

```
J <- c(2, 4, 4)
poststrat <- as.data.frame(array(NA, c(prod(J), length(J)+1)))
colnames(poststrat) <- c("sex", "age", "eth", "N")
count <- 0
for (i1 in 1:J[1]){
  for (i2 in 1:J[2]){
    for (i3 in 1:J[3]){
      count <- count + 1
      poststrat[count, 1:3] <- c(i1, i2, i3)
    }
  }
}
```

Next we make up numbers for the populations of the cells and enter these into the table:

```
p_sex <- c(0.52, 0.48)
p_age <- c(0.2, 0.25, 0.3, 0.25)
p_eth <- c(0.7, 0.1, 0.1, 0.1)
for (j in 1:prod(J)){
  poststrat$N[j] <- 250e6 * p_sex[poststrat[j,1]] * p_age[poststrat[j,2]] *
  p_eth[poststrat[j,3]]
}
```

In this particular case, we have assumed a population of 250 million adults, 52% female and 48% male; an age distribution that is 20% from 18–29, 25% from 30–44, 30% from 45–64, and 25% aged 65+; and an ethnic distribution that is 70% white and 10% each of the other three groups. We have also implicitly assumed that sex, age, and ethnicity are statistically independent in the population. This assumption is wrong—for example, women are more likely than men to survive into old age—and it would not be made if we were directly using Census data. In a fake-data simulation

[2]Code for this example is in the folder Poststrat.

it would be possible to remove the independence assumption by simply changing those values of N inserted into the table; see Exercise 17.2.

We then hypothesize a nonresponse pattern in which women, older people, and whites are more likely to respond than men, younger people, and minorities.

```
p_response_baseline <- 0.1
p_response_sex <- c(1, 0.8)
p_response_age <- c(1, 1.2, 1.6, 2.5)
p_response_eth <- c(1, 0.8, 0.7, 0.6)
p_response <- rep(NA, prod(J))
for (j in 1:prod(J)){
  p_response[j] <- p_response_baseline * p_response_sex[poststrat[j,1]] *
  p_response_age[poststrat[j,2]] * p_response_eth[poststrat[j,3]]
}
```

Again, this is a particularly simple multiplicative model with no interactions; there would be no difficulty in simulating a more complicated model of nonresponse.

We then sample from the assumed population with the assumed nonresponse probabilities:

```
n <- 1000
people <- sample(prod(J), n, replace=TRUE, prob=poststrat$N*p_response)
  ## For respondent i, people[i] is that person's poststrat cell,
  ## some number between 1 and 32
n_cell <- rep(NA, prod(J))
for (j in 1:prod(J)){
  n_cell[j] <- sum(people==j)
}
print(cbind(poststrat, n_cell/n, poststrat$N/sum(poststrat$N)))
```

We shall assume the survey responses come from a logistic regression with these coefficients:

```
coef_intercept <- 0.6
coef_sex <- c(0, -0.2)
coef_age <- c(0, -0.2, -0.3, -0.4)
coef_eth <- c(0, 0.6, 0.3, 0.3)
```

Hence the probabilities are:

```
prob_yes <- rep(NA, prod(J))
for (j in 1:prod(J)){
  prob_yes[j] <- invlogit(coef_intercept + coef_sex[poststrat[j,1]] +
    coef_age[poststrat[j,2]] + coef_eth[poststrat[j,3]])
}
```

We can then simulate the fake data:

```
y <- rbinom(n, 1, prob_yes[people])
```

### Performing regression and poststratification

Having simulated data in this artificial world, we can conduct the statistical analysis on the simulated data and then poststratify.

First we fit a logistic regression, predicting the survey response given sex, age, and ethnicity, with no interactions:

```
sex <- poststrat[people,1]
age <- poststrat[people,2]
eth <- poststrat[people,3]
fake <- data.frame(y, sex, age, eth)
fit <- stan_glm(y ~ factor(sex) + factor(age) + factor(eth),
  family=binomial(link="logit"), data=fake)
print(fit)
```

We use posterior_epred to estimate the proportion of Yes responses in the $2 \times 4 \times 4$ cells:

```
pred_sim <- posterior_epred(fit, newdata=as.data.frame(poststrat))
pred_est <- colMeans(pred_sim)
print(cbind(poststrat, prob_yes, pred_est))
```

Finally, we poststratify:

```
poststrat_est <- sum(poststrat$N*pred_est)/sum(poststrat$N)
round(poststrat_est, 2)
```

The above gives us the point estimate; to get inferential uncertainty, we can work with the matrix of posterior simulations:

```
poststrat_sim <- pred_sim %*% poststrat$N / sum(poststrat$N)
round(c(mean(poststrat_sim), sd(poststrat_sim)), 3)
```

## 17.3 Models for missingness

Example: Social Indicators Survey

Missing data arise in almost all serious statistical analyses. In this chapter we discuss a variety of methods to handle missing data, including some relatively simple approaches that can often yield reasonable results. We use as a running example the Social Indicators Survey, a telephone survey of New York City families conducted by the Columbia University School of Social Work. Nonresponse in this survey is a distraction to our main goal of studying trends in attitudes and economic conditions, and we would like to simply clean the dataset so it could be analyzed as if there were no missingness. After some background in this section and the next, we discuss in Section 17.5 our general approach of random *imputation*, the insertion of modeled values in place of missing observations in a dataset. Section 17.6 discusses situations where not just the missing observations but also the missing-data process itself must be modeled in order to perform imputations correctly.

### Missing-data codes in R

In R, missing values are indicated by NA's. For example, to see some of the data from five respondents (arbitrarily picking rows 91–95 of the data file) in the Social Indicators Survey, we type:[3]

```
SIS[91:95, c("male", "white", "educ_r", "earnings")]
```

and get,

```
   male white educ_r earnings
91    1     0      3       NA
92    0     1      2      135
93    0     0      2       NA
94    1     1      3        3
95    1     0      1        0
```

In regression (as well as many of its other procedures), R automatically excludes all cases in which the outcome or any of the inputs are missing; this can limit the amount of information available in the analysis, especially if the model includes many inputs with potential missingness. This approach is called a complete-case analysis, and we discuss some of its weaknesses below.

### Missingness mechanisms

It is helpful to understand why data are missing. We consider four general "missingness mechanisms," moving from the easiest to most difficult to handle:

[3]Data and code for this example are in the folder Imputation.

1. *Missingness completely at random.* A variable is *missing completely at random* if the probability of missingness is the same for all units, for example, if each survey respondent decides whether to answer the earnings question by rolling a die and refusing to answer if a 6 shows up. If data are missing completely at random, then throwing out cases with missing data does not bias your inferences.
2. *Missingness at random.* Missingness is not typically *completely* at random, as can be seen from the observed data and pattern of missingness. For example, the different nonresponse rates for whites and blacks (see Exercise 17.3) indicate that the earnings question in the Social Indicators Survey is not missing completely at random.
   A more general assumption, *missing at random*, is that the probability that a variable is missing depends only on available information. Thus, if sex, race, education, and age are recorded for all the people in the survey, then `earnings` is missing at random if the probability of nonresponse to this question depends only on these other, fully recorded variables. It is often reasonable to model this process as a logistic regression, where the outcome variable equals 1 for observed cases and 0 for missing.
   When an outcome variable is missing at random, it is acceptable to exclude the missing cases (that is, to treat them as NA's), as long as the regression adjusts for all the variables that affect the probability of missingness. Thus, any model for earnings would have to include predictors for ethnicity to avoid nonresponse bias.
   Missingness at random is the same sort of assumption as ignorability, which we discuss later in the context of causal inference. Both require that sufficient information has been collected that we can "ignore" the assignment mechanism (assignment to treatment, assignment to nonresponse).
3. *Missingness that depends on unobserved predictors.* Missingness is no longer "at random" if it depends on information that has not been recorded, and this information also predicts the missing values. For example, suppose that "surly" people are less likely to respond to the earnings question, surliness is predictive of earnings, and "surliness" is unobserved. Or, suppose that people with college degrees are less likely to reveal their earnings, having a college degree is predictive of earnings, and there is also some nonresponse to the education question. Then, once again, earnings are not missing at random.
   A familiar example from medical studies is that if a particular treatment causes discomfort, a patient is more likely to drop out of the study. This missingness is not at random (unless "discomfort" is measured and observed for all patients).
   If missingness is not at random, it must be explicitly modeled, or else you must accept some bias in your inferences.
4. *Missingness that depends on the missing value itself.* The most difficult situation is when missingness can depend on the (potentially missing) variable itself. For example, suppose that people with higher earnings are less likely to reveal them. In the extreme case (for example, all people earning more than $100 000 refuse to respond), this is called *censoring*, but even the probabilistic case causes difficulty.
   Censoring and related missing-data mechanisms can be modeled (as discussed in Section 17.6) or else mitigated by including more predictors in the missing-data model and thus bringing it closer to missing at random. For example, whites and people with college degrees tend to have higher-than-average incomes, so adjusting for these predictors will somewhat—but probably only somewhat—correct for the higher rate of nonresponse among higher-income people. More generally, such predictions can be challenging in that the nature of the missing-data mechanism may force these predictive models to extrapolate beyond the range of the observed data.

### General impossibility of proving that data are missing at random

As discussed above, missingness at random is conceptually easy to handle—simply include as regression inputs all variables that affect the probability of missingness, and then build a good predictive model. Unfortunately, we generally cannot be sure whether data really are missing at random, or whether the missingness depends on unobserved predictors or the missing data themselves. The fundamental difficulty is that these potential "lurking variables" are unobserved—by definition—and so we can never rule them out. We generally must make assumptions, or check with reference to other studies (for example, surveys in which extensive follow-ups are done in order to ascertain the earnings of nonrespondents).

In practice, we typically try to include as many predictors as possible in a model, so that the assumption of missingness at random is reasonable. For example, it may be a strong assumption that nonresponse to the earnings question depends only on sex, race, and education—but this is a lot more plausible than assuming that the probability of nonresponse is constant, or that it depends only on one of these predictors.

## 17.4 Simple approaches for handling missing data

### Missing-data methods that discard data

Many missing data approaches simplify the problem by throwing away data. We discuss in this section how these approaches may lead to biased estimates (one of these methods tries to directly address this issue). In addition, throwing away data can lead to estimates with larger standard errors due to reduced sample size.

### Complete-case analysis

A direct approach to missing data is to exclude them. In the regression context, this usually means *complete-case analysis*: excluding all units for which the outcome or any of the inputs are missing. In R, this is done automatically for classical regressions (data points with any missingness in the predictors or outcome are ignored by the regression).

Two problems arise with complete-case analysis:

1. If the units with missing values differ systematically from the completely observed cases, this could bias the complete-case analysis.
2. If many variables are included in a model, there may be very few complete cases, so that most of the data would be discarded for the sake of a simple analysis.

### Available-case analysis

Another simple approach is *available-case analysis*, where different aspects of a problem are studied with different subsets of the data. For example, in the 2001 Social Indicators Survey, all 1501 respondents stated their education level, but 16% refused to state their earnings. We could thus summarize the distribution of education levels of New Yorkers using all the responses and the distribution of earnings using the 84% of respondents who answered that question. This approach has the problem that different analyses will be based on different subsets of the data and thus will not necessarily be consistent with each other. In addition, as with complete-case analysis, if the nonrespondents differ systematically from the respondents, this will bias the available-case summaries. For example in the Social Indicators Survey, 90% of African Americans but only 81% of whites report their earnings, so the "earnings" summary represents a different population than the "education" summary.

Available-case analysis also arises when a researcher simply excludes a variable or set of variables from the analysis because of their missing-data rates (sometimes called "complete-variables analyses"). In a causal inference context (as with many prediction contexts), this may lead to omission of a variable that is necessary to satisfy the assumptions necessary for desired (causal) interpretations.

### Nonresponse weighting

As discussed previously, complete-case analysis can yield biased estimates because the sample of observations that have no missing data might not be representative of the full sample. Is there a way of reweighting this sample so that representativeness is restored?

Suppose, for instance, that only one variable has missing data. We could build a model to predict the nonresponse in that variable using all the other variables. The inverse of predicted probabilities of response from this model could then be used as survey weights to make the complete-case sample representative (along the dimensions measured by the other predictors) of the full sample. This method becomes more complicated when there is more than one variable with missing data. Moreover, as with any such weighting scheme, standard errors can become erratic if predicted probabilities are close to 0 or 1.

### Simple missing-data approaches that retain all the data

Rather than removing variables or observations with missing data, another approach is to fill in or "impute" missing values. A variety of imputation approaches can be used that range from extremely simple to rather complex. These methods keep the full sample size, which can be advantageous for bias and precision; however, they can yield different kinds of bias, as detailed in this section.

Whenever a single imputation strategy is used, the standard errors of estimates tend to be too low. The intuition here is that we have substantial uncertainty about the missing values, but by choosing a single imputation we in essence pretend that we know the true value with certainty.

**Mean imputation.** Perhaps the easiest way to impute is to replace each missing value with the mean of the observed values for that variable. Unfortunately, this strategy can severely distort the distribution for this variable, leading to complications with summary measures including, notably, underestimates of the standard deviation. Moreover, mean imputation distorts relationships between variables by pulling estimates of the correlation toward zero.

**Last value carried forward.** In evaluations of interventions where pre-treatment measures of the outcome variable are also recorded, a strategy that is sometimes used is to replace missing outcome values with the pre-treatment measure. This is often thought to be a conservative approach (that is, one that would lead to underestimates of the true treatment effect). However, there are situations in which this strategy can be anti-conservative. For instance, consider a randomized evaluation of an intervention that targets couples at high risk of HIV infection. From the regression-to-the-mean phenomenon (see Section 12.3), we might expect a reduction in risky behavior even in the absence of the randomized experiment; therefore, carrying the last value forward will result in values that look worse than they truly are. Differential rates of missing data across the treatment and control groups will result in biased treatment effect estimates.

**Using information from related observations.** Suppose we are missing data regarding the income of fathers of children in a dataset. Why not fill these values in with the mother's report of the values? This is a plausible strategy, although these imputations may propagate measurement error. We must consider ways in which the reporting person might misrepresent the measurement for the person about whom he or she is providing information.

**Imputation based on logical rules.** Sometimes we can impute using logical rules: for example, the Social Indicators Survey includes a question on "number of months worked in the previous year," which all 1501 respondents answered. Of the people who refused to answer the earnings question, 10 reported working zero months during the previous year, and thus we could impute zero earnings to them. This type of imputation strategy does not rely on particularly strong assumptions since, in effect, the missing-data mechanism is known.

## 17.5 Understanding multiple imputation

Rather than replacing each missing value in a dataset with one randomly imputed value, it may make sense to replace each with several imputed values that reflect our uncertainty about our imputation model. For example, if we impute using a regression model we may want our imputations to reflect not only sampling variability (as random imputation should) but also our uncertainty about the regression coefficients in the model. If these coefficients themselves are modeled, we can draw a new set of missing-value imputations for each draw from the distribution of the coefficients.

*Multiple imputation* does this by creating several (say, five) imputed values for each missing value, each of which is predicted from a slightly different model and each of which also reflects sampling variability. How do we analyze these data? The simple idea is to use each set of imputed values to form (along with the observed data) a *completed* dataset. Within each completed dataset a standard analysis can be run. Then inferences can be combined across datasets.

### Combining inferences from multiple imputations

Before getting to the construction of the imputations we consider how they will be used.

For instance, suppose we want to make inferences about a regression coefficient, $\beta$. We obtain estimates $\hat{\beta}_m$ in each of the $M$ datasets as well as standard errors, $\mathrm{se}_1, \dots, \mathrm{se}_M$. To obtain an overall point estimate, we then simply average over the estimates from the separate imputed datasets; thus, $\hat{\beta} = \frac{1}{M}\sum_{m=1}^{M}\hat{\beta}_m$. We then can combine variation within and between imputations to get a standard error for this estimate:

$$\mathrm{se}_\beta = \sqrt{W + \left(1 + \frac{1}{M}\right)B},$$

where the within-imputation variance is estimated by $W = \frac{1}{M}\sum_{m=1}^{M}\mathrm{se}_m^2$ and the between-imputation variance is estimated by $B = \frac{1}{M-1}\sum_{m=1}^{M}(\hat{\beta}_m - \hat{\beta})^2$.

### Simple random imputation

Example: Social Indicators Survey

In order to understand missing-data imputation, we start with the relatively simple setting in which missingness is confined to a single variable, $y$, with a set of variables $X$ that are observed on all units. We shall consider the case of imputing missing earnings in the Social Indicators Survey.[4]

The simplest approach is to impute missing values of earnings based on the observed data for this variable. We can write this as an R function:

```
random_imp <- function(a) {
  missing <- is.na(a)
  n_missing <- sum(missing)
  a_obs <- a[!missing]
  imputed <- a
  imputed[missing] <- sample(a_obs, n_missing)
  imputed
}
```

To see how this function works, take a small dataset and evaluate the function line by line. We use `random_imp()` to create a *completed data* vector of earnings:

```
SIS$earnings_imp <- random_imp(SIS$earnings)
```

imputing into the missing values of the original `earnings` variable. This approach does not make much sense—it ignores the useful information from all the other questions asked of these survey responses—but these simple random imputations can be a convenient starting point. A better approach is to fit a regression to the observed cases and then use that to predict the missing cases, as we show next.

[4]Data and code for this example are in the folder `Imputation`.

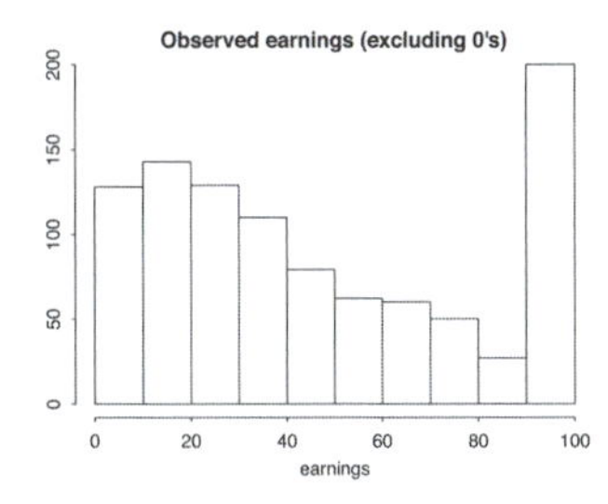


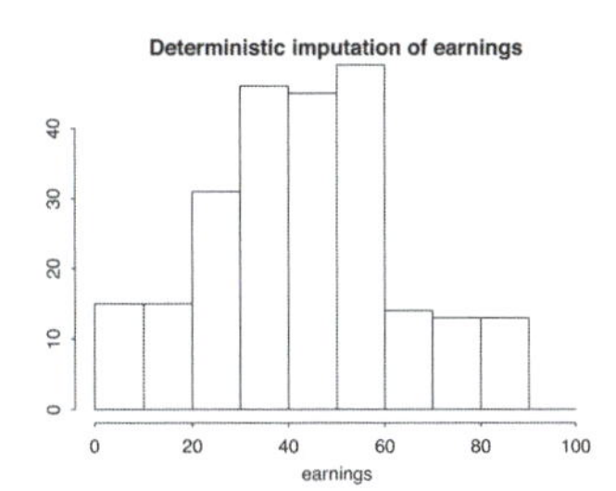


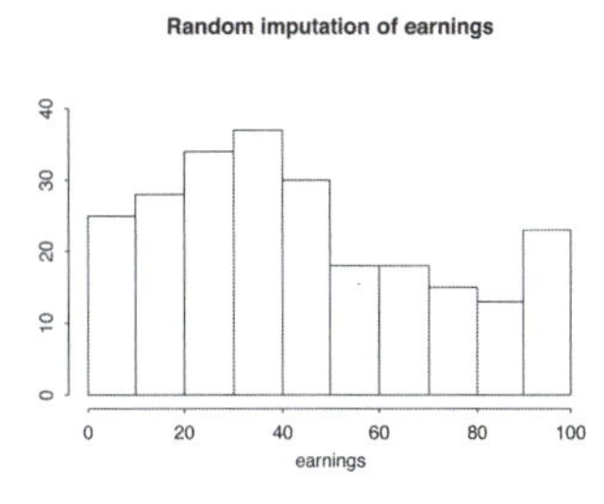


Figure 17.6 *Histogram of earnings (in thousands of dollars) in the Social Indicators Survey: (a) for the 988 respondents who answered the question and had positive earnings, (b) deterministic imputations for the 241 missing values from a regression model, (c) random imputations from that model. All values are topcoded at 100, with zero values excluded.*

## Zero coding and topcoding

We begin with some practicalities of the measurement scale. We shall fit the regression model to those respondents whose earnings were observed and positive (since, as noted earlier, the respondents with zero earnings can be identified from their zero responses to the "months worked" question). In addition, we shall "topcode" all earnings at \$100 000—that is, all responses above this value will be set to \$100 000—before running the regression. Figure 17.6a shows the distribution of positive earnings after topcoding.

```
topcode <- function(a, top){
  ifelse(a>top, top, a)
}
SIS$earnings_top <- topcode(SIS$earnings, 100)    # earnings are in $thousands
hist(SIS$earnings_top[SIS$earnings>0])
```

Topcoding reduces the sensitivity of the results to the highest values, which in this survey go up to the millions. By topcoding, we lose information, but the main use of earnings in this survey is to categorize families into income quantiles, for which purpose topcoding at \$100 000 has no effect.

Similarly, we topcoded number of hours worked per week at 40 hours. The purpose of topcoding was not to correct the data—we have no particular reason to disbelieve the high responses—but rather to perform a simple transformation to improve the predictive power of the regression model.

## The problem with using regression predictions for deterministic imputation

A simple and general imputation procedure that uses individual-level information uses a regression to the nonzero values of earnings. We first fit a regression to positive values of earnings:

```
fit_imp_1 <- stan_glm(earnings ~ male + over65 + white + immig + educ_r +
  workmos + workhrs_top + any_ssi + any_welfare + any_charity,
  data=SIS, subset=earnings>0)
```

We shall describe these predictors shortly, but first we go through the steps needed to create deterministic and then random imputations. We first get point predictions for all the data:

```
SIS_predictors <- SIS[, c("male", "over65", "white", "immig", "educ_r",
  "workmos", "workhrs_top", "any_ssi", "any_welfare", "any_charity")]
pred_1 <- predict(fit_imp_1, newdata=SIS_predictors)
```

To get predictions for the entire data vector, we must include the data frame for the predictors, `SIS_predictors`, in the `posterior_predict()` call. We needed to create a data frame just with the predictors, excluding the earnings and interest variables, which have missing values (`NA` in R) and would cause the `predict` function to skip these cases. As discussed below, we can include predictors with missing data only if we first impute provisional values for these variables.

Next we write a little function to create a completed dataset by imputing the predictions into the missing values:

```
impute <- function(a, a_impute){
  ifelse(is.na(a), a_impute, a)
}
```

and use this to impute missing earnings:

```
SIS$earnings_imp_1 <- impute(SIS$earnings, pred_1)
```

### Transformations

For the purpose of predicting incomes in the low and middle range, where we are most interested in this application, we work on the square root scale of income, topcoded to 100 (in thousands of dollars); as discussed in Section 12.5, we would expect a linear prediction model to fit better on that compressed scale. Here is the imputation procedure:

```
fit_imp_2 <- stan_glm(sqrt(earnings_top) ~ male + over65 + white + immig + educ_r +
  workmos + workhrs_top + any_ssi + any_welfare + any_charity,
  data=SIS, subset=earnings>0)
print(fit_imp_2)
pred_2_sqrt <- predict(fit_imp_2, newdata=SIS_predictors)
pred_2 <- topcode(pred_2_sqrt^2, 100)
SIS$earnings_imp_2 <- impute(SIS$earnings_top, pred_2)
```

Figure 17.6b shows the resulting deterministic imputations:

```
hist(SIS$earnings_imp_2[is.na(SIS$earnings)])
```

From this graph, it appears that most of the nonrespondents have incomes in the middle range (compare to Figure 17.6a). Actually, the central tendency of Figure 17.6b is an artifact of the deterministic imputation procedure. One way to see this is through the regression model: its $R^2$ is 0.44, which means that the explained variance from the regression is only 44% of the total variance. Equivalently, the explained standard deviation is $\sqrt{0.44} = 0.66 = 66\%$ of the data standard deviation. Hence, the predicted values from the regression will tend to be less variable than the original data. If we were to use the resulting deterministic imputations, we would be falsely implying that most of these nonrespondents had incomes in the middle of the scale.

### Random regression imputation

We can put the uncertainty back into the imputations by adding the prediction error into the regression, as discussed in Section 9.2. For this example, this involves creating a vector of random predicted values for the 241 missing cases—here we simply grab the first row from the matrix of simulated predicions—and then squaring, as before, to return to the original dollar scale:

```
pred_4_sqrt <- posterior_predict(fit_imp_2, newdata=SIS_predictors, draws=1)
pred_4 <- topcode(pred_4_sqrt^2, 100)
SIS$earnings_imp_4 <- impute(SIS$earnings_top, pred_4)
```

Figure 17.6c shows the resulting imputed values from a single simulation draw. Compared to Figure 17.6b, these random imputations are more appropriately spread across the range of the population.

The new imputations certainly do not look perfect—in particular, there still seem to be too few imputations at the topcoded value of $100 000—suggesting that the linear model on the square root scale, with normal errors, is not quite appropriate for these data. (This makes sense given the spike in the data from the topcoding.) The results look much better than the deterministic imputations, however.

Figure 17.7 illustrates the deterministic and random imputations in another way. Figure 17.7a

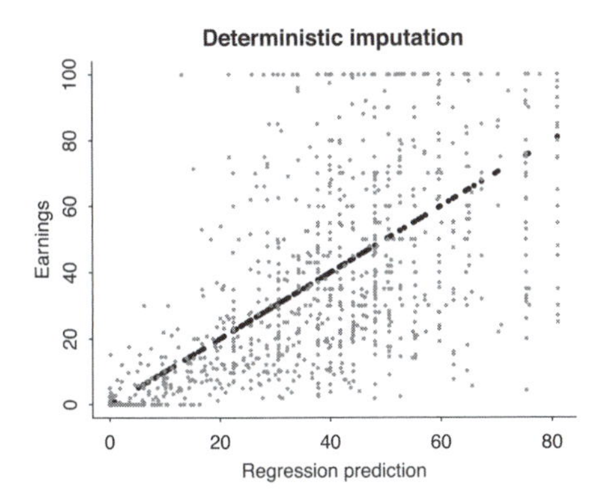


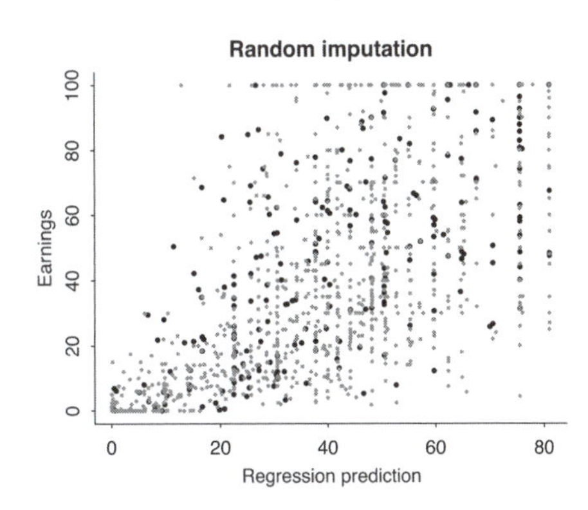


Figure 17.7 *Gray dots show observed data, and black dots show deterministic and random imputations for the 241 missing values of earnings in the Social Indicators Survey. (a) The deterministic imputations are exactly at the regression predictions and ignore predictive uncertainty. (b) In contrast, the random imputations are more variable and better capture the range of earnings in the data. See also Figure 17.6.*

shows the deterministic imputations as a function of the predicted earnings from the regression model. By the definition of the imputation procedure, the values are identical and so the points fall along the identity line. Figure 17.7b shows the random imputations, which follow a generally increasing pattern but with scatter derived from the unexplained variance in the model. The increase in variance as a function of predicted value arises from fitting the model on the square root scale and squaring at the end.

### Predictors used in the imputation model

We fit a regression of earnings on sex, age, ethnicity, nationality, education, the number of months worked in the previous year and hours worked per week, and indicators for whether the respondent's family receives each of three forms of income support (from disability payments, welfare, and private charities).

It might seem strange to model earnings given information on income support—which is, in part, a consequence of earnings—but for the purposes of imputation, this is acceptable. The goal here is not causal inference but simply accurate prediction, and it is acceptable to use any inputs in the imputation model to achieve this goal.

### Two-stage modeling to impute a variable that can be positive or zero

In the Social Indicators Survey, we only need to impute the positive values of earnings: the "hours worked" and "months worked" questions were answered by everyone in the survey, and these variables are a perfect predictor of whether the value of earnings (more precisely, employment income) is positive. For the missing cases of `earnings`, we can impute 0 if `workhrs` = 0 and `workmos` = 0, and impute a continuous positive value when either of these is positive. This imputation process is what was described above, with the regression based on $n = 988$ data points and displayed in Figure 17.7. The survey as a whole included 1501 families, of whom 272 reported working zero hours and months and were thus known to have zero earnings. Of the 1229 people reporting positive working hours or months, 988 responded to the earnings question and 241 did not.

Now suppose that the `workhrs` and `workmos` variables were *not* available, so that we could not immediately identify the cases with zero earnings.We would then impute missing responses to the earnings question in two steps: first, imputing an indicator for whether earnings are positive, and, second, imputing the continuous positive values of earnings.

Mathematically, we would impute earnings $y$ given regression predictors $X$ in a two-step process, defining

$$y = I^y y^{\text{pos}},$$

where $I^y = 1$ if $y > 0$ and 0 otherwise, and $y^{\text{pos}} = y$ if $y > 0$. The first model is a logistic regression for $I^y$:

$$\Pr(I_i^y = 1) = \text{logit}^{-1}(X_i \alpha),$$

and the second part is a linear regression for the square root of $y^{\text{pos}}$:

$$\sqrt{y_i^{\text{pos}}} = X_i \beta + \text{error}.$$

The first model is fit to all the data for which $y$ is observed, and the second model is fit to all the data for which $y$ is observed and positive.

We illustrate with the earnings example. First we fit the two models:

```
fit_positive <- stan_glm((earnings>0) ~ male + over65 + white + immig + educ_r +
  any_ssi + any_welfare + any_charity, data=SIS, family=binomial(link=logit))
print(fit_positive)
fit_positive_sqrt <- stan_glm(sqrt(earnings_top) ~ male + over65 + white + immig +
  educ_r + any_ssi + any_welfare + any_charity, data=SIS, subset=earnings>0)
print(fit_positive_sqrt)
```

Then we impute an indicator for whether the missing earnings are positive:

```
pred_sign <- posterior_predict(fit_positive, newdata=SIS_predictors, draws=1)
pred_pos_sqrt <- posterior_predict(fit_positive_sqrt, newdata=SIS_predictors, draws=1)
```

and then impute the earnings themselves:

```
pred_pos <- topcode(pred_pos_sqrt^2, 100)
SIS$earnings_imp <- impute(SIS$earnings, pred_sign*pred_pos)
```

## Matching and hot-deck imputation

A different way to impute is through *matching*: for each unit with a missing $y$, find a unit with similar values of $X$ in the observed data and take its $y$ value. This approach is also sometimes called "hot-deck" imputation (in contrast to "cold deck" methods, where the imputations come from a previously collected data source). Matching imputation can be combined with regression by defining "similarity" as closeness in the regression predictor, $X\beta$. Matching can be viewed as a nonparametric or local version of regression and can also be useful in some settings where setting up a regression model can be challenging.

For example, the New York City Department of Health has the task of assigning risk factors to all new HIV cases. The risk factors are assessed from a reading of each patient's medical file, but for a large fraction of the cases, not enough information is available to determine the risk factors. For each of these "unresolved" cases, we proposed taking a random imputation from the risk factors of the five closest resolved cases, where "closest" is defined based on a scoring function that penalizes differences in sex, age, the clinic where the HIV test was conducted, and other information that is available on all or most cases.

More generally, one could estimate a *propensity score* that predicts the missingness of a variable conditional on several other variables that are fully observed, and then match on this score (see Section 20.7) to impute missing values.

## Multiple imputation of several missing variables

It is common to have missing data in several variables in an analysis, in which case one cannot simply set up a model for a single partially observed variable $y$ given a set of fully observed $X$ variables. In fact, even in the Social Indicators Survey example, some of the predictor variables (ethnicity, interest income, and the indicators for income supplements) had missing values in the data, which we crudely imputed before running the regression for the imputations. More generally, we must think of the dataset as a multivariate outcome, any components of which can be missing.

### Routine multivariate imputation

The direct approach to imputing missing data in several variables is to fit a multivariate model to all the variables that have missingness, thus generalizing the approach of Section 17.5 to allow the outcome $Y$ as well as the predictors $X$ to be a vector. The difficulty of this approach is that it requires a lot of effort to set up a reasonable multivariate regression model, and so in practice an off-the-shelf model is typically used, most commonly the multivariate normal or $t$ distribution for continuous outcomes, and a multinomial distribution for discrete outcomes. Software exists to fit such models automatically, so that one can conceivably "press a button" and impute missing data. These imputations are only as good as the model, and so they need to be checked in some way—but this automatic approach is easy enough that it can be a good place to start.

### Iterative regression imputation

A different way to generalize the univariate methods of the previous section is to apply them iteratively to the variables with missingness in the data. If the variables with missingness are a matrix $Y$ with columns $Y_{(1)}, \dots, Y_{(K)}$ and the fully observed predictors are $X$, this entails first imputing all the missing $Y$ values using some crude approach (for example, choosing imputed values for each variable by randomly selecting from the observed outcomes of that variable); and then imputing $Y_{(1)}$ given $Y_{(2)}, \dots, Y_{(K)}$ and $X$; imputing $Y_{(2)}$ given $Y_{(1)}, Y_{(3)}, \dots, Y_{(K)}$ and $X$ (using the newly imputed values for $Y_{(1)}$), and so forth, randomly imputing each variable and looping through until approximate convergence.

For example, the Social Indicators Survey asks about several sources of income. It would be helpful to use these to help impute each other since they have non-overlapping patterns of missingness. We illustrate for the simple case of imputing missing data for two variables—interest income and earnings—using the same fully observed predictors used to impute earnings in the previous section.

We create random imputations to get the process started:

```
SIS$interest_imp <- random_imp(SIS$interest)
SIS$earnings_imp <- random_imp(SIS$earnings)
```

and then we write a loop to iteratively impute. For simplicity in programming, we set up the function on the original (non-square-root) scale of the data and we take the first simulation draw from each fitted model:

```
n_loop <- 10
for (loop in 1:n_loop){
  fit1 <- stan_glm(earnings ~ interest_imp + male + over65 + white + immig + educ_r +
    workmos + workhrs_top + any_ssi + any_welfare + any_charity, data=SIS)
  SIS_predictors$interest_imp <- SIS$interest_imp
  pred1 <- posterior_predict(fit1, newdata=SIS_predictors, draws=1)
  SIS$earnings_imp <- impute(SIS$earnings, pred1)

  fit2 <- stan_glm(interest ~ earnings_imp + male + over65 + white + immig + educ_r +
    workmos + workhrs_top + any_ssi + any_welfare + any_charity, data=SIS)
  SIS_predictors$earnings_imp <- SIS$earnings_imp
  pred2 <- posterior_predict(fit2, newdata=SIS_predictors, draws=1)
  SIS$interest_imp <- impute(SIS$interest, pred2)
}
```

This yields some negative imputations, but here we are just illustrating the general principles of iterative imputation. The code could be easily elaborated to handle topcoding, transformations, and

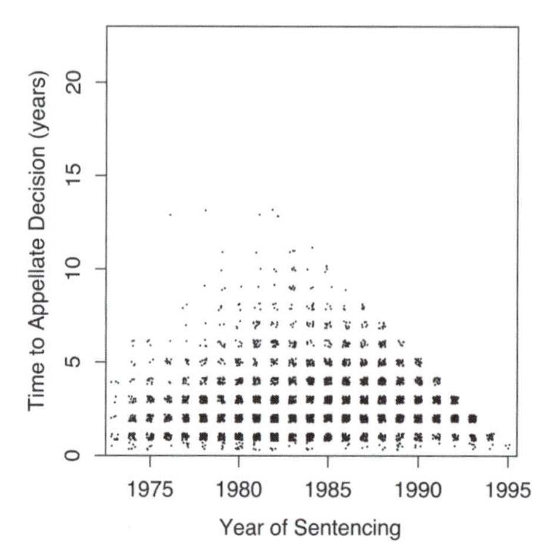


Figure 17.8 *Delays in state appeals courts for death penalty cases, plotted versus year of sentencing (jittered to allow individual data points to be visible). We only have results up to the year 1995. The data show a steady increase in delay times for the first decade, but after that, the censoring makes the graph difficult to interpret directly.*

two-stage modeling for variables that could be zero or positive; see Exercise 17.8. These operations should be done within the imputation loop, not merely tacked on at the end. The above is not the most efficient code; we show it here to demonstrate the general concept of iterative imputation.

Iterative regression imputation has the advantage that, compared to the full multivariate model, the set of separate regression models (one for each variable, $Y_{(k)}$) is easier to understand, thus allowing the imputer to potentially fit a reasonable model at each step. Moreover, it is easier in this setting to allow for interactions (difficult to do using most joint model specifications).

The disadvantage of the iterative approach is that the researcher has to be more careful in this setting to ensure that the separate regression models are consistent with each other. For instance, it would not make sense to use income when imputing age but then to later ignore age when imputing income.

Moreover, even if such inconsistencies are avoided, the resulting specification will not in general correspond to any joint probability model for all of the variables being imputed. It is an open research project to develop methods to diagnose problems with multivariate imputations, by analogy to the existing methods such as residual plots for finding problems in regressions. In the meantime, it makes sense to examine histograms and scatterplots of observed and imputed data to check that the imputations are reasonable.

In any case, the imputations are only as good as the model—but this is the case for any method of handling missing data. As discussed earlier, simple approaches using data exclusion or deterministic imputation also come with their own strong implicit models. Ultimately, with imperfect data, there is no way forward without making assumptions.

## 17.6 Nonignorable missing-data models

Example: Death sentences

Finally, for some missing-data problems, it is best to explicitly model the missingness mechanism. For example, in the study of death penalty appeals described in Section 15.3, we were interested in the duration of the appeals process for individual cases. A challenge in fitting these data was that more than half the cases had unknown dispositions—they were still in the appeals process at the time of data collection in 1995. Figure 17.8 shows the data.

Simply excluding the undecided cases would be a mistake, since the very fact they are undecided can convey information, depending on the year of the decision. For instance, a case that was sentenced in 1980 and still undecided in 1995 is likely to ultimately be upheld—that it was still under review after fifteen years implies that it has probably been upheld in the early stages of the review process. In contrast, if a death sentence was applied in 1993, the fact that it was still under review in 1995 conveys little information.

The data-inclusion model looks like,

$$y_i = \begin{cases} z_i & \text{if } z_i \leq 1995 - t_i \\ \text{censored} & \text{otherwise,} \end{cases}$$

where $y_i$ is the observed waiting time for case $i$, $z_i$ is the ultimate waiting time, and $t_i$ is the year of sentencing. We shall not analyze these data further here; we have introduced this example just to illustrate the complexities that arise in realistic censoring situations. The actual analysis for this problem is more complicated because death sentences have three stages of review, and cases can be waiting at any of these stages.

## 17.7 Bibliographic note

Little and Rubin (2002) provide an overview of methods for analysis with missing data. For more on multiple imputation in particular, see Rubin (1987, 1996). "Missing at random" and related concepts were formalized by Rubin (1976). A simple discrete-data example appears in Rubin, Stern, and Vehovar (1995).

For routine imputation of missing data, Schafer (1997) presents a method based on the multivariate normal distribution, Liu (1995) uses the $t$ distribution, and Van Buuren, Boshuizen, and Knook (1999) use interlocking regressions. Abayomi, Gelman, and Levy (2008) discuss methods for checking the fit of imputation models, and Troxel, Ma, and Heitjan (2004) present a method to assess sensitivity of inferences to missing-data assumptions.

Software for routine imputation in R and SAS has been developed by Van Buuren and Oudshoom (2011); Raghunathan, Van Hoewyk, and Solenberger (2001); Raghunathan, Solenberger, and Van Hoewyk (2002); and Su et al. (2011). Kropko et al. (2014) compare some of those approaches.

Specialized imputation models have been developed for particular problems, with multilevel models used to adjust for discrete predictors. Some examples include Clogg et al. (1991) and Belin et al. (1993). See also David et al. (1986).

Meng (1994), Fay (1996), Rubin (1996), and Clayton et al. (1998) discuss situations in which the standard rules for combining multiple imputations have problems. Barnard and Meng (1994) and Robins and Wang (2000) propose alternative variance estimators and reference distributions.

For more on the Social Indicators Survey, see Garfinkel and Meyers (1999). The death-sentencing example is discussed by Gelman, Liebman, et al. (2004) and Gelman (2004a); see also Finkelstein et al. (2006).

For more on the Xbox survey, see Wang et al. (2015) and Gelman, Goel, et al. (2016). For some background on regression and poststratification, see Gelman and Little (1997), Gelman (2007), and Lax and Phillips (2009).

## 17.8 Exercises

17.1 *Regression and poststratification*: Section 10.4 presents some models predicting weight from height and other variables using survey data in the folder `Earnings`. But these data are not representative of the population. In particular, 62% of the respondents in this survey are women, as compared to only 52% of the general adult population. We also know the approximate distribution of heights in the adult population: normal with mean 63.7 inches and standard deviation 2.7 inches for women, and normal with mean 69.1 inches and standard deviation 2.9 inches for men.

(a) Use poststratification to estimate the average weight in the general population, as follows: (i) fit a regression of linear weight on height and sex; (ii) use `posterior_epred` to make predictions for men and women for each integer value of height from 50 through 80 inches; (iii) poststratify using a discrete approximation to the normal distribution for heights given sex, and the known proportion of men and women in the population. Your result should be a set of simulation draws representing the population average weight. Give the median and mad sd of this distribution: this represents your estimate and uncertainty about the population average weight.

(b) Repeat the above steps, this time including the `height:female` interaction in your fitted model before poststratifying.

(c) Repeat (a) and (b), this time performing a regression of log(weight) but still with the goal of estimating average weight in the population, so you will need to exponentiate your predictions in step (ii) before poststratifying.

17.2 *Regression and poststratification with fake data*: Repeat the fake-data simulation of Section 17.2, replacing the made-up population numbers and the assumption that sex, age, and ethnicity are statistically independent in the population, instead using Census numbers for the distribution of sex $\times$ age $\times$ education of adults in the United States. You can keep the same 32 demographic categories; just use the actual numbers.

(a) Compare your newly created vector `poststrat$N` to the vector `poststrat$N` constructed in Section 17.2.

(b) You have done a fake-data simulation, so you can compare inferences to the true values. Your baseline is the true population mean of the outcome under your model, so compute that first.

(c) Compute the sample mean in your simulated data, $y$, and the corresponding standard error based on the (incorrect) binomial model corresponding to independent sampling with equal probabilities. Compare these to the true population mean.

(d) Examine the posterior estimate and standard error obtained by poststratification

17.3 *Response rates*: Based on the summaries at the very end of Section 17.4, compare the response rates for the earnings question in the Social Indicators Survey for whites and blacks.

17.4 *Bias in deterministic imputation*: Suppose you are imputing missing responses for income in a social survey of American households, using for the imputation a regression model given demographic variables. Which of the following two statements is basically true?

(a) If you impute income deterministically using a fitted regression model (that is, imputing using $X\beta$ rather than $X\beta + \epsilon$), you will tend to impute too many people as *rich* or *poor*: A deterministic procedure overstates your certainty, making you more likely to impute extreme values.

(b) If you impute income deterministically using a fitted regression model (that is, imputing using $X\beta$ rather than $X\beta + \epsilon$), you will tend to impute too many people as *middle class*: By not using the error term, you'll impute too many values in the middle of the distribution.

17.5 *Using simulation to evaluate missing-data procedures*: Take a complete dataset (with no missingness) of interest to you with two variables, $x$ and $y$. Call this the "full data."

(a) Write a program in R to cause approximately half of the values of $x$ to be missing. Design this missingness mechanism to be at random but *not* completely at random; that is, the probability that $x$ is missing should depend on $y$. Call this new dataset, with missingness in $x$, the "available data."

(b) Perform the regression of $x$ on $y$ (that is, with $y$ as predictor and $x$ as outcome) using complete-case analysis (that is, using only the data for which both variables are observed) and show that it is consistent with the regression on the full data.

(c) Perform the complete-case regression of $y$ on $x$ and show that it is *not* consistent with the corresponding regression on the full data.

(d) Using just the available data, fit a model in R for $x$ given $y$, and use this model to randomly impute the missing $x$ data. Perform the regression of $y$ on $x$ using this imputed dataset and compare to your results from (c).

17.6 *Using simulation to evaluate missing-data procedures*: Continuing the template of the previous exercise:

(a) Construct the missingness in such a way that the complete-case analysis and the analysis of imputed data give similar results.

(b) Construct the missingness in such a way that the complete-case analysis and the analysis of imputed data give dramatically different results.

(c) Explain the aspects of the missingness model that leads to the complete-case and imputed-data analyses being similar or different.

17.7 *Imputation for continuous and binary data*: Write a program in R to create a fake dataset with three continuous variables and two variables that take on the values 0 and 1. Write a program in R to cause approximately 20% of values of each variable to be missing. Design this missingness mechanism so that the probability of a variable being missing depends on the other four variables in the dataset. Write a program in R to impute the missing values using iterative regression.

17.8 *Multiple imputation*: Use iterative regression to impute missing data for all the income components (earnings, retirement, interest, public assistance, and other sources) in the Social Indicators Survey (data at folder `Imputation`).

(a) Perform the imputations using untransformed linear regressions.

(b) Use appropriate models, given that these variables are never negative and are often exactly zero.

17.9 *Imputation compared to complete-case analysis*: Find survey data and set up a regression analysis on a topic of interest to you.

(a) Use iterative regression to impute missing data so you can perform the regression on all the survey respondents.

(b) Compare to the results of a complete-case analysis.

17.10 *Nonignorable missing data*: The folder `Congress` has election outcomes and incumbency for U.S. congressional election races in the 1900s.

(a) Construct three bad imputation procedures and one good imputation procedure for votes for the two parties in the uncontested elections.

(b) Define clearly how to interpret these imputations. (These election outcomes are not actually "missing"—it is known that they were uncontested.)

(c) Fit the model to the completed dataset under each of the imputation procedures from (a) and compare the results.

We continue this example in Exercise 19.10.

17.11 *Working through your own example*: Continuing the example from the final exercises of the earlier chapters, perform multiple imputation for missing data (in earlier analyses, you may have just excluded cases with missingness) and then compare one of your earlier regression results to what you obtain with the imputed datasets.

# Part 5: Causal inference

# Chapter 18

# Causal inference and randomized experiments

So far, we have been interpreting regressions predictively: given the values of several inputs, the fitted model allows us to predict $y$, typically considering the $n$ data points as a simple random sample from a hypothetical infinite "superpopulation" or probability distribution. Then we can make comparisons across different combinations of values for these inputs. This section of the book considers *causal inference*, which concerns what *would happen* to an outcome $y$ as a result of a treatment, intervention, or exposure $z$, given pre-treatment information $x$. This chapter introduces the notation and ideas of causal inference in the context of randomized experiments, which allow clean inference for average causal effects and serve as a starting point for understanding the tools and challenges of causal estimation.

## 18.1 Basics of causal inference

For most statisticians, ourselves included, causal effects are conceptualized as a comparison between different potential outcomes of what might have occurred under different scenarios. This comparison could be between a factual state (what did happen) and one or more counterfactual states (representing what might have happened), or it could be a comparison among various counterfactuals.

### Running example

xample: ypothetical ietary xperiment

In this chapter, we use an example of a hypothetical small experiment on the effect of a dietary intervention. By using artificial data, we will be able to demonstrate key ideas of causal inference. Unobserved potential outcomes are a key aspect of causal inference, and with fake data we can look at all these potential outcomes and better understand how they are combined to define different causal summaries.

Increasingly over the past two decades, consumption of omega-3 fatty acids has been promoted as an effective strategy for addressing a variety of medical conditions ranging from coronary heart disease to rheumatoid arthritis to anxiety and depression. Suppose you had done some reading on the subject and were concerned about your own blood pressure levels but were confused by the conflicting evidence regarding the effect of omega-3 fish oil supplements on blood pressure. So you decided to investigate the link yourself. You found eight friends who agreed to be part of an informal study on the relationship between fish oil supplements and systolic blood pressure. Four of the friends were placed in the "fish oil supplement" treatment group. Members of this group agreed to consume 3 grams of fish oil supplements per day for one year while otherwise maintaining their current diets. The other four friends agreed to simply maintain their current diets free from fish oil supplements for the same year. At the end of the study period, blood pressure was measured for each of the eight participants. To simplify the discussion (and recognizing that classifications of hypertension vary), we will henceforth consider only systolic blood pressure, considering levels of 160 mmHg and above to represent "high blood pressure." We also assume that study participants actually take up the treatment they are assigned. Before we examine the study design and data more closely, however, let us step back and be explicit about what we are trying to estimate.

### Potential outcomes, counterfactuals, and causal effects

To formalize our approach, we use notation for the treatment variable that takes one of two values. In our running example, $z = 0$ denotes "no fish oil supplements ingested," and $z = 1$ denotes "3 grams per day of fish oil supplements ingested." To think about what it means for fish oil supplements (relative to no supplements) to *cause* lower blood pressure, we need to consider two possible outcomes: the blood pressure that would result if the person had no supplement, $y_i^0$, and the blood pressure that would result if the person had received the prescribed supplement, $y_i^1$. These possible outcomes are commonly referred to in the causal inference literature as *potential outcomes*. For the people in the experiment, the "factual state" is represented by the potential outcome that corresponds to the treatment actually received ($y_i^1$ for those who received the treatment and $y_i^0$ for those who did not). The "counterfactual state" is represented by the potential outcome that corresponds to the treatment not received ($y_i^0$ for those who received the treatment and $y_i^1$ for those who did not). The counterfactual state is thus not observed and only could be if we went back in time and had the study participant choose the opposite treatment. For the people not in the experiment, both states are counterfactual and we are interested in what might happen under either scenario.

The observed outcome for the people in the experiment is simply the potential outcome that was revealed due to treatment assignment and thus can be considered to be a function of both potential outcomes, $y_i = y_i^0(1 - z_i) + y_i^1 z_i$. The causal effect of supplements versus no supplements for person $i$ can be expressed by a mathematical statement such as $\tau_i = y_i^1 - y_i^0$. Causal effects can also be expressed as nonlinear functions of the potential outcomes. In this book we focus on linear functions because they are conceptually and mathematically simpler and because linear models work well in many settings, especially after appropriate transformations.

Why go through all the effort of defining potential outcomes and counterfactual states? Consider the clarity that is gained with this approach. Suppose the $j^{\text{th}}$ individual, Audrey, was observed pre-treatment to have a blood pressure of 140 mmHg. She received supplements for a year at which time her systolic blood was measured at 135; thus $y_j^1 = 135$. (We assume throughout that measurement error is small enough compared to natural variation that it can be ignored.) Does the pre/post decline from 140 to 135 provide evidence that fish oil supplements *caused* a reduction in Audrey's blood pressure? The only way we could make this claim would be if we knew that *if* Audrey had not received the supplements, then her blood pressure *would have been* higher, for instance if we knew that $y_j^0 = 140$ for her. If, instead, Audrey would have had the same blood pressure either way—that is, if $y_j^0 = 135$ for Audrey as well—then the causal effect of supplements on her blood pressure would be 0.

### The fundamental problem of causal inference

The problem inherent in determining the effect for any given individual, however, is that we can never observe both potential outcomes $y_j^0$ *and* $y_j^1$. In the language of this example, we cannot observe the blood pressure that would have resulted *both* if Audrey had taken the supplements *and* if she had not, thus the causal effect is impossible to directly measure. This is commonly referred to as the *fundamental problem of causal inference*.

To see this in a broader context, consider the eight-person study you implemented. Figure 18.1 displays the data the researcher could observe. Even though we observe outcomes under each of the two treatment regimes, we cannot observe any given person's outcome under both regimes. Therefore we cannot determine any individual causal effect without further assumptions.

If treatments are randomly assigned, we can estimate an average causal effect, but even then this will not tell us about the effect on any given person. We put a lot of effort into defining and

| Unit $i$ | Female, $x_{1i}$ | Age, $x_{2i}$ | Treatment, $z_i$ | Potential outcomes if $z_i = 0$, $y_i^0$ | Potential outcomes if $z_i = 1$, $y_i^1$ | Observed outcome, $y_i$ |
|---|---|---|---|---|---|---|
| Audrey | 1 | 40 | 0 | 140 | ? | 140 |
| Anna | 1 | 40 | 0 | 140 | ? | 140 |
| Bob | 0 | 50 | 0 | 150 | ? | 150 |
| Bill | 0 | 50 | 0 | 150 | ? | 150 |
| Caitlin | 1 | 60 | 1 | ? | 155 | 155 |
| Cara | 1 | 60 | 1 | ? | 155 | 155 |
| Dave | 0 | 70 | 1 | ? | 160 | 160 |
| Doug | 0 | 70 | 1 | ? | 160 | 160 |

Figure 18.1 *Hypothetical causal inference data for the effect of fish oil supplements on systolic blood pressure. For each person, we as researchers can only observe the potential outcome corresponding to the treatment actually received. Therefore we cannot directly observe either the individual-level or group-level causal effects. In this example, a simple difference of average outcomes across the two groups,* 157.5 − 145, *would lead to the estimate that supplements produce a 12.5 mmHg increase in systolic blood pressure. This is a poor estimate because the treatment and control groups here are highly imbalanced.*

estimating various average causal effects, but these only apply directly to individuals if we are willing to assume that effects are constant across people. One reason it can be important to include treatment interactions in causal regression models is because this is a way of estimating variation in effects.

## Close substitutes

But what if we had Audrey's pre-study blood pressure? Could we use that as a substitute for her $y_j^0$? This is the basic idea behind a "pre-post" or before-after study design. The problem is that Audrey's blood pressure before the study starts, $y_j^{\text{before}}$, is not necessarily an accurate reflection of what her blood pressure would be without the supplements one year later. For all we know, Audrey might have had other life changes during that study year: she might have lost her job or started a new one, she might have gotten married or divorced, she might have taken up running or discovered a love of baby back ribs. Any of these life changes could have affected her blood pressure irrespective of whether she received the omega-3 supplements or not, making $y_j^{\text{before}}$ a poor substitute for $y_j^0$. In Section 19.3, we discuss before-after studies in the context of a particular example.

Things would get still more complicated if the study were implemented so that Audrey received the supplements in the first year but not the second. If the supplements turned out to have a long-lasting effect, then the results from the second year would be affected by the treatment from the first year. We might attempt to avoid this contamination by allowing for a wash-out period between study segments (in our example to attempt to make sure the body is no longer being affected by the supplements), but this design still suffers from the defect of implicitly attempting to substitute measurements from one time period as estimates of potential outcomes from another.

There are ways of adding rigor to these types of pre-post designs. For instance, one could include many participants, randomize the order of receipt of treatment, and take multiple measurements on each person. Such studies fall under the general classification of *crossover trials*.

When considering a study that attempts to make causal inferences by substituting values in this way, keep in mind the strong assumptions often implicit in such strategies, which we can now formalize by explicitly requiring that $y_j^0$ and $y_j^1$ correspond to the same point in time. Otherwise, differences between these might be due to factors other than just the treatment.

One might object to the formulation of the fundamental problem of causal inference by noting situations where it appears one can actually measure both $y_i^0$ and $y_i^1$ on the same unit. Consider, for example, drinking tea one evening and milk another evening, and then measuring the amount of sleep each time. A careful consideration of this example reveals the implicit assumption that there are no

| Unit $i$ | Female, $x_{1i}$ | Age, $x_{2i}$ | Treatment, $z_i$ | Potential outcomes if $z_i = 0$, $y_i^0$ | Potential outcomes if $z_i = 1$, $y_i^1$ | Observed outcome, $y_i$ |
|---|---|---|---|---|---|---|
| Audrey | 1 | 40 | 0 | **140** | 135 | 140 |
| Anna | 1 | 40 | 0 | **140** | 135 | 140 |
| Bob | 0 | 50 | 0 | **150** | 140 | 150 |
| Bill | 0 | 50 | 0 | **150** | 140 | 150 |
| Caitlin | 1 | 60 | 1 | 160 | **155** | 155 |
| Cara | 1 | 60 | 1 | 160 | **155** | 155 |
| Dave | 0 | 70 | 1 | 170 | **160** | 160 |
| Doug | 0 | 70 | 1 | 170 | **160** | 160 |

Figure 18.2 *Hypothetical causal inference scenario for the effect of fish oil supplements on systolic blood pressure. Based on these numbers, there is a treatment effect of* $-10$ *for men and a treatment effect of* $-5$ *for women. For each row, the observation in bold is the one actually seen, and the numbers not in bold are unobserved; we only see them here because this is a theoretical simulation where we construct all potential outcomes for all people in the study. A naive comparison of outcomes across the two groups* ($157.5 - 145$) *would lead to the erroneous conclusion that supplements produce a 12.5 mmHg average* increase *in systolic blood pressure.*

systematic differences between days that could also affect sleep. An additional assumption is that applying the treatment on one day has no effect on the outcome on another day.

More pristine examples can generally be found in the physical sciences. For instance, imagine dividing a piece of plastic into two parts and then exposing each piece to a corrosive chemical. In this case, the hidden assumption is that pieces are identical in how they would respond with and without treatment, that is, $y_1^0 = y_2^0$ and $y_1^1 = y_2^1$.

#### More than two treatment levels, continuous treatments, and multiple treatment factors

Going beyond a simple treatment-and-control setting, multiple treatment effects can be defined relative to a baseline level. With random assignment, this simply follows general principles of regression modeling.

If treatment levels are numeric, the treatment level can be considered as a continuous input variable. To conceptualize randomization with a continuous treatment, think of spinning a spinner that can land on any of the potential levels of the treatment assignment. As with regression inputs in general, it can make sense to fit more complicated models as suggested by theory or supported by data. A linear model—which estimates the average effect on $y$ for each additional unit of $z$—is a natural starting point for effects that are believed to be monotonically increasing or decreasing functions of the treatment level.

With several discrete treatments that are unordered, as in a comparison of three different sorts of psychotherapy, we can move to multilevel modeling, with the group index indicating the treatment assigned to each unit, and a second-level model on the group coefficients, or treatment effects.

Additionally, multiple treatments can be administered in combination. For instance, depressed individuals could be randomly assigned to receive nothing, drugs, counseling sessions, or both drugs and counseling sessions. These combinations could be modeled as two treatments and their interaction or as four distinct treatments.

## 18.2 Average causal effects

The counterfactual or potential outcomes framework adds clarity to the meaning of causal effects, but it also highlights the challenge inherent in estimating them. In experiments in which only one treatment is applied to each individual, it will not be possible to estimate individual-level causal effects, $y_i^1 - y_i^0$. So how do we proceed? Can we gain any traction by collecting data on many individuals, some who took the supplements and some who did not?

Consider again our hypothetical eight-person study. Suppose Figure 18.2 were a display of the data that could be seen if we were somehow able to apply both the treatment and the control to each person. Under this state of omniscience, we would get to see both $y_i^1$ and $y_i^0$ for each study participant, even though the researcher in any real study sees at most one of these for each participant. Given all these potential outcomes, we can directly compute the treatment effect for each person:

$$\text{individual treatment effect: } \tau_i = y_i^1 - y_i^0.$$

For the example shown in Figure 18.2, the treatment effect for each of the men is a reduction in systolic blood pressure of 10 mmHg; the effect for each of the women is a reduction of 5 mmHg.

The value of framing this as a causal inference question, rather than simply comparing observed outcomes, is that when there are differences between treatment and control groups (as there will be in practice), some adjustment will be needed to estimate the causal effect—and for this it helps to understand what underlying summary is being estimated.

The causal effect, $y_i^1 - y_i^0$, can in general vary by person; hence, any definition of average causal effect will depend on what group of people is being averaged over. This is similar to the choice of population distribution in poststratification, as discussed in Section 17.1. Here we work things out in our simple example with just eight individuals.

### Sample average treatment effect (SATE)

For the data shown in Figure 18.2, the average treatment effect across all eight units in this example is a 7.5 mm Hg reduction in systolic blood pressure. This estimand, called the *sample average treatment effect*, or SATE, can be calculated by averaging all the $y^1$'s in the sample, $\frac{1}{n}\sum_{i=1}^n y_i^1$, and subtracting the average of the $y^0$'s in the sample, $\frac{1}{n}\sum_{i=1}^n y_i^0$. Equivalently, and perhaps more intuitively, we can simply average the individual-level causal effects,

$$\tau_{\text{SATE}} = \frac{1}{n}\sum_{i=1}^{n}(y_i^1 - y_i^0).$$

### Conditional average treatment effect (CATE)

We can also calculate the average treatment effect for well-defined subsets such as men (for which the sample average treatment effect in this example is $-10$, as can be seen by averaging the relevant numbers in Figure 18.2), women (sample average treatment effect of $-5$), or, for instance, 50-year-olds (sample average treatment effect of $-10$). These estimands are sometimes referred to as *conditional average treatment effects* (CATEs) and can also take more complicated forms such as expectations (average predictions) from linear regression models.

### Population average treatment effect (PATE)

Researchers often want inferences about some *population* of interest rather than simply the study sample. Realistically, the people (or, more generally, experimental units) in a social or medical experiment are typically not a random or representative sample of any known or well-understood population of inference. In fact, often the more well-tailored a sample is for making causal inferences for a given sample, the less likely the sample is to be a random sample from, or representative of, a known population.

However, the *population average treatment effect* (PATE) is often a goal of causal inference. For a population of size $N$, we can define the PATE as,

$$\tau_{\text{PATE}} = \frac{1}{N}\sum_{i=1}^{N}(y_i^1 - y_i^0).$$

To know the PATE, our omniscience would need to extend to seeing both potential outcomes for the full population of interest. In essence, in this inferential problem, there are then two types of missing data: (1) the missing counterfactual values within our sample, and (2) both potential outcomes for those observations not in our sample.

If our study sample is a random sample of the population of interest, then any unbiased estimate of the SATE will also be an unbiased estimate of PATE. More generally, when we fit regression models, the estimate of the PATE will depend on the assumed distribution of the regression predictors in the population of interest. For example, if the estimated treatment effect is larger among women than men, then the PATE will depend on the proportion of men and women in the population.

In the absence of a random sample, *estimating* the PATE requires a model of the treatment effect given pre-treatment predictors that will allow us to extrapolate from experimental units to the general population, in the same way that poststratification allows us to generalize from imperfect samples as discussed in Section 17.1. We discuss such methods in the section on generalization.

### Problems with self-selection into treatment groups

Let's return to our goal of trying to estimate an average treatment effect. The starting point is the comparison of treatment and control groups, and we can run into trouble if these two groups are not sufficiently similar or *balanced.* In Figure 18.1, we see that the people who received treatment were on average older than the controls. This difference could have occurred just by chance or perhaps because those who agreed to take the supplements were more concerned about their blood pressure and the study offered them a chance to try out the supplements for free, while those who agreed to be in the no-supplements group did not care if the supplements might benefit their health.

What is the implication of this type of sorting or self-selection into treatment groups? In our hypothetical example, we know all the counterfactual outcomes, and we have already seen from Figure 18.2 that the true sample average treatment effect is $-7.5$. However, what happens if we perform the most straightforward analysis and simply compare the average outcomes for those who chose to be in the treatment group to the average outcomes for those who chose to be in the control group? The difference in means between these two groups in Figure 18.1 yields a naive estimated treatment effect estimate of $-12.5$. What went wrong?

Applied researchers will intuitively know the answer to this question. The groups whose outcomes we were comparing differed in their pre-treatment characteristics. For example, although we had the same ratio of men to women across the two groups, those in the control group were quite a bit younger on average than those in the treatment group. This difference matters because age is also predictive of the outcome.

Perhaps more interestingly though, even if the observed characteristics had been exactly the same across the two groups, if the potential outcomes remained as they are in Figure 18.2, clearly the simple comparison would have given the wrong answer. In short, *the potential outcomes encapsulate all of the necessary information regarding what we need to be similar across the two groups.* Since we can't directly adjust for these, we use our pre-treatment variables as a proxy. But we never know if these variables are sufficient to act as proxies.

### Using design and analysis to address imbalance and lack of overlap between treatment and control groups

In practice, we can never ensure that treatment and control groups are balanced on all relevant pre-treatment characteristics. However, there are statistical approaches that may bring us closer. At the design stage, we can use *randomization* to ensure that treatment and control groups are balanced in expectation, and we can use *blocking* to reduce the variation in any imbalance. At the analysis stage, we can *adjust* for pre-treatment variables to correct for differences between the two groups to reduce bias in our estimate of the sample average treatment effect. We can further adjust for differences between sample and population if our goal is to estimate the population average treatment effect.

| Unit $i$ | Female, $x_{1i}$ | Age, $x_{2i}$ | Treatment, $z_i$ | Potential outcomes if $z_i = 0$, $y_i^0$ | Potential outcomes if $z_i = 1$, $y_i^1$ | Observed outcome, $y_i$ |
|---|---|---|---|---|---|---|
| Audrey | 1 | 40 | 0 | **140** | 135 | 140 |
| Anna | 1 | 40 | 1 | 140 | **135** | 135 |
| Bob | 0 | 50 | 0 | **150** | 140 | 150 |
| Bill | 0 | 50 | 1 | 150 | **140** | 140 |
| Caitlin | 1 | 60 | 0 | **160** | 155 | 160 |
| Cara | 1 | 60 | 1 | 160 | **155** | 155 |
| Dave | 0 | 70 | 0 | **170** | 160 | 170 |
| Doug | 0 | 70 | 1 | 170 | **160** | 160 |

Figure 18.3 *Hypothetical causal inference data for the effect of a fish oil supplements on systolic blood pressure from an idealized randomized experiment that happens to achieve perfect balance. There is a treatment effect of* $-10$ *for men and a treatment effect of* $-5$ *for women. For each row, the observation in bold is the one actually seen, and the numbers not in bold are not observed; we only see them here because this is a theoretical simulation where we know all potential outcomes for all people in the study. In this case, the randomization got lucky and the simple difference in means,* 155 − 147.5, *happens to recover the true sample average treatment effect (SATE) of* $-7.5$. *Compare, however, to Figure 18.4, which illustrates how the same design can give a much different result in this small sample.*

## 18.3 Randomized experiments

Randomly assigning units to treatment and control groups ensures that there are no differences in expectation in the distribution of potential outcomes between groups receiving different treatments (or treatment and control). We can randomly assign units to treatments (or treatments to units) by flipping a coin, drawing names from a hat, or, more generally, assigning random numbers to units. In this day and age we usually let a computer software package create random assignments. For instance in our example the following command in R could be used to create a randomly ordered vector z of four ones and four zeroes:

```
z <- sample(c(0,0,0,0,1,1,1,1), 8)
```

### Completely randomized experiments

In a *completely randomized experiment*, the probability of being assigned to the treatment is the same for each unit in the sample. Let us go back in time and randomly assign your eight friends to treatment conditions.

**An idealized outcome from a completely randomized experiment.** Figure 18.3 displays the same units as in Figures 18.1 and 18.2, but now the treatment assignment and observed outcome reflect an idealized random assignment. As with Figure 18.2, we are omniscient, and thus both potential outcomes are revealed for each unit. We characterize the random assignment as idealized because it happens to have led to perfectly balanced distributions of potential outcomes.

In this case, given the perfect balance across groups, it makes sense that a simple difference in means exactly recovers the true treatment effect of $-7.5$. Equivalently, the average causal effect of the treatment corresponds to the coefficient $\tau$ in the regression, $y_i = \alpha + \tau z_i + \text{error}_i$, so we could estimate the treatment effect using a regression of the outcome on the treatment assignment, as discussed in the next chapter.

**A less idealized randomized result.** What happens if our randomization yields treatment assignments that are not so perfectly balanced? Consider, for example, the treatment assignment in Figure 18.4. In this case, the random assignment led to imbalance between treatment and control groups, with younger participants disproportionately more likely to have received treatment. More to the point, the $y_i^0$'s are 140, 140, 150, and 170 in the control group and 150, 160, 160, and 170 in the treatment

| | | | | Potential outcomes | | Observed |
|---|---|---|---|---|---|---|
| Unit $i$ | Female, $x_{1i}$ | Age, $x_{2i}$ | Treatment, $z_i$ | if $z_i = 0$, $y_i^0$ | if $z_i = 1$, $y_i^1$ | outcome, $y_i$ |
| Audrey | 1 | 40 | 1 | 140 | **135** | 135 |
| Anna | 1 | 40 | 1 | 140 | **135** | 135 |
| Bob | 0 | 50 | 1 | 150 | **140** | 140 |
| Bill | 0 | 50 | 0 | **150** | 140 | 150 |
| Caitlin | 1 | 60 | 0 | **160** | 155 | 160 |
| Cara | 1 | 60 | 0 | **160** | 155 | 160 |
| Dave | 0 | 70 | 0 | **170** | 160 | 170 |
| Doug | 0 | 70 | 1 | 170 | **160** | 160 |

Figure 18.4 *Hypothetical causal inference data for the effect of fish oil supplements on systolic blood pressure from a randomized experiment that happened to have a less perfectly balanced assignment, compared to that shown in Figure 18.3. The difference in means is again an unbiased estimate of the treatment effect under this completely randomized design, but it is subject to sampling variation, and for this particular realization the estimate,* 142.5 – 160 = –17.5, *happens to be far from the underlying and unobservable sample average treatment effect (SATE) of* –7.5.

group. This implies that even if the treatment had no effect at all, we'd see a big difference between treatment and control groups in the measured outcome.

For this particular treatment assignment, the simple difference between average treatment and control measurements comes to −17.5, which is quite a bit off from the true sample average treatment effect of −7.5. Randomization ensures balance on average but not in any given sample, and imbalance can be large when sample size is small.

## 18.4 Sampling distributions, randomization distributions, and bias in estimation

An unbiased estimate leads us to the right answer on average. As discussed in Section 4.2, in classical statistical inference we think about using samples to estimate population quantities. Properties of a statistical procedure are reflected in the distribution of the estimate over repeated samples from that population. That is, we can envision taking an infinite number of samples from the population and for each sample calculating the estimate. The distribution of these estimates is the sampling distribution.

The estimate is *unbiased* if the mean of this sampling distribution is equal to the estimand. The estimate is *efficient* if the sampling distribution has small variance. How can we conceptualize a sampling distribution when we are estimating the average treatment effect from the sample itself?

As an alternative way to conceptualize the inherent variability in our estimate, imagine a different way to think about the uncertainty. First of all, we can simplify matters by considering all covariates and potential outcomes to be fixed (this is a representation common both to the survey sampling world and the randomization-based inference framework). Then imagine randomly allocating observations to treatment groups again and again. Each new allocation will imply a different set of observed outcomes (since observed outcomes are a function of both potential outcomes and treatment assignment). Suppose that with each re-randomization the difference in mean outcomes between the treatment and control groups is calculated. For the $k^{\text{th}}$ sample, we could write this as $d^k = \frac{1}{n_1}\sum_{i;z_i^k=1} y_i^k - \frac{1}{n_0}\sum_{i;z_i^k=0} y_i^k$. The set of these estimates represents the *randomization distribution* for this estimate, which plays the same role as the sampling distribution discussed in Section 4.1. If the estimate is unbiased, then the average of all of these estimates (the mean of the randomization distribution) equals the true sample average treatment effect; that is, $\text{E}(d_k) = \tau_{\text{SATE}}$.

When considering the population average treatment effect, we can conceptualize drawing a random sample of size $n_1$ from a population to receive the treatment, drawing a random sample of size $n_0$ from the same population to receive the control, and taking the difference between the average response in each as an estimate of the population average treatment effect. If this were done over and over, the estimates would form a sampling distribution for this estimate. For the estimate to be unbiased, we would need the mean of this sampling distribution to be centered on the true $\tau_{\text{PATE}}$. A stricter criterion that would also lead to an unbiased treatment effect estimate would require the sampling distributions of the mean of $y$ for the control and treated units to be centered on the population average of $y^0$ and the population average of $y^1$, respectively.

Even if an estimate is unbiased, this does not guarantee that the estimate from any particular random assignment will be close to the true value of the estimand, particularly when the sample is small. The sampling or randomization distribution may be wide, reflecting unlucky, but still possible, randomizations such as the one represented in Figure 18.4. We can increase the chances that our estimate is closer to the true value of the estimand through designs that reduce the potential for imbalance and thus decrease the variance of the treatment effect estimates. The most important aspect of such designs is to collect pre-treatment information that can then be used at the design stage to assign treatments in a balanced way and can be used in the analysis stage to adjust for imbalance between treatment and control groups. Exercise 18.6 provides the opportunity to simulate such randomization distributions.

## 18.5 Using additional information in experimental design

In many experimental settings, pre-treatment information is available that can be used in the experimental design to yield more precise and less biased estimates of treatment effects.

Consider again the data in Figure 18.3 that resulted from the idealized assignment with perfect balance. Let us look more closely at what made that configuration so successful at recovering the true sample average treatment effect. There are in essence four different types of participants in the study. One way to characterize these four types would be as women aged 40, men aged 50, women aged 60, and men aged 70. A more definitive characterization would reference the four unique pairs of potential outcomes, $y_i^0$ and $y_i^1$: 140 and 135, 150 and 140, 160 and 155, and 170 and 160. Let's focus on the observable characteristics, however, since a researcher could act on those.

In experimental design, these groups are called *blocks*. As described above, given that there are equal numbers of control and treated units within each block, each treated unit in the block gets to act as the perfect comparison for the control unit and vice versa. In this idealized example, there is not even sampling variability: the potential outcomes are exactly equal, not simply drawn from the same distribution. Therefore, the distribution of $y^0$'s in the treatment group is exactly the same as the distribution of $y^0$'s in the control group. The same is true for the $y^1$'s.

### Randomized blocks experiments

What would have happened if the ratio of treated to control units were not equal across blocks? Consider the data displayed in Figure 18.5 arising from a design involving eight extra recruits from among your friends. The number of treated units is no longer equal to the number of control units within each block. In fact, the ratio of treated to control units now varies across blocks. In each of the top two blocks in the table, there are three controls and one treated unit. The bottom two blocks contain one control and three treated units. Because of this, the overall distribution of $y^0$'s in the treatment group is quite different from the overall distribution of $y^0$'s for the controls. This level of imbalance in potential outcomes between the treatment and control groups might occur in such a small sample with a completely randomized design, although it would be unlikely (see Exercise 18.6 at the end of this chapter to quantify this). In this scenario, the simple difference in means is not close to the true sample average treatment effect, either for this sample or across repeated samples.

How did the data from this version of the hypothetical study arise? That is, what was the study design? Suppose the older participants had objected to a completely randomized design because they thought that they should be given preference regarding access to the supplements. Wanting to address these concerns but still feeling committed to randomization, you could have decided to

| | | | | Potential outcomes | | Observed |
|---|---|---|---|---|---|---|
| Unit $i$ | Female, $x_{1i}$ | Age, $x_{2i}$ | Treatment, $z_i$ | if $z_i = 0$, $y_i^0$ | if $z_i = 1$, $y_i^1$ | outcome, $y_i$ |
| Audrey | 1 | 40 | 0 | **140** | 135 | 140 |
| Abigail | 1 | 40 | 0 | **140** | 135 | 140 |
| Arielle | 1 | 40 | 0 | **140** | 135 | 140 |
| Anna | 1 | 40 | 1 | 140 | **135** | 135 |
| Bob | 0 | 50 | 0 | **150** | 140 | 150 |
| Bill | 0 | 50 | 0 | **150** | 140 | 150 |
| Burt | 0 | 50 | 0 | **150** | 140 | 150 |
| Brad | 0 | 50 | 1 | 150 | **140** | 140 |
| Caitlin | 1 | 60 | 0 | **160** | 155 | 160 |
| Cara | 1 | 60 | 1 | 160 | **155** | 155 |
| Cassie | 1 | 60 | 1 | 160 | **155** | 155 |
| Cindy | 1 | 60 | 1 | 160 | **155** | 155 |
| Dave | 0 | 70 | 0 | **170** | 160 | 170 |
| Doug | 0 | 70 | 1 | 170 | **160** | 160 |
| Dylan | 0 | 70 | 1 | 170 | **160** | 160 |
| Derik | 0 | 70 | 1 | 170 | **160** | 160 |

Figure 18.5 *Hypothetical causal inference data for the effect of fish oil supplements on systolic blood pressure from a* randomized block *experiment. Compare to Figures 18.3 and 18.4, which show the same pre-treatment predictors and the same potential outcomes but with different patterns of treatment assignments. With the randomized block design shown here, a simple difference in means, 152.5 − 150 = 2.5, is no longer an unbiased estimate of the true SATE of −7.5, but an appropriately adjusted estimate will be unbiased.*

give the eight oldest participants preference by allowing them a higher probability of receiving the supplements. Within each age group you might also have considered it important to ensure there were at least one male and one female in each treatment condition. These considerations led to a design in which randomization was performed within each group defined by age and sex (or first letter of the first name, as it so happens). In the first two blocks (Audrey through Brad) that contain the younger participants, the probability of receiving the supplements is 0.25 under this design, whereas in the last two blocks, with the older participants, this probability is 0.75.

Now that we know that observations were randomized within blocks, we can estimate average treatment effects, *taking into account the design in the analysis*. In practice, this means that we calculate treatment effects conditional on the blocks within which the randomization was performed, and then average over the blocks to represent the sample, or the general population. We can do this in any one of several ways.

The simplest approach conceptually starts with calculating a separate treatment effect for each block, $\hat{\tau}_j$. Since this design implies that a separate (tiny) randomized experiment was implemented within each of these blocks of units, each of these is a valid (if potentially imprecise) treatment effect estimate for that block. To get an estimate of the sample or population average treatment effect (SATE or PATE), we could just average these estimates with weights proportional to the number of units in each, specifically $\sum_j n_j \hat{\tau}_j / \sum_j n_j$ for the SATE or $\sum_j N_j \hat{\tau}_j / \sum_j N_j$ for the PATE.

Another approach is to fit a linear regression of blood pressure on the treatment variable and indicators for the three (of the four) blocks (denoted by $b_k$),

$$y_i = \alpha + \tau_{\mathrm{RB}} z_i + \gamma_1 b_{1i} + \gamma_2 b_{2i} + \gamma_3 b_{3i} + \mathrm{error}_i,$$

where the coefficient of interest is $\tau_{\mathrm{RB}}$. We might be able to sharpen this estimate by adding sex as a predictor. In theory, we could also recover block-specific treatment effect estimates by including block × treatment interactions in the model. We give R code for these approaches in Sections 19.2–19.4.

**Motivations for randomization in blocks.** In this example, the randomized block experimental design was chosen in part to increase people's willingness to participate in the study. Given that the older participants might have had more to gain from taking the supplements, this could be thought of as a design motivated by both logistical and ethical concerns. Another logistical constraint that often motivates use of randomized block experiments is location. When an experiment is being implemented across several different geographic sites, it can be simpler to implement the randomization separately within site.

Another motivation for a randomized block design is that it can reduce the uncertainty in the estimate of the treatment effect. This can be particularly important when measurements are noisy or sample size is small.

**Defining blocks in practice.** How do we choose blocks when our goal is to increase statistical efficiency? The goal when creating blocks is to minimize the variation of each type of potential outcome, $y^0$ and $y^1$, within the block. In our contrived example we have access to data on both potential outcomes for each person and thus in theory could identify blocks within which there is no variation in the values of either potential outcome across participants. In practice researchers only have access to observed pre-treatment variables when making decisions regarding how to define blocks. The randomized block experiment in this section is an example with blocks defined by age. So, to the extent possible, the predictors used to define the blocks should be those that are believed to be predictive of the outcome based on either theory or on results from previous studies. The more predictive the blocking variable, the bigger the precision gains at the end of the day.

### Matched pairs experiments

Suppose there is an important characteristic of your friends who participated in this study that you neglected to reveal at the outset. The 16 friends are actually a collection of 8 identical twins! We can identify these pairs because in each case the siblings have names that start with the same letter of the alphabet. This feature of the sample presents a new design possibility: a matched pairs randomized experiment. This design is a special case of a randomized block design in which each matched pair represents a 2-unit block. The randomization that occurs simply assigns one member of each pair to receive the treatment and the other to receive the control.

If the members of the pair are closely matched to each other, this design can yield substantial efficiency gains. In fact, the data presented in Figure 18.4 would be much more likely to arise from a matched pairs randomized design than from a completely randomized design, and we have discussed the idealized nature of this design that arises from the fact that it can be characterized as sets of identically matched pairs.

In practice it can be difficult to find pairs that are closely matched even on their observable characteristics, let alone their potential outcomes, which are unobserved at the time of treatment assignment. If the members of the pair do not have similar potential outcomes, the paired design will not reduce variance in the estimated treatment effect.

The most effective paired or blocked designs can arise when pairs or groupings arise naturally in the data, such as multiple children in a family or multiple students in a school. In general an effective strategy can be to match based on a wide variety of characteristics using a dimension-reduction strategy such as Mahalanobis or propensity score matching, as discussed in Chapter 20.

### Group or cluster-randomized experiments

Sometimes it is difficult or problematic to randomize treatment assignments to individual units, and so researchers instead randomize groups or clusters to receive the treatment. This design choice might be motivated by how a treatment is implemented. For instance, group therapy strategies typically are implemented at the group level—that is, with every person in a group receiving the same treatment. Similarly, schoolwide reform interventions by definition require assignment at the school level.

A decision to assign treatments at the group level can be driven by cost or logistical concerns. It might be more cost effective to provide free flu shots to a random subset of health clinics, for example, than to have professionals go to every clinic and then randomly assign individuals to receive shots.

Assignment at the clinic level would also avoid creating ill will among potential study participants being deprived of a service that others in the same location are able to receive. Cluster-randomized experiments are also used to avoid spillover or contagion effects (which can also be considered as violations of the stable unit treatment value assumption or SUTVA; see page 353).

When treatments have been assigned at the group level, the simplest approach to analysis is to conceptualize the groups as the level of analysis and use aggregated measures of the response variables as the outcome in the analysis. The example discussed later in this chapter does exactly this. An alternative is to use multilevel regression to model the outcomes at the individual level while including treatment assignment as a group-level predictor.

### Using design-relevant information in subsequent data analysis

Blocking and pairing in experimental design are effective to the extent that units within the same block or pair are similar to each other in their pre-treatment characteristics, so that splitting each block or pair ensures some level of pre-treatment balance between treatment and control groups. In contrast, clustered designs are typically less efficient than complete randomization. In any of these designs, the grouping should be accounted for in the analysis to remove bias and to get appropriate uncertainty bounds for estimated treatment effects.

It is best to include pre-treatment information in the design if this is feasible. But in any case, whether or not this information is accounted for in the design, it should be included in the analysis by including relevant pre-treatment information as predictors. We demonstrate in Chapter 19 with an education experiment on elementary school classrooms that used a paired design and two pre-treatment predictors: grade and pre-treatment test score.

## 18.6 Properties, assumptions, and limitations of randomized experiments

By examining the consequences of our design choices, we have revealed some desirable properties of the randomized experiment. When the design yields no average differences between groups, a simple difference in means will yield an unbiased estimate of the sample average treatment effect, and a regression on treatment indicator, also adjusting for pre-treatment predictors, will also be unbiased but can do even better by reducing variance. This section provides a more detailed discussion of the properties implied by randomization for completely randomized, randomized block, and matched pairs experiments.

### Ignorability

We first discuss how the ignorability assumption differs across different randomized study designs.

**Completely randomized experiments.** In a completely randomized design, the treatment assignment is a random variable that is independent of the potential outcomes, a statement that can be written formally as,

$$z \perp y^0, y^1. \tag{18.1}$$

The implication of this independence is that, under repeated randomizations, there will be no differences, on average, in the potential outcomes, comparing treatment and control groups. This property is commonly referred to as *ignorability* in the statistics literature (though it is a special case of a more general assumption also called ignorability, as we shall see shortly). Ignorability does not imply that the groups are perfectly balanced. Rather, it implies that there is no imbalance *on average* across repeated randomizations. If we were to redo the randomized assignment, we'd be equally likely to see imbalance in one direction as in another. Said another way, ignorability implies that the value of someone's potential outcomes does not provide any information about his or her treatment group assignment. For instance, someone who would have low blood pressure under either regime would not be any more likely to end up in the treatment group.

More formally, the independence between the treatment indicator $z$ and the potential outcomes that exists under this assumption implies the unbiasedness property discussed in Section 18.4. To see why, consider the following equivalencies:

$$\mathrm{E}(y|z=1) = \mathrm{E}(y^1|z=1) = \mathrm{E}(y^1),$$
$$\mathrm{E}(y|z=0) = \mathrm{E}(y^0|z=0) = \mathrm{E}(y^0).$$

The first line says that the average of the outcomes for those assigned to treatment is equal to the average of the treatment potential outcomes for those assigned to treatment which is equal to the average of all the treatment potential outcomes. The first equivalency in this statement is achieved by definition; observed outcomes for those in the treatment group are $y_i^1$'s. The second equivalency holds due to ignorability. Since $y^1$ and $z$ are independent, conditioning on $z$ makes no difference in expectation. Parallel arguments can be made for the second statement in this pair. Together, these imply that the observed difference in means across treatment and control groups will be an unbiased estimate of the true treatment effect.

In this discussion, we do not specify if the expectation is over the sample or the population. To consider the former, think about the randomization distribution discussed above. These properties imply that it will be centered on the true sample average treatment effect (SATE). If the data are also a random sample from the population, then these assumptions additionally imply that the sampling distribution will be centered on the true population average treatment effect (PATE).

Another important property of randomized experiments is that they create independence between treatment assignment and *all* variables $x$ that occur before treatment assignment, so that $x \perp z$. In general, the relevant cutoff time is when the treatment is *assigned*, not when it is *implemented*, because people can adjust their behavior based on the anticipation of a treatment to which they have been assigned. To put it another way, if the act of assignment is potentially part of the treatment, then the study should be analyzed as such.

Pre-treatment variables include those that do not change value over time or variables such as participant age, which change over time but cannot be affected by a treatment. We can think of the potential outcomes as a subset of these pre-treatment variables because we can conceptualize them as existing before the treatment is even implemented, as shown in Figure 19.1. The treatment assignment identifies which of the potential outcomes will be observed.

This property is what motivates researchers who implement randomized experiments to check whether the randomization worked well by comparing means or distributions of observed pre-treatment covariates across randomized treatment groups. It is also important because it justifies use of subgroup difference in means (and regression analogs of such estimates) as unbiased estimates of conditional treatment effects. It also justifies inclusion of pre-treatment variables as predictors in models that estimate treatment effects using data from randomized experiments (more below).

**Randomized blocks experiments.** When we randomize within blocks, the independence between potential outcomes and the treatment variable holds only conditional on block membership, $w$:

$$z \perp y^0, y^1 \,|\, w. \tag{18.2}$$

This more general version of the randomization property is also known as ignorability (sometimes called conditional ignorability, but the word "conditional" here is redundant), as will any such statement of conditional independence between the potential outcomes and the treatment assignment as long as the data being conditioned on (here $w$) are fully observed. Intuitively, we can think of a randomized block experiment as a collection of experiments, each of which is performed within the members of a specific block. This means that all those in the same block (those who look the same with regard to their $w$ variable) have the same probability of being assigned to the treatment group.

The implication of this property is that it is necessary to condition on the blocks when estimating treatment effects as described in the previous section. If the probability of receiving the treatment is the same across all the blocks, we can ignore this conditioning and still get an unbiased treatment effect estimate, but to achieve the efficiency gains of this design the conditioning is still necessary.

**Matched pairs experiments.** Recall that the paired design is the special case of randomized blocks in which each block contains exactly two units. Therefore, we can formalize the randomization property for matched pairs in the same way as in (18.2), where $w$ now indexes the pairs. Given that there are only two units in each block of the paired design and exactly one must receive the treatment, the probability of receiving the treatment is, by definition, the same across blocks. A simple difference in means will constitute an unbiased treatment effect estimate in this setting. However, this strategy will miss out on the potential efficiency gains of the design.

The most straightforward approach to treatment effect estimation that accounts for the matched pairs design is to calculate pair-specific differences in outcomes, $d_j = y_j^T - y_j^C$ (where $y_j^T$ is the outcome of the treated unit in pair $j$ and $y_j^C$ is the outcome of the control unit in pair $j$), and then calculate an average of those $K$ differences, $\bar{d} = \frac{1}{K}\sum_{k=1}^{K} d_j$. Including a predictor for each pair makes such models noisy to estimate using simple regression; data from such designs can often be more efficiently fit using multilevel models.

We will revisit the more general version of ignorability from (18.2) in Chapter 20 because the analysis strategies in that chapter rely on this assumption. Thus another reason for having a strong conceptual understanding of the properties of randomized block experiments is that this will help us to understand the similar assumptions that are often made when adjusting for differences between treatment and control groups in observational studies, in which case there is no randomization to ensure that the model holds, highlighting the value of being explicit about assumptions.

## Efficiency

Another design property was revealed in our discussion above. The more similar the potential outcomes are across treatment groups, the *closer* our estimate will be on average to the value of the estimand. Said another way, the more similar are the units being compared across treatment groups, the smaller the variance of the randomization or sampling distribution will be.

**Randomized blocks experiments.** Ideally, the randomized block experiment creates subgroups (blocks) within which the $y_i^0$'s are more similar to each other and the $y_i^1$'s are more similar to each other across treatment groups. This makes it possible to get sharper estimates of block-specific treatment effects, which can then be combined using a weighted average into an estimate of the average treatment effect across the entire experimental sample or population. A linear regression using block indicators as predictors achieves a similar result. This block-specific homogeneity is why the randomized block experiment yields estimates that have smaller standard errors on average than estimates from completely randomized experiments of the same sample size.

**Increasing precision by adjusting for pre-treatment variables.** Another strategy for achieving efficiency gains in treatment effect estimates using data from a randomized experiment is to adjust for pre-treatment variables that are predictive of the outcome. This requires regression modeling, but models that are fit to data from a randomized experiment tend to be robust to deviations from the parametric assumptions inherent in linear regression. (See Section 19.6 for a discussion of why you should not include regression predictors that use information occurring *after* the treatment.)

Consider the data from the idealized randomized experiment displayed in Figure 18.3. If we estimate the treatment effect for these data using a regression only on the treatment indicator (so, effectively, a difference-in-means estimate) the standard error is 8.8. However, if we include sex and age as predictors, the standard error drops to 1.2! What's going on?

This randomization (that is, this random draw from the randomization distribution) in particular yielded exquisite balance, but under the completely randomized design the balance can vary simply by chance. The standard error in the model without additional predictors reflects the fact that we might not be so lucky the next time and thus our estimate might end up much farther from the truth. If, however, this balance had resulted from a randomized block or matched pairs experiment, then we would not expect the treatment effect estimates to vary nearly so much across repeated randomizations. Exercise 18.6 illustrates this point via simulation. Using a model that accounts for this design by conditioning on the blocking variables in our model would allow us to capture these efficiency gains.

When we condition on pre-treatment variables in the absence of such an intentional design, we are capitalizing on the association between the potential outcomes and these variables to reduce the variance of our treatment effect estimate. With a pre-treatment variable that is highly predictive of the response, this association creates homogeneity with regard to potential outcomes for observations with similar values of that pre-treatment variable. It is as if nature created a randomized block experiment and we are taking advantage of it.

### Stable unit treatment value assumption (SUTVA): no interference among units and no hidden versions of the treatment

Our definition of potential outcomes from the start of this chapter implicitly encapsulates yet another assumption! Each person's potential outcome is defined in terms of only his or her own treatment assignment, $y_i^{z_i}$. One could imagine, alternatively, defining each potential outcome in terms of the *collection* of treatment assignments, $z$, across all study participants, $y_i^{z}$. In other words, person $i$'s outcome would be a function not only of her own treatment assignment, but also the treatment assignments of others in the sample. This might make conceptual sense in situations where the outcome of one person could potentially be affected by the treatment of others, but this level of generality would quickly become intractable. Even with our small study with two levels of the treatment, there are $2^8 = 256$ different possible allocations of treatments the 2 treatments to these 8 people. We clearly do not have enough data to inform 256 potential outcomes for each person. For this reason, researchers often simply hope that such interference among units, or spillover, does not exist, or else they model spillover in some way. For more recent advances and alternatives, see the bibliographic note at the end of the chapter.

We can formalize this *stable unit treatment value assumption* (SUTVA) as,

$$\text{SUTVA:}\;\; y_i^{z} = y_i^{z'} \text{ if } z_i = z_i',$$

so that the potential outcome for unit $i$ depends only on the treatment. Thus if SUTVA holds, $y_i^{z} = y_i^{z_i}$, and we are back where we started with the number of potential outcomes equal to the number of treatments.

SUTVA also implies that there are no hidden versions of treatments. That is, all of the units receive the same well-defined treatment. If not, the effect of $z$ is not well defined because, in effect, $z_i$ doesn't mean the same thing as $z_j$. Typically researchers focus on the no-interference aspect of SUTVA, perhaps reasoning that the definition of the treatment can always be broadened sufficiently to reflect various versions of treatments. However, this can be a dangerous path leading eventually to treatment effects that are virtually uninterpretable. It is best to try to ensure that treatment definitions and implementations are well defined across experimental units.

Examples of potential SUTVA violations abound. An experiment testing the effect of a new fertilizer by randomly assigning adjacent plots to treatment or control is a classic example. Fertilizer from one plot might leach into an adjacent plot assigned to receive no fertilizer and thus affect the yield in that control plot. Vaccines that reduce the probability of a contagious disease within a school, business, or community could easily lead to violation of SUTVA if the vaccine is actually effective. Consider an experiment that recruited families from the same public housing complex and randomized them to receive a voucher to move to a better neighborhood or not. This could suffer from interference if a given family moving might influence (positively or negatively) the well-being of another family that happened to be randomized to not receive the voucher.

Tensions may exist between what is optimal for an intervention to succeed and what is optimal to evaluate the magnitude of its impact. A program developer might reasonably *hope* for interference among units to increase the potential impact of a treatment. For instance, consider a behavioral intervention whose goal is to reduce bullying among students. If it were possible to introduce the intervention to just a small fraction of the student body of the school and yet, by some diffusion mechanism, create impacts on all the students, that would be a positive social externality. However, if we tried to estimate the magnitude of the effect of such an intervention with an experiment that randomized at the student level, it would be virtually impossible to interpret the estimated effects.

If we are planning a study such as this in which SUTVA is not likely to hold due to interference, one option is to assign the treatment at the level of a group beyond which interference is not likely to occur. For instance, there is a long tradition in education research of assigning treatments at the classroom or school level and then randomizing at this level of aggregation. In this way the students might be most likely to transfer knowledge or behaviors among themselves are all receiving the same treatment. With this type of design the most conservative approach is to interpret results only at the group level. However, more sophisticated approaches now exist that attempt to directly model the impact of this interference or diffusion (see bibliographic note at the end of the chapter).

### External validity: Difficulty of extrapolating to new individuals and situations

Randomized experiments are widely regarded as an ideal research design for identifying average causal effects. The ability of this design to recover causal effects averaging over the sample studied is often referred to as *internal validity*. But those who agree to participate in a study can differ in important ways from the population of interest. In addition, the effects being measured in an experimental setting may differ "in the wild." To the extent that the sample is not representative of the population, and that the environment in the study does not reflect the outside world, we say that the design lacks *external validity*. External validity is an issue with nonexperimental studies as well. However the treatments have been assigned or chosen, there are statistical approaches that can help to generalize effect estimates to new populations, as we discuss in Section 19.4.

### How the experiment affects the participants

Randomized experiments pose other challenges or "threats to validity," a number of which revolve around the interaction between the study design and patient or researcher behavior.

For instance, it is possible that simply participating in a study will change one's behavior, a phenomenon that has been called the "Hawthorne effect," in reference to a series of experiments conducted in Western Electric's Hawthorne Works plant in the 1920s. The studies, designed to estimate the effect of light on productivity, purportedly revealed that workers in both experimental groups increased their productivity not based on their exposure to light but rather simply because they knew they were being observed. This interpretation of these data has since been challenged, but the potential for this sort of effect surely still exists. This type of participant reaction can prevent the researcher from estimating the effect of the treatment that would have occurred naturally. As an example, in the set of hypothetical studies just described, if the researcher had decided to collect daily food diaries for each participant during the intervention year, such scrutiny might have led participants to alter their diets in ways that would affect subsequent blood pressure measurements. Relatedly, one may wonder if it is possible for participants to maintain their current diet once they have been alerted to the potential relationship between omega-3 and blood pressure (among other health issues). Members of the control group might (either consciously or inadvertently) start to include more fish in their diet, for example, thus possibly attenuating the effects of the supplements by altering the intended counterfactual condition. These types of expectancy effects have been documented not only in human studies but in studies with animal participants (even rats!).

It is not only the patients' expectations and psychology around receiving the treatment and participating in the study that could be problematic. The professionals administering the treatment or intervention (doctors, teachers, social workers) or the researchers gathering the outcome data might also be influenced by knowing who received which treatment. If so, they could inadvertently bias the results. For instance, an assessor who knows that a given study participant has been taking an anti-depressant medication for two months might be more likely to rate his depression level as less severe than in the absence of such knowledge.

This sort of problem has led medical researchers (in particular) to favor *double-blind* research designs in which neither the patients nor the researchers (and in some cases even the professionals administering the treatment) have knowledge regarding which people are receiving which treatment. This strategy won't work in all settings (people are bound to know whether or not they received a six-week job training intervention!) and doesn't cure all problems but can address some concerns. These designs also lend credibility to estimates of complier effects using instrumental variables, as we discuss in Section 21.1.

### Missing data and noncompliance

A different set of concerns that are also present in observational studies are those of missing data and noncompliance with treatment assignment. Since pre-treatment variables are independent of the treatment assignment, missing pre-treatment variables can be ignored without affecting the internal validity. A simple difference in means can be used on the full sample. Or we can run a conditional analysis on the complete-cases sample that deletes observations missing values for the predictors in the model and still achieve an unbiased treatment effect estimate for this subsample. On the other hand, either of these strategies could increase standard errors and the latter might focus inference on a less interesting segment of the analysis sample.

Missing outcome data are common, however, and pose greater challenges than missing pre-treatment data. That's because they are often more likely to occur for those in the control group than those in the treatment group because those in the treatment group are more likely to be emotionally invested in the study (and thus will be more likely to show up for the test or assessment or to fill out the required survey). Even if the missingness rates are the same across groups, if those participants for whom we are missing outcome data in the control group are different from those for whom we are missing outcome data in the treatment group (that is, the distributions of their potential outcomes are different), and the missing outcome data are ignored (that is, a complete-case analysis is used), then the benefits of the initial randomization will be destroyed. If analyses are performed on the sample of complete cases, then the resulting treatment effect estimates will be biased. We discuss this issue more thoroughly in Chapters 20 and 21.

Noncompliance with treatment assignment describes a situation in which, for instance, patients choose not to take their medication or to participate in an intervention to which they were randomly assigned. In our hypothetical omega-3 fatty acids study, the onus was on the participants to remember to take the supplements as well as to continue to choose to keep taking them. Furthermore, participants assigned to maintain their usual diet could have decided to start taking the supplements anyway. Human beings are capricious and any participant at any time could have failed to comply with their treatment assignment. This sort of noncompliance creates a disconnect between the effect of assignment and the effect of actually taking the treatment, which makes it more challenging to estimate the latter effect. We discuss this issue and a potential solution in more detail in Section 21.2.

## 18.7 Bibliographic note

You can get a sense of different perspectives on causal inference from the recent textbooks by Kennedy (2008), Angrist and Pischke (2009), Imbens and Rubin (2015), VanderWeele (2015), and Hernan and Robins (2020).

Campbell and Stanley (1963) is an early presentation of causal inference in experiments and observational studies from a social science perspective; see also Achen (1986). Shadish, Cook, and Campbell (2002), as well as earlier incarnations of that text, provide accessible and intuitive explanations for threats to validity in causal inference, including interactions between study design and patient or researcher behavior.

The stable unit treatment value assumption (SUTVA) was defined by Rubin (1978); see also Sobel (2006) for a discussion of the implications of SUTVA violations in the context of a public policy intervention and evaluation. VanderWeele and Hernan (2013) discuss aspects of inference and generalization when there are hidden versions of the treatment. Ainsley, Dyke, and Jenkyn (1995) and Besag and Higdon (1999) discuss spatial models for interference between units in agricultural experiments. Further developments in models of interference appear in Hong and Raudenbush (2006), Rosenbaum (2007), Hudgens and Halloran (2008), and Aronow and Samii (2017). Gelman (2004c) discusses treatment interactions in before-after studies.

Several strategies have been proposed for implementing pairing in the context of individual-level experimental designs with many pre-treatment variables; examples include Morris (1979), Hill, Rubin, and Thomas (2000), and Greevy et al. (2004). Pair matching in the context of cluster-randomized experiments has been discussed recently by several authors including Zhang et al. (2012).

A classic reference on placebo effects is Beecher (1955); a more recent review can be found in Meissner et al. (2011). Claims about the original "Hawthorne effect" have been questioned; see Franke and Kaul (1978), Jones (1992), and Levitt and List (2011).

## 18.8 Exercises

18.1 *Designing an experiment*: Suppose you are interested in the effect of the presence of vending machines in schools on childhood obesity. What controlled experiment would you want to do (in a world without ethical, logistical, or financial constraints) to evaluate this question?

18.2 *Designing an experiment with ethical constraints*: Suppose you are interested in the effect of smoking on lung cancer. What controlled experiment could you plausibly perform (in the real world) to evaluate this effect?

18.3 *Mapping a causal question to a hypothetical controlled experiment*: Suppose you are a consultant for a researcher who is interested in investigating the effects of teacher quality on student test scores. Use the strategy of mapping this question to a controlled experiment to help define the question more clearly. Write a memo to the researcher asking for needed clarifications to this study proposal.

18.4 *Average treatment effects*: The table below describes a hypothetical experiment on 8 people. Each row of the table gives a participant and her pre-treatment predictor $x$, treatment indicator $z$, and potential outcomes $y^0$ and $y^1$.

| | $x$ | $z$ | $y^0$ | $y^1$ |
|---|---|---|---|---|
| Anna | 3 | 0 | 5 | 5 |
| Beth | 5 | 0 | 8 | 10 |
| Cari | 2 | 1 | 5 | 3 |
| Dora | 8 | 0 | 12 | 13 |
| Edna | 5 | 0 | 4 | 2 |
| Fala | 10 | 1 | 8 | 9 |
| Geri | 2 | 1 | 4 | 1 |
| Hana | 11 | 1 | 9 | 13 |

(a) Give the average treatment effect in the population, the average treatment effect among the treated, and the estimated treatment effect based on a simple comparison of treatment and control.

(b) Simulate a new completely randomized experiment on these 8 people; that is, resample $z$ at random with the constraint that equal numbers get the treatment and the control. Report your new randomization and give the corresponding answers for (a).

18.5 *Potential outcomes*: The table below describes a hypothetical experiment on 2400 people. Each row of the table specifies a category of person, as defined by his or her pre-treatment predictor $x$, treatment indicator $z$, and potential outcomes $y^0$, $y^1$. For simplicity, we assume unrealistically that all the people in this experiment fit into these eight categories.

| Category | # people in category | $x$ | $z$ | $y^0$ | $y^1$ |
|---|---|---|---|---|---|
| 1 | 300 | 0 | 0 | 4 | 6 |
| 2 | 300 | 1 | 0 | 4 | 6 |
| 3 | 500 | 0 | 1 | 4 | 6 |
| 4 | 500 | 1 | 1 | 4 | 6 |
| 5 | 200 | 0 | 0 | 10 | 12 |
| 6 | 200 | 1 | 0 | 10 | 12 |
| 7 | 200 | 0 | 1 | 10 | 12 |
| 8 | 200 | 1 | 1 | 10 | 12 |

In making the table we are assuming omniscience, so that we know both $y^0$ and $y^1$ for all observations. But the (non-omniscient) investigator would only observe $x$, $z$, and $y^z$ for each unit. For example, a person in category 1 would have $x=0, z=0, y=4$, and a person in category 3 would have $x=0, z=1, y=6$.

(a) What is the average treatment effect in this population of 2400 people?

(b) Another summary is the mean of $y$ for those who received the treatment minus the mean of $y$ for those who did not. What is the relation between this summary and the average treatment effect (ATE)?

(c) Is it plausible to believe that these data came from a completely randomized experiment? Defend your answer.

(d) For these data, is it plausible to believe that treatments were assigned using randomized blocks conditional on given $x$? Defend your answer.

18.6 *Sampling distributions under randomization*: Use the covariate and potential outcome data from Figure 18.1 as a starting point for considering the randomization distributions of four different designs by creating simulations in R.
Comment on the relative bias and efficiency for each of the following designs:

- Completely randomized design,
- Randomized design blocked by the oldest four participants versus the youngest four,
- Matched pairs design,

using each of the following estimates:

- Difference in means,
- Regression on treatment indicator and age,
- Regression on treatment indicator, age, and sex,
- Regression on treatment indicator, age, sex, and treatment × sex interaction.

In addition, you can consider the matched pairs estimate for the matched pairs experiment.

xample: esame treet xperiment

18.7 *Before-after comparisons*: The folder Sesame contains data from an experiment in which a randomly selected group of children was encouraged to watch the television program Sesame Street and the randomly selected control group was not.

(a) The goal of the experiment was to estimate the effect on child cognitive development of watching more Sesame Street. In the experiment, encouragement but not actual watching was randomized. Briefly explain why you think this was done. Think of practical as well as statistical reasons.

(b) Suppose that the investigators instead had decided to test the effectiveness of the program simply by examining how test scores changed from before the intervention to after. What assumption would be required for this to be an appropriate causal inference? Use data on just the control group from this study to examine how realistic this assumption would have been.

We return to this example in Chapter 21.

18.8 *Evaluating an encouragement design*: Return to the Sesame Street example from the previous exercise. Did encouragement (the variable `viewenc`) lead to an increase in post-test scores for letters (`postlet`) and numbers (`postnumb`)? Fit an appropriate model to address this question.

18.9 *Regression and causal inference*: Write the functional form of the regression that would yield the data in Figure 18.3.

18.10 *Imbalance from a randomized design*: Calculate how unlikely it is that you would see the level of imbalance displayed in Figure 18.4 or worse in the context of a completely randomized experiment. To do so you will first have to create a reasonable definition of imbalance. You can calculate this probability using mathematical computations or a simulation in R.

18.11 *Conditional and marginal effects*: Show how the law of total probability (or law of iterated expectation) allows us to combine conditional estimates to get a marginal estimate.

18.12 *Simulating potential outcomes*: In this exercise, you will simulate an intervention study with a pre-determined average treatment effect. The goal is for you to understand the potential outcome framework, and the properties of completely randomized experiments through simulation.
The setting for our hypothetical study is a class in which students take two quizzes. After quiz 1 but before quiz 2, the instructor randomly assigns half the class to attend an extra tutoring session. The other half of the class does not receive any additional help. Consider the half of the class that receives tutoring as the treated group. The goal is to estimate the effect of the extra tutoring session on average test scores for the retake of quiz 1. Assume that the stable unit treatment value assumption is satisfied.

(a) Simulating all observed and potentially observed data (omniscient mode). For this section, you are omniscient and thus know the potential outcomes for everyone. Simulate a dataset consistent with the following assumptions.

i. The average treatment effect on all the students, $\tau$, equals 5.
ii. The population size, $N$, is 1000.
iii. Scores on quiz 1 approximately follow a normal distribution with mean of 65 and standard deviation of 3.
iv. The potential outcomes for quiz 2 should be linearly related to the pre-treatment quiz score. In particular they should take the form,

$$y^0 = \beta_0 + \beta_1 x + 0 + \epsilon^0,$$
$$y^1 = \beta_0 + \beta_1 x + \tau + \epsilon^1,$$

where the intercept $\beta_0 = 10$ and the slope $\beta_1 = 1.1$. Draw the errors $\epsilon^0$ and $\epsilon^1$ independently from normal distributions with mean 0 and standard deviations 1.

(b) Calculating and interpreting average treatment effects (omniscient mode). Answer the following questions based on the data-generating process or using your simulated data.

i. What is your interpretation of $\tau$?
ii. Calculate the sample average treatment effect (SATE) for your simulated dataset.
iii. Why is SATE different from $\tau$?
iv. How would you interpret the intercept in the data-generating process for $y^0$ and $y^1$?
v. How would you interpret $\beta_1$?
vi. Plot the response surface versus $x$. What does this plot reveal?

(c) Random assignment (researcher mode). For the remaining parts of this exercise, you are a mere researcher! Return your goggle of omniscience and use only the observed data available to the researcher; that is, you do not have access to the counterfactual outcomes for each student.

Using the same simulated dataset generated above, randomly assign students to treatment and control groups. Then, create the observed dataset, which will include pre-treatment scores, treatment assignment, and observed $y$.

(d) Difference in means (researcher mode).

i. Estimate SATE using a difference in means.

ii. Is this estimate close to the true SATE? Divide the difference between SATE and estimated SATE by the standard deviation of the observed outcome, $y$.

iii. Why is $\widehat{\text{SATE}}$ different from SATE and $\tau$?

(e) Researcher view: linear regression.

i. Now you will use linear regression to estimate SATE for the observed data created as above. With this setup, you will begin to better understand some fundamental assumptions crucial for the R homework assignments in the following chapters.

ii. What is gained by estimating the average treatment effect using linear regression instead of the mean difference estimate from above?

iii. What assumptions do we need to make in order to believe this estimate? Given how you generated the data, do you believe these assumptions have been satisfied?

18.13 *Varying treatment effect*: Now you will explore a situation where treatment effect varies by a known covariate.

(a) Simulate data from the following response surface and plot both curves,

$$y^0 = \beta_0^0 + \beta_1^0 x + \epsilon^0,$$
$$y^1 = \beta_0^0 + \beta_1^0 x + \epsilon^1,$$

where $\beta_0^0 = 35$, $\beta_1^0 = 0.6$, $\beta_0^1 = 15$, $\beta_1^1 = 1$, and $\epsilon^0$ and $\epsilon^1$ are drawn independently from normal distributions with mean 0 and standard deviation 1.

(b) Comment on your findings. In particular, there is no longer a $\tau$ included in the data-generating process. Is there still a sample average treatment effect? What is it? How do you interpret the average treatment effect in this setting?

(c) Is the treatment effect the same for all students? If not, is there a pattern to the way it varies? Why should you care about variation in the treatment effect?

(d) Plot the response surface versus $x$. What does this plot reveal, particularly as compared to the plot from the previous question?

18.14 *Completely randomized experiment*: In a completely randomized experiment, the number of treatment and control units is decided in advance, and each unit has the same probability of being assigned to receive the treatment. This exercise extends Exercise 18.12 by comparing the same estimates with respect to both bias and efficiency.
You will operationalize these properties by appealing to randomization-based inference. Therefore you will create a randomization distribution (similar to a sampling distribution) for each estimate. This distribution only considers the variability in estimates that would manifest as a result of the randomness in who is assigned to receive the treatment.

(a) Simulate the dataset from Exercise 18.12. You will use this as your observed data.

(b) Now create a randomization distribution for each estimate by following these steps:

i. Create a vector of length 10 000 for each estimate to save these estimates across simulated datasets. These could be named, for instance, `dm_res` and `ols_res`.

ii. For the $r^{\text{th}}$ simulated draw, generate a new treatment assignment vector using the `rbinom()` or `sample()` function in R. Use it to generate a new outcome vector, recalling that $y = (1 - z)y^0 + zy^1$.

iii. Apply each of the estimates (difference in means and linear regression with the one covariate) to the newly generated dataset. Save each estimate as the $r^{\text{th}}$ item in the results vector.

iv. Repeat 10 000 times.

(c) We will use these estimates to create a Monte Carlo estimate of a randomization distribution for each of these estimates for the sample average treatment effect.

(d) Plot the randomization distribution for each of the two estimates: difference in means and regression. Either overlay the plots (with different colors for each) or make sure the $x$-axis range on both plots is the same. Also add vertical lines (using different colors) for the SATE and the mean of the randomization distribution.

(e) Use these distributions to calculate the bias of each of these two methods. What is the difference between methods with respect to bias?

(f) Use these distributions to calculate the efficiency of each of these two methods. What is the difference between methods with respect to efficiency?

(g) Re-run the simulation by first generating data as in Exercise 18.12 but with a zero coefficient for the pre-treatment score. Does the zero coefficient lead to a different bias and efficiency estimate compared to when the coefficient for pre-treatment score was at 1.1 from before?

18.15 *Randomized block design with interactions*: In a randomized block design, randomization occurs separately within blocks, and the ratio of treatment to control observations is allowed to vary across blocks. And, as always, the underlying treatment effect can vary across blocks. For this exercise, you will simulate datasets for a randomized block design that includes sex as a blocking variable. You will then estimate the average treatment with two estimates: one that accounts for the blocking structure and one that does not. You will compare the bias and efficiency of these estimates. We will walk you through this in steps.

(a) First simulate the blocking variable and potential outcomes. In particular:

- Generate sex as a blocking variable, with 30% female and 70% male.
- Generate $y^0$ and $y^1$ with the following features:
  - The intercept of the regression of $y$ on treatment indicator, block indicator, and their interaction, is 70.
  - The residual standard deviation is 1.
  - Treatment effect varies by block: the average treatment effect is 7 for women and 3 for men. Further assume that men and women have no average difference under the control.

(b) Calculate the sample average treatment effect (SATE) overall and for each block.

(c) Now create a function for assigning the treatment. In particular, within each block create different assignment probabilities:

$$\Pr(z = 1 \mid \text{male}) = 0.6,$$
$$\Pr(z = 1 \mid \text{female}) = 0.4.$$

(d) Generate the treatment and create a vector for the observed outcomes implied by that treatment. We will use this to create a randomization distribution for two different estimates for the SATE. Take 100 000 draws from that distribution.

(e) Plot the Monte Carlo estimate of the randomization distribution for each of the two estimates: difference in means and regression. Either overlay the plots (with different colors for each) or make sure the $x$-axis on both plots is the same.

(f) Calculate the bias and efficiency of each estimate. Also calculate the root mean squared error.

(g) Consider the estimate that ignores the blocking. Why is it biased? Is the efficiency meaningful here? Why did we ask you to calculate the root mean squared error?

(h) Describe one possible real-life scenario where treatment assignment probabilities or treatment effects vary across levels of a covariate.

(i) How could you use a regression to estimate the treatment effects separately by group? Calculate estimates for our original sample and treatment assignment.

18.16 *Population average treatment effect*: Suppose you wanted to redo Exercise 18.14 but using the population average treatment effect rather than the sample average treatment effect. Does your preferred mode of inference depend at all on the estimand that you care most about?

18.17 *Sampling distributions instead of randomization distributions*:

(a) We could have also evaluated the properties of the estimates in Exercise 18.14 using sampling distributions that take into account uncertainty in all of the variables in the data-generating process. Simulate sampling distributions (with 100 000 draws) for the data-generating process and associated estimates from above.

(b) Create histograms of the sampling distributions just as you did above for the randomization distributions.

(c) Use the sampling distribution to evaluate the properties of the estimates relative to the population average treatment effect.

(d) What is the difference between a sampling distribution and a randomization distribution?

18.18 *Working through your own example*: Continuing the example from the final exercises of the earlier chapters, frame a substantive question in terms of the effect of a binary treatment. For this example, explain what are the outcome variable $y$, the treatment variable $z$, the pre-treatment variables $x$, and the potential outcomes $y^0$ and $y^1$.

## Chapter 19

# Causal inference using regression on the treatment variable

In the usual regression context, predictive inference relates to comparisons *between* units, whereas causal inference addresses comparisons of different treatments if applied to the *same* units. More generally, causal inference can be viewed as a special case of prediction in which the goal is to predict what *would have happened* under different treatment options. Causal interpretations of regression coefficients can only be justified by relying on much stronger assumptions than are needed for predictive inference. As discussed in the previous chapter, controlled experiments are ideal settings for using regression to estimate a treatment effect because the design of data collection guarantees that treatment assignment is independent of the potential outcomes, conditional on the information used in the design. This chapter illustrates the use of regression in the setting of controlled experiments, going through issues of adjustment for pre-treatment predictors, interactions, and pitfalls that can arise when building a regression using experimental data and interpreting coefficients causally.

## 19.1 Pre-treatment covariates, treatments, and potential outcomes

In this chapter, we consider a scenario in which there is a sample of items or *study units*, each of which has been assigned to receive either Treatment 1 or Treatment 0. Often these are labeled as the *treatment group* and the *control group*, but in some settings these can represent two distinct treatments. As shown in Figure 19.1, there is typically (but not necessarily) some background information on each observation, then there is the treatment assignment, and then, after some period of time (allowing for the treatment to have its effect), an outcome is measured. We thus can have at least three sorts of measurements on each item $i$:

- Pre-treatment measurements, also called *covariates*, $x_i$. As noted above, these are not strictly required for causal inference but in practice can be essential for checking and adjusting for pre-treatment differences between treatment and control groups, estimating treatment interactions, and making inferences about average treatment effects in subpopulations and in the general population.
- The treatment $z_i$, which equals 1 for treated units and 0 for controls.
- The outcome measurement, $y_i$, which we label as $y_i^1$ for units that have been exposed to the treatment and $y_i^0$ for units that received the control. As discussed in detail in Section 18.1, $y_i^0$ and $y_i^1$ represent *potential outcomes* under different possible treatment assignments, and $y_i$ refers to which of these is actually observed.

Here are some examples in this framework:

- Units are high school students, $x$ is college admissions test score on the first try, $z$ is 1 for students who are assigned to get coaching or 0 otherwise, $y$ is test score on the second try, six months later.
- Units are male or female musicians auditioning for jobs at orchestras, $x$ is their sex (for example, $x = 1$ for men and 0 for women), $z = 1$ if the audition is performed behind a screen or 0 if the

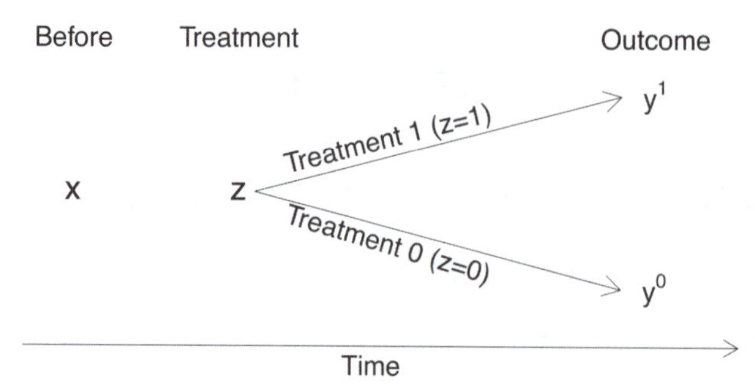


Figure 19.1 *Basic framework for causal inference. The treatment effect is $y^1 - y^0$, but we can never observe both* potential outcomes $y^1$ *and* $y^0$ *on the same item. For any given unit, the unobserved outcome is called the* counterfactual.

evaluators can see the musician audition, and $y$ is the outcome, defined as 1 if the applicant gets a job offer and 0 otherwise. The key point here is that sex is a background variable, not the treatment. The treatment, which is the condition that can be manipulated, is whether a screen is used.

More generally, there can be multiple pre-treatment measurements, multiple treatment levels, and multiple outcome measurements.

## 19.2 Example: the effect of showing children an educational television show

Example: Electric Company experiment

We illustrate the use of regression to estimate a causal effect in the context of an educational experiment that was undertaken around 1970 on a set of 192 elementary school classes.[1] The goal was to measure the effect of a new educational television program, The Electric Company, on children's reading ability. We discussed this study briefly on page 6. Selected classes of children in grades 1–4 were randomized into treated and control groups. At the beginning and the end of the school year, students in all the classes were given a reading test, and the average test score within each class was recorded. Unfortunately, we do not have data on individual students, and so our entire analysis will be at the classroom level; that is, we treat the classes as the observational units in this study.

### Displaying the data two different ways

Figure 19.2 displays the distribution of the outcome, average post-treatment test scores, in the control and treatment group for each grade. Recall that the experimental treatment was applied to classes, not to schools, and so we treat the average test score in each class as a single measurement. Rather than try to cleverly put all the data on a single plot, we arrange them on a $4 \times 2$ grid, using a common scale for all the graphs to facilitate comparisons among grades and between treatment and control. We also extend the axis all the way to zero, which is not strictly necessary, in the interest of clarity of presentation. In this example, as in many others, we are not concerned with the exact counts in the histogram; thus, we simplify the display by eliminating $y$-axes, and we similarly clarify the $x$-axis by removing tick marks and using minimal labeling.

As discussed in Section 2.3, all graphs are comparisons. Figure 19.2 most directly allows comparisons between grades. This comparison is useful—if for no other reason than to ground ourselves and confirm that scores are higher in the higher grades—but we are more interested in the comparison of treatment to control within each grade.

Thus, it might be more helpful to arrange the histograms as shown in Figure 19.3, with treatment

[1]Data and code for this example are in the folder `ElectricCompany`.

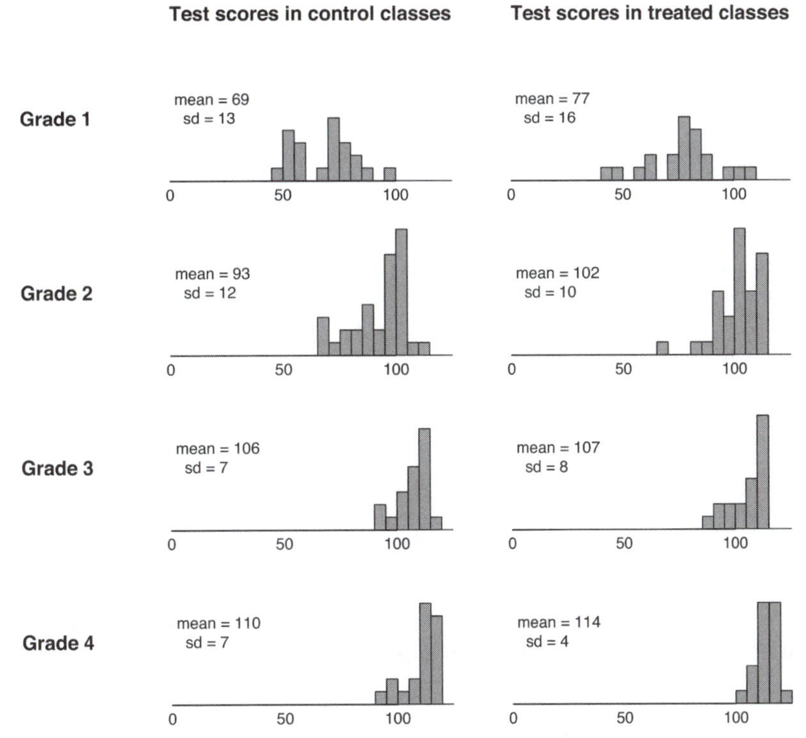


Figure 19.2 *Post-treatment test scores from an experiment measuring the effect of an educational television program, The Electric Company, on children's reading abilities. The experiment was applied on a total of 192 classrooms in four grades. At the end of the experiment, the average reading test score in each classroom was recorded.*

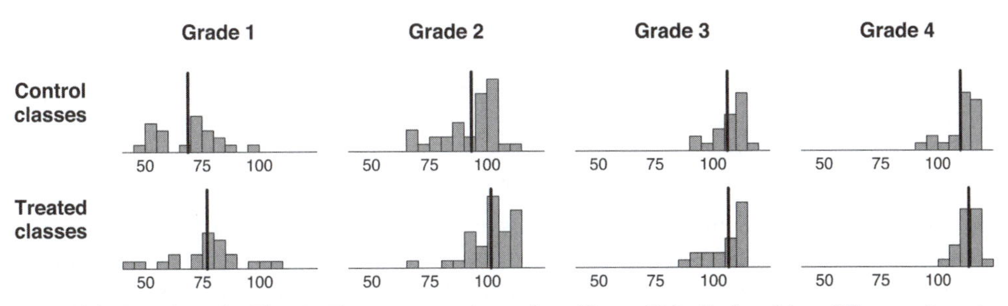


Figure 19.3 *Data from the Electric Company experiment, from Figure 19.2, displayed in a different orientation to allow easier comparison between treated and control groups in each grade. For each histogram, the average is indicated by a vertical line.*

and control aligned for each grade. With four histograms arranged horizontally on a page, we need to save space, and so we restrict the $x$-axes to the combined range of the data. We also indicate the average value in each group with a vertical line to allow easier comparisons of control to treatment in each grade.

Visual comparisons across treatment groups within grades suggest that watching The Electric Company may have led to small increases in average test scores, particularly for the lower grades. This plot also reveals an apparent ceiling with respect to the reading assessment used. There is not much room for improvement for those classes that scored well in the later grades. This feature of the assessment makes it difficult to know whether the relatively small estimated effects at the higher grades might have been larger if the assessments had included more difficult items to allow for more room for improvement among the more advanced classes.

### Paired comparisons design

The experiment was performed in two cities, Fresno and Youngstown. For each city and grade, the experimenters selected a small number of schools (10–20) and, within each school, they selected the two poorest reading classes of that grade. For each pair, one of these classes was randomly assigned to continue with its regular reading course and the other was assigned to view the TV program.

This is an example of a *matched pairs* design (which in turn is a special case of a *randomized block* design, with exactly two units within each block). Recall that the idea behind this is that there are characteristics of the school—both observable and, potentially, unobservable—that are predictive of future student outcomes that we would like to adjust for explicitly by forcing balance through our design.

For simplicity we shall analyze this experiment *as if the treatment assignment had been completely randomized within each grade*. It would be better to use a multilevel model to account for the pairing in the design, but that goes beyond the scope of this book.

### Simple difference estimate (equivalently, regression on an indicator for treatment), appropriate for a completely randomized experiment with no pre-treatment variables

We start by estimating a single treatment effect using the simplest possible estimate, a regression of post-test on treatment indicator, which would be the appropriate analysis had the data come from a completely randomized experiment with no available pre-treatment information.

When treatments are assigned completely at random, we can think of the treatment and control groups as two random samples from a common population. The population average under each treatment, $\text{avg}(y^0)$ and $\text{avg}(y^1)$, can then be estimated by the sample average, and the population average difference between treatment and control, $\text{avg}(y^1) - \text{avg}(y^0)$—that is, the average causal effect—can be estimated by the difference in sample averages, $\bar{y}_1 - \bar{y}_0$.

Equivalently, the average causal effect of the treatment corresponds to the coefficient $\theta$ in the regression, $y_i = \alpha + \theta z_i + \text{error}_i$; as discussed in Section 7.3, linear regression on an indicator variable is a comparison of averages. In R:

```
stan_glm(post_test ~ treatment, data=electric)
```

Applied to the Electric Company data, this yields an estimate of 5.7 with a standard error of 2.5. This estimate is a starting point, but we should be able to do better because it makes sense that the effects of the television show could vary by grade. Indeed this appears to be the case in the visual comparisons of the histograms in Figure 19.2.

### Separate analysis within each grade

Given the large variation in test scores from grade to grade, it makes sense to take the next step and perform a separate regression analysis on each grade's data. This is equivalent to fitting a model in which treatment effects vary by grade—that is, an interaction between treatment and grade indicators—and where the residual variance can be different from grade to grade as well.

In R we fit this as four separate models:

```
fit_1 <- as.list(rep(NA, 4))
for (k in 1:4)
  fit_1[[k]] <- stan_glm(post_test ~ treatment, data=electric, subset=(grade==k))
```

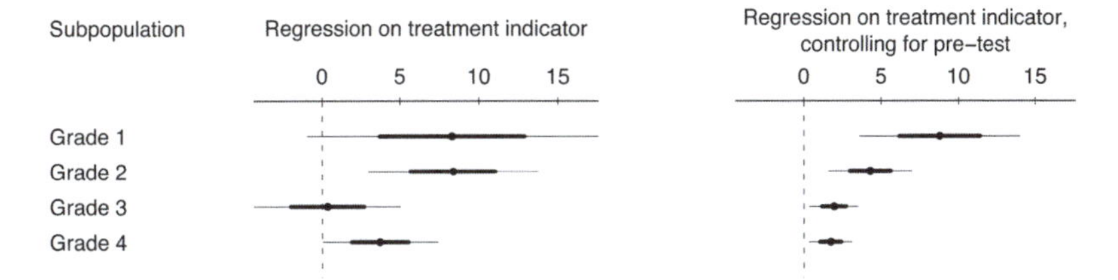


Figure 19.4 *Estimates, 50%, and 95% intervals for the effect of watching The Electric Company (see data in Figures 19.2 and 19.5) as estimated in two ways: (a) from a regression on treatment alone, and second,(b) also adjusting for pre-test data. In both cases, the coefficient for treatment is the estimated causal effect. Including pre-test data as a predictor increases the precision of the estimates.*
*Displaying these coefficients and intervals as a graph facilitates comparisons across grades and across estimation strategies (adjusting for pre-test or not). For instance, the plot highlights how adjusting for pre-test scores increases precision and reveals decreasing effects of the program for the higher grades, a pattern that would be more difficult to see in a table of numbers.*
*Sample sizes are approximately the same in each of the grades. The estimates for higher grades have lower standard errors because the residual standard deviations of the regressions are lower in these grades; see Figure 19.5.*

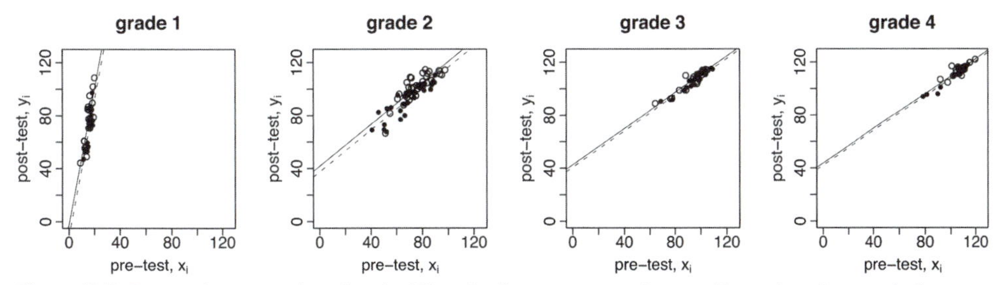


Figure 19.5 *Pre-test/post-test data for the Electric Company experiment. Treated and control classes are indicated by circles and dots, respectively, and the solid and dotted lines represent parallel regression lines fit to the treatment and control groups, respectively. The solid lines are slightly higher than the dotted lines, indicating slightly positive estimated treatment effects. Compare to Figure 19.2, which displays only the post-test data.*

The resulting estimates and uncertainty intervals for the Electric Company experiment are graphed in Figure 19.4a. The treatment appears to be generally effective, perhaps more so in the low grades, but it is hard to be sure, given the large standard errors of estimation. Sample sizes are approximately the same in each of the grades, but the estimates for higher grades have lower standard errors because the residual standard deviations of the regressions are lower in these grades; see Figure 19.5.

The separate analysis for each grade is appropriate, given that this is a block design in which classrooms were randomly assigned to treatments within each grade. But in this example, even if there had been a completely randomized design with no blocking by grade, we still would have recommended separate analyses, or an interaction of treatment effect with grade. The effect varies enough by grade that it would not make sense to try to estimate a single treatment effect averaging over the four grades.

## 19.3 Including pre-treatment predictors

### Adjusting for pre-test to get more precise estimates

In the Electric Company experiment, a pre-test was given in each class at the beginning of the school year, before the treatment was applied, and we can use this information to improve our treatment effect estimates, using a regression model such as, $y_i = \alpha + \theta z_i + \beta x_i + \text{error}_i$, where $z_i$ indicates the treatment (1 if Electric Company and 0 if control) and $x_i$ denotes the average pre-test scores of the students in classroom $i$.

Figure 19.5 shows the before-after data for the Electric Company experiment. For grades 2–4, the same test was given to the students at the beginning and the end of the year, and so it is no surprise that all the classes improved whether treated or not. For grade 1, the pre-test was a subset of the longer test, which explains why the pre-test scores for grade 1 are so low. We can also see that the distribution of post-test scores for each grade is similar to the next grade's pre-test scores, which makes sense.

As before, we fit the model separately for each grade:

```
fit_2 <- as.list(rep(NA, 4))
for (k in 1:4)
  fit_2[[k]] <- stan_glm(post_test ~ treatment + pre_test, data=electric,
    subset=(grade==k))
```

We display the fitted lines along with the data in Figure 19.5. For each grade, the difference between the regression lines for the two groups represents the estimated treatment effect as a function of pre-treatment score. Since we have not included any interaction in the model, the treatment effect within each grade is assumed constant over all levels of the pre-treatment score.

In the regression,

$$y_i = \alpha + \theta z_i + \beta x_i + \text{error}_i, \tag{19.1}$$

the coefficient $\theta$ still represents the average treatment effect in the grade, but adjusting for pre-treatment score, $x_i$, can reduce the uncertainty in the estimate. More generally, the regression can adjust for multiple pre-treatment predictors, in which case the model has the form $y_i = \alpha + \tau z_i + X_i\beta + \text{error}_i$, where $X_i$ and $\beta$ now are vectors of predictors and coefficients, respectively; alternatively $\alpha$ could be removed from the equation and considered as a constant term in the linear predictor $X\beta$.

## Benefits of adjusting for pre-treatment score

Under a clean randomization, adjusting for pre-treatment predictors in this way does not change what we are estimating. However, if the predictor has a strong association with the outcome it can help to bring each estimate closer (on average) to the truth, and if the randomization was less than pristine, the addition of predictors to the equation may help us adjust for *systematically* unbalanced characteristics across groups. Thus, this strategy has the potential to adjust for both random and systematic differences between the treatment and control groups (that is, to reduce both variance and bias), as long as these differences are characterized by differences in the pre-test.

The estimates for the Electric Company study appear in of Figure 19.4b. It now appears that the treatment is effective on average for each of the grades, although, consistent with our earlier visual inspection of the data, the effects seem larger in the lower grades.

To get a sense of what we get by adjusting for a pre-treatment predictor, suppose that in a particular grade the average pre-test score is $\Delta_x$ points higher for the treatment than the control group. Such a difference would not necessarily represent a failure of assumptions; it could just be chance variation that happened to occur in this particular randomization. In any case, *not* adjusting for this pre-treatment imbalance would be a mistake: scores on pre-test and post-test are positively correlated, and so the unadjusted comparison would tend to overestimate the treatment effect by an amount $b\Delta_x$, in this case. Performing the regression automatically performs this adjustment on the estimate of $\theta$.

Is this a problem of "bias" or "variance"? It depends on how you look at it. With a simple randomized design, the pre-test difference between treatment and control, and thus the estimation error induced by not adjusting for pre-test, is random and has expected value zero. Averaging over the sampling procedure, the unadjusted estimate is unbiased but has large variance. However, for any given sample, we can observe which group has higher average pre-test, and by how much. Once we've seen that pre-test scores are higher (or lower) in the treatment group, we can make an adjustment—and if we do not make this adjustment, we have in practice introduced a bias in the sense that our estimated treatment effect is too high by a certain amount. It would not be appropriate to use an unadjusted estimate, if the pre-test data are readily available. Whether this predictable correction counts as bias or variance depends on whether we are conditioning on the particular treatment assignment that occurred.

This reasoning applies not just to pre-test but to any pre-treatment variables that help to predict the outcome. In practice, we can neither collect nor analyze everything, and our models (linear and otherwise) are themselves only approximations, so we can only try to reduce our errors of estimation, not bring them all the way to zero.

Crucially, when fitting such a regression it is only appropriate to adjust for *pre*-treatment predictors, or, more generally, predictors that would not be affected by the treatment. We discuss this point in more detail in Section 19.6.

### Problems with simple before-after comparisons

Given that we have pre-test and post-test measurements, why not simply summarize the treatment effect by their difference? Why bother with a controlled experiment at all? The problem with the simple before-after estimate is that, when estimating causal effects we are interested in the difference between treatment and control conditions, not in the simple improvement from pre-test to post-test. The improvement is not a causal effect (except under the assumption, unreasonable in this case, that under the control there would be no change in reading ability during the school year).

There are real-world settings in which there is no control, and then strong assumptions do need to be made to draw any causal inferences. But when a control group is available, it should be used. In a regression context, the treatment effect is the coefficient on the treatment indicator, and if there is a treatment group but no control group, then the treatment variable, $z_i$, equals 1 for all units, and there is no way to compute the regression coefficient of $z$ without strong modeling assumptions that cannot be checked with the data at hand.

### Gain scores: a special case of regression in which the coefficient for pre-test is fixed at 1

More generally, an alternative way to specify a model that controls for pre-test measures is to use these measures to transform the response variable. A simple approach is to subtract the pre-test score, $x_i$, from the outcome score, $y_i$, thereby creating a "gain score," $g_i$. Then this score can be regressed on the treatment indicator (and other predictors if desired), $g_i = \alpha + \tau z_i + \text{error}_i$. In the simple case with no other predictors, the regression estimate is simply $\hat{\tau} = \bar{g}^T - \bar{g}^C$, the average difference of gain scores in the treatment and control groups.

Using gain scores is most effective if the pre-treatment score is comparable to the post-treatment measure. For instance, in our Electric Company example it would not make sense to create gain scores for the classes in grade 1 since their pre-test measure was based on only a subset of the full test. Also, as with other ways of adjusting for predictors, it is never appropriate to create a gain score by subtracting a variable that was measured after the treatment assignment.

One perspective on the analysis of gain scores is that it implicitly makes an unnecessary assumption, namely, that $\beta = 1$ in model (19.1). To see this, note the algebraic equivalence between $y_i = \alpha + \tau z_i + x_i + \text{error}_i$ and $y_i - x_i = \alpha + \tau z_i + \text{error}_i$. On the other hand, if this assumption is close to being true, then $\tau$ may be estimated more precisely. One way to resolve this concern about misspecification would simply be to include the pre-test score as a predictor as well, $g_i = \alpha + \tau z_i + \gamma x_i + \text{error}_i$. However, in this case, $\hat{\tau}$, the estimate of the coefficient for $z$, is equivalent to the estimated coefficient from the original model, $y_i = \alpha + \tau z_i + \beta x_i + \text{error}_i$; see Exercise 19.3.

Sometimes gain score models are motivated by concern that the pre-treatment score may have been measured with error. In this case we would prefer to adjust for the latent "true score" rather than the observed pre-treatment measurement. One can set up a regression adjusting for an unmeasured variable using a multilevel model; this goes beyond the scope of this book. For here, we merely note that there are settings where performing a simple regression using the gain score can be a useful

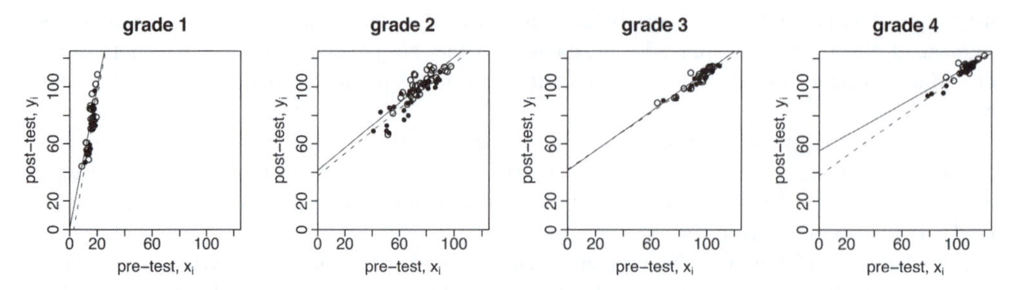


Figure 19.6 *Pre-test/post-test data for the Electric Company experiment. Treated and control classes are indicated by circles and dots, respectively, and the solid and dotted lines represent separate regression lines fit to the treatment and control groups, respectively—that is, the model interacts treatment with pre-test score. For each grade, the difference between the solid and dotted lines represents the estimated treatment effect as a function of pre-test score. Compare to Figure 19.5, which displays the same data but with parallel regression lines in each graph. The non-parallel lines in the current figure represent interactions in the fitted model.*

approximation to that latent-variable regression. In a randomized experiment such as the Electric Company study this is less of a concern because the treatment assignment is independent of all pre-treatment variables that have not been already accounted for in the design.

Another motivation for use of gain scores is the desire to interpret effects on changes in the outcome rather than the effect on the outcome on its own. Compare this interpretation to the interpretation of a treatment effect estimate from a model that adjusts for the pre-test; in this case we could interpret an effect on the outcome for those with the same value of the pre-test. The difference between these interpretations is subtle.

## 19.4 Varying treatment effects, interactions, and poststratification

Once we include pre-test in the model, it is natural to interact it with the treatment effect. The treatment effect is then allowed to vary with the level of the pre-test. Figure 19.6 shows the Electric Company data with separate intercepts and slopes estimated for the treatment and control groups. As with Figure 19.5, for each grade the difference between the regression lines is the estimated treatment effect as a function of pre-test score.

We illustrate in detail for grade 4. First we fit the simple model including only the treatment indicator:

```
stan_glm(post_test ~ treatment, data=electric, subset=(grade==4))
```

which yields,

```
            Median MAD_SD
(Intercept) 110.4    1.2
treatment     3.7    1.8

Auxiliary parameter(s):
      Median MAD_SD
sigma 6.0    0.7
```

The estimated treatment effect is 3.7 with a standard error of 1.8, displayed as the Grade 4 entry in Figure 19.4a.

As discussed in the previous section, we can reduce the standard error of the estimate by adjusting for pre-test:

```
stan_glm(post_test ~ treatment + pre_test, data=electric, subset=(grade==4))
```

which yields,

```
            Median MAD_SD
(Intercept) 41.9    4.3
treatment    1.7    0.7
pre_test     0.7    0.0

Auxiliary parameter(s):
      Median MAD_SD
sigma 2.2    0.2
```

The new estimated treatment effect is 1.7 with a standard error of 0.7 and is displayed as the Grade 4 entry in Figure 19.4a. Compared to the model fit just before, the residual standard deviation has declined a lot, from 6.0 to 2.2, telling us that pre-test adds a lot of information as a linear predictor in the model.

We next include the interaction of treatment with pre-test:

```
stan_glm(post_test ~ treatment + pre_test + treatment:pre_test,
  data=electric, subset=(grade==4))
```

yielding,

```
                   Median MAD_SD
(Intercept)        39.41   4.90
treatment          11.61   7.98
pre_test            0.68   0.05
treatment:pre_test -0.09   0.07

Auxiliary parameter(s):
      Median MAD_SD
sigma 2.17   0.25
```

The estimated treatment effect is now $11.61 - 0.09x$, which is difficult to interpret without knowing the range of $x$. From Figure 19.6 we see that, in grade 4, the pre-test scores range from approximately 80 to 120; in this zone, the estimated treatment effect varies from $11.61 - 0.09 * 80 = 4.4$ for classes with pre-test scores of 80 to $11.61 - 0.09 * 120 = 0.8$ for classes with pre-test scores of 120. This range represents the *variation* in estimated treatment effects as a function of pre-test score, *not* uncertainty in the estimated effect. Centering $x$ before including it in the model allows the treatment coefficient to represent that treatment effect for classes with the mean pre-test score for the sample.

To get a sense of the uncertainty, we can plot the 20 random simulation draws of the estimated treatment effect:

```
fit_4 <- stan_glm(post_test ~ treatment + pre_test + treatment:pre_test,
  data=electric, subset=(grade==4))
sims_4 <- as.matrix(fit_4)
n_sims <- nrow(sims_4)
plot(c(80, 120), c(-5, 10), xlab="pre-test", ylab="treatment effect",
  main="treatment effect in grade 4", type="n")
abline(0, 0, lty=2)
subset <- sample(n_sims, 20)
for (s in subset){
  curve(sims_4[s,2] + sims_4[s,4]*x, lwd=0.5, col="gray", add=TRUE)
}
```

This produces the graph shown in Figure 19.7.

Finally, we can estimate a mean treatment effect by averaging over the values of $x$ in the data. If we write the regression model as $y_i = \alpha + \tau_1 z_i + \beta x_i + \tau_2 z_i x_i + \text{error}_i$, then the treatment effect is $\tau_1 + \tau_2 x$, and the summary treatment effect in the sample is $\frac{1}{n}\sum_{i=1}^{n}(\tau_1 + \tau_2 x_i)$, averaging over the $n$ fourth-grade classrooms in the data. We can compute the estimated average treatment effect as follows:

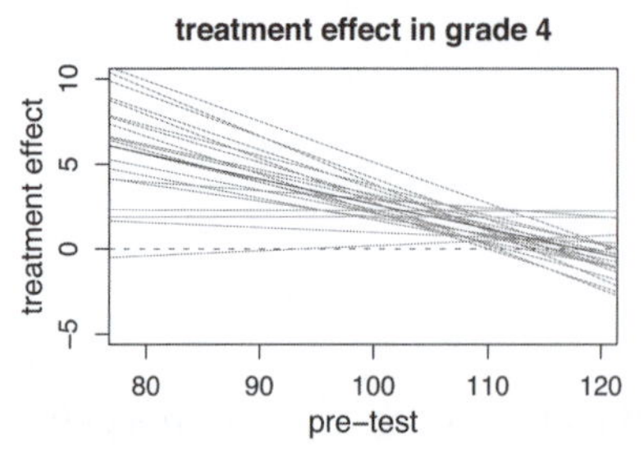


Figure 19.7 *Uncertainty about the effect of viewing The Electric Company (compared to the control) for fourth-graders. Compare to the rightmost plot in Figure 19.6. The dark line here—the estimated treatment effect as a function of pre-test score—is the difference between the two regression lines in the grade 4 plot in Figure 19.6. The gray lines represent 20 random draws from the uncertainty distribution of the treatment effect.*

```
n_sims <- nrow(sims_4)
effect <- array(NA, c(n_sims, sum(grade==4)))
for (s in 1:n_sims){
  effect[s,] <- sims_4[s,2] + sims_4[s,4]*pre_test[grade==4]
}
avg_effect <- rowMeans(effect)
```

The `rowMeans()` function averages over the grade 4 classrooms, and the result of this computation, `avg_effect`, is a vector of length `n_sims` representing the uncertainty in the average treatment effect. We can summarize with the posterior median and mad sd (see Section 5.3):

```
print(c(median(avg_effect), mad(avg_effect)))
```

The result is 1.8 with a standard deviation of 0.7—similar to the result from the model adjusting for pre-test but with no interactions. In general, for a linear regression model, the estimate obtained by including the interaction, and then averaging over the data, reduces to the estimate with no interaction. The motivation for including the interaction is thus to get a better idea of how the treatment effect varies with pre-treatment predictors, not to simply estimate an average effect.

Identification of treatment interactions is also important when we want to generalize experimental results to a broader population. If treatment effects vary with pre-treatment characteristics and the distribution of these characteristics varies between the experimental sample and the population of interest, the average treatment effects will typically be different. If these characteristics are observable, however, we should be able to extrapolate the one estimate to the other.

## Poststratification of conditional treatment effects gives an average treatment effect

In survey sampling, *stratification* refers to the procedure of dividing the population into disjoint subsets (strata), sampling separately within each stratum, and then combining the stratum samples to get a population estimate. Poststratification is the analysis of an unstratified sample, breaking the data into strata and reweighting as would have been done had the survey actually been stratified. Stratification can adjust for potential differences between sample and population using the survey design; poststratification makes such adjustments in the data analysis.

We discussed this in the general regression context in Section 17.1 where the challenge was to align sample and population. Here we apply the idea to causal inference, where the challenge is to align treatment and control groups.

For example, suppose we have treatment variable $z$ and pre-treatment control variables $x_1, x_2$, with regression predictors $z, x_1, x_2$ and interactions $zx_1$ and $zx_2$, so the model is $y = \beta_0 + \beta_1 z + \beta_2 x_1 + \beta_3 x_2 + \beta_4 z x_1 + \beta_5 z x_2 + \text{error}$. The estimated treatment effect is then $\beta_1 + \beta_4 x_1 + \beta_5 x_2$, and its average, in a linear regression, is simply $\beta_1 + \beta_4 \mu_1 + \beta_5 \mu_2$, where $\mu_1$ and $\mu_2$ are the averages of $x_1$ and $x_2$ in the population. These population averages might be available from another source, or

| Predictor | Estimate ± s.e. |
|---|---|
| (Intercept) | 1.6 ± 1.8 |
| Value of incentive | 0.12 ± 0.05 |
| Prepayment | 4.0 ± 2.1 |
| Gift | −5.4 ± 2.2 |
| Burden | 2.9 ± 1.6 |

Figure 19.8 *Fitted regression estimating the effects of monetary incentives on survey response rates. Each data point in the regression corresponds to a different survey that had multiple incentive conditions, with the outcome being the difference in percentage response rates comparing high to low (or zero) incentives, and the predictors corresponding to various conditions of the survey. In a mistaken naive interpretation, each of these coefficients would be taken to represent the causal effect on response rate of some aspect of the survey.*

else they could be estimated using the averages of $x_1$ and $x_2$ in the data at hand. Standard errors for summaries such as $\beta_1 + \beta_4\mu_1 + \beta_5\mu_2$ can be determined analytically, but it is easier to simply compute them using simulations obtained using `stan_glm` as discussed in earlier chapters.

Modeling interactions is important when we care about differences in the treatment effect for different groups, and poststratification then arises naturally if a population average estimate is of interest.

## 19.5 Challenges of interpreting regression coefficients as treatment effects

ample: ects of entives surveys

It can be tempting to take the coefficients of a fitted regression model and give them a causal interpretation, but this can be a mistake—even if the data come from randomized experiments. We illustrate with an example of a meta-analysis of studies of incentives in sample surveys.[2]

A team of researchers performed a literature search and found 39 randomized experiments on the effects of incentives on survey response rates. In each experiment, respondents were randomly assigned to two or more conditions (for example, no incentive, an incentive of $2, or an incentive of $5). For each survey, the zero or lowest incentive condition was considered the baseline, and for each other condition, the difference in observed response rate compared to the baseline was recorded. A summary dataset was then constructed, with one data point corresponding to each non-baseline condition on each survey: the 39 surveys included a total of 101 experimental conditions, hence 62 data points. A regression was then fit, predicting change in response rate compared to baseline given several predictors: the difference in dollar value of the incentive compared to baseline, and indicators for whether the incentive was prepaid (given before the survey was conducted), whether the incentive was in gift form (with the default condition being cash), and whether the survey was assessed as requiring high burden of effort on respondents.

The result of the regression is shown in Figure 19.8. Given that the treatments were assigned randomly to survey respondents, it might seem reasonable at first to interpret the coefficients causally:

- Just the act of assigning any incentive at all is estimated to increase response rate by 1.6 percentage points, but this estimate has a high uncertainty, with standard error 1.8 percentage points.
- Each additional dollar of incentive (here inflation-adjusted to 1999, the year that the meta-analysis was conducted) is estimated to increase response rate by 0.12 ± 0.05 percentage points.
- An incentive is estimated to be better prepaid than postpaid, by 4.0 ± 2.1 percentage points.
- Gifts are estimated to be less effective than cash, by an amount of −5.4 ± 2.2 percentage points.
- Incentives are estimated to be 2.9 ± 1.6 percentage points more effective in high-burden surveys.

Cautions need to be attached to these estimates: they only apply to the sorts of surveys in the

[2]Data and code for this example are in the folder `Incentives`.

meta-analysis, which might not be representative of future surveys of interest, and it could be a mistake to extrapolate beyond the data. For example, if an extra $1 of incentives increases the response rate by 0.12 percentage points, we should not naively expect another $100 to add 12 percentage points, as such a conclusion would lean far too heavily on the linearity assumption in the model.

Here, though, we want to discuss a different issue, which is that the above coefficients should not be directly interpreted as causal effects, *even for the surveys and experimental conditions in the data*. Given that the treatments in the survey experiments were randomly assigned, it may at first seem paradoxical to deny the causal interpretation. The resolution is that, although the incentive conditions were assigned randomly *within* each experiment, the differences in the conditions were not assigned at random *between* experiments. The difference in response rate comparing two incentive conditions within a survey is an unbiased estimate of the effect of that particular implementation of the incentive for that particular survey—but when comparing incentives implemented in different surveys, what we have is an observational study.

For example, the coefficient of Prepayment in Figure 19.8 represents a difference of effects on response rates of prepaid as compared to postpaid incentives. But it is possible that surveys with prepaid incentives happened to be in conditions more susceptible to effects of incentives, compared to the surveys in the meta-analysis where postpaid incentives were tried. Perhaps the surveys with prepayment were simply more professionally conducted. We can't really know, given just the information given in this regression. And these issues arise for all the coefficients in the model. We are in the familiar setting with observational data of interpreting coefficients in the context of potential background variables and interactions that are not yet in the model.

These concerns are not just theoretical, as can be seen from a careful examination of the coefficients in Figure 19.8. Consider the large negative coefficient for Gift in that regression. What happens if, instead of just looking at its sign and statistical significance and concluding that gift incentives are worse than cash, we try to take the numerical estimate seriously? Consider a low-value postpaid incentive, perhaps a mug worth $3, in a low-burden survey. The fitted model yields an estimated effect of $1.6 + 0.12 * 3 - 5.4 = -3.4$. It is hardly plausible that this incentive would be expected to cause this large reduction in response rates; rather, it seems more natural to suppose that, for some reason, incentives happened on average to be less effective in the surveys where gifts were tried. Because the incentive conditions were not randomly assigned to surveys, we have to be open to this sort of non-causal interpretation.

## 19.6 Do not adjust for post-treatment variables

As illustrated in the examples of this chapter, we recommend adjusting for pre-treatment covariates when estimating causal effects in experiments and observational studies. However, it is generally not a good idea to adjust for variables measured *after* the treatment. By "adjusting for" a variable, we specifically are referring to including it as a regression input. As we discuss in Section 21.1, information on post-treatment variables can be included in the more complicated framework of instrumental variables or in more general mediator strategies.

In this section we explain why naively adjusting for a post-treatment variable can bias the estimate of the treatment effect, *even when the treatment has been randomly assigned to study participants*. In the next section we describe the related difficulty of using regression on "mediators" or "intermediate outcomes" (variables measured post-treatment but generally prior to the primary outcome of interest) to estimate so-called mediating effects.

Example: Hypothetical experiment with intermediate outcomes

Consider a hypothetical study of a treatment that incorporates a variety of social services including high-quality child care and home visits by trained professionals. We label $y$ as the child's IQ score measured 2 years after the treatment regime has been completed, $q$ as a continuous parenting quality measure (ranging from 0 to 1) measured one year after treatment completion, $z$ as the *randomly assigned* binary treatment, and $x$ as the pre-treatment measure reflecting whether both parents have a high school education (in general this could be a vector of pre-treatment predictors). The goal here

| Unit, $i$ | Pre-treatment covariate, $x_i$ | Treatment, $z_i$ | Observed intermediate outcome, $q_i$ | Potential intermediate outcomes, $q_i^0$ | $q_i^1$ | Final outcome, $y_i$ |
|---|---|---|---|---|---|---|
| 1 | 0 | 0 | 0.5 | **0.5** | *0.7* | $y_1$ |
| 2 | 0 | 1 | 0.5 | *0.3* | **0.5** | $y_2$ |
| ⋮ | ⋮ | ⋮ | ⋮ | ⋮ | ⋮ | ⋮ |

Figure 19.9 *Hypothetical example illustrating the problems with regressions that adjust for a continuous intermediate outcome. If we adjust for $q$ when regressing $y$ on $z$, we will be essentially making comparisons between units such as 1 and 2 above, which differ in $z$ but are identical in $q$. The trouble is that such units are not comparable, as can be seen by looking at the parenting quality potential outcomes, $q^0$ and $q^1$ (which can never both be observed, but which we can imagine for the purposes of understanding this comparison). Unit 1, which received the control, has higher parenting quality potential outcomes than unit 2, which received the treatment. Matching on the observed $q$ thus inherently leads to misleading comparisons as reflected in the parenting quality potential outcomes. The coefficient $\tau$ in regression (19.2) thus in general represents an inappropriate comparison of units that fundamentally differ. See Figure 19.10 for a similar example with a discrete intermediate outcome.*

is to measure the effect of $z$ on $y$, and we shall explain why it is not a good idea to adjust for the intermediate outcome, $q$, when estimating this effect.

Due to the randomization we know that unconditional ignorability holds, that is, $y^0, y^1 \perp z$. Thus a model for $y$ given $z$ alone would serve to unbiasedly estimate the treatment effect:

$$\text{regression estimating the treatment effect: } y = \tau z + \epsilon.$$

The coefficient on $z$ in this equation is equivalent to the difference in mean outcomes across treatment groups. The model has no intercept because the treatment effect is defined relative to the control group so by definition it is zero when $z = 0$.

As discussed above, we also know that when $z$ is randomized, ignorability also holds conditional on $x$, that is, $y^0, y^1 \perp z \mid x$. Therefore a model for $y$ given $z$ and $x$—excluding $q$—is straightforward, with the coefficient on $z$ representing the total effect of the treatment on the child's cognitive outcome:

$$\text{regression estimating the treatment effect: } y = \tau z + \beta x + \epsilon.$$

The difficulty comes if the intermediate outcome, $q$, is added to this model. Adding $q$ as a predictor could improve the model fit, explaining much of the variation in $y$:

$$\text{regression including intermediate outcome: } y = \tau^* z + \beta^* x + \delta^* q + \epsilon^*. \tag{19.2}$$

We add the asterisks here because adding a new predictor changes the meaning of each of the parameters. Unfortunately, the new coefficient $\tau^*$ does *not*, in general, estimate the effect of $z$. Formally this is because it is not in general true that $y^0, y^1 \perp z \mid x, q$ if $q$ was measured post-treatment, and in particular if $q$ was affected by the treatment.

For instance, suppose the true model for $q$ given $z$ and $x$ is a linear regression:

$$q = 0.3 + 0.2z + \gamma x + \text{error}, \tag{19.3}$$

with independent errors. By saying this is the true model, we are implying that $\mathrm{E}(q(0)|x) = 0.3 + \gamma x$ and $\mathrm{E}(q(1)|x) = 0.3 + \gamma x + 0.2$. Further suppose that the pre-treatment variable $x$ has been standardized to have mean 0. The parenting quality variable then has an average of 0.3 for the controls and 0.5 for the treated parents. Thus, $z$ increases parenting quality by 0.2 points on this scale that runs from 0 to 1.

Figure 19.9 illustrates the problem with adjusting for an intermediate variable that has been affected by the treatment. The coefficient of $z$ in (19.2) corresponds to a comparison of units that are identical in $x$ and $q$ but differ in $z$. The trouble is, they will then automatically differ in their *potential outcomes*, $q^0$ and $q^1$. For example, consider two families, both with $q=0.5$ and $x=0$, but one with $z=0$ and one with $z=1$. Under the (simplifying) assumption that the effect of $z$ is to increase $q$ by exactly 0.2 (as in the assumed model (19.3)), the first family has potential outcomes $q^0=0.5, q^1=0.7$, and the second family has potential outcomes $q^0=0.3, q^1=0.5$. Given two families with the same observed intermediate outcome $q$, the one that received the treatment has lower underlying parenting skills. Thus, in the regression of $y$ on $(x, z, q)$, the coefficient of $z$ represents a comparison of families that differ fundamentally in their underlying characteristics. This is an inevitable consequence of adjusting for an intermediate outcome that has been affected by the treatment.

This reasoning suggests a strategy of estimating treatment effects conditional on the potential outcomes—in this example, including both $q^0$ and $q^1$, along with $z$ and $x$, in the regression. The practical difficulty here (as usual) is that we observe at most one potential outcome for each observation, and thus such a regression would require imputation (prediction) of $q^0$ or $q^1$ for each case (perhaps using pre-treatment variables as proxies for $q^0$ and $q^1$), and correspondingly strong assumptions.

## 19.7 Intermediate outcomes and causal paths

Randomized experimentation is often described as a "black box" approach to causal inference. We see what goes into the box (treatments) and we see what comes out (outcomes), and we can make inferences about the relation between these inputs and outputs, without the need to see what happens *inside* the box. This section discusses some difficulties that arise from using naive techniques to try to ascertain the role of post-treatment *mediating* variables in the causal path between treatment and outcomes, as part of a well-intentioned attempt to peer inside the black box.

### Hypothetical example of a binary intermediate outcome

Continuing the hypothetical experiment on child care, suppose that the randomly assigned treatment increases children's IQ after three years by an average of 10 points (compared to the outcome under usual care). In comparison to the previous section, now we would like to understand to what extent these positive results *were the result of* improved parenting practices. This question is sometimes phrased as: "What is the 'direct' effect of the treatment, net of the effect of parenting?" Does the experiment allow us to easily evaluate this question? The short answer is no—at least not without making further assumptions.

Yet it is not unusual to see such a question addressed by simply running a regression of the outcome on the randomized treatment indicator variable along with the (post-treatment) mediating variable "parenting" added to the equation. In this model, the coefficient on the treatment indicator represents a comparison between those assigned to treatment and control, within subgroups defined by post-treatment parenting practices. Let us consider what is estimated by such a regression in this example.

For simplicity, assume that parenting quality is now measured by a simple categorization: "good" or "poor." The comparison of these two groups can be misleading, because parents who demonstrate good parenting practices after the treatment is applied are likely to be different, on average, from the parents who would have been classified as having good parenting practices even in the absence of the treatment. Such comparisons in essence lose the advantages imparted by the randomization, and it becomes unclear what they represent. Said another way, in general this approach of adjusting for an intermediate outcome will lead to biased estimates of the average treatment effect.

| | Parenting quality after assigned to | | Child's IQ score after assigned to | | Proportion |
|---|---|---|---|---|---|
| Parenting potential | control | treat | control | treat | of sample |
| Poor parenting either way | Poor | Poor | 60 | 70 | 0.1 |
| Good parenting if treated | Poor | Good | 65 | 80 | 0.7 |
| Good parenting either way | Good | Good | 90 | 100 | 0.2 |

Figure 19.10 *Hypothetical example illustrating the problems with regressions that adjust for intermediate outcomes. The table shows, for three categories of parents, their potential parenting behaviors and the potential outcomes for their children under the control and treatment conditions. The proportion of the sample falling into each category is also provided. In actual data, we would not know which category was appropriate for each individual parent—it is the fundamental problem of causal inference that we can observe at most one treatment condition for each person—but this theoretical setup is helpful for understanding the properties of statistical estimates. See Figure 19.9 for a similar example with a continuous intermediate outcome.*

### Regression adjusting for intermediate outcomes cannot, in general, estimate "mediating" effects

Some researchers who perform these analyses will say that these models are still useful, making the claim that, if the estimate of the coefficient on the treatment variable is (statistically indistinguishable from) zero after including the mediating variable, then we have learned that the entire effect of the treatment acts through the mediating variable. Similarly, if including the intermediate outcome as a predictor results in the treatment effect being cut in half, users of this method might claim that half of the effect of the treatment acts through better parenting practices or, equivalently, that the effect of treatment net the effect of parenting is half the total value. These sorts of conclusions are generally *not* appropriate, however, as we illustrate with a hypothetical example.

**Hypothetical scenario with direct and indirect effects.** Figure 19.10 displays potential outcomes of the children of the three different kinds of parents in our sample: those who will demonstrate poor parenting practices with or without the intervention, those whose parenting will get better if they receive the intervention, and those who will exhibit good parenting practices with or without the intervention. We can think of these categories as reflecting parenting *potential*. For simplicity, we have defined the model deterministically, with no individual variation within the three categories of family. We have also ruled out the existence of parents whose parenting quality is adversely affected by the intervention.

Here the effect of the intervention is 10 IQ points for children whose parents' parenting practices were unaffected by the treatment. For those parents who would improve their parenting due to the intervention, the children would see a 15-point improvement. In some sense, philosophically, it is difficult (some would say impossible) to even define questions such as "What percentage of the treatment effect can be attributed to improved parenting practices?',' since treatment effects, and fractions attributable to various causes, can vary across people. For instance, if families that were treated have good parenting only due to exposure to the treatment, how can we ever separate what part of the effect is attributable to better parenting from the effects due to other aspects of the program? If we could assume that the effect on children not due to parenting practices stays constant at 10 points over different types of people, then we might say that, at least for those with the potential to have their parenting improved by the intervention, this improved parenting accounts for about $(15 - 10)/15 = 1/3$ of the effect. But this would be a strong assumption. Realistically we can only talk about average effects, but then the question always arises of who we are averaging over.

**A regression adjusting for the intermediate outcome does not generally work.** However, if one were to try to estimate the effect of this treatment using a regression of the outcome on the randomized treatment indicator and a measure of observed parenting behavior, the coefficient on the treatment indicator would be $-1.5$, falsely implying that the treatment has some sort of negative "direct effect" on IQ scores.

To understand what is happening here, recall that this coefficient is based on comparisons of treated and control groups within subpopulations defined by *observed* parenting behavior. Consider, for instance, the comparison between treated and control groups within those observed to have poor parenting behavior. The group of parents who did not receive the treatment and are observed to have poor parenting behavior is a mixture of those who would have exhibited poor parenting either way and those who exhibited poor parenting simply because they did not get the treatment. Those in the treatment group who exhibited poor parenting are all those who would have exhibited poor parenting either way. Those whose poor parenting is not changed by the intervention have children with lower test scores on average—under either treatment condition—than those whose parenting would have been affected by the intervention.

The regression adjusting for the intermediate outcome thus implicitly compares unlike groups of people and underestimates the treatment effect, because the treatment group in this comparison is made up of lower-performing children, on average. A similar phenomenon occurs when we make comparisons across treatment groups among those who exhibit good parenting. Those in the treatment group who demonstrate good parenting are a mixture of two groups (good parenting if treated and good parenting either way), whereas the control group is simply made up of the parents with the highest-performing children (good parenting either way). This estimate does not reflect the effect of the intervention in the absence of the effect of parenting. It does not estimate any causal effect. It is simply a mixture of some nonexperimental comparisons.

This example is an oversimplification, but the basic principles hold in more complicated settings. In short, randomization allows us to calculate causal effects of the variable randomized, but not other variables, unless a whole new set of assumptions is made. Moreover, the benefits of the randomization for treatment effect estimation are generally destroyed by including post-treatment variables if they are included directly as regression predictors.

### What can theoretically be estimated: principal stratification

We noted earlier that questions such as "What proportion of the treatment effect works through variable A?" are in some sense, inherently unanswerable. What can we learn about the role of intermediate outcomes or mediating variables? As we discussed in the context of Figure 19.10, treatment effects can vary depending on the extent to which the mediating variable (in this example, parenting practices) is affected by the treatment. The key theoretical step here is to divide the population into categories based on their potential outcomes for the mediating variable—what would happen under each of the two treatment conditions. In statistical parlance, these categories are sometimes called *principal strata*. The problem is that the principal strata are generally unobserved. It is theoretically possible to statistically infer principal-stratum categories based on covariates, especially if the treatment was randomized—because then at least we know that the distribution of principal strata is the same across the randomized groups. In practice, however, this reduces to making the same kinds of ignorability assumptions as are made in typical observational studies.

Principal strata are important because they can define, even if only theoretically, the categories of people for whom the treatment effect can be estimated from available data. For example, if treatment effects were nonzero only for the study participants whose parenting practices had been changed, and if we could reasonably exclude other causal pathways, even stronger conclusions could be drawn regarding the role of this mediating variable. We discuss this scenario of *instrumental variables* in greater detail in Section 21.1.

### Intermediate outcomes in the context of observational studies

If trying to adjust directly for mediating variables is problematic in the context of controlled experiments, it should come as no surprise that it generally is also problematic for observational studies. The concern is nonignorability—systematic differences between groups defined conditional on the post-treatment intermediate outcome. In the example above, if we could adjust for the true parenting potential designations, the regression would yield the correct estimate for the treatment effect if we are willing to assume constant effects across groups (or willing to posit a model for how effects change across groups). One conceivably can obtain the same result by adjusting for covariates that adequately proxy this information.

Example: Arsenic in Bangladesh

As an example where the issues discussed in this and the previous section come into play, consider one of the logistic regressions from Chapter 13:

$$\text{Pr(switch)} = \text{logit}^{-1}(-0.21 - 0.90 * \text{dist100} + 0.47 * \text{arsenic} + 0.17 * \text{educ4}),$$

predicting the probability that a household switches nonignorability drinking-water wells as a function of distance to the nearest safe well, arsenic level of the current well, and education of head of household.

This model can simply be considered as data description, but it is natural to try to interpret it causally: being further from a safe well makes one less likely to switch, having a higher arsenic level makes switching more likely, and having more education makes one more likely to switch. Each of these coefficients is interpreted with the other two inputs held constant—and this is what we want to do, in isolating the "effects" (as crudely interpreted) of each variable. For example, households that are farther from safe wells turn out to be more likely to have high arsenic levels, and in studying the "effect" of distance, we would indeed like to compare households that are otherwise similar, including in their arsenic level. This fits with a psychological or decision-theoretic model in which these variables affect the perceived costs and benefits of the switching decision, as outlined in Section 15.7.

However, in the well-switching example as in many regression problems, additional assumptions beyond the data are required to justify the convenient interpretation of multiple regression coefficients as causal effects—what would happen to $y$ if a particular input were changed, with all others held constant—and it is rarely appropriate to give more than one coefficient such an interpretation, and then only after careful consideration of ignorability. Similarly, we cannot learn about causal pathways from observational data without strong assumptions.

For example, a careful estimate of the effect of a potential intervention (for example, digging new, safe wells near existing high-arsenic households) should include, if not an actual experiment, a model of what would happen in the particular households being affected, which returns us to the principles of observational studies discussed earlier in this chapter.

## 19.8 Bibliographic note

Neyman (1923) introduced the potential outcome notation, and the fundamental problem of causal inference was formally introduced by Rubin (1974, 1978) who also coined the term "ignorability." Related earlier work includes Haavelmo (1943), Roy (1951), and Cox (1958). Heckman (1990), Manski (1990), and Dawid (2000) offer other perspectives on potential outcomes.

Rosenbaum (2002b) and Imbens (2004) present overviews of inference for observational studies. Leamer (1978, 1983) explores some challenges of relying on regression models for answering causal questions.

The study on blinding of orchestra auditions is described by Goldin and Rouse (2000), some of whose conclusions were overstated, as discussed by Pallesen (2019). The Electric Company experiment is described by Ball and Bogatz (1972) and Ball et al. (1972). Rubin (2000, 2004) discusses direct and indirect effects for multilevel designs. The example of survey incentives comes from Singer et al. (1999) and Gelman, Stevens, and Chan (2003).

We do not attempt here to review the vast literature on structural equation modeling; Morgan and Winship (2014) connect to other approaches for causal inference. Rosenbaum (1984) provides a helpful discussion of the dangers, outlined in Section 19.7, involved in trying to adjust for post-treatment outcomes; see also Montgomery, Nyhan, and Torres (2018). The term "principal stratification" was introduced by Frangakis and Rubin (2002); examples of its application include Frangakis et al. (2003) and Barnard et al. (2003). Similar ideas appear in Robins (1989, 1994).

## 19.9 Exercises

19.1 *External validity*: Comment on the external validity of the Electric Company example.

19.2 *Compliance*: You are consulting for a researcher who has performed a randomized trial where the treatment was a series of 26 weekly therapy sessions, the control was no therapy, and the outcome was self-report of emotional state one year later. However, most people in the treatment group did not attend every therapy session, and there was a good deal of variation in the number of therapy sessions actually attended. The researcher is concerned that her results represent "watered down" estimates because of this variation and suggests adding in another predictor to the model: number of therapy sessions attended. What would you advise her?

19.3 *Gain scores*: In the discussion of gain-score models in Section 19.3, we noted that if we include the pre-treatment measure of the outcome in a gain score model, the coefficient on the treatment indicator will be the same as if we had just run a standard regression of the outcome on the treatment indicator and the pre-treatment measure. Show why this is true.

19.4 *Pre-test and post-test*: 100 students are given a pre-test, then a treatment or control is randomly assigned to each, then they get a post-test. Given the following regression model:

$$\text{post\_test} = a + b * \text{pre\_test} + \theta * z + \text{error},$$

where $z = 1$ for treated units and 0 for controls. Further suppose that `pre_test` has mean 40 and standard deviation 15. Suppose $b = 0.7$ and $\theta = 10$ and the mean for `post_test` is 50 for the students in the control group. Further suppose that the residual standard deviation of the regression is 10.

(a) Determine $a$.

(b) What is the standard deviation of the post-test scores for the students in the control group?

(c) What are the mean and standard deviation of the post-test scores in the treatment group?

19.5 *Causal inference using logistic regression*: Suppose you have fit a model,

```
fit <- stan_glm(y ~ z + age + z:age, family=binomial(link="logit"), data=mydata)
```

with binary outcome $y$, treatment indicator $z$, and age measured in years. Give R code to produce an estimate and standard error of average treatment effect in a large population, given a vector `n_pop` of length 82 that has the number of people in the population at each age from 18 through 99.

19.6 *Sketching the regression model for causal inference*: Assume that linear regression is appropriate for the regression of an outcome, $y$, on treatment indicator, $z$, and a single confounding covariate, $x$. With pen on paper, sketch hypothetical data (plotting $y$ versus $x$, with treated and control units indicated by circles and dots, respectively) and regression lines (for treatment and control group) that represent each of the following situations:

(a) No treatment effect,

(b) Constant treatment effect,

(c) Treatment effect increasing with $x$.

19.7 *Linearity assumptions and causal inference*: Consider a study with an outcome, $y$, a treatment indicator, $z$, and a single confounding covariate, $x$. Draw a scatterplot of treatment and control observations that demonstrates each of the following:

(a) A scenario where the difference in means estimate would not capture the true treatment effect but a regression of $y$ on $x$ and $z$ would yield the correct estimate.

(b) A scenario where a linear regression would yield the wrong estimate but a nonlinear regression would yield the correct estimate.

ample: ow feed periment

19.8 *Messy randomization*: The folder Cows contains data from an agricultural experiment that was conducted on 50 cows to estimate the effect of a feed additive on 6 outcomes related to the amount of milk fat produced by each cow.
Four diets (treatments) were considered, corresponding to different levels of the additive, and three variables were recorded before treatment assignment: lactation number (seasons of lactation), age, and initial weight of the cow.
Cows were initially assigned to treatments completely at random, and then the distributions of the three covariates were checked for balance across the treatment groups; several randomizations were tried, and the one that produced the "best" balance with respect to the three covariates was chosen. The treatment depends only on fully observed covariates and not on unrecorded variables such as the physical appearances of the cows or the times at which the cows entered the study, because the decisions of whether to re-randomize are not explained.
We shall consider different estimates of the effect of additive on the mean daily milk fat produced.

(a) Consider the simple regression of mean daily milk fat on the level of additive. Compute the estimated treatment effect and standard error, and explain why this is not a completely appropriate analysis given the randomization used.
(b) Add more predictors to the model. Explain your choice of which variables to include. Compare your estimated treatment effect to the result from (a).
(c) Repeat (b), this time considering additive level as a categorical predictor with four levels. Make a plot showing the estimate (and standard error) of the treatment effect at each level, and also showing the inference from the model fit in part (b).

19.9 *Causal inference based on data from individual choices*: Our lives involve tradeoffs between monetary cost and physical risk, in decisions ranging from how large a car to drive, to choices of health care, to purchases of safety equipment. Economists have estimated how people implicitly trade off dollars and danger by comparing choices of jobs that are similar in many ways but have different risks and salaries. This can be approximated by fitting regression models predicting salary given the probability of death on the job (and other characteristics of the job). The idea is that a riskier job should be compensated with a higher salary, with the slope of the regression line corresponding to the "value of a statistical life."

(a) Set up this problem as an individual choice model, as in Section 15.7. What are an individual's options, value function, and parameters?
(b) Discuss the assumptions involved in assigning a causal interpretation to these regression models.

See Dorman and Hagstrom (1998), Viscusi and Aldy (2003), and Costa and Kahn (2004) for different perspectives of economists on assessing the value of a life, and Lin et al. (1999) for a discussion in the context of the risks from exposure to radon gas.

xample: cumbency dvantage

19.10 *Estimating causal effects*: The folder Congress has election outcomes and incumbency for U.S. congressional election races in the 1900s.

(a) Take data from a particular year, $t$, and estimate the effect of incumbency by fitting a regression of $v_{i,t}$, the Democratic share of the two-party vote in district $i$, on $v_{i,t-2}$ (the outcome in the previous election, two years earlier), $I_{it}$ (the incumbency status in district $i$ in election $t$, coded as 1 for Democratic incumbents, 0 for open seats, $-1$ for Republican incumbents), and $P_{it}$ (the incumbent *party*, coded as 1 if the sitting congressmember is a Democrat and $-1$ if he or she is a Republican). In your analysis, include only the districts where the congressional election was contested in both years (if you are interested in missing-data imputation for these elections, see Exercise 17.10), and do not pick a year ending in 2. District lines in the United States are redrawn every 10 years, and district election outcomes $v_{it}$ and $v_{i,t-2}$ are not comparable across redistrictings, for example, from 1970 to 1972.

(b) Plot the fitted model and the data, and discuss the political interpretation of the estimated coefficients.

(c) What assumptions are needed for this regression to give a valid estimate of the causal effect of incumbency? In answering this question, define clearly what is meant by incumbency as a "treatment variable."

See Erikson (1971); Gelman and King (1990); Cox and Katz (1996); Levitt and Wolfram (1997); Ansolabehere, Snyder, and Stewart (2000); Ansolabehere and Snyder (2004); and Gelman and Huang (2008) for further work and references on this topic.

19.11 *Adjusting for post-treatment variables*: Work out the algebra in Section 19.6 in the context of an applied problem of interest to you, giving reasonable values for the parameters in the model. Adjusting for the post-treatment variable should give you a biased estimate of the causal effect in some special case. Identify what is the special case in the context of your example.

19.12 *Working through your own example*: Continuing the example from the final exercises of the earlier chapters, estimate the causal effect you defined in Exercise 18.18 using a regression of $y$ on the treatment indicator $z$, at least one pre-treatment predictor $x$, and their interaction. Plot the data and the fitted regression lines for treatment and control, and discuss the assumptions that are required for this to be a good estimate of the causal effect of interest.

# Chapter 20

# Observational studies with all confounders assumed to be measured

In Chapters 18 and 19, we introduced a statistical formalization of causal effects using potential outcomes, focusing on the estimation of average causal effects and interactions using data from controlled experiments. In practice, logistic, ethical, or financial constraints can make it difficult or impossible to externally assign treatments, and simple estimates of the treatment effect based on differences or regressions can be biased when selection into treatment and control group is not random. To estimate effects when there is imbalance and lack of overlap between treatment and control groups, you should include as regression predictors all the confounders that explain this selection. The present chapter discusses methods for causal inference in the presence of systematic pre-treatment differences between treatment and control groups. A key difficulty is that there can be many pre-treatment variables with mismatch, hence the need for adjustment on many variables.

## 20.1 The challenge of causal inference

To remind the reader of the challenges presented when we are not able to randomly assign study participants to treatments, we present two simple examples in which predictive comparisons with observational data do not yield appropriate causal inferences.

### Hypothetical example of zero causal effect but positive predictive comparison

ample: balance tween atment d control ups

Consider a hypothetical medical experiment in which 100 patients receive the treatment and 100 receive the control condition. In this scenario, the causal effect represents a comparison between what would have happened to a given patient had he or she received the treatment compared to what would have happened under control. We first suppose that the treatment would have no effect on the health status of any given patient, compared with what would have happened under the control. That is, the *causal effect* of the treatment is zero.

Now let us further suppose that treated and control groups systematically differ, with healthier patients receiving the treatment and sicker patients receiving the control. This scenario is illustrated in Figure 20.1, where the distribution of *previous* health status is different for the two groups. This scenario leads to a positive *predictive comparison* between the treatment and control groups, even though the causal effect is zero. This sort of discrepancy between the predictive comparison and the causal effect is sometimes called selection bias, because different sorts of participants are selected for, or select themselves into, different treatments.

### Hypothetical example of positive causal effect but zero positive predictive comparison

Conversely, it is possible for a truly nonzero treatment effect to be erased in the predictive comparison. Figure 20.2 illustrates. In this scenario, the treatment has a positive effect for all patients, whatever

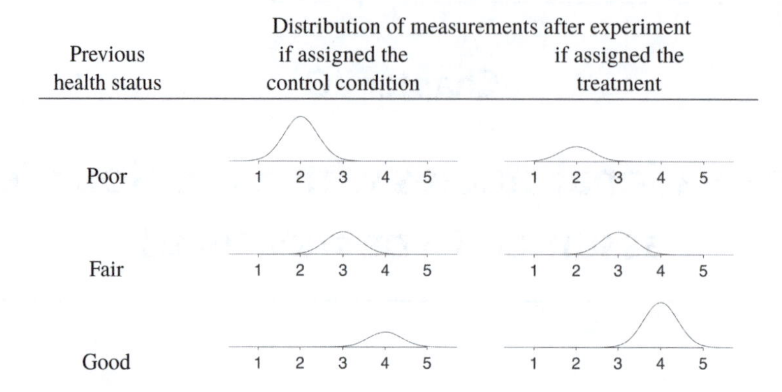


Figure 20.1 *Hypothetical scenario of* zero causal effect *of treatment: for any value of previous health status, the distributions of potential outcomes are* identical *under control and treatment. The heights of these distributions reflect the relative frequency of treatment receipt. Therefore, the* predictive comparison *between treatment and control is positive, because more of the healthier patients receive the treatment and more of the sicker patients receive the control condition.*

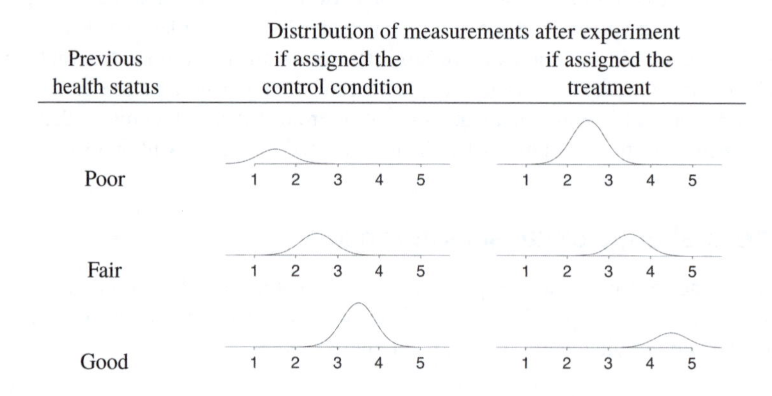


Figure 20.2 *Hypothetical scenario of* positive causal effect *of treatment: for any value of previous health status, the distributions of potential outcomes are centered at* higher *values for the treatment group than for the control group. Once again, the heights of these distributions reflect the relative frequency of treatment receipt. Therefore, the* predictive comparison *between treatment and control can be zero, because more of the sicker patients receive the treatment and more of the healthier patients receive the control condition. Compare to Figure 20.1.*

their previous health status, as displayed by outcome distributions that for the treatment group are centered one point to the right of the corresponding (same previous health status) distributions in the control group. So, for any given unit, we would expect the outcome to be better under treatment than control. However, suppose that this time, sicker patients are given the treatment and healthier patients are assigned to the control condition, as illustrated by the different heights of these distributions. It is then possible to see equal average outcomes of patients in the two groups, with sick patients who received the treatment canceling out healthy patients who received the control.

Previous health status plays an important role in both these scenarios because it is related both to treatment assignment and future health status. If a valid causal estimate is desired, simple comparisons of average outcomes across groups that ignore this variable will be misleading because the effect of the treatment will be "confounded" with the effect of previous health status. For this reason, such predictors are sometimes called *confounding covariates*.

### Adding regression predictors; "omitted" or "lurking" variables

The preceding examples illustrate how a simple predictive comparison is not necessarily an appropriate estimate of a causal effect. In these examples, however, there is a simple solution, which is to compare treated and control units conditional on previous health status. Intuitively, the simplest way to do this is to compare the averages of the current health status measurements across treatment groups only within each previous health status category. We discuss this kind of subclassification strategy for observational studies below.

Another way to estimate the causal effect in this scenario is to regress the outcome on two inputs: the treatment indicator and previous health status. If health status is the only confounding covariate—that is, the only variable that predicts both the treatment and the outcome—and if the regression model is properly specified, then the coefficient of the treatment indicator corresponds to the average causal effect in the sample.

In general, then, causal effects can be estimated using regression if the model includes all confounding covariates (predictors that can affect treatment assignment or the outcome) and if the model is correct. If the confounding covariates are all observed (as in this example), then accurate estimation comes down to proper modeling and the extent to which the model is forced to extrapolate beyond the support of the data. If the confounding covariates are not observed (for example, if we suspect that healthier patients received the treatment, but no accurate measure of previous health status is included in the model), then they are "omitted" or "lurking" variables that complicate the quest to estimate causal effects.

We consider these issues in more detail below, but first we give a mathematical derivation that may provide some intuition.

### Omitted variable bias

We can quantify the bias incurred by excluding a confounding covariate in the context where a simple linear regression model is appropriate and there is only one confounding covariate. First, suppose the correct specification is,

$$y_i = \beta_0 + \beta_1 z_i + \beta_2 x_i + \epsilon_i, \tag{20.1}$$

where $z_i$ is the treatment and $x_i$ is the covariate for unit $i$.

If instead the confounding covariate, $x_i$, is ignored, one can fit the model,

$$y_i = \beta_0^* + \beta_1^* z_i + \epsilon_i^*.$$

What is the relation between these models? To understand, it helps to define a third regression,

$$x_i = \gamma_0 + \gamma_1 z_i + \nu_i.$$

If we substitute this representation of $x$ into the original, correct, equation and rearrange terms, we get

$$y_i = \beta_0 + \beta_2\gamma_0 + (\beta_1 + \beta_2\gamma_1)z_i + \epsilon_i + \beta_2\nu_i. \tag{20.2}$$

Equating the coefficients of $z$ in (20.1) and (20.2) yields

$$\beta_1^* = \beta_1 + \beta_2\gamma_1.$$

This correspondence helps demonstrate the definition of a confounding covariate. If there is no association between the treatment and the possible confounder (that is, $\gamma_1 = 0$) or if there is no association between the outcome and the confounder (that is, $\beta_2 = 0$), then the variable is not a confounder because there will be no bias ($\beta_2\gamma_1 = 0$).

This expression is commonly presented in regression texts to describe the bias that can be incurred if a model is specified incorrectly, and it provides some intuition for the types of problems that can arise when we fail to account for all confounders. However, this term is difficult to interpret unless we are estimating a causal effect. Moreover this bias calculation relies on the model specification, which includes assumptions such as additivity and linearity.

## 20.2 Using regression to estimate a causal effect from observational data

### Observational studies and confounding covariates

In this book we use the term *observational study* to refer to any nonexperimental research design. This study could be prospective or retrospective, and it may or may not involve direct manipulation of the treatment or potential causal variable of interest. Observational studies often have the advantage of being more practical to conduct and may more accurately reflect how the treatment is likely to be administered in practice or the population that might be likely to be exposed to it. However, in observational studies, treatment exposure is observed rather than manipulated (for example, comparisons of smokers to nonsmokers), and it is not reasonable to consider the observed data as reflecting a random allocation across treatment groups.

Thus, in an observational study, there can be systematic differences between groups of units that receive different treatments with respect to key covariates, $x$, that can affect the outcome, $y$. Such covariates that are associated with the treatment and the potential outcomes are typically called *confounders* or *confounding covariates* because if we observe differences in average outcomes across these groups, we can't separately attribute these differences to the treatment or the confounders—the effect of the treatment is thus "confounded" by these variables. We discussed these kinds of systematic differences in Section 18.4, where we went a step further to define the source of the bias more precisely as imbalance in the potential outcomes across the treatment groups.

In this chapter, we explore both the dangers of and possibilities for using observational data to infer causal effects. The approaches discussed all involve direct or indirect attempts to address imbalance and lack of overlap in potential outcomes by adjusting for potential confounding covariates that act, in essence, as proxies for the potential outcomes. These approaches include regression adjustments, stratification, matching, and weighting, and combinations of these. Many of these methods will make use of the estimated propensity score, an important one-number summary of the covariates. All these strategies rest on the crucial, but untestable, assumption that all confounding covariates have been measured. A variety of alternative approaches to causal inference in the absence of randomized experiments that rely on different structural assumptions are described in the following chapter.

### A return to the Electric Company example

Example: Electric Company experiment

We begin with an observational study that was embedded in the Electric Company experiment described in Section 19.2.[1] As it turns out, once the treatments had been assigned in this experiment, the teacher for each class assigned to the treatment group had the choice of *replacing* or *supplementing* the regular reading program with the television show. That is, all the classes in the treatment group watched the show, but some watched it instead of the regular reading program and others received it in addition. This procedural detail reveals that the treatment for the randomized experiment is more subtle than described earlier. As implemented, the experiment estimated the effect of *making the program available*, either as a supplement or replacement for the current curriculum. We now consider something slightly different: the effect of using the show to complement versus substitute for the existing curriculum.

### Informal discussion of the structural assumption of ignorability

Given our inferential goal, it would be naive to simply compare outcomes across the children assigned to these two new treatment options—Replace or Supplement—and expect this to estimate the treatment effect. This is an example of the concern that prompts the advice "correlation is not causation." If we compare outcomes across these groups, we can't be sure that the differences are

[1]Data and code for this example are in the folder `ElectricCompany`.

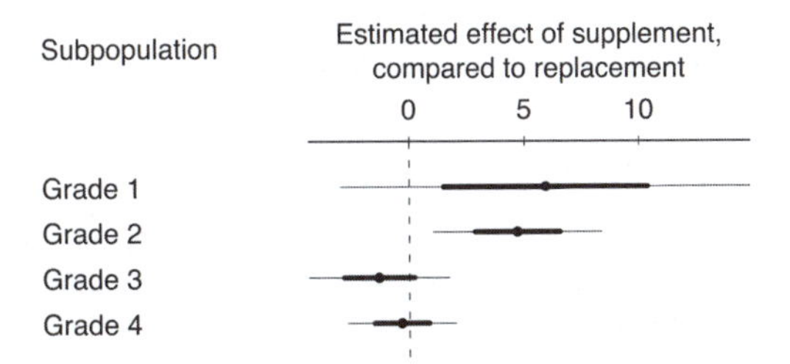


Figure 20.3 *Estimates, 50%, and 95% intervals for the effect of The Electric Company as a supplement rather than a replacement, as estimated by a regression on the supplement/replacement indicator also adjusting for pre-test data. For each grade, the regression is performed only on the treated classes; this is an observational study embedded in an experiment.*

attributable to the different treatment as opposed to the characteristics of the observations that led to selection into the treatment option.

What if we could envision a conditional random assignment similar to the randomized block designs discussed in the previous chapter? For instance suppose that the probability of assignment to the Replace or Supplement groups was determined by the average pre-test scores in that classroom, plus, potentially, some "noise," or variables unrelated to the potential outcomes. To conceptualize this, imagine grouping classrooms into bins defined by ranges of the pre-test scores and performing random assignment within each bin (or block), where the probability that a classroom is assigned to treatment increases with higher classroom pre-test scores. Random assignment conditional on average pre-test score satisfies a form of the ignorability assumption discussed in Section 18.6. We shall review this formalization in the following section and discuss the implications for regression analysis.

In the Electric Company example, ignorability of the Replace or Supplement decision implies that pre-test score is the only confounding covariate. This is a strong assumption in the absence of any evidence on how decisions regarding variety of treatment were actually made. In reality it is likely that teachers made decisions based on factors relating to the children's temperament and motivation—including attributes not captured by classroom average pre-test scores—children's home environment, their own teaching ability and experience, and practical considerations such as scheduling and curricular or assessment requirements. Nonetheless, the ignorability assumption can be a useful starting point, especially in a setting such as this where pre-test score is such a strong predictor of post-test score.

## Setting up a regression for causal inference

We know from Chapters 18 and 19 that when the probability of assignment to the treatment varies with the level of a pre-treatment variable, it is important to account for that variable when estimating the treatment effect. Perhaps the easiest way to do this with a continuous covariate is by including it as a regression predictor. For our example, we add to our data frame a variable called supp that equals 0 for the replacement form of the treatment, 1 for the supplement, and NA for the controls; this last drops the control observations from the analysis. We then estimate the effect of the supplement, as compared to the replacement, for each grade as a regression of post-test on this indicator and pre-test:

```
for (k in 1:4) {
  fit_supp <- stan_glm(post_test ~ supp + pre_test, data=electric,
    subset = (grade==k) & (!is.na(supp)))
}
```

The estimates are graphed in Figure 20.3. The uncertainties are high enough that the comparison is inconclusive except in grade 2, but on the whole the pattern is consistent with the reasonable hypothesis that supplementing is more effective than replacing in the lower grades.

#### Plausibility of regression assumptions

Even if the ignorability assumption appears to be justified, this does not mean that simple regression of our outcomes on confounding covariates and a treatment indicator is the best modeling approach for estimating treatment effects. We still need to be concerned with the regression assumptions listed in Section 11.1. In addition, exploration of two related properties of the data, balance and overlap (also called "common support"), can help us to better understand the extent to which our model may be robust to misspecification. Imbalance and lack of complete overlap force stronger reliance on our modeling than if covariate distributions were the same across treatment groups (as occurs in a completely randomized experiment). We discuss these properties further in Section 20.4.

## 20.3 Assumption of ignorable treatment assignment in an observational study

This section formalizes the ignorability assumption that will be assumed throughout the rest of the chapter. In observational studies, unlike completely randomized studies, it is typically implausible to assume independence between the potential outcomes and the treatment indicator. Recall that this unconditional version of the ignorability assumption can be formalized by the statement,

$$y^0, y^1 \perp z.$$

which says that the distribution of the potential outcomes, $(y^0, y^1)$, is the same across levels of the treatment variable, $z$.

Instead, the key structural assumption underlying all the methods discussed in this chapter is that, *conditional on the covariates, $x$, used in the analysis*, the distribution of potential outcomes is the same across levels of the treatment variable, $z$. We use the term *structural* for this assumption to convey that it can be described as independent of any particular parametric model for the data. This assumption can be formalized by the conditional independence statement,

$$y^0, y^1 \perp z \,|\, x.$$

Therefore, as opposed to making the same assumption as the completely randomized experiment, the key assumption underlying the estimate in the observational study is that the treatment has (or varieties of the treatment have) been assigned at random *conditional* on the inputs in the regression analysis (in this case, pre-test score) with respect to the potential outcomes.

Colloquially, we can think of this assumption as a requirement, for a hypothetical group of observations who all have the same values for the vector of covariates $x$, that they have in essence been randomly assigned to treatment and control groups. More specifically, for this condition to hold, assignment must be random *with respect to the potential outcomes*. That is, the distribution of the potential outcomes, $(y^0, y^1)$, should be the same across levels of the treatment variable, $z$, *once we condition on confounding covariates*, $x$.

This assumption looks almost exactly the same as the assumption we described in the previous chapter for the randomized block experiment. What's different in this scenario? In the randomized block experiment, units are randomly assigned to treatment conditions within strata defined by the blocking variables, $w$. This design ensures that, within blocks defined by $w$, the distribution of the potential outcomes is the same across treatment groups—just as is true for all pre-treatment variables (or variables unaffected by the treatment). Recall that potential outcomes are conceptualized as existing before the treatment even occurs. In the randomized block experiment, the blocking variables are the only confounding covariates *by design*. Technically these are confounders only to the extent that the probability of treatment varies across blocks and the blocks are themselves predictive of the outcome. Otherwise there are no confounders.

In the observational studies we consider in this chapter, however, no actual randomized assignment has taken place. Our hope is that the units in each block (defined by values of the confounding covariates) have the property of balanced potential outcomes that *would have* resulted had they actually been randomized. Crucially, however, we must make the leap of faith that we have conditioned on the appropriate set of confounders, $x$, such that the distribution of potential outcomes for observations who have the same level of these confounders is the same across treatment groups. This group of confounders might be sufficient to achieve this independence for one set of potential outcomes (for instance, achievement scores in an educational intervention with rich measurement of pre-intervention cognitive ability) but not another set of potential outcomes (for instance, behavioral outcomes in the same evaluation). A loose way to think about the ignorability assumption is that it requires adjustment for all confounding covariates: the pre-treatment variables that are associated both with the treatment and the outcome of interest.

As with the randomized block experiment, this assumption is called ignorability of the treatment assignment in the statistics literature. It is also called *selection on observables* or the *conditional independence assumption* in econometrics. The same assumption is often referred to as *all confounders measured* or *exchangeability* in the epidemiology literature. Failure to satisfy the assumption is sometimes referred to as *hidden bias* or *omitted variable bias*, a term that is also used more generally in statistics outside the causal context.

The term *ignorability* reflects that this assumption allows the researcher to *ignore* the model for the treatment assignment as long as analyses regarding the causal effects condition on the predictors needed to satisfy it. If ignorability holds, causal inference does not, in theory, require modeling the treatment assignment mechanism.

In general, one can never prove that the treatment assignment process in an observational study is ignorable—it is always possible that treatment exposure depends on relevant information that has not been recorded. In a study of teaching methods, this information could reflect characteristics of the teacher or school that predict both treatment assignment and post-treatment test scores. Thus, if we interpret the estimates in Figure 20.3 as causal effects, we do so with the understanding that we would prefer to have further pre-treatment information, especially on the teachers, in order to be more confident in ignorability. If we believe that treatment assignments depend on information not included in the model, then we should choose a different analysis strategy. We discuss some options in the next chapter.

## Implications of ignorability for analysis

Recall that analyses of data resulting from completely randomized experiments need not condition on any pre-treatment variables to be unbiased—this is why we can use a simple difference in means to estimate causal effects. Analyses of such data can benefit from conditioning on pre-treatment variables, however, by achieving more precise estimates through a reduction in unexplained variation in the response variable.

Randomized experiments that block or match satisfy ignorability conditional on those design variables used to block or match (this assumes that the blocking variables are associated with the outcome). One should therefore include these blocking or matching variables when estimating causal effects, both for concerns of bias and efficiency.

The same holds for observational studies that satisfy ignorability. If the probability of treatment varies with a covariate that also predicts the outcome (a confounder), then estimation of treatment effects must condition on this confounding covariates in order to be unbiased. If a variable is related to the outcome but not the treatment, then we can include it to increase efficiency.

To understand why, recall that the ignorability assumption implies that the distributions of potential outcomes are only equivalent across treatment groups within subgroups defined by the confounding covariates, $x$. As opposed to the similar unconditional statements from the previous chapters, this version of the ignorability assumption allows for identities conditional on the confounding covariates:

$$\mathrm{E}(y \mid z = 1, x) = \mathrm{E}(y^1 \mid z = 1, x) = \mathrm{E}(y^1 \mid x),$$
$$\mathrm{E}(y \mid z = 0, x) = \mathrm{E}(y^0 \mid z = 0, x) = \mathrm{E}(y^0 \mid x).$$

For instance, if $x$ denotes sex (1 for female and 0 for male) and it is our only confounding covariate, these equations would say that the average outcome among women who were treated would be equal to the average potential outcome, $y^1$, for them. Similarly, the average outcome among women who were not treated would be equal to the average potential outcome, $y^0$. The second equality in each line holds by definition. These same properties would hold for men as well.

The conditional quantities just discussed are related to the marginal quantities needed to construct the average treatment effect by considering the distribution of $x$. If, as in the example in the previous paragraph, $x$ represents sex, then the average treatment effect would be calculated as the difference between two marginal effects, each of which represents a weighted average of the two conditional effects. That is,

$$\mathrm{E}(y^1 - y^0) = \mathrm{E}(y^1) - \mathrm{E}(y^0),$$

where,

$$\mathrm{E}(y^1) = \mathrm{E}(y^1 \mid z = 1, x = 1) \Pr(x = 1) + \mathrm{E}(y^1 \mid z = 1, x = 0) \Pr(x = 0),$$
$$\mathrm{E}(y^0) = \mathrm{E}(y^0 \mid z = 0, x = 1) \Pr(x = 1) + \mathrm{E}(y^0 \mid z = 0, x = 0) \Pr(x = 0).$$

This averaging gets more complicated for continuous or multidimensional $x$. Different estimation strategies discussed in this chapter have different ways of addressing this.

**Identifying causal effects with subclassification.** As we have seen, identifying causal effects when assuming ignorability requires some way of estimating conditional expectations such as $\mathrm{E}(y \mid z = 1, x)$ and $\mathrm{E}(y \mid z = 0, x)$. This should be easy in a setting with only one binary confounder, $x$. For instance, if $x$ denotes sex (because sex is our only confounding covariate), then we could simply estimate each of the six distinct pieces of the equations above. We can estimate $\mathrm{E}(y^1 \mid x = 1)$ using $\bar{y}_1^{x=1} = \frac{\sum_{i=1}^n y_i z_i x_i}{\sum_{i=1}^n z_i x_i}$, the sample mean for women assigned to the treatment group; the other expectations can be estimated similarly. We can estimate $\Pr(x = 1)$ using the sample proportion of women, $\frac{1}{n}\sum_{i=1}^n x_i$.

It is straightforward to use these averages to get estimates of $\mathrm{E}(y^1)$ and $\mathrm{E}(y^0)$ simply by plugging in estimates of each of the quantities above. The difference between these is an estimate of the average treatment effect.

**Identifying causal effects with regression.** This strategy will get more complicated as $x$ gets more complex. For instance, if we had two binary confounders, we could avoid making additional assumptions by considering for four subgroups corresponding to the cells of the contingency table defined by the two variables (that is, all combinations of the levels of the two variables). This would require having both treated and control units in each of these four subgroups. More variables, or more levels for each variable, or continuous variables, would increase the complexity of this enterprise quickly and make it increasingly less likely that we would have enough data to support this kind of nonparametric analysis.

Once the number and type of confounders gets more complicated, perhaps the simplest parametric model that we can fit to estimate these expectations is linear regression. For instance, if we assume that the treatment effect, $\tau$, is constant (or at least additive) we might posit that $\mathrm{E}(y^z|x) = \beta_0 + X\beta + \tau z$. If ignorability is satisfied and this model holds, we simply need to regress the outcome on the treatment indicator and confounders. The estimated coefficient on $z$ from this fit, $\hat{\tau}$, can be conceptualized as a weighted version of all of the conditional effect estimates. However, fitting a model to estimate these quantities is not without potential weaknesses. The two most obvious concerns are imbalance and lack of complete overlap, which we discuss in the next section.

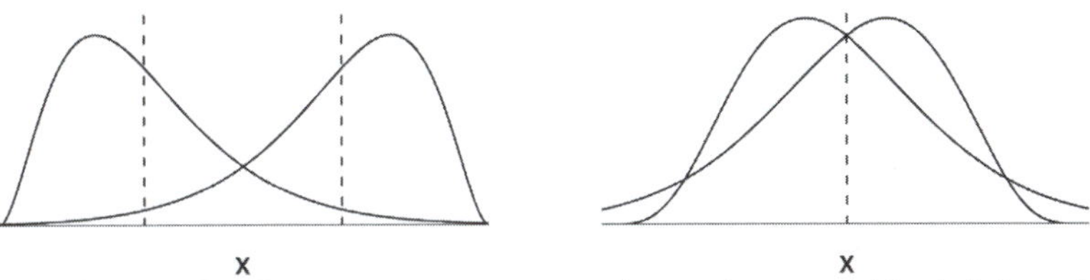


Figure 20.4 *Imbalance in distributions across treatment and control groups. (a) In the left panel, the groups differ in their averages (dotted vertical lines) but cover the same range of x. (b) The right panel shows a more subtle form of imbalance, in which the groups have the same average but differ in their distributions.*

## 20.4 Imbalance and lack of complete overlap

A repeated theme in Chapters 18 and 19 was that causal inference is cleanest when the units receiving the treatment are comparable to those receiving the control. However, in an observational study, the treatment and control groups are likely to be different in multiple ways. If these differences across groups are with respect to *unobserved* confounders, then ignorability cannot be satisfied and the methods in this chapter are not appropriate. If these differences are with respect to *observed* confounders, then we can try to create comparability using the approaches discussed in this chapter. To better understand the tradeoffs between these approaches, it will help to have a clearer understanding of two sorts of departures from comparability—*imbalance* and *lack of complete overlap*. Imbalance and lack of complete overlap are issues for causal inference even if ignorability holds because they force us to rely more heavily on model specification and less on direct support from the data.

### Imbalance and model sensitivity

xample: nbalance nd lack of omplete verlap

Imbalance with measured confounders occurs when the distributions of confounders differ for the treatment and control groups. This could manifest, for instance, as differences in means or standard deviations of a covariate between treatment and control groups. More generally, any differences in covariate distributions across groups can be referred to as lack of balance across groups. When treatment and control groups suffer from *imbalance*, the simple comparison of group averages, $\bar{y}_1-\bar{y}_0$, is not, in general, a good estimate of the average treatment effect. Instead, some analysis must be performed to adjust for the pre-treatment differences between the groups.

Figure 20.4 shows two examples of imbalance with respect to a single covariate, $x$. In Figure 20.4a, the groups have different means (dotted vertical lines) and different skews. In Figure 20.4b, groups have the same mean but different skews. In both examples, the standard deviations are the same across groups. Differences in spread are another common manifestation of imbalance.

Imbalance creates problems by forcing us to rely more on the correctness of our model than we would have to if the samples were balanced. To see this, consider what happens when we try to make inferences about the effect of a treatment variable (for instance, a new reading program) on test score, $y$, while adjusting for a crucial confounding covariate, pre-test score, $x$. Suppose that the true treatment effect is $\theta$ and the relationship between the response variable, $y$, and the sole confounding covariate, $x$, is quadratic, as indicated by the following regressions:

$$\begin{aligned}\text{treated:}\, y_i &= \beta_0 + \beta_1 x_i + \beta_2 x_i^2 + \theta + \text{error}_i \\ \text{controls:}\, y_i &= \beta_0 + \beta_1 x_i + \beta_2 x_i^2 + \text{error}_i .\end{aligned}$$

Averaging over each group separately, solving the second equation for $\beta_0$, plugging back into the first, and solving for $\theta$ yields,

$$\theta = \bar{y}_1 - \bar{y}_0 - \beta_1(\bar{x}_1 - \bar{x}_0) - \beta_2(\overline{x_1^2} - \overline{x_0^2}), \tag{20.3}$$

where $\bar{y}_1$ and $\bar{y}_0$ denote the average of the outcome test scores in the treatment and control groups,

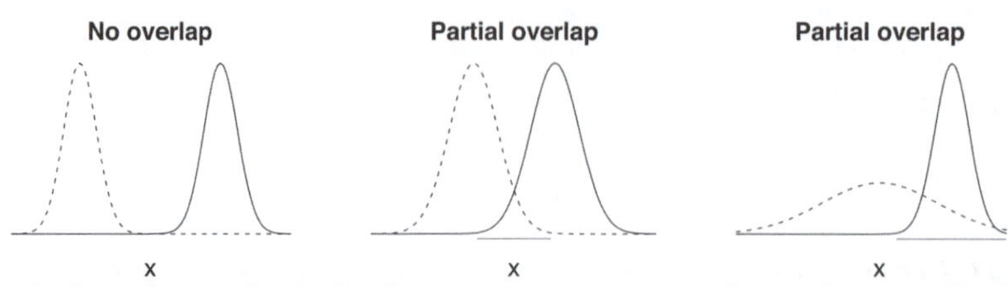


Figure 20.5 *Lack of complete overlap in distributions across treatment and control groups. Dashed lines indicate distributions for the control group; solid lines indicate distributions for the treatment group. (a) Two distributions with no overlap; (b) two distributions with partial overlap; (c) a scenario in which the* range *of one distribution is a subset of the range of the other.*

respectively, $\bar{x}_1$ and $\bar{x}_0$ represent average pre-test scores for treatment and control groups, respectively, and $\overline{x_1^2}$ and $\overline{x_0^2}$ represent these averages for squared pre-test scores. Ignoring $x$ and simply using the raw treatment/control comparison, $\bar{y}_1 - \bar{y}_0$, will yield a poor estimate of the treatment effect. It will be off by the amount $\beta_1(\bar{x}_1 - \bar{x}_0) + \beta_2(\overline{x_1^2} - \overline{x_0^2})$, which corresponds to systematic pre-treatment differences between groups 0 and 1. The magnitude of this bias depends on how different the distribution of $x$ is across treatment and control groups (specifically with regard to variance in this case) and how large $\beta_1$ and $\beta_2$ are. The closer the distributions of pre-test scores are across treatment and control groups, the smaller this bias will be.

Moreover, a linear model regression using $x$ as a predictor would also yield the wrong answer; it will be off by the amount $\beta_2(\overline{x_1^2} - \overline{x_0^2})$. The closer the distributions of pre-test scores are across treatment and control groups, the smaller $(\overline{x_1^2} - \overline{x_0^2})$ will be, and the less worried we need to be about correctly specifying the form of this relationship. More generally, the curve could deviate from linearity in any number of ways. One option is to use a nonparametric fitting algorithm. Alternatively, we can use approaches that first match or weight to create balance across treatment and control groups. These may help reduce the problems from incorrect specification of this model.

## Lack of complete overlap

*Overlap* or *common support* describes the extent to which the support of the covariate data is the same between the treatment and control groups. We use the term "overlap," rather than "common support," though both appear in the literature. There is complete overlap when there exist both treatment and control units in all neighborhoods of the covariate space. Lack of complete overlap in the confounders creates problems, because in that setting there are treatment observations for which we have no *empirical counterfactuals* (that is, control observations with the same covariate distribution) or vice versa. Since we rely on these empirical counterfactual units to inform counterfactual outcomes, when treatment and control groups do not completely overlap, the data are inherently limited in what they can tell us about treatment effects in the regions of nonoverlap.

Figure 20.5 displays several scenarios of lack of complete overlap with respect to one confounder. It becomes increasingly difficult to visualize overlap as the dimension of the confounder space gets larger. Areas with no overlap represent conditions under which we may not want to make causal inferences. Any inferences in those areas would rely on modeling assumptions in place of direct support from the data. Adhering to this structure would imply that in the setting of Figure 20.5a, it would be impossible to make purely data-based causal inferences about any of the observations. Figure 20.5b shows a scenario in which data-based inferences are only possible for the region of overlap, which is underscored on the plot. In Figure 20.5c, causal inferences are possible for the full treatment group but only for a subset of the control group (again indicated by the underscored region). There is an important correspondence between areas of overlap and the estimand of interest, and it can make sense to choose your inferential goal based on the support in the data.

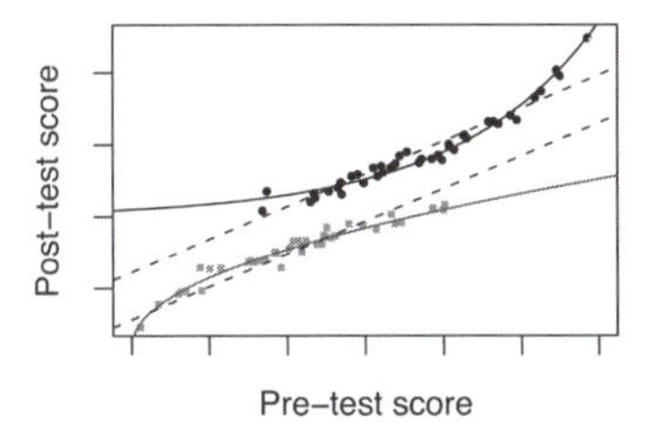


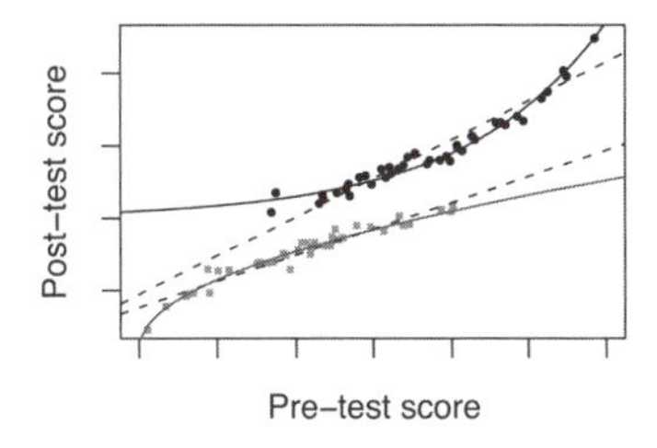


Figure 20.6 *Hypothetical before/after data demonstrating potential problems in using linear regression for causal inference. The dark dots correspond to the children who received the educational supplement; the lighter dots correspond to the children who did not receive the supplement. The dark solid line represents the true relationship between the potential outcome for treatment receipt and the pre-test, E($y^1$|pre-test). The light solid line represents the true relationship between the potential outcome for the control condition and the pre-test, E($y^0$|pre-test). The causal effect at any level of the pre-test is simply the vertical distance between the two solid lines. Each average causal effect is an average across the relevant subset of these individual-level causal effects. The dashed lines are regression lines fit to the observed data. The dashed model fit shown in the right panel allows for an interaction between receiving the supplement and pre-test scores.*

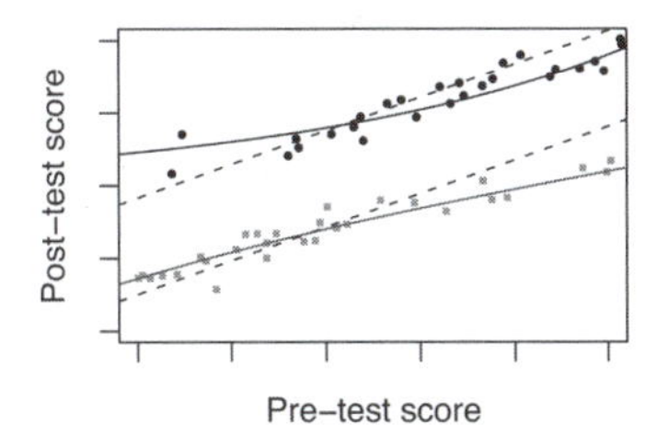


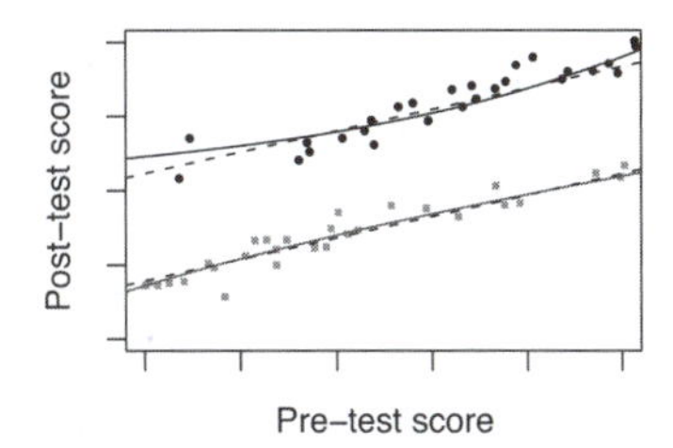


Figure 20.7 *These plots are restricted to observations from the example in 20.6 in the region where there is overlap in terms of the pre-treatment test score across treatment and control groups. The left panel shows only the portion of the plot in Figure 20.6 where there is overlap. The right panel shows new regression lines fit only using observations in this overlapping region.*

## Model extrapolation

xample: ectric mpany xperiment

A traditional model fitted to data without complete overlap is forced to extrapolate beyond the support of the data. For example, suppose we are interested in the effect on test scores of a supplementary educational activity, such as viewing The Electric Company, that was not randomly assigned. Let's assume, as we did above, however, that only one predictor, pre-intervention test score, is necessary to satisfy ignorability—that is, there is only one confounding covariate. Suppose further that students who participate in the supplementary activity tend to have higher mean pre-test scores, on average, than those who do not participate. One manifestation of this hypothetical scenario is illustrated in Figure 20.6. The dark curve represents the true relation between pre-test scores ($x$-axis) and post-test scores ($y$-axis) for those who receive the supplement, $\mathrm{E}(y^1 \mid x)$. The lighter curve represents the true relation between pre-test scores and post-test scores for those who do not receive the supplement, $\mathrm{E}(y^0 \mid x)$. Thus these lines display the response surface.

These lines are not parallel. Since the treatment effect at each level of the average pre-test score is represented as the vertical distance between the two lines, the non-parallel lines imply that the effect of the treatment varies across observations with different average pre-test scores. The average treatment effect for any collection of classrooms would be an average of such pre-test-specific treatment effects.

Estimated linear regression lines for these data are superimposed as dotted lines on these plots. The linear model has problems fitting the true nonlinear regression relation—a problem that is compounded by the lack of overlap in average pre-test scores across the two groups. Because there are no control classrooms with high test scores and virtually no treatment classrooms with low test scores, these linear models, to create counterfactual predictions, are forced to extrapolate over portions of

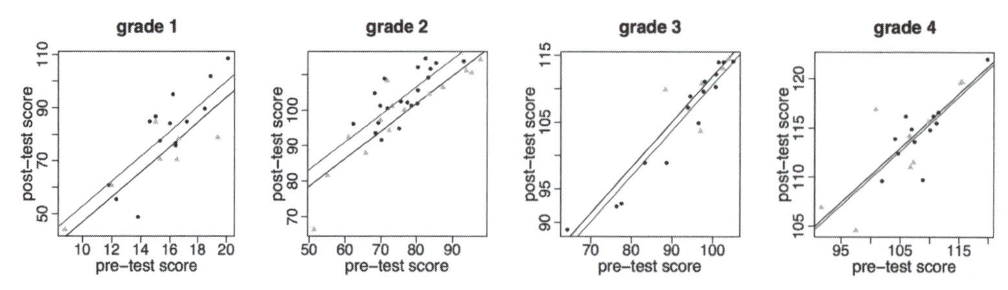


Figure 20.8 *Pre-test/post-test data examining the overlap in pre-test scores across treatment groups as well as the extent to which models are being extrapolated to regions where there is no support in the data. Classrooms that viewed The Electric Company as a supplement are represented by the dark points and regression line; classrooms that viewed the show as a replacement are represented by the lighter points and regression line. No interactions were included when estimating the regression lines.*

the space where there are no data to support them. These two problems combine to create, in this case, a substantial underestimate of the true average treatment effect. Allowing for an interaction, as illustrated in Figure 20.6b, does not solve the problem; the linear extrapolation for the control group is somewhat better than in the original (noninteracted) model, however, the extrapolations of the linear fit for the treatment group are even farther from the true response surface where there is no support from the data.

In the region of pre-test scores where there are observations from both treatment groups, however, even the incorrectly specified linear regression lines applied to the full sample do not provide such a bad fit to the data. And no model extrapolation is required, so diagnosing this lack of fit would be possible. This is demonstrated in Figure 20.7a by restricting the plot from Figure 20.6a to the area of overlap. Furthermore, if the regression lines are fit only using this restricted sample, they fit well in this region, as is illustrated in Figure 20.7b. Some of the strategies discussed in the propensity score sections of this chapter capitalize on this idea of limiting analyses to observations within the region of complete overlap.

### Examining overlap in the Electric Company embedded observational study

For the Electric Company data, we can use plots similar to those in Figures 20.6 and 20.7 to assess the appropriateness of the modeling assumptions and the extent to which we are relying on unsupported model extrapolations. For the most part, Figure 20.8 reveals a reasonable amount of overlap in pre-test scores across treatment groups within each grade. Grade 3, however, has some classrooms that received the supplement whose average pre-test scores that are lower than any of the classrooms that received the replacement. It might be appropriate to decide that no counterfactual classrooms exist in our data for these classrooms and thus the data cannot support causal inferences for these classrooms. The sample sizes for each grade make it difficult to come to any firm conclusions one way or another, however. Therefore, in order to trust a causal interpretation we must feel confident in the (probably relatively minor) degree of model extrapolation relied upon by these estimates. All this is in addition to believing the ignorability assumption.

## 20.5 Example: evaluating a child care program

Example: Child care experiment

We illustrate the concepts of imbalance and lack of overlap (and in later sections the analysis strategies subclassification and matching) with data collected on the development of nearly 4500 children born in the 1980s. A subset of 290 of these children received special services in the first few years of life, including high-quality child care (five full days a week) in the second and third years of life as part of a formal intervention, the Infant Health and Development Program (IHDP). These children were

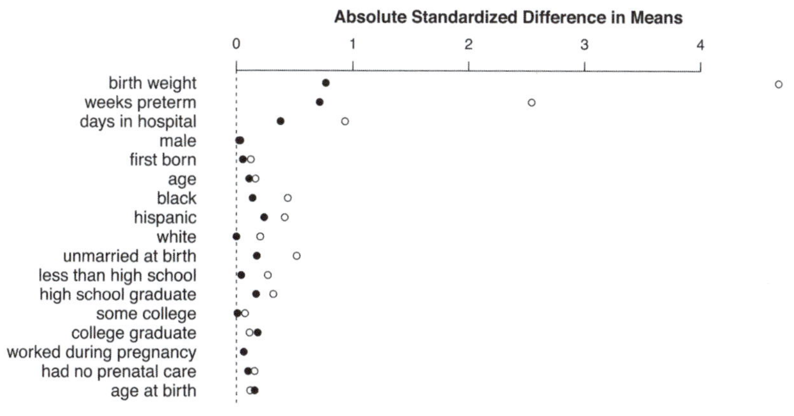


Figure 20.9 *Imbalance in averages of confounding covariates across treatment groups. Open circles represent absolute differences in means for the unmatched groups standardized by the standard deviation in the treatment group. Solid circles represent absolute differences in means for groups matched without replacement standardized by the standard deviation in the treatment group. Matching methods are described later in the chapter.*

targeted because they were born prematurely, had low birth weight (less than or equal to 2500 grams), and lived in the eight cities where the intervention took place. Children in the sample who did not receive the intervention exhibited a more representative range of birth timing and birth weight.[2]

We want to evaluate the impact of this intervention on the children's subsequent cognitive outcomes by comparing the outcomes for children in the intervention group to the outcomes in a comparison group of 4091 children who did not participate in the program. The outcome of interest is test score at age 3; this test is similar to an IQ measure, so we simplistically refer to these scores as IQ scores from now on.

## Missing data

Incomplete data arise in virtually all observational studies. For this sample dataset, we imputed missing data once, using a model-based random imputation (see Chapter 17 for a general discussion of this approach). We excluded the most severely low-birth-weight children (those at or below 1500 grams) from the sample because they are so different from the comparison sample. For these reasons, results presented here do not exactly match the published analyses for this IHDP comparison, which used imputations for the missing values.

## Examining imbalance for several covariates

To illustrate the ways in which the treated and comparison groups differ, the open circles in Figure 20.9 display the absolute standardized differences in mean values (absolute differences in means divided by the pooled within-group standard deviations for the treatment group) for a set of confounding covariates that we think predict both program participation and subsequent test scores. Many of these differences are large!

Displaying the confounders in this way and plotting absolute standardized differences in means—rather than displaying a table of numbers—facilitates comparisons across predictors and methods. For instance, we can see that there are important differences between the treatment and control groups with respect to birth weight, weeks preterm, and days in hospital.

From Figure 20.10, which displays a scatterplot and regression lines of test scores on birth weight (separately by treatment group), we see that average birth weights differ in the two groups (imbalance),

[2]Data and code for this example are in the folder `Childcare`.

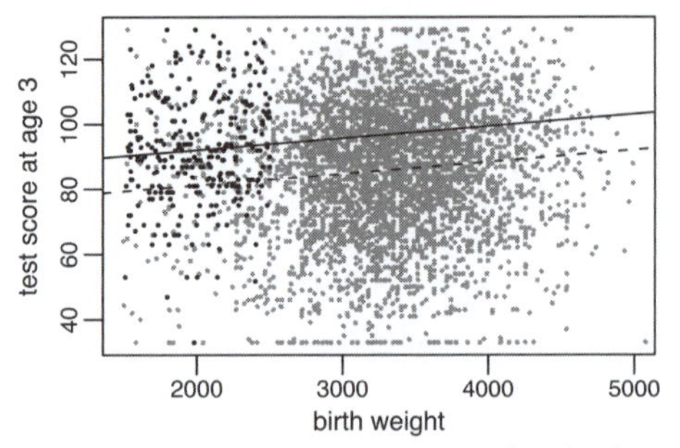


Figure 20.10 *Data from children who received an intervention targeting low-birth-weight, premature children (black dots), and data from a comparison group of children (gray dots). Test scores at age 3 are plotted against birth weight. The solid line and dotted lines are regressions fit to the black and gray points, respectively.*

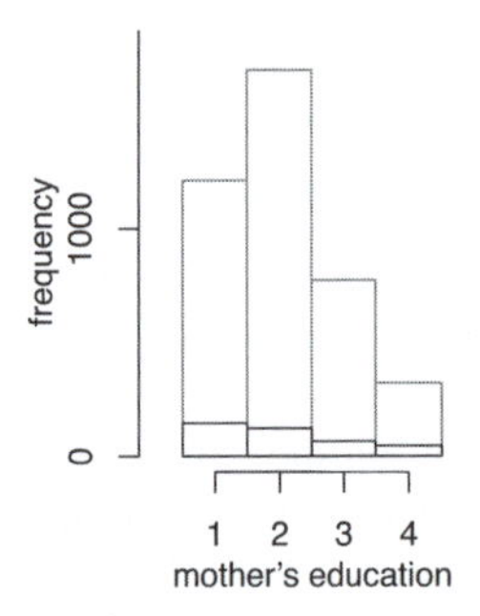


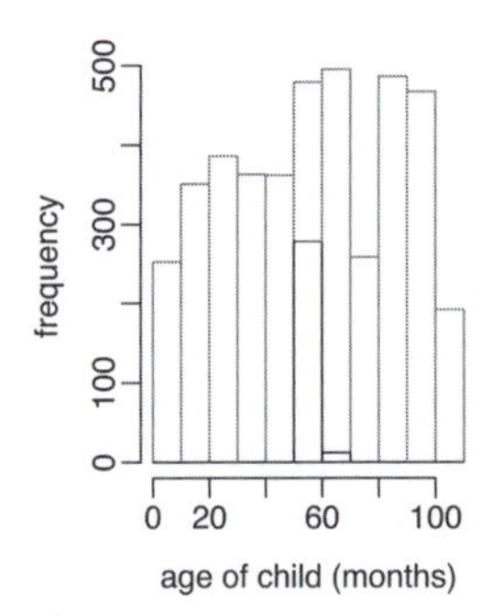


Figure 20.11 *Comparisons of the treatment (black histogram bars) and control (gray histogram bars) groups for the child-intervention study, with respect to two of the pre-treatment variables. There is lack of complete overlap for child age, but the averages are similar across groups. In contrast, mother's education shows complete overlap, but imbalance exists in that the distributions differ for the two groups.*

and also there are many control observations (gray dots) with birth weights far outside the range of birth weights in the treatment population (black dots). This is an example of *lack of complete overlap* in this predictor across groups. If birth weight is a confounding covariate, that is, a covariate we need to adjust for to achieve ignorability, Figure 20.10 demonstrates that if we want to make inferences about the effect of the program on children with birth weights above 2500 grams, we will have to rely on model extrapolations that may be inappropriate. Alternatively, we may want to restrict our sample to observations that satisfy overlap.

## Imbalance is not the same as lack of overlap

As we have discussed, imbalance does not necessarily imply lack of complete overlap; conversely, lack of complete overlap does not necessarily result in imbalance in the sense of different average values in the two groups. Typically, lack of overlap is a more serious problem, corresponding to a lack of data that limits the causal conclusions that can be made without uncheckable modeling assumptions. Figure 20.11 highlights the distinction between balance and overlap.

Figure 20.11a demonstrates complete overlap across groups in terms of mother's education. Each category includes observations in each treatment group. However, the percentages falling in each category (and the overall average, were we to code these categories as 1–4) differ when comparing treatment and control groups—thus there is clearly imbalance.

Figure 20.11b shows balance in mean values but without complete overlap. As the histograms show, the averages of children's ages differ little across treatment groups, but the vast majority of

| Mother's education | Treatment effect estimate ± s.e. | Sample size treated | Sample size controls |
|---|---|---|---|
| Not a high school graduate | 9.3 ± 1.5 | 126 | 1232 |
| High school graduate | 4.1 ± 1.9 | 82 | 1738 |
| Some college | 7.9 ± 2.4 | 48 | 789 |
| College graduate | 4.6 ± 2.3 | 34 | 332 |

Figure 20.12 *Estimates ± standard errors of the effect on children's test scores of a child care intervention, for each of four subclasses formed by mother's educational attainment. The study was of premature infants with low birth weight, most of whom were born to mothers with low levels of education.*

control children have ages that are not represented in the treatment group. Thus, there is a lack of complete overlap across groups for this variable. More specifically, there appears to be sufficient overlap with regard to this variable in terms of the treatment observations, but not in terms of the control observations. If we believe age to be a crucial confounding covariate, we probably would not want to make inferences about the full set of controls in this sample. For these data, children's ages were measured at roughly the time of the IHDP assessment. Since the sample of children from National Longitudinal Study of Youth includes those who were born several years before or after the IHDP children, this predictor in effect proxies for time cohort.

## 20.6 Subclassification and average treatment effects

Assuming we are willing to trust the ignorability assumption, are there approaches we can use to estimate treatment effect that make few if any parametric assumptions? As discussed briefly in 20.3, subclassification is a simple option, possibly followed by poststratification to estimate population average treatment effects.

### Subclassification as a simple nonparametric approach

Developmental psychologists know that a mother's educational attainment is an important predictor of her child's test scores. Education level is also traditionally associated with participation in interventions such as this program for children with low birth weights. Let us make the (unreasonable) assumption for the moment that this is the only confounding covariate (that is, the only predictor associated with both participation in this program and test scores). How would we want to estimate causal effects? In this case, a simple solution would be to estimate the difference in mean test scores within each subclass defined by mother's education. These averages, as well as the associated standard error and sample size in each subclass, are displayed in Figure 20.12. These point to positive effects for all participants (although with some uncertainty), with by far the largest estimates for the children whose mothers had not graduated from high school.

There is overlap on this variable between the treatment and control groups, as is evidenced by the sample sizes for treated and control observations within each subclass in Figure 20.12. If there were a subclass with observations only from one group, we would not be able to make inferences for this type of person. Also, if there were a subclass with only a small number of observations in either the treatment group or the control group, we would probably be wary of making inferences for these children as well.

To get an estimate of the average treatment effect, the subclass-specific estimates could be combined using a weighted average where the weights are defined by the number of children in each subclass:

$$\text{Est avg treatment effect} = \frac{9.3 * 1358 + 4.1 * 1820 + 7.9 * 837 + 4.6 * 366}{1358 + 1820 + 837 + 366} = 6.5, \qquad (20.4)$$

with a standard error of $\sqrt{\frac{1.5^2 * 1358^2 + 1.9^2 * 1738^2 + 2.4^2 * 789^2 + 2.3^2 * 366^2}{(1358+1820+837+366)^2}} = 1.01$. A regression fit with indicators for each education attainment level and interactions between the treatment and these subclasses would yield similar results.

This strategy has the advantage of imposing overlap and, moreover, forcing the control sample to have roughly the same covariate distribution as the treated sample. This reduces reliance on the type of model extrapolations discussed previously. Moreover, one can choose to avoid modeling altogether after subclassifying, and simply use a difference in averages across treatment and control groups to perform inferences, thus avoiding parametric assumptions about the relation between the response and confounding covariates.

One drawback of subclassification, however, is that when adjusting for a continuous variable, some information may be lost when discretizing the variable. A more substantial drawback is that it is difficult to adjust for many variables at once, which can make it difficult to satisfy ignorability.

### Whom are we making inferences about?

Figure 20.12 demonstrates how treatment effects can vary over different subpopulations. Previously in Figure 20.6 we saw another example of heterogeneous treatment effects. One implication of the existence of treatment effects that vary across different types of people is that we should think about what types of treatment effects we are measuring to answer a particular research question. In our high-quality child care example, the intervention was designed for the special needs of low-birth-weight, premature children—not for typical children—hence there is less interest in its effect on comparison children who would not have participated. Thus from a policy perspective we could argue that we should focus on estimating the treatment effect averaged just over these children.

This is an example of an alternative to the average treatment effect called the average effect of the treatment on the treated (ATT). This estimand can make sense in situations where the treatment would only ever be administered to certain types of people, so that it makes sense to focus on the average effect for people who fit that profile.

For instance in our child care example, suppose for simplicity that the only difference between groups was that those who received the high-quality child care had disproportionately low birth weight (that is, less than 2500 grams) and that most of those children who did not receive that high-quality child care were born with typical weight (that is, above 2500 grams). In reality, *all* those in the treatment group had low birth weight, but for the moment we want to avoid conflating the issue of complete overlap with the choice of estimand.

Further suppose that those children with low birth weight had much more to gain from high-quality child care. Perhaps even those children with low birth weight would gain 10 IQ points on average from the high-quality child care whereas those with typical birth weight would receive no benefit from the child care. In this scenario the average treatment effect on the treated would certainly be larger than the average treatment effect in the general population.

To obtain an estimate of the ATT, or the overall effect for those who participated in the high quality child care in our example, the subclass-specific estimates could be combined using a weighted average where the weights are defined by *the number of children in each subclass who participated in the program*:

$$\text{Estimated effect on the treated} = \frac{9.3 * 126 + 4.1 * 82 + 7.9 * 48 + 4.6 * 34}{126 + 82 + 48 + 34} = 7.0, \tag{20.5}$$

with a standard error of $\sqrt{\frac{1.5^2 * 126^2 + 1.9^2 * 82^2 + 2.4^2 * 48^2 + 2.3^2 * 34^2}{(126+82+48+34)^2}} = 0.9$.

Had we weighted instead by the number of *control* children in each subclass, we would have obtained an estimate of the average treatment effect on controls (ATC) of 6.4. As it turns out, the average treatment effect on treated and controls (ATT and ATC) are not dramatically different from each other, but this will not always be the case in practice.

Why would we expect the average treatment effect on the treated and controls to be different from each other? If the observations in the treatment and control groups are different from each other in terms of pre-treatment characteristics, and those characteristics are associated with different treatment effects, then we would expect these estimands to be different. The effect of the intervention might vary, for instance, across children with different initial birth weights, and since we know that the mix of children's birth weights differs in treatment and comparison groups, the average effects across these groups could also differ.

In reality, the children who received the high-quality child care were more disadvantaged on average across many dimensions (they may also have been differentially advantaged in other ways). We'd like to average over the portion of our analysis sample that is representative of the group about whom we are most interested in making inferences.

How can we decide which estimand is most relevant? This particular intervention was designed for the special needs of low-birth-weight, premature children—not for typical children—and there is little interest in its effect on children who would not even have been eligible to participate. Therefore the average treatment effect on the treated would be more relevant to the scientific question that drove the researchers who ran the study.

Another consideration when choosing between estimands is that there may not exist sufficient support in the data to estimate a particular estimand. For instance, we saw in Figure 20.10 that there are so many control observations with no counterfactual observations in the treatment group with regard to birth weight that these data are likely inappropriate for drawing inferences about the control group either directly (the effect of the treatment on the controls) or as part of an average effect across the entire sample.

### Poststratification and average treatment effects

Another framing of this issue appeals to the idea of poststratification. We can think of the estimate of the effect of the treatment on the treated as a poststratified version of the estimate of the average causal effect. As the methods we discuss in this section rely on more and more covariates, it can be more attractive to apply methods that estimate the effect of the treatment on the treated while avoiding explicit stratification on all potential treatment effect modifiers, as we discuss next.

## 20.7 Propensity score matching for the child care example

*Matching* refers to any of a variety of procedures that restructure the original sample in preparation for a statistical analysis. The goal of this restructuring in a causal inference setting is to create an analysis sample that *looks like* it was created from a randomized experiment. Therefore, we would like the matched groups to exhibit balance and overlap with respect to the pre-treatment variables that we consider to be confounders. Restructuring may remove some observations or may upweight the contribution of others. For instance, if the goal of the analysis is to estimate the effect of the treatment on the treated, for each treatment observation we might find a control observation that we deem most similar to that observation by some metric, and retain that unit as its *match*. Unmatched control units would be discarded. If ignorability is satisfied conditional on the specified covariates and if we can achieve balance and overlap with respect to them, then we should be able to estimate causal effects without bias.

Matching is sometimes considered an alternative to fitting a model to adjust for potential confounders. We prefer to think of matching as a first step in the analysis that allows for reduced reliance on parametric assumptions of the model we end up fitting to estimate the treatment effect. Just as with a randomized experiment, the intuition is that if sufficient overlap exists and we can create balance between our treatment and control groups, then, even if we misspecify the model used to estimate the treatment effect, we should still get a reasonable estimate of the treatment effect. We witnessed this phenomenon in the comparison between Figures 20.6 and 20.7.

We introduce this topic by walking through a specific implementation of a form of matching called propensity score matching. Our goal is to estimate the effect of the child care intervention introduced in Section 20.5. We can think of propensity score matching as proceeding through a series of five steps, which we outline here. Various methods have been developed in R and other software packages to automate some of these steps, but we want the reader to understand the logic of each step before using such methods. Relatedly, to maintain focus on the necessary logical steps, we present relatively simple implementations of each step. Extensions and potential complications are discussed in the following section of this chapter.

### Step 1: Defining the confounders and estimand

The first step in our analysis is to define the confounders and the estimand.

**Defining the confounders.** How should we choose which covariates to consider confounders and what are the implications of this choice? Confounders are the covariates that predict both treatment assignment (or receipt) and the outcome. These are the covariates that we need to condition on to satisfy ignorability. Recall that if ignorability holds, then we can estimate treatment effects without bias if we appropriately adjust these covariates in our analysis. We discuss in more detail below what it means to "appropriately adjust" for these covariates. This first step focuses simply on identifying which covariates in our data we consider to be confounders.

Choosing which covariates are confounders can be challenging in a dataset with many covariates. Often researchers choose these variables based on related research; however, more data-driven approaches exist as well. In our child care example there are not many covariates available, so we will use them all. We discuss more general concerns with choosing which covariates to adjust for in Section 20.8.

**What are we estimating?** We discussed above that the estimand of greatest interest in this example is arguably the *effect of the treatment on the treated.* Choice of estimand will have implications for our matching strategy. In particular, we will want our analysis sample to be representative of the treatment group. Therefore, when matching we will keep our treatment group intact but find matches for each treatment group member from among the comparison group. In essence, we will be restructuring our comparison group to *look like* our treatment group in terms of the distributions of observed covariates.

If we were interested in estimating the effect of the treatment on the controls, we would instead keep the comparison group intact and find matches for each control unit from among the treated units. Since the group that stays intact is the one that we will be making inferences about (for example, we are making inferences about the treatment group when we estimate the effect of the treatment on the treated), we sometimes refer to that group as the *inferential group*.

### Step 2: Estimating the propensity score

The next step in creating matches is to fit a model to estimate the probability of receiving the exposure or treatment.

We can use the estimated propensity score as a one-number summary of our covariates; we'll briefly discuss the theory surrounding this in Section 20.8. Given that the goal of matching is to create groups that have similar distributions of confounding covariates, a natural starting point would be a logistic regression that includes these covariates:

```
ps_fit_1 <- stan_glm(treat ~ bw + bwg + hispanic + black + b.marr + lths + hs +
  ltcoll + work.dur + prenatal + sex + first + preterm + momage + dayskidh +
  income, family=binomial(link="logit"), data=cc2, algorithm="optimizing")
```

We will evaluate the appropriateness of this model in later steps. For now we extract the predictions:

```
pscores <- apply(posterior_linpred(ps_fit_1), 2, mean)
```

We will use this estimated score in the next step to find matches for these treated units from among the controls. Above we use `posterior_linpred` to get predictions on the scale of the linear predictor, that is, $X^{\text{new}}\beta$. If we wanted predictions on the probability scale, we would use posterior_epred. In this example, similar results would arise from using either approach. We discuss additional approaches to propensity score estimation in Section 20.8.

### Step 3: Matching to restructure the data

Using the propensity score as a summary allows us to relatively easily find matches for each treated unit from among the untreated units. For each treated unit, we can simply choose the untreated unit with the closest estimated propensity score as the "match."

**Matching without replacement.** In this example we use the estimated propensity scores to create matched pairs using an R function called matching. We begin by implementing matching *without* replacement:

```
matches <- matching(z=cc2$treat, score=pscores, replace=FALSE)
```

This version of the algorithm will not use any comparison unit more than once as a match. We can use the output from the matching command to create a new matched dataset that holds the treated observations and only those control observations that were chosen as matches.

```
matched <- cc2[matches$match.ind,]
```

However, the results from most matching methods won't translate so seamlessly into a new "matched" dataset. Instead, they will yield weights that can be applied to our full dataset to create a *virtual* matched dataset. Matching without replacement is a simple example where all these "weights" simply equal 0 or 1.

**Matching with replacement.** Suppose instead that we had implemented matching *with* replacement:

```
matches_wr <- matching(z=cc2$treat, score=pscores, replace=TRUE)
wts_wr <- matches_wr$cnts
```

This version of the algorithm chooses as a match the comparison unit with the closest estimated propensity score to the treatment unit, *even if that comparison unit was previously chosen as a match for another treatment unit.* Thus, any given comparison unit might be chosen as a match more than once. The cnts output keeps track of how many times each unit should contribute to subsequent analyses: once for each treated unit and the number of times used as a match for each comparison unit. These counts allow us to restructure the data through weights.

Matching with replacement should yield better matches on average compared to matching without replacement. The hope is that it will lead to better balance and less biased treatment effect estimates. However, it can result in over-using certain units or ignoring other close matches, thus failing to capitalize on important information in the data. Given the smaller number of unique observations this also might increase the variance of our treatment effect estimates. Consequently the choice between these matching methods is often framed as a bias-variance tradeoff.

### Step 4: Diagnostics for balance and overlap

*Balance* can be assessed for the matched sample by comparing the distribution of each potential confounder across treatment and matched control groups. Figure 20.9 displays a set of comparisons. The open circles correspond to balance before matching. The solid circles correspond to the differences in matched groups resulting from matching without replacement based on the propensity score equation above. In that plot balance is operationalized as the standardized absolute differences in means. We calculate this measure by dividing the absolute difference in means by the standard deviation of the given variable for those observations in the treatment group (more generally, for those in the inferential group). This standardization allows us to better understand the practical importance of each difference, similar to the argument made in certain disciplines for presenting results in terms of "standardized coefficients" or "effect sizes."

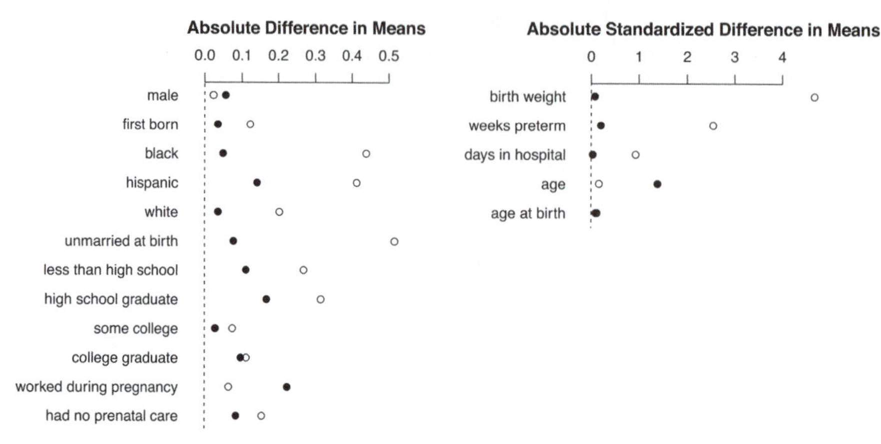


Figure 20.13 *Imbalance in averages of confounding covariates across treatment groups. The left panel focuses on binary covariates and displays absolute differences in means. Open circles correspond to the unmatched groups. Solid circles correspond to groups matched with replacement. The right panel focuses on continuous variables and displays absolute differences in means standardized by the standard deviation in the treatment group. Open circles correspond to the unmatched groups. Solid circles correspond to groups matched with replacement.*

In contrast, Figure 20.13 displays pre-match (open circles) and post-match (solid circles) balance resulting from matching *with* replacement (the propensity score model used for this plot is described in the next subsection). Moreover, this figure displays separate balance plots for continuous and binary variables. Balance for continuous variables is displayed as standardized absolute difference in means, as above. Since the standard deviation for binary variables is determined by exclusively the mean it could be misleading to standardize. Therefore, we present absolute difference in means without standardization for binary variables.

```
bal_wr <- balance(rawdat=cc2[,covs], cc2$treat,
  matched=matches_20.9_wr$cnts, estimand="ATT")
plot.balance(bal_wr, longcovnames=cov_names, which.cov="cont")
plot.balance(bal_wr, longcovnames=cov_names, which.cov="binary")
```

In practice, this diagnostic is used as a way of evaluating the adequacy of the propensity score model fit in step 2. If we have succeeded in creating groups that look similar in terms of their observed covariates, we feel more secure that our subsequent analyses adequately adjust for these inputs.

It is common at this stage to assess whether we have sufficient *overlap* in our covariate distributions to believe that we can reasonably make inferences about our treatment group. Assuming our model yields a decent estimate of the propensity score (and assuming ignorability holds), a quick way to gauge the extent to which overlap is achievable via matching is to plot overlapping histograms with respect to the estimated propensity score. Figure 20.14 displays these histograms both for the unmatched sample and the matched sample chosen through matching with replacement. We plot the propensity scores on the logit scale to better display variation at the extremes, which correspond to probabilities near 0 and 1.

These plots appear to demonstrate both decreased imbalance and increased overlap. However, if overlap did not exist for some treatment units before restructuring, it cannot possibly exist afterwards. For this reason some researchers often examine only pre-match overlap plots. On the other hand, pre-match plots with treatment and control groups that have very different distributions of propensity scores can make it difficult to properly understand overlap because the plot devotes too much visual space to portions of the control distribution that will likely disappear after restructuring. Thus post-match plots may provide a clearer illustration of overlap.

Overlap in the histograms for the matched groups do not ensure that the distributions of *every*

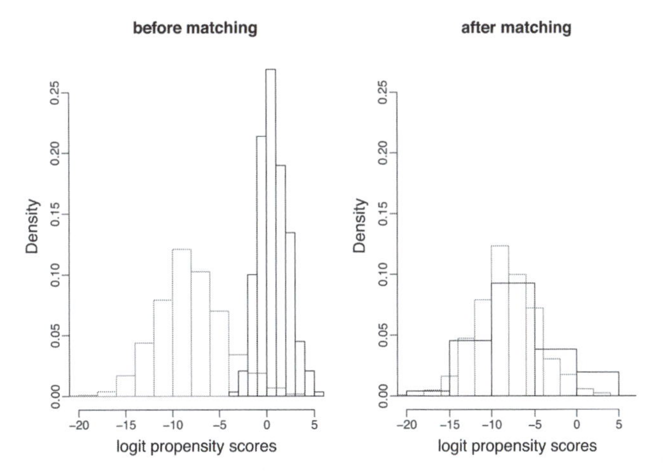


Figure 20.14 *(a) Distribution of logit propensity scores for treated (dark lines) and control groups (gray lines)* before *matching. The 12 observations with scores below −6 were omitted from the plot to facilitate focus on the relevant portion of the plot. (b) Distributions of logit propensity scores for treated (dark lines) and control groups (gray lines)* after matching.

*predictor* included in the model will have common support or be balanced. However, to the extent that the propensity score is modeled well and includes all the confounders, such plots can be used to help assess the extent to which empirical counterfactuals exist for all our inferential units.

In our example, a more specific check,

```
sum(pscores[cc2$treat==1] > max(pscores[cc2$treat==0]))
```

reveals that 12 treated units have propensity scores outside the range of the control propensity scores. Therefore, we might want to consider restricting the treated sample to observations with propensity scores that enjoy common support.

### Before Step 5: Repeat steps 2–4 until adequate balance is achieved

Balance achieved after our first attempt at matching is often not as close as we would like. There are two potential reasons for imbalance. The first is that there is insufficient overlap in the data to ever achieve a set of matches that will yield adequate balance. The second is that we did a poor job of accurately estimating propensity scores and we need to use a different model. We address these concerns in turn.

If you aren't happy with the balance achieved from your initial propensity score estimates and matching strategy, then you don't want to proceed yet to step 5. Instead, you can try to achieve better balance again using a different approach. In general this either involves changing the way you estimate the propensity score or changing how you use the propensity score to restructure your data, for instance, by using a different matching method.

**Changing the propensity score model.** In our example, we evaluated several different model fits by iterating between finding matches using estimated propensity scores from a particular fitted model and evaluating the balance achieved in that matched sample. Strategies for adjusting the propensity score model to attempt to address imbalance for particular variables included adding interactions between that variable and other potential confounders and including transformed versions of some of the continuous variables. Typically, creating better balance for one variable (or set of variables) resulted in greater imbalance for others. In particular, a tradeoff appears to exist between finding balance for birth weight, weeks preterm, and days in hospital, versus finding balance for mother's ethnicity education variables. In the end, after fitting several dozen models, we chose the one that seemed to provide the best balance that seemed achievable overall. It took the form,

```
ps_fit_2 <- stan_glm(treat ~ bw + preterm + dayskidh + sex + hispanic + b.marr +
  lths + hs + ltcoll + work.dur + prenatal + momage + income + bwT + pretermT +
  income + black:dayskidT + b.marr:bw + b.marr:preterm +  b.marr:dayskidT +
  bw:income, data=cc2, family=binomial(link="logit")
```

This model is messy, and to the extent that it is overfitting the data, its predictions and then the resulting matches can be noisy and may be misleading with regard to our assessment of overlap. As with regression fitting more generally, it can make sense to stabilize the estimate by using regularization techniques appropriate for regressions with many predictors, but here we are focusing on the steps of matching and so we just accept this particular fit as is.

### Step 5: Estimating a treatment effect using the restructured data

After choosing a propensity score model and matching algorithm based on the balance of the matches, we could simply estimate the effect of the treatment on the treated in this sample using a weighted difference in means. A better strategy, is to fit a regression model including the treatment indicator and all potential confounders as predictors using the restructured data, as reflected in the integer counts.

In our example we could fit the following regression model using the weights provided by the matching-with-replacement function to restructure our data:

```
reg_ps2_design <- svydesign(ids=~1, weights=~matches2_wr$cnts, data=cc2)
reg_ps2 <- svyglm(ppvtr.36 ~ treat + bw + bwg + hispanic + black + b.marr +
  lths + hs + ltcoll + work.dur + prenatal + sex + first + preterm + momage +
  dayskidh + income, design=reg_ps2_design, data=cc2)
```

We use the survey package in R to account for the weighting structure.

Fitting this model gives us an additional chance to adjust for differences in covariate distributions that typically remain between the groups (to decrease bias and increase efficiency). The overlap and balance created by the matching should make this model more robust to potential model misspecification; that is, even if this model isn't quite right (for example, excluding a key interaction, or assuming linearity when the underlying relation is strongly nonlinear) our coefficient estimate should still be close to correct, conditional on ignorability being satisfied. Our estimate of the effect of the treatment on the treated from the data created through matching without replacement is $10.1 \pm 1.5$. The estimate in the data created through matching with replacement is $10.6 \pm 2.3$. That is, if we trust the causal interpretation, we would say that the children who were assigned to participate in the IHDP intervention had test scores that were, on average, about 10 points higher than they would have been had they not had exposure to the intervention.

These estimates can be compared to the standard regression estimate of $11.4 \pm 1.2$ based on the full data without matching. Recall, however, that the standard regression estimate targets an estimand that is similar to the average treatment effect. Thus, the differences between these estimates could be driven both by the difference in target estimands and by the matching being a more robust estimation procedure.

The standard errors presented for the analyses fitted to matched samples are still not technically correct. First, matching induces correlation among the matched observations. The regression model, however, if correctly specified, should account for this by including the variables used to match. Second, the fact that the propensity score has been estimated from the data is not reflected in our calculations. This issue has no perfect solution to date and is currently under investigation by researchers in this field.

### Geographic information

We have excluded some important information from these analyses. We have access to indicators reflecting the state where each child resides. Given the tremendous variation in test scores and child care quality across states, it seems prudent to adjust for this variable as well. Variation in quality of child care is important because it reflects one of the most important alternatives that can be chosen by the parents in the control group. If we redo the propensity score matching by including state indicators in both the propensity score model and final regression model, we get estimates of $8.0 \pm 2.0$ and $6.7 \pm 2.4$ for the matching without replacement and matching with replacement strategies, respectively. Extending the regression analysis on the full data to include state indicators changes the estimate only from $11.4 \pm 1.2$ to $11.3 \pm 1.4$.

### Experimental benchmark by which to evaluate our estimates

In reality, the IHDP intervention was evaluated using a randomized experiment. In the preceding example, we simply replaced the true experimental control group with a comparison group pulled from the National Longitudinal Survey of Youth to create a hypothetical "constructed observational study." The advantage of this example as an illustration of propensity score matching is that we can compare the estimates obtained from the constructed observational study to the estimates found using the original randomized experiment. For this sample, the experimental estimate is 7.4. The estimates based on propensity score adjustment were close to this once we include geographic information, while simple unadjusted regression estimates were substantially farther.

Finding treatment effect estimates that are close to the experimental benchmark doesn't present a clear referendum on the performance of any method, however. Why? Because ignorability is still an approximation. To reinforce this point, recall that subclassification on mother's education alone yields an estimated treatment effect of 7.0, which happens to be close to the experimental benchmark. However, this does not imply that subclassifying on one variable is generally the best strategy overall. In this example, failure to adjust for all confounding covariates leads to many biases (some negative and some positive—the geographic variables complicate this picture), and unadjusted differences in average outcomes yield estimates that are lower than the experimental benchmark. Adjusting for one variable appears to work well for this example because the biases caused by the imbalances in the other variables just happen to cancel.

## 20.8 Restructuring to create balanced treatment and control groups

We now expand our discussion of restructuring to consider more general issues surrounding the choice of confounders, the choice and estimation of distance metrics, balance and overlap, and approaches to matching and weighting. We organize the steps as in the previous section.

### Step 1: Estimands and confounders

Moving from a research question to statistical analyses requires making many decisions. In this section we discuss additional issues with regard to choosing estimands and the covariates considered to be confounders.

**Defining the population of interest.** Ideally, the estimand targeted by the analysis should be chosen based on the research goals, in particular, deciding whom we want to make inferences about. For example, if the goal is to understand the effect of a social program, and data have been collected for both program participants and non-participants, it is common to focus on the effect of the treatment on the treated, so as to estimate the effect of the program for those who might actually choose to participate in it. On the other hand, if the analysis sample is representative of the population of interest, all of whom might have been exposed to the treatment, then the average treatment effect might be more appropriate.

More generally, when we fit a regression model with interactions, we are estimating the treatment effect as a function of pre-treatment predictors, and any summary of this estimate implies some distribution of who in the population was given this treatment (for a retrospective analysis) or who might take it in the future (for making future decisions). This is mathematically the same problem as the generalization-to-population problem discussed in Section 17.1 on poststratification.

In practice, however, data constraints can make it difficult to estimate the ideal estimand. If the goal is to estimate the effect of the treatment on the treated but there are treated observations with no empirical counterfactuals, then it might make sense to switch to a different estimand that focuses on the controls or on a subset of the treatment group for whom sufficient overlap exists.

**Choosing covariates.** In observational studies we adjust for covariates in an attempt to better approximate ignorability. In some situations there aren't many covariates available to include in an analysis. In other situations there might be more covariates than observations. In many contexts there are more covariates than can be included in any given analysis without raising concerns about overfit. Regularized regression and multilevel models can be used to adjust for large numbers of pre-treatment predictors even when data are sparse. Here, though, we assume simple regression methods, and so we need to consider which covariates to include and which to exclude.

First, as discussed in Section 19.6, it is not appropriate to adjust for post-treatment variables; this can induce bias even if ignorability is satisfied.

A second point is more subtle. There is a false intuition that, when adjusting for covariates, every covariate added to an analysis will decrease bias. This simply is not true. Properly adjusting for *all* confounding covariates (if that were possible) should eliminate bias, but understanding the bias that remains from adjusting for only a subset of the confounders is more complicated. As an example, recall that adjusting for only mother's education in the child care analysis yielded an estimate that was close to the experimental benchmark even though many confounders were excluded from the analysis. Adding more covariates *increased* the bias relative to that initial analysis.

Given this, how can we make decisions in a scenario where we are not convinced that all confounders have been included? One strategy to reduce risk is to avoid covariates that are strongly related to the treatment variable but not strongly related to the outcome. The classic example of a covariate that should not just be includes as an additional predictor is an instrumental variable, as discussed in Section 21.1. If ignorability has not been satisfied, adjusting for such variables can inflate whatever bias is present. More generally, it makes sense to privilege inclusion of covariates that are strongly related to the outcome. Assessing the strength of this relationship can be justified based on theoretical or empirical grounds. Theoretical justifications alone may overlook associations not previously considered, while empirical justifications based solely on marginal associations or overly simplistic models that make parametric assumptions may perpetuate some of the problems we are trying to avoid.

### Step 2: Calculating distance metrics: finding observations with different treatments that are similar in their pre-treatment characteristics

In the previous section we provided an example of using the propensity score as a distance metric. Here we formally motivate and define the propensity score, discuss estimation issues, and briefly discuss alternative distance metrics.

**Motivation for finding a distance metric.** Matching on one continuous confounding covariate is an intuitive idea. For instance, if targeting the effect of the treatment on the treated when we knew that a pre-test score was the only confounding covariate, we could simply choose for each treated unit the untreated unit that had the closest value of the pre-test. Matching on one binary and one continuous variable still seems doable. A simple approach would be to stratify within subgroups defined by the binary variable and then match on the continuous variable within each subgroup.

Matching on two continuous variables immediately becomes more tricky. One strategy might be to first create a subset of potential matches within a certain distance with regard to one of the variables and then within those choose the observation that was closest with regard to the second variable. But that approach raises issues about which variable to match on first and how large the initial window should be. If it's tricky to match on two continuous variables, what happens when we have 20, 50, or 200 confounding covariates?

One way to simplify the issue of matching or subclassifying on many confounding covariates at once is to define a univariate distance metric between observations as a function of the observed covariates. A starting point is the *Euclidean distance* between predictor vectors $X_i$ and $X_j$, $d_{ij}^{\text{Euclidean}} = \sum_{k=1}^{K}(X_{ik} - X_{jk})^2$, where $K$ is the dimensionality, that is, the number of covariates in the predictive model. Euclidean distance is problematic because it is not scale invariant and it ignores the correlation structure of the covariates. *Mahalanobis distance* addresses both of these shortcomings using a matrix transformation: $d_{ij}^{\text{Mahalanobis}} = (X_i - X_j)^t \Sigma^{-1}(X_i - X_j)$, where $\Sigma$ is the covariance matrix of the data, or perhaps the population of interest. A criticism of both these distances is that they implicitly over-privilege higher-order interactions between variables. Said another way, each defines proximity based on what are arguably specialized neighborhoods of the covariate space.

**The propensity score.** The *propensity score* emerged as an alternative approach several decades ago, and the absolute difference between propensity scores remains the most popular distance metric for matching in causal inference to this day. The propensity score for the $i^{\text{th}}$ observation is defined as the probability that unit $i$ receives the treatment given pre-treatment covariates, $x_i$. More generally, with a binary treatment, the propensity score can be defined as $\Pr(z = 1|x) = \mathrm{E}(z|x)$. This equivalence highlights that the propensity score model can be conceptualized as a regression (broadly defined as a conditional expectation) with a binary response. Thus, in theory, propensity scores can be estimated using standard models such as logistic or probit regression, where the response is the treatment indicator and the predictors are all the potential confounders. After using the fitted model to estimate propensity scores, matches can be found by choosing for each treatment observation the control observation with the closest estimated propensity score.

There are many ways to create a one-number summary of the covariates. Why choose this one? It turns out that if the covariates that are included in the propensity score model are sufficient to satisfy ignorability, then appropriate conditioning on the propensity score (for instance by matching, subclassifying, or weighting on functions of it) is sufficient to estimate unbiased treatment effect estimates. Formally, we can say that if $y^0, y^1 \perp x$, then $y^0, y^1 \perp e(x)$, where $e(x)$ denotes the propensity score. Thus,

$$\mathrm{E}(y^0 \,|\, x, z = 0) = \mathrm{E}(y^0 \,|\, x) = \mathrm{E}(y^0 \,|\, e(x)),$$
$$\mathrm{E}(y^1 \,|\, x, z = 1) = \mathrm{E}(y^1 \,|\, x) = \mathrm{E}(y^1 \,|\, e(x)).$$

These identities hold because if $x$ is sufficient to satisfy ignorability then $e(x)$ is also sufficient to satisfy ignorability.

**Estimating the propensity score.** There are many estimation strategies for the propensity score ranging from parametric to nonparametric. Some considerations are discussed here.

*Multicollinearity.* Students learning about regression are often urged to avoid multicollinearity. However, since our focus is only on estimating the treatment effect, and the goal of our matching is to create independence between the covariates and the treatment indicator, we are not particularly concerned about issues of multicollinearity among our covariates. That said, if we have a large number of covariates relative to our sample size, we can easily run into related estimation problems such as overfitting, as discussed below.

On the other hand, if you are trying to adjust for covariates that are highly correlated with your treatment, you might want to consider whether that variable really should be considered as part of the definition of the treatment. For instance, consider estimating the effect of being a single parent. Single-parent status is strongly associated with household income. Because of this, one might subset the analysis sample based on income, which is roughly equivalent to including in the model an interaction between treatment and income (or discretized income). If insufficient overlap existed for the high-income group, one could decide to avoid making inferences for this group and be explicit about this change in estimand when reporting results.

*Overfitting.* For many years, overfitting was not perceived to be a problem in fitting the propensity score, particularly if the goal is to estimate the sample average treatment effect (SATE). Researchers were often advised to use approaches like stepwise regression to fit their propensity score models, particularly if many covariates were involved. However, overfitting can lead to problems. To understand why, consider a model or algorithm that perfectly predicts for each observation in our sample whether it received the treatment or control. Perfect prediction sounds like a good thing, right? However, this model would estimate propensity scores equal to 1 for all treated and equal to 0 for all controls. This is a declaration that no overlap exists. Moreover, this model would not be useful for predicting treatment assignment in a different sample. Therefore, it is important to ensure that your model does not suffer this fate.

Alternatively, or in addition, we can fit a model designed to avoid overfitting. The estimation strategy used in Section 20.7 based on `stan_glm` should help since it relies on a Bayesian model that partially pools the coefficient estimates toward 0. This is an example of a regularization strategy, but many other options exist. We discuss further in Section 22.4.

*Predictive accuracy.* There is sometimes a temptation when fitting propensity score models to think of prediction as a goal in itself. But that would be a mistake. To see why, consider that a completely randomized experiment is often an excellent design but corresponds to a propensity score model with no predictive power at all. On the other hand, poor predictive power of a propensity score model in a real-world observational study could instead indicate that important pre-treatment confounders have not yet been included in the model, which could be a major concern. Looking at the predictive power of the fitted model does not distinguish between these crucially different scenarios.

**Alternative distance metrics.** As mentioned briefly above, the propensity score is not the only way to determine proximity between treatment and control units. Moreover, the propensity score is optimized for finding balance on average across groups, not for finding close pair matches. Consider a simple example where the propensity score is estimated using a logistic regression equation based on two covariates, age (in years) and income (in thousands of dollars), as in

$$\text{logit}(p) = \beta_0 + 0.4 * \text{age} + 0.4 * \text{income}.$$

In this example, we could end up with a Mahalanobis treated observation and a control observation that are identified as matches with identical propensity scores even though one of them is 30 years old and makes \$60 000 and the other is 60 years old and makes \$30 000. A more flexible model for the propensity score might alleviate some of these problems but will have the challenges with respect to overfitting described above.

Alternatively, a metric like Mahalanobis matching would privilege matches in the same part of covariate space. However, when the number of covariates grows large, this space will be sparse and it will be more difficult to find close matches. More recent approaches to matching rely on algorithms that directly optimize balance or other criteria specified by the user. We describe some possibilities in the references at the end of the chapter.

### Step 3: Restructuring the data

Once we have defined a measure of distance between units, we need to decide on a strategy for restructuring. Options include subclassification, weighting, and any number of matching algorithms. We describe a few of the basic choices involved in such algorithms here.

**Options for matching algorithms.** In the previous section, we described the simplest form of matching: one-to-one (or nearest neighbor). That is, for each member of the inferential group we find exactly one match from among the other group. If there is a large comparison group, one-to-one matching can waste a great deal of information.

An alternative is to match a pre-specified number of controls to each member of the inferential group; this is sometimes referred to as $k$-to-1 matching. This can help use more observations in the sample but lacks flexibility with regard to finding close matches. For instance, suppose you have 200 treated observations and 1200 control observations. At first glance it might seem like there would be plenty of controls to find 3, 4, or even 5 matches from among the control reservoir for each treated observation. However, depending on the overlap across groups it might instead be the case that some treated units easily have 10 close matches and some only have 1 or 2, or even none.

Caliper matching addresses this issue of differential numbers of "good" matches because it allows for a variable number of matches for each observation in the inferential group. The basic idea is to define a symmetric window of a specified width around each observation in the inferential group and accept all control units within this window as matches for that observation. One approach is to weight these matches equally. However, it might make sense to give more weight to closer matches. So sometimes the contribution of these matches is weighted by a function, often called a kernel, that downweights matches in the window that are farthest from the inferential unit to which they are being matched. Either way this can be a helpful approach to use more of your data.

**Choice of matching algorithm.** When matching without replacement, the order in which units are matched can affect which units end up getting chosen for the matched comparison group. The simplest option for programming such a matching algorithm is to first order the inferential units in some way. For instance, units could be randomly ordered or, alternatively, ordered from largest to smallest propensity scores. Then for each observation in turn, the algorithm would find the comparison unit that is closest to that inferential observation with respect to the given distance metric. This type of strategy is referred to as a greedy algorithm because it does not consider how the choices at one step affect the choices down the line. The consequence is that the resulting matches may not result in the best balance overall. An alternative is to implement a matching algorithm that jointly considers the consequences to all units. For instance "full matching" and "optimal matching" algorithms optimize over the full set of matches to minimize the total distance between units.

### Interpreting matching as a particular weighting scheme

Recall that we can think of matching as a special case of weighting. For instance, in one-to-one matching each observation is implicitly assigned a weight of either 1 (retained in the matched sample) or 0 (not retained in the matched sample). In most software packages for matching, this weighting becomes explicit. More complicated matching methods map to more complicated weights. For instance, in $k$-to-1 matching (without replacement) used to estimate the average treatment effect for the treated group, each treated observation has an implicit weight of 1 and each control observation has an implicit weight of $1/k$. In caliper matching, each control observation would accrue weights based on its relative frequency of its use as a match across treated units. For instance, if a unit was used as one of three matches for one treated unit, one of five matches for another unit, and one of six matches for a third treated unit, it would receive a weight of $1/3 + 1/5 + 1/6 = 0.7$. If kernels are used, these fractions would be further adjusted by the kernel weights.

### Weighting based on estimated inverse probability of treatment assignment

Functions of the propensity scores can also be used more explicitly as weights in the absence of any matching. The basic idea, similar to ideas from survey sampling, is to weight the sample to be representative of a pseudo-population that is unconfounded and represents the population of interest. In practice this target population is often simply set to the sample of units in the study (to estimate the sample average treatment effect), or just the treatment group (to estimate the average effect among the treated units), or just the control group (to estimate the average effect among the controls). More generally one could aim to estimate the treatment effect for a larger population or different population in an attempt to generalize the results more broadly, in the same way that we use poststratification to generalize results to a new population, as discussed in Section 17.1.

| Sample size | LBW | Child care | Potential outcomes ($z_i = 0$) | ($z_i = 1$) | Observed outcome | Prop score | Normed weight | Pseudo-pop |
|---|---|---|---|---|---|---|---|---|
| $n$ | $x_{1i}$ | $z_i$ | $y_i^0$ | $y_i^1$ | $y_i$ | $p_i$ | $\tilde{w}_i$ | $\tilde{n}$ |
| 80 | 0 | 0 | **110** | 120 | 110 | 0.2 | 0.625 | 50 |
| 50 | 1 | 0 | **90** | 90 | 90 | 0.5 | 1 | 50 |
| 20 | 0 | 1 | 110 | **120** | 120 | 0.2 | 2.5 | 50 |
| 50 | 1 | 1 | 90 | **90** | 90 | 0.5 | 1 | 50 |

Figure 20.15 *Hypothetical causal inference data for the effect of high-quality child care on IQ scores. There is a treatment effect of 0 for low-birth-weight children and a treatment effect of 10 IQ points for typical-birth-weight children. The average treatment effect is 5. Each row represents one of four categories of children as defined by their birth weight status (low=1) and their treatment assignment (1 for high-quality child care). The number of children in each of these categories in the observational study is displayed in the first column. A naive comparison of average outcomes across the two groups (98.6 and 102.3) would lead to the erroneous conclusion that child care caused a small decrease in IQ scores. However, reweighting the data implicitly creates a pseudo-population with sample sizes displayed in the last column of the table. A comparison of mean outcomes within this pseudo-population leads to an estimated average treatment effect of 5.*

Consider the weighting used to estimate an average treatment effect. Treated units receive weights of $1/\hat{p}$, the inverse of the estimated probabilities that they ended up in their particular (treated) sample. These weight the treated observations to be approximately representative of the full sample and by extension the population of which that sample is representative. By the same reasoning, the control units receive weights of $1/(1-\hat{p})$; they are weighted by the inverse of the estimated probability that they ended up in their sample (the control group). Typically these weights are normed so that they average to 1; this can be accomplished by simply dividing by the average of the unnormed weights. This is the same principle that motivates the method from survey sampling where observations in a sample receive weights equal to the inverse of their estimated probabilities of selection.

Now consider the weighting that is used to estimate the average treatment effect among the treatment. The goal is to weight the sample so that it is representative of the treatment group. The treatment group observations would be assigned weights of 1 since they are already representative of themselves. The control observations would be assigned (unnormed) weights equal to $\hat{p}_i/(1-\hat{p}_i)$. This strategy is reversed for estimating the average effect among the controls. The control group observations would be assigned weights of 1 since they are already representative of themselves. The treatment group observations would be assigned (unnormed) weights equal to $(1-\hat{p}_i)/\hat{p}_i$.

For any of these estimands, just as with matching, the researcher should perform diagnostics to assess balance and overlap in the reweighted analysis sample.

Example: Illustration of inverse estimated probability weighting

*Hypothetical example.* We illustrate the impact of weighting for estimating an average treatment effect using the hypothetical observational study data in Figure 20.15. In these data, $x_i$ is an indicator for whether the child had low birth weight (less than 2500 grams). For the sake of simplicity this is the only confounder. The estimated propensity score for low-birth-weight children is 0.5, because 50 out of 100 of these children were treated. The estimated propensity score for typical-weight children is 0.2, because 20 out of 100 of these children were treated.

Using the inverse weighting scheme to estimated the sample average treatment effect as discussed above yields the normed weights displayed in the table. In this simple example these weights will yield a restructured dataset that is balanced in the covariates and potential outcomes. Thus the weighted estimate will be unbiased (in this case exactly equal to) the true sample average treatment effect of 5. Exercise 20.3 will give you a chance to calculate weights for estimating the average treatment effect among the treated and control groups, check for balance, and understand how well the weighted estimates of these estimands approximate the truth.

*Child care example.* What happens if we implement estimated inverse probability weighting in the IHDP analysis to estimate the average treatment effect among the treated? Given the number of covariates involved, we estimate propensity scores using a logistic regression model (as with

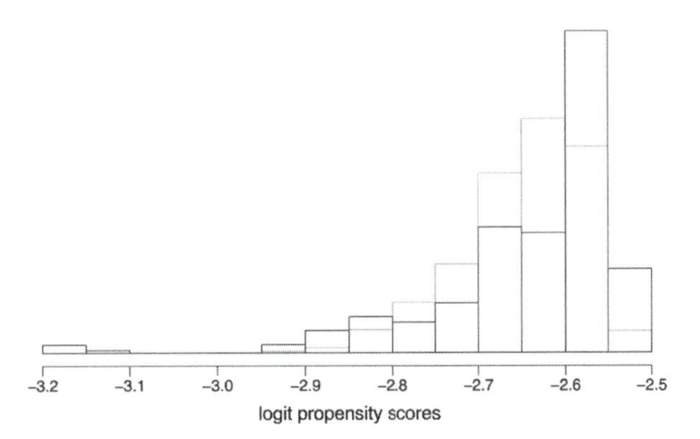


Figure 20.16 *Distribution of logit propensity scores for treated (dark lines) and control groups (gray lines) before matching for an example where the overlap in propensity scores between the treatment and control groups suggest a good propensity score model fit, but the propensity score model excludes some important pre-treatment predictors. This model has low predictive power, and overlap on this propensity score should not be taken as a sign of good overlap of treatment and control groups in the multivariate space of covariates.*

matching). Then the weights are incorporated in a linear regression of the outcome on an indicator for treatment assignment and the observed confounders. This yields a treatment effect estimate of 8.6, which is close to the experimental benchmark.

These strategies have the advantage (in terms of precision) of retaining the full sample. However, the weights may have wide variability and may be sensitive to model specification, which could lead to instability. Therefore, these strategies work best when care is taken to create stable weights and to use robust models to estimate the weights. A simple fix is to cap the weights such that none can exceed a certain value, for instance, 10% of the sample size of the inferential group. Such methods are beyond the scope of this book.

### Step 4: Diagnostics: Balance and overlap

Here we discuss some additional considerations regarding balance and overlap.

**Don't use overlap to understand model fit.** Sometimes there is an inclination to use overlap plots as a way of diagnosing model fit. Researchers search for models that lead to overlap in the propensity score. However, if we haven't fit an adequate model to estimate the propensity scores, then we might be misled by such a plot. As an example of how this might lead one astray, consider that a terribly fit model with no predictive power should lead to plots of propensity scores with complete overlap (or nearly so). Clearly this overlap does not imply that we have estimated useful propensity scores.

That is why we advocate for checking overlap in propensity scores only after using diagnostics such as balance as an initial check of the propensity score model. All these concepts are inherently linked. If the treatment and control groups don't overlap sufficiently, then matching will not be able to create good balance regardless of the propensity score model or matching method used. On the other hand, if we don't have a good model for the propensity score, then our overlap plots of the propensity score will not yield accurate information, as we illustrate in Figure 20.16. One way to address this issue is to examine overlap directly for the most important confounders, as in Figure 20.11.

**Overlap and overfit.** One implication of model overfit is that overlap plots may appear to demonstrate lack of overlap even when overlap exists. What can we do?

Overfit can be diagnosed in several ways. One approach is to use cross validation to assess whether a propensity score model has good out-of-sample prediction properties. The optimal metric for this approach (for example, balance versus classification rates for the binary treatment variable) is still a topic of debate. A less formal strategy is to fit your preferred model to an analysis sample with the original data augmented with a treatment assignment mechanism that is completely randomly generated (with marginal probability of being treated equal to the original analysis sample). Since our simulated dataset mimics a completely randomized experiment, if overlap plots reveal lack of overlap in this scenario we know we have a problem with our model.

**Overlap and the estimand (again).** What happens if there are observations in the inferential group for which there are no observations with similar propensity scores in the comparison group? For instance, suppose we are interested in the effect of the treatment on the treated but there are some treated observations with propensity scores far from the propensity scores of all the control observations. One option is to accept some lack of comparability (and corresponding level of imbalance) in covariates. Another option is to eliminate the problematic treated observations. If the latter choice is made, it is important to be clear about the change in the population about whom inferences will now generalize. It can also be helpful to profile the observations that are omitted from the analysis.

**Overlap in confounders not covariates.** Researchers often err on the side of including more variables than they think are strictly necessary among their potential confounders. However, if the propensity score model includes covariates that aren't actually confounders because they don't predict the outcome, then it might appear that overlap is violated even if it actually holds with respect to the true confounders. Said another way, it is only necessary to have common support with respect to confounders, not all covariates.

**Beyond balance in means.** In Section 20.4 above, we discussed the relationship between imbalance in covariate distributions and bias, illustrating with a simple example of a response surface that was quadratic in $x$. We can safeguard against the failure of our analysis model to capture nonlinearity in the response surface by balancing other aspects of the distributions, not simply means. This will not be an issue for binary variables whose distributions are entirely defined by the mean. A step in the right direction would be to consider balance diagnostics for continuous variables such as the ratio of standard deviations across treatment and control groups. We display these for the continuous variables in our data:

```
          unmatched MwoR  MwR
bw             0.50 0.87 0.55
preterm        0.95 0.80 1.20
dayskidh       2.07 0.96 2.50
momage         1.86 1.75 1.96
income         0.27 0.19 1.11
age            0.07 0.07 0.08
```

Matching without replacement in the child care example made some progress toward the goal of equal standard deviations. Four out of the five variables in the original (unmatched) sample had ratios of standard deviations in the treatment and control groups that were far from 1, whereas both matching methods yield ratios for the variables for birth weight and the number of days children were in the hospital much closer to 1. Neither matching method made much progress in balancing the spread of the distributions for child's age or mother's age. We hope that the true response surface does not include a strong nonlinear term for these variables.

Balancing the standard deviation of $x$ is equivalent to balancing the mean of $(x - \bar{x})^2$. Some software packages that evaluate this criterion do so by including squared versions of the continuous variables in the balance-of-means check.

Balancing standard deviations makes most sense when the variables in question are reasonably symmetrically distributed. A more comprehensive goal, more difficult to achieve, would be to attempt to balance the full distribution of each continuous (or ordered categorical) variable. This could be checked using plots or through summaries of the distance between empirical distributions.

These measures all reflect the distance between marginal (univariate) distributions of each variable. For maximum robustness to model misspecification we should try to balance the *joint* distribution of the covariates. For instance, nonadditivity (interactions) in the true response surface can be addressed by balancing some measure of associations between variables. Thus, some software packages evaluate this criterion by checking the balance of means of the product of each pair of potential confounders.

### Before Step 5: Iterating across Steps 2–4

Researchers are often advised to keep fitting propensity score models until "sufficient" balance is achieved. However, it's not clear exactly what constitutes sufficient balance, and so researchers will fit many models and then try to decide which is best. Why might this cause problems?

One reason is that this may lead to a greater chance of obtaining results that are idiosyncratic to specific combinations of propensity score estimation strategies, matching methods, and post-estimation models. To the extent that the decisions to pursue each model and strategy are not made explicit this becomes increasingly problematic. Moreover such practice may make it difficult for other researchers to reproduce the original results.

An additional concern is the possibility of peeking at the treatment effect estimate, allowing a researchers to choose results that are consistent with his or her expectations or hopes about the direction and magnitude of the effect. Or a researcher may lean toward propensity score strategies that yield treatment effect estimates with small $p$-values.

A more careful strategy is to define in advance a set of criteria for acceptable balance and to avoid estimating the treatment effect until a final propensity score model has been chosen. Various matching and weighting algorithms now exist to automate and optimize this type of strategy; see the bibliographic note at the end of this chapter. A related approach is to present results from *all* models that satisfy a given balance threshold.

### Step 5: Estimating the treatment effect using restructured data

Once the dataset has been restructured, it can be analyzed in the same way you would analyze a controlled experiment. For instance, you can estimate the causal effect by fitting a regression to the outcome conditional on a treatment indicator and pre-treatment variables (potential confounders).

**A warning about matched pairs.** Although many matching algorithms form matched pairs, the overall goal of propensity score matching is not to ensure that each *pair* of matched observations is similar in terms of all their covariate values. Rather, the pair matching is a means to an end to create matched *groups* that are similar in their covariate distributions. The members of each pair are typically too different from each other for there to be any real benefit from including the pairing as a predictor in the analysis.

## 20.9 Additional considerations with observational studies

Sometimes the term *observational study* refers to a situation in which a specific intervention was offered in an uncontrolled way to a population or in which a population was exposed to a well-defined treatment but with exposure not determined by any experimenter. As we discuss in this chapter and the next, the primary characteristic that distinguishes causal inference in these settings from causal inference in controlled experiments is the inability to identify effects without assumptions about how treatments were assigned or chosen.

Observational studies can also refer more broadly to settings where no explicit intervention has been performed, in which case it can be more descriptive to speak of exposed and unexposed groups rather than treatment and control. Another concern that the treatment or exposure in the data can differ from potential policies, interventions, or exposures for new people or cases. We also should be aware of the temporal ordering of the measured confounders, treatment, and outcome variables, and the challenge of separately identifying effects of multiple treatment factors. When thinking about these issues for your problem, we recommend that, when attempting causal inference using observational data, you formalize a controlled experiment as a conceptual comparison.

### Defining the treatment or exposure variable

We have defined a causal effect as a comparison between counterfactual states. This requires being able to conceive of the treatment or cause as being *manipulable*, such that each unit could potentially experience any level of the treatment variable (for instance, both treatment and control or treatment A and treatment B for binary variables). For instance, consider the oft-studied "effect" of height on earnings. This effect is ill-defined without reference to the particular treatment that could change one's height. Otherwise, what does it mean to define a potential outcome for a person that would occur *if* he or she had been shorter or taller? A hypothetical swapping of genes, for example, would have different effects than a change in the fetal environment, which in turn is different from changing a child's diet or introducing medication such as growth hormone.

As another example, consider the effect of single motherhood on children's outcomes. We might be able to envision several different kinds of interventions that could change a mother's marital status either before or after birth: changes in tax laws, participation in a marriage encouragement program for unwed parents, new child support enforcement policies, divorce laws, and so on. These potential "treatments" vary in the timing of marriage relative to birth and even the strength of the marriages that might result, and consequently might be expected to have different effects on the children involved. Therefore, this conceptual mapping to a hypothetical intervention can be important for choice of study design, analysis, and interpretation of results.

As a related example, consider a study that examines children who were randomly assigned to families for adoption. Does this allow for fair comparisons across conditions such as being raised in one-parent versus two-parent households? In any case, this is a different kind of treatment altogether than considering whether a couple should get married. There is no attempt to compare *parents* who are similar to each other except in terms of marital status; instead it is the *children* who are similar on average at the outset. The treatment in question then has to do with the child's placement in a family. This addresses an interesting although perhaps less policy-relevant question (at least in terms of policies that affect incentives for marriage formation or dissolution).

### Understanding the counterfactual state

We have repeatedly emphasized that causal inference requires a comparison of at least two conditions (for example, treatment/control or exposed/unexposed). Discussion often focuses on the treatment received without sufficient consideration for what was experience by those who don't receive the treatment. For example, consider a study of the effect on weight loss of a new diet. The treatment (following the diet) may be clear but the control is not. Is it to continue eating as before? to try a different diet? to exercise more? Different control conditions imply different counterfactual states and imply different causal effects.

A more subtle example is embedded in the IHDP child care study discussed above. Recall that the study took place in eight distinct sites around the country. In each location the child care options in the absence of the intervention were different with respect to both the quality of care and type of care available. Thus, if we see differences in the magnitude of effects across sites, it might be due to differences in treatment implementation, differences in characteristics of participants, or differences in the counterfactual. This requires careful thought even in the case of a randomized experiment, and it represents a violation of the stable unit treatment value assumption as discussed on page 353.

### Temporal ordering of variables

Causal inference implicitly rests on an assumption, so basic that we rarely bother to articulate it, that causal effects are measured *after* a treatment has been assigned and administered. Outcomes may be measured five minutes or five years after, but they must reflect post-treatment phenomena; careful thought is often given regarding optimal timing for measuring outcomes to ensure for instance that enough time has elapsed for an outcome to manifest or to understand whether effects fade over time.

Furthermore, we discussed in Section 19.6 the importance of only adjusting for covariates that reflect phenomena that occur *before* the treatment was administered. When implementing a randomized experiment or a prospective observational design, understanding the temporal ordering of these three types of variables is straightforward.

In a retrospective observational study—that is, a study that uses data that have already been collected—the temporal ordering of the variables may be less clear. The most obvious problematic example is when study data come from a cross-sectional design (that is, when data on all variables were collected at the same time). It is possible in such a design that some of the questions refer to events in the past (previous schooling, jobs, health, relationship patterns) such that careful selection of variables may still be able to reflect the proper time ordering. This type of questioning about past events typically yields less reliable information, however. More desirable is a longitudinal design in which the researcher can select variables from different waves of the study.

A final warning is in order. If the participants in a study know that a treatment *will be* administered in the future and they know or suspect what treatment group they will be in, they might alter their behavior *even before* the treatment is formally administered. In that case, the measurements taken prior to the treatment cannot be considered to be independent of the treatment and must be handled with the same precaution as other post-treatment variables.

### Multiple treatment factors

It is difficult to directly interpret more than one input variable causally in an observational study. Suppose we are interested in estimating the effect of low self-esteem and stressful events on the level of depression. To estimate both effects at once, we would have to imagine a manipulation of self-esteem that leaves stressful events unchanged and a manipulation of stressful events that leaves self-esteem unchanged. In observational examples we have seen, it is generally difficult to envision both these interventions happening independently. For instance, in this example are we willing to believe that having low self-esteem wouldn't lead to a choice of engagement in events that might be differentially stressful compared to how it otherwise might have been? Similarly, if one engages in a high-stress event, might that not have an impact on self-esteem?

If we knew that self-esteem was measured before the measure of stressful events in time or logical sequence, then we can estimate the effect of stressful events adjusting for self-esteem but not the reverse, because of the problem with adjusting for post-treatment variables, as discussed in Section 19.6. An additional challenge here is to define and measure self-esteem and stress in a way relevant to the questions at hand.

### Studying causes of effects

A related practice when studying a social problem (for example, poverty), is to run regressions to predict that outcome with many inputs included to see which was the strongest predictor. As opposed to the way we have tried to frame causal questions thus far in this chapter, as the effect of causes, this is a strategy that searches for the causes of an effect. We think of such questions as a part of model checking and hypothesis generation. Looking for the most important cause of an outcome is a confusing framing for a research question because one can always find an earlier cause that affected the "cause" you determine to be the strongest from your data. This phenomenon is sometimes called the infinite regress of causation. We further discuss the search for causes in Section 21.5.

### Thought experiment: what would be an ideal controlled experiment?

If you find yourself confused about what can be estimated and how the various aspects of your study should be defined, a simple strategy is to try to formalize a controlled experiment you would have liked to have performed to answer your causal question. A perfect mapping rarely exists between this experimental ideal and your data, so often you will be forced instead to figure out, given the data you have, what experiment could be thought to have generated such data.

For instance, if you were interested in the effect of breastfeeding on children's cognitive outcomes, what controlled experiment would you want to perform in the absence of any practical, legal, or moral barriers? We could imagine randomizing mothers to either breastfeed their children exclusively or bottle-feed them formula exclusively. We would have to consider how to handle those who do not adhere to their treatment assignment, such as mothers and children who are not able to breastfeed, and children who are allergic to standard formula. Moreover, what if we want to separately estimate the physiological effects of the breast milk from the potential psychological implications (to both mother and child) of nursing at the breast and the more extended physical contact that is often associated with breastfeeding? In essence, then, we can think of breastfeeding as representing several concurrent treatments. Perhaps we would want to create a third treatment group of mothers who feed their babies with bottles of expressed breast milk. This exercise of considering the controlled experiment helps to clarify what the true nature of the intervention is that we are using our treatment variable to represent.

Finally, thinking about hypothetical experiments can help with problems of trying to establish a causal link between two variables when neither has temporal priority and when they may have been simultaneously determined. For instance, consider a regression of crime rates in each of 50 states using a cross section of data, where the goal is to determine the effect of the number of police officers while adjusting for the social, demographic, and economic features of each state as well as characteristics of the state (such as the crime rate) that might affect decisions to increase the size of the police force. The problem is that it may be difficult (if not impossible) to disentangle the effect of the size of the police force on crime from the effect of the crime rate on the size of the police force.

If one is interested in figuring out policies that can affect crime rates, it might be more helpful to conceptualize both "number of police officers" and "crime rate" as outcome variables. Then one could imagine different treatments (policies) that could affect these outcomes. For example, the number of police officers could be affected by a bond issue to raise money for hiring new police, or a change in the retirement age, or a reallocation of resources within local and state law enforcement agencies. These different treatments could have different effects on the crime rate.

## 20.10 Bibliographic note

Modeling strategies exist that rely on ignorability but loosen the relatively strict functional form imposed by linear regression. Early examples include Hahn (1998) and Heckman, Ichimura, and Todd (1998). More recent approaches using machine learning include Hill (2011), Wager and Athey (2018), and Hahn, Murray, and Carvalho (2019), and Carnegie, Dorie, and Hill (2019). In addition to potentially correcting for biases arising from imbalance, these nonlinear models include the possibility of deep interactions and thus open the door to more serious consideration of varying treatment effects. Recent causal inference data analysis challenges have attempted to construct testing grounds to compare the performance of nonparametric causal models, as compared to more traditional propensity score approaches; see Dorie et al. (2019).

Imbalance and lack of complete overlap have been discussed in many places; an early reference is Cochran and Rubin (1973). The intervention for low-birth-weight children is described by Brooks-Gunn, Liaw, and Klebanov (1992) and Hill, Brooks-Gunn, and Waldfogel (2003). More recent advances in discussions of overlap can be found in Crump et al. (2009); Li, Thomas, and Li (2019); and Hill and Su (2013). The last of these publications distinguishes between overlap in all covariates and overlap in the confounders; the latter is called *common causal support*.

Cochran (1968) discusses subclassification and its connection to regression. The local average treatment effect was introduced by Imbens and Angrist (1994). Cochran and Rubin (1973), Rubin (1973), Rubin (1979), Rubin and Thomas (2000), Rubin (2006), and Abadie and Imbens (2011) discuss the use of matching followed by regression for causal inference. Dehejia (2003) discusses an example of the interpretation of a treatment effect with interactions.

Propensity scores were introduced by Rosenbaum and Rubin (1983a, 1984, 1985). Examples across several fields include Lavori, Keller, and Endicott (1995), Lechner (1999), Hill, Waldfogel, and Brooks-Gunn (2002), Vikram et al. (2003), and O'Keefe (2004). Rosenbaum (1989) and Hansen (2004) discuss full matching. Diamond and Sekhon (2013) present an iterative matching algorithm that has analogies to genetic adaptation. Drake (1993) discusses robustness of treatment effect estimates to misspecification of the propensity score model. Joffe and Rosenbaum (1999), Imbens (2000), and Imai and van Dyk (2004) generalize the propensity score beyond binary treatments. Rubin and Stuart (2005) extends to matching with multiple control groups. Stuart (2010) reviews matching methods, and Imbens (2004) provides a review of methods for estimating causal effects assuming ignorability using matching and other approaches.

Use of propensity scores as weights is discussed by Rosenbaum (1987), Hirano, Imbens, and Ridder (2003), and Frolich (2004), among others. This work has been extended to a "doubly robust" framework by Robins and Rotnitzky (1995), Robins, Rotnitzsky, and Zhao (1995), and Robins and Ritov (1997). Interval estimation for treatment effect estimates obtained via propensity score matching is discussed by Hill and Reiter (2006).

As far as we are aware, LaLonde (1986) was the first to use a constructed observational study as a testing ground for nonexperimental methods. Other examples include Friedlander and Robins (1995); Heckman, Ichimura, and Todd (1997); Dehejia and Wahba (1999); Michalopoulos, Bloom, and Hill (2004); and Agodini and Dynarski (2004). Dehejia (2005a, b), in response to Smith and Todd (2005), provides useful guidance regarding appropriate uses of propensity scores (the need to think hard about ignorability and to specify propensity score models that are specific to any given dataset). The constructed observational analysis presented in this chapter is based on a more complete analysis presented in Hill, Reiter, and Zanutto (2004). Shadish, Clark, and Steiner (2008) present an extension of this type of evaluation strategy that addresses some of the limitations of typical constructed observational studies.

Wainer, Palmer, and Bradlow (1998) provide a friendly introduction to selection bias. Heckman (1979) and Diggle and Kenward (1994) are influential works on selection models in econometrics and biostatistics, respectively.

Rosenbaum and Rubin (1983b), Rosenbaum (2002a), and Greenland (2005) consider sensitivity of inferences to ignorability assumptions; see also the `rbounds` package in R (Keele, 2015). Carnegie et al. (2016) and Dorie et al. (2016) present more recent sensitivity analysis work that doesn't rely on matching. Scott et al. (2018) extend this framework to the multilevel setting. Ding and VanderWeele (2016) present a bounding approach that relaxes some of the assumptions of previous methods.

## 20.11 Exercises

20.1 *Omitted variable bias*: Work out the formulas for omitted variable bias on page 385 in the context of an applied problem of interest to you, setting reasonable values for all the coefficients.

xample: aLonde tudy

20.2 *Constructed observational studies*: The folder `Lalonde` contains data from an observational study constructed by LaLonde (1986) based on a randomized experiment that evaluated the effect on earnings of a job training program called National Supported Work. The constructed observational study was formed by replacing the randomized control group with a comparison group formed using data from two national public-use surveys: the Current Population Survey (CPS) and the Panel Study of Income Dynamics.

Dehejia and Wahba (1999) used a subsample of these data to evaluate the potential efficacy of propensity score matching. The subsample they chose removes men for whom only one pre-treatment measure of earnings is observed. There is evidence in the economics literature that adjusting for earnings from only one pre-treatment period is insufficient to give a good approximation to ignorability. This exercise replicates some of Dehejia and Wahba's findings based on the CPS comparison group.

(a) Estimate the treatment effect from the experimental data in two ways: (i) a simple difference in means between treated and control units, and (ii) a regression-adjusted estimate (that is, a regression of outcomes on the treatment indicator as well as predictors corresponding to the pre-treatment characteristics measured in the study).

(b) Now use a regression analysis to estimate the causal effect from Dehejia and Wahba's subset of the constructed observational study. Examine the sensitivity of the model to model specification, for instance, by excluding the employed indicator variables or by including interactions. How close are these estimates to the experimental benchmark?

(c) Now estimate the causal effect from the Dehejia and Wahba subset using propensity score matching. Do this by first trying several different specifications for the propensity score model and choosing the one that you judge to yield the best balance on the most important covariates.

Perform this propensity score modeling *without* looking at the estimated treatment effect that would arise from each of the resulting matching procedures.

For the matched dataset you construct using your preferred model, report the estimated treatment effects using the difference-in-means and regression-adjusted methods described in part (a) of this exercise. How close are these estimates to the experimental benchmark (about $1800)?

(d) Assuming that the estimates from (b) and (c) can be interpreted causally, what causal effect does each estimate? What populations are we making inferences about for each of these estimates?

(e) Redo both the regression and the matching exercises, excluding the variable for earnings in 1974, which is two time periods before the start of this study. How important does the earnings-in-1974 variable appear to be in terms of satisfying the ignorability assumption?

20.3 *Understanding the difference between average treatment effect in the treated and control groups*: Create a hypothetical dataset in which the average treatment effect on the treated and controls (ATT and ATC) are clearly different. What are the distinguishing characteristics of this dataset?

20.4 *Exploring the properties of observational data under a linear model*: The goal of this exercise is to learn how to simulate a few different types of observational causal structures and evaluate the properties of different approaches to estimating the treatment effect through linear regression.

(a) Simulate data from 1000 people. First we want to simulate the underlying data from the joint distribution of a continuous pre-treatment predictor or covariate $x$, a binary treatment variable $z$, and continuous potential outcomes $y^0, y^1$. The data-generating process is $p(x, z, y^0, y^1) = p(x)p(z|x)p(y^0, y^1|z, x)$.

i. Start by simulating the values of $x$ from a normal distribution with mean 0 and standard deviation 1.

ii. What role does $x$ play in the data-generating process?

iii. The next step is to simulate $z$ from $p(z|x) = \text{binomial}(p)$, where the vector of probabilities can vary across observations. Come up with a strategy for generating the vector $z$ conditional on $x$ that forces you to create be explicit about how these probabilities are conditional on $x$. An inverse logit function is one approach but there are others. Make sure that $x$ is strongly associated with $z$ and that the vector of probabilities used to draw $z$ does not go below 0.05 or above 0.95 for the values of $x$ in the data.

iv. The last step is to simulate $y^0, y^1$ given $x$ and $z$. Construct a model for simulating the two potential outcomes with appropriate conditioning on $z$ and $x$ with the following stipulations:

- Make sure that $\text{E}(y^1|x) - \text{E}(y^0|x) = 5$.
- Make sure that $x$ has a linear and strong relationship with the outcome.
- Finally, set your error term to have a standard deviation of 1 and allow the residual standard deviation to be different for the same person across potential outcomes.

v. Create a data frame containing your simulated $x$, $z$, $y^0$, $y^1$, and save it for use later.
vi. Calculate the sample average treatment effect and save it for use later.
vii. Create a data frame with the observed data, $x$, $z$, $y$.

(b) Play the role of the applied researcher. Pretend someone handed you the observed data generated above and asked you to estimate the treatment effect. You will try two approaches: difference in means and regression.

i. Estimate the treatment effect using a difference in mean outcomes across treatment groups.
ii. Estimate the treatment effect using a regression of the outcome on the treatment indicator and covariate.
iii. Create a scatterplot of observed $y$ vs. $x$ using red dots for treatment and blue for control observations. If you were the researcher, would you be comfortable using linear regression with no interaction in this setting?

(c) Play the role of the statistician studying the properties of estimates.

i. Create a scatterplot of both potential outcomes vs. $x$, using red dots for values of $y^1$ and blue dots for $y^0$. Does linear regression with no interaction seem like a reasonable model to estimate causal effects for the observed data? Why or why not?
ii. Find the bias of each of the estimates calculated by the researcher in (b) relative to the sample average treatment effect computed using all the potential outcomes.
iii. Think harder about the practical importance of the bias by dividing this estimate by the standard deviation of the observed outcome $y$.
iv. Find the bias of each of the estimates by creating a randomization distribution for each. When creating randomization distributions, remember to be careful to keep the original sample the same, only varying treatment assignment and the observed outcome.

20.5 *Simulating observational data under a nonlinear model*: Now we'll explore what happens when we fit a wrong model in an observational study in a simple problem with a single pre-treatment predictor.

(a) Simulate the data. Create an R function called `sim_nlin()` that does the following:

i. The predictor $x$ should be drawn from a uniform distribution between 0 and 2.
ii. Treatment assignment should be drawn from the following binomial where $\mathrm{E}(z|x) = p = \mathrm{logit}^{-1}(-2 + x^2)$. Save the $p$ vector for use later.
iii. The response surface (model for $y^0$, $y^1$) should be drawn from the following distributions: $y^0 = x + \epsilon_0$, $y^1 = 2x + 3x^2 + \epsilon_1$, where the error terms are independent and normally distributed, each with mean 0 and standard deviation 1.
iv. Make sure the returned dataset has a column for the probability of treatment assignment as well.
v. Simulate a dataset called `data_nlin` with sample size 1000.

(b) Make the following plots:

i. Overlaid histograms of the probability of assignment.
ii. A scatterplot of observed $y$ vs. $x$ using red dots for treatment and blue for control observations.
iii. A scatterplot of both potential outcomes vs. $x$, using red dots for values of $y^1$ and blue dots for $y^0$. Does linear regression with no interactions seem like a reasonable model to estimate causal effects for the observed data? Why or why not?

(c) Create randomization distributions to investigate the properties of each of three estimates with respect to the sample average treatment effect: (1) difference in means, (2) linear regression of the outcome on the treatment indicator and $x$, (3) linear regression of the outcome on the treatment indicator, $x$, and $x^2$.

(d) Calculate the standardized bias (bias divided by the standard deviation of $y$) of these estimates relative to the sample average treatment effect.

20.6 *Observational data with multiple covariates*:

(a) Simulate from $p(x_1, x_2, x_3, z, y^0, y^1) = p(x_1, x_2, x_3)p(z|x_1, x_2, x_3)p(y^0, y^1|z, x_1, x_2, x_3)$.

i. Simulate the predictors $x_1, x_2, x_3$ to be independent of each other.

ii. Make sure that the probability of being treated is between 0.05 and 0.95 for each person and that there is a reasonable amount of overlap across the treatment and control groups.

iii. Generate the response surface as in the following:

$$y^0 = x_1 + x_2 + x_3 + \epsilon_0$$
$$y^1 = x_1 + x_2 + x_3 + 5 + \epsilon_1,$$

with independent errors normally distributed with mean 0 and standard deviation 1.

(b) Create randomization distributions for (1) a regression that adjusts for only one of the three covariates and (2) a regression estimate that adjusts for all three covariates. Evaluate the standardized bias of these estimates relative to the sample average treatment effect (SATE).

(c) Repeat the above but $x_1$, $x_2$, and $x_3$ generated with a dependence structure.

(d) Repeat the above but with a nonlinear response surface.

20.7 *Propensity score matching*: Perform your own propensity score analysis using the IHDP childcare data in the folder `Childcare` discussed in Section 20.5. Follow all of the steps and report back in your findings.

20.8 *Weighted estimate of treatment effect*: Perform an inverse estimated probability of treatment weighting (IPTW) analysis for the IHPD problem. Make sure to check balance and overlap.

20.9 *Propensity score matching*: Decide on a balance metric for the covariates in the IHPD problem. Try out a variety of propensity score approaches, varying the propensity score models and method for restructuring, including weighting by inverse estimated probability of treatment. Find at least eight approaches that satisfy your balance criteria. Comment on the variation in your estimates.

20.10 *Inverse estimated probability of treatment weighting and other estimates*: Figure 20.15 displays hypothetical data from an observational study with one confounder.

(a) Determine the weights (normed and unnormed) for estimating the average treatment effect for the treated units (ATT) and the sample size in the ATT pseudo-population.

(b) Check for balance in the covariate and potential outcomes in the reweighted sample.

(c) Calculate the ATT in this example using the potential outcomes.

(d) Estimate the ATT in this example using the reweighted data.

(e) Show the equivalency between weighting for ATT in this example and the stratification weights discussion in Section 20.6.

(f) Repeat all of the above for the average treatment effect for the control units (ATC).

20.11 *Observational studies and hypothetical experiments*: Consider an applied problem of interest to you with a causal effect that has been estimated using an observational study. Think about possible hypothetical experiments that could be performed to estimate different aspects of this causal effect. Consider how the effect might vary across the population and across different implementations of the treatment.

20.12 *Working through your own example*: Continuing the example from the final exercises of the earlier chapters, consider a treatment effect that can only be estimated using an observational study. Using your data, assess issues of imbalance and lack of overlap. Use matching if necessary to get comparable treatment and control groups, then perform a regression analysis adjusting for matching variables or the propensity score and estimate the treatment effect. Graph the data and fitted model and assess your assumptions.

# Chapter 21

# Additional topics in causal inference

The previous chapters described causal inference strategies that assume ignorability of exposure or treatment assignment. It is reasonable to be concerned about this assumption, however. After all, when are we really confident that we have measured *all* confounders? This chapter explores several alternative causal inference strategies that rely on slightly different sets of assumptions that may be more plausible in certain settings. We also discuss the connection between statistical causal inference (estimates of the effects of specified treatments or exposures) and causal explorations or searches for causes of patterns in observed data.

## 21.1 Estimating causal effects indirectly using instrumental variables

In some situations when the argument for ignorability of the treatment assignment seems weak, there may exist another variable that does appear to be randomly assigned or can be considered as such. If this variable, called the *instrument*, $z$, is predictive of the treatment, $T$, then we *may* be able to use it to isolate a particular kind of targeted causal estimand. The instrument should only affect the treatment assignment but not have a direct effect on the outcome, an *exclusion restriction* that we explain more precisely below.

### Example: a randomized-encouragement design

ample: same eet periment

Suppose we want to estimate the effect of watching the educational television program Sesame Street on recognition of letters in the alphabet. We might consider implementing a controlled experiment where the participants are preschool children, the treatment of interest is watching Sesame Street, the control condition is not watching, and the outcome is the score on a test of letter recognition.

However, it is not so easy for an experimenter to force children to watch a TV show or to refrain from watching a show that is currently being broadcast. Thus *watching* could not be randomized. Instead, when this study was actually performed, what was randomized was *encouragement* to watch the show—this is called a *randomized encouragement design*.

A simple comparison of outcomes across randomized groups in this study will yield an estimate of the effect of *encouraging* these children to watch the show, not an estimate of the effect of actually viewing the show. The former estimand is often referred to as the *intent-to-treat* (ITT) effect. However, we may be able to take advantage of the randomization to estimate the effect of watching for at least some of the people in the study by using the randomized encouragement as an *instrument* that induces variation in the treatment variable of interest.

### Compliance as an intermediate potential outcome

The researchers in the Sesame Street study recorded four viewing categories: (1) rarely watched, (2) watched once or twice a week, (3) watched 3–5 times a week, and (4) watched more than 5 times a week on average. Since there is not a category for "never watched," for the purposes of this illustration we treat the lowest viewing category ("rarely watched") as if it were equivalent to "never watched."[1]

We begin by describing a framework that will help us to understand the estimand targeted by this approach as well as the required assumptions. In the Sesame Street example, this requires considering the children's compliance behavior: the extent to which their viewing behavior matched what they were encouraged to do.

A critical feature of this type of study is that only some of the children's viewing patterns were affected by the encouragement. Those children whose viewing patterns could be altered by encouragement are the only participants in the study for whom the design provides information on counterfactuals with regard to viewing behavior since under different experimental conditions they might have been observed either viewing or not viewing. Therefore, a comparison of the potential outcomes for these children makes sense. Following the conventions of the statistics literature on instrumental variables, we shall label these children *compliers*. These are the only children for whom we will make inferences about the effect of watching Sesame Street and the local average treatment effect for them is referred to as the *complier average causal effect* (CACE).

What other types of children exist in the study? We know that there were children who were encouraged to watch but did not; we might plausibly assume that these children also would not have watched if not encouraged. A child who would not watch regardless of his assignment is labeled a *never-taker*. We cannot directly estimate the effect of viewing for these children, since in this context they would never be observed watching the show. Similarly, for the children who watched Sesame Street even though not encouraged, we might plausibly assume that if they had been encouraged they would have watched as well. Again, these children cannot shed light on the effect of viewing, since all of them are regular viewers of the program: there is no direct way of comparing outcomes between viewing conditions. Those children who would watch whether encouraged or not, are labeled as *always-takers*. The assumptions made above about profiles of the never-takers and always-takers ruled out children who would always do the opposite of what they were told, so-called *defiers*; this assumption will be formalized below.

### Assumptions for instrumental variables estimation

Instrumental variable analyses rely on several key assumptions, one combination of which we discuss in this section in the context of a simple example with binary treatment and binary instrument:

- Ignorability of the instrument.
- Monotonicity.
- Nonzero association between instrument and treatment variable.
- Exclusion restriction.

We discuss each of these assumptions below. The model additionally assumes no interference between units (the stable unit treatment value assumption) as with most other causal analyses, an issue we have already discussed at the end of Section 19.2.

**Ignorability of the instrument.** The first assumption in the list above is *ignorability of the instrument* with respect to the potential outcomes (both for the primary outcome of interest and the treatment variable). Formally, we can write this as $y^0, y^1 \perp z$. This is trivially satisfied in a randomized experiment (assuming the randomization was pristine), assuming, as always, that any design features are reflected in the analysis (for example, block indicators would need to be included in the analysis of data arising from a randomized block design). In the absence of a controlled experiment or natural experiment, this property may be more difficult to satisfy and often requires conditioning on other pre-treatment variables (potential confounders). To the extent that this assumption is difficult to justify, any advantage over a traditional observational study (relying on ignorability of the treatment assignment mechanism) is generally lost.

[1]Data and code for this example are in the folder Sesame.

**Monotonicity.** In defining never-takers and always-takers, we assumed that there were no children who would watch if they were not encouraged but who would *not* watch if they *were* encouraged; that is, we assumed that there were no defiers. Formally this is called the *monotonicity assumption*, and it will not necessarily hold in practice, though there are many situations in which it is defensible. In particular, in some studies it would be impossible for study participants in the non-encouraged group to gain access to the treatment of interest. In those situations, there are no defiers or always-takers.

**Nonzero association between instrument and treatment variable.** To demonstrate how we can use the instrument to obtain a causal estimate of the treatment effect in our example, first consider that about 90% of those encouraged watched the show regularly; by comparison, only 55% of those not encouraged watched the show regularly. Therefore, if we are interested in the effect of actually viewing the show, we should focus on the 35% of the treatment population who decided to watch the show because they were encouraged but who otherwise would not have watched the show. If the instrument (encouragement) did not affect regular watching, then we could not proceed. Fortunately, nonzero association between the instrument and the treatment is an assumption that can be directly checked from the data.

**Exclusion restriction.** To estimate the effect of viewing for those children whose viewing behavior would have been affected by the encouragement (the induced watchers), we must make another important assumption, called the *exclusion restriction*. This assumption says for those children whose behavior would not have been changed by the encouragement (never-takers and always-takers) there is no effect of encouragement on outcomes. So for the never-takers (children who would not have watched either way), for instance, we assume encouragement to watch did not affect their outcomes. And for the always-takers (children who would have watched either way), we assume encouragement to watch did not affect their outcomes. Technically, the assumptions regarding always-takers and never-takers represent distinct exclusion restrictions. In this simple framework, however, the analysis suffers if either assumption is violated. When using more complicated estimation strategies, it can be helpful to consider these assumptions separately, as it may be possible to weaken one or or both.

It is not difficult to tell a story that violates the exclusion restriction. Consider, for instance, the conscientious parents who do not let their children watch television and are concerned with providing their children with a good start educationally. The materials used to encourage them to have their children watch Sesame Street for its educational benefits might instead have motivated them to purchase other types of educational materials for their children or to read to them more often.

To illustrate the instrumental variables approach, however, we proceed as if the exclusion restriction were true (or at least approximately true). In this case, if we think about individual-level effects, the answer becomes relatively straightforward.

### Derivation of instrumental variables estimation with complete data (including unobserved potential outcomes)

Figure 21.1 illustrates with hypothetical data displaying for each study participant not only the observed data (encouragement and viewing status as well as observed outcome test score) but also the unobserved categorization, $c_i$, into always-taker, never-taker, or complier, based on potential watching behavior as well as the counterfactual test outcomes (the potential outcome corresponding to the treatment not received). Here, potential outcomes are the outcomes we would have observed under either *encouragement* option. Because of the exclusion restriction, for always-takers and never-takers the potential outcomes are unaffected by the encouragement (really they need not be *exactly* the same, just distributionally the same, but this simplifies the exposition).

The true intent-to-treat effect for these 20 observations is then an average of the effects for the 8 induced watchers, along with 12 zeroes corresponding to the encouragement effects for the always-takers and never-takers:

| Unit, $i$ | Potential viewing outcomes, $T_i^0$ | $T_i^1$ | Compliance indicator, $c_i$ | Encouragement indicator, $z_i$ | Potential test outcomes, $y_i^0$ | $y_i^1$ | Encouragement effect, $y_i^1 - y_i^0$ |
|---|---|---|---|---|---|---|---|
| 1 | **0** | 1 | (complier) | 0 | **67** | 76 | 9 |
| 2 | **0** | 1 | (complier) | 0 | **72** | 80 | 8 |
| 3 | **0** | 1 | (complier) | 0 | **74** | 81 | 7 |
| 4 | **0** | 1 | (complier) | 0 | **68** | 78 | 10 |
| 5 | **0** | 0 | (never-taker) | 0 | **68** | 68 | 0 |
| 6 | **0** | 0 | (never-taker) | 0 | **70** | 70 | 0 |
| 7 | **1** | 1 | (always-taker) | 0 | **76** | 76 | 0 |
| 8 | **1** | 1 | (always-taker) | 0 | **74** | 74 | 0 |
| 9 | **1** | 1 | (always-taker) | 0 | **80** | 80 | 0 |
| 10 | **1** | 1 | (always-taker) | 0 | **82** | 82 | 0 |
| 11 | 0 | **1** | (complier) | 1 | 67 | **76** | 9 |
| 12 | 0 | **1** | (complier) | 1 | 72 | **80** | 8 |
| 13 | 0 | **1** | (complier) | 1 | 74 | **81** | 7 |
| 14 | 0 | **1** | (complier) | 1 | 68 | **78** | 10 |
| 15 | 0 | **0** | (never-taker) | 1 | 68 | **68** | 0 |
| 16 | 0 | **0** | (never-taker) | 1 | 70 | **70** | 0 |
| 17 | 1 | **1** | (always-taker) | 1 | 76 | **76** | 0 |
| 18 | 1 | **1** | (always-taker) | 1 | 74 | **74** | 0 |
| 19 | 1 | **1** | (always-taker) | 1 | 80 | **80** | 0 |
| 20 | 1 | **1** | (always-taker) | 1 | 82 | **82** | 0 |

Figure 21.1 *Hypothetical complete data in a randomized encouragement design. Units have been ordered for convenience. For each unit, the students are encouraged to watch Sesame Street ($z_i = 1$) or not ($z_i = 0$). This reveals which of the potential viewing outcomes ($T_i^0, T_i^1$) and which of the potential test outcomes ($y_i^0, y_i^1$) we get to observe. The observed outcomes are displayed in boldface. Here, potential outcomes are what we would observe under either encouragement option. The exclusion restriction forces the potential outcomes to be the same for those whose viewing would not be affected by the encouragement. The effect of watching for the "compliers" is equivalent to the intent-to-treat effect (encouragement effect over the whole sample) divided by the proportion induced to view, 3.4/0.4 = 8.5. Researchers cannot calculate this effect directly because they cannot see both potential outcomes. Moreover, true compliance class is known for only some individuals and even then only under the monotonicity assumption.*

$$
\begin{aligned}
\text{ITT} &= \frac{9+8+7+10+9+8+7+10+0+\cdots+0}{20} \\
&= 8.5 * \frac{8}{20} + 0 * \frac{12}{20} \\
&= 8.5 * 0.4 \\
&= 3.4.
\end{aligned}
$$

The effect of watching Sesame Street for the compliers is 8.5 points on the letter recognition test. This is equal to the intent-to-treat effect estimate of 3.4 divided by the proportion of compliers, 8/20 = 0.40.

## Deconstructing the complier average causal effect

To better understand the implications of the assumptions we have made to identify this effect, let us back up for a minute and consider a more general formulation. We start by conceptualizing the intent-to-treat effect as a weighted average of four different ITT effects—one for each of the four types of compliance classifications. Informally we can write,

$$\text{ITT} = \text{ITT}_{c=\text{complier}}\Pr(c = \text{complier}) + \text{ITT}_{c=\text{never-taker}}\Pr(c = \text{never-taker}) + \\ + \text{ITT}_{c=\text{always-taker}}\Pr(c = \text{always-taker}) + \text{ITT}_{c=\text{defier}}\Pr(c = \text{defier}),$$

where $\text{ITT}_{c=\text{complier}}$ denotes the effect of the *instrument* on the outcomes for the compliers, and the other ITT effects in the expression are similarly defined.

The exclusion restriction sets $\text{ITT}_{c=\text{never-taker}}$ and $\text{ITT}_{c=\text{always-taker}}$ to 0 and the monotonicity assumption sets $\Pr(c = \text{defier})$ to 0, simplifying the expression to,

$$\text{ITT} = \text{ITT}_{c=\text{complier}}\Pr(c = \text{complier}),$$

which is easily rearranged to look like our complier average causal effect estimand,

$$\text{ITT}_{c=\text{complier}} = \text{CACE} = \frac{\text{ITT}}{\Pr(c = \text{complier})} = \frac{\text{ITT}}{\text{E}(T(z=1) - T(z=0))}, \tag{21.1}$$

where the expression in the denominator can be conceived of as the ITT effect of the instrument on the treatment variable, $T$. This expression provides another way of understanding the assumption that the instrument must have an effect on the treatment; that is, the denominator of the CACE estimand cannot be zero.

**Violations of ignorability.** The ignorability assumption is arguably the most essential to the instrumental variables framework presented here because it is required to unbiasedly estimate both the numerator and denominator of the estimand in (21.1). Violations of this assumption could lead to either positive or negative bias. This bias will be exacerbated in situations when the estimated proportion of compliers is small because dividing the biased ITT estimate by a small number that is less than 1 is equivalent to multiplying it by a big number.

**Violations of the exclusion restriction.** What happens when the exclusion restriction is violated? Consider a scenario in which the effect of the instrument on the never-takers is $\alpha$. In this case, our expanded ITT formula reduces to,

$$\text{ITT} = \text{CACE} * \Pr(c = \text{complier}) + \alpha * \Pr(c = \text{never-taker}),$$

and straightforward algebraic manipulations reveal,

$$\frac{\text{ITT}}{\Pr(c = \text{complier})} = \text{CACE} + \alpha\frac{\Pr(c = \text{never-taker})}{\Pr(c = \text{complier})},$$

with the term on the far right representing the bias. This bias increases with the size of the effect of the instrument on the never-takers, $\alpha$. Also, this effect is either shrunk or magnified depending on the ratio of never-takers to compliers. Whenever the percentage of never-takers is greater than the percentage of compliers, the bias captured in the never-taker effect, $\alpha$, will be amplified. This suggests that *weak instruments*—those that are not strongly predictive of the treatment—will be highly vulnerable to violations of the exclusion restriction.

**Violations of the monotonicity assumption.** If the monotonicity assumption is violated, then $\Pr(c = \text{defier}) \neq 0$ and consequently the equivalence between $\Pr(c = \text{complier})$ and $\text{E}(T(1) - T(0))$ is lost. In this case, the calculations to derive the bias are slightly more complicated, so we omit the steps here; we present the result that the bias in this case looks like,

$$\text{bias} = \delta * \left(\text{ITT}_{c=\text{complier}} + \text{ITT}_{c=\text{defier}}\right),$$

where

$$\delta = \frac{\Pr(c = \text{defier})}{\Pr(c = \text{complier}) - \Pr(c = \text{defier})}.$$

Somewhat oddly, this bias is driven by the *difference* between the intent-to-treat effects for the compliers and defiers. The ITT effect for a defier represents $y(T = 0) - y(T = 1)$, since defiers, by definition, are the people who choose the opposite of their treatment assignment; whereas for compliers, who always do what they are told, the ITT effect represents $y(T = 1) - y(T = 0)$. Thus, we can reframe this bias cancellation property by noting that the bias will disappear if the effect of the treatment (watching) on the outcome is the same for compliers and defiers. Any difference between the causal effect of the treatment on the outcomes will be magnified if the proportion of defiers is high relative to the proportion of compliers. Again, this points to the potential dangers of a weak instrument when there is a chance that an assumption will be violated.

### Local average treatment effect (LATE) versus intent-to-treat effect (ITT)

As we have discussed, the instrumental variables strategy here does not estimate an overall effect of watching Sesame Street across everyone in the study, or even an effect for all those treated. The complier average causal effect (CACE) estimate applies only to those children whose treatment receipt is dictated by their randomized instrument assignment and is a special case of what is commonly called a *local average treatment effect* (LATE) by economists.

Some researchers argue that intent-to-treat effects are more interesting from a policy perspective because they accurately reflect that not all targeted individuals will participate in the intended program. However, the intent-to-treat effect only parallels a true policy effect if in the subsequent policy implementation the compliance rate remains unchanged. We recommend estimating both the intent-to-treat effect and the complier average causal effect to maximize what we can learn about the intervention.

Compliers are a latent subpopulation, so it is not possible to directly intervene on them. Also, if the instrument is weak, so that $\mathrm{E}(T(z = 1) - T(z = 0))$ is small, then any statement about compliers is only directly relevant to a small fraction of the population. For this reason it can make sense to perform additional statistical modeling with the aim of inference for effects within other subgroups of the population.

### Instrumental variables estimate: Sesame Street

We can calculate an estimate of the effect of watching Sesame Street for the compliers with the actual data using the same principles.

We first estimate the percentage of children actually induced to watch Sesame Street by the intervention, which is the coefficient on the instrument (encouraged), in the following regression:

```
itt_zt <- stan_glm(watched ~ encouraged, data=sesame)
```

The estimated coefficient of encouraged here is 0.36 (which, in this regression with a single binary predictor, is simply the proportion of compliers in the data).

We then compute the intent-to-treat estimate, obtained in this case using the regression of the outcome (in the data, labeled postlet, that is, the *post*-treatment measurement of the *letter* recognition task) on the instrument:

```
itt_zy <- stan_glm(postlet ~ encouraged, data=sesame)
```

The estimated coefficient of encouraged in this regression is 2.9, which we then "inflate" by dividing by the percentage of children affected by the intervention:

```
wald_est <- coef(itt_zy)["encouraged"] / coef(itt_zt)["encouraged"]
```

The estimated effect of regularly viewing Sesame Street is thus $2.9/0.36 = 8$ points on the letter recognition test. This ratio is sometimes called a *Wald estimate*. We discuss the standard error of the estimate below.

## 21.2 Instrumental variables in a regression framework

Instrumental variables models and estimates can also be derived by focusing on the mean structure of the regression model, allowing us to more easily extend the basic concepts discussed in the previous section. A general instrumental variables model with (potentially continuous) instrument, $z$, and (potentially continuous) treatment, $T$, can be written as,

$$\begin{aligned} y_i &= \beta_0 + \beta_1 T_i + \epsilon_i, \\ T_i &= \gamma_0 + \gamma_1 z_i + \nu_i. \end{aligned} \tag{21.2}$$

The assumptions can now be expressed slightly differently. The first set of assumptions is that $z_i$ is uncorrelated with both $\epsilon_i$ and $\nu_i$; these translate informally into the ignorability assumption and exclusion restriction (expressed informally as, "the instrument only affects the outcome through its effect on the treatment"). The requirement that the correlation between $z_i$ and $T_i$ must be nonzero is explicit in both formulations. We next address how this framework identifies the effect of $T$ on $y$.

### Identifiability with instrumental variables

Generally speaking, *identifiability* refers to whether the data contain sufficient information for unique estimation of a given parameter or set of parameters in a particular model.

What if we did not impose the exclusion restriction for our basic model? The model (ignoring covariate information, and switching to mathematical notation for simplicity and generalizability) can be written as,

$$\begin{aligned} y &= \beta_0 + \beta_1 T + \beta_2 z + \text{error}, \\ T &= \gamma_0 + \gamma_1 z + \text{error}, \end{aligned} \tag{21.3}$$

where $y$ is the response variable, $z$ is the instrument, and $T$ is the treatment of interest. Our goal is to estimate $\beta_1$, the treatment effect. The difficulty is that $T$ has not been randomly assigned; it is observational and, in general, can be correlated with the error in the first equation. Thus we cannot simply estimate $\beta_1$ by fitting a regression of $y$ on $T$ and $z$.

However, as described in the previous section, we can estimate $\beta_1$ using instrumental variables. We derive the estimate here algebraically, in order to highlight the assumptions needed for identifiability.

Substituting the formula for $T$ into the formula for $y$ yields,

$$\begin{aligned} y &= \beta_0 + \beta_1 T + \beta_2 z + \text{error} \\ &= \beta_0 + \beta_1(\gamma_0 + \gamma_1 z) + \beta_2 z + \text{error} \\ &= (\beta_0 + \beta_1 \gamma_0) + (\beta_1 \gamma_1 + \beta_2) z + \text{error}. \end{aligned} \tag{21.4}$$

We now show how to estimate $\beta_1$, the effect of interest, using the slope of this regression, along with the regressions (21.3) and the exclusion restriction.

The first step is to express (21.4) in the form,

$$y = \delta_0 + \delta_1 z + \text{error}.$$

From this equation we need $\delta_1$, which can be estimated from a simple regression of $y$ on $z$. We can now solve for $\beta_1$ in the following equation:

$$\delta_1 = \beta_1 \gamma_1 + \beta_2,$$

which we can rearrange to get,

$$\beta_1 = (\delta_1 - \beta_2)/\gamma_1. \tag{21.5}$$

We can directly estimate the denominator of this expression, $\gamma_1$, from the regression of $T$ on $z$ in (21.3)—this is not a problem since we are assuming that the instrument, $z$, is randomized.

The only challenge that remains in estimating $\beta_1$ from (21.5) is to estimate $\beta_2$, which in general cannot simply be estimated from the top equation of (21.3) since, as already noted, the error in that equation can be correlated with $T$. However, under the exclusion restriction, we know that $\beta_2$ is zero, and so $\beta_1 = \delta_1/\gamma_1$, leaving us with the standard instrumental variables estimate.

**Other models.** There are other ways to achieve identifiability in this two-equation setting. Approaches such as selection correction models rely on functional form specifications to identify the causal effects even in the absence of an instrument. For example, a probit specification could be used for the regression of $T$ on $z$. The resulting estimates of treatment effects are often unstable if a true instrument is not included as well.

### Two-stage least squares: Sesame Street

The Wald estimate discussed in the previous section can be used with this formulation of the model as well. We now describe a more general estimation strategy, *two-stage least squares* (TSLS).

Example: Sesame Street experiment

To illustrate we return to our Sesame Street example. The first step is to regress the "treatment" variable—an indicator for regular watching (watched)—on the randomized instrument, encouragement to watch (encouraged). This is the same model as our ITT estimate of the effect of encouragement on watching above but we run again here and relabel to indicate that in this framework it is our first-stage model. Then we plug predicted values of watched into the equation predicting the letter recognition outcome, y:

```
fit_2a <- stan_glm(watched ~ encouraged, data=sesame)
sesame$watched_hat <- fit_2a$fitted
fit_2b <- stan_glm(postlet ~ watched_hat, data=sesame)
```

The result is,

```
            Median MAD_SD
(Intercept) 20.7    3.8
watched_hat  7.8    4.9

Auxiliary parameter(s):
      Median MAD_SD
sigma 13.4    0.6
```

Here the coefficient on watched_hat is the effect of watching Sesame Street on letter recognition for those who would watch if encouraged but not otherwise (compliers). This two-stage estimation strategy is especially useful for more complicated versions of the model, for instance, when multiple instruments are included.

This second-stage regression does not give the correct standard error, however, as we discuss below.

### Adjusting for covariates in an instrumental variables framework

It turns out that the randomization for this particular experiment took place within sites and settings; it is therefore appropriate to adjust for these covariates in estimating the treatment effect. Additionally, pre-test scores are available that are highly predictive of post-test scores. Our preferred model would adjust for all of these predictors. We can calculate the same ratio (intent-to-treat effect divided by effect of encouragement on viewing) as before using models that include these additional predictors but pulling out only the coefficients on encouraged for the ratio.

Here we equivalently perform this analysis using two-stage least squares:

```
fit_3a <- stan_glm(watched ~ encouraged + prelet + as.factor(site) + setting,
  data=sesame)
watched_hat_3 <- fit_3a$fitted
```

```
fit_3b <- stan_glm(postlet ~ watched_hat_3 + prelet + as.factor(site) + setting,
  data=sesame)
```

yielding,

```
                   Median MAD_SD
(Intercept)         1.3    4.7
watched_hat_3      14.1    3.9
prelet              0.7    0.1
as.factor(site)2  8.3    1.9
as.factor(site)3 -4.0    1.8
as.factor(site)4  0.9    2.5
as.factor(site)5  2.6    2.8
setting             1.6    1.5

Auxiliary parameter(s):
      Median MAD_SD
sigma 9.7    0.5
```

The estimated effect of watching Sesame Street on the compliers is about 14 points on the letter recognition test. Again, we do not trust this standard error and will discuss later how to appropriately adjust it for the two stages of estimation.

Since the randomization took place within each combination of site (five categories) and setting (two categories), it would be appropriate to interact these variables in our equations. Moreover, it would probably be interesting to estimate variation of effects across sites and settings. However, for simplicity of illustration (and also due to the complication that one site × setting combination has no observations) we only include main effects for this discussion.

### Standard errors for instrumental variables estimates

The second step of two-stage regression yields the instrumental variables estimate, but the standard-error calculation is complicated because we cannot simply look at the second regression in isolation. We show here how to adjust the standard error to account for the uncertainty in both stages of the model. We illustrate with the model we have just fitted.

The regression of compliance on treatment and other covariates (model fit_3b) is unchanged. We next compute the standard deviation of the adjusted residuals, $r_i^{\text{adj}} = y_i - X_i^{\text{adj}}\hat{\beta}$, where $X^{\text{adj}}$ is the predictor matrix from fit_3b but with the column of predicted treatment values replaced by observed treatment values:

```
X_adj <- X <- model.matrix(fit_3b)
X_adj[,"watched_hat_3"] <- sesame$watched
n <- nrow(X)
p <- ncol(X)
RMSE1 <- sqrt(sum((sesame$postlet - X %*% coef(fit_3b))^2)/(n-p))
RMSE2 <- sqrt(sum((sesame$postlet - X_adj %*% coef(fit_3b))^2)/(n-p))
```

Finally, we compute the adjusted standard error for the two-stage regression estimate by taking the standard error from fit_3b and scaling by the adjusted residual standard deviation, divided by the residual standard deviation from fit_3b itself:

```
se_adj <- summary(fit_3b)$coef["watched_hat_3",2] * RMSE1 / RMSE2
```

So the adjusted standard errors are calculated as the square roots of the diagonal elements of $(X^tX)^{-1}\hat{\sigma}^2_{\text{TSLS}}$ rather than $(X^tX)^{-1}\hat{\sigma}^2$, where $\hat{\sigma}$ is the residual standard deviation from fit_3b, and $\hat{\sigma}_{\text{TSLS}}$ is calculated using the residuals from a model predicting the outcome from watched using the two-stage least squares estimate of the coefficient, not the coefficient that would have been obtained in a least squares regression of the outcome on watched. In this example, the resulting standard error for our example, 4.1, is slightly larger than the unadjusted standard error of 3.9.

### Performing two-stage least squares automatically using brms

We have illustrated the key concepts in our instrumental variables discussion using basic R commands with which you were already familiar so that the steps were transparent. It is also possible to perform a Bayesian version of the two-stage least squares model directly using brms.

Here is how to estimate the effect of regularly watching Sesame Street on post-treatment letter recognition scores using encouragement as an instrument:

```
library("brms")
f1 <- bf(watched ~ encour)
f2 <- bf(postlet ~ watched)
IV_brm_a <- brm(f1 + f2, data=sesame)
print(IV_brm_a)
```

The resulting estimate is the coefficient postlet_watched, which is the instrumental variables estimate of the effect of watched on postlet; the estimate is 8.0 with standard error 4.7.

To incorporate other pre-treatment variables as controls, we must include them in both stages of the regression model; for example,

```
f1 <- bf(watched ~ encour + prelet + setting + factor(site))
f2 <- bf(postlet ~ watched + prelet + setting + factor(site))
IV_brm_b <- brm(f1 + f2, data=sesame)
```

The resulting estimate is the coefficient postlet_watched, which is the instrumental variables estimate of the effect of watched on postlet; the estimate is 14.1 with standard error 4.0.

### More than one treatment variable; more than one instrument

In the experiment discussed in Section 20.7, the children randomly assigned to the intervention group received several services ("treatments") that the children in the control group did not receive, most notably, access to high-quality child care and home visits from trained professionals. Children assigned to the intervention group did not make full use of these services. Simply conceptualized, some children participated in the child care while some did not, and some children received home visits while others did not. Can we use the randomization to treatment or control groups as an instrument for these two treatments? The answer is no.

Similar arguments as those used in Section 21.2 can be given to demonstrate that a single instrument cannot be used to identify more than one treatment variable. As a general rule, we need to use at least as many instruments as treatment variables in order for all the causal estimates to be identifiable. A model in which the number of instruments is equal to the number of treatment variables is called a just-identified model. A model in which the number of instruments exceeds the number of treatment variables is called an overidentified model. We discuss issues with overidentified models in the section below on weak instruments.

### Continuous treatment variables or instruments

When using two-stage least squares, the models we have discussed can easily be extended to accommodate continuous treatment variables and instruments, although at the cost of complicating the interpretation of the causal effects.

Researchers must be careful, however, in the context of binary instruments and continuous treatment variables. A binary instrument cannot without further assumptions identify a continuous treatment or dosage effect. If we map this back to a randomized experiment, the randomization assigns someone only to be encouraged or not. This is equivalent to a setting with many different treatments (one at each dosage level) but only one instrument; therefore effects for all these treatments are not identifiable without further assumptions. To identify such dosage effects without making further parametric assumptions, one would need to randomly assign encouragement levels corresponding to the different dosages or levels of participation.

### Have we really avoided the ignorability assumption?

We have motivated instrumental variables using the cleanest setting, within a controlled, randomized experiment. The drawback of illustrating instrumental variables using this example is that it de-emphasizes one of the most important assumptions of the instrumental variables model, *ignorability of the instrument*. In the context of a randomized experiment, this assumption should be trivially satisfied (assuming the randomization was pristine). However, in practice an instrumental variables strategy may also be reasonable in the context of a *natural experiment*, that is, an observational study context in which a "randomized" variable or instrument appears to have occurred naturally. Examples of this include:

- The weather in New York as an instrument for estimating the effect of supply of fish on their price;
- The sexes of the first two children in a family (in an analysis of people who have at least two children) as an instrument for estimating the effect of number of children on labor supply.

In these examples we have traded one ignorability assumption (ignorability of the treatment variable) for another (ignorability of the instrument) that we believe to be more plausible. Additionally, we must assume the exclusion restriction.

In the absence of an experiment or a natural experiment, instrumental variables strategies are more suspect. While we cannot confirm the existence of ignorability in the absence of planned or naturally occurring randomization, we may be able to rule it out by performing the types of balance checks we did in the propensity score matching context. Or we may believe that ignorability holds only conditional on observed covariates in which case we should adjust for these in our analysis. Broadly speaking, if the ignorability assumption is not highly plausible, the expected gains from performing an instrumental variables analysis are not likely to outweigh the potential for bias.

### Plausibility of exclusion restriction

One way to assess the plausibility of the exclusion restriction is to calculate an estimate within a sample that would not be expected to be affected by the instrument. For instance, researchers estimated the effect of military service on earnings (and other outcomes) using, as an instrument, the lottery number for young men eligible for the draft during the Vietnam War. This number was assigned randomly and strongly affected the probability of military service. It was hoped that the lottery number would only have an effect on earnings for those who served in the military only because they were drafted (as determined by a low enough lottery number). Satisfaction of the exclusion restriction is not certain, however, because, for instance, men with low lottery numbers may have altered their educational plans so as to avoid or postpone military service. So the researchers also ran their instrumental variables model for a sample of men who were assigned numbers so late that the war ended before they ever had to serve. This showed no clear relation between lottery number and earnings, which provides some support for the exclusion restriction.

### Weak instruments

Our assumptions require that the instrument have nonzero correlation with the treatment variable. This is the only assumption that can be directly tested. Thus the first-stage model (which predicts the treatment using the instrument) should be examined closely to assess the strength of the instrument. As discussed in Section 21.1, a weak instrument can exacerbate the bias that can result from failure to satisfy the ignorability or monotonicity assumptions or the exclusion restriction. If a weak instrument leads to a small proportion of compliers, this increases the potential for bias if one of these assumptions is violated.

In addition there is the concern that the compliers—the group about which the data are informative regarding causal effects—may not be particularly representative of the full population of interest. This is a form of sampling bias that is subtle because we cannot directly observe whether individuals are compliers; we can only use statistical methods to infer characteristics of compliers and noncompliers in the sample and the population.

It is also important to ensure that the association between the instrument and treatment is *positive*. If the association is not in the expected direction, this might be the result of a mixture of two different mechanisms, the expected process and one operating in the opposite direction, which could in turn imply a violation of the monotonicity assumption.

Finally, if the relationship between the instrument and the treatment variable is not sufficiently strong, two-stage least squares can produce biased estimates even if all of the instrumental variables assumptions hold.

### Structural equation models

A goal in many areas of social science is to infer causal relations among many variables, a generally difficult problem (as discussed in Section 19.7). *Structural equation modeling* is a family of methods of multivariate data analysis that are sometimes used for causal inference. The same statistical model is also used to estimate latent factors in noncausal regression settings with multiple inputs and sometimes multiple outcome variables.

In the causal setting, structural equation modeling relies on conditional independence assumptions in order to identify causal effects, and the resulting inferences can be sensitive to strong parametric assumptions such as linear relationships and multivariate normality of errors. Instrumental variables can be considered as a special case of a structural equation model where certain dependencies are set to be exactly zero. As we have just discussed, even in a relatively simple instrumental variables model, the assumptions needed to identify causal effects are difficult to satisfy and largely untestable. A structural equation model that tries to estimate many causal effects at once multiplies the number of assumptions required with each desired effect so that it quickly becomes difficult to justify all of them. Therefore, we do not discuss the use of structural equation models for causal inference in any greater detail here. We have no objection to complicated models, but we are cautious about attempting to estimate complex causal structures from observational data.

## 21.3 Regression discontinuity: known assignment mechanism but no overlap

Often the assumption of ignorability is not plausible; that is, it does not make sense to assume that treatment assignment depends only on observed pre-treatment predictors. However, we can design observational studies for which the assignment mechanism is entirely known (as with a controlled experiment) but for which no explicit randomization is involved.

Example: Hypothetical school tutoring experiment

As an example, suppose we want to evaluate the effect of an after-school tutoring program for fourth graders, but the school administrators want to prioritize the children who are struggling the most academically. One way to balance the desire for equity but also provide a reasonable design for evaluating the effectiveness of the program would be to assign children to receive the program or not based on their scores on a pre-test. For instance, if funding were available for 100 students to participate and 1000 students were enrolled in fourth grade in the district, the program administrators would simply choose the 100 students with the *lowest* scores on the pre-test to participate in the tutoring program. Those with higher scores would not be allowed access.

Suppose that the 100 lowest pre-test scores were all below 60 and the rest of the scores were above 60. In the purest form of this design, the assignment to the treatment, $z$, is completely deterministic, specifically: $\Pr(z = 1 \mid \text{pre-test} < 60) = 1$ and $\Pr(z = 1 \mid \text{pre-test} > 60) = 0$. Another way of framing this property is that the pre-test score is our only confounding covariate. This sounds like a randomized block experiment, doesn't it? A randomized block experiment with one blocking variable also has a known assignment mechanism with one confounding covariate. What's the difference?

In a randomized block experiment, observations are randomly assigned to treatment groups within subclasses defined by the blocking groups. Since the probability of assignment depends only on this variable, it is our only confounder. Within each block the potential outcomes will be balanced (in distribution) across treatment groups. If the probability of assignment to the treatment varies substantially across blocks, the blocking variable (our only confounder) will be imbalanced across treatment groups. However, there is a requirement with a randomized block design that the probability of assignment to treatment can never be 0 or 1. The implication of this design feature is that there will always be overlap across treatment groups with respect to this crucial confounder.

Contrast this scenario with that of the regression discontinuity design. By construction the pure form of this design leads to a situation in which there is *absolutely no overlap* across treatment groups with respect to our only confounder, the *assignment variable* or "forcing variable," the pre-test score; there must also be substantial imbalance in these distributions due to the lack of overlap. These features stand in stark contrast with the scenarios in the previous chapter in which we worked so hard to assure balance and overlap. With a pure regression discontinuity design there is no hope of achieving either balance or overlap. How then do we proceed to make causal inferences?

One way to conceptualize making progress is by considering the students whose scores lie just below and just above the cutoff of 60. For instance, if the standard deviation on the test were 15, would we really think that students who received scores of 57 were different in important ways than those who received scores of 63? More precisely, would we expect that students in this range of pre-test scores would have much different potential test score outcomes? If not, then maybe we could consider the data from students with scores in this range to be nearly equivalent to data that were generated by a controlled experiment.

In practice we might not trust this assumption too closely and would do well to adjust for our pre-test as well, and we would want to adjust for pre-test in any case to increase the efficiency of the estimate. Moreover we could restrict the estimand to apply just to the observations right at the cutoff. Although we don't have overlap, all we need to do is to built a sufficiently robust model on either side of the cutoff to get good predictions of the outcome if not treated ($y^0$) and the outcome if treated ($y^1$) at the threshold. If we limit our analysis to a narrow enough range of the data, a smooth model can make sense. At the same time, when we limit the range of pre-tests, we limit the information we can use to make inferences. We could use a broader window of scores, but the wider the range of pre-test scores considered, the more we will have to worry about the model extrapolations necessary to perform causal inference. So a tension exists between the range of values we are willing to consider for the assignment variable and the size of the analysis sample.

### Example: The effect of an educational program on test scores in Chile

xample: hile :hools xperiment

We illustrate some of the key features of regression discontinuity analysis using data that arose based on a policy decision. No one prospectively designed this study to address the research question at hand; rather, researchers retrospectively realized that the data generated from the policy mimics such a design. In the regression discontinuity framework, these data can be used to estimate the effect of an educational program in Chile on subsequent school-level test scores.[2]

The Chilean government introduced the "900 schools program" (P-900) in 1990 in an effort to increase the performance of struggling public schools by providing resources in four different areas. In 1990 and 1991 the focus was on infrastructure improvements and instructional materials. Starting in 1992 the focus shifted to teacher training and after-school tutoring. In an effort to target the neediest schools, the government provided these resources only to schools whose mean fourth-grade test scores (averaged both across students and across separate reading and math tests) were lower than a set cutoff.

There are three features of this program assignment that make this discontinuity design more complicated than the simple example presented above. First, the cutoff varied by region of the country,

[2]Data and code for this example are in the folder ChileSchools.

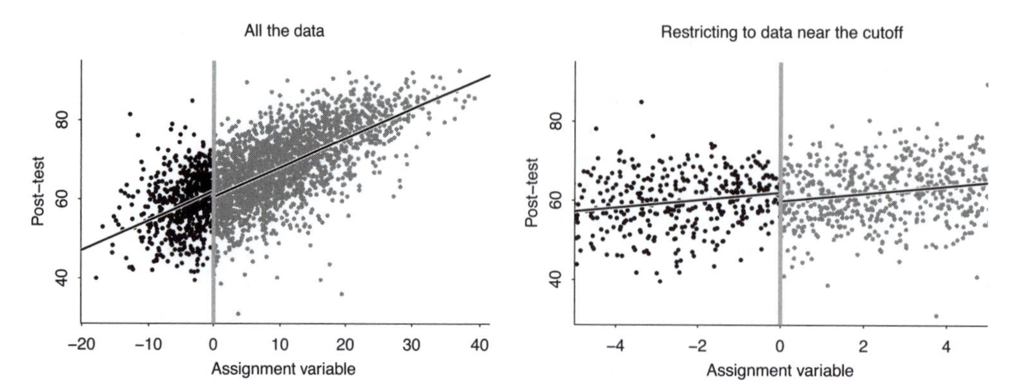


Figure 21.2 *Example of a regression discontinuity analysis: the intent-to-treat effect of exposure to P-900 on schools in Chile. (a) School-level data and regression lines predicting the outcome (reading test score in 1992) given the assignment variable. (b) Subset of data close to the cutoff, representing schools that had a reasonable chance of receiving or not receiving the assignment. For both graphs, black dots represent schools that were below the cutoff (so they received the intention to treat) and gray dots indicate schools above the cutoff. Lines show the fitted regression model in each case with a discontinuity at the threshold.*

so we have defined our assignment, or forcing, variable to be the positive or negative distance of each school's average score from the cutoff for its region.

Second, the precise cutoff was not recorded and must be deduced from the data. Here we use one of the cutoff rules introduced by the researchers who first analyzed these data using a regression discontinuity design.

The third challenge is that assignment rules do not seem to have been strictly followed: there are schools that did not receive the treatment while having average test scores that fall below the average test scores of the treated schools. In fact, while fewer than 2% of the schools above the derived cutoff received the program, only about 68% of those below the cutoff received it. This is an example of a "fuzzy" regression discontinuity. We will discuss further options for analyzing such designs later, but for now will just conceptualize the estimand as an intent-to-treat effect. That is, we estimate the effect of being on one side of the cutoff or the other, regardless of whether the P-900 resources were actually received. We shall refer to this characteristic as *eligibility*.

Figure 21.2a displays school-level data for the outcome (reading test score in 1992) and the assignment variable (the average of earlier test scores in the school, minus the threshold score for that school's region). By construction, the cutoff for the assignment variable is at zero for all schools. The graph also shows the fitted linear regression of outcome given the continuous assignment variable and an indicator for the discontinuity, which in this case equals 1 when the assignment variable is negative and 0 when it is positive, hence a positive discontinuity coefficient in this example.

Figure 21.2b restricts the data to a range within 5 points of the cutoff, with a regression fit to this subset of the data. Restricting the analysis to data near the cutoff reduces the imbalance between treatment and control groups.

**Basic regression discontinuity analysis.** At this point, the simplest model that adjusts for pre-treatment differences between treatment and control group is,

$$y_i = \beta_0 + \tau * \text{eligible}_i + \beta_1 * \text{rule2}_i + \text{error}_i, \tag{21.6}$$

where $y_i$ is the average reading test score in school $i$ in 1992, $\text{eligible}_i$ is the indicator for being below the cutoff ($z$ in our more general notation), and $\text{rule2}_i$ is the assignment variable. As shown in Figure 21.2, $\texttt{eligible}_i = 1$ when $\texttt{rule2}_i < 0$.

Here is the result of the regression with a subsetting condition set up to only include the schools that were within 5 points of the threshold:

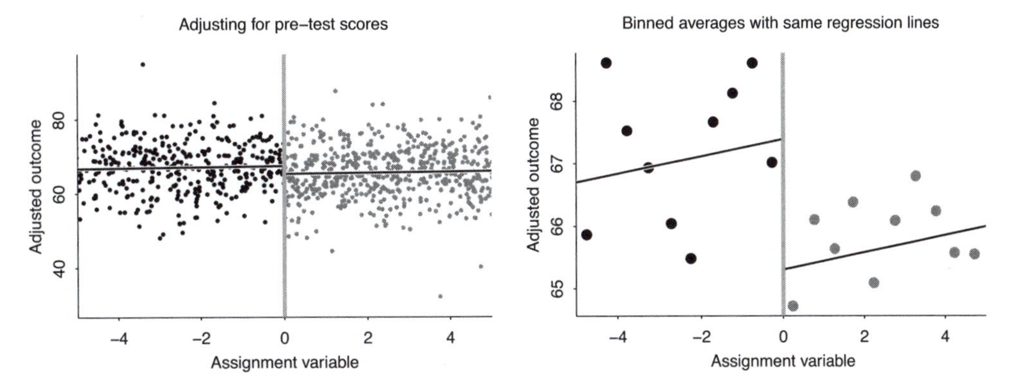


Figure 21.3 *Continuing from Figure 21.2, plots for the regression discontinuity analysis of reading test scores in Chile, this time also adjusting for pre-treatment measurements, reading and mathematics test scores in 1988. (a) Adjusted outcome, plotted vs. the assignment variable in a range close to the cutoff. (b) Binned averages of the data. For both plots, black dots represent schools or averages of schools that received the intention to treat, and gray dots represent those that did not. The discontinuity is more apparent in the binned graph because the range on the $y$-axis is much smaller.*

```
stan_glm(read92 ~ eligible + rule2, data=chile, subset = abs(rule2)<5)

            Median MAD_SD
(Intercept) 59.8    0.5
eligible     2.2    1.0
rule2        0.9    0.2

Auxiliary parameter(s):
      Median MAD_SD
sigma 7.2    0.2
```

The fit is displayed by the line in Figure 21.2b. The coefficient of `rule2` is the slope of the line, and the estimated effect on eligibility for the P-900 program on reading test scores is 2.2 points, which corresponds the discontinuity at the cutoff (comparing `eligibility` = 1 to `eligibility` = 0) in that graph.

**Including other pre-treatment variables as regression predictors.** The challenge of any regression discontinuity is the lack of overlap between treatment and control groups. Including the assignment variable as a pre-treatment predictor, as in (21.6) above, is a start, but, as with any observational study, we should be able to do better by including other pre-treatment characteristics that capture differences between schools.

In this example, we have two clearly relevant pre-treatment variables: reading and mathematics scores in 1988, which suggests the following regression:

$$y_i = \beta_0 + \tau * \text{eligible}_i + \beta_1 * \text{rule2}_i + \beta_2 * \text{reading pre-test}_i + \beta_2 * \text{math pre-test}_i + \text{error}_i. \quad (21.7)$$

Now we are adjusting not just for the assignment variable but also for two other relevant predictors:

```
stan_glm(read92 ~ eligible + rule2 + read88 + math88, data=chile, subset = abs(rule2)<5)

            Median MAD_SD
(Intercept) 23.4    4.4
eligible     2.1    0.9
rule2        0.1    0.2
read88       0.6    0.1
math88       0.2    0.1
```

```
Auxiliary parameter(s):
      Median MAD_SD
sigma 6.9    0.2
```

After adjustment for prior reading and math scores, the coefficient for the assignment variable `rule2` is small and highly uncertain, but we keep it in the model because of concern about bias from lack of overlap if we were not adjusting for this predictor.

The effect of being eligible for the P-900 program as estimated by the coefficient on `eligible` is an increase in reading test scores of about 2 points on average. To display the fitted model (21.7), we construct an adjusted outcome measurement as follows:

$$y_i^{\text{adj}} = y_i - \hat{\beta}_2 * (\text{reading pre-test}_i - \overline{\text{reading pre-test}}) - \hat{\beta}_3 * (\text{math pre-test}_i - \overline{\text{math pre-test}}).$$

With this adjustment, our new model reduces to the simple regression discontinuity (21.6) with $y^{\text{rep}}$ in place of $y$.

Figure 21.3a plots these adjusted test scores vs. the assignment variable, `rule2`, along with the fitted discontinuity model with an estimated drop of 2.1 points moving from the left side of the graph (with `eligible` = 1) to the right side (`eligible` = 0). To make this pattern clearer, Figure 21.3b shows binned averages of the data along with the same fitted line.

One downside of the regression discontinuity analysis is that the estimated effect applies only for those schools close to each threshold. The upside is that the assumptions (formalized below) can be more reasonable than a naive regression analysis on these data that attempts to estimate an average treatment effect across the entire sample.

**Including interactions.** As with causal regressions more generally, it can make sense to consider interactions of the treatment with pre-treatment predictors that have large estimated effects. In this case, we can interact the treatment with the reading pre-test score. As discussed in Section 12.2, it can help to center and scale predictors before including their interactions, so we define standardized pre-test scores:

```
chile$z_read88 <- (chile$read88 - mean(chile$read88))/sd(chile$read88)
chile$z_math88 <- (chile$math88 - mean(chile$math88))/sd(chile$math88)
```

Then we fit the model interacting treatment with reading pre-test:

```
stan_glm(read92 ~ eligible + rule2 + z_read88 + z_math88 + eligible:z_read88,
  data=chile, subset = abs(rule2)<5)

                  Median MAD_SD
(Intercept)       65.6    0.8
eligible           1.5    1.5
rule2              0.1    0.2
z_read88           6.3    0.9
z_math88           1.4    0.7
eligible:z_read88 -0.6    1.4

Auxiliary parameter(s):
      Median MAD_SD
sigma 6.9    0.2
```

In this case, the available data are clearly too noisy to estimate this interaction. This relates to general concerns discussed in Chapter 20 of adjusting for many covariates in an observational study.

**Regression fit to all the data.** Alternatively, we could fit the model to the whole dataset, which includes schools with average test scores that range from 20 points below their cutoff to 40 points above; see Figure 21.2a. Here are results from a linear fit without interactions:

```
stan_glm(read92 ~ eligible + rule2 + read88 + math88, data=chile)

            Median MAD_SD
(Intercept) 21.6    2.4
eligible     1.9    0.4
rule2        0.0    0.1
read88       0.7    0.0
math88       0.1    0.0

Auxiliary parameter(s):
      Median MAD_SD
sigma 6.7    0.1
```

The coefficient on `eligible` is similar to the earlier analyses above but is estimated much more precisely, which makes sense given that we are now fitting to three times as many data points. However, it is unclear how much we can trust a regression fit to all these data for the purpose of causal inference. If the goal is to estimate an average treatment effect for the entire sample, we would need to extrapolate the model for $y^1$ in neighborhoods of the covariate space where there are no treated observations and the model for $y^0$ in neighborhoods where there are no control observations.

A more realistic goal might be to use the full sample to fit models for each group that accurately predict $y^1$ and $y^0$ just at the threshold, the typical goal of regression discontinuity. In any case, a reasonable first step is to make sure that the model at least fits well for the observed data. That is typically easier to achieve if we use more flexible models and when we focus on data closer to the threshold. After all, how much influence do we want observations far from the threshold to have on predictions just at the threshold?

### Concerns and assumptions

We can learn from natural experiments. In a natural experiment, the exposure is assigned externally so we don't have to worry about selection bias. Many natural experiments have the form of a discontinuity. If so, there's a potentially serious concern with lack of overlap of the exposed and control groups, inference will be sensitive to our model for the outcome given the assignment variable, so we should take that aspect of the model seriously. But the assignment variable should not be the only concern. The exposed and control groups can differ in other pre-treatment characteristics. As in any observational study, you should do your best to adjust for differences between treated and control groups. And, as with causal inferences in general, we recommend performing the simple comparison and then seeing how that changes as adjustments are included in the model.

What assumptions are required to believe that regression discontinuity estimates identify causal effects? We will walk through a few settings as a way of understanding how regression discontinuity allows us to relax our assumptions relative to more standard observational studies.

Regression discontinuity analysis works particularly well when the assignment variable has a strong logical and empirical connection to the outcome variable, for example pre-test and post-test scores in an education study. In such cases, if you adjust for the assignment variable and other relevant covariates, you don't need to be so concerned about imbalance or lack of overlap on other pre-treatment variables.

In other settings, such as some examples where the treatment assignment depends on geography, the assignment variable is not particularly predictive of the outcome, and little if anything is gained by focusing on the discontinuity rather than simply considering the problem as a generic observational study. Indeed, performing a regression discontinuity analysis can lead to mistaken conclusions if you mistakenly think that the discontinuity implies that you don't need to worry about adjusting for other pre-treatment variables.

If our goal were to estimate the average causal effect for everyone within some range of the cutoff and we were to simply use a difference in means to estimate the effect of being eligible for the treatment (i.e. either above or below the threshold, $C$), we would need to assume unconditional ignorability in that range. We could formalize this as $y^1, y^0 \perp z$ for $x \in (C - a, C + a)$, where $C$ represents the threshold for assignment to treatment or control.

In practice, though, we always want to condition on the assignment variable $x$ by incorporating it as a predictor in our estimation model. If we still wanted to estimate the effect for the full analysis sample, we could weaken the ignorability assumption to $y^1, y^0 \perp z \mid x$ for $x \in (C - a, C + a)$. However, we would then require an appropriate model for $\mathrm{E}(y^0 \mid x)$ and $\mathrm{E}(y^1 \mid x)$ in this range, even without any overlap between treatment and control groups to support this estimation from the data.

Most commonly with regression discontinuity, the focus is on the causal effect at the threshold, $\mathrm{E}(y^1, y^0 \mid x = C)$. The ignorability assumption remains the same: $y^1, y^0 \perp z \mid x$ for $x \in (C-a, C+a)$. This is sometimes expressed as an assumption that no confounders vary discontinuously across the threshold. Now we only need to require that we can appropriately model the relationship between the potential outcomes on the side of the threshold supported by the data, as in $\mathrm{E}(y^0 \mid z = 0, x)$ and $\mathrm{E}(y^1 \mid z = 1, x)$.

## Fuzzy regression discontinuity

What happens when the discontinuity is not starkly defined, that is, when treated and control observations fall on the opposite side of the threshold from where they were assigned? This is sometimes called a "fuzzy" regression discontinuity, as opposed to the "sharp" or "clean" discontinuity discussed thus far.

As a simple example, consider a situation where the decision whether to promote children to the next grade is supposed to be made solely based upon results from a standardized test. Theoretically, this should create a situation with no overlap in these test scores across the children forced to repeat their grade and those promoted to the next grade (the treatment and control groups). In reality, though, students may not comply with their treatment assignment. Some children may be promoted despite failure to exceed the threshold based on any of several reasons, including parental pressure on school administrators, a teacher who advocates for the child, or an exemption from the policy due to classification as learning disabled. Other students may be retained even though they earned a score above the threshold, due perhaps to social or behavioral issues.

Violation of the strict discontinuity rule creates partial overlap between the treatment and control groups in terms of the variable that should be the sole confounding covariate, the assignment variable, promotion test scores. Given our concern about the lack of overlap, this may at first glance seem to be a preferable scenario! Unfortunately, this overlap arises from deviations from the stated assignment mechanism. If the reasons for these deviations are well defined and measurable, then ignorability can be maintained by adjusting for the appropriate child, parent, or school characteristics. Similarly, if the reasons for these deviations are independent of the potential outcomes of interest, there is no need for concern. More plausibly, however, such inferences could be compromised by failure to adjust for important omitted confounders.

## Connection to instrumental variables

An alternative strategy is to approach the causal effect estimation from an instrumental variables perspective. In certain scenarios we can extend our conceptualization of a sharp regression discontinuity design from that of a simple randomized experiment (for units within a narrow margin of the threshold) to that of a randomized experiment with noncompliance. In this framework the indicator for whether an observation falls above or below the threshold, $z$ ("eligibility" in the Chile example) acts as the instrument. The treatment variable of interest, $T$, is the indicator for whether the program or treatment was actually received (P-900 in the Chile example). For this strategy to be successful, all the standard instrumental variables assumptions must hold. It is far from true that instrumental variables estimation will always clean up a fuzzy regression discontinuity. Let's walk through these assumptions in turn.

*Ignorability of the instrument* is a bit more complicated in the regression discontinuity setting (as compared to a randomized experiment at least) because of the necessity of either choosing an appropriately narrow bandwidth or specifying an appropriate model to properly adjust for the assignment variable.

The interpretation of the *exclusion restriction* is similar here to a traditional instrumental variables setting. Never-takers (those who will never receive the treatment regardless of their placement with regard to the cutoff) and always-takers (those who will always receive the treatment regardless of their placement with regard to the cutoff) must have the same outcomes regardless of whether they end up above or below the cutoff (that is, regardless of their value for the instrument). More colloquially, we can express this assumption as a requirement that the instrument (falling on either side of the cutoff) can only affect the outcome through its effect on the treatment.

To satisfy *monotonicity*, we need to exclude the possibility of defiers. That is, there cannot be observations that would receive the treatment if they fall on the side of the cutoff that makes them eligible but would not receive it otherwise.

Finally, *the instrument must be predictive of the treatment variable*. If the cutoff that determines "eligibility" is unrelated to the probability that observations receive the treatment, then we cannot make progress.

As with standard instrumental variables estimation, if the required assumptions hold, the procedure will estimate the effect of the treatment only for the compliers—that is, the people (or, more generally, experimental units) who *would have* adhered to their assignment no matter which side of the threshold they fell on—and only those compliers in the chosen bandwidth. In our grade retention example these are the students who would have been retained in grade had their test scores fallen short of the cutoff but who would have been promoted if their test scores had exceeded the threshold. Never-takers are those students who would have been promoted no matter their test score and always-takers are those who would be held back regardless. As with the ignorability assumption, the exclusion restriction may become more or less plausible as we vary the margin around the threshold that defines the subset of students that we choose to consider for conducting our analyses.

### Returning to the Chile schools example

Our example with the P-900 program in Chile also represents a fuzzy discontinuity design. In the restricted sample, about 5% of schools that weren't defined as eligible received the P-900 program nonetheless. Moreover, about 39% of those who were defined as eligible did not receive the program.

So can we use instrumental variables to estimate the effect of actually receiving the P-900 program? In addition to the ignorability assumption that requires the schools close to the threshold to, in effect, be randomly assigned to being above or below that threshold, other assumptions would be required. To satisfy the exclusion restriction we would need to believe that lying above or below the threshold has no effect on outcomes except through its effect on whether schools received P-900. Another way of framing the assumption is that schools that were "never-takers" (here, those schools who would not receive the program no matter which side of the threshold they fell on) and "always-takers" (here, schools that would receive the program no matter which side of the threshold they fell on) would not have had different test scores if they fell above or below the threshold. For a narrow range of schools this may be plausible.

The monotonicity assumption would require that there are no schools that would not receive P-900 if they fell in the eligible range of test scores but would receive it if they fell in the non-eligible range. We know that there is a strong correlation between the instrument (falling above or below the derived eligibility cutoff) and the treatment (receiving P-900).

If we perform an instrumental variables analysis on our subset of schools close to the threshold we obtain an estimate of the effect of participating in the P-900 program of about 5 points. This estimate applies to the schools that would participate in P-900 if they were eligible (below the cutoff) and not otherwise (the compliers). But we do not know enough about education policy in Chile to feel completely secure in the assumptions required for these results to be valid, particularly with regard to the exclusion restriction.

## 21.4 Identification using variation within or between groups

### Comparisons within groups using varying-intercept ("fixed effects") models

Another set of strategies capitalizes on situations in which repeated observations within groups can provide a means for adjusting for unobserved characteristics of these groups. If comparisons are made across the observations within a group, implicitly such comparisons hold constant all characteristics intrinsic to the group or individual that do not vary across members of the group.

Example: Hypothetical study of birth weight

For example, suppose you want to examine the effect of low birth weight, $z$, on a health outcome, $y$. One difficulty in establishing a causal effect is that children with low birth weight are also typically disadvantaged in genetic endowments and socioeconomic characteristics of the family, some of which may not be easy or possible to measure. Rather than trying to directly adjust for all of these characteristics, one could instead *implicitly* adjust for them by comparing outcomes across children within a pair of twins. So we may be able to consider birth weight to be randomly assigned (ignorable) *within* twin pairs. In essence, then, each twin acts as a counterfactual for his or her sibling. Theoretically, if there is enough variation in birth weight, within sets of twins, we can estimate the effect of birth weight on subsequent outcomes. A regression model that is sometimes used to approximate this conceptual comparison simply allows for each group (here, the twin pair) to have its own intercept. This model can be expressed as,

$$y_{ij} = \beta_0 + \tau z_{ij} + \alpha_i + \epsilon_{ij}.$$

To fit this model, one could regress outcomes on birth weight (the "treatment" variable, $z$) along with an indicator variable for each pair of twins (keeping one pair as a baseline category to avoid collinearity). This model can also be fit using differences. With pairs (that is, groups of size 2), for example, we would replace each pair of observations with their difference:

$$y_{i2} - y_{i1} = \tau(z_{i2} - z_{i1}) + \epsilon_{i2} - \epsilon_{i1}.$$

With more than two observations per group, a more general strategy would be to subtract the group-specific means of each variable from each individual-level value, as in,

$$y_{ij} - \bar{y}_i = \tau(z_{ij} - \bar{z}_i) + \epsilon_{ij} - \bar{\epsilon}_i,$$

where $z$ now denotes the treatment in this more general model. The `plm` package in `R` will fit this model (using the "within" option) and will produce standard errors that adjust for the fact that the group-level means are estimated.

This varying-intercepts model is sometimes called a fixed-effects model, a term we avoid because it is used in various ambiguous ways in the statistics literature. We discuss varying-intercept and varying-slopes models in more detail in our book on multilevel modeling. Two reasons why hierarchical or multilevel modeling can be useful for varying intercepts and slopes are: (1) when the number of observations per group is small, the simple unregularized linear regression estimates can be noisy, and (2) with a multilevel model, we can include group-level predictors and allow coefficients to vary by group at the same time, which cannot be done using classical regression. For now, though, we will continue with the simple classical estimate to get across the general point of varying-intercepts models, with the understanding that you could go back later and fit the model better using multilevel regression, at which point you could also learn what pitfalls to avoid when using that approach.

Causal inference using multilevel modeling is not completely straightforward, because the simplest assumption for a multilevel regression is that the varying intercepts are independent of the other predictors in the model, including the treatment. From a causal standpoint, however, we are always guarding against failure to adjust for variables we think may be associated with both the treatment (one of our predictors) and the outcome. Recall that the motivation for fitting a model with group-specific intercepts was, in essence, to adjust for group-specific unobserved confounders. Confounders, by definition, are associated with both the treatment and outcome variables. Therefore it is important when fitting such models to either include the within-group means of the treatment variable as a group-level predictor, or to include enough relevant group-level predictors so that confounding is not a concern.

**The causal predictor needs to vary within group.** This study design can create comparisons across different levels of the treatment variable only if it varies *within group*. In examples where the treatment is dichotomous, it is possible for a substantial portion of the sample to exhibit no variation at all in the causal variable within groups. For instance, suppose a varying-intercepts model is used to estimate the effect of maternal employment on child cognitive outcomes by including indicators for each family (set of siblings) in a regression of outcomes on an indicator for whether the mother worked in the first year of the child's life. In some families, the mother may not have varied her employment status across children. Therefore, no inferences about the effect of maternal employment status can be made for these families. We can only estimate effects for the type of family where the mother varied her employment choice across the children (for example, working after her first child was born but staying home from work after the second).

**Assumptions.** What assumptions need to hold for varying-intercept regression to identify a causal effect? The structural assumption required can be conceived of as an extension of the standard ignorability assumption that additionally conditions on the group indicators,

$$y^0, y^1 \perp z \,|\, \alpha.$$

We can map this to a situation in which we have a separate randomized experiment within each group. Another framing is that ignorability holds once we condition on (potentially unmeasured) group-specific confounders that are common to all members of the group.

In addition to the ignorability assumption, the varying-intercepts model makes all the assumptions of a standard regression model. In the simple version of the model without covariates, this may not translate to strong assumptions about linearity and additivity (depending on the form of the treatment variable). We also assume that the observations are (conditionally) independent of each other. While this may not seem plausible given the grouped structure of the data, recall that we can now conceive of each group member's observation as simply the distance of this value from the group mean. Therefore, if the mechanism driving the similarity in responses within a group was the shared mean, then subtracting the mean of the data should eliminate the problem. Depending on context, more complicated correlation structures can be fit, but that goes beyond the scope of this book.

**Covariates.** The researcher might also want to adjust for pre-treatment covariates, $x$, to improve the plausibility of the ignorability assumption (to account for the fact that the treatment may not be strictly randomly assigned even within each group). Formally this extends the ignorability assumption to

$$y^0, y^1 \perp z \,|\, \alpha, x.$$

Conditioning on such confounders may come at the cost of additional parametric assumptions, but this is generally worth the tradeoff.

In our birth-weight example it is difficult to find child-specific predictors that vary across children within a pair but can still be considered pre-treatment. However, the haphazard nature of the "assignment" to birth weight may obviate the need for such controls; that is, ignorability seems plausible even without conditioning on other variables.

In the maternal-employment example, however, we are less likely to believe that the choice to enter the labor market is independent of child potential outcomes (or, said another way, is independent of child and family characteristics that might also influence child developmental outcomes). Therefore the analysis would likely be stronger if researchers could include child covariates measured before the mother returned to work, such as birth weight or number of days in hospital as well as immutable characteristics such as sex. As we discuss in more detail next, however, researchers should be wary of conditioning on confounding covariates that may have been affected by the treatment.

### Within-person controls

Perhaps the most common use of varying intercepts when estimating causal effects occurs in the context of panel data, in which the same individuals are surveyed over time, typically with a fair amount of overlap in the questions asked at each time point. For instance, consider examining the effect of marriage on men's earnings by analyzing data that follows men over time and tracks marital status, earnings, and predictors (potential confounding covariates) such as race, educational attainment, number of children, and occupation. Researchers have justified the use of varying-intercept regressions in this setting by arguing that it allows for within-person comparisons that implicitly adjust for unobserved characteristics of each individual that do not vary over time. However when these models also condition on confounding covariates that could be affected by the treatment, they end up adjusting for post-treatment variables, which, as we know from Section 19.6, can lead to bias.

In the above example it is hard to rule out the possibility that educational attainment, number of children, and occupation were affected by whether the study participant had gotten married at a previous (or current) time point. A prudent choice might be to omit any covariate from the model that has the potential to be affected by the treatment. Unfortunately, this might render ignorability completely implausible. The bibliographic note at the end of the chapter points to more advanced approaches to adjust for such variables using more complex assumptions related to ignorability.

### Comparisons within and between groups: difference-in-differences estimation

Almost all causal strategies make use of comparisons across groups: one or more groups that were exposed to a treatment, and one or more groups that were not. *Difference-in-difference* strategies additionally make use of another source of variation in outcomes, typically time, to help adjust for potential (observed and unobserved) differences across these groups. For example, consider estimating the effect of a newly-introduced school busing program on housing prices in a school district where some neighborhoods were affected by the program and others were not. A simple comparison of housing prices for all houses across affected and unaffected areas several years after the busing program went into effect might seem tempting, but it would not be appropriate because these neighborhoods might be different in other ways that are related to housing prices. For instance, the busing program might have been implemented in neighborhoods that were more disadvantaged than the other neighborhoods in the school district and thus already had lower housing prices. A simple before-after comparison of housing prices would also typically be inappropriate because other changes that occurred during this time period (for example, a recession or economic boom) might also be influencing housing prices.

A difference-in-differences approach would instead capitalize on both sources of variation: the difference in housing prices across time and the difference in prices across exposed and unexposed neighborhoods. Perhaps the most intuitive way to think about this model is that it makes comparisons between exposed and unexposed groups with respect to the differences in means over time (housing price trajectories). For example, suppose the housing prices in the unexposed areas increased on average by \$20 000 between the pre-intervention and post-intervention time periods but that they increased by only \$10 000 in the exposed areas over this same period of time. If we can assume that the growth in housing prices in the unexposed areas reflects what would have happened in the exposed areas had the busing program never occurred, then we could say that the estimated effect of the busing for the houses in the exposed areas was a decrease in house price of \$10 000.

### Regression framework

The calculation we just performed simply involved four means. We could estimate the same effect in a regression framework by including each combination of observation and time point as a separate row in our dataset and fitting the model,

$$y_i = \beta_0 + \beta_1 z_i + \beta_2 P_i + \tau z_i P_i + \epsilon_i, \tag{21.8}$$

where $z$ denotes the treatment and $P$ denotes the time period, with $P = 0$ before exposure and $P = 1$ after exposure. In this model the difference-in-differences estimand is the coefficient on the interaction term. Intuitively this makes sense because the interaction term represents the difference in the slope coefficients on time across the treatment groups where the slope coefficients on time each represent a difference in means (across time points). Equivalently, we could say that the interaction term represents the difference in slope coefficients on treatment across the time periods.

In our housing example, we saw prices on the same sample of houses at two different times. If we include each house twice in our analysis we know that our regression assumption of independent observations will be violated. A simple fix is to instead model the differences:

$$d_i = \alpha + \tau P_i + \epsilon_i, \tag{21.9}$$

where the estimand, $\tau$, has the same interpretation as before. This simplification will not work if our observational units are not the same across the two time points.

Why would we bother to fit a regression if this coefficient simply represents the difference between two mean differences? Why not do a simple calculation with four means? First, the regression approach allows us to easily estimate the standard error on our treatment effect estimate. Furthermore, this framework will extend easily to allow complications such as including covariates in the model.

### Assumptions

We discussed the required structural assumption above colloquially as a requirement that the change in test scores for the control group needs to represent what would have happened to the treatment group had they not been exposed to the treatment. We can formalize the assumption by defining a new quantity that we refer to as a potential change (rather than a potential outcome). In particular, define $d^0 = y^0 - y_{P=0}$, to reflect the potential change in outcomes (appraised house price) between the pre-exposure and post-exposure time periods if the unit were never exposed to the treatment. Here $y_{P=0}$ denotes the baseline (pre-exposure time period) measure of the outcome variable, which in other contexts we would simply think of as an important confounding covariate.

Our use of the term "outcome" and the notation $y$ to denote both post- and pre-exposure versions of this variable is admittedly a bit confusing. Until this point we have used $y$ to refer to post-treatment outcomes, but here we use $y$ to refer to the variable we think that the treatment might affect, even during time periods before the treatment can have affected it. The benefit of this choice is that it may help us to better understand the regression models used in this context in which the response will sometimes reflect a true post-exposure outcome and will sometimes reflect a pre-exposure version of the same variable. Also, in the notation for potential changes, this choice makes it clear that we aren't just subtracting the value of some important confounder—we are subtracting the same measure at a different time point.

Mapping this to our earlier informal discussion of the assumptions and estimand, we can say that if $d^0 \perp T$ we can identify the effect for the treated, $\mathrm{E}(d^1 - d^0 \mid T = 1)$.

To define the effect of the treatment on the controls, we need to first define the potential change under treatment assignment, $d^1 = y^1 - y_{P=0}$. Then to identify the effect of the treatment on the controls, $\mathrm{E}(d^1 - d^0 \mid T = 0)$, we would need to assume $d^1 \perp T$. To estimate the average treatment effect $\mathrm{E}(d^1 - d^0)$ we would need to assume $d^0, d^1 \perp T$.

**Adjusting for potential confounding covariates.** Without introducing confounding covariates, however, we have no way of distinguishing between average treatment effects for the control group and average treatment effects for the treated. Said another way, our current assumptions define them to be the same (just as in a completely randomized experiment these effects are the same). We can loosen this restriction and create more plausible ignorability assumptions by conditioning on other confounding covariates, $x$. This also allows us to create a more plausible ignorability assumption, $d^0, d^1 \perp T \mid x$.

If we think that there is heterogeneity in our treatment effect and we wanted to explicitly target the effect for the treated or the effect for the controls, we could combine the difference-in-differences framework with a method (such as one of the ones discussed in the previous chapter) that allows for the targeting of such estimands. For instance, we could run a change score model on a reduced sample consisting of the treatment units and the control units that have been matched to them.

**Comparison with standard ignorability assumptions.** The assumption needed with this strategy appears to be weaker than the unconditional ignorability assumption, because rather than assuming that potential outcomes are the same across treatment groups, one only has to assume that the potential *changes* in outcomes over time are the same across groups (for example, exposed and unexposed neighborhoods). Therefore, we need only believe that the difference in housing prices over time would be the same across the two types of neighborhoods, not that the average post-program potential housing prices if exposed or unexposed would be the same.

In studies where we observe the same units at both time periods, however, the pre-exposure outcome can be thought of simply as an important confounding covariate. If we had access to such information, we would certainly want to adjust for that covariate, possibly using a regression model and propensity scores. In that case we would be relying on a related assumption to identify the causal effect: $y^0, y^1 \perp T \mid y_{P=0}$. This says that we would expect that among those with the same housing price during the pre-exposure time period, there is no difference in the distributions of potential outcomes in treated and untreated groups. The differences between this assumption and the difference-in-differences assumption are subtle.

**Different observations at each time point.** Thus far we have been discussing a special case within the difference-in-differences framework in which the same units are observed at both time points. As we have just noted, in this setting the advantages of the difference-in-differences strategy are less apparent because an alternative model would be to include an indicator for treatment exposure but then simply regress on the pre-treatment version of the outcome variable.

One strength of the difference-in-differences framework is that it accommodates situations in which the observations are *not* the same across time periods simply by using the initial regression framework introduced in (21.8). This usage requires the additional assumption, however, that the units sampled across the two time points are both random samples from the same distribution.

When might this assumption be violated? Consider the earlier example with busing and housing prices. Suppose that instead of using appraisal values to obtain housing prices of every house in each neighborhood, the researchers used sale prices from houses that had actually sold in each time period. Then if the busing also had an impact on the ability of owners to sell their house, the sample of houses that sold in the post-exposure time period would not be representative of the sample of houses that sold in the pre-exposure time period and this assumption would be violated. The ability to condition on covariates could weaken this assumption, but the cautions discussed below with regard to such extensions of the model must be considered.

This design is often presented as if it provides a weaker set of assumptions than traditional observational studies, as it a quasi-experiment, not just a study where we have to hope we've measured all confounders. However, looking carefully at the assumptions, it is unclear if the assumption of randomly assigned *changes* in potential outcome is truly more plausible than the assumption of randomly assigned potential outcomes for those with the same value of the pre-treatment variable (and of course both approaches could allow for conditioning on additional covariates). In either case, we need not assume actual random manipulation of treatment assignment for either assumption to hold, only results that would be consistent with such manipulation.

**Do not condition on post-treatment outcomes.** Once again, to make the (new) ignorability assumption (as well as the additional assumption required when we have different observational units across time points) more plausible, it can be desirable to condition on potential confounders. However, for standard difference-in-differences models in which one of the differences is across time, this model will implicitly condition on post-treatment variables (similar to the varying-intercepts specification with panel data). If these predictors can be reasonably assumed to be unchanged by the treatment, then this may be reasonable. However, as discussed in Section 19.7, it is otherwise inappropriate to adjust for post-treatment variables by adding them as regression predictors.

When we are working with panel data, a better strategy presents itself: we can fit the model in (21.9) with the addition of pre-treatment covariates. This amounts to running a regression of the change in outcomes on the treatment variable and covariate values from the pre-exposure time period.

**Regression discontinuity versus difference-in-differences.** There may be a temptation to use a difference-in-differences approach in situations where the data arise from a prospective or retrospective regression discontinuity design. After all, don't such designs often lead to situations in which there is an exposed and an unexposed group and measures of a response of interest both before and after the treatment group is exposed? Why wouldn't this be ideal?

As it turns out, this design is particularly *poorly* suited to a difference-in-differences approach because, by design, a simple application of difference-in-differences is likely to suffer from bias arising from regression to the mean. For example, consider our hypothetical regression discontinuity example with grade retention imposed based on student test scores falling below a given threshold. The regression discontinuity design is capitalizing on the fact that, for students with scores close to the threshold, falling above or below the cutoff is for all intents and purposes randomly assigned. For instance you might imagine that students with test scores in that range all have scores that come from a probability distribution centered at the threshold value.

If we accept this premise, we realize that if nothing changed for those students over time, that is, *if there were no treatment effect*, then the next time we administered a similar test there would be a high probability that the students who scored in the lower part of this distribution (below the cutoff) the first time would be likely to score higher the next time and that the students who score relatively high (above the cutoff) the first time would be likely to score lower the next time. Therefore, if there were truly no treatment effect then we would still (repeatedly across samples) estimate a *negative* treatment effect due to regression to the mean.

The regression discontinuity analysis avoids this problem by conditioning on the assignment variable. This sets up direct comparisons between units with the same value of the assignment variable (confounder) and highlights the role of specification of the models for $\mathrm{E}(y^1|x)$ and $\mathrm{E}(y^0|x)$.

## 21.5 Causes of effects and effects of causes

Statistical methods for causal and policy analysis are more focused on "effects of causes" than on "causes of effects." Causal researchers generally advise against studying the causes of any particular outcome because of the difficulty of identifying all of the mechanisms that might contribute to an observed phenomenon. For instance, consider what it would mean to pursue a question such as "What are the causes of poverty?" The list would be very long! Moreover, it would be difficult to determine at what point these mechanisms would be defined relative to the point in time when poverty is assessed. For any identified cause, "Cause A," we could likely identify something that happened before it, "Cause B," that caused it, and a "Cause C" before that ...

Here we argue that the search for causes can be understood within traditional statistical frameworks as a part of model checking and hypothesis generation. Moreover, while it can make sense to ask *questions* about the causes of effects, ultimately the *answers* to these questions will be in terms of effects of causes.

**Example: cancer clusters.** A map shows an unexpected number of cancers in some small location, and this raises questions: What is going on? Why are there so many cancers in this place? These questions might be addressed via the identification of some carcinogenic agent (for example, a local environmental exposure) or, less interestingly, some latent variable (some systematic difference between people living in and not living in this area), or perhaps some statistical explanation (for example, a multiple-comparisons argument demonstrating that an apparent anomalous count in a local region is no anomaly at all, after accounting for all the possible clusters that could have been found). Until it is explained in some way, the question, "Why are there so many cancers in that location?" refers to a data pattern that does not fit some pre-existing model. Researchers are then motivated to look for an explanation that leads to a deeper understanding and ultimately identifies a carcinogenic agent or some other reason for the observed pattern.

The cancer-cluster problem is typical of a lot of scientific reasoning. An anomaly is observed and it needs to be explained. To borrow concepts from the philosophy of science, the resolution of the anomaly may be an entirely new paradigm or a revision of an existing research program.

The purpose of the present discussion is to place "What caused this?" questions within a statistical framework of causal inference. We argue that a question such as "Why are there so many cancers in this place?" can be viewed not directly as a question of causal inference, but rather indirectly as an identification of a problem with an existing statistical model, motivating the development of more sophisticated statistical models to directly address causation in terms of counterfactuals and potential outcomes.

## Types of questions

We distinguish between two broad classes of causal queries:

1. *Effects of causes.* What might happen if we do $z$? What is the effect of some manipulation, for example, the effect of job training on poverty status, the effect of smoking on health, the effect of schooling on earnings, the effect of campaigns on election outcomes, and so forth?
2. *Causes of effects.* What causes $y$? Why do more attractive people earn more money? Why does *per capita* income vary so much by country? Why do many poor people vote for Republicans and rich people vote for Democrats? Why do some economies collapse while others thrive?

When methodologists write about causal inference, they generally focus on the effects of causes. We are taught to answer questions of the type "What is the effect of $z$?", rather than "What caused $y$?" As we have discussed in the preceding chapters, potential outcomes can be framed in terms of manipulations: if $z$ were changed by one unit, how much would $y$ be expected to change? But "What caused this?" questions are important too. They are a natural way to think, and in many ways, these causal questions motivate the research, including experiments and observational studies, that we use to estimate particular causal effects.

How can we incorporate "What caused this?" questions into a statistical framework that is centered around "What is the effect of this?" causal inference? A potentially useful frame of this issue is as follows: "What is the effect?" causal inference is about estimation; "What caused this?" questions are more about model checking and hypothesis generation. Therefore we do not try to *answer* "What caused this?" questions; rather, our exploration of these questions *motivates* "What is the effect of?" questions that can be studied using statistical tools such as the ones we have discussed in these chapters: experiments and observational studies.

Example: Incumbency advantage

**Example: political campaigns.** Consider a "causes of effects" question. Why do incumbents running for reelection to Congress get so much more funding than challengers? Many possible answers have been suggested, including the idea that people like to support a winner, that incumbents have higher name recognition, that certain people give money in exchange for political favors, and that incumbents are generally of higher political "quality" than challengers and get more support of all types. Various studies could be performed to evaluate these different hypotheses, all of which could be true to different extents and in some interacting ways.

We can frame our approach to such questions as model checking. It goes like this: what we see is some pattern in the world that needs an explanation. For example, if we ask, "Why do incumbents get more contributions than challengers?", we are comparing to an implicit model in which incumbency has correlation with fundraising success. If we gather some data on funding, compare incumbents to challengers, and find the difference is large and precisely estimated, then we are comparing to the implicit model that there is variation but it is not related to incumbency status. If we get some measure for candidate quality (for example, previous elective office and other qualifications) and still see a big difference between the funds given to incumbents and challengers, then it seems we need more explanation. And so forth.

Here is an example told to us by a statistician working in a business environment:

xample: ausation in narketing esearch

> A lot of real world problems can be framed as "causes of effects" and, arguably, it's irresponsible to ignore them. . . . Take for example P&G who spend a huge amount of money on marketing and advertising activities. The money is spread out over many vehicles, such as television, radio, newspaper, supermarket coupons, events, emails, display ads, search engine, etc. The key performance metric is sales amount.
>
> If sales amounts suddenly drop, then the executives will want to know what caused the drop. This is the classic reverse causality question. Many possible hypotheses could be generated . . . TV advertisements were not effective, coupons weren't distributed on time, emails suffered a deliverability problem, etc.
>
> The same question can be posed as a set of "effects of causes" problem. We now start with a list of treatments. We will divide the country into regions, and vary the so-called marketing mix, that is, the distribution of spending across the many vehicles. This generates variations in the spend patterns by vehicle, which with some assumptions allows us to estimate the effect of each of the constituent vehicles on the overall sales performance.

This is the pattern: a quantitative analyst notices an anomaly, a pattern that cannot be explained by current models. The "causes of effects" question is: Why did this happen? And this can motivate improved modeling and data collection.

### Relation to formal models of causal inference

Now let us put some mathematical structure on the problem using the potential outcome framework. Let $y_i$ denote the outcome of interest for unit $i$, say, an indicator whether individual $i$ developed cancer. We may also observe characteristics of units, individuals in this case, that are known to, or expected to, be related to the outcome, in this case, related to the probability of developing cancer. Denote those by $x_i$. Finally, there is a characteristic of the units, denoted by $u_i$, that the researcher feels should not affect the outcome, and so one would expect that in populations homogeneous in $w_i$, there should be no association between the outcome and $u_i$:

$$y_i \perp u_i \mid x_i.$$

This attribute $u_i$ may be the location of individuals. It may be a discrete characteristic, say, female versus male, or an indicator for a subpopulation. The key is that the researcher interprets this variable as an attribute that should not be correlated with the outcome in homogeneous subpopulations.

However, suppose the data reject this hypothesis and show a substantial association between $u_i$ and the outcome, $y_i$, even after adjusting for differences in the other attributes, $x_i$. This implies that our conditional independence model must be altered in some way.

Such a finding is consistent with two alternative models. First, it may be that there is a cause, its value for unit $i$ denoted by $z_i$, such that the potential outcome given the cause is not associated with $u_i$, possibly after conditioning on $x_i$. Let $y_i(z)$ denote the potential outcomes in the potential outcome framework, and $y_i(z_i)$ the realized outcome. Formally we would hypothesize that, for all $z$,

$$y_i(z) \perp u_i \mid x_i.$$

Thus, the observed association between $y_i$ and $u_i$ is the result of a causal effect of $z_i$ on $u_i$, and an association between the cause $z_i$ and the attributes $x_i$. In the cancer example, there may be a carcinogenic agent that is more common in the area with high cancer rates.

The second possible explanation that is still consistent with no causal link between $u_i$ and $y_i$ is that the researcher omitted an important attribute, say, $v_i$. Given this attribute and the original attributes $x_i$, the association between $u_i$ and $y_i$ would disappear:

$$y_i \perp u_i \mid v_i, x_i.$$

For example, it may be that individuals in this area have a different genetic background that makes them more susceptible to the particular cancer.

Both these alternative models could in principle provide satisfactory explanations for the anomaly of the strong association between $u_i$ and the outcome. Whether these models do so in practice, and which of these models do, and in fact whether the original association is even viewed as an anomaly, depends on the context and the researcher. Consider the finding that taller individuals command higher wages in the labor market. Standard economic models suggest that wages should be related to productivity, rather than height, and such an association might therefore be viewed as an anomaly. It may be that childhood nutrition affects both adult height and components of productivity. One could view that as an explanation of the first type, with childhood nutrition as the cause that could be manipulated to affect the outcome. One could also view health as a characteristic of adult individuals that is a natural predictor of productivity.

As stressed before, there are generally not unique answers to these questions. In the case of the association between height and earnings, one researcher might find that health is the omitted attribute, and another researcher might find that childhood nutrition is the omitted cause. Both could be right, and which answer is more useful will depend on the context. The point is that the finding that height itself is correlated with earnings is the starting point for an analysis that explores causes of earnings, that is, alternative models for earnings determination, that would reproduce the correlation between height and earnings without the implication that intervening in height would change earnings.

### What does this mean for statistical practice?

A key theme in this discussion is the distinction between causal *statements* and causal *questions*. A reverse causal question does not in general have a well-defined answer, even in a setting where all possible data are made available. But this does not mean that such questions are valueless or that they fall outside the realm of statistics. A reverse question places a focus on an anomaly—an aspect of the data unlikely to be reproducible by the current (possibly implicit) model—and points toward possible directions of model improvement.

It has been (correctly) said that one of the main virtues of the potential outcome framework is that it motivates us to be explicit about interventions and outcomes in forward causal inference. Similarly, one of the main virtues of reverse causal thinking is that it motivates us to be explicit about what we consider to be problems with our model.

In terms of graphical models, the anomalies also suggest that the current model is inadequate. In combination with the data, the model suggests the presence of arrows that do not agree with our prior understanding. The implication is that one needs to build a more general model involving additional variables that would eliminate the arrow between the attribute and the outcome.

By formalizing reverse causal reasoning within the process of data analysis, we hope to make a step toward connecting our statistical reasoning to the ways that we naturally think and talk about causality, to be able to talk about reverse causal questions in a way that is complementary to, rather than outside of, the mainstream formalisms of statistics and econometrics.

Just as we view graphical exploratory data analysis as a form of model checking (which may be implicit), we similarly hold that reverse causal questions arise from anomalies—aspects of our data that are not readily explained—and that the search for causal explanations is, in statistical terms, an attempt to improve our models so as to reproduce the patterns we see in the world.

## 21.6 Bibliographic note

Instrumental variables formulations date back to work in the economics literature by Tinbergen (1930) and Haavelmo (1943). Angrist and Krueger (2001) present an upbeat applied review of instrumental variables. Imbens (2004) provides a review of statistical methods for causal inference that is a little less enthusiastic about instrumental variables. Woolridge (2001, chapter 5) presents instrumental variables from a classical econometric perspective. The "always-taker," "complier," and "never-taker" categorizations here are adapted from Angrist, Imbens, and Rubin (1996), who reframe the classic econometric presentation of instrumental variables in statistical language and clarify the assumptions and the implications when the assumptions are not satisfied. For a discussion of all of the methods discussed in this chapter from an econometric standpoint, see Angrist and Krueger (1999).

The Vietnam War draft lottery example comes from several papers including Angrist (1990). The weather and fish price example comes from Angrist, Graddy, and Imbens (2000). The sex of child example comes from Angrist and Evans (1998).

Imbens and Rubin (1997) discuss a Bayesian approach to instrumental variables in the context of a randomized experiment with noncompliance. Hirano et al. (2000) extend this framework to include covariates. Barnard et al. (2003) describe further extensions that additionally accommodate missing outcome and covariate data. For discussions of prior distributions for instrumental variable models, see Dreze (1976), Maddala (1976), Kleibergen and Zivot (2003), and Hoogerheide, Kleibergen, and van Dijk (2007). Glickman and Normand (2000) derive an instrumental variables estimate using a latent-data model; see also Carroll et al. (2004).

For a discussion of use of instrumental variables to estimate bounds for the average treatment effect (as opposed to the local average treatment effect), see Robins (1989), Manski (1990), and Balke and Pearl (1997). Robins (1994) discusses estimation issues.

Sarsons (2015) discusses problems with unexamined assumptions in instrumental variables analysis, in the context of a previously fitted model of income and political conflict that used rainfall as an instrument even though the exclusion restriction was violated in this case.

The focus of our discussion of weak instruments is different from much of the econometrics literature on the topic, which focuses on the fact that in an overidentified instrumental variables model (that is, one with more instruments than causal variables), if the instruments are not sufficiently predictive (as measured by a test of the hypothesis that the coefficients in the first-stage model are all equal to zero) then the resulting effect estimates may be biased. Given the difficulty of meeting the assumptions required of even one instrument, our advice is to stick to situations in which there is one instrument and one treatment variable (or at least the same number of instruments as treatment variables).

For more on the Sesame Street experiment, see Bogatz and Ball (1971) and Murphy (1991).

Regression discontinuity analysis is described by Thistlethwaite and Campbell (1960) and Hahn, Todd, and van der Klaauw (2001). Gelman and Zelizer (2015) and Gelman and Imbens (2017) explore some potential pitfalls when regression discontinuity analysis is applied in an automatic manner where it is not appropriate. The example of the P-900 program in Chile is adapted from Chay, McEwan, and Urquiola (2005), and the example of children's promotion in school is drawn from Jacob and Lefgren (2004).

Ashenfelter, Zimmerman, and Levine (2003) discuss "fixed effects" and difference-in-differences methods for causal inference. The twins and birth weight example is based on a paper by Almond, Chay, and Lee (2005). Another interesting twins example examining the returns from education on earnings can be found in Ashenfelter and Krueger (1994). Aaronson (1998) provides a further example of the application of these approaches. The busing and housing prices example is from Bogart and Cromwell (2000). Card and Krueger (1994) discuss a classic example of a difference-in-differences model that uses panel data. The `plm` package is described by Croissant and Millo (2008). Hill (2013) discusses some of the complications that can arise when estimating causal effects in settings with grouped data.

For more on structural models and causal inference, see Morgan and Winship (2014) and Pearl

(2000). Sobel (1990, 1998) discusses the assumptions needed for structural modeling, and Gelman (2011b) considers different approaches to causal inference.

Section 21.5 is taken from Gelman and Imbens (2013), and the quote on page 447 is from Kaiser Fung. For another discussion of causes of effects and effects of causes, see Dawid, Musio, and Fienberg (2016).

## 21.7 Exercises

21.1 *Instrumental variables*: The following study is performed at a university. Students are sent emails encouraging them to click on a university website. Each student is randomly assigned to one of two sites: a site with encouragement to vote in the upcoming student government election, and a neutral site with study tips. The students are then followed up to see if they voted. Define $y = 1$ if the student voted or 0 otherwise; define $u = 1$ if the student was assigned to the encouragement site or 0 if he or she was assigned to the neutral site; define $v = 1$ if the student actually accessed the site (which can be checked using unique identifiers) or 0 if he or she never clicked on the link.

(a) From which of the following regressions or pair of regressions can we compute the instrumental variables estimate of the effect of accessing the site on voting?

- Regression of $y$ on $u$.
- Regression of $y$ on $v$.
- Regression of $y$ on $u$ and $v$.
- Regression of $y$ on $u$, and the regression of $v$ on $u$.
- Regression of $y$ on $v$, and the regression of $v$ on $u$.

(b) What assumptions are required for the instrumental variables estimate to be reasonable in this case? Do these assumptions seem plausible here?

21.2 *Instrumental variables, simulating a population*: The goal of this exercise is to simulate data consistent with the assumptions of the instrumental variables procedure described in Section 21.1. You will also evaluate the properties of different approaches to estimating the complier average causal effect (CACE).

To help conceptualize the type of data that might be consistent with the instrumental variable assumptions, consider a hypothetical randomized encouragement design. In particular, imagine a study in which 1000 students entering any undergraduate degree program in the sciences in a major university are randomly assigned to one of two conditions. One group is encouraged via an email from the chair of their department to participate in a one-week math boot camp just before the start of their first semester. Students in the other group are also allowed to participate but receive no special encouragement. These students would have had to discover on their own the existence of the program on the university website. The running variable is test score on the final exam for a required math course, and the outcome variable $y$ that you will simulate below represents the *difference* between that score and the threshold for passing. Thus a negative value for a student reflects the fact that the student would not be eligible for that course.

Generate data for a sample of 1000 individuals consistent with the IV assumptions discussed in this chapter. Follow the guidelines below.

(a) Simulate compliance status. Assume that 25% of individuals are compliers, 60% are never-takers, and 15% are always-takers. Generate $d^0$ and $d^1$ vectors to reflect this. You can also generate a vector indicating compliance type, $c$, if that is helpful.

(b) Which compliance group has been omitted from consideration? What assumption does that imply?

(c) Simulate the potential outcomes in a way that meets the following criteria:

- The exclusion restriction is satisfied.

- The average effect of $z$ on $y$ for the compliers is 4.
- The average $y(z = 0)$ for never-takers is 0; the average $y(z = 0)$ for compliers is 3; the average $y(z = 0)$ for always-takers is 6.
- The residual standard deviation is 1 for everyone in the sample (generated independently for each potential outcome).

(d) Calculate the sample average treatment effect (SATE; average effect of $z$ on $y$ for the people in the experiment) for each of the compliance groups.

(e) What is another name for the SATE for the compliers?

(f) Calculate the intent-to-treat (ITT) effect using your simulated data.

(g) Put $T^0$, $T^1$, $y^0$, $y^1$ into one dataset called `dat_full`. You can also include a variable, $c$, indicating compliance group if you created one.

21.3 *Instrumental variables, playing the role of the experimenter*: Now switch to the role of the researcher. Pretend that you ran the experiment described in the previous exercise. Generate a binary indicator $z$ for to indicate who was to be encouraged to participate in the program. The probability of encouragement should be 0.5.

21.4 *Instrumental variables, understanding which potential outcome manifests as an observed outcome*: Use `dat_full` from the previous exercise to create a dataset that the researcher would actually get to see given the $z$ you just generated. This observed dataset should only have $T$, $z$, and $y$ in it. Call it `dat_obs`.

21.5 *Instrumental variables, inference*: Given the simulated `dat_obs`:

(a) Estimate the percentage of compliers, never-takers, and always-takers, assuming that there are no defiers.

(b) Perform the naive regression estimate of the effect of the treatment on the outcome. What is another name for this type of analysis?

(c) Estimate the intent-to-treat on the treated (ITT) effect.

(d) Estimate the complier average causal effect (CACE) by dividing the ITT estimate by the estimated percentage of compliers in the sample.

(e) Estimate the CACE by fitting two-stage regression as in Section 21.1.

(f) Estimate the CACE and its standard error using `brms`.

21.6 *Instrumental variables, assumptions*: Continuing Exercises 21.2–21.5:

(a) Describe the assumptions required for the instrumental variables procedure to give a good estimate of the treatment effect. We have generated data that satisfy these assumptions. Suppose instead you were handed data from the study described above. Comment on the plausibility of each of the required assumptions in that setting.

(b) Suppose that the data-generating process above included a covariate that predicted both $z$ and $y$. Which of the assumptions described in (a) would that change and how?

(c) Suppose that the above directions were amended as follows: "The average of $y^0$ for never-takers is 0; the average of $y^0$ for compliers is 3; the average of $y^0$ for always-takers is 6; the average of $y^1$ for never-takers is 2." Which of the assumptions of instrumental variables would that violate?

(d) How would you alter the simulation to violate the monotonicity assumption?

(e) How might one alter the administration of this program to preclude the existence of always-takers? Does this raise ethical questions?

21.7 *Instrumental variables, evaluating statistical properties when the assumptions hold*: Simulate a sampling distribution for any of the estimators used in Exercise 21.5. Which of these estimators is unbiased? For each, also report the standard deviation of the sampling distribution and compare to the standard error computed in Exercise 21.5.

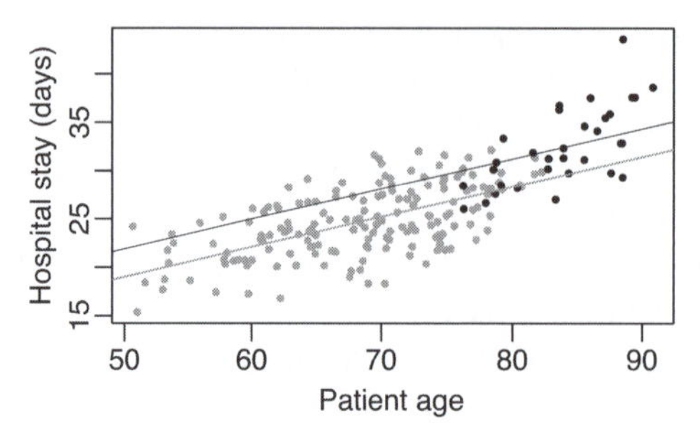


Figure 21.4 *Hypothetical data of length of hospital stay and age of patients, with separate points and regression lines plotted for each treatment condition: the new procedure in gray and the old procedure in black.*

21.8 *Instrumental variables, evaluating statistical properties when the assumptions are violated*: Simulate a sampling distribution for any of the estimators used in Exercise 21.5 in one of the worlds implied in Exercise 21.6. What happens to the properties of these estimators when the required assumptions do not hold?

21.9 *Regression discontinuity*: Take the Chile schools example from Section 21.3 and perform a series of analyses using different subsetting ranges. Plot the estimate ± standard error as a function of the subsetting range. As the width of the range increases, the standard error should go down, but there should be more of a concern about the use of the estimate for causal inference, given the lack of overlap.

21.10 *Regression discontinuity*: Suppose you are trying to evaluate the effect of a new procedure for coronary bypass surgery that is supposed to help with the postoperative healing process. The new procedure is risky, however, and is rarely performed in patients who are over 80 years old. Data for this (hypothetical) example are displayed in Figure 21.4.

(a) Does this seem like an appropriate setting in which to implement a regression discontinuity analysis?

(b) The folder `Bypass` contains data for this example: `stay` is the length of hospital stay after surgery, `age` is the age of the patient, and `new` is the indicator for whether the new surgical procedure was used. Preoperative disease severity (`severity`) was unobserved by the researchers, but we have access to it for illustrative purposes. Can you find any evidence using these data that the regression discontinuity design is inappropriate?

(c) Estimate the treatment effect using a regression discontinuity estimate (ignoring) severity. Estimate the treatment effect in any way you like, taking advantage of the information in severity. Explain the discrepancy between these estimates.

21.11 *Regression discontinuity, setting up an artificial world*: This assignment simulates hypothetical data collected on women who gave birth at any one of several hospitals in disadvantaged neighborhoods in New York City in 2010. We are envisioning a government policy that makes health care available for pregnant women, new mothers, and their children though 2 years post-birth. This program is only available for women in households with income below \$20 000 at the time they gave birth. The general question of interest is whether this program increases a measure of child health at age 3. You will generate data for a sample of 1000 individuals.

For this assignment we make the unrealistic assumption that everyone who is eligible for the program participates and no one participates who is not eligible. This is an example of a "clean" or "sharp" regression discontinuity design.

(a) Simulate the "assignment variable" (sometimes referred to as the running variable, forcing variable, or rating), income, in units of thousands of dollars. Call the variable `income`. Try to create a distribution that mimics the key features of the data displayed in Figure 21.5.

(b) Create an indicator for program eligibility for this sample. Call this variable `eligible`.

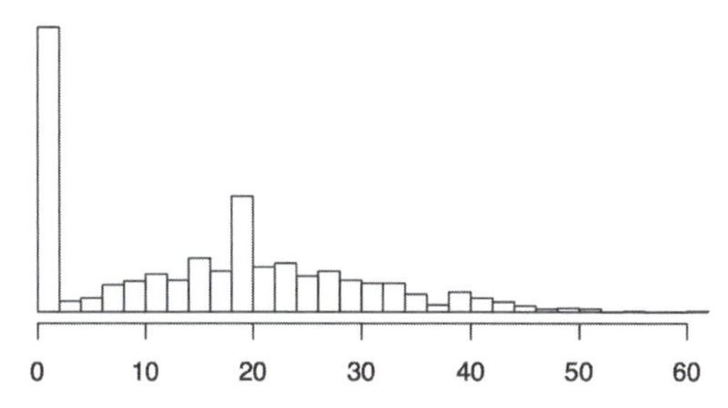


Figure 21.5 *Distribution of women's incomes (in tens of thousands of dollars) in the hypothetical regression discontinuity study in Exercise 21.11.*

(c) Simulate outcomes for World A. Generate the potential outcomes for health assuming linear models for both $\mathrm{E}(y^0|x)$ and $\mathrm{E}(y^1|x)$. This health measure should have a minimum possible score of 0 and maximum possible score of 30. The expected treatment effect for everyone should be 4; in other words, $\mathrm{E}(y^1 - y^0|x)$ should be 4 at all levels of $x$. The residual standard deviation of each potential outcome should be 1.

Save two datasets: (1) `fullA` should have the assignment variable and both potential outcomes and (2) `obsA` should have the assignment variable, the eligibility variable, and the observed outcome.

(d) Simulate outcomes for World B. Generate the potential outcomes for health assuming a linear model for $\mathrm{E}(y^0|x)$ and a quadratic model for $\mathrm{E}(y^1|x)$. The treatment effect at the threshold (the level of $x$ that determines eligibility) should be 4. The residual standard deviation of each potential outcome should be 1. Creating this data-generating process may be facilitated by using a transformed version of your income variable that subtracts out the threshold value.

Save two datasets: (1) `fullB` should have the assignment variable and both potential outcomes and (2) `obsB` should have the assignment variable, the eligibility variable, and the observed outcome.

21.12 *Regression discontinuity analysis*: Now you will act as a researcher and analyze the data created in the previous exercise.

(a) Plot your data! Make two scatterplots of observed health vs. income, one corresponding to each world. In each, plot eligible participants in red and non-eligible participants in blue.

(b) Estimate the treatment effect for World A and World B using all the data. Now we will estimate effects in a number of different ways. Each model should include reported income and eligible as predictors. In each case use the model fit to report the estimate of the effect of the program at the threshold level of income.

i. Fit a linear model to the full dataset. Do not include an interaction.
ii. Fit a linear model to the full dataset, including an interaction between income and eligible.
iii. Fit a model that is quadratic in income and includes an interaction between both income terms and eligible; that is, allow the shape of the relationship to vary between treatment and control groups.

(c) Fit the same models using the data from World B.

(d) Fit the same models using the data from World A but restricting the sample to those with incomes between \$16 000 and \$24 000.

(e) Fit the same models using the data from World B but restricting the sample to those with incomes between \$16 000 and \$24 000.

(f) Comment on the differences in the estimates from the previous four parts of this exercise.

(g) Provide a causal interpretation of your favorite estimate above.

21.13 *Regression discontinuity assumptions*: What are the three most important assumptions for the causal estimates in the previous exercise?

21.14 *Thinking harder about regression discontinuity*: Consider the scenario of Exercise 21.11.

(a) A colleague now points out that some women may have incentives in these settings to misreport their actual income. Plot a histogram of reported income and look for anything that might support such a claim. What assumption is called into question if women are truly misreporting in this manner?

(b) Another colleague points out that several other government programs such as food supplements and Head Start have the same income threshold for eligibility. How might this knowledge impact your interpretation of your results?

21.15 *Intermediate outcomes*: In Exercise 19.10, you estimated the effect of incumbency on votes for Congress. Now consider an additional variable: money raised by the congressional candidates. Assume this variable has been coded in some reasonable way to be positive in districts where the Democrat has raised more money and negative in districts where the Republican has raised more.

(a) Explain why it is inappropriate to include money as an additional input variable to "improve" the estimate of incumbency advantage in the regression in Exercise 19.10.

(b) Suppose you are interested in estimating the effect of money on the election outcome. Set this up as a causal inference problem (that is, define the treatments and potential outcomes).

(c) Explain why it is inappropriate to simply estimate the effect of money using instrumental variables, with incumbency as the instrument. Which of the instrumental variables assumptions would be reasonable in this example and which would be implausible?

(d) How could you estimate the effect of money on congressional election outcomes?

See Campbell (2002) and Gerber (2004) for more on this topic.

21.16 *Difference-in-differences estimation*: Consider the Electric Company example from Chapter 19, estimating a separate treatment effect in each grade. What is the difference-in-differences estimate here? Explain why the regression estimate used in that chapter is a better choice.

21.17 *Causes of effects and effects of causes*: Apply the ideas of Section 21.5 to an applied problem of interest for you. Consider reverse causal questions and then potential forward causal questions, along with associated experiments or experiments that could be used to estimate relevant causal effects.

21.18 *Working through your own example*: Continuing the example from the final exercises of the earlier chapters, consider a causal problem that it would make sense to estimate using instrumental variables. Perform the instrumental variables estimate, compare to the estimated causal effect using direct regression on the treatment variable (in both cases including relevant pre-treatment predictors in your regressions), and discuss the different assumptions of these different approaches.

# Part 6: What comes next?

# Chapter 22

# Advanced regression and multilevel models

Going forward, there are various ways in which we find it useful in applied work to push against the boundaries of linear regression and generalized linear models. Consider this concluding chapter as an introduction to various methods that would require much more space to cover in detail.

## 22.1 Expressing the models so far in a common framework

### Linear regression with one predictor

We start by reviewing what we have learned so far, starting with the basic linear regression, $y = a + bx + \text{error}$, or, more formally,

$$y_i = a + bx_i + \epsilon_i, \text{ for } i = 1, \ldots, n. \tag{22.1}$$

Our first extension to (22.1) was to give a probability distribution for the errors:

$$\epsilon_i \sim \mathrm{N}(0, \sigma^2), \text{ for } i = 1, \ldots, n : \tag{22.2}$$

independent errors with mean 0 and standard deviation $\sigma$. As discussed in Section 10.7, this model is also written as $y_i \sim \mathrm{N}(a + bx_i, \sigma^2)$ with the understanding that the errors are independent unless otherwise specified, or as $y \sim \mathrm{N}(a + bx, \sigma^2 I)$, where $x$ and $y$ are vectors of length $n$, and $I$ is the $n \times n$ identity matrix.

As discussed in Chapters 7–9, much can be done even with the basic model, starting with least squares estimation (which is equivalent to maximum likelihood for the model with independent, equal-variance, normally distributed errors), point estimation, and prediction; inclusion of prior information (which can also be viewed as regularization or stabilization of inferences); and forecasting including predictive uncertainty.

### Linear regression with multiple predictors

The next step is linear regression with multiple predictors, which can be viewed in two ways:

- Most directly, as presented in Chapters 10 and 12, we replace $a + bx$ in the basic formula by the expression $X\beta$, where $X$ is a matrix of predictors (including the "constant term" as a vector of 1's), as pictured in Figure 10.8.
- Alternatively, we can think of $X\beta$ as a combined *linear predictor*, which we can call $u$, so that $u_i = X_i\beta$, and then $y = u + \text{error}$. An advantage of this latter approach is that we can use this derived quantity, the linear predictor $u$, in further models or graphical explorations.

When there is more than one control variable, one approach is to plot on the $x$-axis the linear predictor created from all the control variables with coefficients estimated from their regression models. For example, with a regression model of the form $y_i = \beta_0 + \beta_1 X_{i1} + \beta_2 X_{i2} + \beta_3 X_{i3} + \text{error}$, one can plot $y_i$ versus $\beta_2 X_{i2} + \beta_3 X_{i3}$ with different symbols for different values of $X_{i1}$. In that plot one would plot dotted lines of $y = c + x$, for $c = \beta_0 + \beta_1 x_1$ for the different values for $x_1$, to illustrate the expected relationship. Figure 2.5 shows an example with one predictor that plays the role of "treatment" and other "background" predictors which are combined in the $x$-axis.

### Nonlinear models

Nonlinearity can be incorporated into the regression framework in different ways. Most simply, we can include nonlinear transformations within regression predictors, for example $y = \beta_0+\beta_1 x+\beta_2 x^2+\text{error}$ or $y = \beta_0 + \beta_1 \sin(x) + \beta_2 \sin(2x) + \text{error}$. This formulation is limited in that it only works when all but the linear parameters are known. One cannot, for example, use linear regression to estimate the parameter $\beta_2$ in the model $y = \beta_0 + \beta_1 \sin(\beta_2 x) + \text{error}$ or $y = \beta_0 + \beta_1 \exp(-\beta_2 x) + \text{error}$; for such problems we need a more general approach as discussed in Section 22.6.

Chapters 13–15 introduce logistic regression and generalized linear models, in which a linear predictor, $X\beta$, is transformed onto the space of data prediction using some nonlinear link function. We estimate these models using maximum likelihood or Bayesian inference, methods that can also be used for more general prediction models.

## 22.2 Incomplete data

In Chapter 17 we discussed the use of regression models for imputing missing data. Such methods can be helpful, especially in cleaning up small amounts of missingness so as to create a "rectangular" dataset (a matrix of $k$ observations on each of $n$ units). There are problems, however, where it is worth the effort to set up a model specifically to handle the missingness process.

Consider an experiment comparing time until death of cancer patients randomly assigned to treatment or control. The challenge in analyzing such data is that we only learn the length of life for patients who die before the study is over; for the others, the full length of life is, in statistics jargon, "censored."

Such data can be included in a regression model by including the censoring into the likelihood function. For example, in the peacekeeping example from page 7, the units of analysis are countries at civil war that received or did not receive United Nations peacekeeping, the background variable $x$ is a "badness measure" that captures pre-existing conditions in the country, and the outcome is the time until resumption of fighting. As shown in Figure 1.4, many of the cases are censored, meaning that there was no return to civil war during the period in which data were collected.

Censored data arise in *survival analysis*, where commonly used statistical methods include graphs and nonparametric analyses of survival times, and also extensions of maximum likelihood or Bayesian regression that are estimated by including the censoring into the likelihood. The same ideas also work for other missing-data patterns such as rounded data, or observations that are stochastically missing (for example, a survey of young children in which some ages are reported in months and others are rounded to the nearest year or half-year).

Another form of incomplete data is *measurement error*. In the linear regression $y = a+bx+\text{error}$, measurement error in $y$ does not pose a problem, as it folds right into the error term. Suppose we would like to estimate the "underlying" model $y = a + bx + \epsilon$ but $y$ is itself measured with error, so that what we actually observe is $y^* = y + \eta$. In that case, we can combine the two models to get $y^* = a + bx + \epsilon + \eta$, and if the two error terms are independent and each with mean 0, this reduces to a linear regression of $y^*$ on $x$, which can be fit directly; see Exercise 22.1.

It is more challenging to estimate a regression model when there is measurement error in the predictors. Suppose again that we want to estimate $y = a + bx + \epsilon$, but this time it is $x$ that is measured with error, with observations $x^*$ for which $x^* = x + \eta$. It might at first seem that we could shake this out and write, $y = a + b(x^* - \eta) + \epsilon$, thus $y = a + bx^* + \epsilon - b\eta$, implying that we could just regress $y$ on $x^*$, considering $(\epsilon - b\eta)$ as the new error term—but this won't work because this new error term will be correlated with the regression predictor. Regressing $y$ on $x^*$ will *not* return the desired slope $b$, not even in expectation and not even in the asymptotic limit of large sample size; see Exercise 22.2.

To correctly fit a regression with measurement error in $x$, one must know, or estimate, the variance of the measurement errors $\eta$. Having done that, one can either take the raw estimate from the regression of $y$ on $x^*$ and apply a bias correction, or else directly fit the full "simultaneous-equation model" (so called because the models for $y$ and $x^*$ are being estimated together) using a marginal likelihood or Bayesian approach. The mathematics of such models are the same as for instrumental variables (see Section 21.1), which makes sense, as instrumental variables are applied when regression is performed on an "instrument" rather than directly on the causal treatment variable of interest.

## 22.3 Correlated errors and multivariate models

Another direction in which regression models can be generalized is to consider correlated or structured errors. Examples include time series, spatial correlation, networks, and factor analysis. From a modeling standpoint, a common feature of many of these models is that they can be expressed as linear regressions and generalized linear models but with error terms that are dependent in some way, with the amount of dependence typically characterized by some hyperparameters. Estimation proceeds by likelihood or Bayesian methods as before, with the hyperparameters either pre-specified or estimated together with the regression coefficients.

We demonstrate with the simple example of a first-order autoregression on a time series. Start with the model $y_t = X_t\beta + \epsilon_t$, $t = 1, \dots, T$, and then give the errors the following model: $\epsilon_t \sim \mathrm{N}(\rho\,\epsilon_{t-1}, (1 - \rho^2)\,\sigma^2)$, with the residual variance of this autoregression set to $(1 - \rho^2)\,\sigma^2$, because then the math works out so the marginal variance becomes $\sigma^2$ for each of the error terms $\epsilon_t$. It is also necessary that the autocorrelation parameter $\rho$ be less than 1 in absolute value. The model as just stated can also be expressed as a multivariate regression, $y \sim \mathrm{N}(X\beta, \Sigma)$, where $\Sigma$ is a covariance matrix with elements $\Sigma_{ij} = \rho^{|i-j|}\sigma^2$, and one can simultaneously estimate the parameters $\beta, \sigma, \rho$ using maximum likelihood or Bayesian inference.

Modeling and inference for more complicated time series, spatial, etc., models proceeds the same way, in that any linear dependence structure for normally distributed errors can be expressed as a multivariate normal distribution on the errors.

## 22.4 Regularization for models with many predictors

There are various settings where we want to fit regressions with large numbers of predictors. In causal inference for observational studies, including more pre-treatment predictors can make the assumption of ignorability more plausible, thus giving us more confidence that we can interpret the coefficient on the treatment variable as a causal effect. In missing-data imputation, including more predictors can help give more reasonable imputed values for the missing observations. And if regression is simply being used for prediction or explanation of variation, as in the mesquite-bushes example in Section 12.6 or the arsenic example in Section 13.7, then the inclusion of more information can lead to more accurate predictions and better understanding.

The challenge is to ensure stability in estimation. If we increase the number of predictors $k$ while keeping the number of data points $n$ fixed, then estimates using least squares, maximum likelihood, or Bayesian inference with noninformative priors, can become increasingly noisy. Adding too many predictors can make estimates and predictions less accurate.

The traditional solution to the noise problem is to reduce the number of predictors, but this has the problem of discarding information. One way around the problem is to combine predictors in a structured way, rather than simply throwing them into the regression. For example, suppose that in the election forecasting model of Section 7.1, we had 10 different measures of economic performance. Instead of fitting a least squares regression with 10 predictors, we could first combine the measures into a single omnibus predictor (for example, by first putting each measure of economic performance on a standardized −1 to 1 scale and then averaging the 10 scores) followed by a linear regression with one predictor. This two-step procedure removes our ability to use the data to estimate the best weighting for the 10 factors, but in this case we would prefer it to a big regression as the latter would be too unstable.

In other settings, large numbers of predictors come in without there being any natural way to combine them via preprocessing, and so techniques have been developed for *regularizing* or stabilizing the predictions and inferences, by selecting only a subset of the input variables, or by pulling coefficients toward zero, or some combination of the two. Some of these methods go by the names ridge regression, lasso, and horseshoe.

Using just the techniques in this book, we can perform some regularization using Bayesian regression with informative priors; for example, if you think the coefficient for a particular variable is likely to be less than 0.1 in absolute value, you can assign it a prior distribution with mean 0 and standard deviation 0.1 in `stan_glm`. When sample size is small and the number of predictors is large, it will make sense to assign strong priors to most of the predictors to keep their coefficients controlled. In Chapter 11 we briefly mentioned the regularized horseshoe prior which is useful when we expect only a small number of coefficients to have a large magnitude but without knowing which ones we expect to be large, and the lasso prior, which partially pools coefficient estimates toward zero. The appropriate scale of priors will depend on context; as discussed in Section 9.1, `stan_glm` works by default on a unit scale setting priors corresponding to normalized predictors. We can think of more elaborate variable-selection and regularization procedures as adaptations or approximations of Bayesian regression where aspects of the prior distributions are fit from the data.

## 22.5 Multilevel or hierarchical models

Examples of hierarchical data include students within schools, or survey respondents with states: in the most simple two-level model, observational units are clustered within groups. What counts as a group depends on context. For example, in a survey in which multiple measurements are conducted on each person, the observational unit is the measurement, and the person is the group.

Structures with more than two levels can be *nested* (for example, students within schools within school districts, or respondents within Congressional districts within states) or *non-nested* (for example, students classified by school and by grade level, or election outcomes in each of 50 states for each of 10 years).

In hierarchical structures (sometimes called *multilevel* to avoid the implication of nesting that can come with the term *hierarchical*), it can make sense to allow regression coefficients to vary by group, and such variation can already be included in the simple regression framework. For example, if you want to fit the model $y = a + bx +$ error with a different intercept for respondents in each of the 50 states, just write `stan_glm(y ~ x + factor(state))`, and if you want to allow both intercepts and slopes to vary by state, just use the formula, `y ~ x + factor(state) + x:factor(state)`.

The challenge comes when the number of observations per group is small, so that simply including group indicators in a least squares regression gives unacceptably noisy estimates. Multilevel regression is a method of partially pooling varying coefficients, equivalent to Bayesian regression where the variation in the data is used to estimate prior distribution on the variation of intercepts and slopes.

## 22.6 Nonlinear models, a demonstration using Stan

Sometimes we want to fit a nonlinear model that cannot be expressed in the framework of linear predictors. For example, $y = a(1 + b_1x_1 + b_2x_2)/(1 + c_1x_1 + c_2x_2) +$ error, or $y = (a_1 e^{-b_1 x} + a_2 e^{-b_2 x}) * e^{\text{error}}$. Such models can be fit with maximum likelihood or Bayesian methods using Stan.

In this setting, we will set up a simple problem, display the data, fit a simple linear regression,

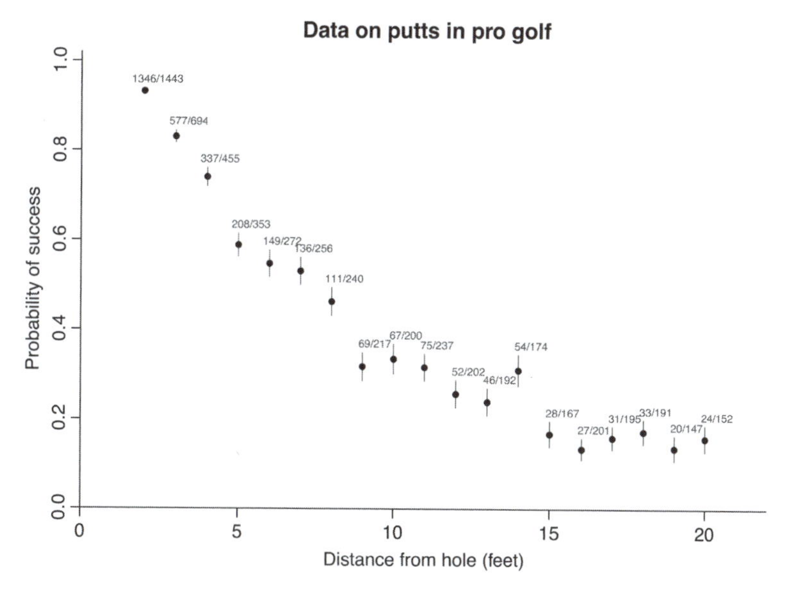


Figure 22.1 *Golf data. We would like to fit a model predicting the probability of the ball going in, as a function of distance from the hole. The error bars associated with each point j are simple binomial standard deviations,* $\sqrt{\hat{p}_j(1-\hat{p}_j)/n_j}$, *where* $\hat{p}_j = y_j/n_j$ *is the success rate for putts taken at distance* $x_j$.

then fit a custom-designed nonlinear regression in Stan. There is no space in this book to give a full introduction to Stan, but we will demonstrate its use here to give a sense of our workflow.

### Data from golf putts

Example: Golf putting

Figure 22.1 shows data on the proportion of successful golf putts as a function of distance from the hole.[1] Unsurprisingly, the probability of making the shot declines as a function of distance.

Can we model this? Given what we've seen so far in this book, the natural starting point would be logistic regression:

$$y_j \sim \text{binomial}(n_j, \text{logit}^{-1}(a + bx_j)), \text{ for } j = 1, \ldots, J.$$

In R, this can be fit using,

```
stan_glm(cbind(y, n-y) ~ x, family=binomial(link="logit"), data=golf)
```

And here is the result:

```
            Median MAD_SD
(Intercept)  2.23   0.06
x           -0.26   0.01
```

Figure 22.2 shows the fitted model along with the data.

### Logistic regression using Stan

Before moving to our nonlinear model, we fit the above logistic regression directly in Stan. We do this as a way of introducing Stan in a simple example where we already know the model.

Here is our Stan model, which we save in a file called `golf_logistic.stan`:

[1] Data and code for this example are in the folder `Golf`.

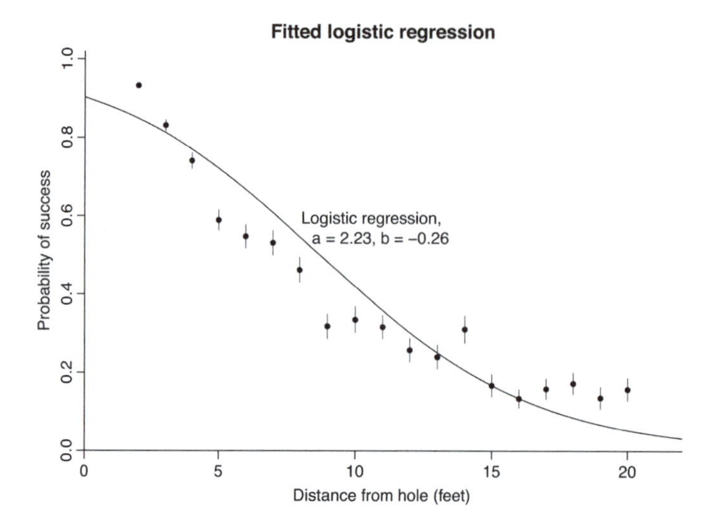


Figure 22.2: *Logistic regression fit to the golf-putting data from Figure 22.1.*

```
data {
  int J;
  int n[J];
  vector[J] x;
  int y[J];
}
parameters {
  real a;
  real b;
}
model {
  y ~ binomial_logit(n, a + b*x);
}
```

Without getting into details of the syntax, the Stan code maps pretty clearly onto the statistical model being fit. No prior distributions have been declared for the parameters $a$ and $b$ so by default they have been assigned uniform priors.

Here is the code for fitting the model:

```
golf <- read.table("golf.txt", header=TRUE, skip=2)
golf_data <- list(x=golf$x, y=golf$y, n=golf$n, J=nrow(golf))
fit_logistic <- stan("golf_logistic.stan", data=golf_data)
```

And here is the result:

```
     mean se_mean   sd    25%    50%    75% n_eff Rhat
a    2.23    0.00 0.06   2.19   2.23   2.27  1135    1
b   -0.26    0.00 0.01  -0.26  -0.26  -0.25  1101    1
```

Stan has computed the posterior means ± standard deviations of $a$ and $b$ to be $2.23 \pm 0.06$ and $-0.26 \pm 0.01$, respectively. The Monte Carlo standard error of the mean of each of these parameters is 0 (to two decimal places), indicating that the simulations have run long enough to estimate the posterior means precisely. The posterior quantiles give a sense of the uncertainty in the parameters, with 50% posterior intervals of [2.19, 2.27] and [−0.26, −0.25] for $a$ and $b$, respectively. Finally, the effective sample sizes of 1100 and $\widehat{R}$ near 1 tell us that the simulations from Stan's four simulated chains have mixed well.

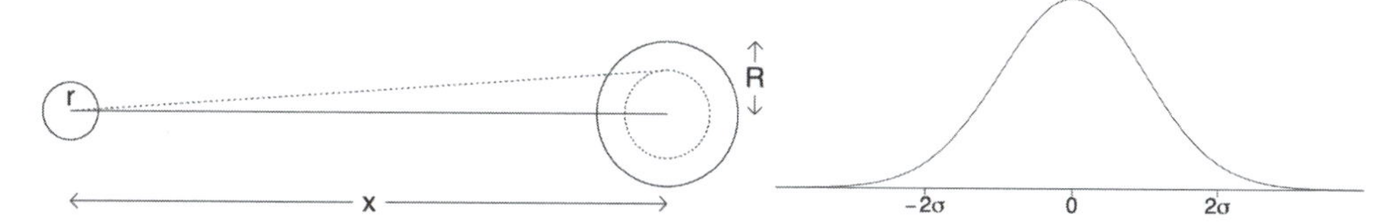


Figure 22.3 *Simple mathematical model of golf putting: (a) Geometry showing the range of angles for which the ball will go in the hole; (b) Assumed normal distribution for the angle of the shot. As the distance x to the hole increases, the probability of a successful shot decreases.*

### Fitting a nonlinear model from scratch using Stan

As an alternative to logistic regression, we build a model from first principles and fit it to the data. Figure 22.3a shows a simplified sketch of a golf shot. The dotted line represents the angle within which the ball of radius $r$ must be hit so that it falls within the hole of radius $R$. This threshold angle is $\sin^{-1}((R-r)/x)$.

The next step is to model human error. We assume that the golfer is attempting to hit the ball completely straight but that many small factors interfere with this goal, so that the actual angle follows a normal distribution centered at 0 with some standard deviation $\sigma$.

The probability that the ball goes in the hole is then the probability that the angle is less than the threshold; that is, $2\Phi(\sin^{-1}((R-r)/x)) - 1$, where $\Phi$ is the cumulative normal distribution function.

Our model then has two parts:

$$
\begin{aligned}
y_j &\sim \text{Binomial}(n_j, p_j), \\
p_j &= 2\Phi(\sin^{-1}((R-r)/x)) - 1, \text{ for } j = 1, \dots, J.
\end{aligned}
$$

We can fit this model using the following program which we save in the file golf_trig.stan:

```
data {
  int J;
  int n[J];
  vector[J] x;
  int y[J];
  real r;
  real R;
}
parameters {
  real<lower=0> sigma;
}
model {
  vector[J] p = 2*Phi(asin((R-r) ./ x) / sigma) - 1;
  y ~ binomial(n, p);
}
```

The ./ in the definition of $p$ represents elementwise division, so that the line of code is mathematically equivalent to computing `p[j] = 2*Phi(asin((R-r)/x[j])/sigma) - 1` for each element `p[j]` of the vector p.

Fitting the model is easy: the data $J, n, x, y$ have already been set up, and so we just need to define $r$ and $R$ (the golf ball and hole have diameters 1.68 and 4.25 inches, respectively), and call Stan:

```
r <- (1.68/2)/12
R <- (4.25/2)/12
golf_data <- list(x=golf$x, y=golf$y, n=golf$n, J=nrow(golf), r=r, R=R)
fit_trig <- stan("golf_trig.stan", data=golf_data)
```

Here is the result:

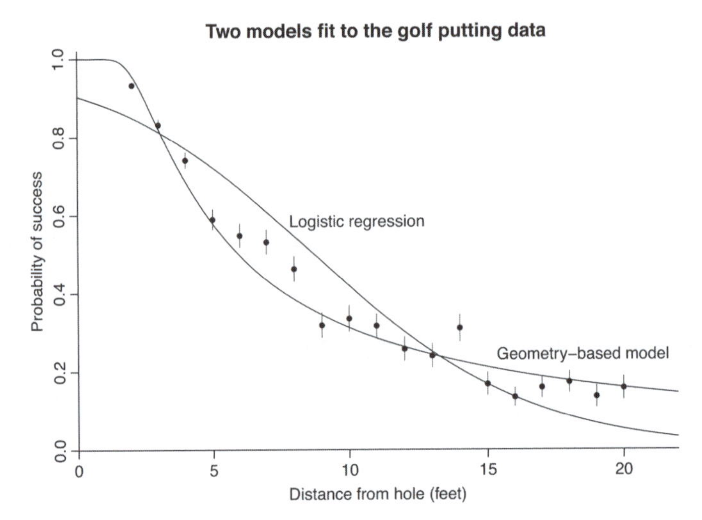


Figure 22.4 *Logistic regression and custom geometry-based model fit to the golf-putting data. The custom model fits better, which is particularly impressive, given that it contains only one parameter and the logistic regression has two.*

```
         mean se_mean     sd    25%    50%    75% n_eff   Rhat
sigma  0.0267  0.0000 0.0004 0.0264 0.0266 0.0269  1416 1.0015
```

The model has a single parameter, $\sigma$. From the output, we find that Stan has computed the posterior mean of $\sigma$ to be 0.0267 (multiplying this by $180/\pi$, this comes to 1.5 degrees). The Monte Carlo standard error of the mean is 0 (to four decimal places), indicating that the simulations have run long enough to estimate the posterior mean precisely. The posterior standard deviation is calculated at 0.0004 (that is, 0.02 degrees), indicating that $\sigma$ itself has been estimated with high precision, which makes sense given the large number of data points and the simplicity of the model. The precise posterior distribution of $\sigma$ can also be seen from the narrow range of the posterior quantiles. Finally, the effective sample size of 1400 and $\widehat{R}$ near 1 tell us that the simulations from Stan's four simulated chains have mixed well.

### Comparing the two models

Figure 22.4 plots the data and the fitted model (here using the posterior median of $\sigma$ but in this case the uncertainty is so narrow that any reasonable posterior summary would give essentially the same result), along with the logistic regression fitted earlier. The custom nonlinear model fits the data much better. This is not to say that the model is perfect—any experience with golf will reveal that the angle is not the only factor determining whether the ball goes in the hole—but it seems like a useful start, and it is good to know that we can fit nonlinear models by just coding them up in Stan.

## 22.7 Nonparametric regression and machine learning

So far, we have considered nonlinear models in which the functional form is specified by the user. In *nonparametric regression*, the regression curve $E(y|x)$ is not constrained to follow any particular parametric form, instead adapting to the data.

The unavoidable concern with nonparametric modeling is *overfitting*: a model with no constraints can be fit arbitrarily close to data, yielding a low prediction error on observed data but noisy inferences and poor performance on new cases. Effective nonparametric modeling works via "soft constraints," which enforce some level of smoothness on the regression curve, penalize its roughness, or use some sort of cross validation (see Section 11.8) to control or regularize the inference.

The term *machine learning* is used to describe various nonparametric regression algorithms. One distinction between machine learning and regression is that the former is focused more on prediction of individual cases, while the latter is associated more with parameter estimation.

Nonparametric regression and machine learning methods have *tuning parameters* that govern the scale at which they adapt to the data. Sometimes the tuning parameter is set from the outside; in other cases the tuning parameter is itself estimated from the data using a method such as hierarchical modeling or cross validation.

### Loess

Perhaps the simplest general-purpose nonparametric regression algorithm is locally weighted regression, also called *loess*, in which $\mathrm{E}(y|x)$ at each data point $x_i$ is estimated using a weighted regression that gives higher weights to the data points with predictors that are close to $x_i$ and lower weights to more distant points. Loess is easy to understand and to use, but the procedure is not so flexible, in part because it is a procedure that is applied to data but does not have any underlying statistical model.

The tuning parameter of loess controls the severity of the local weighting, the fraction $f$ of points included in each local regression. At one extreme, at $f \to 0$, all the weight for the local regression goes to the point at the value $x_i$ exactly, and the fitted curve at this point goes through $y_i$ (or through the local average if there are multiple data points $i$ with the same value of $x$). The resulting regression curve goes through all the data or local averages. At the other extreme, $f = 1$, all points are weighted equally, and loess reduces to simple linear regression. For intermediate values of $f$, the loess curve represents some tradeoff between smoothness and fidelity to the data.

### Splines and Gaussian processes

*Splines* are a model-based approach in which a nonlinear regression curve is expressed as the sum of many localized pieces, with the entire fit constructed to maximize the smoothness of these individual terms. The tuning parameter in a spline model determines the degree of local smoothness in the estimated regression curve.

A fitted spline can be expressed as a linear regression in which the predictors are particular pre-specified nonlinear functions, called *basis functions*, of the input variables in the model. In the scenario with only one input, $x$, we can write $y = \sum_{h=1}^{H} \beta_h b_h(x) + \text{error}$, with the basis functions $b_h$ being locally smooth. In higher dimensions, splines can be constructed as additive models, with a separate spline for each input variable and interactions if desired, or the basis functions can be multivariate, in which case a single spline is fit to the entire vector of predictors. An advantage of splines is that they can easily be incorporated into our general framework by simply adding regression predictors. When data are sparse, it can be important to include informative priors on these new coefficients.

Another way to penalize roughness in nonparametric regression is by setting up a prior distribution that rewards smoothness in the underlying regression function. In the context of a regression model, $y = g(x) + \text{error}$, a *Gaussian process* is a prior distribution on $g$ in which, for any vector $x = (x_1, \dots, x_n)$, $g(x)$ is modeled as coming from a multivariate normal distribution with correlation matrix depending on the distances of the points $x$ from each other: For any points $x_i, x_j$, the correlation between $g(x_i)$ and $g(x_j)$ can be written as a function of $d_{ij} = ||x_i - x_j||$ so that the correlation approaches 1 as $d_{ij} \to 0$, and the correlation goes to 0 as $d_{ij} \to \infty$. The scale of the correlation matrix—how fast the correlations decline as a function of the distance $d_{ij}$—serves as a tuning parameter for the model. Gaussian processes are a flexible class of models that can work well in multiple dimensions. When $n$ is large, however, in its general form the fitted Gaussian process has dimensionality equal to the number of data points and so computations can be slow, hence much effort has gone into efficient approximate algorithms for fitting these and similar models.

### Tree models and BART

*Tree models* for prediction are typically set up using binary splits of predictors. If we let $T$ denote a binary tree, $T$ consists of the tree structure and the rule corresponding to each split. For a prediction tree with $b$ leaves, we label $M = (\mu_1, \ldots, \mu_b)$ as the predicted values corresponding to these leaves. Thus, $T, M$ corresponds to a prediction rule for $\mathrm{E}(y|x)$, which we write as $g(x, T, M)$, where $x$ represents the vector of inputs that are used in the model. The tree $T$ corresponds to some set of nested partitions of the space of $x$. The inputs $x$ can be discrete or continuous; if the latter, the locations of the partitions are considered as part of the information encoded in $T$.

A *Bayesian additive regression tree (BART)* consists of two pieces: a sum over $J$ trees and a regularization prior. The model for continuous outcomes can be written as $y = \sum_{j=1}^{J} g(x, T_j, M_j) +$ error. For the normal model, one would assume independent, equal-variance normally distributed errors by default. For continuous outcomes restricted to be positive, it would be most natural to apply the predictive model to log $y$; and for other sorts of data, link functions can be used by analogy to generalized linear models, for example, $\Pr(y = 1) = \mathrm{logit}^{-1}(\sum_{j=1}^{J} g(x, T_j, M_j))$ if the outcome variable $y$ is binary.

The motivation for the sum-of-trees model is to represent patterns in the prediction that cannot be easily captured by a single tree. Any particular tree can be a terrible representation of $\mathrm{E}(y|x)$, but the sum of a hundred or so little trees can have enough resolution to fit the data well.

The Bayesian part of BART is the prior distribution on the trees $T_j$ and the vectors $M_j$ of predictive values. Given the flexibility of the sum-of-trees model and the lack of identification, the prior is essential, regularizing the overall fit so the model does not overfit and also limiting the contribution of each tree to the overall sum. The prior favors smaller trees and predictive components that are near zero, thus keeping the fit smooth except to the extent that the data force otherwise.

### Machine learning meta-algorithms

Many other techniques have been developed for statistical prediction and machine learning, with the common goal being to include as much information as possible into the predictors and allowing flexibility in the functional form, while avoiding overfitting. In principle, one can assess overfitting by comparing within-sample to out-of-sample prediction errors, using cross validation when only a single dataset is available; see Section 11.8. Cross validation does not solve all overfitting problems—if a model is itself fit to minimize cross validation error on an existing dataset, then it could still be overfitting, just at a slightly higher level. As a result, there is still a need for theoretical understanding and practical experience when considering methods with large numbers of free parameters.

Some important ideas in machine learning include ensemble learning (averaging several different prediction models in a way that should outperform any individual component), deep learning (an approach in which nonlinear predictive models are applied in a sequence of latent or hidden layers to form a network that can fit data more flexibly than any of these single models alone), and genetic algorithms (in which a set of prediction models are set up and then allowed to "mutate," "breed," and "evolve," following a series of open-ended rules which, by trial and error, can lead to prediction algorithms that would never have been constructed by a human from scratch).

These are called meta-algorithms because they adapt existing classes of prediction methods, and they are based on the idea of automatically building flexible regression models from many small, interchangeable components. As such, these methods are well suited to parallel implementations, which can be a necessity when data sizes become large. Conversely, big-data machine learning methods are typically set up in such generality that they can in practice *require* a large sample to work well. One would not, for example, imagine fitting such a model to the election forecasting example in Section 7.1. Linear and logistic regression models make strong assumptions that allow us

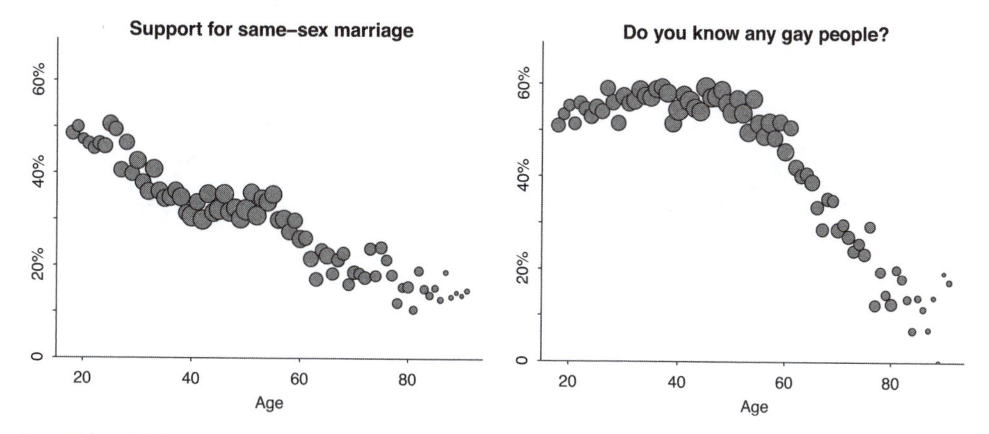


Figure 22.5 *(a) Support for same-sex marriage and (b) percentage of people who say they know at least one gay person, as a function of age, from a national survey taken in 2004. The patterns are nonlinear with no clear functional forms, hence the motivation to fit using a nonparametric regression. The area of each circle is proportional to the number of respondents at each age. There are fewer respondents at the highest ages, hence the noisier appearance of the data above age 80 in each graph.*

to summarize small datasets while getting a handle on predictive and inferential uncertainty, and Bayesian inference can allow even stronger statements, as long as relevant information can be supplied in the prior distribution. In contrast, machine learning meta-algorithms supply very little structure, with the aim of maximum flexibility for problems where enough data are available to support it.

### Demonstration: political attitudes as a function of age

xample: ttitudes on ay rights

We demonstrate nonparametric regression using the examples of support for same-sex marriage, and percentage of people who say they know someone gay, as a function of age, using responses from a national political survey from 2004.[2] Figure 22.5 shows the raw data aggregated by age: the patterns are nonlinear and do not appear to follow any simple functional forms. The complicated patterns arise from the combination of several factors, including steady generational trends in attitudes toward gay rights along with variation in partisanship at shorter time scales corresponding to voters who came of age during different political eras from the New Deal through the Clinton presidency. The point of nonlinear fits for these responses is not to directly make any causal discoveries but rather to summarize the patterns in the data.

Figure 22.6 shows support for same-sex marriage and the percentage who know any gay people as a function of age, estimated separately using loess and splines, using current default settings in R's `loess` and `stan_gamm4` functions, respectively. The estimates are broadly similar, with the largest differences near the endpoints, where the loess extrapolations look more like straight lines, while the splines are pulled toward constants.

## 22.8 Computational efficiency

As datasets become larger and more complex and realistic, computation becomes more expensive for several reasons. The time required to compute a model is roughly proportional to the number of observations multiplied by the number of predictors. In addition, big data are typically messy data, so that we need to include more predictors when adjusting for differences between sample and population, and differences between treatment and control groups for causal inference. And, beyond

[2]Data and code for this example are in the folder `Gay`.

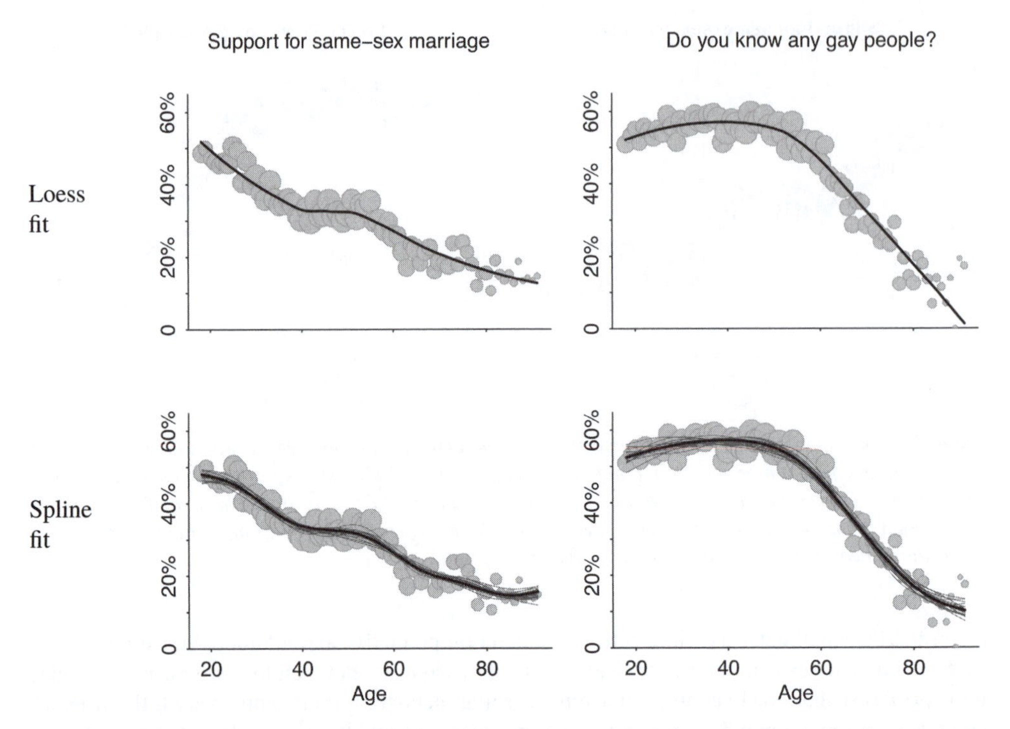


Figure 22.6 *Estimated support for same-sex marriage, and percentage of people who say they know someone gay, as a function of age, based on the data in Figure 22.5, fit separately using two different nonparametric regression models: loess and splines. The fits are similar but differ at the boundaries of the data, which is where the choice of model becomes more decisive.*

this, when data are richer, departures from linearity can become more apparent in any particular application, motivating nonparametric regression, which comes with its own computational costs.

Further discussion of these issues of *scalability* is beyond the scope of this book, but we find it helpful when thinking of advanced regression models to be aware at some level of the tradeoffs between realism of the model (which facilitates accurate prediction for individual cases, generalization to larger populations, and causal inference) and computational cost. A central research goal in statistics and machine learning is to push against the efficient frontier so as to allow more elaborate models to be fit to larger, more complex datasets in less time.

Most of the examples in this book are small, with no more than a few thousand data points and only a few predictors, in which case we can simply run `stan_glm` at its default settings and wait a few seconds for reliable results. But when the number of data points or predictors is large, computation can be slow, and we can use alternative, approximate algorithms.

### The efficiency of models fit in Stan

Before discussing diagnostics and an alternative fitting algorithm, we briefly describe the workings of `stan_glm`. Consider a model with $D$ parameters; for example, in a simple linear regression, $y = a + bx + \text{error}$, the dimensionality is $D = 3$, corresponding to $a$, $b$, and $\sigma$, the standard deviation of the errors. Let $\theta$ represent the $D$-dimensional vector of parameters. Stan fits models using a stochastic algorithm called Hamiltonian Monte Carlo[3] (HMC) that is programmed to take a guided

[3]The algorithm is named after Hamiltonian dynamics, a model of physics that is used to construct the steps of the computation, and Monte Carlo, the district of Monaco that is associated with casinos and random algorithms more generally.

random walk through parameter space. Starting at some initial point $\theta^0$, the procedure goes through points $\theta^t, t = 1, 2, \dots,$, with the algorithm constructed so that, when it has run long enough, the set of iterations $\theta$ looks like a random sample from the target distribution, in this case the posterior distribution of $\theta$.

We will not go into the details of Hamiltonian Monte Carlo here, but we need to understand some things about it in order not tp get tripped up when fitting regressions in difficult problems.

For our purposes here, the two most important aspects of HMC are that is it *iterative*—it does not jump directly to the solution; instead, it proceeds step by step, only fully converging asymptotically—and it is *stochastic*: running it multiple times will give different results. These two issues complicate our analysis, as we must, first, make sure that the algorithm has run long enough to have reached approximate convergence, and, second, appropriately acknowledge the variation in the results. In typical regression settings, Stan handles both these tasks automatically, so mostly what we need to learn here is how to understand its output.

### Convergence and effective sample size

At the time of this writing, `stan_glm` in its default setting runs four randomly initialized chains in parallel, each with 1000 warm-up iterations followed by 1000 iterations that are saved, yielding a total of 4000 simulated draws of the vector of parameters in the model. Running several chains is a form of replication that allows us to compare the results from different chains. If the results are in agreement we have more confidence that sampling is working well. If the results are in disagreement we say that the chains have not mixed well and that sampling has not reached approximate convergence. Hamiltonian Monte Carlo works well for all the models and datasets used in this book, but it is useful to be aware of the available diagnostics that are needed for harder problems.

Figure 8.2a shows the fitted regression line to the economy and elections data, and Figure 8.2b shows 50 Bayesian posterior draws obtained from Stan.[4] We can check the convergence diagnostics after fitting the model:

```
M1 <- stan_glm(vote ~ growth, data=hibbs)
summary(M1)
```

The relevant diagnostic part of the output is,

```
              mcse Rhat n_eff
(Intercept)   0.0  1.0  3194
growth        0.0  1.0  3567
sigma         0.0  1.0  2743
```

`Rhat` denotes the split-$\widehat{R}$ convergence diagnostic, which compares the results from different chains, reaching the value 1 at convergence and being higher than 1 for chains that have not fully mixed. `n_eff` denotes the effective sample size of the Hamiltonian Monte Carlo run, which due to its iterative nature can be less or more than the efficiency of independent draws. Usually `n_eff` > 400 is sufficient for reliable diagnostics and accurate parameter estimates. `mcse` stands for Monte Carlo standard error, which is the additional uncertainty due to using stochastic algorithm (not to be confused with the uncertainty presented by the posterior distribution). Monte Carlo standard error is negligible in all the examples in this book.

### Parallel processing

We often do our computing on computers that contain multiple processors. To take advantage of this, execute this command after loading in the `rstanarm` package:

```
options(mc.cores = parallel::detectCores())
```

[4]Data and code for this example are in the folder `ElectionsEconomy`.

This sets up Stan to automatically run in parallel using multiple processors that can give an automatic factor of 4 speedup on many laptops. Further improvements in efficiency of Stan using parallel computing are under development.

### Mode-based approximations

Bayesian inference using `stan_glm` is slower than maximum likelihood using `glm`. In case of linear regression with normal prior, the speed difference is small. In case of generalized linear models, speed can be a concern with large datasets, for regressions with many predictors, and when the computation is embedded within some larger algorithm where real-time processing is required.

What, then, do we do in settings where speed is a concern but we want to retain the advantages of Bayesian inference: the ability to include prior information and to propagate uncertainty using simulation? Various algorithms are available that approximate the full Bayesian calculations but at lower computational cost.

Here we discuss one such approach: the normal approximation centered at the posterior mode. The idea is to take the Bayesian posterior distribution, and instead of sampling from it, which can be slow when the number of regression coefficients is large, we use an optimization algorithm to find its mode, and then use the curvature of the posterior density to construct a normal approximation which we use to take random draws. This optimization is as fast as maximum likelihood—indeed, it can be faster, in that the prior distribution can make the optimization problem more stable—and can be a good choice for problems where full Bayesian inference is too slow.

To perform the normal approximation, just add the argument `algorithm="optimizing"` to the `stan_glm` call, and then everything else proceeds as before.

We illustrate with a simple fake-data example.[5] First we create predictors, assume true values of the logistic regression coefficients, simulate data, and save into a data frame:

```
n <- 1e4
k <- 100
x <- rnorm(n)
xn <- matrix(rnorm(n*(k-1)), nrow=n)
a <- 2
b <- 3
sigma <- 1
y <- as.numeric(a + b*x + sigma*rnorm(n) > 0)
fake <- data.frame(x, xn, y)
```

We then fit the logistic regression in three different ways:

```
fit_1 <- glm(y ~ ., data=fake, family=binomial())
fit_2 <- stan_glm(y ~ ., data=fake, family=binomial(), algorithm="optimizing")
fit_3 <- stan_glm(y ~ ., data=fake, family=binomial())
```

The algorithms used to create `fit_2` and `fit_3` are different, but both return point estimates and simulations that can be used for probabilistic estimation and prediction, as discussed in the earlier chapters.

The normal approximation is not perfect, especially for small datasets where there is more uncertainty in the inference or for more elaborate priors such as the regularized horseshoe mentioned in Chapter 11, so we recommend using `stan_glm` in its usual, non-optimizing, setting where possible. For more advanced models, going beyond simple regressions and generalized linear models, further approximations are possible, but this goes beyond the scope of this book.

[5] Code for this example is in the folder `Scalability`.

## 22.9 Bibliographic note

The material in this chapter will be expanded in our forthcoming book on advanced regression and multilevel models, which includes an expanded version of the second half of Gelman and Hill (2007).

Cox and Oakes (1984) is a classic text on hazard regression and survival analysis. For more on measurement error models, see Fuller (1987) and Carroll et al. (2006). Some references on models for structured data include Box and Jenkins (1976), Brillinger (1981) and Pole, West, and Harrison (1994) for time series; Cressie (1993) and Banerjee, Gelfand, and Carlin (2003) for spatial models; and Hoff, Raftery, and Handcock (2002) and Hoff (2005) for network models.

Ridge regression was introduced by Hoerl and Kennard (1970), lasso by Tibshirani (1996), and horseshoe regression by Carvalho, Polson, and Scott (2010). Other relevant papers on variable selection and regularization for regression include Hoeting et al., (1999); Chipman, George, and McCulloch (2001); and Piironen and Vehtari (2017b, c).

Three classic texts on multilevel models are Snijders and Bosker (1999), Kreft and De Leeuw (1998), and Raudenbush and Bryk (2002).

The golf data come from Berry (1995) and the model from Gelman and Nolan (2002); see Gelman (2019b) for an expansion of the model.

The original form of loess appeared in Cleveland (1979). Wahba (1978) and Rasmussen and Williams (2006) are important references on splines and Gaussian processes, respectively; see also Hastie and Tibshirani (1990). BART was introduced by Chipman, George, and McCulloch (2010); see Hill, Linero, and Murray (2020) for a recent review. Bishop (2006) and Hastie, Tibshirani, and Friedman (2009) are early overviews of machine learning methods in statistics.

The convergence diagnostic $\widehat{R}$ was introduced by Gelman and Rubin (1992), and further developments appear in Gelman et al. (2013) and Vehtari et al. (2019).

## 22.10 Exercises

22.1 *Measurement error in $y$*: Simulate data $(x, y)_i$, $i = 1, \dots, n$ from a linear regression model, $y = a + bx +$ error, but suppose that the outcome $y$ is not observed directly, but instead we observe $v = y +$ error, with independent measurement errors with mean zero. Use simulations to understand the statistical properties of the observed-data regression of $v$ on $x$, compared to the desired regression of $y$ on $x$.

22.2 *Measurement error in $x$*: Simulate data $(x, y)_i$, $i = 1, \dots, n$ from a linear regression model, $y = a + bx +$ error, but suppose that the predictor $x$ is not observed directly, but instead we observe $u = x +$ error, with independent measurement errors with mean zero. Use simulations to understand the statistical properties of the observed-data regression of $y$ on $u$, compared to the desired regression of $y$ on $x$.

22.3 *Nonlinear modeling*: The folder `Golf` contains the dataset used in Section 22.6 and also an additional, larger dataset on golf putting, including putts of up to 75 feet in length.

(a) Fit the two models in Section 22.6 to this new data. Display the data and fitted models, and comment on the fit.

(b) Expand the geometry-based model to allow for the possibility that the putt can be hit too short or too long. Tinker with the model as necessary to get a reasonable fit to the larger dataset.

22.4 *Nonlinear modeling*: In the absence of air resistance, a falling object has acceleration equal to the gravitational constant $g$, and so a dropped object will fall a distance $\frac{1}{2}gt^2$ during time $t$. For this exercise you will conduct an experiment to estimate the gravitational constant. Using a yardstick, mark a set of heights on a wall, for example, 3, 4, 5, 6, and 7 feet off the ground. At each height, drop a ball twice and measure the time it takes to fall. Using these data and a measurement error model, fit a model in Stan to estimate $g$.

22.5 *Smoothing and sample size*: Take a random sample of 1/10 of the data from the Gay folder and re-fit the loess and spline models to estimate the support for same-sex marriage and percentage of people who say they know at least one gay person. Compare to the estimates from the full data shown in Figure 22.6.

22.6 *Smoothing and sample size*: Repeating the previous exercise with different fractions of data subsampled, explore the behavior of loess and splines at their default settings as a function of sample size.

22.7 *Nonparametric modeling of a continuous outcome*: Find data of interest to you with a continuous outcome and one continuous predictor and at least 100 data points. Fit loess, spline, and BART and plot these along with the data. Discuss how these fitted models differ. If the three fits are essentially identical, add some new (fake) points to the dataset to create differences between the three fitted models.

22.8 *Nonparametric modeling of a binary outcome*: Repeat the previous exercise, but with a binary outcome. This could be a different example or simply a discretized version of the outcome variable in the previous exercise, as long as the discretization is of substantive interest and is not just an arbitrary cutpoint.

22.9 *Visualizing a fitted nonparametric model in multiple dimensions*: Find data of interest to you with a continuous outcome and two continuous predictors and at least 100 data points. Fit loess, spline, and BART. Consider different displays of the data and fitted model, such as three-dimensional plots or separate plots of $y$ vs. $x_1$ and $y$ vs. $x_2$.

22.10 *Computational efficiency*: Perform fake-data experiments to study the speed of optimization and full Bayesian inference for logistic regression, following the example at the end of Section 22.8. Make a graph on the log-log scale showing compute time as a function of sample size for a fixed number of predictors.

22.11 *Working through your own example*: Continuing the example from the final exercises of the earlier chapters, consider a prediction problem where the linear regressions and generalized linear models would not fit the data well. Construct a predictive model using one of the more advanced methods discussed in this chapter, and use both graphical methods and cross-validation compare the result to a simpler linear regression or generalized linear model fit.

# Appendixes

# Appendix A

# Computing in R

Computing is central to our conception of statistics, both in practice—we can do many more things in statistics if we do not restrict ourselves to problems with analytical solutions—and in teaching, as we use simulation and graphics to understand data, models, and the statistical properties of model fits. In this appendix, we explain how to get started with R and Stan and how to access the data and examples used in the book. We then give a brief introduction to R. There are many R tutorials online that can provide additional information; the introduction here is focused on some of the data processing, analysis, and programming tasks that arise in applied regression. We also provide pointers to useful packages for data analysis in R and Stan.

## A.1 Downloading and installing R and Stan

We do our computing in the open-source package R, a command-based statistical software environment that you can download and operate on your own computer, using RStudio, a free graphical user interface for R. We fit Bayesian regressions in `rstanarm`, an R package that calls Stan, an open-source program written in C++. The `rstanarm` package includes a library of already compiled Stan models for fitting linear and logistic regression, generalized linear models, and a selection of other models that can be called using `stan_glm` and other functions. See `www.mc-stan.org/rstanarm` for documentation and vignettes for modeling examples. Stan itself is more general, allowing the user to program and fit arbitrary Bayesian models.

### R and RStudio

To set up R, go to `www.r-project.org` and click on the "download R" link. This will take you to a list of mirror sites. Choose any of these. Now click on the link under "Download and Install R" at the top of the page for your operating system (Linux, Mac, or Windows) and download the binaries corresponding to your system. Follow all the instructions. Default settings should be fine.

Go to `www.rstudio.com`, click on "Download RStudio," click on "Download RStudio Desktop," and click on the installer for your platform (Windows, Mac, etc.). Then a download will occur.

When the download is done, click on the RStudio icon to start an R session.

To check that R is working, go to the Console window within RStudio. There should be a ">" prompt. Type `2+5` and hit Enter. You should get `[1] 7`. From now on, when we say "type" _____, we mean type _____ into the R console and hit Enter.

You will also want to install some R *packages*, in particular, `ggplot2` (for building graphs), `knitr` (which allows you to process certain R documentation), `bayesplot` (which has convenient functions for displaying fitted models), `rstanarm` (which has functions for fitting Bayesian regression models using simulation), `rstan` (which allows you to fit more general models in Stan), `loo` (fast leave-one-out cross validation; see Section 11.8), `survey` (which contains functions for classical survey analysis, some of which we use in working with propensity scores in Chapter 20), `arm` (which includes functions we have written for applied regression modeling, some of which are used in

Chapter 21), and rprojroot (which has been used in the code for this book to make it easier to work with many folders). Some of the example code also uses other packages which you can see at the beginning of each code file. You can install these packages from RStudio by clicking on the Tools tab and then on Install Packages and then entering the names of the packages and clicking Install. When you want to use a package, you load it as needed, for example by typing library("brms").

### Stan and rstanarm

If your only use for Stan is to fit the regression models in this book, it will be enough to install rstanarm as discussed above, and then begin any R session with,

```
library("rstanarm")
```

We almost always do our computing by writing scripts, rather than simply typing commands into the R console window; thus, we recommend simply typing the above line at the beginning of any of your R scripts that include calls to stan_glm.

For more flexible Bayesian modeling you will need to use Stan itself, for which downloads, documentation, and other information are available at www.mc-stan.org. Follow the instructions to set up rstan, the R interface to Stan. This will automatically install Stan itself on your computer, which in turn requires a C++ compiler; again, the necessary instructions are at the rstan set-up page.

Examples to get started are in the rstan and rstanarm documentation. We list some useful functions in Section A.8.

## A.2 Accessing data and code for the examples in the book

The data and code used in the examples and to make the figures in this book are available on Github, linked from www.stat.columbia.edu/~gelman/regression. You can copy all the files to your own computer by clicking on the Clone or Download button and then clicking on Download ZIP, or by using a direct link. If you are familiar with git version control, you can also clone the repository. For quick viewing of code and example results, they are also available as web pages at this directory.

## A.3 The basics

Computing is central to modern statistics at all levels, from basic to advanced. If you already know how to program, great. If not, consider this a start.

Try a few things, typing these one line at a time and looking at the results on the console:

```
1/3
sqrt(2)
curve(x^2 + 5, from=-2, to=2)
```

These will return 0.3333333, 1.414214, and a new graphics window plotting the curve $y = x^2 + 5$.

Finally, quit your R session by closing the RStudio window.

### Calling functions and getting help

Reopen R and play around, using the assignment function ("<-"). To start, type the following lines into your script.R file and copy-and-paste them into the R window:

```
a <- 3
print(a)
b <- 10
a + b
a*b
```

```
exp(a)
10^a
log(b)
log10(b)
a^b
round(3.435, 0)
round(3.435, 1)
round(3.435, 2)
```

R is based on *functions*, which include mathematical operations (`exp`, `log`, `sqrt`, and so forth) and lots of other routines (`print`, `round`, ...).

The function `c()` concatenates numbers together into a vector. For example, type `c(4,10,-1,2.4)` in the R console or type the following:

```
x <- c(4,10,-1,2.4)
print(x)
```

The function `seq` creates an equally-spaced sequence of numbers; for example, `seq(4,54,10)` returns the sequence, 4, 14, 24, 34, 44, 54. The `seq` function works with non-integers as well: try `seq(0,1,0.1)` or `seq(2,-5,-0.4)`. For integers, `a:b` is shorthand for `seq(a,b,1)` if `b>a`, or `seq(a,b,-1)` if `b<a`. Let's try a few more commands:

```
c(1, 3, 5)
1:5
c(1:5, 1, 3, 5)
c(1:5, 10:20)
seq(-1, 9, 2)
```

You can get help on any function using "?" in R. For example, type `?seq`. This should open a window with a help file for `seq`. R help files typically have more information than you'll know what to do with, but if you scroll to the bottom of the page you'll find some examples that you can cut and paste into your console. Whenever you are trying out a new function, we recommend using "?" to view the help file and running the examples at the bottom to see what happens.

### Sampling and random numbers

Here's how to get a random number, uniformly distributed between 0 and 100:

```
runif(1, 0, 100)
```

And now 50 more random numbers:

```
runif(50, 0, 100)
```

Suppose we want to pick one of three colors with equal probability:

```
color <- c("blue", "red", "green")
sample(color, 1)
```

Suppose we want to sample with unequal probabilities:

```
color <- c("blue", "red", "green")
p <- c(0.5, 0.3, 0.2)
sample(color, 1, prob=p)
```

Or we can do it all in one line, which is more compact but less readable:

```
sample(c("blue","red","green"), 1, prob=c(0.5,0.3,0.2))
```

### Data types

**Numeric data.** In R, numbers are stored as *numeric* data. This includes many of the examples above as well as special constants such as `pi`.

**Big and small numbers.** R recognizes scientific notation. A million can be typed in as 1000000 or 1e6, but not as 1,000,000. (R is particular about certain things. Capitalization matters, "," doesn't belong in numbers, and spaces usually aren't important.) Scientific notation also works for small numbers: 1e-6 is 0.000001 and 4.3e-6 is 0.0000043.

**Infinity.** Type these into R, one line at a time, and see what happens:

```
1/0
-1/0
exp(1000)
exp(-1000)
1/Inf
Inf + Inf
-Inf - Inf
0/0
Inf - Inf
```

Those last two operations return NaN (Not a Number); type ?Inf for more on the topic. In general we try to avoid working with infinity, but it is convenient to have Inf for those times when we accidentally divide by 0 or perform some other illegal mathematical operation.

**Missing data.** In R, NA is a special keyword that represents missing data. For more information, type ?NA. Try these commands in R:

```
NA
2 + NA
NA - NA
NA / NA
NA * NA
c(NA, NA, NA)
c(1, NA, 3)
10 * c(1, NA, 3)
NA / 0
NA + Inf
is.na(NA)
is.na(c(1, NA, 3))
```

The is.na function tests whether or not the argument is NA. The last line operates on each element of the vector and returns a vector with three values that indicate whether the corresponding input is NA.

**Character strings.** Let's sample a random color and a random number and put them together:

```
color <- sample(c("blue","red","green"), 1, prob=c(0.5,0.3,0.2))
number <- runif(1, 0, 100)
paste(color, number)
```

Here's something prettier:

```
paste(color, round(number,0))
```

## TRUE, FALSE, and ifelse

Try typing these:

```
2 + 3 == 4
2 + 3 == 5
1 < 2
2 < 1
```

In R, the expressions ==, <, > are *comparisons* and return a logical value, `TRUE` or `FALSE` as appropriate. Other comparisons include <= (less than or equal), >= (greater than or equal), and != (not equal).

Comparisons can be used in combination with the `ifelse` function. The first argument takes a logical statement, the second argument is an expression to be evaluated if the statement is true, and the third argument is evaluated if the statement is false. Suppose we want to pick a random number between 0 and 100 and then choose the color red if the number is below 30 or blue otherwise:

```
number <- runif(1, 0, 100)
color <- ifelse(number<30, "red", "blue")
```

## Loops

A key aspect of computer programming is *looping*—that is, setting up a series of commands to be performed over and over. Start by trying out the simplest possible loop:

```
for (i in 1:10){
  print("hello")
}
```

Or:

```
for (i in 1:10){
  print(i)
}
```

Or:

```
for (i in 1:10){
  print(paste("hello", i))
}
```

The curly braces define what is repeated in the loop. The spaces and line breaks are not necessary—one could just as well do `for(i in 1:10)print(paste("hello",i))`—but they improve readability.

Here's a loop of random colors:

```
for (i in 1:10){
  number <- runif(1, 0, 100)
  color <- ifelse(number<30, "red", "blue")
  print(color)
}
```

## Working with vectors

In R, a vector is a list of items. These items can include numerics, characters, or logicals. A single value is actually represented as a vector with one element. Here are some vectors:

- (1, 2, 3, 4, 5)
- (3, 4, 1, 1, 1)
- ("A", "B", "C")

Here's the R code to create these:

```
x <- 1:5
y <- c(3, 4, 1, 1, 1)
z <- c("A", "B", "C")
```

And here's a random vector of 5 random numbers between 0 and 100:

```
u <- runif(5, 0, 100)
```

Mathematical operations on vectors are done componentwise. Take a look:

```
x
y
x + y
1000*x + u
```

There are scalar operations on vectors:

```
1 + x
2 * x
x / 3
x^4
```

We can summarize vectors in various ways, including the sum and the average (called the "mean" in statistics jargon):

```
sum(x)
mean(x)
```

We can also compute weighted averages if we know the weights. We illustrate with a vector of three elements:

```
x <- c(100, 200, 600)
w1 <- c(1/3, 1/3, 1/3)
w2 <- c(0.5, 0.2, 0.3)
```

In the above code, the vector of weights `w1` has the effect of counting each of the three items equally; vector `w2` counts the first item more. Here are the weighted averages:

```
sum(w1*x)
sum(w2*x)
```

Or suppose we want to weight in proportion to population:

```
N <- c(310e6, 112e6, 34e6)
sum(N*x)/sum(N)
```

Or, equivalently,

```
N <- c(310e6, 112e6, 34e6)
w <- N/sum(N)
sum(w*x)
```

The `cumsum` function does the cumulative sum. Try this:

```
a <- c(1, 1, 1, 1, 1)
cumsum(a)
a <- c(2, 4, 6, 8, 10)
cumsum(a)
```

## Subscripting

Vectors can be indexed by using brackets, "`[ ]`". Within the brackets we can put in a vector of elements we are interested in either as a vector of numbers or a logical vector. When using a vector of numbers, the vector can be of arbitrary length, but when indexing using a logical vector, the length of the vector must match the length of the vector you are indexing. Try these:

```
a <- c("A", "B", "C", "D", "E", "F", "G", "H", "I", "J")
a[1]
a[2]
a[4:6]
a[c(1,3,5)]
a[c(8,1:3,2)]
a[c(FALSE, FALSE, FALSE, TRUE, TRUE, TRUE, FALSE, FALSE, FALSE, FALSE)]
```

As we have seen in some of the previous examples, we can perform mathematical operations on vectors. These vectors have to be the same length, however. If the vectors are not the same length, we can subset the vectors so they are compatible. Try these:

```
x <- c(1, 1, 1, 2, 2)
y <- c(2, 4, 6)
x[1:3] + y
x[3:5] * y
y[3]^x[4]
x + y
```

The last line runs but produces a warning. These warnings should not be ignored since it isn't guaranteed that R would carry out the operation as you intended.

## A.4 Reading, writing, and looking at data

### Your working directory

Choose a *working directory* on your computer where you will do your R work. Suppose your working directory is `c:/myfiles/stat/`. Then you should put all your data files in this directory, and all the files and graphs you save in R will appear here too. To set your working directory in RStudio, click on the Session tab and then on Set Working Directory and then on Choose Directory and then navigate from there.

RStudio has several subwindows: a text editor, the R console, a graphics window, and a help window. It's generally best to type your commands into the RStudio text editor, select the lines you want to run, and run them by pressing Ctrl-Enter. When your session is over, you can save the contents of the text editor into a plain-text file with a name such as `todays_work.R` that you can save in your working directory.

Now go to the R console and type `getwd()`. This shows your R working directory, which you can change by typing `setwd("c:/myfiles/stat/")`, or whatever you would like to use. In RStudio you can also choose the working directory using menu `Session -> Set Working Directory`. You can type `getwd()` again to confirm that you are in the right place.

The code for the book uses the `rprojroot` package, which makes it easy to run code from different subfolders without need to change the working directory to a specific folder. It is sufficient to set the working directory to the main demo folder or any of the subfolders.

### Reading data

Let's read some data into R. The file `heads.csv` has data from a coin-flipping experiment done in a previous class. Each student flipped a coin 10 times, and we have a count of the number of students who saw exactly 0, 1, 2, ..., 10 heads. The data file is in the folder `Coins` at our data and code repository. Start by going to this location, finding the directory, downloading the file, and saving it as `heads.csv` in your working directory (for example, `c:/myfiles/stat/`).

Now read the file into R:

```
heads <- read.csv("heads.csv")
```

Typing the name of any object in R displays it. So type heads and look at what comes out.

If you have tabular data separated by spaces or tabs, you can just use the `read.table` function. Here is an example of the file `mile.txt` from the folder `Mile`:

```
mile <- read.table("mile.txt", header=TRUE)
mile[1:5,]
```

The `header=TRUE` argument is appropriate here because the first line of the file mile.txt is a "header," that is, a list of column names. If the file had just data with no header, we would simply call `read.table("mile.txt")` with no header argument.

### Writing data

You can save data into a file using `write` instead of `read`. For example, to write the R object heads into a comma-separated file `output1.csv`, we would type `write.csv(heads,"output1.csv")`. To write it into a space-separated file `output2.txt`, it's just `write.table(heads,"output2.txt")`.

### Examining data

At this point, you should have two variables in your R environment, heads and `mile`. You can see what you have in your session by clicking on the Environment tab in the RStudio window.

**Data frames, vectors, and subscripting.** Most of the functions used to read data return a structure called a *data frame*. You can see an example by typing `class(heads)`. Each column of a data frame is a vector. We can access the first column of heads by typing `heads[,1]`. Data frames are indexed using two vectors inside "[" and "]"; the two vectors are separated by a comma. The first vector indicates which rows you are interested in, and the second vector indicates what columns you are interested in. For example, `heads[6,1]` shows the number of heads observed and `heads[6,2]` shows the number of students that observed that number of heads. Leaving it blank is shorthand for including all. Try:

```
heads[6,]
heads[1:3,]
heads[,1]
heads[,1:2]
heads[,]
```

To find the number of columns in a data frame, use the function `length`. To find the number of rows, use `nrow`. We can also find the names of the columns by using `names`. Try these:

```
length(heads)
nrow(heads)
names(heads)
```

## A.5 Making graphs

### Graphing data

Example: Mile run

**Scatterplots.** A scatterplot shows the relation between two variables. The data frame called `mile` in the folder `Mile` contains the world record times in the mile run since 1900 as four columns, `yr`, `month`, `min`, and `sec`. For convenience we create the following derived quantities in R:

```
mile$year <- mile$yr + mile$month/12
mile$seconds <- mile$min*60 + mile$sec
```

Figure A.1 plots the world record time (in seconds) against time. We can create scatterplots in R using `plot`. Here is the code to make the basic graph:

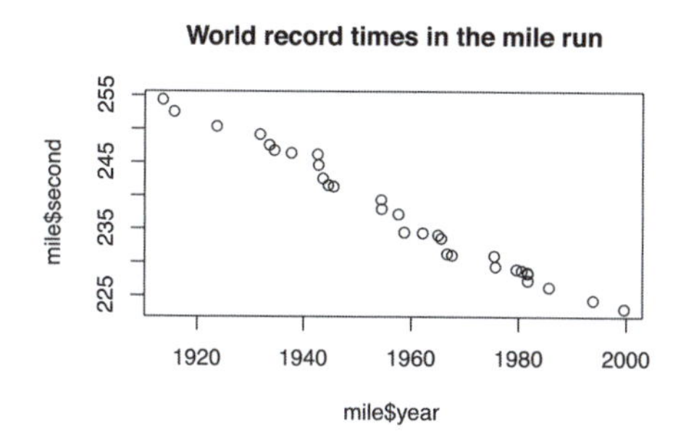


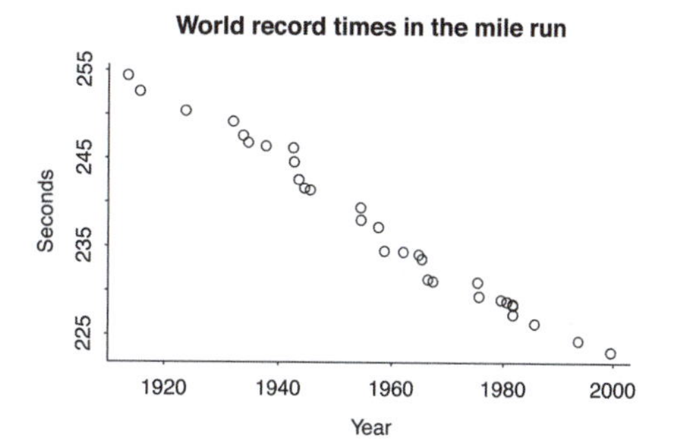


Figure A.1 *World record times (in minutes) in the mile run since 1900. The left plot was made with the basic function call:* plot(mile$year, mile$seconds, main="World record times in the mile run"). *The right plot was made with more formatting:* par(mar=c(3,3,3,1), mgp=c(2,.5,0), tck=-.01); plot(mile$year, mile$seconds, bty="l", main="World record times in the mile run", xlab="Year", ylab="Seconds").

```
plot(mile$year, mile$seconds, main="World record times in the mile run")
```

And here is a slightly prettier version:

```
par(mar=c(3,3,3,1), mgp=c(2,.5,0), tck=-.01)
plot(mile$year, mile$seconds, bty="l",
     main="World record times in the mile run", xlab="Year", ylab="Seconds")
```

The `par` function sets graphical parameters, in this case reducing the blank border around the graph, placing the labels closer to the axes, and reducing the size of the tick marks, compared to the default settings in R. In addition, in the call to `plot`, we have set the box type to `l`, which makes an "L-shaped" box for the graph rather than fully enclosing it.

We want to focus on the basics, so for the rest of this section we will show simple `plot` calls that won't make such pretty graphs. But we thought it would be helpful to show one example of a pretty graph, hence the code just shown above. The folder `Mile` also includes an example of making a pretty graph using the `ggplot2` package.

**Fitting a line to data.** We can fit a linear trend of world record times as follows:

```
fit <- stan_glm(seconds ~ year, data=mile)
print(fit)
```

Here is the result:

```
            Median MAD_SD
(Intercept) 1006.1   23.3
year          -0.4    0.0

Auxiliary parameter(s):
      Median MAD_SD
sigma 1.4    0.2
```

Interpretation of these results is discussed elsewhere in the book. All you need to know now is that the estimated coefficients are 1006.1 and –0.4; that is, the fitted regression line is $y = 1006.1 - 0.4 * \text{year}$. It will help to have more significant digits on this slope, so we type `print(fit, digits=2)` to get:

```
            Median   MAD_SD
(Intercept) 1006.15   23.33
year          -0.39    0.01

Auxiliary parameter(s):
      Median MAD_SD
sigma 1.42   0.19
```

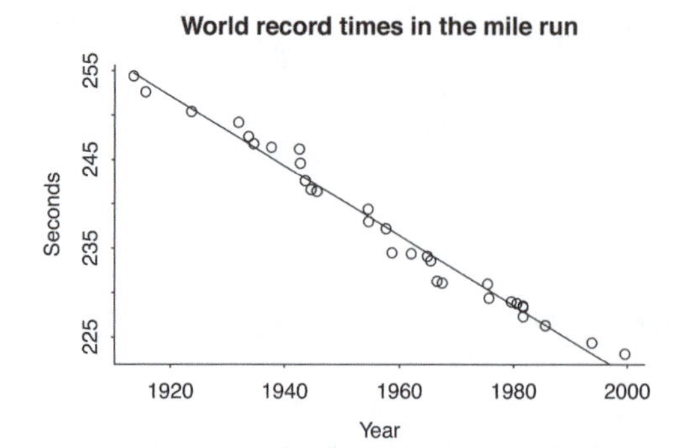


Figure A.2 *Fitted line predicting the world record time in the mile run given year, overlaying the data shown in Figure A.1b, plotted by adding the line,* curve(coef(fit)[1] + coef(fit)[2]*x, add=TRUE), *to that code.*

The estimated line is $y = 1006.15 - 0.39 * \text{year}$.

We can add the straight line to the scatterplot by adding the following line after the call to `plot`:

```
curve(1006.14 - 0.39*x, add=TRUE)
```

The first argument is the equation of the line as a function of $x$. The second argument, add=TRUE, tells R to draw the line onto the existing scatterplot. More cleanly, we can extract the parameters for the curve from the fitted regression:

```
curve(coef(fit)[1] + coef(fit)[2]*x, add=TRUE)
```

The result is displayed in Figure A.2.

### Multiple graphs on a page

Visualizations can be much more powerful using *small multiples*: repeated graphs on a similar theme. There are various ways to put multiple graphs on a page in R; one way uses the `par` function with its `mfrow` option, which tells R to lay out graphs in a grid, row by row.

We illustrate in Figure A.3 with a simple example plotting random numbers:

```
par(mfrow=c(2,4))
for (i in 1:2){
  for (j in 1:4){
    x <- rnorm(10)
    y <- rnorm(10)
    plot(x, y, main=paste("Row", i, "Column", j))
  }
}
```

The actual code we used to make the graphs also includes instructions to label the graphs, put them on a common scale, and size them to better fit on the page, but the above code gives the basic idea.

## A.6 Working with messy data

### Reading in survey data, one question at a time

Example: Height and weight

Data on the heights, weights, and earnings of a random sample of Americans are available from the Work, Family, and Well-Being Survey conducted by Catherine Ross in 1990. We downloaded the data file, `06666-0001-Data.txt`, and the codebook, `06666-0001-Codebook.txt`, from the Inter-university Consortium for Political and Social Research. Information on the survey is at `dx.doi.org/10.3886/ICPSR06666` and can be downloaded from the ICPSR.

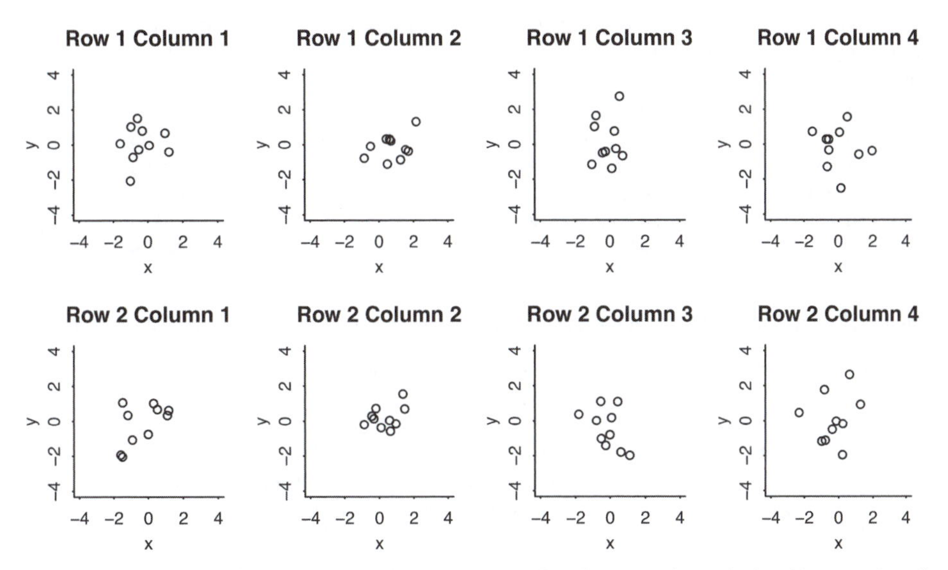


Figure A.3 *Example of a grid of graphs of random numbers produced in R. Each graph plots 10 pairs of numbers randomly sampled from the 2-dimensional normal distribution, and the display has been formatted to make 20 plots visible on a single page with common scales on the x and y-axes and customized plot titles.*

We saved the files under the names `wfw90.dat` and `wfwcodebook.txt` in the `Earnings` folder.

Figure A.4 shows the first 10 lines of the data, and Figure A.5 shows the relevant portion of the codebook. Our first step is to save the data file `wfwcodebook.txt` in our working directory. We then want to extract the responses to the questions of interest. To do this we first create a simple function to read columns of data, making use of R's function `read.fwf` (read in fixed width format).

Copy the following into the R console and then you will be able to use our function for reading one variable at a time from the survey. The following code is a bit tricky, so you're not expected to understand it; you can just copy it in and use it.

```
read.columns <- function (filename, columns) {
  start <- min(columns)
  length <- max(columns) - start + 1
  if (start==1)
    return(read.fwf(filename, widths=length))
  else
    return(read.fwf(filename, widths=c(start-1, length))[,2])
}
```

The data do not come in a convenient comma-separated or tab-separated format, so we use the function `read.columns` to read the coded responses, one question at a time:

```
height_feet <- read.columns("wfw90.dat", 144)
height_inches <- read.columns("wfw90.dat", 145:146)
weight <- read.columns("wfw90.dat", 147:149)
earn_exact <- read.columns("wfw90.dat", 203:208)
earn2 <- read.columns("wfw90.dat", 209:210)
sex <- read.columns("wfw90.dat", 219)
```

## Cleaning data within R

We now must put the data together in a useful form, doing the following for each variable of interest:

1. Look at the data.

```
100022  31659123     121222113121432 22  2 3 411179797979797 1  4 503100100...
100081  486 2122     111222141122221222  2 1 997979797979797 1  4  01 25 25...
100091   1371123     1232122111113111314 1 0                          30100...
100101  15684222     133122121113232 22  1 0                          10 40...
100111  25371122     122222111111421222  2 2 4 6979797979797 1  2 853 30 95...
100202   2389013     1111221412   22 314 2 0                         100100...
100281   7884021     2232132422   42 22  2 0                          80 60...
100351  15684221     233223242112212 32  2 0                          75100...
100571  88341221     243233321113452 12  2 0                          90 98...
100641  15684223     122112211113432 22  2 0                         100 75...
```

Figure A.4 *First 10 lines of the file* wfw90.dat, *which has data from the Work, Family, and Well-Being Survey.*

```
    HEIGHT      144-146   F3.0      Q.46 HEIGHT IN INCHES
    WEIGHT      147-149   F3.0      Q.47 WEIGHT
. . .
    EARN1       203-208   F6.0      Q.61  PERSONAL INCOME - EXACT AMOUNT
    EARN2       209-210   F2.0      Q.61  PERSONAL INCOME - APPROXIMATION
    SEX         219       F1.0      Q.63  GENDER OF RESPONDENT
. . .
    HEIGHT
    46. What is your height without shoes on?
        __________ ft.  __________in.
    WEIGHT
    47. What is your weight without clothing?
        __________ lbs.
. . .
    61a. During 1989, what was your personal income from your own wages,
    salary, or other sources, before taxes?
    EARN1
    $ __________--> (SKIP TO Q-62a)
    DON'T KNOW . . . 98
    REFUSED  . . . . 99
```

Figure A.5 *Selected rows of the file* wfwcodebook.txt, *which first identifies the columns in the data corresponding to each survey question and then gives the question wordings.*

2. Identify errors or missing data.
3. Transform or combine raw data into summaries of interest.

We start with `table(sex)`, which simply yields:

```
sex
  1    2
749 1282
```

No problems. But we prefer to have a more descriptive name, so we define a new variable, `male`:

```
male <- 2 - sex
```

This *indicator variable* equals 0 for men and 1 for women.

We next look at height, typing: `table(height_feet, height_inches)`. Here is the result:

```
           height_inches
height_feet   0   1   2   3   4   5   6   7   8   9  10  11  98  99
          4   0   0   0   0   0   0   0   0   0   1   3  17   0   0
          5  66  56 144 173 250 155 247 127 174 105 145  90   0   0
          6 129  59  46  20   8   5   1   0   0   0   1   0   0   0
          7   0   0   0   0   0   0   0   1   0   0   0   0   0   0
          9   0   0   0   0   0   0   0   0   0   0   0   0   2   6
```

Most of the data look fine, but there are some people with 9 feet and 98 or 99 inches (missing data codes) and one person who is 7 feet 7 inches tall (probably a data error). We recode as missing:

```
height_inches[height_inches>11] <- NA
height_feet[height_feet>=7] <- NA
```

And then we define a combined height variable:

```
height <- 12*height_feet + height_inches
```

Next, we type `table(weight)` and get the following:

```
weight
 80  85  87  89  90  92  93  95  96  98  99 100 102 103 104 105 106 107 108 110
  1   1   1   1   1   1   1   2   2   3   1  12   5   4   3  16   1   5   7  46
111 112 113 114 115 116 117 118 119 120 121 122 123 124 125 126 127 128 129 130
  4  15   5   5  42   5   4  21   4  72   4  14  20  11  61  11   3  25   8 106
131 132 133 134 135 136 137 138 139 140 141 142 143 144 145 146 147 148 149 150
  4  16   9   5  85   9  10  15   4  94   1  12   2   4  74   2   7   8   5 121
151 152 153 154 155 156 157 158 159 160 161 162 163 164 165 166 167 168 169 170
  2   4   8   7  49   3   6  14   1  88   1   8   4   9  65   2   2   8   2  81
171 172 173 174 175 176 178 180 181 182 183 184 185 186 187 188 189 190 192 193
  2  10   4   2  58   5   4  78   3   4   2   4  62   1   5   1   3  46   1   3
194 195 196 197 198 199 200 201 202 203 205 206 207 208 209 210 211 212 214 215
  4  26   3   2   3   2  57   1   2   2  11   2   3   2   3  36   1   2   3  10
217 218 219 220 221 222 223 225 228 230 231 235 237 240 241 244 248 250 255 256
  1   1   1  21   2   1   1  13   2  17   1   3   1  13   1   1   1  10   3   1
260 265 268 270 275 280 295 312 342 998 999
  2   3   1   3   1   3   1   1   1   6  36
```

Everything looks fine until the end. The numbers 998 and 999 must be missing data, which we duly code as such:

```
weight[weight>500] <- NA
```

Coding the earnings responses is more complicated. The variable `earn_exact` contains responses (in dollars per year) for those who answered the question. Typing `table(is.na(earn_exact))` reveals that 1380 people answered the question (that is, `is.na(earn_exact)` is `FALSE` for these respondents), and 651 did not answer (`is.na` is `TRUE`). These nonrespondents were asked a second, discrete, earnings question, `earn2`, which gives earnings in round numbers (in thousands of dollars per year) for people who were willing to answer in this way. A careful look at the codebook reveals a set of nonreponse codes requring recoding of the data; details appear in the file `earnings_setup.R` in the `Earnings` folder. In particular, a code of 90 represents nonresponse or refusal to answer, and a code of 1 corresponds to people who did not supply an exact earnings value but did say it was more than \$100 000. For these people we perform a simple imputation using the median of the exact earnings responses that exceeded 100 000:

```
earn_approx[earn2>=90] <- NA
earn_approx[earn2==1] <- median(earn_exact[earn_exact>100000], na.rm=TRUE)/1000
```

We divided by 1000 because `earn_approx` is on the scale of thousands of dollars. We then create a combined earnings variable:

```
earn <- ifelse(is.na(earn_exact), 1000*earn_approx, earn_exact)
```

The new `earn` variable still has 211 missing values (out of 2031 respondents in total) and is imperfect in various ways, but we have to make some choices when working with real data.

### Looking at the data

If you stare at the table of responses to the weight question you can see more. People typically round to the nearest 5 or 10 pounds, and so we see a lot of weights reported as 100, 105, 110, and so forth, but not so many in between. Beyond this, people appear to like round numbers: 57 people report weights of 200 pounds, compared to only 46 and 36 people reporting 190 and 210, respectively.

Similarly, if we go back to reported heights, we see some evidence that the reported numbers do not correspond exactly to physical heights: 129 people report heights of exactly 6 feet, compared to 90 people at 5 feet 11 inches and 59 people at 6 feet 1 inch. Who are these people? Let's look at the breakdown of height and sex:

```
table(male, height)
```

Here's the result:

```
       height
male  57  58  59  60  61  62  63  64  65  66  67  68  69  70  71  72  73  74  75
   0   1   3  17  63  54 140 170 236 142 200  84  93  28  26   8   4   3   2   1
   1   0   0   0   3   2   4   3  14  13  47  43  81  77 119  82 125  56  44  19
     height
male  76  77  78  82
   0   0   0   0   0
   1   8   5   1   1
```

The extra 6-footers (72 inches tall) are just about all men. But there appear to be too many women of exactly 5 feet tall and an excess of both men and women who report being exactly 5 feet 6 inches.

## A.7 Some R programming

### Writing your own functions

You can write your own function. Most of the functions we will be writing will take in one or more vectors and return a vector. Below is an example of a simple function that triples the value provided:

```
triple <- function(x) {3*x}
```

For example, to call this function, type `triple(c(1,2,3,5))`. This function has one argument, `x`. The body of the function is within the curly braces and the arguments of the function are available for use within the braces. In our example function, we multiply `x` by 3 and we return it back to the user. If we wanted to have more than one argument, we could do something like this:

```
new_function <- function(x, y, a, b) {
  a*x+b*y
}
```

### Optimization

Example: Simple optimization

**Finding the peak of a parabola.** Figure A.6 shows the parabola $y = 15 + 10x - 2x^2$. As you can see, the peak is at $x = 2.5$. How can we find this solution systematically? Finding the maximum of a function is called an *optimization* problem. Here's how we do it in R and Stan.

1. Plot the function, which you can do using `curve(15+10*x-2*x^2,from=___,to=___)`, entering numbers in the blanks. If necessary, you can play around with the "from" and "to" arguments until the maximum appears on the graph.
2. Write it as a Stan program called `parabola.stan`:[1]

[1]Code for this example is in the folder `Parabola`.

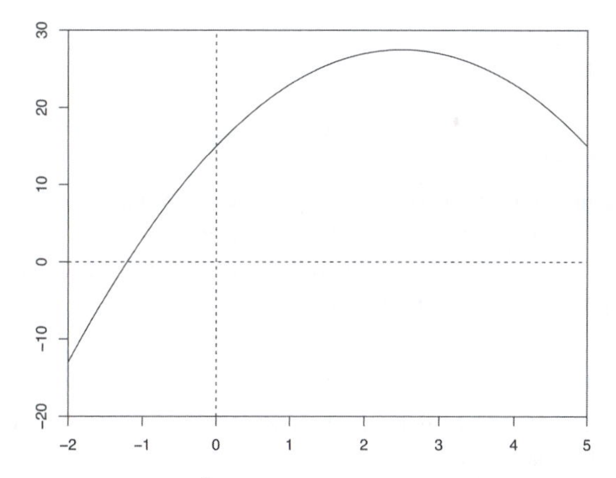


Figure A.6 *The parabola* $y = 15 + 10x - 2x^2$, *plotted in R using* curve(15 + 10*x - 2*x^2, from=-2, to=5). *The maximum is at* $x = 2.5$.

```
parameters {
  real x;
}
model {
  target += 15 + 10*x - 2*x^2;
}
```

3. In R, compile the Stan function and optimize it:

```
library("rstan")
model <- stan_model("parabola.stan")
fit <- optimizing(model)
print(fit)
```

This yields:

```
$par
  x
2.5

$value
[1] 27.5

$return_code
[1] 0
```

The output labeled `par` is the value of $x$ at which the function is optimized, and the output labeled `value` is the target function at the optimum. A `return_code` of 0 corresponds to the optimizer running with no problem, and you can ignore `theta_tilde`.

4. Check the solution on the graph to see that it makes sense.

Example: Restaurant pricing

**Restaurant pricing.** Suppose you own a small restaurant and are trying to decide how much to charge for dinner. For simplicity, suppose that dinner will have a single price, and that your marginal cost per dinner is \$11. From a marketing survey, you have estimated that, if you charge \$$x$ for dinner, the average number of customers per night you will get is $5000/x^2$. How much should you charge, if your goal is to maximize expected net profit?

Your expected net profit is the expected number of customers times net profit per customer, that is, $(5000/x^2) * (x - 11)$. To get a sense of where this is maximized, we can first make a graph:

```
net_profit <- function(x){
  (5000/x^2)*(x-11)
}
curve(net_profit(x), from=10, to=100, xlab="Price of dinner",
  ylab="Net profit per night")
```

From a visual inspection of this curve, the peak appears to be around $x = 20$. There are two ways we can more precisely determine where net profit is maximized.

First, the brute-force approach. The optimum is clearly somewhere between 10 and 100, so let's just compute the net profit at a grid of points in this range:

```
x <- seq(10, 100, 0.1)
y <- net_profit(x)
```

The maximum value of net profit is then simply `max(y)`, which equals 113.6, and we can find the value of $x$ where profit is maximized:

```
x[y==max(y)]
```

It turns out to be 22, which is `x[121]`, the 121st element of $x$. Above, we are subscripting the vector `x` with a logical vector `y==max(y)`, which is a vector of 900 `FALSE`'s and one `TRUE` at the maximum.

We can also directly optimize using Stan. First we write the function as a Stan program and save it in the file `restaurant.stan`:[2]

```
parameters {
  real<lower=0,upper=100> x;
}
model {
  target += (5000/x^2)*(x-11);
}
```

Then, in R:

```
resto <- stan_model("restaurant.stan")
fit <- optimizing(resto)
print(fit)
```

The result:

```
$par
       x
22.0

$value
[1] 113.6

$return_code
[1] 0
```

The function returns 113.6 at the input value $x = 22.0$. If the return code had not been zero, that would have indicated a problem with the optimization.

## A.8 Working with rstanarm fit objects

### Cleaner output

By default, `stan` and `stan_glm` print intermediate output into the R console. Including the argument `refresh=0` suppresses this output.

[2]Code for this example is in the folder `Restaurant`.

### Extracting posterior summaries, simulation draws, and predictions

We have collected here some useful functions for obtaining information or manipulating fitted models from stan_glm—these are called stanreg objects. As an example we use the children's test score model introduced in Chapter 10.

```
fit <- stan_glm(kid_score ~ mom_hs + mom_iq, data=kidiq)
```

Most simply, print(fit) displays the parameter estimates and associated uncertainties:

```
            Median MAD_SD
(Intercept) 25.7    5.9
mom_hs       6.0    2.4
mom_iq       0.6    0.1

Auxiliary parameter(s):
      Median MAD_SD
sigma 18.2    0.6
```

The coefficients are always displayed first, with auxiliary parameters at the end. In linear regression, the only auxiliary parameter is the residual standard deviation, $\sigma$. Logistic regression has no auxiliary parameters at all. Other generalized linear models can have various auxiliary parameters such as overdispersions and cutpoints.

Markov chain Monte Carlo diagnostics are provided by summary(fit):

```
Diagnostics:
              mcse Rhat n_eff
(Intercept)   0.1  1.0  5021
mom_hs        0.0  1.0  4663
mom_iq        0.0  1.0  4557
sigma         0.0  1.0  5115
mean_PPD      0.0  1.0  4443
log-posterior 0.0  1.0  1926
```

Bayesian $R^2$ (see Section 11.6) can be extracted using bayes_R2(fit), which returns posterior draws of Bayesian $R^2$. A point summary can be obtained, for example, using median(bayes_R2(fit)).

To check the priors used for the inference, use prior_summary(fit). Default priors can be replaced when calling stan_glm. For more information, type vignette("priors", package="rstanarm") in the R console.

When fitted models are used within programs, it can be helpful to extract the estimated vector of coefficients and residual standard deviation using coef(fit) and sigma(fit).

You can extract the estimated residual standard deviation using se(fit), which can be helpful when quickly plotting inferences (as in Figure 19.4); just remember that the estimates and standard errors do *not* fully capture inferential uncertainty, as they do not encode posterior correlation of the parameters. When propagating uncertainty in coefficients, it is best to extract the posterior simulations directly, which can be done using as.matrix(fit) or as.data.frame(fit).

As discussed in Sections 9.2 and 13.3, predictions can be obtained with different levels of uncertainty: predict(fit) returns point predictions, and posterior_linpred(fit), posterior_epred(fit), and posterior_predict(fit) return draws from the posterior distribution of the linear predictor, the expected value of data, and the posterior predictive distribution, respectively. If you call these predictive functions without supplying a newdata argument, they are automatically applied to the existing data, so that predict returns a vector of $n$ point predictions, while posterior_linpred(fit), posterior_epred(fit), and posterior_predict(fit) each return a $n_{\text{sims}} \times n$ matrix of simulations. If you supply a newdata argument that is a data frame with $\tilde{n}$ new observations—for that data frame you just need to supply the predictors for these new data—then predict returns a vector of length $\tilde{n}$, and posterior_linpred(fit), posterior_epred(fit), and posterior_predict(fit) each return a $n_{\text{sims}} \times \tilde{n}$ matrix.

To find more functions that accept a stanreg object, type methods(class=class(fit)) from the R console.

### R packages for visualizing and post-processing stanreg objects

The shinystan package (mc-stan.org/shinystan) provides a graphical user interface for interactive Markov chain Monte Carlo (MCMC) diagnostics and plots and tables helpful for analyzing a posterior sample. Simply run launch_shinystan(fit) after model fitting.

The bayesplot package (mc-stan.org/bayesplot) provides plotting functions for posterior analysis, model checking, and MCMC diagnostics using ggplot2 graphics. In the examples in the book, we have used the functions mcmc_areas and mcmc_scatter to make marginal posterior, and bivariate marginal posterior plots. In Chapter 11 we used ppc_hist, ppc_dens_overlay, and ppc_stat to make posterior predictive checking plots. The posterior and tidybayes packages provide further functions for posterior analysis, data manipulation, and graphics for Bayesian inference.

### Fitting Stan models in R using brms

The brms package (github.com/paul-buerkner/brms) uses an extended version of the formula syntax used by rstanarm. Unlike rstanarm, brms does not have pre-compiled models, which allows more flexibility in the formula syntax, but on the other hand requires C++ compilation, making interactive use slower. The models used in this book also work with brms, just by replacing stan_glm with brm. The default prior distributions for brms are slightly different, but this does not cause major differences for our examples. Using the function make_stancode, the brms package can be used to produce Stan model code, which can make it easier to learn the Stan language and start writing new models which can then be modified by editing that code.[3]

## A.9 Bibliographic note

The books by Becker, Chambers, and Wilks (1988); Venables and Ripley (2002); Wickham (2014); and Wickham and Grolemund (2017) provide thoughtful overviews of R, and its predecessor S, from different perspectives. R is currently maintained by R Core Team (2019).

Stan is introduced by Gelman, Lee, and Guo (2015) from a users' perspective and Carpenter et al. (2017) from a developers' perspective. Much more documentation on R and Stan is available at www.r-project.org and www.mc-stan.org. Stan is currently maintained by Stan Development Team (2020).

The following additonal free software has been used in the examples: rstanarm (Goodrich et al., 2019), brms (Buerkner, 2017, 2018), bayesplot (Gabry and Mahr, 2019), shinystan (Gabry, 2018), loo (Vehtari et al., 2019), arm (Gelman and Su, 2018), MASS (Venables and Ripley, 2002), survey (Lumley, 2019), BART (McCulloch et al., 2019), ggplot2 (Wickham, 2016), knitr (Xie, 2015, 2019), and rprojroot (Mueller, 2018); see also posterior (Buerkner et al., 2020) and tidybayes (Kay, 2020).

The Work, Family, and Well-Being Survey is described by Ross (1990).

[3]Code to illustrate this is in the folder DifferentSoftware.

# Appendix B

# 10 quick tips to improve your regression modeling

## B.1 Think about variation and replication

Variation is central to regression modeling, and not just in the error term. If a regression is fit to different datasets, we can expect the relations between variables, and thus coefficients and causal effects, to vary. Fitting the same model to different datasets—the technique called the "secret weapon" in Section 10.9—can give a sense of variation across problems, which in many settings is more relevant to applications than the standard errors from a single study.

Replication ideally implies performing all the steps of a study from the start, not just increasing the sample size and collecting more data within an existing setting. Repeating an entire experiment can be seen as a way of capturing the variation corresponding to various aspects of data collection and measurement, not just the variation seen within a single study. And this is all in addition to the advantages of a fresh perspective and an avoidance of forking paths in data coding and analysis.

In some fields, such as psychology and cell biology, it can be easy and inexpensive to replicate an experiment from scratch. In mostly observational sciences, such as economics and political science, replication can be more difficult—we cannot re-run the international economy and political system so as to observe 10 more recessions or 20 more civil wars. For such problems, replication will need to be more indirect, for example analyzing local economic or political activity within different countries.

## B.2 Forget about statistical significance

Forget about $p$-values, and forget about whether your confidence intervals exclude zero.

We say this for three reasons. First, if you discretize results based on significance tests, you are throwing away information. Measures of significance such as $p$-values are noisy, and it is misleading to treat an experiment as a success or failure based on a significance test. Second, in the sorts of problems we work on, there are no true zeroes. For example, religious attendance is associated with attitudes on economic as well as social issues, and both these correlations vary by state. And it does not interest us, for example, to test a model in which social class affects vote choice through party identification but not along a direct path. More generally, no true populations are identical, and anything that plausibly could have an effect will not have an effect that is exactly zero. Third, comparisons and effects vary by context, so there is typically little reason to focus on whether a confidence interval excludes zero, as if that would tell us something useful about future differences.

## B.3 Graph the relevant and not the irrelevant

### Graphing the fitted model

Graphing the data is fine (see Chapter 2), but it is also useful to graph the estimated model itself; see lots of examples of regression lines and curves throughout this book. A table of regression coefficients does not give you the same sense as graphs of the model. This point should seem obvious but can be obscured in statistics textbooks that focus so strongly on plots for raw data and for regression diagnostics, forgetting the simple plots that help us understand a model.

### Make many graphs

Real data are full of complexity, and regression models can be hard to understand. Try different visualizations of your data, and look at your model from different angles. Grids of plots can be helpful in visualizing many dimensions, as in Figure 2.10, and a series of graphs can tell a story in a way that would not be possible with a single image; see Section 2.4 for an example. Letting go of the search for the single perfect graph liberates you to learn more from your data and to understand and explain your findings better.

### Don't graph the irrelevant

Are you sure you really want to make those quantile-quantile plots, influence diagrams, and all the other things that spew out of a standard regression package? What are you going to do with all that? Just forget about it and focus on something more important. A quick rule: any graph you show, be prepared to explain.

## B.4 Interpret regression coefficients as comparisons

Regression coefficients are commonly called "effects," but this terminology can be misleading. From the data alone, a regression only tells us about comparisons between individuals, not about changes within individuals.

Taken as a data description, a linear regression coefficient is the modeled average difference in the outcome, comparing two individuals that differ in one predictor, while being at the same levels of all the other predictors. In the special case of a single binary predictor, the coefficient is a simple difference: the average of $y$ for individuals with $x = 1$, minus the average with $x = 0$. For a continuous predictor, we should either scale it so that a difference of 1 unit is of interest, or we should multiply the coefficient by a reasonable change in that predictor.

There are several benefits to thinking of regressions as comparisons. First, the interpretation as a comparison is always available: it is a description of the model and does not require any causal assumptions. Second, we can consider more complicated regressions as built up from simpler models, starting with simple comparisons and adding adjustments. Third, the comparative interpretation also works in the special case of causal inference, where we can consider comparisons between the same individual receiving two different levels of a treatment.

Causal inference can be considered as an application of statistical modeling in which predictions are being made about potential outcomes and where we often summarize inferences as average causal effects, which represent average predictive comparisons under the model. In the special case of an ignorable treatment assignment with no interaction, the average treatment effect is the same as the coefficient of the treatment indicator; more generally, we need to work through the model's predictions to construct an average causal effect.

## B.5 Understand statistical methods using fake-data simulation

Simulating fake data can take more effort than fitting models to real data. For example, to simulate from a linear model, you need to pick reasonable values of the coefficients and residual standard deviation, and you also need to specify all the predictors $x$ for your fake data. Depending on the context, you might want some structure in these predictors, as demonstrated in the simulated midterm and final exams in Sections 6.5 and 16.6 and the simulated poststratification in Section 17.2.

The effort of creating and simulating from a fake world has several payoffs. First, the decisions made in constructing this world can be clarifying: how large a treatment effect could we realistically expect to see, how large are the interactions we want to consider, what might be the correlation between pre-test and post-test, and so forth. Asking these questions requires contact with the application in a way that can increase our understanding. Second, fake-data simulation is a general way to study the properties of statistical methods under repeated sampling. Put the simulation and inference into a loop and you can see how close the model's estimates and predictions are to the assumed true values. Here it can make sense to simulate from a process that includes features not included in the model you will use to fit the data—but, again, this can be a good thing in that it forces you to consider assumptions that might be violated. Third, fake-data simulation is a way to debug your code. With large samples or small data variance, your fitted model should be able to recover the true parameters; if it can't, you may have a coding problem or a conceptual problem, where your model is not doing what you think it is doing. It can help in such settings to plot the simulated data overlaid with the assumed and fitted models. Finally, fake-data simulation, or its analytical equivalent, is necessary if you want to design a new study and collect new data with some reasonable expectation of what you might find.

## B.6 Fit many models

Think of a series of models, starting with the too-simple and continuing through to the hopelessly messy. Generally it's a good idea to start simple. Or start complex if you'd like, but prepare to quickly drop things out and move to the simpler model to help understand what's going on. Working with simple models is not a research goal—in the problems we work on, we usually find complicated models more believable—but rather a technique to help understand the fitting process.

A corollary is the need to be able to fit models relatively quickly. Realistically, you don't know what model you want, so it's rarely a good idea to run the computer overnight fitting a single model. At least, wait until you've developed some understanding by fitting many models.

When fitting multiple models, you should keep track of what models you have fit. This is important for the purpose of understanding your data and also to protect yourself from the biases that can arise when you have many possible ways of analyzing your data (many researcher degrees of freedom or forking paths). In such settings, it's best to record all that you've done and report results from all relevant models, rather than to pick just one and then overinterpret a story from it.

## B.7 Set up a computational workflow

With a bit of work, you can make your computations faster and more reliable. This sounds like computational advice but is really about statistical workflow: if you can fit models faster, you can fit more models and better understand both data and model.

### Data subsetting

Related to the "fit many models" approach are simple approximations that speed the computations. Computers are getting faster and faster—but models are getting more and more complicated! And so these general tricks might remain important. A simple and general trick is to break a large dataset into subsets and analyze each subset separately. For example, perform separate analyses within each region of a country, and then display in one plot the estimates and uncertainties corresponding to the different regions.

An advantage of working with data subsets is that computation is faster, allowing you to explore the data by trying out more models. In addition, separate analyses, when well chosen, can reveal variation that is of interest.

There are two disadvantages of working with data subsets. First, it can be inconvenient to partition the data, perform separate analyses, and summarize the results. Second, the separate analyses may not be as accurate as would be obtained by putting all the data together in a single analysis. Moving forward, one can use multilevel modeling to get some of the advantages of subsetting without losing inferential efficiency or computational stability.

### Fake-data and predictive simulation

When computations get stuck, or a model does not fit the data, it is usually not clear at first if this is a problem with the data, the model, or the computation. Fake-data and predictive simulation are effective ways of diagnosing problems. First use fake-data simulation to check that your computer program does what it is supposed to do, then use predictive simulation to compare the data to the fitted model's predictions.

## B.8 Use transformations

Consider transforming just about every variable in sight:

- Logarithms of all-positive variables (primarily because this leads to multiplicative models on the original scale, which often makes sense).
- Standardizing based on the scale or potential range of the data (so that coefficients can be more directly interpreted and scaled); an alternative is to present results in scaled and unscaled forms.

Plots of raw data and residuals can also be informative when considering transformations, as with the log transformation for arsenic levels in Section 14.5.

In addition to univariate transformations, consider interactions and predictors created by combining inputs (for example, adding several related survey responses to create a total score). The goal is to create models that *could* make sense (and can then be fit and compared to data) and that include all relevant information.

## B.9 Do causal inference in a targeted way, not as a byproduct of a large regression

Don't assume that a comparison or regression coefficient can be interpreted causally. If you are interested in causal inference, consider your treatment variable carefully and use the tools of Chapters 18–21 to address the challenges of balance and overlap when comparing treated and control units to estimate a causal effect and its variation across the population. Even if you are using a natural experiment or identification strategy, it is important to compare treatment and control groups and adjust for pre-treatment differences between them.

When considering several causal questions, it can be tempting to set up a single large regression to answer them all at once; however, in observational settings (including experiments in which certain conditions of interest are observational) this is not appropriate, as we discuss in Sections 19.5–19.7.

## B.10 Learn methods through live examples

We have demonstrated the concepts, methods, and tools in this book through examples that are of interest to us, many of which came from our applied research. Consider these as a starting point: when learning these and more complicated ideas yourself, apply them to problems that you care about, gather data on these examples, and develop statistical understanding by simulating and graphing data from models that make sense to you. Know your data, your measurements, and your data-collection procedure. Be aware of your population of interest and the larger goals of your data collection and analysis. Understand the magnitudes of your regression coefficients, not just their signs. You will need this understanding to interpret your findings and catch things that go wrong.

# References


Aaronson, D. (1998). Using sibling data to estimate the impact of neighborhoods on children's educational outcomes. *Journal of Human Resources* **33**, 915–946.

Abadie, A., and Imbens, G. W. (2011). Bias-corrected matching estimators for average treatment effects. *Journal of Business and Economic Statistics* **29**, 1–11.

Abayomi, K., Gelman, A., and Levy, M. (2008). Diagnostics for multivariate imputations. *Applied Statistics* **57**, 273–291.

Achen, C. (1986). *Statistical Analysis of Quasi-Experiments*. Berkeley: University of California Press.

Agodini, R., and Dynarski, M. (2004). Are experiments the only option? A look at dropout prevention programs. *Review of Economics and Statistics* **86**, 180–194.

Agresti, A. (2002). *Categorical Data Analysis*, second edition. New York: Wiley.

Agresti, A., and Coull, B. A. (1998). Approximate is better than exact for interval estimation of binomial proportions. *American Statistician* **52**, 119–126.

Ainsley, A. E., Dyke, G. V., and Jenkyn, J. F. (1995). Inter-plot interference and nearest-neighbour analysis of field experiments. *Journal of Agricultural Science* **125**, 1–9.

Akaike, H. (1973). Information theory and an extension of the maximum likelihood principle. In *Proceedings of the Second International Symposium on Information Theory*, ed. B. N. Petrov and F. Csaki, 267–281. Budapest: Akademiai Kiado. Reprinted in *Breakthroughs in Statistics*, ed. S. Kotz, 610–624. New York: Springer-Verlag, 1992.

Albert, A., and Anderson, J. A. (1984). On the existence of maximum likelihood estimates in logistic regression models. *Biometrika* **71**, 1–10.

Almond, D., Chay, K. Y., and Lee, D. S. (2005). The costs of low birth weight. *Quarterly Journal of Economics* **120**, 1031–1083.

Amemiya, T. (1981). Qualitative response models: A survey. *Journal of Economic Literature* **19**, 481–536.

Anderson, D. A. (1988). Some models for overdispersed binomial data. *Australian Journal of Statistics* **30**, 125–148.

Angrist, J. D. (1990). Lifetime earnings and the Vietnam era draft lottery: Evidence from Social Security administrative records. *American Economic Review* **80**, 313–336.

Angrist J. D., and Evans, W. N. (1998). Children and their parents' labor supply: Evidence from exogenous variation in family size. *American Economic Review* **88**, 450–477.

Angrist, J. D., Graddy, K., and Imbens, G. W. (2000). The interpretation of instrumental variables estimators in simultaneous equations models with an application to the demand for fish. *Review of Economic Studies* **67**, 499–527.

Angrist, J. D., Imbens, G. W., and Rubin, D. B. (1996). Identification of causal effects using instrumental variables. *Journal of the American Statistical Association* **91**, 444–455.

Angrist, J. D., and Krueger, A. (1999). Empirical strategies in labor economics. In *Handbook of Labor Economics*, volume 3A, ed. O. Ashenfelter and D. Card, 1278–1366. Amsterdam: North-Holland.

Angrist, J. D., and Krueger, A. (2001). Instrumental variables and the search for identification: From supply and demand to natural experiments. *Journal of Economic Perspectives* **15**, 69–85.

Angrist, J. D., and Pischke, J. S. (2009). *Mostly Harmless Econometrics*. Princeton University Press.

Ansolabehere, S., and Snyder, J. M. (2004). Using term limits to estimate incumbency advantages when officeholders retire strategically. *Legislative Studies Quarterly* **29**, 487–515.

Ansolabehere, S., Snyder, J. M., and Stewart, C. (2000). Old voters, new voters, and the personal vote: Using redistricting to measure the incumbency advantage. *American Journal of Political Science* **44**, 17–34.

Aronow, P. M., and Samii, C. (2017). Estimating average causal effects under general interference, with application to a social network experiment. *Annals of Applied Statistics* **11**, 1912–1947.

Ashenfelter, O., and Krueger, A. (1994). Estimates of the economic return to schooling from a new sample of twins. *American Economic Review* **84**, 1157–1173.

Ashenfelter, O., Zimmerman, P., and Levine, D. (2003). *Statistics and Econometrics: Methods and Applications*. New York: Wiley.

Assmann, S. F., Pocock, S. J., Enos, L. E., and Kasten, L. E. (2000). Subgroup analysis and other (mis)uses of baseline data in clinical trials. *Lancet* **355**, 1064–1069.

Atkinson, A. C. (1985). *Plots, Transformations, and Regression*. Oxford University Press.

Balke, A., and Pearl, J. (1997). Bounds on treatment effects from studies with imperfect compliance. *Journal of the American Statistical Association* **92**, 1172–1176.

Ball, S., and Bogatz, G. A. (1972). Reading with television: An evaluation of the Electric Company. Report PR-72-2. Princeton, N.J.: Educational Testing Service.

Ball, S., Bogatz, G. A., Kazarow, K. M., and Rubin, D. B. (1972). Reading with television: A follow-up evaluation of the Electric Company. Report PR-74-15. Princeton, N.J.: Educational Testing Service.

Banerjee, S., Carlin, B. P., and Gelfand, A. E. (2004). *Hierarchical Modeling and Analysis for Spatial Data*. London: Chapman and Hall.

Barnard, J., Frangakis, C., Hill, J. L., and Rubin, D. B. (2003). A principal stratification approach to broken randomized experiments: A case study of vouchers in New York City (with discussion). *Journal of the American Statistical Association* **98**, 299–311.

Barnard, J., and Meng, X. L. (1994). Exploring cross-match estimators with multiply-imputed data sets. *Proceedings of the American Statistical Association, Section on Survey Research Methods*. Alexandria, Va.: American Statistical Association.

Becker, R. A., Chambers, J. M., and Wilks, A. R. (1988). *The New S Language: A Programming Environment for Data Analysis and Graphics*. Pacific Grove, Calif.: Wadsworth.

Beecher, H. K. (1955). The powerful placebo. *Journal of the American Medical Association* **159**, 1602–1616.

Belin, T. R., Diffendal, G. J., Mack, S., Rubin, D. B., Schafer, J. L., and Zaslavsky, A. M. (1993). Hierarchical logistic regression models for imputation of unresolved enumeration status in undercount estimation (with discussion). *Journal of the American Statistical Association* **88**, 1149–1166.

Bennett, C. M., Baird, A. A., Miller, M. B., and Wolford, G. L. (2009). Neural correlates of interspecies perspective taking in the post-mortem Atlantic salmon: An argument for multiple comparisons correction. Poster presented at the 15th Annual Meeting of the Organization for Human Brain Mapping, San Francisco.

Berk, R. A. (2004). *Regression Analysis: A Constructive Critique*. Thousand Oaks, Calif.: Sage.

Bernardo, J. M., and Smith, A. F. M. (1994). *Bayesian Theory*. New York: Wiley.

Berry, D. (1995). *Statistics: A Bayesian Perspective*. Belmont, Calif.: Duxbury.

Bertin, J. (1967). *Semiology of Graphics*. Translated by W. J. Berg. Madison: University of Wisconsin Press.

Besag, J., and Higdon, D. (1999). Bayesian analysis of agricultural field experiments (with discussion). *Journal of the Royal Statistical Society B* **61**, 691–746.

Bishop, C. (2006). *Pattern Recognition and Machine Learning*. New York: Springer.

Blalock, H. M. (1961). Evaluating the relative importance of variables. *American Sociological Review* **26**, 866–874.

Bogart, W. T., and Cromwell, B. A. (2000). How much is a neighborhood school worth? *Journal of Urban Economics* **47**, 280–306.

Bogatz, G. A., and Ball, S. (1971). *The Second Year of Sesame Street: A Continuing Evaluation*, two volumes. Princeton, N.J.: Educational Testing Service.

Box, G. E. P., and Cox, D. R. (1964). An analysis of transformations (with discussion). *Journal of the Royal Statistical Society B* **26**, 211–252.

Box, G. E. P., Hunter, J. S., and Hunter, W. G. (2005). *Statistics for Experimenters*, second edition. New York: Wiley.

Box, G. E. P., and Jenkins, G. M. (1976). *Time Series Analysis: Forecasting and Control*, second edition. San Francisco: Holden-Day.

Bradley, R. A., and Terry, M. E. (1952). The rank analysis of incomplete block designs: I. The method of paired comparisons. *Biometrika* **39**, 324–345.

Brainard, J., and Burmaster, D. E. (1992). Bivariate distributions for height and weight of men and women in the United States. *Risk Analysis* **12**, 267–275.

Brillinger, D. R. (1981). *Time Series: Data Analysis and Theory*, expanded edition. San Francisco: Holden-Day.

Bring, J. (1994). How to standardize regression coefficients. *American Statistician* **48**, 209–213.

Brooks-Gunn, J., Liaw, F. R., and Klebanov, P. K. (1992). Effects of early intervention on cognitive function of low birth weight preterm infants. *Journal of Pediatrics* **120**, 350–359.

Browner, W. S., and Newman, T. B. (1987). Are all significant P values created equal? *Journal of the American Medical Association* **257**, 2459–2463.

Buja, A., Asimov, D., Hurley, C., and McDonald, J. A. (1988). Elements of a viewing pipeline for data analysis. In *Dynamic Graphics for Statistics*, ed. W. S. Cleveland and M. E. McGill, 277–308. Belmont, Calif.: Wadsworth.

Buja, A., Cook, D., Hofmann, H., Lawrence, M., Lee, E. K., Swayne, D., and Wickham, H. (2009). Statistical inference for exploratory data analysis and model diagnostics. *Philosophical Transactions of the Royal Society A* **367**, 4361–4383.

Buerkner, P. C. (2017). brms: An R package for Bayesian multilevel models using Stan. *Journal of Statistical Software* **80**, 1–28.

Buerkner, P. C. (2018). Advanced Bayesian multilevel modeling with the R package brms. *R Journal* **10**, 395–411.

Buerkner, P. C., and Vehtari, A. (2019). Approximate leave-future-out cross-validation for time series models. `arxiv.org/abs/1902.06281`

Buerkner, P. C., Gabry, J., and Kay, M. (2020). posterior: Tools for working with posterior distributions. `https://github.com/jgabry/posterior`

Buerkner, P. C., and Vehtari, A. (2020). Efficient leave-one-out cross-validation for Bayesian non-factorized normal and Student-*t* models. `arxiv.org/abs/1810.10559`

Bush, R. R., and Mosteller, F. (1955). *Stochastic Models for Learning*. New York: Wiley.

Button, K. S., Ioannidis, J. P. A., Mokrysz, C., Nosek, B. A., Flint, J., Robinson, E. S. J., and Munafo, M. R. (2013). Power failure: Why small sample size undermines the reliability of neuroscience. *Nature Reviews Neuroscience* **14**, 365–376.

Campbell, D. T., and Stanley, J. C. (1963). *Experimental and Quasi-Experimental Designs for Research*. Chicago: Rand McNally.

Campbell, J. E. (2002). Is the House incumbency advantage mostly a campaign finance advantage? Department of Political Science, State University of New York at Buffalo.

Card, D., and Krueger, A. (1994). Minimum wages and employment: A case study of the fast-food industry in New Jersey and Pennsylvania. *American Economic Review* **84**, 772–784.

Carlin, J. B., and Forbes, A. (2004). *Linear Models and Regression*. Melbourne: Biostatistics Collaboration of Australia.

Carnegie, N., Dorie, V., and Hill, J. L. (2019). Examining treatment effect heterogeneity using BART. *Observational Studies* **5**, 21–35.

Carnegie, N., Harada, M, and Hill, J. L. (2016). Assessing sensitivity to unmeasured confounding using a simulated potential confounder. *Journal of Research on Educational Effectiveness* **9**, 395–420.

Carpenter, B., Gelman, A., Hoffman, M., Lee, D., Goodrich, B., Betancourt, M., Brubaker, M., Guo, J., Li, P., and Riddell, A. (2017). Stan: A probabilistic programming language. *Journal of Statistical Software* **76** (1).

Carroll, R. J., and Ruppert, D. (1981). On prediction and the power transformation family. *Biometrika* **68**, 609–615.

Carroll, R. J., Ruppert, D., Crainiceanu, C. M., Tosteson, T. D., and Karagas, M. R. (2004). Nonlinear and nonparametric regression and instrumental variables. *Journal of the American Statistical Association* **99**, 736–750.

Carroll, R. J., Ruppert, D., Stefanski, L. A., and Crainiceanu, C. M. (2006). *Measurement Error in Nonlinear Models: A Modern Perspective*. London: CRC Press.

Carvalho, C. M., Polson, N. G., and Scott, J. G. (2010). The horseshoe estimator for sparse signals. *Biometrika* **97**, 465–480.

Case, A., and Deaton, A. (2015). Rising morbidity and mortality in midlife among white non-Hispanic Americans in the 21st century. *Proceedings of the National Academy of Sciences* **112**, 15078–15083.

Case, A., and Deaton, A. (2016). Reply to Schmid, Snyder, and Gelman and Auerbach: Correlates of the increase in white non-Hispanic midlife mortality in the 21st century. *Proceedings of the National Academy of Sciences* **113**, E818–819.

Casella, A., Gelman, A., and Palfrey, T. (2006). An experimental study of storable votes. *Games and Economic Behavior* **57**, 123–154.

Chambers, J. M., Cleveland, W. S., Kleiner, B., and Tukey, P. A. (1983). *Graphical Methods for Data Analysis*. Pacific Grove, Calif.: Wadsworth.

Chapman, R. (1973). The concept of exposure. *Accident Analysis and Prevention* **5**, 95–110.

Chay, K. Y., McEwan, P. J., and Urquiola, M. (2005). The central role of noise in evaluating interventions that use test scores to rank schools. *American Economic Review* **95**, 1237–1258.

Chipman, H., George, E. I., and McCulloch, R. E. (2001). The practical implementation of Bayesian model selection (with discussion). In *Model Selection*, ed. P. Lahiri, 67–116. Institute of Mathematical Statistics Lecture Notes 38.

Chipman, H., George, E. I., and McCulloch, R. E. (2010). BART: Bayesian additive regression trees. *Annals of Applied Statistics* **4**, 266–298.

Clayton, D. G., Dunn, G., Pickles, A., and Spiegelhalter, D. (1998). Analysis of longitudinal binary data from multi-phase sampling. *Journal of the Royal Statistical Society B* **60**, 71–87.

Cleveland, W. S. (1979). Robust locally weighted regression and smoothing scatterplots. *Journal of the American Statistical Association* **74**, 829–836.

Cleveland, W. S. (1985). *The Elements of Graphing Data*. Pacific Grove, Calif.: Wadsworth.

Cleveland, W. S. (1993). *Visualizing Data*. Summit, N.J.: Hobart Press.

Clinton, J., Jackman, S., and Rivers, D. (2004). The statistical analysis of roll call data. *American Political Science Review* **98**, 355–370.

Clogg, C. C., Rubin, D. B., Schenker, N., Schultz, B., and Wideman, L. (1991). Multiple imputation of industry and occupation codes in Census public-use samples using Bayesian logistic regression. *Journal of the American Statistical Association* **86**, 68–78.

Cochran, W. G. (1968). The effectiveness of adjustment by subclassification in removing bias in observational studies. *Biometrics* **24**, 205–213.

Cochran, W. G. (1977). *Sampling Techniques*, third edition. New York: Wiley.

Cochran, W. G., and Rubin, D. B. (1973). Controlling bias in observational studies: A review. *Sankhya A* **35**, 417–446.

Cook, S. R., Gelman, A., and Rubin, D. B. (2006). Bayesian model validation. *Journal of Computational and Graphical Statistics* **15**, 675–692.

Cortez, P., and Silva, A. (2008). Using data mining to predict secondary school student performance. In *Proceedings of 5th Future Business Technology Conference*, ed. A. Brito and J. Teixeira, 5–12.

Costa, D. L., and Kahn, M. E. (2004). Changes in the value of life, 1940–1980. *Journal of Risk and Uncertainty* **29**, 159–180.

Cox, D. R. (1958). *Planning of Experiments*. New York: Wiley.

Cox, D. R., and Oakes, D. (1984). *Analysis of Survival Data*. New York: Wiley.

Cox, G. W., and Katz, J. (1996). Why did the incumbency advantage grow? *American Journal of Political Science* **40**, 478–497.

Cramer, C. S. (2003). *Logit Models from Economics and Other Fields*. Cambridge University Press.

Cressie, N. A. C. (1993). *Statistics for Spatial Data*, second edition. New York: Wiley.

Croissant, Y., and Millo, G. (2008). Panel data econometrics in R: The plm package. *Journal of Statistical Software* **27**, 1–43.

Crump, R. K., Hotz, V. J., Imbens, G. W., and Mitnik, O. A. (2009). Dealing with limited overlap in estimation of average treatment effects. *Biometrika* **96**, 187–199.

David, M. H., Little, R. J. A., Samuhel, M. E., and Triest, R. K. (1986). Alternative methods for CPS income imputation. *Journal of the American Statistical Association* **81**, 29–41.

Dawid, A. P. (2000). Causal inference without counterfactuals (with discussion). *Journal of the American Statistical Association* **95**, 407–448.

Dawid, A. P., Musio, M., and Fienberg, S. E. (2016). From statistical evidence to evidence of causality. *Bayesian Analysis* **11**, 725–752.

de Groot, A. D. (1956). The meaning of the concept of significance in studies of various types. *Nederlands Tijdschrift voor de Psychologie en Haar Grensgebieden* **11**, 398–409. Translated 2013 by E. J. Wagenmakers, D. Borsboom, J. Verhagen, R. Kievit, M. Bakker, A. Cramer, D. Matzke, D. Mellenbergh, and H. L. J. van der Maas. dl.dropboxusercontent.com/u/1018886/Temp/DeGrootv3.pdf

Dehejia, R. (2003). Was there a Riverside miracle? A framework for evaluating multi-site programs. *Journal of Business and Economic Statistics* **21**, 1–11.

Dehejia, R. (2005a). Practical propensity score matching: A reply to Smith and Todd. *Journal of Econometrics* **125**, 355–364.

Dehejia, R. (2005b). Does matching overcome LaLonde's critique of nonexperimental estimators? A postscript. Technical report, Department of Economics, Columbia University.

Dehejia, R., and Wahba, S. (1999). Causal effects in non-experimental studies: Re-evaluating the evaluation of training programs. *Journal of the American Statistical Association* **94**, 1053–1062.

Diamond, A., and Sekhon, J. S. (2013). Genetic matching for estimating causal effects. *Review of Economics and Statistics* **95**, 932–945.

Diggle, P., and Kenward, M. G. (1994). Informative drop-out in longitudinal data analysis. *Journal of the Royal Statistical Society C* **43**, 49–73.

Ding, P., and VanderWeele, T. J. (2016). Sensitivity analysis without assumptions. *Epidemiology* **27**, 368–377.

Dobson, A. (2001). *An Introduction to Generalized Linear Models*, second edition. London: CRC Press.

Dorie, V., Harada, M., Carnegie, N., and Hill, J. L. (2016). A flexible, interpretable framework for assessing sensitivity to unmeasured confounding. *Statistics in Medicine* **35**, 3453–3470.

Dorie, V., Hill, J. L., Shalit, U., Scott, M., and Cervone, D. (2019). Automated versus do-it-yourself methods for causal inference: Lessons learned from a data analysis competition (with discussion). *Statistical Science* **34**, 43–99.

Dorman, P., and Hagstrom, P. (1998). Wage compensation for dangerous work revisited. *Industrial and Labor Relations Review* **52**, 116–135.

Drake, C. (1993). Effects of misspecification of the propensity score on estimators of treatment effect. *Biometrics* **49**, 1231–1236.

Dreze, J. H. (1976). Bayesian limited information analysis of the simultaneous equations model. *Econometrica* **44**, 1045–1075.

Efron, B. (1979). Bootstrap methods: Another look at the jackknife. *Annals of Statistics* **7**, 1–26.

Efron, B. (1982). Maximum likelihood and decision theory. *Annals of Statistics* **10**, 340–356.

Efron, B., and Tibshirani, R. J. (1993). *An Introduction to the Bootstrap*. London: Chapman and Hall.

Ehrenberg, A. S. C. (1978). *Data Reduction: Analysing and Interpreting Statistical Data*. New York: Wiley.

Erikson, R. S. (1971). The advantage of incumbency in congressional elections. *Polity* **3**, 395–405.

Erikson, R. S., and Romero, D. W. (1990). Candidate equilibrium and the behavioral model of the vote. *American Political Science Review* **4**, 1103–1126.

Fair, R. C. (1978). The effect of economic events on votes for President. *Review of Economics and Statistics* **60**, 159–173.

Fay, R. E. (1996). Alternative paradigms for the analysis of imputed survey data. *Journal of the American Statistical Association* **91**, 490–498.

Felton, J., Mitchell, J., and Stinson, M. (2003). Web-based student evaluations of professors: The relations between perceived quality, easineness, and sexiness. *Assessment and Evaluation in Higher Education* **29**, 91–108.

Fienberg, S. E. (1977). *The Analysis of Cross-Classified Categorical Data*. Cambridge, Mass.: MIT Press.

Finkelstein, M. O., Levin, B., McKeague, I. M., and Tsai, W. Y. (2006). A note on the censoring problem in empirical case-outcome studies. *Journal of Empirical Legal Studies* **3**, 375–395.

Firth, D. (1993). Bias reduction of maximum likelihood estimates. *Biometrika* **80**, 27–38.

Fortna, V. P. (2008). *Does Peacekeeping Work? Shaping Belligerents' Choices after Civil War*. Princeton University Press.

Fox, J. (2002). *An R and S-Plus Companion to Applied Regression*. Thousand Oaks, Calif.: Sage.

Francis, G. (2013). Replication, statistical consistency, and publication bias (with discussion). *Journal of Mathematical Psychology* **57**, 153–169.

Frangakis, C. E., Brookmeyer, R. S., Varadhan, R., Mahboobeh, S., Vlahov, D., and Strathdee, S. A. (2003). Methodology for evaluating a partially controlled longitudinal treatment using principal stratification, with application to a needle exchange program. *Journal of the American Statistical Association* **99**, 239–249.

Frangakis, C. E., and Rubin, D. B. (2002). Principal stratification in causal inference. *Biometrics* **58**, 21–29.

Franke, R. H., and Kaul, J. D. (1978). The Hawthorne experiments: First statistical interpretation. *American Sociological Review* **43**, 623–643.

Friedlander, D., and Robins, P. K. (1995). Evaluating program evaluations—new evidence on commonly used nonexperimental methods. *American Economic Review* **85**, 923–937.

Friendly, M., and Kwan, E. (2003). Effect ordering for data displays. *Computational Statistics and Data Analysis* **43**, 509–539.

Frolich, M. (2004). Finite-sample properties of propensity-score matching and weighting estimators. *Review of Economics and Statistics* **86**, 77–90.

Fuller, W. A. (1987). *Measurement Error Models*. New York: Wiley.

Gabry, J. (2018). shinystan: Interactive visual and numerical diagnostics and posterior analysis for Bayesian models. R package version 2.5.0. `cran.r-project.org/package=shinystan`

Gabry, J., and Mahr, T. (2019). bayesplot: Plotting for Bayesian models. R package version 1.7.0. `mc-stan.org/bayesplot`

Gabry, J., Simpson, D., Vehtari, A., Betancourt, M., and Gelman, A. (2019). Visualization in Bayesian workflow (with discussion). *Journal of the Royal Statistical Society A* **182**, 389–402.

Garfinkel, I., and Meyers, M. K. (1999). A tale of many cities: The New York City Social Indicators Survey. School of Social Work, Columbia University.

Geisser, S., and Eddy, W. F. (1979). A predictive approach to model selection. *Journal of the American Statistical Association* **74**, 153–160.

Gelman, A. (2004a). Exploratory data analysis for complex models (with discussion). *Journal of Computational and Graphical Statistics* **13**, 755–787.

Gelman, A. (2004b). 55,000 residents desperately need your help! *Chance* **17** (2), 28–31.

Gelman, A. (2004c). Treatment effects in before-after data. In *Applied Bayesian Modeling and Causal Inference from Incomplete Data Perspectives*, ed. A. Gelman and X. L. Meng, chapter 18. New York: Wiley.

Gelman, A. (2007). Struggles with survey weighting and regression modeling (with discussion). *Statistical Science* **22**, 153–188.

Gelman, A. (2009a). Debunking the so-called Human Development Index of U.S. states. Statistical Modeling, Causal Inference, and Social Science, 20 May. `statmodeling.stat.columbia.edu/2009/05/20/debunking_the_s/`

Gelman, A. (2009b). Rich people are more likely to be Republican but not more likely to be conservative. Statistical Modeling, Causal Inference, and Social Science, 12 Aug. `statmodeling.stat.columbia.edu/2009/08/12/rich_people_are/`

Gelman, A. (2009c). Healthcare spending and life expectancy: A comparison of graphs. Statistical Modeling, Causal Inference, and Social Science, 30 Dec. `statmodeling.stat.columbia.edu/2009/12/30/healthcare_spen/`

Gelman, A. (2011a). The pervasive twoishness of statistics; in particular, the "sampling distribution" and the "likelihood" are two different models, and that's a good thing. Statistical Modeling, Causal Inference, and Social Science, 20 Jun. `statmodeling.stat.columbia.edu/2011/06/20/the_sampling_di_1/`

Gelman, A., (2011b). Causality and statistical learning. *American Journal of Sociology* **117**, 955–966.

Gelman, A. (2013). Childhood intervention and earnings. *Symposium*, 3 Nov. `www.symposium-magazine.com/childhood-intervention-and-earnings/`

Gelman, A., (2014). Experimental reasoning in social science. In *Field Experiments and their Critics*, ed. Dawn Teele, 185–195. New Haven, Conn.: Yale University Press.

Gelman, A. (2015a). The connection between varying treatment effects and the crisis of unreplicable research: A Bayesian perspective. *Journal of Management* **41**, 632–643.

Gelman, A. (2015b). What's the most important thing in statistics that's not in the textbooks? Statistical

Modeling, Causal Inference, and Social Science, 28 Apr. statmodeling.stat.columbia.edu/2015/04/28/whats-important-thing-statistics-thats-not-textbooks/

Gelman, A. (2015c). First, second, and third order bias corrections. Statistical Modeling, Causal Inference, and Social Science, 18 Nov. statmodeling.stat.columbia.edu/2015/11/18/first-second-and-third-order-bias-corrections-also-my-ugly-r-code-for-the-mortality-rate-graphs/

Gelman, A. (2015d). Statistics and the crisis of scientific replication. *Significance* **12** (3), 23–25.

Gelman, A. (2016a). Why this gun control study might be too good to be true. Statistical Modeling, Causal Inference, and Social Science, 11 Mar. statmodeling.stat.columbia.edu/2016/03/11/why-this-gun-control-study-might-be-too-good-to-be-true/

Gelman, A. (2016b). Kalesan, Fagan, and Galea respond to criticism of their paper on gun laws and deaths. Statistical Modeling, Causal Inference, and Social Science, 17 Mar. statmodeling.stat.columbia.edu/2016/03/17/kalesan-fagan-and-galea-respond-to-criticism-of-their-paper-on-gun-laws-and/

Gelman, A. (2016c). What has happened down here is the winds have changed. Statistical Modeling, Causal Inference, and Social Science, 21 Sep. statmodeling.stat.columbia.edu/2016/09/21/what-has-happened-down-here-is-the-winds-have-changed/

Gelman, A. (2017). Easier-to-download graphs of age-adjusted mortality trends by sex, ethnicity, and age group. Statistical Modeling, Causal Inference, and Social Science, 29 Mar. statmodeling.stat.columbia.edu/2017/03/29/easier-download-graphs-age-adjusted-mortality-trends-sex-ethnicity-age-group/

Gelman, A. (2018). The failure of null hypothesis significance testing when studying incremental changes, and what to do about it. *Personality and Social Psychology Bulletin* **44**, 16–23.

Gelman, A. (2019a). Post-hoc power using observed estimate of effect size is too noisy to be useful. *Annals of Surgery* **270** (2), e64.

Gelman, A. (2019b). Model building and expansion for golf putting. *Stan Case Studies* **6**. mc-stan.org/users/documentation/case-studies/golf.html

Gelman, A., and Auerbach, J. (2016). Age-aggregation bias in mortality trends. *Proceedings of the National Academy of Sciences* **113**, E816–817.

Gelman, A., and Carlin, J. B. (2014). Beyond power calculations: Assessing Type S (sign) and Type M (magnitude) errors. *Perspectives on Psychological Science* **9**, 641–651.

Gelman, A., Carlin, J. B., and Nallamothu, B. K. (2019). Objective Randomised Blinded Investigation With Optimal Medical Therapy of Angioplasty in Stable Angina (ORBITA) and coronary stents: A case study in the analysis and reporting of clinical trials. *American Heart Journal* **214**, 54–59.

Gelman, A., Carlin, J. B., Stern, H. S., Dunson, D. B., Vehtari, A., and Rubin, D. B. (2013). *Bayesian Data Analysis*, third edition. London: CRC Press.

Gelman, A., Fagan, J., and Kiss, A. (2007). An analysis of the NYPD's stop-and-frisk policy in the context of claims of racial bias. *Journal of the American Statistical Association* **102**, 813–823.

Gelman, A., Goegebeur, Y., Tuerlinckx, F., and Van Mechelen, I. (2000). Diagnostic checks for discrete-data regression models using posterior predictive simulations. *Applied Statistics* **49**, 247–268.

Gelman, A., Goel, S., Rivers, D., and Rothschild, D. (2016). The mythical swing voter. *Quarterly Journal of Political Science* **11**, 103–130.

Gelman, A., Goodrich, B., Gabry, J., and Vehtari, A. (2019). R-squared for Bayesian regression models. *American Statistician* **73**, 307–309.

Gelman, A., and Greenland, S. (2019). Are confidence intervals better termed "uncertainty intervals"? *British Medical Journal* **366**, l5381.

Gelman, A., and Hill, J. L. (2007). *Data Analysis Using Regression and Multilevel/Hierarchical Models*. Cambridge University Press.

Gelman, A., and Huang, Z. (2008). Estimating incumbency advantage and its variation, as an example of a before/after study (with discussion). *Journal of the American Statistical Association* **103**, 437–451.

Gelman, A., Hwang, J., and Vehtari, A. (2014). Understanding predictive information criteria for Bayesian models. *Statistics and Computing* **24**, 997–1016.

Gelman, A., and Imbens, G. (2013). Why ask why? Forward causal inference and reverse causal questions. NBER Working Paper No. 19614. www.nber.org/papers/w19614

Gelman, A., and Imbens, G. (2017). Why high-order polynomials should not be used in regression discontinuity designs. *Journal of Business and Economic Statistics* **37**, 225–456.

Gelman, A., Jakulin, A., Pittau, M. G., and Su, Y. S. (2008). A weakly informative default prior distribution for logistic and other regression models. *Annals of Applied Statistics* **2**, 1360–1383.

Gelman, A., Katz, J. N., and Bafumi, J. (2004). Standard voting power indexes don't work: An empirical analysis. *British Journal of Political Science* **34**, 657–674.

Gelman, A., and King, G. (1990). Estimating incumbency advantage without bias. *American Journal of Political Science* **34**, 1142–1164.

Gelman, A., and King, G. (1993). Why are American presidential election campaign polls so variable when votes are so predictable? *British Journal of Political Science* **23**, 409–451.

Gelman, A., and King, G. (1994). A unified model for evaluating electoral systems and redistricting plans. *American Journal of Political Science* **38**, 514–554.

Gelman, A., King, G., and Boscardin, W. J. (1998). Estimating the probability of events that have never occurred: When is your vote decisive? *Journal of the American Statistical Association* **93**, 1–9.

Gelman, A., Lee, D., and Guo, J. (2015). Stan: A probabilistic programming language for Bayesian inference and optimization. *Journal of Educational and Behavioral Statistics* **40**, 530–543.

Gelman, A., Liebman, J., West, V., and Kiss, A. (2004). A broken system: The persistent pattern of reversals of death sentences in the United States. *Journal of Empirical Legal Studies* **1**, 209–261.

Gelman, A., and Little, T. C. (1997). Poststratification into many categories using hierarchical logistic regression. *Survey Methodology* **23**, 127–135.

Gelman, A., and Loken, E. (2014). The statistical crisis in science. *American Scientist* **102**, 460–465.

Gelman, A., Meng, X. L., and Stern, H. S. (1996). Posterior predictive assessment of model fitness via realized discrepancies (with discussion). *Statistica Sinica* **6**, 733–807.

Gelman, A., and Nolan, D. (2002). A probability model for golf putting. *Teaching Statistics* **24**, 93–95.

Gelman, A., and Nolan, D. (2017). *Teaching Statistics: A Bag of Tricks*, second edition. Oxford University Press.

Gelman, A., and Pardoe, I. (2007). Average predictive comparisons for models with nonlinearity, interactions, and variance components. *Sociological Methodology* **37**, 23–51.

Gelman, A., Park, D., Shor, B., and Cortina, J. (2009). *Red State, Blue State, Rich State, Poor State: Why Americans Vote the Way They Do*, second edition. Princeton University Press.

Gelman, A., Pasarica, C., and Dodhia, R. (2002). Let's practice what we preach: Using graphs instead of tables. *American Statistician* **56**, 121–130.

Gelman, A., and Rubin, D. B. (1992). Inference from iterative simulation using multiple sequences (with discussion). *Statistical Science* **7**, 457–511.

Gelman, A., Silver, N., and Edlin, A. (2012). What is the probability your vote will make a difference? *Economic Inquiry* **50**, 321–326.

Gelman, A., Simpson, D., and Betancourt, M. (2017). The prior can often only be understood in the context of the likelihood. *Entropy* **19**, 555.

Gelman, A., Skardhamar, T., and Aaltonen, M. (2017). Type M error might explain Weisburd's paradox. *Journal of Quantitative Criminology*.

Gelman, A., and Stern, H. S. (2006). The difference between "statistically significant" and "not significant" is not itself significant. *American Statistician* **60**, 328–331.

Gelman, A., Stevens, M., and Chan, V. (2003). Regression models for decision making: A cost-benefit analysis of incentives in telephone surveys. *Journal of Business and Economic Statistics* **21**, 213–225.

Gelman, A., and Su, Y. S. (2018). arm: Data analysis using regression and multilevel/hierarchical models. R package version 1.10-1. `cran.r-project.org/package=arm`

Gelman, A., Trevisani, M., Lu, H., and van Geen, A. (2004). Direct data manipulation for local decision analysis, as applied to the problem of arsenic in drinking water from tube wells in Bangladesh. *Risk Analysis* **24**, 1597–1612.

Gelman, A., and Tuerlinckx, F. (2000). Type S error rates for classical and Bayesian single and multiple comparison procedures. *Computational Statistics* **15**, 373–390.

Gelman, A., and Unwin, A. (2013). Infovis and statistical graphics: Different goals, different looks (with discussion). *Journal Journal of Computational and Graphical Statistics* **22**, 2–49.

Gelman, A., and Vehtari, A. (2014). Bootstrap averaging: Examples where it works and where it doesn't work. *Journal of the American Statistical Association* **109**, 1015–1016.

Gelman, A., and Weakliem, D. (2009). Of beauty, sex, and power: Statistical challenges in estimating small effects. *American Scientist* **97**, 310–316.

Gelman, A., and Zelizer, A. (2015). Evidence on the deleterious impact of sustained use of polynomial regression on causal inference. *Research and Politics* **2**, 1–7.

Gerber, A. (2004). Does campaign spending work? *American Behavioral Scientist* **47**, 541–574.

Gertler, P., Heckman, J., Pinto, R., Zanolini, A., Vermeerch, C., Walker, S., Chang, S. M., and Grantham-McGregor, S. (2013). Labor market returns to early childhood stimulation: A 20-year followup to an experimental intervention in Jamaica. Institute for Research on Labor and Employment working paper #142-13.

Geweke, J. (2004). Getting it right: Joint distribution tests of posterior simulators. *Journal of the American Statistical Association* **99**, 799–804.

Glickman, M. E., and Normand, S. L. (2000). The derivation of a latent threshold instrumental variables model. *Statistica Sinica* **10**, 517–544.

Goldin, C., and Rouse, C. (2000). Orchestrating impartiality: The impact of "blind" auditions on female musicians. *American Economic Review* **90**, 715–741.

Goodrich, B., Gabry, J., Ali, I., and Brilleman, S. (2019). rstanarm: Bayesian applied regression modeling via Stan. R package version 2.19.2. `mc-stan.org/rstanarm`

Greenland, S. (2005). Multiple bias modelling for analysis of observational data (with discussion). *Journal of the Royal Statistical Society A* **168**, 267–306.

Greenland, S. (2019). Valid p-values behave exactly as they should: Some misleading criticisms of p-values and their resolution with s-values. *American Statistician* **73** (supl), 106–114.

Greenland, S., Maclure, M., Schlesselman, J. J., Poole, C., and Morgenstern, H. (1991). Standardized regression coefficients: A further critique and a review of alternatives. *Epidemiology* **2**, 387–392.

Greenland, S., Schlessman, J. J., and Criqui, M. H. (1986). The fallacy of employing standardized regression coefficients and correlations as measures of effect. *American Journal of Epidemiology* **123**, 203–208.

Greevy, R., Lu, B., Silber, J. H., and Rosenbaum, P. (2004). Optimal multivariate matching before randomization. *Biostatistics* **5**, 263–275.

Groves, R., Fowler, F. J., Couper, M. P., Lepkowski, J. M., Singer, E., and Tourangeau, R. (2009). *Survey Methodology*, second edition. New York: Wiley.

Haavelmo, T. (1943). The statistical implications of a system of simultaneous equations. *Econometrica* **11**, 1–12.

Hahn, J. (1998). On the role of the propensity score in efficient semiparametric estimation of average treatment effects. *Econometrica* **66**, 315–331.

Hahn, J., Todd, P. E., and van der Klaauw, W. (2001). Identification and estimation of treatment effects with a regression-discontinuity design. *Econometrica* **69**, 201–209.

Hahn, P. R., Murray, J. S., and Carvalho, C. (2019). Bayesian regression tree models for causal inference: Regularization, confounding, and heterogeneous effects. `arxiv.org/abs/1706.09523`

Hamermesh, D. S., and Parker, A. M. (2005). Beauty in the classroom: Instructors' pulchritude and putative pedagogical productivity. *Economics of Education Review* **24**, 369–376.

Hansen, B. B. (2004). Full matching in an observational study of coaching for the SAT. *Journal of the American Statistical Association* **99**, 609–619.

Harrell, F. (2001). *Regression Modeling Strategies*. New York: Springer-Verlag.

Hastie, T. J., and Tibshirani, R. J. (1990). *Generalized Additive Models*. New York: Chapman and Hall.

Hastie, T. J., Tibshirani, R. J., and Friedman, J. (2009). *The Elements of Statistical Learning: Data Mining, Inference, and Prediction*, second edition. New York: Springer-Verlag.

Hauer, E., Ng, J. C. N., and Lovell, J. (1988). Estimation of safety at signalized intersections. *Transportation Research Record* **1185**, 48–61. Washington, D.C.: National Research Council.

Healy, K. (2018). *Data Visualization: A Practical Introduction*. Princeton University Press.

Healy, M. J. R. (1990). Measuring importance. *Statistics in Medicine* **9**, 633–637.

Heckman, J. (1979). Sample selection bias as a specification error. *Econometrica* **47**, 153–161.

Heckman, J. (1990). Varieties of selection bias. *American Economic Review, Papers and Proceedings* **80**, 313–318.

Heckman, J. J., Ichimura, H., and Todd, P. E. (1998). Matching as an econometric evaluation estimator. *Review of Economic Studies* **65**, 261–294.

Heeringa, S. G., West, B. T., and Berglund, P. A. (2017). *Applied Survey Data Analysis*. London: CRC Press.

Heinze, G., and Schemper, M. (2003). A solution to the problem of separation in logistic regression. *Statistics in Medicine* **12**, 2409–2419.

Heitjan, D. F., Moskowitz, A. J., and Whang, W. (1999). Problems with interval estimates of the incremental cost-effectiveness ratio. *Medical Decision Making* **19**, 9–15.

Hernan, M. A., and Robins, J. M. (2020). *Causal Inference*. London: CRC Press.

Hibbs, D. (2000). Bread and peace voting in U.S. presidential elections. *Public Choice* **104**, 149–180.

Hibbs, D. (2012). Obama's reelection prospects under "Bread and Peace" voting in the 2012 U.S. presidential election. *PS: Political Science and Politics* **45**, 635–639.

Hibbs, D., Rivers, D., and Vasilatos, N. (1982). On the demand for economic outcomes: Macroeconomic performance and mass political support in the United States. *Journal of Politics* **44**, 426–462.

Hill, J. L. (2011). Bayesian nonparametric modeling for causal inference. *Journal of Computational and Graphical Statistics* **20**, 217–240.

Hill, J. L. (2013). Multilevel models and causal inference. In *The SAGE Handbook of Multilevel Modeling*, ed. M. Scott, J. Simonoff, and B. Marx. Thousand Oaks, Calif.: Sage.

Hill, J. L., Brooks-Gunn, J., and Waldfogel, J. (2003). Sustained effects of high participation in an early intervention for low-birth-weight premature infants. *Developmental Psychology* **39**, 730–744.

Hill, J. L., Linero, A., and Murray, J. (2020). Bayesian additive regression trees: A review and look forward. *Annual Review of Statistics and Its Application* **7**.

Hill, J. L., and Reiter, J. (2006). Interval estimation for treatment effects using propensity score matching. *Statistics in Medicine* **25**, 2230–2256.

Hill, J. L., Reiter, J., and Zanutto, E. (2004). A comparison of experimental and observational data analyses. In *Applied Bayesian and Causal Inference from an Incomplete Data Perspective*, ed. A. Gelman and X. L. Meng. New York: Wiley.

Hill, J. L., Rubin, D. B., and Thomas, N. (2000). The design of the New York school choice scholarship program evaluation. In *Research Design: Donald Campbell's Legacy*, ed. L. Bickman. Thousand Oaks, Calif.: Sage.

Hill, J. L., and Su, Y. (2013). Assessing lack of common support in causal inference using Bayesian nonparametrics: implications for evaluating the effect of breastfeeding on children's cognitive outcomes. *Annals of Applied Statistics* **7**, 1386–1420.

Hill, J. L., Waldfogel, J., and Brooks-Gunn, J. (2002). Assessing the differential impacts of high-quality child care: A new approach for exploiting post-treatment variables. *Journal of Policy Analysis and Management* **21**, 601–627.

Hill, J. L., Waldfogel, J., Brooks-Gunn, J., and Han, W. J. (2005). Maternal employment and child development: A fresh look using newer methods. *Developmental Psychology* **41**, 833–850.

Hirano, K., Imbens, G. W., and Ridder, G. (2003). Efficient estimation of average treatment effects using the estimated propensity score. *Econometrica* **71**, 1161–1189.

Hirano, K., Imbens, G. W., Rubin, D. B., and Zhou, A. (2000). Assessing the effect of an influenza vaccine in an encouragement design. *Biostatistics* **1**, 69–88.

Hoenig, J. M., and Heisey, D. M. (2001). The abuse of power: The pervasive fallacy of power calculations for data analysis. *American Statistician* **55**, 19–24.

Hoerl, A. E., and Kennard, R. W. (1970). Ridge regression: Biased estimation for nonorthogonal problems. *Technometrics* **12**, 55–67.

Hoeting, J., Madigan, D., Raftery, A. E., and Volinsky, C. (1999). Bayesian model averaging (with discussion). *Statistical Science* **14**, 382–417.

Hoff, P. D. (2005). Bilinear mixed-effects models for dyadic data. *Journal of the American Statistical Association* **100**, 286–295.

Hoff, P. D., Raftery, A. E., and Handcock, M. S. (2002). Latent space approaches to social network analysis. *Journal of the American Statistical Association* **97**, 1090–1098.

Hong, G., and Raudenbush, S. W. (2006). Evaluating kindergarten retention policy: A case study of causal inference for multilevel observational data. *Journal of the American Statistical Association* **475**, 901–910.

Hoogerheide, L., Kleibergen, F., and van Dijk, H. K. (2007). Natural conjugate priors for the instrumental variables regression model applied to the Angrist-Krueger data. *Journal of Econometrics* **138**, 63–103.

Hudgens, M. G., and Halloran, M. E. (2008). Toward causal inference with interference. *Journal of the American Statistical Association* **482**, 832–842.

Hullman, J., Resnick, P., and Adar, E. (2015). Hypothetical outcome plots outperform error bars and violin plots for inferences about reliability of variable ordering. *PLOS One* **10** (11): e0142444.

Imai, K., and van Dyk, D. A. (2003). A Bayesian analysis of the multinomial probit model using marginal data augmentation. *Journal of Econometrics* **124**, 311–334.

Imai, K., and van Dyk, D. A. (2004). Causal inference with general treatment regimes: Generalizing the propensity score. *Journal of the American Statistical Association* **99**, 854–866.

Imbens, G. W. (2000). The role of the propensity score in estimating dose-response functions. *Biometrika* **87**, 706–710.

Imbens, G. W. (2004). Nonparametric estimation of average treatment effects under exogeneity: A review. *Review of Economics and Statistics* **86**, 4–29.

Imbens, G. W., and Angrist, J. (1994). Identification and estimation of local average treatment effects. *Econometrica* **62**, 467–475.

Imbens, G. W., and Rubin, D. B. (1997). Bayesian inference for causal effects in randomized experiments with noncompliance. *Annals of Statistics* **25**, 305–327.

Imbens, G., W., and Rubin, D. B. (2015). *Causal Inference for Statistics, Social, and Biomedical Sciences: An Introduction*. Cambridge University Press.

Jacob, B. A., and Lefgren, L. (2004). Remedial education and student achievement: A regression-discontinuity analysis. *Review of Economics and Statistics* **86**, 226–244.

J. F. (2015). Home advantage in basketball: As sweet as ever. *Economist* Game theory blog, 24 May. `www.economist.com/blogs/gametheory/2015/06/home-advantage-basketball`

Joffe, M. M., and Rosenbaum, P. R. (1999). Propensity scores. *American Journal of Epidemiology* **150**, 327–333.

Jones, S. R. G. (1992). Was there a Hawthorne effect? *American Journal of Sociology* **98**, 451–468.

Kahneman, D., and Tversky, A. (1973). On the psychology of prediction. *Psychological Review* **80**, 237–251.

Kalesan, B., Mobily, M., Keiser, O., Fagan, J., and Galea, S. (2016). Firearm legislation and firearm mortality in the USA: A cross-sectional, state-level study. *Lancet* **387**, 1847–1855.

Kay, M. (2020). tidybayes: Tidy data and geoms for Bayesian models. `mjskay.github.io/tidybayes`

Keele, L. J. (2015). Rbounds: An R package for sensitivity analysis with matched data. `lukekeele.com/rbounds-an-r-package-for-sensitivity-analysis-with-matched-data/`

Kennedy, P. (2008). *A Guide to Econometrics*, sixth edition. New York: Wiley.

Kleibergen, F., and Zivot, E. (2003). Bayesian and classical approaches to instrumental variable regression. *Journal of Econometrics* **114**, 29–72.

Krantz, D. H. (1999). The null hypothesis testing controversy in psychology. *Journal of the American Statistical Association* **94**, 1372–1381.

Krantz-Kent, R. (2018). Television, capturing America's attention at prime time and beyond. *Beyond the Numbers: Special Studies and Research* **7**, 14. Washington, D.C.: U.S. Bureau of Labor Statistics. `www.bls.gov/opub/btn/volume-7/television-capturing-americas-attention.htm`

Kreft, I., and De Leeuw, J. (1998). *Introducing Multilevel Modeling*. London: Sage.

Kropko, J., Goodrich, B., Gelman, A., and Hill, J. L. (2014). Multiple imputation for continuous and categorical data: Comparing joint and conditional approaches. *Political Analysis* **22**, 497–519.

LaLonde, R. J. (1986). Evaluating the econometric evaluations of training programs using experimental data. *American Economic Review* **76**, 604–620.

Landwehr, J. M., Pregibon, D., and Shoemaker, A. C. (1984). Graphical methods for assessing logistic regression models. *Journal of the American Statistical Association* **79**, 61–83.

Lange, K. L., Little, R. J. A., and Taylor, J. M. G. (1989). Robust statistical modeling using the $t$ distribution. *Journal of the American Statistical Association* **84**, 881–896.

Lavori, P. W., Keller, M. B., and Endicott, J. (1995). Improving the aggregate performance of psychiatric diagnostic methods when not all subjects receive the standard test. *Statistics in Medicine* **14**, 1913–1925.

Lax, J., and Phillips, J. (2009). How should we estimate public opinion in the states? *American Journal of Political Science* **53**, 107–121.

Leamer, E. (1978). *Specification Searches: Ad Hoc Inference with Nonexperimental Data*. New York: Wiley.

Leamer, E. (1983). Let's take the con out of econometrics. *American Economic Review* **73**, 31–43.

Lechner, M. (1999). Earnings and employment effects of continuous off-the-job training in East Germany after unification. *Journal of Business and Economic Statistics* **17**, 74–90.

Lenth, R. V. (2001). Some practical guidelines for effective sample size determination. *American Statistician* **55**, 187–193.

Lesaffre, E., and Albert, A. (1989). Partial separation in logistic discrimination. *Journal of the Royal Statistical Society B* **51**, 109–116.

Levitt, S. D., and List, J. A. (2011). Was there really a Hawthorne effect at the Hawthorne plant? An analysis of the original illumination experiments. *American Economic Journal: Applied Economics* **3**, 224–238.

Levitt, S. D., and Wolfram, C. D. (1997). Decomposing the sources of incumbency advantage in the U.S. House. *Legislative Studies Quarterly* **22**, 45–60.

Li, F., Thomas, L. E., and Li, F. (2019). Addressing extreme propensity scores via the overlap weights. *American Journal of Epidemiology* **188**, 250–257.

Liang, K. Y., and McCullagh, P. (1993). Case studies in binary dispersion. *Biometrics* **49**, 623–630.

Lin, C. Y., Gelman, A., Price, P. N., and Krantz, D. H. (1999). Analysis of local decisions using hierarchical modeling, applied to home radon measurement and remediation (with discussion). *Statistical Science* **14**, 305–337.

Lira, P. I., Ashworth, A., and Morris, S. S. (1998). Effect of zinc supplementation on the morbidity, immune function, and growth of low-birth-weight, full-term infants in northeast Brazil. *American Journal of Clinical Nutrition* **68**, 418S–424S.

Little, R. J. A., and Rubin, D. B. (2002). *Statistical Analysis with Missing Data*, second edition. New York: Wiley.

Liu, C. (1995). Missing data imputation using the multivariate $t$ distribution. *Journal of Multivariate Analysis* **48**, 198–206.

Liu, C. (2004). Robit regression: A simple robust alternative to logistic and probit regression. In *Applied Bayesian Modeling and Causal Inference from Incomplete-Data Perspectives*, ed. A. Gelman and X. L. Meng, 227–238. London: Wiley.

Lohr, S. (2009). *Sampling: Design and Analysis*, second edition. Pacific Grove, Calif.: Duxbury.

Loken, E. (2004). Multimodality in mixture models and factor models. In *Applied Bayesian Modeling and Causal Inference from Incomplete-Data Perspectives*, ed. A. Gelman and X. L. Meng, 203–213. London: Wiley.

Lord, F. M. (1967). A paradox in the interpretation of group comparisons. *Psychological Bulletin* **68**, 304–305.

Lord, F. M. (1969). Statistical adjustments when comparing preexisting groups. *Psychological Bulletin* **72**, 336–337.

Lumley, T. (2019). survey: Analysis of complex survey samples. R package version 3.35-1.

Maddala, G. S. (1976). Weak priors and sharp posteriors in simultaneous equation models. *Econometrica* **44**, 345–351.

Maddala, G. S. (1983). *Limited Dependent and Qualitative Variables in Econometrics*. Cambridge University Press.

Mallows, C. L. (1973). Some comments on $C_p$. *Technometrics* **15**, 661–675.

Manski, C. F. (1990). Nonparametric bounds on treatment effects. *American Economic Reviews, Papers and Proceedings* **80**, 319–323.

McCullagh, P. (1980). Regression models for ordinal data (with discussion). *Journal of the Royal Statistical Society B* **42**, 109–142.

McCullagh, P., and Nelder, J. A. (1989). *Generalized Linear Models*, second edition. London: Chapman and Hall.

McCulloch, R. E., Sparapani, R., Gramacy, R., Spanbauer, C., and Pratola, M. (2019). BART: Bayesian additive regression trees. R package version 2.5. `cran.r-project.org/package=BART`

McDonald, G. C., and Schwing, R. C. (1973). Instabilities of regression estimates relating air pollution to mortality. *Technometrics* **15**, 463–482.

McElreath, R. (2020). *Statistical Rethinking: A Bayesian Course with Examples in R and Stan*, second edition. London: CRC Press.

McFadden, D. (1973). Conditional logit analysis of qualitative choice behavior. In *Frontiers in Econometrics*, ed. P. Zarembka, 105–142. New York: Academic Press.

McShane, B. B., and Gal, D. (2017). Statistical significance and the dichotomization of evidence. *Journal of the American Statistical Association* **112**, 885–895.

McShane, B. B., Gal, D., Gelman, A., Robert, C., and Tackett, J. L. (2019). Abandon statistical significance. *American Statistician* **73** (sup1), 235–245.

Meehl, P. E. (1967). Theory-testing in psychology and physics: A methodological paradox. *Philosophy of Science* **34**, 103–115.

Meehl, P. E. (1978). Theoretical risks and tabular asterisks: Sir Karl, Sir Ronald, and the slow progress of soft psychology. *Journal of Consulting and Clinical Psychology* **46**, 806–834.

Meehl, P. E. (1990). Why summaries of research on psychological theories are often uninterpretable. *Psychological Reports* **66**, 195–244.

Meissner, K., Bingel, U., Colloca, L., Wager, T. D., Watson, A., and Flaten, M. A. (2011). The placebo effect: Advances from different methodological approaches. *Journal of Neuroscience* **31**, 16117–16124.

Meng, X. L. (1994). Multiple-imputation inferences with uncongenial sources of input (with discussion). *Statistical Science* **9**, 538–573.

Michalopoulos, C., Bloom, H. S., and Hill, C. J. (2004). Can propensity score methods match the findings from a random assignment evaluation of mandatory welfare-to-work programs? *Review of Economics and Statistics* **86**, 156–179.

Montgomery, D. C. (1986). *Design and Analysis of Experiments*, second edition. New York: Wiley.

Montgomery, J. M., Nyhan, B., and Torres, M. (2018). How conditioning on post-treatment variables can ruin your experiment and what to do about it. *American Journal of Political Science* **62**, 760–775.

Morgan, S. L., and Winship, C. (2014). *Counterfactuals and Causal Inference: Methods and Principles for Social Research*, second edition. Cambridge University Press.

Morris, C. (1979). Measurement issues in the second generation of social experiments: The health insurance study. *Journal of Econometrics* **11**, 117–129.

Mosteller, F. (1951). Remarks on the method of paired comparisons. *Psychometrika* **16**, 3–9, 203–206, 207–218.

Mosteller, F., and Tukey, J. W. (1977). *Data Analysis and Regression*. Reading, Mass.: Addison-Wesley.

Mueller, K. (2018). rprojroot: Finding files in project subdirectories. R package version 1.3-2. `cran.r-project.org/web/packages/rprojroot/vignettes/rprojroot.html`

Muller, O., Becher, H., van Zweeden, A. B., Ye, Y., Diallo, D. A., Konate, A. T., Gbangou, A., Kouate, B., and Gareene, M. (2001). Effect of zinc supplementation on malaria and other causes of morbidity in west African children: Randomized double blind placebo controlled trial. *British Medical Journal* **322**, 1567–1573.

Mulligan, C. B., and Hunter, C. G. (2003). The empirical frequency of a pivotal vote. *Public Choice* **116**, 31–54.

Murphy, R. T. (1991). Educational effectiveness of Sesame Street: A review of the first twenty years of research, 1969–1989. Report RR-91-55. Princeton, N.J.: Educational Testing Service.

Murrell, P. (2005). *R Graphics*. London: CRC Press.

Nelder, J. A., and Wedderburn, R. W. M. (1972). Generalized linear models. *Journal of the Royal Statistical Society A* **135**, 370–384.

Neter, J., Kutner, M. H., Nachtsheim, C. J., and Wasserman, W. (1996). *Applied Linear Statistical Models*, fourth edition. Burr Ridge, Ill.: Richard D. Irwin, Inc.

Newton, M. A., Kendziorski, C. M., Richmond, C. S., Blattner, F. R., and Tsui, K. W. (2001). On differential variability of expression ratios: Improving statistical inference about gene expression changes from microarray data. *Journal of Computational Biology* **8**, 37–52.

Neyman, J. (1923). On the application of probability theory to agricultural experiments. Essay on principles. Section 9. Translated and edited by D. M. Dabrowska and T. P. Speed. *Statistical Science* **5**, 463–480 (1990).

Nosek, B. A., Spies, J. R., and Motyl, M. (2012). Scientific utopia II. Restructuring incentives and practices to promote truth over publishability. *Perspectives on Psychological Science* **7**, 615–631.

O'Keefe, S. (2004). Job creation in California's enterprise zones: A comparison using a propensity score matching model. *Journal of Urban Economics* **55**, 131–150.

Open Science Collaboration (2015). Estimating the reproducibility of psychological science. *Science* **349** aac4716, 1–8.

Orben, A., and Przybylski, A. K. (2019). The association between adolescent well-being and digital technology use. *Nature Human Behaviour* **3**, 173–182.

Pagano, M., and Anoke, S. (2013). Mommy's baby, Daddy's maybe: A closer look at regression to the mean. *Chance* **26** (3), 4–9.

Pallesen, J. (2019). Blind auditions and gender discrimination. `jsmp.dk/posts/2019-05-12-blindauditions/`

Pardoe, I., and Cook, R. D. (2002). A graphical method for assessing the fit of a logistic regression model. *American Statistician* **56**, 263–272.

Pardoe, I., and Simonton, D. K. (2008). Applying discrete choice models to predict Academy Award winners. *Journal of the Royal Statistical Society A* **171**, 375–394.

Pearl, J. (2000). *Causality*. Cambridge University Press.

Pearson, K., and Lee, A. (1903). On the laws of inheritance in man. I. Inheritance of physical characters. *Biometrika* **2**, 357–462.

Persico, N., Postlewaite, A., and Silverman, D. (2004). The effect of adolescent experience on labor market outcomes: The case of height. *Journal of Political Economy* **112**, 1019–1053.

Piironen, J., Paasiniemi, M., and Vehtari, A. (2020). Projective inference in high-dimensional problems: Prediction and feature selection. *Electronic Journal of Statistics*.

Piironen, J., and Vehtari, A. (2017a). Comparison of Bayesian predictive methods for model selection. *Statistics and Computing* **27**, 711–735.

Piironen, J., and Vehtari, A. (2017b). On the hyperprior choice for the global shrinkage parameter in the horseshoe prior. *Proceedings of Machine Learning Research* **54**, 905–913.

Piironen, J., and Vehtari, A. (2017c). Sparsity information and regularization in the horseshoe and other shrinkage priors. *Electronic Journal of Statistics* **11** 5018–5051.

PlatypeanArchcow (2009). Map of Human Development Index (Creative Commons License). `https://commons.wikimedia.org/wiki/File:US_states_HDI.png`

Pole, A., West, M., and Harrison, J. (1994). *Applied Bayesian Forecasting and Time Series Analysis*. London: Chapman and Hall.

R Core Team (2019). R: A language and environment for statistical computing. R Foundation for Statistical Computing, Vienna, Austria. `www.R-project.org/`

Raghunathan, T. E., Solenberger, P. W., and Van Hoewyk, J. (2002). IVEware. `www.isr.umich.edu/src/smp/ive/`

Raghunathan, T. E., Van Hoewyk, J., and Solenberger, P. W. (2001). A multivariate technique for multiply imputing missing values using a sequence of regression models. *Survey Methodology* **27**, 85–95.

Ramsey, F. L., and Schafer, D. W. (2001). *The Statistical Sleuth*, second edition. Pacific Grove, Calif.: Duxbury.

Rasmussen, C. E., and Williams, C. K. I. (2006). *Gaussian Processes for Machine Learning*. Cambridge, Mass.: MIT Press.

Raudenbush, S. W., and Bryk, A. S. (2002). *Hierarchical Linear Models*, second edition. Thousand Oaks, Calif.: Sage.

Ripley, B. D. (1988). *Statistical Inference for Spatial Processes*. Cambridge University Press.

Robins, J. M. (1989). The analysis of randomized and non-randomized AIDS treatment trials using a new approach to causal inference in longitudinal studies. *Health Services Research Methodology: A Focus on AIDS*, ed. L. Sechrest, H. Freeman, and A. Mulley, 113–159. Washington, D.C.: U.S. Public Health Service, National Center for Health Services Research.

Robins, J. M. (1994). Correcting for non-compliance in randomized trials using structural nested mean models. *Communications in Statistics* **23**, 2379–2412.

Robins, J. M., and Ritov, Y. (1997). Towards a curse of dimensionality appropriate (CODA) asymptotic theory for semi-parametric models. *Statistics in Medicine* **16**, 285–319.

Robins, J. M., and Rotnitzky, A. (1995). Semiparametric efficiency in multivariate regression models with missing data. *Journal of the American Statistical Association* **90**, 122–129.

Robins, J. M., Rotnitzky, A., and Zhao, L. P. (1995). Analysis of semiparametric regression models for repeated outcomes in the presence of missing data. *Journal of the American Statistical Association* **90**, 106–121.

Robins, J. M., and Wang, N. (2000). Inference for imputation estimators. *Biometrika* **87**, 113–124.

Rodu, B., and Plurphanswat, N. (2018). Re: Electronic cigarette use and progression from experimentation to established smoking. *Pediatrics* comment, 15 Mar. pediatrics.aappublications.org/content/141/4/e20173594/tab-e-letters

Rosado, J. L., Lopez, P., Munoz, E., Martinez, H., and Allen, L. H. (1997). Zinc supplementation reduced morbidity, but neither zinc nor iron supplementation affected-growth or body composition of Mexican preschoolers. *American Journal of Clinical Nutrition* **65**, 13–19.

Rosenbaum, P. R. (1984). The consequences of adjustment for a concomitant variable that has been affected by the treatment. *Journal of the Royal Statistical Society A* **147**, 656–666.

Rosenbaum, P. R. (1987). Model-based direct adjustment. *Journal of the American Statistical Association* **82**, 387–394.

Rosenbaum, P. R. (1989). Optimal matching in observational studies. *Journal of the American Statistical Association* **84**, 1024–1032.

Rosenbaum, P. R. (2002a). Covariance adjustment in randomized experiments and observational studies. *Statistical Science* **17**, 286–327.

Rosenbaum, P. R. (2002b). *Observational Studies*. New York: Springer-Verlag.

Rosenbaum, P. R. (2007). Interference between units in randomized experiments. *Journal of the American Statistical Association* **102**, 191–200.

Rosenbaum, P. R., and Rubin, D. B. (1983a). The central role of the propensity score in observational studies for causal effects. *Biometrika* **70**, 41–55.

Rosenbaum, P. R., and Rubin, D. B. (1983b). Assessing sensitivity to an unobserved binary covariate in an observational study with binary outcome. *Journal of the Royal Statistical Society B* **45**, 212–218.

Rosenbaum, P. R., and Rubin, D. B. (1984). Reducing bias in observational studies using subclassification on the propensity score. *Journal of the American Statistical Association* **79**, 516–524.

Rosenbaum, P. R., and Rubin, D. B. (1985). Constructing a control group using multivariate matched sampling methods that incorporate the propensity score. *American Statistician* **39**, 33–38.

Rosenstone, S. J. (1983). *Forecasting Presidential Elections*. New Haven, Conn.: Yale University Press.

Ross, C. E. (1990). Work, family, and well-being in the United States. Survey data available from Inter-university Consortium for Political and Social Research, Ann Arbor, Mich.

Roy, A. D. (1951). Some thoughts on the distribution of earnings. *Oxford Economic Papers* **3**, 135–146.

Rubin, D. B. (1973). The use of matched sampling and regression adjustment to remove bias in observational studies. *Biometrics* **29**, 185–203.

Rubin, D. B. (1974). Estimating causal effects of treatments in randomized and nonrandomized studies. *Journal of Educational Psychology* **66**, 688–701.

Rubin, D. B. (1976). Inference and missing data. *Biometrika* **63**, 581–592.

Rubin, D. B. (1978). Bayesian inference for causal effects: The role of randomization. *Annals of Statistics* **6**, 34–58.

Rubin, D. B. (1979). Using multivariate matched sampling and regression adjustment to control bias in observational studies. *Journal of the American Statistical Association* **74**, 318–328.

Rubin, D. B. (1981). Estimation in parallel randomized experiments. *Journal of Educational Statistics* **6**, 377–401.

Rubin, D. B. (1984). Bayesianly justifiable and relevant frequency calculations for the applied statistician. *Annals of Statistics* **12**, 1151–1172.

Rubin, D. B. (1987). *Multiple Imputation for Nonresponse in Surveys.* New York: Wiley.

Rubin, D. B. (1996). Multiple imputation after 18+ years (with discussion). *Journal of the American Statistical Association* **91**, 473–520.

Rubin, D. B. (2000). Statistical issues in the estimation of the causal effects of smoking due to the conduct of the tobacco industry. In *Statistical Science in the Courtroom*, ed. J. L. Gastwirth, 322–350. New York: Springer-Verlag.

Rubin, D. B. (2004). Direct and indirect causal effects via potential outcomes (with discussion). *Scandinavian Journal of Statistics* **31**, 161–201.

Rubin, D. B. (2006). *Matched Sampling for Causal Effects*. Cambridge University Press.

Rubin, D. B., Stern, H. S, and Vehovar, V. (1995). Handling "Don't Know" survey responses: The case of the Slovenian plebiscite. *Journal of the American Statistical Association* **90**, 822–828.

Rubin, D. B., and Stuart, E. A. (2005). Matching with multiple control groups and adjusting for group differences. In *Proceedings of the American Statistical Association, Section on Health Policy Statistics*. Alexandria, Va.: American Statistical Association.

Rubin, D. B., and Thomas, N. (2000). Combining propensity score matching with additional adjustments for prognostic covariates. *Journal of the American Statistical Association* **95**, 573–585.

Ruel, M. T., Rivera, J. A., Santizo, M. C., Lonnerdal, B., and Brown, K. H. (1997). Impact of zinc supplementation on morbidity from diarrhea and respiratory infections among rural Guatemalan children. *Pediatrics* **99**, 808–813.

Sarsons, H. (2015). Rainfall and conflict: A cautionary tale. *Journal of Development Economics* **115**, 62–72.

Schafer, J. L. (1997). *Analysis of Incomplete Multivariate Data*. London: Chapman and Hall.

Schmid, C. H. (2016). Increased mortality for white middle-aged Americans not fully explained by causes suggested. *Proceedings of the National Academy of Sciences* **113**, E814.

Schmidt-Nielsen, K. (1984). *Scaling: Why is Animal Size So Important?* Cambridge University Press.

Scott, M., Diakow, R., Hill, J. L., and Middleton, J. (2018). Potential for bias inflation with grouped data: A comparison of estimators and a sensitivity analysis strategy. *Observational Studies* **4**, 111–149.

Shadish, W. R., Clark, M. H., and Steiner, P. M. (2008). Can nonrandomized experiments yield accurate answers? A randomized experiment comparing random and nonrandom assignments. *Journal of the American Statistical Association* **103**, 1334–1344.

Shadish, W. R., Cook, T., and Campbell, S. (2002). *Experimental and Quasi-Experimental Designs*. Boston: Houghton Mifflin.

Shirani-Mehr, H., Rothschild, D., Goel, S., and Gelman, A. (2018). Disentangling bias and variance in election polls. *Journal of the American Statistical Association* **113**, 607–614.

Shirley, K., and Gelman, A. (2015). Hierarchical models for estimating state and demographic trends in U.S. death penalty public opinion. *Journal of the Royal Statistical Society A* **178**, 1–28.

Simmons, J., Nelson, L., and Simonsohn, U. (2011). False-positive psychology: Undisclosed flexibility in data collection and analysis allow presenting anything as significant. *Psychological Science* **22**, 1359–1366.

Singer, E., Van Hoewyk, J., Gebler, N., Raghunathan, T., and McGonagle, K. (1999). The effects of incentives on response rates in interviewer-mediated surveys. *Journal of Official Statistics* **15**, 217–230.

Smith, J., and Todd, P. E. (2005). Does matching overcome LaLonde's critique of nonexperimental estimators? (with discussion). *Journal of Econometrics* **120**, 305–375.

Snedecor, G. W., and Cochran, W. G. (1989). *Statistical Methods*, eighth edition. Ames: Iowa State University Press.

Snijders, T. A. B., and Bosker, R. J. (1999). *Multilevel Analysis*. London: Sage.

Sobel, M. E. (1990). Effect analysis and causation in linear structural equation models. *Psychometrika* **55**, 495–515.

Sobel, M. E. (1998). Causal inference in statistical models of the process of socioeconomic achievement: A case study. *Sociological Methods and Research* **27**, 318–348.

Sobel, M. E. (2006). What do randomized studies of housing mobility reveal? Causal inference in the face of interference. *Journal of the American Statistical Association* **101**, 1398–1407.

Spiegelhalter, D. J., Best, N. G., Carlin, B. P., and van der Linde, A. (2002). Bayesian measures of model complexity and fit (with discussion). *Journal of the Royal Statistical Society B* **64**, 583–639.

Spitzer, E. (1999). The New York City Police Department's "stop and frisk" practices. Office of the New York State Attorney General. `ag.ny.gov/sites/default/files/pdfs/bureaus/civil_rights/stp_frsk.pdf`

Stan Development Team (2020). *Stan Modeling Language User's Guide and Reference Manual*. `mc-stan.org/`

Stigler, S. M. (1977). Do robust estimators work with real data? (with discussion). *Annals of Statistics* **5**, 1055–1098.

Stigler, S. M. (1983). Discussion of "Parametric empirical Bayes inference: Theory and applications," by C. Morris. *Journal of the American Statistical Association* **78**, 62–63.

Stigler, S. M. (1986). *The History of Statistics*. Cambridge, Mass.: Harvard University Press.

Stone, M. (1974). Cross-validatory choice and assessment of statistical predictions (with discussion). *Journal of the Royal Statistical Society B* **36**, 111–147.

Stone, M. (1977). An asymptotic equivalence of choice of model cross-validation and Akaike's criterion. *Journal of the Royal Statistical Society B* **36**, 44–47.

Stuart, E. A. (2010). Matching methods for causal inference: A review and a look forward. *Statistical Science* **25**, 1–21.

Su, Y. S., Gelman, A., Hill, J. L., and Yajima, M. (2011). Multiple imputation with diagnostics (mi) in R: Opening windows into the black box. *Journal of Statistical Software* **45** (2).

Swartz, T. B., and Arce, A. (2014). New insights involving the home team advantage. *International Journal of Sports Science & Coaching* **9**, 681–692.

Talts, S., Betancourt, M., Simpson, D., Vehtari, A., and Gelman, A. (2018). Validating Bayesian inference algorithms with simulation-based calibration. `arxiv.org/abs/1804.06788`

Thistlethwaite, D., and Campbell, D. (1960). Regression-discontinuity analysis: An alternative to the ex post facto experiment. *Journal of Educational Psychology* **51**, 309–17.

Thurstone, L. L. (1927a). A law of comparative judgment. *Psychological Review* **34**, 273–286.

Thurstone, L. L. (1927b). The method of paired comparison for social values. *Journal of Abnormal and Social Psychology* **21**, 384–400.

Tibshirani, R. (1996). Regression shrinkage and selection via the lasso. *Journal of the Royal Statistical Society B* **58**, 267–288.

Tinbergen, J. (1930). Determination and interpretation of supply curves: An example. *Zeitschrift fur Nationalokonomie*. Reprinted in *The Foundations of Econometrics*, eds. D. Hendry and M. Morgan, 233–245. Cambridge University Press.

Tobin, J. (1958). Estimation of relationships for limited dependent variables. *Econometrica* **26**, 24–36.

Troxel, A. B., Ma, G., and Heitjan, D. F. (2004). An index of sensitivity to nonignorability. *Statistica Sinica* **14**, 1221–1237.

Tufte, E. R. (1983). *The Visual Display of Quantitative Information*. Cheshire, Conn.: Graphics Press.

Tufte, E. R. (1990). *Envisioning Information*. Cheshire, Conn.: Graphics Press.

Tukey, J. W. (1977). *Exploratory Data Analysis*. Reading, Mass.: Addison-Wesley.

Unwin, A., Volinsky, C., and Winkler, S. (2003). Parallel coordinates for exploratory modelling analysis. *Computational Statistics and Data Analysis* **43**, 553–564.

Urbanek, S. (2004). *Exploratory Model Analysis: An Interactive Graphical Framework for Model Comparison and Selection*. Norderstedt: Books on Demand.

Van Buuren, S., Boshuizen, H. C., and Knook, D. L. (1999). Multiple imputation of missing blood pressure covariates in survival analysis. *Statistics in Medicine* **18**, 681–694.

Van Buuren, S., and Oudshoom, C. G. M. (2011). MICE: Multivariate imputation by chained equations in R. *Journal of Statistical Software* **45** (3).

VanderWeele, T. J. (2015). *Explanation in Causal Inference: Methods for Mediation and Interaction*. Cambridge University Press.

VanderWeele, T. J., and Hernan, M. (2013). Causal inference under multiple versions of treatment. *Journal of Causal Inference* **1**, 1–20.

van Geen, A., Zheng, Y., Versteeg, R., et al. (2003). Spatial variability of arsenic in 6000 tube wells in a 25 km$^2$ area of Bangladesh. *Water Resources Research* **39**, 1140.

Vehtari, A., Gabry, J., Yao, Y., and Gelman, A. (2019). loo: Efficient leave-one-out cross-validation and WAIC for Bayesian models. R package version 2.1.0. `mc-stan.org/loo`

Vehtari, A., Gelman, A., and Gabry, J. (2017). Practical Bayesian model evaluation using leave-one-out cross-validation and WAIC. *Statistics and Computing* **27**, 1413–1432.

Vehtari, A., Gelman, A., Simpson, D., Carpenter, B., and Buerkner, P. C. (2019). Rank-normalization, folding, and localization: An improved R-hat for assessing convergence of MCMC. `arxiv.org/abs/1903.08008`

Vehtari, A., and Ojanen, J. (2012). A survey of Bayesian predictive methods for model assessment, selection and comparison. *Statistics Surveys* **6**, 142–228.

Venables, W. N., and Ripley, B. D. (2002). *Modern Applied Statistics with S*, fourth edition. New York: Springer-Verlag.

Vikram, H. R., Buenconsejo, J., Hasbun, R., and Quagliarello, V. J. (2003). Impact of valve surgery on 6-month mortality in adults with complicated, left-sided native valve endocarditis: A propensity analysis. *Journal of the American Medical Association* **290**, 3207–3214.

Viscusi, W. K., and Aldy, J. E. (2003). The value of a statistical life: A critical review of market estimates throughout the world. *Journal of Risk and Uncertainty* **27**, 5–76.

von Mises, R. (1957). *Probability, Statistics, and Truth*, second edition. New York: Dover. Reprint.

Vul, E., Harris, C., Winkielman, P., and Pashler, H. (2009). Puzzlingly high correlations in fMRI studies of emotion, personality, and social cognition. *Perspectives on Psychological Science* **4**, 274–290.

Wachsmuth, A. W., Wilkinson, L., and Dallal, G. E. (2003). Galton's bend: A previously undiscovered nonlinearity in Galton's family stature regression data. *American Statistician* **57**, 190–192.

Wager, S., and Athey, S. (2018). Estimation and inference of heterogeneous treatment effects using random forests. *Journal of the American Statistical Association* **113**, 1228–1242.

Wahba, G. (1978). Improper priors, spline smoothing and the problem of guarding against model errors in regression. *Journal of the Royal Statistical Society B* **40**, 364–372.

Wainer, H. (1984). How to display data badly. *American Statistician* **38**, 137–147.

Wainer, H. (1997). *Visual Revelations*. New York: Springer-Verlag.

Wainer, H., Palmer, S., and Bradlow, E. T. (1998). A selection of selection anomalies. *Chance* **11** (2), 3–7.

Walker, S. H., and Duncan, D. B. (1967). Estimation of the probability of an event as a function of several independent variables. *Biometrika* **54**, 167–178.

Wallis, W. A., and Friedman, M. (1942). The empirical derivation of indifference functions. In *Studies in Mathematical Economics and Econometrics in Memory of Henry Schultz*, ed. O. Lange, F. McIntyre, and T. O. Yntema, 267–300. University of Chicago Press.

Wang, W., Rothschild, D., Goel, S., and Gelman, A. (2015). Forecasting elections with non-representative polls. *International Journal of Forecasting* **31**, 980–991.

Wattenberg, L. (2007). Where all boys end up nowadays. Baby Name Wizard blog, 19 Jul. `www.babynamewizard.com/archives/2007/7/where-all-boys-end-up-nowadays`

Weisburd, D., Petrosino, A., and Mason, G. (1993). Design sensitivity in criminal justice experiments. *Crime and Justice* **17**, 337–379.

Wickham, H. (2006). Exploratory model analysis with R and GGobi. Technical report. `had.co.nz/model-vis/2007-jsm.pdf`

Wickham, H. (2014). *Advanced R*. London: CRC Press.

Wickham, H. (2016). *ggplot2: Elegant Graphics for Data Analysis*. New York: Springer-Verlag. `ggplot2.tidyverse.org`

Wickham, H., and Grolemund, G. (2017). *R for Data Science*. Sebastopol, Calif.: O'Reilly.

Wiens, B. L. (1999). When log-normal and gamma models give different results: A case study. *American Statistician* **53**, 89–93.

Wilkinson, L. (2005). *The Grammar of Graphics*, second edition. New York: Springer-Verlag.

Wlezien, C., and Erikson, R. S. (2004). The fundamentals, the polls, and the presidential vote. *PS: Political Science and Politics* **37**, 747–751.

Wlezien, C., and Erikson, R. S. (2005). Post-election reflections on our pre-election predictions. *PS: Political Science and Politics* **38**, 25–26.

Woolridge, J. M. (2001). *Econometric Analysis of Cross Section and Panel Data*. Cambridge, Mass.: MIT Press.

Xie, Y. (2015). *Dynamic Documents with R and knitr*, second edition. London: CRC Press.

Xie, Y. (2019). knitr: A general-purpose package for dynamic report generation in R. R package version 1.25. `yihui.org/knitr/`

Yao, Y., Vehtari, A., Simpson, D., and Gelman, A. (2018). Using stacking to average Bayesian predictive distributions (with discussion). *Bayesian Analysis* **13**, 917–1003.

Yates, F. (1967). A fresh look at the basic principles of the design and analysis of experiments. *Proceedings of the Fifth Berkeley Symposium on Mathematical Statistics and Probability* **4**, 777–790. Berkeley: University of California Press.

Zellner, A. (1976). Bayesian and non-Bayesian analysis of the regression model with multivariate Student-*t* error terms. *Journal of the American Statistical Association* **71**, 400–405.

Zhang, K., Traskin, M., and Small, D. S. (2012). A powerful and robust test statistic for randomization inference in group-randomized trials with matched pairs of groups. *Biometrics* **68**, 75–84.

Zorn, C. (2005). A solution to separation in binary response models. *Political Analysis* **13**, 157–170.

# Author Index

*Note:* with references having more than three authors, the names of all authors appear in this index, but only the first authors appear in the in-text citations.

# Subject Index